T0138240

Ecology and Evolution of Poeciliid Fishes

Ecology and Evolution of Poeciliid Fishes

Edited by Jonathan P. Evans,
Andrea Pilastro, and Ingo Schlupp

The University of Chicago Press
Chicago and London

JONATHAN P. EVANS is associate professor at the University of Western Australia's Centre for Evolutionary Biology. ANDREA PILASTRO is professor of zoology at the University of Padova. INGO SCHLUPP is associate professor of zoology at the University of Oklahoma and adjunct professor of integrative biology at the University of Texas.

The University of Chicago Press, Chicago 60637
The University of Chicago Press, Ltd., London
© 2011 by The University of Chicago
All rights reserved. Published 2010
Printed in the United States of America

20 19 18 17 16 15 14 13 12 11   1 2 3 4 5

ISBN-13: 978-0-226-22274-5 (cloth)
ISBN-10: 0-226-22274-8 (cloth)

A glossary for this book appears online at www.press.uchicago.edu/books/evans/.

Library of Congress Cataloging-in-Publication Data

Ecology and evolution of poeciliid fishes / edited by Jonathan P. Evans, Andrea Pilastro, and Ingo Schlupp.
        p. cm.
  Includes bibliographical references and index.
   ISBN-13: 978-0-226-22274-5 (cloth : alkaline paper)
   ISBN-10: 0-226-22274-8 (cloth : alkaline paper)   1. Poeciliidae.
2. Poeciliidae—Evolution.   3. Poeciliidae—Ecology.   I. Evans, Jonathan P. editor.   II. Pilastro, Andrea, editor.   III. Schlupp, Ingo, editor.
   QL638.P73E28 2011
   597´.667—dc22
                                      2011004668

♾ The paper used in this publication meets the minimum requirements of the American National Standard for Information Sciences—Permanence of Paper for Printed Library Materials, ANSI Z39.48-1992.

Contents

Part IV. Sexual selection

Part V. Genetics

Part VI. Conclusions

Foreword

B Y ALMOST any measure, the Poeciliidae is a fascinating family of fishes of a significance to scientific study greatly out of proportion to their individual sizes (small) or family diversity (medium). Possibly no other family of fishes has attracted so much interest for so many diverse reasons. The scientific areas of investigation that have drawn heavily upon these fishes make up a lengthy list that includes (in no particular order) population and community ecology, life-history evolution, sex determination, viviparity, unisexuality, basic inheritance studies, sexual selection, behavior, social systems, ecological genetics, phenotypic plasticity, parasitology, exotic-species problems, conservation, biogeography, sex linkage, sensory systems, speciation, physiology, immunology, oncology, and . . . the list goes on. Few families have contributed so broadly to scientific investigation of the natural world.

Poeciliids also play a strong role in very practical, species- and ecosystem-level management. Some are the focus of endangered-species recovery efforts, others have gone locally extinct at the hands of humanity, and in other cases they are harmful and destructive exotic invasives, even placing other species of rare poeciliids in harm's way. Some poeciliids are erroneously used in mosquito control and then become major pests; others have been more benignly transplanted outside their native ranges, fortunately with modest effects.

In addition to their scientific importance, poeciliids are of tremendous interest to aquarists, have developed a large and devoted following, and are the focus of lay publications, hobbyist societies, and conferences. They are raised in captivity in huge numbers for the aquarium trade, as well as bred and traded privately among devotees. In some cases, rare species are maintained in captivity by aquarists. Thus, they are an economically important group of fishes. For all these reasons, poeciliids deserve close scientific scrutiny.

It has been 22 years since Franklin Snelson and I edited a volume on these species, *Ecology and Evolution of Livebearing Fishes (Poeciliidae)*, a book now out of print. At that time it had been more than 25 years since Donn Rosen and Reeve Bailey in 1963 published their seminal and classic review of the structure, zoogeography, and systematics of poeciliids, and we felt in 1989 that a collection of chapters on the ecology, reproductive biology, genetics, life-history evolution, systematics, and conservation and impacts of poeciliids was overdue. Consequently, we persuaded some of the top researchers in the world to contribute to that volume, and because of their hard work and knowledge base, the family gained appropriate attention as subjects of scientific study and renewed efforts in the laboratory and the field. Twenty-two years later we are well due for another assessment, and I am delighted to see the volume that follows.

I am fascinated to see how far poeciliid science has come in the intervening 22 years, and this volume is a huge leap forward. The level of sophistication in poeciliid studies has increased tremendously, as has our detailed knowledge of these fishes. The chapters that follow demonstrate in no uncertain terms that this is a fundamentally important group of fishes for understanding how nature operates on levels from molecules to ecosystems. These species are treasure troves of scientific inquiry and continue to enlighten us at so many levels. I am personally gratified that so much attention is still being paid to this group of species.

When all is said and done, I think that, cold scientific assessments aside, these are simply fascinating fishes that feed our curiosities! Any organism that can have several (sometimes many) broods of offspring developing simultaneously and fed by the mother is well worth our attention. To understand viviparity, superfetation, measurable responses to selection in just a few generations, species interactions, unisexuality, complex color and inheritance patterns, and the dozens of other topics represented in this book is intellectually satisfying and personally exciting. I believe these fishes will continue to capture our attention and imagination for decades to come, and there is little danger of exhausting the deep supply of newly revealed secrets any time soon.

We still have much to learn from this endlessly fascinating family, and this book is a major step in that direction.

The real danger, of course, is in losing populations and species of poeciliids in the catastrophic global declines of nature that we are experiencing; addressing that tragedy is our most urgent calling. We cannot continue to harvest scientific wealth from species that no longer exist, and consequently we must deal immediately with their conservation in situ. If we fail in that effort, then books like this one and the previous volumes simply will become testimonials to our scientific brilliance and our simultaneous failures as caretakers.

Gary K. Meffe

Preface

FEW VERTEBRATE GROUPS have made such significant inroads into multiple fields of research as poeciliid fishes. The value of this family as evolutionary and ecological models is unprecedented and unrivaled among vertebrates and indeed surpasses that of many invertebrate systems as models in genetics, ecology, life-history evolution, and sexual selection. The first synthesis of the emerging body of literature on evolutionary ecology in poeciliid fishes resulted in the publication of Gary Meffe and Franklin Snelson Jr.'s (1989b) edited volume *Ecology and Evolution of Livebearing Fishes (Poeciliidae)*. Our motivation to compile a successor to this volume was simple: for over two decades, poeciliid biologists have relied heavily on Meffe and Snelson's volume, but sadly this has long been out of print, and in the many years since its publication there has been a tremendous amount of new work, encompassing both existing and new topics. Our motivation here is to provide a single reference source that documents these exciting developments. Throughout we have sought reviews that critically evaluate the literature and outline gaps in our knowledge. To this end we have brought together 53 researchers working in North America, continental Europe, Great Britain, Mexico, and Australia, covering subjects that broadly fall into five main research areas, which reflect the subdivisions of this book: (I) reproductive biology and life history, (II) evolutionary ecology, (III) behavior and cognition, (IV) sexual selection, and (V) genetics.

Unlike its predecessor, our volume does not include a chapter on phylogeny and taxonomy. This area has always been in flux and has seen rapid change over the past years, leading to several detailed phylogenetic hypotheses within smaller taxonomic groups. A very broadly based phylogenetic hypothesis addressing the question of the outgroup of all poeciliids and the general zoogeography was proposed recently by Hrbek et al. (2007). This is one of the latest phylogenetic hypotheses that builds on classical work by Rosen and Bailey (1963), Parenti (1989), and many more and is the one most frequently referred to throughout the present volume. The latest hypotheses, however, are more complex than those outlined by the older literature and many questions remain open. Especially important will be attempts to reconcile biogeography and phylogenetics. The current data suggest multiple waves of colonization of Middle America, but not enough detail is known. Another topic that needs additional attention is the phylogenetic position of *Tomeurus gracilis*, the only poeciliid that is not livebearing. This species was thought to be ancestral, but this view has been challenged recently by Hrbek et al. (2007), who suggest that *T. gracilis* has secondarily lost this trait. This has important implications for studies using character mapping. Clearly, more detailed studies are needed not only to understand the bigger picture but also to unravel relationships within and among genera. The list of named species within the poeciliids is still growing, as new species are being described. Clearly, widely distributed species such as *Poecilia mexicana* are likely candidates for finding cryptic species. Describing the diversity within the poeciliids is unfortunately a race against the extinction of species, just as documenting the biogeography of species is increasingly hampered by artificial introductions of species into foreign habitats.

The present volume begins with a section devoted to

reproductive biology and life-history evolution, commencing with an overview of poeciliid reproductive biology (**chapter 1**) by Hartmut Greven. This chapter provides a comprehensive overview of the structural adaptations employed during reproduction and embryonic development, including those involved in maternal-offspring interactions. Such maternal-fetal interactions are explored in detail in **chapter 2**, by Edie Marsh-Matthews, who summarizes recent developments in our understanding of the role of maternal contributions to offspring. Continuing these themes, in **chapter 3** Marcelo Pires and colleagues focus on the adaptive basis and evolutionary origins of such traits, shedding light on what drives their evolutionary diversification across the family. In **chapter 4** Jerald Johnson and Justin Bagley employ a comparative framework to infer links between life-history evolution and agents of natural selection, primarily focusing on studies that document intraspecific life-history diversification. Part I concludes (**chapter 5**) with Ingo Schlupp and Rüdiger Riesch's review of one of the more unusual aspects of poeciliid biology: unisexual reproduction, a mode of reproduction that was thought to be impossible in vertebrates until Laura and Carl Hubbs described sperm-dependent, clonal reproduction in *Poecilia formosa* in 1932.

Part II, on evolutionary ecology, commences with Gregory Grether and Gita Kolluru's review (**chapter 6**) examining the role of resource availability as an agent of natural selection, arguing that such effects, although largely overlooked in the literature, may have profound evolutionary implications. In **chapter 7**, Seth Coleman reviews a growing body of literature that investigates the use of specific sensory modalities in poeciliid behavior and the genetics and physiology of poeciliid sensory receptors, in particular of the visual system. Joanne Cable (**chapter 8**) then reviews the literature on poeciliid parasites—a rich and fascinating area—and explores the evolutionary interactions of hosts and parasites and highlights the importance of parasites in both wild and captive poeciliids. In **chapter 9** Joel Trexler and colleagues develop a new theory that integrates community assembly and parental care in order to predict the distribution of livebearing fishes. Next, Gil Rosenthal and Francisco García de León (**chapter 10**) explore the mechanistic and ecological factors that influence reproductive isolation, speciation, and hybridization. In **chapter 11** Michael Tobler and Martin Plath review the ways in which poeciliids have adapted to extreme environments such as hydrogen sulfide–rich and cave habitats. Part II concludes with Craig Stockwell and Sujan Henkanaththegedara's review of the emerging field of evolutionary conservation biology, an approach geared toward conserving species in the context of ecological and evolutionary processes (**chapter 12**).

Part III focuses on behavior and cognition and commences with a chapter by Jens Krause and colleagues (**chapter 13**), who consider the factors driving shoal composition and grouping behavior and explore recent methodological advances for quantifying such interactions. The next two chapters recognize the increasing use of poeciliid fishes as models for understanding learning and cognition. In **chapter 14** Mike Webster and Kevin Laland review the myriad ways in which poeciliid fishes learn and innovate in order to extract and exploit information from conspecifics or their environment. Next, in **chapter 15** Angelo Bisazza considers the complex cognitive abilities of poeciliids, including memory, spatial organization, numerical abilities, and lateralization of cognitive function. To conclude part III, Jennifer Kelley and Culum Brown look at predation risk and decision making in poeciliid prey, evaluating the evidence that predation risk is an important factor driving the evolution of prey cognition (**chapter 16**).

Part IV turns to sexual selection, starting with two chapters that review the two broad mechanisms of this evolutionary process: intra- and intersexual selection. In **chapter 17** Oscar Rios-Cardenas and Molly Morris explore how both mechanisms of sexual selection operate *before* mating, through mate choice and intrasexual competition, and how they interact to influence male mating success. In **chapter 18** Jonathan Evans and Andrea Pilastro focus on both mechanisms as agents of sexual selection *after* mating, in the form of sperm competition and cryptic female choice, and examine how both processes influence patterns of male reproductive success. The next chapter (**chapter 19**), by Anne Magurran, focuses on sexual coercion—a prominent feature of poeciliid mating systems—and explores its costs for females, the ecological settings in which it occurs, and the reasons for individual differences in the responses to coercion. In **chapter 20**, in their review of communication networks and sexual selection, Matt Druen and Lee Dugatkin explore how socially acquired information (eavesdropping) expands our understanding of mate choice and aggression. Part IV concludes with a chapter on genital evolution by Brian Langerhans (**chapter 21**), who evaluates several hypotheses proposed to account for the extraordinary diversity of male genital form in poeciliid fishes.

Part V combines papers on a broad array of subjects involving genetics. One of the extraordinary features of poeciliids is the high degree of phenotypic variation among populations and species. Since the publication of Meffe and Snelson's book, DNA markers have almost completely supplanted allozyme markers. In their opening chapter (**chapter 22**), Felix Breden and Anna Lindholm survey what we know in the DNA era about genetic structure of natural populations of poeciliids, looking for the signatures that sexual selection, natural selection, migration, and/or drift leave on the population phenotypic and genetic variation.

In **chapter 23** Robert Brooks and Erik Postma focus on guppies as an evolutionary model for understanding the inheritance of color pattern genes, the genetic basis for their expression, and the processes thought to underlie the extreme polymorphism in these traits. In **chapter 24** Manfred Schartl and colleagues provide an overview of sex determination in poeciliids, highlighting the fascinating diversity of sex-determining mechanisms in this group. This is followed by Mark McMullan and Cock van Oosterhout's chapter (**chapter 25**) on the major histocompatibility complex (MHC), a large multigene family involved in immunocompetence. Using data from the guppy and other poeciliid species, they investigate the role of natural selection (parasite resistance), sexual selection, and selection on linked mutations that hitchhike with the MHC alleles in the maintenance of MHC polymorphism. Finally, in **chapter 26** Manfred Schartl and Svenja Meierjohann focus on how poeciliids are used in cancer research. The workhorse of this research is still the classical Gordon-Kosswig melanoma system, but this chapter goes well beyond traditional genetics and highlights recent advances in understanding skin cancer.

The final chapter of the book, **chapter 27**, by John Endler, is an integrative commentary that critically evaluates the state of the field while highlighting fruitful directions for future research. We are delighted that John has agreed to provide such a commentary, and we can think of no one more suitable to stimulate the next generation of evolutionary and ecological biologists working on this fascinating family of fishes. If our volume serves its purpose, and as John himself hints in the final chapter, we will no doubt see a third volume on the subject in years to come.

J. P. Evans (University of
Western Australia, Perth, Australia)

A. Pilastro (University of Padua, Padua, Italy)

I. Schlupp (University of Oklahoma, Norman, USA)

Acknowledgments

OUR MAIN THANKS go to the contributors of this volume, who made enormous efforts in synthesizing diverse and extensive bodies of literature into highly concise and readable chapters. We thank them all for responding swiftly to our questions and for complying with our numerous requests. We are very grateful to Gary Meffe for writing a foreword to this book. We are also particularly grateful to the following reviewers, including several authors of the current volume, for their detailed reviews of one or more chapters: Farrah Bashey, Paul Bentzen, Angelo Bisazza, Felix Breden, Culum Brown, Molly Cummings, Tariq Ezaz, Bennett Galef, Cameron Ghalambor, Jean-Guy Godin, Hartmut Greven, Sue Healy, David Hosken, Anne Houde, Kimberly Hughes, Michael Jennions, Jerry Johnson, Jennifer Kelley, Jens Krause, Joachim Kurtz, Brian Langerhans, Robin Liley, Constantino Macias Garcia, Anne Magurran, Edie Marsh-Matthews, Mariana Mateos, Brian Mautz, Peter McGregor, Manfred Milinski, Bob Montgomery, Tim Mousseau, Bryan Neff, Francesc Piferrer, Trevor Pitcher, Martin Plath, Bart Pollux, Russell Rader, Aldemaro Romero, Gil Rosenthal, Locke Rowe, Eric Schultz, Ole Seehausen, Stephen Stearns, Colette St. Mary, Michael Tobler, Cock van Oosterhout, Jean-Nicolas Volff, Ashley Ward, Gary A. Wellborn, Christoph Winkler, Bob Wong, and two anonymous reviewers. Their efforts in ensuring the scientific validity of this volume are sincerely appreciated. We also thank the whole production team at the University of Chicago Press, but especially Christie Henry, Michael Koplow, and Pamela Bruton, for their constant encouragement and support throughout the editorial process. We reserve very special thanks to Alessandro Devigili for his painstaking work in preparing the index and references and for helping to synthesize an enormous collection of files into a single volume, and to Clelia Gasparini for her invaluable help with proofreading and indexing. Finally, we thank our partners, Jen, Francesca, and Andrea, for their constant encouragement (and tolerance!) throughout.

Part I

Reproductive biology
and life history

Chapter 1 Gonads, genitals, and reproductive biology

Hartmut Greven

1.1 Introduction

ALL POECILIIDAE, or Poeciliinae (Parenti 1981; Hrbek et al. 2007), give birth to competent fry, and their mode of reproduction is traditionally named viviparity. Only one rarely studied species, *Tomeurus gracilis*, appears to be more plastic in this respect, primarily laying eggs with embryos at an early stage of development (e.g., Rosen & Bailey 1963). The evolution, adaptive value, and benefits of viviparity and associated costs, most of which accrue to the gravid mother, have been reviewed repeatedly elsewhere (Pires et al., **chapter 3**; Marsh-Matthews, **chapter 2**) and are therefore not considered in this chapter.

Structural adaptations to viviparity are diverse, but their treatment in the recent literature is largely limited to a few common species. Other than the copulatory organ, which is frequently used for species classification, the structures involved in reproduction are poorly described in many poeciliids. However, some generalizations are possible, and in this chapter I briefly summarize this information. I focus on primary sex characteristics, that is, the gonads, and some secondary sex characteristics, such as genital ducts and external genitals. I also point out structural adaptations involved in maternal-fetal exchange, touch upon parturition and embryonic development, and point out gaps in our knowledge.

1.2 Development

1.2.1 Gonads and gonadal ducts

Determination of gonadal sex (i.e., primary sex determination) in gonochoristic poeciliids, studied in a few species only, involves the formation of either an ovary or a testis and development of intragonadal ducts from the primarily bipotential gonad (e.g., *Xiphophorus* [formerly *Platypoecilus*] *maculatus*: Wolf 1931; Schreibman et al. 1982; *Poecilia reticulata* [formerly *Lebistes reticulatus*] [guppy]: Weishaupt 1925; Goodrich et al. 1934; Dildine 1936; Anteunis 1959; *Xiphophorus hellerii*: Essenberg 1923; *Gambusia affinis*: Koya et al. 2003). Primordial germ cells (PGCs) are specified by determinants found in the germinal plasma that can be identified in very early developmental stages. During embryonic development PGCs migrate from their extragonadal position in the genital ridge on each side of the dorsal mesentery to form a pair of gonadal primordia. These primordia consist of somatic cells, which are derived from the peritoneal wall, and germ cells, which are derived from PGCs. Proliferation of germ cells gives rise to a stem-cell population of oogonia or spermatogonia.

The first sign of prenatal sex differentiation of the still paired primordia is the larger size or larger number of germ cells distributed in the prospective ovary or the larger number of somatic cells, especially in the hilar region, of the prospective testis, with germ cells located at the periphery. Differentiation into ovarian or testicular tissue can take place before birth (Goodrich et al. 1934; Dildine 1936; *G. affinis*: Nakamura et al. 1998) or after birth (e.g., Schreibman et al. 1982), although this may depend on the species.

Before fusion, somatic cells form an intragonadal duct in each primordium. After fusion, intragonadal ducts of the female unite to form a single ovarian cavity (fig. 1.1), which continues in a short extragonadal gonoduct. In sexually mature females the gonoduct ends blindly in the connective tissue between the urethra and hindgut and opens

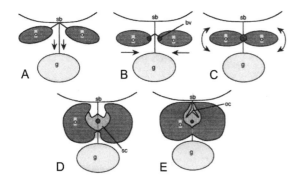

Figure 1.1 Development of the ovary in the western mosquitofish. (A) The paired gonadal primordia hang from the dorsal coelomic wall. (B) Primordia have approached just after birth. (C) Separate primordia fuse at the hilar region. (D) The lateral side of the dorsal stroma elongates upward to the coelomic wall, and the outer part of the ovary bends dorsally. (E) Elongated lateral sides of dorsal somatic cells fuse at the top of the ovary to form the ovarian cavity. bv = blood vessel; g = gut; o = gonadal primordium; oc = ovarian cavity; sb = swim bladder; sc = somatic cell cluster. From Koya et al. 2003. (Figure reproduced with permission of the Council of the Academic Societies, Japan, and Dr. Y. Koya.)

into the urogenital sinus, where the urethra also drains (see section 1.4.3).

In the male the two primordia fuse incompletely, forming a bilobed testis. Branches (efferent ducts) of the paired intragonadal ducts (testicular ducts) extend toward the periphery, where nests of spermatogonia accumulate. Posteriorly the testicular ducts merge to form a common, short, extragonadal vas deferens, which enters the urogenital sinus. The epithelium lining the ovarian cavity and that forming the spermatocysts (see section 1.3.1) are referred to as germinal epithelia, consisting of somatic cells and of oogonia or spermatogonia, respectively.

Extragonadal ducts arise from an anterior (the posterior parts of the gonadal primordium consist of somatic cells only) and a posterior primordium of peritoneal origin, which join together with the gonad later (see also Anteunis 1959). Although not studied in detail, the mature gonoduct appears to consist of two structurally different portions, an anterior aglandular and a posterior secretory portion (M. Uribe, pers. comm.).

Development of the extragonadal ducts is likely to be under the control of gonadal hormones. As shown histochemically, steroidogenesis already occurs in the gonads of newborns (*Poecilia* [formerly *Mollienesia*] *latipinna* [black molly]: Hurk 1974; *X. maculatus*: Schreibman et al. 1982).

In some species gonads contain oocytes surrounded by follicle cells before birth. During their subsequent reversal, the testes appear to be hermaphroditic. Temporary feminization (juvenile hermaphroditism) is not consistently observed among poeciliids (Goodrich et al. 1934; Dildine 1936), and the presence or absence of this phenomenon

may differ among populations, for example, in *G. affinis* (Koya et al. 2003).

Generally, sex differentiation is controlled genetically (for a review, see Devlin & Nagahama 2002; Schartl et al., **chapter 24**), but in some species it is influenced by environmental factors during early stages of gonadal differentiation, for example, pH variation in *X. hellerii* (Rubin 1985). Furthermore, higher temperatures usually result in the production of more males, while lower temperatures result in relatively more females (*Poeciliopsis lucida*: Sullivan & Schultz 1986; Schultz 1993; *Poecilia* [formerly *Limia*] *melanogaster*: Römer & Beisenherz 1996; *Cnesterodon decemmaculatus*: Johnen 2006). Thus, by actively selecting a certain temperature range, females may control the predominant sex of their offspring (Greven, unpublished data). It is likely that temperature-dependent sex determination is more widespread in poeciliids than currently documented. The stage of gonadal differentiation at which environmental sex determination occurs and the length of time the female has to be exposed to the sex-determining factors have yet to be determined.

The time to sexual maturity varies greatly among poeciliids. Maturation and size may be controlled genetically (e.g., *Xiphophorus* spp.; summarized in Kallman 2005), but these factors can also be influenced by the social environment (Snelson 1989). Males are sexually active before the gonopodium is completely developed and before the production of spermatozeugmata (unencapsulated sperm bundles) has begun (e.g., the mosquitofish *G. affinis*: Bisazza et al. 1996; the guppy *P. reticulata*: Evans et al. 2002b).

1.2.2 External genitals

External genitals are secondary sex characteristics. The development of the sexually dimorphic anal fin, which transforms into a copulatory organ (gonopodium) in the male, and of its (internal) axial and appendicular skeletal support (suspensorium) has been studied in species with bilaterally symmetrical "armed" gonopodia (see section 1.3.4 below and, e.g., Essenberg 1923; Turner 1941a/1941b, 1942; Hopper 1949; Rosa-Molinar 2005). At birth, both structures (anal fin and suspensorium) are identical in males and females. Unfortunately, studies on the development of structural adaptations of external genitals in females are not available.

In the guppy, transformation of the anal fin into a gonopodium takes place after differentiation of the testis, and developmental stages of the gonopodium correlate with the degree of maturation and steroid histochemistry of the testis (e.g., Schreibman et al. 1982). This transformation primarily affects anal-fin rays 3, 4, and 5 and occurs according to the following specific temporal sequence, as

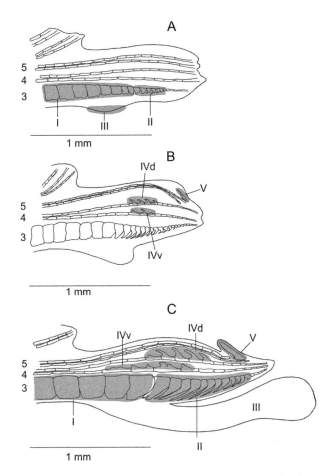

Figure 1.2 Development of the gonopodium of *Poecilia reticulata*. (A, B) The transforming anal fin and (C) mature gonopodium; 3, 4, 5 = anal-fin rays. Accessory structures (shaded) are numbered in the order in which they appear during development: I = thickening of ray 3; II = formation of spines ventrally to ray 3; III = formation of the hood; IVd = spines on the dorsal branch of ray 4; IVv = spines on the ventral branch of ray 4; V = formation of the terminal hook on ray 5. Redrawn from Hopper 1949.

depicted in fig. 1.2: (I) thickening of ray 3; (II) formation of spines ventrally to ray 3, and (III) formation of the hood at the level of segment 1–15 of ray 3; (IV) formation of three areas consisting of a series of spines on the dorsal border of the dorsal branch of ray 4 (IVd) and a series of spines along the dorsal border of the ventral branch of ray 4 (IVv); and (V) formation of a terminal hook at the distal end of the dorsal branch of ray 5. At rest the gonopodium points caudally. In bilaterally symmetrical gonopodia, ray 3 lies below rays 4 and 5 (fig. 1.2; Hopper 1949).

Development of the suspensorium involves the gradual dissolution of the anterior hemal spine(s) and elongation of the interhemal spines (gonactinosts) that articulate ventrally with the gonopodial rays through baseosts. Baseosts carry the muscles that move the fin and are anchored to hemal spines 2, 3 and 4 (gonapophyses I, II, III). The gonapophyses enlarge to form posterior outgrowths (uncini), to which

ligaments attach. A tough network of ligaments holds the gonapophyses, gonactinosts, and baseosts together firmly (see Rosen & Bailey 1963).

A region of six vertebrae (11–16) has been identified as the "genital area" in *G. affinis*. Administration of testosterone to late embryos and adult females induces the anterior displacement of this zone. This shift, as well as growth, elongation, and anterior bending of hemal spines of vertebrae 14–16 (gonapophyses) and the resorption of the 13th hemal arch, leads to the permanent anterior translocation of the anal fin and the suspensorium (Rosa-Molinar et al. 1994). This process is probably mediated by external forces exerted by the interosseal and suspensory ligaments (Rosa-Molinar et al. 1998) and is accompanied by the formation of a sexually dimorphic ano-urogenital nerve plexus of the anal-fin musculature (Rosa-Molinar 2005).

Studies on the development of the enlarged internal and superficial muscles that enable the circumduction of the gonopodium do not exist.

1.3 Male reproductive system

The reproductive system of adult males includes the testis; intragonadal ducts (efferent ducts and paired testicular ducts, or main sperm ducts); the single extragonadal duct (vas deferens) running to the genital papilla (fig. 1.3), which opens into the urogenital sinus; the urogenital aperture, a transversal, or horseshoe-shaped, slit at the origin of the gonopodium; and the suspensorium.

1.3.1 Organization of the mature testis

The relatively large testicular lobes are covered by a thin connective tissue capsule (tunica albuginea), which encloses (1) interstitial tissue, that is, fibroblasts, nerve fibers, smooth muscle cells, blood vessels, cells of the immune system, and steroid-secreting Leydig cells (reviewed by Grier et al. 1981; Grier et al. 2005), which are dispersed mainly in the space between the branches of the efferent ducts (Pandey 1969; Fraile et al. 1992); (2) testicular lobules (tubules, according to some authors) that terminate blindly at the periphery of the testis and are bordered by an incomplete layer of "myoid" boundary cells (Grier et al. 1981; Arenas et al. 1995b); and (3) the duct system. Branching efferent ducts, which meet testicular lobules, radiate from the main sperm ducts in the center of each lobe (figs. 1.3 and 1.4A and B).

Spermatogonia are restricted to the blind ends of lobules (restricted type of testis). The germinal epithelium forms spermatocysts within the lobules and does not border on a lumen ("epitheliod" instead of "epithelial" testis: Grier et al. 1981; Grier et al. 2005). The restricted testis type and

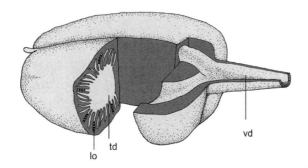

Figure 1.3 Organization of the testis in *Poecilia reticulata*. Lobules (lo) with cysts (black circles); the proximal efferent ducts open into the testicular duct (td). Testicular ducts open into the vas deferens (vd). Redrawn from Billard 1986.

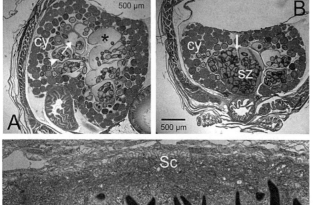

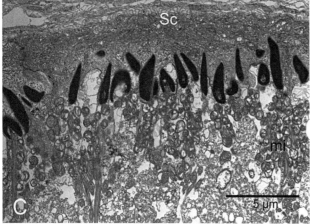

Figure 1.4 Testis and spermatids in *Poecilia* spp. Histological cross sections of (A) the anterior portion of the mature testis of *P. reticulata* with testicular duct (asterisk) and (B) posterior portion with incompletely (arrow) merged ducts; efferent ducts (arrowheads), spermatozeugmata (sz), cysts (cy) with developing sperm. Transmission electron micrograph of late spermatids (C) of *P. melanogaster* with their electron-dense heads embedded in Sertoli cells (Sc).

the epitheliod germinal epithelium are apomorphies of the Atherinomorpha (Parenti & Grier 2004).

New cysts are formed at the distal end of a lobule and contain the synchronously developing stages of spermatogenesis (spermatogonia, spermatocytes, spermatids) and spermiohistogenesis, that is, the final formation of mature spermatozoa from spermatids, during which the elongating

nuclei become tightly associated with Sertoli cells, the somatic portion of the germinal epithelium (fig. 1.4C). During maturation, cysts progress from the distal end down the tubules toward the center of each testis lobe, while the number and size of Sertoli cells increase markedly (Pandey 1969; Grier et al. 1981). Development from meiosis to mature spermatozeugmata (see section 1.3.2) requires at least 35 days (black molly: DeFelice & Rasch 1969).

By the time spermiohistogenesis is complete, cysts are positioned near the efferent-duct system and spermiation begins. During spermiation Sertoli cells transform into secretory efferent-duct cells (Pandey 1969; Grier et al. 2005). The secretions are complex glycoconjugates (Grier et al. 1981; Greven 2005), which hold immotile sperm together to form spermatozeugmata. Each cyst produces a single spermatozeugma, in which sperm heads are external and the tails form the center (Philippi 1908) (fig. 1.4A).

Thus, Sertoli cells support cysts, nourish the germ cells, phagocytose residual bodies that are cast off by maturing spermatids, form spermatozeugmata, transfer and eventually convert metabolites or hormones into germ cells, and establish the blood-testis barrier after meiosis (Marcaillou & Szöllosi 1980; Grier et al. 1981; Bergmann et al. 1984; Billard 1986). In addition, they transport spermatozeugmata, as evident by an increased positive actin immunoreaction in Sertoli cells during spermiogenesis and a strong reaction in efferent-duct cells (Arenas et al. 1995a).

1.3.2 Spermatozeugmata and spermatozoa

Free spermatozeugmata are present in the efferent ducts and are stored in the remaining duct system (fig. 1.4A and B). Storage time might be limited in sexually active males but is considerably longer during the quiescent period in species living in temperate zones (see section 1.3.3).

Strippable spermatozeugmata can vary in size both among individuals of a given species and among species. In ejaculates stripped artificially from male guppies, individual spermatozeugmata measure between 125 and 235 µm in length (Kuckuck & Greven 1997). The number of strippable spermatozeugmata also exhibits considerable variation, ranging from 20 in *Heterandria formosa* to 3000 in *X. hellerii*. Similarly, there is considerable variation in the number of spermatozoa per spermatozeugma among species, ranging from 4000–5500 in *X. hellerii* to up to 34,800 in guppies (Kuckuck & Greven 1997; Greven 2005). Although there can be considerable intraspecific variation in the number of spermatozoa per spermatozeugma (e.g., guppies: 9600–34,800; Kuckuck & Greven 1997), such variation is not systematically explained by male identity (Evans et al. 2003b).

Transfer of spermatozeugmata directly into the female's genital tract or nearby (see section 1.5.1) avoids long exposure to the extreme osmotic stress of the surrounding fresh or brackish water. Despite being internal fertilizers, male poeciliids produce considerable amounts of sperm, exceeding those produced by some oviparous fishes (Billard 1986). Although the ultrastructure of mature poeciliid sperm has been reported in only a handful of species (for review, see Jamieson 1991, 203–207), the pattern is typical for inseminating teleosts (fig. 1.4C). Spermatozoa possess a cone-shaped, elongated head, compressed in the lateral plane and similar in shape to a thick, apically rounded blade; a deep fossa; an elongate midpiece with rows of mitochondria embracing the anterior part of the 9 + 2 flagellum; and glycogen particles. The amount of glycogen is reduced in sperm stored in the male genital tract, indicating some metabolic activity, and in the ovary, perhaps due to aging and utilization by motile sperm (Billard & Jalabert 1973). The two lateral fins of the flagellum, probably present in all poeciliids, may be reminiscent of an aquasperm ancestry, because long fins are typical for spermatozoa of externally fertilizing teleosts. Poeciliid sperm lack an acrosome. In *G. affinis* an elongate acrosome-like vesicle on the apex of the nucleus (pseudoacrosome) has been found, which is devoid of acrosomal enzymes (Baccetti et al. 1989; see fig. 17.6A and B in Jamieson 1991).

Selection via sperm competition and (possibly) cryptic female choice may be responsible for some sperm adaptations (morphology, motility, and swimming velocity) and patterns of sperm expenditure and allocation (for discussion on these topics, see Evans & Pilastro, **chapter 18**).

1.3.3 Testicular cycles

In tropical and subtropical areas (and in captivity) spermatogenesis continues more or less throughout the year, but testes tend to be enlarged and spermatogenesis tends to be more rapid during warmer months (Winemiller 1993). Annual cycles have been studied in species of temperate zones, predominantly in *Gambusia* spp. In these areas, the cycle comprises a spermatogenic period, in which testicular volume and the gonadosomatic index reach their maximum and release of spermatozeugmata is continuous, and a quiescent period, in which the testis is small and, besides a few spermatids, only primary spermatogonia and spermatozeugmata are present. Spermatozeugmata are released at the beginning of the next period of spermatogenesis (Fraile et al. 1992). Mild temperatures, combined with a long photoperiod, induce spermatogenesis during the latter phase of testicular quiescence (Fraile et al. 1993; Fraile et al. 1994).

1.3.4 Diversity of gonopodia and suspensoria

Gonopodia of poeciliids vary in length, in armaments, and in symmetry (Rosen & Gordon 1953; Rosen & Tucker 1961; Rosen & Bailey 1963; Greven 2005) (see figs. 1.5A and C and 1.2C).

Gonopodial length exhibits considerable intra- and interspecific variation, which has been attributed to both sexual and natural selection (e.g., *Poecilia* spp.: Ptacek & Travis 1998; guppies: Brooks & Caithness 1995; Kelly et al. 2000; *Brachyrhaphis episcopi*: Jennions & Kelly 2002; *Gambusia* spp.: Langerhans et al. 2005; see also Langerhans, **chapter 21**; Evans & Pilastro, **chapter 18**).

Generally, gonopodia are defined as "short" (<1/3 of the male's standard length) or "long" (>1/3 of the male's standard length) (e.g., Chambers 1990), but gonopodia of some species are of intermediate length (Greven 2005). Very long gonopodia reach beyond the eye of the male when facing anteriorly (e.g., some *Girardinus* spp., *Phallichthys* spp., *Phalloptychus* spp.: Rosen & Tucker 1961; Chambers 1987).

Gonopodial length is generally correlated with the males' mating tactics (but see Greven 2005 for exceptions). Species with relatively short gonopodia (e.g., *Poecilia, Xiphophorus*) tend to use courtship displays prior to copulation and/or forced copulations. In some species, there is also evidence for intraspecific variability in gonopodium size, which is thought to be related to population differences in the prevalence of forced matings (e.g., in guppies, populations characterized by high levels of forced matings tend to have relatively long gonopodia; Kelly et al. 2000). Furthermore, species with relatively long gonopodia appear to be obligatory nondisplaying species (e.g., *Poeciliopsis* spp., *Heterandria* spp., *Girardinus* spp.). In these species males tend to control the movements of the gonopodium visually (Greven 2005; Greven, unpublished data). The anterior shift in gonopodia and their suspensoria during ontogeny is thought to be an adaptation to facilitate such forced matings in these species (e.g., *C. decemmaculatus, T. gracilis*).

The distal end of the gonopodium appears to play an important role in insemination (see section 1.5.1). The shape of the gonopodial tip exhibits considerable variation, ranging from trowel-like structures (e.g., *Xiphophorus* spp.) to pointed tips in *P. reticulata* (ray 4 forms the apex) and *H. formosa* (see fig. 1.5A–D). Various modifications of the rays are responsible for internal (serrae of ray 4p) and external stabilization (ventral to ray 3 or on the terminal portion of ray 3), and some spines may have sensory functions (ventral spines on ray 3). Holdfast devices ("armaments") such as claws found on rays 4 and 5, hooks on rays 3 and 5, and serrae on rays 4 and 5 (fig. 1.2C) are usually reduced

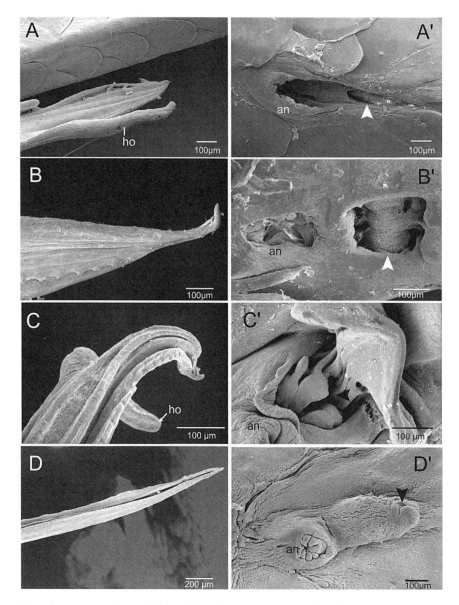

Figure 1.5 Diversity of gonopodial tips (left) and female genital apertures (right). Scanning electron micrographs of (A, A′) *Poecilia reticulata*; (B, B′), *Heterandria formosa*; (C, C′) *Xenophallus umbratilis*; (D, D′) *Poeciliopsis prolifica*. Note the asymmetrical genital aperture; arrowhead = genital aperture; an = anus; ho = hood.

in species with long gonopodia. By contrast, species with short gonopodia exhibit additional devices for positioning (hood, sensory spines), steadying, aligning, and stabilizing the gonopodium (e.g., notched pelvic fins in *Gambusia* spp. and hooked pectoral fins in *Xenodexia* spp.).

Of the various structures that are thought to stabilize the gonopodium as it swings forward (or, more precisely, circumducts; Rosa-Molinar 2005), the external stabilizers ventral to ray 3 are the most variable. These structures range from nonosseous, unpaired palps (hoods, spoons) of varying length and size (e.g., for *Poecilia* spp., see Chambers 1987) to nonosseous paired processes (Girardinini)

and "membranous" structures (*Phallotorynus* spp., Cnesterodontini), etc. (Chambers 1990). Such structures are arguably most bizarre in *T. gracilis* (Rosen & Bailey 1963).

In sexually mature guppies, the large hood extends beyond the tip of the gonopodium (Greven 2005), forming an elongate trough. Its concavity faces dorsally when the gonopodium is in the resting position. The hood is richly innervated and highly vascularized and contains a jelly-like, hyaline core, free of cells. In contrast to the distal segments of the 3–4–5 gonopodial complex, the presence of the hood is not necessary for successful insemination (Clark & Aronson 1951; Clark et al. 1954).

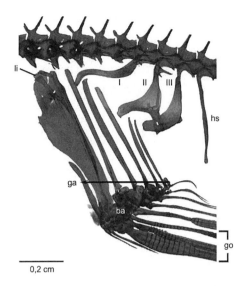

Figure 1.6 Suspensorium of *Phallichthys pittieri* (cleared and stained preparation). ba = baseosts; I–III = gonapophyses with uncini; ga = gonactinosts (partly fused); go = gonopodium; hs = hemal spine; li = ligastyl. From H. Greven and C. Eckstein, unpublished.

"Short" gonopodia are always bilaterally symmetrical, but such symmetry is also evident in some taxa with long gonopodia (e.g., *H. formosa* and some Cnesterodontini with gonopodia that are approximately 70% of the male's standard length; Chambers 1990). When circumducted these bilaterally symmetrical gonopodia form a temporary trough on either side of the body, although in some species preferences for one side may occur (see Hubbs & Hubbs 1945).

Gonopodia of varying degrees of asymmetry have evolved several times independently among poeciliids. In these species, gonopodia form a permanent groove on either the right or left side, but left asymmetrical gonopodia appear to be predominant and are found in about 40% of genera with elongated gonopodia (e.g., *Poeciliopsis, Phallichthys, Carlhubbsia, Xenodexia*; Chambers 1987).

The fully developed suspensorium (fig. 1.6) exhibits considerable variation among taxa. For example, some, but not all, genera have a ligastyl—a relict of the first hemal spine. Likewise, the number of gonapophyses varies among genera, ranging from none (e.g., *T. gracilis*) to two (e.g., *Poecilia* spp.) to four (e.g., *Gambusia* spp.). Species lacking gonapophyses have their anal fins displaced considerably farther forward in relation to the anterior hemal spines (e.g., Cnesterodontini) than those with gonapophyses. Because ray 3 provides mechanical support for rays 4 and 5, the ray 3 baseost acts as the center of gonopodial movement during circumduction. Movements of the gonopodium are effected by a series of large, specialized muscles, which have been described in detail for *X. hellerii* (Rosen & Bailey 1963).

1.4 Female reproductive system

The reproductive system of mature females includes the ovary, with the ovarian cavity extending into the bipartite extraovarian gonoduct; the genital papilla, through which the gonoduct enters the sinus urogenitalis; and the (uro)genital aperture localized between the anus and the origin of the anal fin.

1.4.1 Organization of the mature ovary

The mature ovary is suspended dorsally in the body cavity. Its lumen ends blindly anteriorly and extends posteriorly in the short posterodorsally located gonoduct. The ovarian lumen varies greatly depending on the developmental stages of the oocytes and embryos and is nearly invisible in pregnant females.

The outer surface of the ovary is covered by the thin peritoneal epithelium; the ovarian lumen is lined by the monolayered germinal epithelium, in which oogonia are discontinuously scattered. The ovarian stroma contains connective tissue, some blood vessels, and a few muscles, usually in the dorsal wall, which is free of germ cells (fig. 1.7). The number of muscles increases toward the gonoduct (Philippi 1908).

The germinal epithelium is the continuous source of new follicles. Follicles are composed of the oocyte, that is, an oogonium, which has entered initial phases of meiosis, and the surrounding monolayered follicle (granulosa) cells. They are separated from the surrounding stroma-derived theca by a basement lamina (see fig. 1.8B). Follicle and theca form the follicle complex. Development involves primary oocyte growth (previtellogenesis); secondary oocyte growth (vitellogenesis), that is, accumulation of yolk reserves, mainly of the hepatically derived plasma precursor vitellogenin; and final oocyte maturation. Vitellogenesis affects size, composition, and structure of hepatocytes (e.g., Weis 1972). Atherinomorphs have fluid yolk (fig. 1.8A);

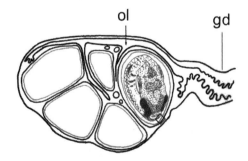

Figure 1.7 Schematic drawing of a poeciliid ovary, midsagittal section. Note the dorsal "roof" free of germ cells. Only one embryo is drawn in the follicle; the square labels the follicular placenta; gd = gonoduct; ol = ovarian lumen.

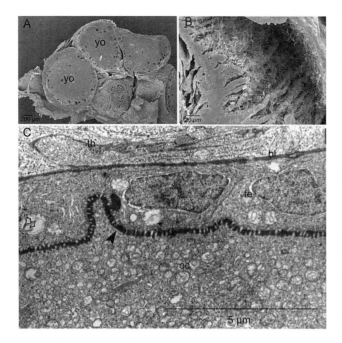

Figure 1.8 The poeciliid ovary. (A) Ovary of *Poecilia reticulata* with a small lumen in the center (asterisk) and large oocytes with confluent yolk (yo) and (B) ovarian lumen; the epithelium houses numerous spermatozoa. (C) Formation of the zona pellucida (arrowhead) in *Heterandria formosa* (transmission electron micrograph); bl = basal lamina; fe = follicle epithelium; oc = oocyte; th = theca folliculi. (Fig. 1.8C courtesy of Dr. R. Riehl.)

that is, yolk globules fuse to form a yolk mass (Parenti & Grier 2004). Immediately before and during vitellogenesis, follicle cells produce steroids (Lambert 1970b).

Oocyte degeneration (atresia) may occur at any stage of development and appears to be especially frequent in young unfertilized females (Fraser & Renton 1940; Rosenthal 1952; Vallowe 1953; Siciliano 1972). Preovulatory atretic follicles, though occasionally named corpora lutea and derived from steroid-producing follicle cells, have no endocrine function (Lambert 1970a). Causes for follicular atresia and inhibition of vitellogenesis include environmental stress, social stress, and poor nutrition (e.g., Dahlgren 1979).

During the primary growth phase, an acellular, smooth zona pellucida (called chorion, zona radiata, egg envelope, or fertilization or vitelline membrane by previous authors) develops around the oocyte, separating it from the follicle epithelium. Its thickness is considerably less than that in oviparous teleosts. This reduction is likely to be due to the need for gaseous exchange and transfer of various molecules (see section 1.6.1) (e.g., Riehl & Greven 1993; Grove & Wourms 1994). The zona pellucida of the embryoparous *T. gracilis* exhibits adhesive filaments, which is typical of eggs of many oviparous taxa (Rosen & Bailey 1963). Zona pellucida material is deposited between the microvilli of the oocyte, creating radial pore canals; microvilli of the

surrounding follicle cells may also penetrate these canals (fig. 1.8C). The intimate contact between oocyte and follicle epithelium facilitates exchange of substances and may stabilize the oocyte within the follicle. In oocytes ready for fertilization, the zona pellucida lacks pore canals (Gravemeier & Greven 2006). A micropyle (a funnel-shaped canal through the zona pellucida), which permits the passage of just a single sperm into the mature oocyte at fertilization, has not been found in any poeciliid.

The diameter of mature oocytes varies considerably (from ca. 0.40 to 3.7 mm) depending on the species and the extent of matrotrophy (Scrimshaw 1946; see also Pires et al., **chapter 3**; Marsh-Matthews, **chapter 2**).

1.4.2 Ovarian cycles

Ovarian activity is clearly related to embryonic development (Pires et al., **chapter 3**). Most species have a "group synchronous ovarian type" (terminology from Wallace & Selman 1981), meaning that in the ovary at least two populations of oocytes can be found: larger oocytes, the "clutch," that become yolked and mature at the time of birth or after parturition of the current brood (Turner 1937) and smaller, more heterogeneous arrested oocytes, which provide the next clutch. Thus, fertilization occurs shortly after parturition or after a lag of several days, depending on the initial size of the oocytes to be fertilized and the duration of the yolk-loading period (e.g., *G. affinis*, *Poecilia* spp., *Xiphophorus* spp.: Tavolga & Rugh 1947; Vallowe 1953; Siciliano 1972). Apparently, ova are not fertilized simultaneously, resulting in various developmental stages among early embryos of a given female (e.g., Hopper 1943; Tavolga 1949). Variations in the yolk-loading period may partly explain the large intra- and interspecific interbrood intervals, which can range from 23 to 75 days (e.g., Snelson et al. 1986), but brood intervals also depend on temperature and photoperiod (Turner 1937; Rosenthal 1952).

Fertilization of the oldest group of oocytes before giving birth to the brood results in superfetation, that is, simultaneously carrying two or more broods at different stages of development (e.g., Turner 1940b, 1940c; Thibault & Schultz 1978; Monaco et al. 1983; Reznick et al. 2002b). Superfetation may be obligatory or facultative. In superfetating species an "asynchronous ovarian type" may occur, in which oocytes of all stages are present in the ovary without a dominant population (e.g., *Poecilia mexicana* and its related gynogenetic forms; Monaco et al. 1978). Interbrood intervals in superfetating species range from 1 to 13 days (Turner 1937).

In *G. affinis* reared under constant ambient light period and temperature (16 h L: 8 h D at 25°C), vitellogenesis of the next clutch was shown to take place before parturition of the current brood (Koya et al. 2000). After parturition, active vitellogenesis of larger oocytes was observed on days 0–3, accompanied by a high hepatosomatic index. Between days 2 and 8 nonsimultaneous maturation and fertilization of ova occurred, and the presence of atretic follicles was noted from days 5 to 10. Embryogenesis in the fertilized ova and slow vitellogenesis of some oocytes for the next clutch began after day 10 (Koya et al. 2000).

Cycles of ovarian activity include estrogen production, and maximal steroid genesis is associated with vitellogenesis (Lambert 1970b). This cycle is correlated with cycles of sexual receptivity and mate attraction in species with male courtship. The period of maximum receptivity, which depends on the presence of mature, fertilizable oocytes, varies among species. These ovarian cycles are also present in virgin females. After being exposed to courting males (without subsequent copulation) the cycles of responsiveness in virgin guppies corresponds with the 20- to 21-day cycle of nonvirgins (Liley & Wishlow 1974). However, in contrast to nonvirgins, virgin females remain responsive to male courtship at all phases of their ovarian cycle (e.g., Liley 1968; Liley & Wishlow 1974). This may account for the long intervals between insemination and the first brood, which are often longer and more variable than those between successive broods (e.g., Rosenthal 1952; Vallowe 1953; Siciliano 1972). Morphological evidence for such cycles in virgin poeciliid females is still absent.

In species inhabiting tropical areas, females may exhibit short but continuous unseasonal reproductive cycles (e.g., *B. episcopi*: Turner 1938). However, in tropical regions characterized by wet and dry seasons, females can exhibit seasonal variation in their ovarian cycles (e.g., Winemiller 1993). In Florida, *Belonesox belizanus* can reproduce year-round, but only during mild winters (Turner & Snelson 1984).

In species inhabiting temperate zones, for example, *Gambusia* spp., annual cycles show five distinct periods: (1) recovery period (fish begin copulatory behavior; ovaries contain oocytes at early stages); (2) prereproductive period (vitellogenesis for the first gestation of the season begins); (3) reproductive period (repeated vitellogenesis, gestations and parturitions at about one-month intervals over four to five successive broods); (4) postreproductive period (last gestation of the season; recruitment of vitellogenic oocytes ceases); (5) degeneration period (after the final parturition, ovaries contain nonvitellogenic oocytes; few spermatozoa in the ovarian cavity) (Koya et al. 1998).

Ovarian recrudescence is initiated by the rise in temperature during spring, while the shorter day length causes ovarian regression during late summer. The critical photoperiod was estimated at about 12.5 h. In nature, vitellogenesis commences when temperatures rise to about 14°C, and

final maturation of oocytes occurs when the temperature reaches about 18°C during spring. In summary, gonadal recrudescence is stimulated by long photoperiods, particularly when combined with warm temperatures (Koya & Kamiya 2000).

The mechanism involved in photoperiod responses appears to be based on a circadian rhythm of sensitivity to light (Nishi 1981).

1.4.3 Gonoduct, genital papilla, and urogenital aperture

At sexual maturity the gonoduct joins the sinus urogenitalis. This may happen spontaneously (Clark et al. 1954), probably under hormonal control, or forcibly during the first copulation. Spontaneous opening must occur in species in which the female's genital papilla is not close to the urogenital aperture and in species in which the outermost tip of the male's gonopodium may only touch the urogenital opening of the female. By contrast, forcible opening may occur in species in which the male's gonopodial tip actually reaches the female's genital papilla (fig. 1.9; see section 1.5.1) and where the gonoduct has still not been opened spontaneously.

In *Gambusia* spp., the relative positions and sizes of the genital papilla, the sinus urogenitalis, and the urogenital aperture differ among species. For example, in *Gambusia gaigei* the genital papilla and urogenital sinus are poorly developed or absent, and the gonoduct and urinary tract terminate more directly on the external surface. In *Gambusia vittata* the urogenital opening is covered by a flap of transparent tissue. Interestingly, in the genus *Gambusia* the shape of the female's genitalia covaries with the shape of the male's gonopodium (Peden 1972a, 1972b). Such comparative studies are missing in other poeciliid genera (see Langerhans, **chapter 21**).

The size and degree of protrusion of the female's genital papilla may also depend on the reproductive cycle. For example, in *Poecilia mexicana* (formerly *Mollienesia sphenops*), the genital papillae are larger and more protruding

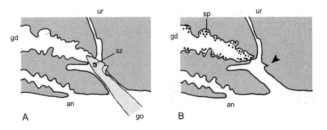

Figure 1.9 Copulation of *Xiphophorus hellerii*. Insertion of the gonopodium into the female genital aperture (A) and wounding (B) of the distal part of the gonoduct (arrowhead); an = anus; gd = gonoduct; go = gonopodium in the sinus urogenitalis; sp = spermatozoa; sz = spermatozeugmata; ur = urethra. Redrawn from Peters and Mäder 1964.

in sexually receptive females than in nonreceptive ones (Parzefall 1973).

The appearance of the urogenital aperture also depends on the age of the female. In fully developed females the appearance varies among individuals and species. The apertures are surrounded by folds or show internal folds and lips that together form the genital pad, or they are situated on the tips of long papillae (fig. 1.5). In *P. mexicana* this pad exhibits many mucous cells and is relatively larger in the cave-dwelling form (Parzefall 1970). Mucous cells may secrete a taste signal that enables males to recognize heterospecific females in darkness (Zeiske 1968). It has been suggested that foliate genital pads trap spermatozeugmata in "external" copulators, but similar pads have also been found in "internal" copulators, which actually insert gonopodia in the female urogenital aperture (see section 1.5.1; Greven 2005).

Constantz (1989) hypothesized that structured genital pads may deflect gonopodia (thereby improving the female's control of paternity) rather than being an adaptation for effective sperm transfer, but this has not yet been substantiated.

Many female poeciliids have a heavily pigmented region of varying size in the anal region. This spot is present either permanently (e.g., the extremely large anal spot in *P. melanogaster*) or are more or less cycling (*P. latipinna*, *G. affinis*). Anal spots appear to be created by melanocytes of the skin, whereas gravid or pregnancy spots are accumulations of melanocytes in the posterior part of the peritoneum (guppy: Greven, unpublished data). The way that males use these cues have yet to be explored, although they appear to have varying functions among different species. For example, in *P. melanogaster* the development of the spot coincides with the time of female sexual attractivity (Farr 1984). In *G. affinis* the spot indicates receptivity and may also serve as a cue for gonopodial orientation (Peden 1973). By contrast, in *P. latipinna* the spot indicates nonreceptivity (Sumner et al. 1994), while in *X. hellerii* the spot is not correlated with receptivity, although in this species males direct more courtship toward females with large spots (Benson 2007).

1.5 Male-female interactions

Males recognize a receptive female by visual and chemical cues. There is evidence for the release of sexual pheromones (probably steroids or steroid conjugates, for example, steroid glucuronides, synthesized in or controlled by the ovary) via the urogenital aperture. Conspecific males perceive the pheromone by taste or olfactorily, either with or without direct oral or nasal contact with the female genital region. Such pheromones may be species specific (Zeiske 1968; Parzefall 1973).

Males are stimulated by the presence of mature virgins and of females immediately after parturition (Parzefall 1973; Crow & Liley 1979) and by water in which postpartum or virgin females have been held previously (Crow & Liley 1979; Brett & Grosse 1982). Ovarian extracts, estradiol, specific glucuronides (Amourique 1965; Colombo et al. 1982; Johansen 1985), and water in which conspecific females (at unknown stages of the gestation cycle) have been held are effective in stimulating male sexual activity in some species (Gandolfi 1969; Thiessen & Sturdivant 1977; Johansen 1985). Release of such cues from the ovary probably requires an open oviduct, as might be present in mature virgins (see section 1.4.3) and shortly before and after parturition. In such cases, opening of the oviduct coincides with the period of female receptivity and the period of pheromone secretion (Crow & Liley 1979).

Other potential mating cues such as the urinary prostaglandin ($PGF_{2\alpha}$) and its metabolite, which are widely used among fish for communication (Stacey & Sorensen 2002), may also function during male-female interactions. Production of the prostaglandins PGE and $PGF_{2\alpha}$, which is relatively high in follicles of mid- and late-gestation females and low in postpartum females, is thought to be related to the induction of ovulation and parturition (guppy: Venkatesh et al. 1992).

1.5.1 Transfer of sperm

During insemination the genital papilla receives spermatozeugmata, which are placed either directly into the gonoduct, into the urogenital sinus ("internal copulators"), or perhaps close to the female's genital aperture ("external copulators"). In each case physical contact between the gonopodial tip and the female's genital region is assumed (Rosen & Gordon 1953) (for functional and evolutionary considerations, see Langerhans, **chapter 21**; Evans & Pilastro, **chapter 18**).

For internal copulators, piercing and subsequent healing of the gonoductal orifice, closure during pregnancy, opening during parturition, and subsequent closure during pregnancy have been reported (e.g., guppy: Weishaupt 1925; *X. hellerii*: Peters & Mäder 1964). In *X. hellerii* the papilla closes within nine days of insemination. However, in guppies mature virgins and pregnant females may have "open" gonoducts, with the gonoductal epithelium tightly apposed (Greven 2005; Greven, unpublished data). Thus, successful inseminations by sneak copulations are not entirely prevented during pregnancy. Open genital papillae are also found in sexually receptive (virgin or recently postpartum) females of *P. mexicana* (Parzefall 1973).

Among internal copulators males typically have "short" gonopodia, and females are often stationary during gonopodial insertion. In these species males also typically have bilateral accessory structures, that is, modified pelvic fins and, in some cases, modified pectoral fins, tactile organs, and holdfasts. In species with holdfasts, the gonopodium may be caught in the female during copulation (e.g., *Xiphophorus* spp.), sometimes injuring the female (fig. 1.9). In such species, copulations tend to be protracted, lasting several seconds in some cases (Clark & Aronson 1951; Clark et al. 1954) and often terminating in a sharp, snaplike break. In some species copulation is followed by a period of postcopulatory jerking (i.e., contractions and hyperactivity after copulation), perhaps a signal to potential rivals (Liley 1966; Greven 2005). By contrast, external depositors typically exhibit "long" gonopodia and deposit spermatozeugmata when the female is in motion during a brief genital contact (Clark & Aronson 1951; Rosen & Tucker 1961).

Gonopodial function has been described in detail in species exhibiting relatively "short," bilaterally symmetrical gonopodia, such as guppies and *Xiphophorus* spp. (Rosen & Gordon 1953; Clark et al. 1954) and *Gambusia* spp. (Peden 1972b) (for details see Langerhans, **chapter 21**).

Circumduction of the gonopodium (e.g., in *X. hellerii*) causes rotation of anal-fin rays 3, 4, and 5, forming a partially enclosed and transitory groove through which spermatozeugmata pass to the gonopodial tip, which is variously supported and stabilized (see section 1.3.4). The mechanism by which the sperm bundles are ejected and reach the female gonopore is uncertain, although Collier (1936) has described an "ejaculatory sphere" connected to ray 5 by ligaments.

Modifications of ray 3 are too large in some species to be inserted in the female genital pore. In *Girardinus metallicus*, for example, the flexible processes on ray 3 may rest against the soft tissue of the females' genital region (folds and protuberances). Other structures, such as the large innervated hood (e.g., guppies), which, rather than being a loosely dangling, "soft and flabby structure" (Clark & Aronson 1951, 61), may serve an adaptive function by sealing the genital aperture to prevent loss of spermatozeugmata, limit the penetration depth (Greven 2005) and protect the tip of the gonopodium during circumduction (Greven, unpublished data).

In swordtails and platys (*Xiphophorus* spp.), the gonopodium is inserted deeply, and mating pairs appear hooked together during copulation (fig. 1.9). Removal of one holdfast reduces insemination efficiency, and removal of both prevents insemination completely (Rosen & Gordon 1953). The most distal part of ray 4 penetrates the genital papilla, whereas the distal claw and hook of the gonopodium are fixed at the posterior wall of the urogenital sinus, causing some lesions. Constantz (1989, 44) hypothesized that "traumatization" of the female in this way may result in an "enforced post-copulatory chastity" in some single-brooded (i.e., nonsuperfetating) poeciliids, thus serving to minimize sperm competition. In superfetating species, where females probably exhibit longer and repeated periods of sexual receptivity, longer and relatively plain gonopodia place spermatozeugmata on to the exterior of the female's genitalia. However, these hypotheses have not been verified and traumatization of the female is not the rule in all single-brooded species.

1.5.2 Sperm storage

Gonoductal fluid breaks up transferred spermatozeugmata in the sinus and in the female's gonoduct (Philippi 1908). Free spermatozoa are motile for a remarkably long time in isotonic physiological solutions (1–2 h; Billard 1986). Spermatozoa migrate upward and are stored for relatively long periods (several months) in epithelial folds of the gonoduct (*C. decemmaculatus*: Philippi 1908), ampulla-like extensions in the ovarian cavity (guppy: Jalabert & Billard 1969), and the caudal part (and, to a lesser extent, the blind dorsocranial end) of the ovary (*X. hellerii*: Paris et al. 1998). Spermatozoa are stored for comparatively shorter periods (perhaps several days or even weeks) in folds of the ovarian epithelium (called *Dellen*, small pockets, or micropockets; see, e.g., Philippi 1908; Fraser & Renton 1940; Kobayashi & Iwamatsu 2002) directly above developing and mature oocytes (fig. 1.10; see also fig. 1.8B).

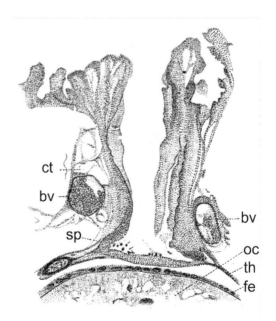

Figure 1.10 Micropocket over an oocyte. ct = connective tissue; oc = oocyte; th = theca folliculi; bv = blood vessel; sp = sperm; fe = follicle epithelium. From Philippi 1908.

Excess and defective sperm are probably endocytized by the epithelium of the dorsal ovarian wall a short time after insemination (Philippi 1908), but endocytosis of low-quality and old spermatozoa should also occur in the ovarian sperm storage sites. However, this has yet to be demonstrated unequivocally, for example, by acid phosphatase cytochemistry.

Sperm storage inevitably reduces the number of and finally abolishes free spermatozoa in the female genital tract (Clark & Aronson 1951). However, sperm remain available to fertilize oocytes and appear to be released more or less continuously from the host cells of the long-term storage site (Jalabert & Billard 1969; Paris et al. 1998). Stored sperm are motionless, suggesting that specific mechanisms must have evolved to inhibit their activity, although it seems likely that stored sperm are nevertheless metabolically active to some extent. Their intimate contact with epithelial cells (Potter & Kramer 2000) and the presence of glycogen in the ovarian epithelium and ovarian sugars (Gardiner 1978) may serve to nourish sperm, which can remain viable and fertile for more than one year. A single insemination is therefore sufficient for producing several successive broods, although brood sizes decline over time (e.g., Clark & Aronson 1951).

Sperm of poeciliids are antigenic. Long-term storage in the ovary is probably due to the ovary's ability to secrete estradiol, which might impede allograft rejection (guppy: Hogarth & Sursham 1972). Apico-lateral zonulae occludentes (tight junctions) between cells of the ovarian epithelium may establish the immunological barrier (Potter & Kramer 2000).

There is no direct evidence that the female actively ejects sperm once they are inside the genital tract. The elimination of excessive, or less viable, spermatozoa must therefore take place by endocytosis, thus suggesting that endocytizing cells must "recognize" such sperm. The most likely sites of sperm elimination are the dorsal wall of the ovary and the long-term storage site (see above) and probably the micropockets. Many spermatozoa occupy these sites along with young and older oocytes after a single insemination (Greven 2005; Greven, unpublished data). If females were able to discriminate sperm from different males at any of these sites, this would influence the outcome of sperm competition. Clearly, this is an aspect of the fertilization process that would be worth exploring in future studies.

1.5.3 Fertilization

Micropockets are the sites of sperm entry into the oocyte; that is, fertilization (karyogamy) occurs within the follicle (Philippi 1908; Fraser & Renton 1940; Kobayashi & Iwamatsu 2002). Micropockets are already associated with oocytes at the beginning of zona pellucida formation. With oocyte growth, the terminal region of micropockets expands and houses an increasing number of spermatozoa. For fertilization, spermatozoa (without an acrosome) have to be activated and have to cross the ovarian epithelium; the underlying connective tissue that includes the theca; the follicular epithelium; and the zona pellucida, in which a micropyle for sperm entry appears to be absent (see above). Despite attempts to explain this phenomenon (Philippi 1908; Fraser & Renton 1940), this process is not yet understood. Micropockets are also the place of rupture for escaping young.

1.6 Mother-offspring interactions

1.6.1 The follicular placenta

The offspring completely develops in the follicle (intrafollicular gestation) and during development maternal and embryonic tissues are closely apposed. As a placenta is "an intimate apposition or fusion of the fetal organs to the maternal (or paternal) tissues for physiological exchange" (Mossman 1937, quoted in Wourms et al. 1988, 39), in the strict sense all viviparous poeciliids exhibit (follicular) placentae (fig. 1.11), albeit with different abilities to transfer nutritive (macro) molecules.

Follicular placentae have evolved repeatedly in poeciliids and within given clades, such as *Poeciliopsis* (e.g., Thibault & Schultz 1978; Reznick et al. 2002b; see Pires et al., **chapter 3**). The transfer of nutritive molecules to the embryo ranges from practically none (lecithotrophic species, which nourish their embryos by yolk deposited prior to fertilization and produce relatively large ova, >2 mm in diameter; embryos lose 30–40% of their initial dry mass during development; lecithotrophy is the plesiomorphic condition) to varying amounts (matrotrophic species, which nourish their embryos by matter transferred from the mother during gestation and produce generally smaller ova; offspring retain or increase their initial dry mass during gestation). Matrotrophy may be facultative, as, for example, in *P. latipinna* (Trexler 1985), or obligatory, as in superfetating species (for literature and a deeper discussion see Pires et al., **chapter 3**; Marsh-Matthews, **chapter 2**).

The maternal component of the placenta is the follicular epithelium, plus its underlying capillary network. The embryonic component is the epithelial embryonic surface, as well as the underlying capillary network. The components are separated from each other by the zona pellucida, which loses its radial canals and becomes gradually thinner during gestation but obviously remains intact until parturition (e.g., Tavolga & Rugh 1947; Jollie & Jollie 1964a; Grove & Wourms 1991). Therefore, barriers for any exchange are

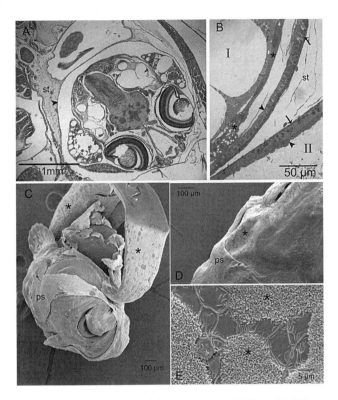

Figure 1.11 Follicular placenta. (A) Embryo in the follicle (guppy); follicle wall (arrowhead); st = stroma. (B) *Poeciliopsis prolifica*; two adjacent follicles (I, II) separated by the stroma (st); note the rich vascularization of the follicle wall (arrows), follicular epithelium (arrowheads), and embryonic surface (asterisks). (C, D) Embryo of *Heterandria formosa* with expanding pericardial sac (ps) and areas of absorptive cells on the body surface (asterisks). (E) Absorptive epithelial cells (asterisks) interspersed with nonabsorptive cells. (A, B) Histological sections; (C, D) scanning electron micrographs.

(1) the maternal capillary endothelium and its basal lamina; (2) the basal lamina of the follicular epithelium and the epithelium itself; (3) the zona pellucida; (4) the embryonic surface epithelium and its basal lamina; and (5) the basal lamina of the embryonic capillary endothelium and the endothelia cells.

Basic functions, such as gas exchange, maintenance of the osmotic environment of the embryo, protection of embryos from immunological rejection, and probably the transfer of small molecules (amino acids) in early phases of development (e.g., in the lecithotrophic guppy: Dépêche 1976; and *Gambusia* sp.: Chambolle 1973) are managed by rather simple squamous epithelia. The follicle epithelium is provided with junctional complexes, including tight junctions (Riehl & Greven 2008), which become more distinct after fertilization (guppy: Jollie & Jollie 1964b; *H. formosa*: Grove & Wourms 1991). After fertilization the follicle complex becomes highly vascularized (e.g., Bailey 1933; Fraser & Renton 1940; Turner 1940b).

The transfer of larger amounts of nutritive molecules requires further modifications. In the highly matrotrophic *H. formosa* follicle cells exhibit apical microvilli and folded basal surfaces. Furthermore, cells exhibit clathrin-coated pits and vesicles apically and along their basal surface. These pits and vesicles are indicative of receptor-mediated endocytosis of specific proteins. These structures suggest controlled uptake of (macro) molecules from the maternal blood and transport of the eventually degraded molecules to the follicular fluid (Grove & Wourms 1994). Turner (1940b) has described vascularized, villi-like extensions covered with secretory cells in the follicular wall of superfetating *Poeciliopsis* spp. The secretory nature of the epithelium has yet to be verified, but follicle-cell structure resembles that in *H. formosa* (fig. 1.11B; Greven, unpublished data).

The most distinctive portions of the embryonic component of the placenta are the pericardial sac and yolk sac, which are both highly vascularized by an extended portal capillary network. In both matrotrophic and lecithotrophic species, it is the pericardial sac and not the yolk sac (Turner 1940b) that expands laterally, envelopes the entire anterior portion of the embryo during maximal development, and forms a short-lived neck strap during regression (Tavolga & Rugh 1947; Tavolga 1949; Kunz 1971). The yolk sac and pericardial sac are pressed against the follicular wall by the greatly enlarged embryonic urinary bladder, perhaps facilitating gas and metabolite exchange (Fraser & Renton 1940; Kunz 1963). Generally, the epithelial covering of the pericardial and yolk sacs is relatively flat and contains osmoregulatory chloride cells, i.e., ionocytes (Dépêche 1973; Grove & Wourms 1991). In the obligatorily matrotrophic *H. formosa* and some *Poeciliopsis* spp. the yolk sac is reduced, and the pericardial sac provides the largest portion of the vascularized embryonic surface. In *H. formosa* the pericardial sac invests the embryonic head almost to the end of gestation (Fraser & Renton 1940; Turner 1940b; Scrimshaw 1944b). In this species (Grove & Wourms 1991) and *Poeciliopsis prolifica* (Greven, unpublished data; see fig. 1.11B) cells specialized for nutrient absorption (probably maternal serum proteins, yolk proteins, and other substances) are structurally similar to the absorptive cells of the follicle epithelium (see above). In *H. formosa* they are distributed over the entire surface in early embryos and are restricted later to the pericardial and yolk sacs (fig. 1.11C–E).

The embryo in the ovarian follicle has to avoid immunological rejection, as antigens foreign to the mother are present in embryos (*X. hellerii*: Hogarth 1968, 1972a, 1972b). When *X. hellerii* embryos with an intact zona pellucida are transplanted in the peritoneal cavity, individuals develop normally, while those without this envelope do not. This finding suggests that the maternally derived zona may act as a barrier against immune rejection (Hogarth 1973), but the tight junctions between cells of the follicle epithelium (see above) are the most likely candidates for an effective immunological barrier.

1.6.2 Gestation

The gestation period begins at fertilization and ends at parturition. Gestation times vary from 20 to 30 days or longer depending on the species, feeding, water temperature, photoperiod (*G. affinis*: Koya et al. 1998), and, at least in the first reproductive cycle, number of mating partners (guppy: Evans & Magurran 2000). Extrinsic ecological factors such as predation risk have also been shown to influence the duration of brood retention in guppies (Dzikowski et al. 2004; Evans et al. 2007).

The increasing volume of the female's abdominal region during gestation reduces the female's sustained swimming ability and, as a possible consequence of this decrease, the ability to obtain food (*G. affinis*: Plaut 2002). Furthermore, the velocity during so-called fast-start swimming declines (*P. reticulata*: Ghalambor et al. 2004), and consequently the risk of predation increases progressively during gestation (*G. affinis*: Britton & Moser 1982; *P. latipinna*: Trexler et al. 1994). Physical limitations of the stomach, which is narrowed by the growing ovary, may constrain the maximum volume of food that can be ingested and, therefore, may result in "indirect" cost (Weeks 1996a).

1.6.3 Embryonic development

Complete tables of early and embryonic development are necessary for accurate comparison of developmental stages in different poeciliids. Such tables have been presented for some species, but these vary in the level of detail (summarized in Haynes 1995). The description of embryonic stages, including organogenesis, of *X. maculatus* and *X. hellerii* (Tavolga & Rugh 1947; Tavolga 1949) is unsurpassed. The amount of time between successive stages varies, and platyfish embryos are almost completely formed four to five days after fertilization, followed by further growth of the existing organs (see also Hopper 1943). In addition, some variation in the developmental stages among the embryos of a given gravid platyfish was noted, probably due to the fact that eggs of a given brood are fertilized over a period of several days (see section 1.4.2). However, initial differences become blurred later, as development proceeds more rapidly in early ontogeny (Tavolga 1949).

Haynes (1995) suggested a simplified, categorical classification applicable to poeciliids exhibiting a variety of developmental modes. His classification includes eleven distinct stages, ranging from the immature oocyte to the mature embryo ready to birth. These stages are characterized largely by external features.

1.6.4 Parturition

Parturition coincides with ovulation in poeciliids (e.g., Turner 1940c). Young from a single brood are born in approximately the same state of development, usually within a few minutes to hours. In superfetating species, only the oldest embryos are delivered. Nothing is known about the factors that signal the female's readiness for birth. Embryos obviously do not actively rupture the follicle. Therefore, birth has to be accomplished by the musculature of the ovary and especially of the gonoduct. During birth the zona pellucida either breaks at the same time as the follicle (Tavolga & Rugh 1947) or within the gonoduct (Philippi 1908). It is not known whether the rupture is achieved mechanically or chemically, for example, by hatching glands. Preterm deliveries due to environmental stress (*G. affinis*: Ishii 1963) or other unknown factors are enclosed by an intact zona pellucida. Females may also interrupt parturition either when disturbed or when affected by other, largely unknown factors (Philippi 1908). Young are born head or tail first (e.g., Philippi 1908; Fraser & Renton 1940). A short time before parturition, the large embryonic urinary bladder is evacuated, which results in the retraction of the residual yolk sac. Therefore, young are normally born without an external yolk sac (e.g., guppy: Kunz 1963).

1.7 Future directions

This chapter identifies numerous gaps in our understanding of poeciliid reproduction and morphology and the need for further comparative and morphological studies on the topics. Among these, studies are required to determine the fate of sperm in the female genital tract that are not used for fertilization, to understand the intra- and interspecific variability of outer genitals in females and males, and to determine whether traumatization of females during copulation plays a significant role in minimizing sperm competition. Many of the comparative studies published at the beginning and in the middle of the last century have not been followed up, for example, by using transmission electron microscopy (TEM), a technique used widely elsewhere. I cite just two examples to illustrate this point. Fertilization in particular is still a mystery, and since the 1940s only a single TEM study (on sperm storage micropockets in guppies) has been performed, and even this study failed to observe fertilization (Kobayashi & Iwamatsu 2002). Furthermore, the follicular placenta has been examined by TEM in just two species: the lecithotrophic guppy, with a comparatively simple placenta, and the highly matrotrophic *H. formosa*, with its elaborated placenta (see section 1.6.1). The guppy was studied at a time when fixation techniques were not perfected (Jollie & Jollie 1964a, 1964b), and therefore, it is worth revisiting this topic. This study, as well as the more modern, descriptive TEM study of the placenta of *H. formosa* (Grove & Wourms 1991, 1994), should be supplemented by experimental work, for example, to demonstrate the possibly ubiquitous uptake of amino acids by the embryonic surface and the allocation and selective uptake of various, more specific proteins in matrotrophic species. Such morphological and experimental studies should be expanded to other genera, for example, *Poeciliopsis* spp., which have complex placentae.

Acknowledgments

I thank the editors for the invitation to contribute to this book, the reviewers and editors for their valuable comments, my wife, Wilma Greven, for drawing figs. 1.2, 1.3, and 1.9, and Marcel Brenner for technical help.

Chapter 2 Matrotrophy

Edie Marsh-Matthews

2.1 Definitions of matrotrophy

MATROTROPHY IS A MODE OF offspring provisioning characterized by postfertilization transfer of nutrients from a mother to her developing embryos (Wourms 1981). Matrotrophy is widespread among livebearing organisms and has evolved independently in groups as divergent as plants (Graham & Wilcox 2000), mussels (Schwartz & Dimock 2001; Korniumshin & Glaubrecht 2003), scorpions (Toolson 1985), leaf beetles (Dobler et al. 1996), and echinoderms (Frick 1998), as well as in most classes of vertebrates (Wourms 1981). Mechanisms and structures supporting nutrient transfer vary widely across livebearing forms, as do the types and amounts of nutrients transferred (Lombardi 1996).

Maternal provisioning of offspring takes many forms and may occur at different periods of development. Among animals, yolk sequestered in eggs prior to fertilization contributes to embryonic nutrition for at least some period of development and is often the sole source (lecithotrophy). In addition to yolk, embryos of livebearing animals may obtain nutrients via matrotrophy, which may be facilitated and enhanced by elaborate placental-type structures (placentotrophy).

Livebearing poeciliids exhibit vast variation in the extent of matrotrophy and placentation (Thibault & Schultz 1978; Wourms 1981). Some species exhibit extensive postfertilization nutrient transfer, which is the source of most embryonic nutrition. These species typically have small eggs with minimal yolk stores (Scrimshaw 1944a, 1946). Embryo mass increases dramatically during gestation, such that the mass of the full-term embryo may be several thousand times that of the mature egg (e.g., *Heterandria formosa*; Scrimshaw 1944a). These species may also have elaborate anatomical modifications to facilitate nutrient transfer (see Greven, **chapter 1**). Wourms (1981) referred to these species as "specialized matrotrophes," and Blackburn (1992) has called this level of nutrient transfer "substantial matrotrophy."

In other livebearing poeciliids, matrotrophy may provide a smaller but still significant contribution to overall embryo nutrition. These species have larger eggs than do substantial matrotrophs (Scrimshaw 1946; Thibault & Schultz 1978; Wourms 1981) but have little or no obvious development of transfer structures. Wourms (1981) referred to such species as "unspecialized matrotrophes" (e.g., *Poeciliopsis occidentalis*, Constantz 1989). In these species, embryo mass remains unchanged during gestation or may increase slightly, despite metabolism of yolk during development.

Many livebearing poeciliids have been considered to be entirely lecithotrophic (Wourms 1981; Wourms et al. 1988; Constantz 1989; Reznick & Miles 1989a, 1989b) with the sole source of embryonic nutrition assumed to be yolk stores sequestered in a large egg. In these species, embryo mass decreases significantly during gestation, at a rate similar to that of an embryo developing outside the mother in oviparous fishes. Wourms (1981) called these species "lecithotrophes." Recent studies (Marsh-Matthews et al. 2001; Marsh-Matthews et al. 2005; Marsh-Matthews et al. 2010), however, have found that nutrient transfer occurs in many species once thought to be strictly lecithotrophic. Blackburn (1992) has called this condition incipient matrotrophy. Marsh-Matthews et al. (2010) suggested that incipient matrotrophy, rather than pure lecithotrophy, may be the plesiomorphic condition for livebearing poeciliids. My aim here is to review postfertilization nutrient transfer in poeciliids, with particular emphasis on incipient matrotrophy and the potential roles it may play in overall offspring provisioning.

2.2 Occurrence of matrotrophy in livebearing poeciliids

Occurrence of matrotrophy has long been recognized for those species in which embryo mass increases or remains the same during gestation (the "matrotrophes" of Wourms 1981). For species that exhibit a decline in embryo mass during development, however, conclusions about the importance of matrotrophy for embryo nutrition have varied among authors as discussed below.

Reznick and Miles (1989b) surveyed the literature and classified 51 species as either lecithotrophic or matrotrophic. Of those, 30 species were classified as lecithotrophic, including all or most species surveyed in the genera *Poecilia, Xiphophorus, Phalloceros, Cnesterodon, Brachyrhaphis, Gambusia, Belonesox, Quintana,* and *Carlhubbsia,* as well as *Priapichthys darienensis* and *Heterandria bimaculata.* Some of the species classified as lecithotrophic by Reznick and Miles (1989b) had been considered to exhibit at least some level of matrotrophy by previous authors. For example, Wourms (1981) had noted that in *Gambusia, Poecilia, Xiphophorus,* and *Belonesox,* embryo mass does not change during gestation and considered these to be unspecialized matrotrophs; Hubbs (1971) and Yan (1986) had concluded that *Gambusia heterochir* embryos must receive nutrients from the mother because embryo size did not change during development; Trexler (1985) had reported matrotrophy in *Poecilia latipinna.*

Since 1989, some species surveyed by Reznick and Miles (1989b) have been reevaluated relative to provisioning mode, and additional species have been assayed. Among those reevaluated, several previously considered to be lecithotrophic have been shown to exhibit postfertilization provisioning, including *Poecilia formosa, Poecilia latipinna, Poecilia mexicana* (Riesch et al. 2009a; Marsh-Matthews et al. 2010), *Poecilia reticulata* (DeMarais & Oldis 2005), *Gambusia affinis* (DeMarais & Oldis 2005; Marsh-Matthews et al. 2005), *Phalloceros caudimaculatus* (Arias & Reznick 2000), and *Cnesterodon decemmaculatus* (Lorier & Berois 1995). Other taxa investigated since 1989 that have been found to be matrotrophic include *Gambusia holbrooki* (Edwards et al. 2006; Marsh-Matthews et al. 2010), *Gambusia clarkhubbsi, Gambusia gaigei, Gambusia*

geiseri, Gambusia nobilis (Marsh-Matthews et al. 2010), *Gambusia puncticulata* (Abney & Rakocinski 2004), *Poeciliopsis paucimaculata, Poeciliopsis latidens, Poeciliopsis baenschi, Poeciliopsis viriosa, Poeciliopsis presidionis* (Reznick et al. 2002b), *Poeciliopsis turrubarensis* (Zúñiga-Vega et al. 2007), and *Xenodexia ctenolepis* (Reznick et al. 2007a).

Of the 51 species listed by Reznick and Miles (1989b), 21 were classified as matrotrophic, including all or most species surveyed in the genera *Priapella, Phalloptychus, Priapichthys, Neoheterandria, Heterandria, Poeciliopsis,* and *Phallichthys,* as well as *Poecilia branneri* and *Gambusia vittata.* Some of these have since been suggested to be lecithotrophic, including several species in the genus *Poeciliopsis* (*P. fasciata, P. infans, P. monacha, P. gracilis, P. catemaco*; Reznick et al. 2002b).

Finally, studies of other taxa not included in Reznick and Miles (1989b) have suggested that several species are lecithotrophic, including *Gambusia hubbsi* (Downhower et al. 2002), *Poecilia butleri, Phallichthys tico, Neoheterandria umbratilis, Priapichthys festae, Poeciliopsis hnilickai,* and *Poeciliopsis scarlii* (Reznick et al. 2002b).

The difficulty in summarizing patterns of maternal provisioning in livebearing poeciliids overall may result from several causes. Among these are use of different methods to assay matrotrophy (see below), spatial and temporal variation in matrotrophic provisioning within taxa (e.g., Trexler 1985 for *Poecilia latipinna*; Schrader & Travis 2005 for *Heterandria formosa*), and lack of a comprehensive survey of this diverse group.

2.3 Matrotrophy and superfetation

Highly matrotrophic poeciliids typically also exhibit superfetation (see also Greven, **chapter 1**; Pires et al., **chapter 3**), that is, the simultaneous development of multiple broods at distinctly different stages of development (Turner 1937; Scrimshaw 1944b). Although these traits are highly correlated (Constantz 1989), their co-occurrence is not absolute: substantial matrotrophy occurs in the absence of superfetation (Arias & Reznick 2000) and superfetation is found in species with low levels of postfertilization nutrient transfer (Thibault & Schultz 1978; Reznick et al. 1996a). The level of superfetation may also vary among populations (e.g., Zúñiga-Vega et al. 2007) or temporally (Thibault & Schultz 1978). Reznick et al. (2007b) suggested that despite the correlation of these life-history traits in livebearing fishes, matrotrophy and superfetation each evolved independently multiple times. The remainder of this chapter will focus on matrotrophy, with only occasional reference to superfetation.

2.4 Methods of matrotrophy assay

2.4.1 Indirect methods

Methods typically used to evaluate matrotrophy are indirect assays of postfertilization provisioning that assess change in embryo mass over the course of gestation rather than directly measure nutrient transfer (box 2.1). Scrimshaw (1945) assessed the extent of matrotrophy in numerous poeciliids by examining the slope of a "dry-weight curve" describing the change in dry embryo mass over the course of gestation. To calculate the slope, Scrimshaw standardized gestation into 10 "relative time units" and plotted relative embryo mass (with mass at fertilization standardized to 1) over time. Use of standardized endpoints and relative embryo mass allowed Scrimshaw to compare species with different gestation times and different sizes of fertilized eggs. To evaluate the extent of matrotrophy, Scrimshaw then compared the calculated slope to a slope expected for oviparous (and hence strictly lecithotrophic) species. Based on published studies for several oviparous species, Scrimshaw (1945) assumed that metabolism of yolk in the absence of postfertilization provisioning will decrease embryo mass by approximately 1/3 over the course of gestation, and he therefore considered any slope for a dry-weight curve exceeding −0.33 (actually −0.033 for his model with 10 relative time units) to indicate matrotrophic provisioning of the developing offspring.

A related method used to assay the occurrence and extent of matrotrophy is the matrotrophy index (MI): the ratio of neonate mass to mass of the egg at fertilization (Wourms et al. 1988; Trexler 1997; Reznick et al. 2002b). In his study of matrotrophy in *Poecilia latipinna*, Trexler (1997) substituted the mass of very early embryos (blastodisc stage) for the size of fertilized eggs to ensure that the initial embryo size reflected the size of a fully yolked egg. In studies using preserved specimens (for which data on developmental endpoints were not available), other authors (e.g., Reznick et al. 2002b) have calculated MI based on construction of a line (or curve; e.g., Pires et al. 2007) describing change in embryo mass over the range of developmental stages represented by the available sample, which then was used to extrapolate sizes of fertilized eggs and neonates. The MI was then calculated as the ratio of the inferred size of the neonate to the inferred size of the fertilized egg.

These indirect methods have been used widely to assess matrotrophy in livebearing poeciliids. For species that exhibit a nonnegative slope for the embryo mass line and for species with a calculated MI > 1, authors have typically concluded that postfertilization provisioning occurs. Conclusions for species that exhibit a negative slope,

or MI < 1.0, however, have been less consistent, and the "cutoff" between concluding that a species exhibits pure lecithotrophy versus some degree of matrotrophy has been ambiguous. For example, Wourms et al. (1988) suggested that embryos of purely lecithotrophic species should lose on average 35% of the initial dry mass during development but cautioned that the loss might vary from 25% to 55%. Reznick et al. (2002b) used MI < 0.7 to indicate lecithotrophy in *Poeciliopsis* spp., but Reznick et al. (2007b) noted that MI < 0.7 was arbitrary and did not preclude some small amount of postfertilization provisioning, and they in fact used MI < 1.0 to indicate lecithotrophy in that later study.

In many cases (e.g., when the only specimens available are preserved), indirect methods such as those described above may be the only tool for assessing matrotrophy. Interpretations should be made with caution (Wourms et al. 1988), however, because recent studies using direct methods (Marsh-Matthews et al. 2001; Marsh-Matthews et al. 2005; Marsh-Matthews et al. 2010) have demonstrated that several species once thought to be strictly lecithotrophic actually exhibit postfertilization transfer of nutrients from mother to embryo (see box 2.1).

2.4.2 Direct methods

Nutrient transfer to developing embryos can be assayed directly using tracers injected (or otherwise inserted) into pregnant females and subsequently detected in embryos. Direct assays of nutrient transfer to developing embryos have been used in taxa as diverse as scorpions (Toolson 1985), skinks (Swain & Jones 1997), and many groups of viviparous fishes (Wourms et al. 1988; MacFarlane & Bowers 1995; Nakamura et al. 2004). Tracer studies in viviparous fishes have typically employed injection of radiolabeled compounds (e.g., Grove & Wourms 1982, 1983; MacFarlane & Bowers 1995), tagged compounds that can be visualized using electron microscopy (e.g., iron dextran; Wourms et al. 1988), or introduction of exogenous proteins that can be detected in the embryo immunologically (e.g., Nakamura et al. 2004). More recently, biologically inert probes (e.g., Fluorospheres™) have been used to investigate potential pathways of nutrient transfer (DeMarais & Oldis 2005).

The use of direct assays to detect and quantify matrotrophy in poeciliids has been limited. Grove and Wourms (1982, 1983) and Wourms et al. (1988) used radiolabeled and tagged compounds to investigate sites of nutrient transfer in the substantial matrotroph *Heterandria formosa*. Wegmann and Götting (1971) used an iron-dextran tracer to examine nutrient transport in *Xiphophorus hellerii*, a species with "moderate" (Wourms et al. 1988, 115) lev-

Box 2.1 Indirect methods of matrotrophy assay: assumptions and a cautionary note

Indirect methods for matrotrophy assay rely on comparisons of embryo mass at different stages of development. For species that do not exhibit superfetation, mass of embryos from broods at different stages of development (i.e., from different females) must be compared. For many poeciliid species, egg or embryo mass is known to vary among females due to factors in addition to gestational stage (e.g., Reznick et al. 1996a). For species that do exhibit superfetation, relative sizes of embryos in different, simultaneously developing broods could potentially be compared (e.g., Lima 2005; Schrader & Travis 2009), and this within-female comparison would seem to provide a more accurate overview of change in embryo size during development. Even this comparison, however, assumes that initial egg size is constant over time, which may not be the case (Schrader & Travis 2005).

Indirect methods typically use preserved specimens, often those that have been stored in alcohol for long periods of time. Storage in alcohol has the potential to extract lipids from both maternal and embryonic tissues. Use of alcohol-preserved specimens assumes that any lipid extraction that takes place will be comparable for all developmental stages of embryos, and this may need to be evaluated.

Another potential problem with the use of indirect measures of matrotrophy relates to construction of lines (or curves) describing change in embryo mass as a function of developmental stage (box-fig. 2.1). Such lines are used to compare slopes for different species (or populations or at different times) and/or extrapolate embryo mass at developmental endpoints (fertilization and parturition) to calculate a matrotrophy index. This approach assumes that the timing of gestation is known or

that the stages compared represent intervals of similar duration throughout development and/or that stages of development display the same timing for the species being compared. If gestational rates are not the same, use of the slope can be misleading. In box-figure 2.1a, the MI based on the ratio of neonate size to egg size is the same for both species, but the shorter gestation time for species 1 results in a steeper slope that would be interpreted as a lower level of postfertilization provisioning. Equal slopes (box-fig. 2.1b) can be equally misleading if initial egg size and gestation time differ between species. In this case, the values for MI differ (0.67 for species 1 and 0.50 for species 2), but the slopes of the embryo mass lines are the same.

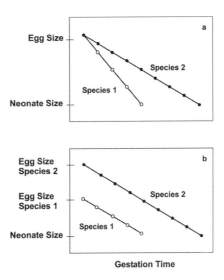

Box-figure 2.1 Hypothetical scenarios leading to misinterpretation of postfertilization provisioning using indirect methods. (a) Embryo mass as a function of development time for two species with identical MI (ratio of neonate size to egg size) but different total gestation time; the slope of the line for the species with the shorter gestation time is steeper. (b) Embryo mass as a function of development time for two species with identical slope of the embryo mass line but different initial egg size and gestation time; the MI for these species differs.

els of postfertilization nutrient transfer. Direct assays also have been used to confirm matrotrophy in species that might otherwise be classified as lecithotrophic using indirect methods. Marsh-Matthews et al. (2001, 2005, 2010) injected radiolabeled amino acids into the caudal peduncle of pregnant *Gambusia* and *Poecilia* females and detected radioactivity in embryos. The radioactivity detected in embryos was a valid indicator of nutrient transfer because assay of embryonic proteins in one study (Marsh-Matthews et al. 2005) showed that tritiated leucine injected into pregnant *Gambusia geiseri* was incorporated into embryonic proteins within two hours of the injection of the mother.

Direct assays confirm that nutrient transfer occurs at the level of the individual, whereas indirect assays often rely on comparison of broods among individuals (box 2.1). Use of radiolabeled nutrients also allows quantification of transfer, as measured by levels of radioactivity present in embryos (corrected for amount of radioactivity injected). These methods, however, usually require euthanizing the mother postinjection and therefore represent "snapshots" of provisioning by individual females. Interpretation of patterns of provisioning over time and space requires comparing multiple females, as with the indirect methods described above.

2.5 Intraspecific variation in matrotrophic provisioning

Matrotrophic provisioning is not a fixed component of maternal provisioning within species but may vary in both space (e.g., Trexler 1985; Reznick et al. 1993; Trexler 1997; Schrader & Travis 2005) and time (e.g., Abney & Rakocinski 2004). Both genetic and environmental factors have been found to affect intraspecific variation in matrotrophy.

2.5.1 Genetic variation

Only a few studies have attempted to detect genetic variation in matrotrophy. Schrader and Travis (2005) investigated pre- and postfertilization provisioning of the substantial matrotroph *Heterandria formosa* using two populations known to exhibit genetically based differences in life history. The matrotrophic component of provisioning differed more between populations than did egg size. In another substantial matrotroph, *Poeciliopsis prolifica*, Pires et al. (2007) detected no difference in matrotrophic provisioning for two populations based on MI measured in field-caught females but found genetic differences in MI (as well as effects of food availability) among populations in common garden experiments using laboratory-raised females.

Lima (2005) and Turcotte et al. (2008) examined matrotrophic variation among hybrid biotypes of the *Poeciliopsis monacha–Poeciliopsis lucida* complex (see Schlupp & Riesch, **chapter 5**). Both parental species exhibit superfetation but differ in matrotrophic provisioning: *P. monacha* appears lecithotrophic, while *P. lucida* is a substantial matrotroph. Among hybrids produced in the laboratory by crossing a female *P. monacha* with a male *P. lucida* (using artificial insemination), Lima (2005) found that some exhibited a decline in embryo mass from early to late in gestation, while others showed an increase. All-female hybrid biotypes (with the ML genotype) from a natural population exhibited a decline in embryo mass during gestation when raised under laboratory conditions. Turcotte et al. (2008) compared the MI in natural populations among the two parental species and the hybrid biotypes with different genomic combinations of the parental species (ML and MML). They found that the MI in both hybrid biotypes resembled that of *P. monacha*, the "lecithotrophic" parent, which led them to suggest that lecithotrophy is dominant to matrotrophy.

2.5.2 Environmentally induced variation

In *H. formosa* (in which most offspring provisioning is matrotrophic), Cheong et al. (1984) found that larger females produced larger offspring in the field. But laboratory studies, in which females were fed to excess, did not show a relationship between maternal size and offspring size. This result led Cheong et al. (1984) to suggest that female resource levels, rather than simply size, affected offspring provisioning and that the results from the field might reflect more efficient foraging by large females. Reznick et al. (1996a) demonstrated that better-fed *H. formosa* produced larger offspring in the laboratory, and Pires et al. (2007) reported production of heavier offspring by high-food females in the laboratory for two populations of *Poeciliopsis prolifica*, another substantial matrotroph.

Feeding history of the mother also affects levels of incipient matrotrophy. Marsh-Matthews and Deaton (2006) maintained *G. geiseri* at high and low feeding levels and found (by direct assay) that better-fed females exhibited higher rates of nutrient transfer than females maintained at low food availability. In addition, they found that the level of nutrient transfer was positively correlated with brood size, regardless of feeding level. Trexler (1997) also found that postfertilization provisioning (as assayed by his index of matrotrophic contribution) increased as a function of brood size in *Poecilia latipinna*. The positive relationship between brood size and matrotrophy may reflect an absolute higher nutrient pool in larger females as suggested by Trexler (1997) and Sakahi and Harada (2001).

2.5.3 Facultative matrotrophy

Trexler (1985, 1997) reported both temporal and spatial variation in matrotrophic provisioning among populations of *Poecilia latipinna* in Florida and suggested that this species exhibits facultative matrotrophy in response to environmental variation in food availability. In his 1997 paper Trexler developed a model (reproduced in fig. 2.1) relating facultative matrotrophy to the long-standing "lecithotrophic-matrotrophic" continuum that had been used to categorize species with respect to provisioning. In addition to acknowledging intraspecific variation in provisioning patterns, the model outlined the relationship among energy sources and timing of provisioning and provided an adaptive context in which to investigate the role of matrotrophy in species that had been considered to be primarily lecithotrophic.

2.6 Nutrient transfer

2.6.1 Sites of nutrient transfer

In placentotrophic poeciliids, nutrient transfer takes place across a follicular placenta (see also Greven, **chapter 1**), which has both embryonic and maternal components (Wourms et al. 1988; Grove & Wourms 1991, 1994). The embryonic portion is derived primarily from the expanded and highly vascularized pericardial sac that grows

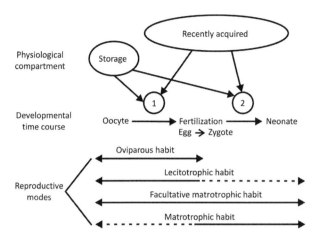

Figure 2.1 Trexler's (1997) conceptual model of facultative matrotrophy. Trexler's original figure legend reads: "A conceptual model of embryo nourishment patterns. Physiological compartment refers to the source of nutrients and energy expended during reproduction by females. Energy expenditure from either compartment may occur during developmental periods 1 or 2, and the size of each pool may vary. Oviparous fish can only contribute during period 1, whereas highly matrotrophic species (those producing eggs with little yolk) contribute primarily during period 2, and lecithotrophic species contribute primarily during period 1. Facultatively matrotrophic species may spread their contributions between periods 1 and 2, or shift their investment primarily to one or the other period, depending on conditions. The maximum absolute volume of the storage compartment is correlated with body size." Redrawn with permission from the Ecological Society of America.

anteriorly to cover the embryo's head (Wourms et al. 1988; Grove & Wourms 1991). Prior to the development of the pericardial placenta, nutrient transfer takes place across the entire embryonic surface, which is covered with microvilli (Wourms et al. 1988; Grove & Wourms 1991); as the placental surface develops, microvilli become confined to the placenta. The maternal component of the follicular placenta consists of follicular epithelium, which becomes vascularized and may develop secretory cells (as reported for *Poeciliopsis* spp. by Turner 1940b). In *H. formosa*, the maternal follicular epithelium appears to be specialized for transport rather than for secretion of materials (Schindler & Hamlett 1993; Grove & Wourms 1994).

Species that lack a well-developed placenta nonetheless have a derivative of the pericardial sac called the neck strap (Turner 1940b). This structure has been assumed by some authors to have a respiratory function (Turner 1940b) but also has been suggested as a site of nutrient transfer (Koya et al. 2000; DeMarais & Oldis 2005). Studies of *Gambusia* spp. provide support for the latter interpretation. DeMarais and Oldis (2005) found that fluorescent microspheres (40 nm) injected into pregnant *Gambusia* subsequently were present in embryos and concentrated in the neck strap region. In another study using fluorescently labeled antibodies and confocal microscopy, DeMarais (unpublished data) examined embryos for the presence and distribution of β-catenin, a protein integral to cadherin-mediated cell adhesion that mediates cell interactions within the ovary

during oocyte development (Cerdà et al. 1999; Luke & DeMarais, unpublished data). DeMarais (pers. comm.) found that β-catenin was present in *Gambusia* embryos and was especially prevalent in the neck strap.

2.6.2 Nutrients transferred

In his 1981 review, Wourms noted the paucity of direct evidence for nutrient transfer from mother to embryo in poeciliid fishes. The only study he cited was that of Wegmann and Götting (1971), which documented transfer of a labeled polysaccharide (iron dextran) from mother to embryos in *Xiphophorus hellerii*. Subsequent work by Grove and Wourms (1982, 1983) used labeled tracers to examine nutrient transfer in *H. formosa* and confirmed transfer of glucose, glycine, iron dextran, and the protein ferritin.

Marsh-Matthews et al. (2001, 2005, 2010) used radiolabeled leucine to investigate matrotrophy in 11 species of poeciliids (*H. formosa*, seven species of *Gambusia*, and three species of *Poecilia*) and demonstrated transfer of this essential nutrient in all species examined. Pilot studies using other radiolabeled nutrients injected into two *Gambusia* species demonstrated mother-to-embryo transfer of phenylalanine (another essential amino acid), tyrosine (nonessential amino acid), and glucose (Marsh-Matthews, unpublished data). Data for lineolic acid (an essential fatty acid) were equivocal; however, only one in four *Gambusia* females exhibited transfer to embryos, and in the one *H. formosa* female injected, evidence of transfer was found only in older and larger embryos (Marsh-Matthews, unpublished data).

Studies of postfertilization provisioning in livebearing teleosts other than poeciliids have documented transfer of simple nutrients (e.g., amino acids, glucose) in numerous groups, including the eelpout *Zoarces viviparous*, several viviparous clinids, the rockfish *Sebastes melanops*, and several species of goodeids (Wourms et al. 1988; Korsgaard 1994). Mother-to-embryo transfer of macromolecules also has been demonstrated, including polysaccharides and proteins in goodeids (Wourms et al. 1988), intact proteins in the embiotocid *Neoditrema ransonneti* (Nakamura et al. 2004), and phospholipids in the rockfish *Sebastes flavidus* (MacFarlane & Bowers 1995). This variety of transferred substances in other viviparous forms compels comprehensive surveys of poeciliids.

To date, studies screening for mother-to-embryo transfer of particular types of nutrients in poeciliids are few, but it is likely that systematic surveys using appropriate (e.g., radioactive, fluorescent, or rare element) tracers will reveal more complexity and perhaps a pattern related to placental development. Literature on placental squamates suggests a correlation between placental complexity and both the amount and the type of substances transferred (Thompson et al. 2000; Thompson & Speake 2006). As a group, poeciliids are ideal for an investigation of nutrient transfer and placental complexity, particularly within the genus *Poeciliopsis*, in which placental development and matrotrophy vary dramatically within the genus (Reznick et al. 2002b; Lawton et al. 2005; O'Neill et al. 2007; Pires et al., **chapter 3**).

2.6.3 Variation in nutrient transfer during gestation

Studies that have examined nutrient transfer over the course of gestation suggest that maternal transfer of nutrients is not constant throughout development. The temporal pattern of postfertilization provisioning, however, appears to vary among species. In *H. formosa*, eggs are extremely small with little yolk (Scrimshaw 1944a, 1946), and embryonic nutrition is derived almost solely from matrotrophic transfer (Cheong et al. 1984). Grove and Wourms (1982) suggested that the rate of nutrient absorption was higher early in development based on incubation of embryos in radiolabeled nutrients (glycine and glucose), although Scrimshaw (1944a) showed that the fastest rate of embryo gain in mass occurred later in development. Thibault and Schlutz (1978) also noted that embryos of placentotrophic species of *Poeciliopsis* (*P. turneri* and *P. prolifica*) showed the greatest gain in mass after the yolk had been consumed and the placenta was well developed. Pires et al. (2007) also concluded that most of maternal provisioning occurs after stage 25 in *Poeciliopsis prolifica*. For the less-specialized matrotroph *Poeciliopsis lucida*, Thibault and Schultz (1978) noted that the pattern of gain in mass was reversed, with rate of mass change greater in younger embryos prior to the disappearance of yolk. They suggested that the rate of transfer in this species might be related to the surface-to-volume ratio of the embryo, which is higher earlier in development.

For species that exhibit incipient matrotrophy, gestational patterns of matrotrophic provisioning also appear to vary among species. Edwards et al. (2006) suggested that maternal provisioning was greatest in midgestation in *G. holbrooki*. Using broods that exhibited a marked developmental gradient, Marsh-Matthews et al. (2005) examined nutrient transfer to individual embryos within broods of two *Gambusia* species. In *G. geiseri*, matrotrophic transfer appeared to increase from early to midgestation, then decrease near the time of parturition; for *G. affinis*, matrotrophic provisioning appeared similar throughout development.

2.7 Roles of matrotrophy in offspring production of livebearing poeciliids

In addition to providing nutrition to embryos, matrotrophic provisioning has the potential to impact many aspects of both maternal and offspring fitness. The most often cited advantage of matrotrophy over pure lecithotrophy is an increase in maternal fecundity (Thibault & Schultz 1978; Trexler 1997; Trexler & DeAngelis 2003). In theory, postfertilization provisioning allows a decrease in the energy invested in individual eggs so that more eggs can be produced from maternal energy available during vitellogenesis (see Trexler et al., **chapter 9**). To the extent that the postfertilization contribution is necessary for development and growth of the embryo, some portion of matrotrophic provisioning would be obligate even for species with low levels of nutrient transfer.

The fecundity advantage of matrotrophy has been suggested by many authors (e.g., Thibault & Schultz 1978; Trexler 1997; Graham & Wilcox 2000). Some comparative studies across species, however, have found more complex variation in the size and number of offspring produced by matrotrophic versus presumably "nonmatrotrophic" species. Reznick et al. (2007b) reported contrasting results in two different genera of livebearing halfbeaks (family Zenarchopteridae): in one genus, the matrotrophic species produced fewer and larger offspring; in the other, the matrotrophic species produced more and smaller offspring.

Reznick et al. (2007b) demonstrated that species with substantial matrotrophy devote a smaller proportion of maternal body mass to reproduction at any given time ("reproductive allocation hypothesis"). They suggested that this reduced allocation in turn reduces locomotory costs (Thibault & Schultz 1978) and predation risks (Plaut 2002) relative to those expected with higher reproductive allocation, particularly when substantial matrotrophy is coupled with superfetation (thus spreading reproductive mass over multiple broods).

Matrotrophy also may allow females to adjust the rate of brood production. In the substantial matrotroph *H. formosa*, Reznick et al. (1996a) found that both the rate of offspring provisioning and the rate of offspring production increased as a function of food availability. Maternal control of interbrood interval also has been reported in poeciliids with incipient matrotrophy (e.g., Evans et al. 2007), and it would be interesting to know if maternal adjustment of matrotrophic provisioning could contribute to maternal control of gestational period.

The protracted period of mother-to-embryo interaction in matrotrophy provides the opportunity for maternal adjustment of offspring size or quality during gestation (e.g., Shine & Downes 1999). In substantial matrotrophs (in which the vast majority of provisioning occurs postfertilization), females on high rations typically produce larger and potentially more fit (Henrich 1988) offspring than those on low rations (*H. formosa*: Cheong et al. 1984; Reznick et al. 1996a; but see Travis et al. 1987; *Poeciliopsis prolifica*: Pires et al. 2007; Banet & Reznick 2008). Postfertilization variation in offspring quality has also been reported in species with incipient matrotrophy. Marsh-Matthews and Deaton (2006) found that better-fed female *G. geiseri* produced the same sized eggs as females on low rations but were more matrotrophic and had larger embryos at all stages of development. Trexler (1997) noted that offspring of more matrotrophic *Poecilia latipinna* mothers tended to have higher lipid stores than offspring of less matrotrophic females.

The period of postfertilization provisioning also may provide the opportunity for cryptic mate choice (Sheldon 2000; Calsbeek & Sinervo 2004; and see Evans & Pilastro, **chapter 18**), via differential provisioning by embryo genotype (Schrader & Travis 2005). In many poeciliids, females are able to store sperm (Medlen 1951; Constantz 1989), and genetic studies have established that multiple paternity within broods is common (see Table 18.1 in Evans & Pilastro, **chapter 18**). With multiple sires, within-brood variation in provisioning could result from selective provisioning of heterozygous offspring (which may have enhanced growth or survival; Feder et al. 1984), of offspring of "preferred" males (Kotiaho et al. 2003), or of offspring of the more "valuable" sex (Clutton-Brock et al. 1981).

Matrotrophy potentially has many roles in livebearing poeciliids, and those roles may vary with the relative contribution that postfertilization nutrient transfer makes to overall provisioning. The primary role of substantial matrotrophy is embryo nutrition. Reznick et al. (1996a) noted that this nutritional commitment may constrain the ability of females to adjust offspring provisioning in response to resource variation in the environment. In species with dual provisioning, females may be more able to adjust the relative contribution of yolk and matrotrophic provisioning (Thibault & Schultz 1978) and vary the matrotrophic component to impact both offspring and maternal fitness. The role of incipient matrotrophy is only beginning to be explored (e.g., Trexler 1997; Marsh-Matthews & Deaton 2006), but given its widespread occurrence in poeciliids (Marsh-Matthews et al. 2010), it should be considered as a potentially important contributor to both maternal and offspring fitness.

2.8 Evolution of matrotrophy

2.8.1 Evolution of matrotrophy from a lecithotrophic ancestor

Matrotrophy has evolved independently from a lecithotrophic ancestor many times within vertebrates (Wourms et al. 1988; Blackburn 1992, 1999, 2005; Reznick et al. 2007b). Although some authors have assumed that within groups that have evolved viviparity, the ancestral livebearer was lecithotrophic (e.g., Wourms et al. 1988; Reznick et al.

2002b; but see Reznick et al. 2007a), others have suggested that viviparity and incipient matrotrophy evolved simultaneously (Blackburn 1992).

Trexler and DeAngelis (2003; see also Trexler et al., **chapter 9**) examined ecological conditions favoring the evolution of matrotrophy from a lecithotrophic ancestor using analytical and simulation models. The models made several important assumptions: (1) offspring size or quality was equal for the two provisioning modes; (2) matrotrophic females produced initially smaller and consequently more eggs for a given brood; (3) given the smaller eggs with less yolk, offspring nutrition for development required maternal transfer of nutrients; and (4) matrotrophic females could potentially produce more eggs than could be brought to term, if resources in the environment became scarce. The model, which they parameterized with data for sailfin mollies (Trexler 1997), predicted that matrotrophy could evolve as an alternative to strict lecithotrophy only if (1) females were able to abort developing embryos and recycle nutrients to remaining embryos in the brood, and (2) resources were sufficiently plentiful and predictable to support prolonged maternal nutrient transfer. The model also predicted that matrotrophic females would be leaner and have a shorter life span.

Marsh-Matthews and Deaton (2006) tested assumptions and predictions of the Trexler-DeAngelis model by manipulating resource availability to female *G. geiseri* and examining nutrient transfer using a direct assay. This species exhibits incipient matrotrophy, which provides an opportunity to evaluate the Trexler-DeAngelis model because factors that favor the "addition" of a matrotrophic component to offspring provisioning (as in the Trexler-DeAngelis model) should also favor an increase in matrotrophic contribution in dual provisioning. Females exposed to a high resource regimen produced more but not smaller eggs, were more matrotrophic, and had embryos that were larger at every stage of development than females on low resources (Marsh-Matthews & Deaton 2006). In addition, more matrotrophic females were leaner (relative to expected fat stores for that feeding level). Overall, most responses of *G. geiseri* to resource variation matched predictions of the Trexler-DeAngelis model, although egg size did not vary as the model assumed, and there was no evidence of differential embryo abortion between treatments. Marsh-Matthews and Deaton (2006) suggested that offspring size in this species has more important maternal fitness consequences than an increase in brood size and that the matrotrophic contribution facilitates the production of larger, instead of more, offspring.

Evidence for embryo abortion in response to resource limitation is mixed. Trexler (1997) found higher rates of abortion and reabsorption of embryos in female *Poecilia latipinna* with lower fat stores following reproduction.

However, Meffe and Vrijenhoek (1981) failed to find evidence of increased embryo abortion in starved *Poecilia* females, and Banet and Reznick (2008) found that low food conditions did not cause embryo abortion in either a substantial matrotroph (*Poeciliopsis prolifica*) or a presumed lecithotroph (*Poeciliopsis monacha*).

2.8.2 Multiple, independent origins of placentotrophy in poeciliids and the role of parent-offspring conflict

Substantial matrotrophy and placentation have evolved independently multiple times in livebearing teleosts, including at least once (and possibly as many as three times) in the family Zenarchopteridae (livebearing halfbeaks; Reznick et al. 2007b) and at least four times within the family Poeciliidae, including three independent origins within the genus *Poeciliopsis* (Reznick et al. 2002b). Reznick et al. (2002b) examined the MI across the genus *Poeciliopsis*, which includes species ranging from presumed lecithotrophs (e.g., *P. monacha*; Wourms 1981) to unspecialized matrotrophs (e.g., *P. occidentalis*; Constantz 1989) to extensive matrotrophs (e.g., *P. turneri*; Wourms 1981). They considered species with MI < 0.7 (or not significantly different from 0.7) to be lecithotrophs, those with 0.7 < MI ≤ 1.6 to have matrotrophic transfer (across a "rudimentary placenta") that supplements yolk during gestation, and those with MI > 5 to be substantial matrotrophs with extensive placentation. Their analyses dated the origin of some placental forms at only 0.75 million years ago, which suggests rapid evolution of this complex structure.

Reznick et al. (2007a) proposed an even more complex scenario for the evolution of substantial matrotrophy within poeciliids based on their finding that *Xenodexia ctenolepis* exhibits both substantial matrotrophy and superfetation. Because genetic analyses show *X. ctenolepis* as the sister taxon to all other members of the family, Reznick et al. (2007a) suggested that substantial matrotrophy and superfetation were likely present in the common ancestor of all poeciliids, in which case those traits would have been lost and reevolved multiple times within the family.

Rapid, independent evolution of substantial matrotrophy and placentotrophy in poeciliids could represent the evolutionary outcome of parent-offspring conflict (Reznick et al. 2002b; Crespi & Semeniuk 2004; Schrader & Travis 2009). Crespi and Semeniuk (2004) proposed the viviparity conflict hypothesis (VCH) to explain the diversity of maternal-fetal interactions in livebearing vertebrates. According to the VCH, although matrotrophy presumably evolved from a lecithotrophic ancestor for ecological reasons, subsequent evolution of substantial matrotrophy resulted from parent-offspring conflict over the rate of maternal provisioning. Conflict over maternal provisioning is assumed to exist because a mother is equally related to all offspring in a brood, but the level of relatedness (*r*) is less than unity (*r* = 1/2 in sexual species); thus, the optimal investment per offspring for the mother should be at a level that maximizes her lifetime reproductive success via maximizing that of all her descendants (Trivers 1974). Optimal nutrient allocation for an individual offspring, however, is greater than the mother's optimum because it is more closely related to itself (*r* = 1) than to full siblings (*r* = 1/2) or to half siblings (*r* = 1/4) in broods with multiple paternity (Trivers 1974). Given this conflict, any trait that allows offspring to manipulate the rate of nutrient transfer from the mother to themselves should be subject to strong selection, and the maternal and embryonic characteristics involved in nutrient transfer should undergo antagonistic coevolution (Crespi & Semeniuk 2004). Characters subject to antagonistic coevolution may show evidence of genomic imprinting and/or positive selection (Summers & Crespi 2005; O'Neill et al. 2007).

Lawton et al. (2005) and O'Neill et al. (2007) examined insulin-like growth factor II (IGF-II) in poeciliids for evidence of genomic imprinting and positive selection. In mammals, IGF-II is involved in growth of the fetus and placenta and displays genomic imprinting of the paternally derived allele (O'Neill et al. 2000). Lawton et al. (2005) examined the expression pattern of IGF-II in two poeciliid fishes with independently derived substantial matrotrophy, *H. formosa* and *Poeciliopsis prolifica*. In both species IGF-II exhibited expression of both maternal and paternal alleles and therefore showed no evidence of genomic imprinting. On the other hand, O'Neill et al. (2007) examined the gene sequence in IGF-II in a variety of teleosts, both egg laying and livebearing, including numerous members of the genus *Poeciliopsis*, and found evidence of positive selection. Additionally, Schrader and Travis (2009) demonstrated that offspring genotype can influence maternal provisioning in *H. formosa*, thereby confirming a key assumption of the VCH. Thus, the current evidence for the VCH to explain the repeated evolution of substantial matrotrophy in poeciliids is mixed, but the group as a whole provides perhaps the best opportunity among viviparous vertebrates to explore the role of parent-offspring conflict because of the vast diversity in offspring provisioning.

2.9 Directions for future research

The widespread occurrence of postfertilization offspring provisioning in poeciliids and the variation throughout the family in the extent of matrotrophy relative to yolk-based nutrition within species together provide vast opportunities to study all aspects of matrotrophy evolution. Basic surveys of provisioning are needed for many species within the family. For those species that appear to lack postfertilization provisioning, direct methods should be applied when possible. The anatomical and physiological bases of nutrient transfer need to be investigated across the range of matrotrophic provisioning, and studies should attempt to identify the molecular bases of transfer. Studies using species that represent the diversity of matrotrophy have the potential to address the integration of genetic, epigenetic, and environmental factors that shape maternal provisioning as a whole. The role of incipient matrotrophy as an integral component of life-history strategies needs to be considered, and the overall provisioning strategies of many incipient matrotrophs will require reevaluation. Poeciliids also provide the range of provisioning strategies necessary to address the evolution of matrotrophy at all levels, from incipient to substantial.

Finally, mounting evidence that incipient matrotrophy, rather than pure lecithotrophy, may be plesiomorphic for livebearing poeciliids demands reevaluation of the evolution of offspring provisioning within the family. Nutrient transfer clearly occurs in the absence of elaborate placental structures, and future studies of placental evolution in the family should consider incipient matrotrophy, rather than lecithotrophy, as the ancestral condition.

Acknowledgments

Many colleagues have provided invaluable discussion of ideas presented here. I particularly thank Dan Blackburn, Melody Brooks, Raelynn Deaton, Alyce DeMarais, William Matthews, Rüdiger Riesch, Ingo Schlupp, and Joel Trexler. Studies in my laboratory were supported by the National Science Foundation (NSF 0096568) and the University of Oklahoma Research Council.

Chapter 3 Variation and evolution of reproductive strategies

Marcelo N. Pires, Amanda I. Banet, Bart J. A. Pollux, and David N. Reznick

3.1 Introduction

THE FAMILY POECILIIDAE (Rosen & Bailey 1963) consists of a well-defined, monophyletic group of nearly 220 species with a fascinating heterogeneity in life-history traits. Reznick and Miles (1989a) made one of the first systematic attempts to gather information from a widely scattered literature on poeciliid life histories. They focused on two important female reproductive traits: (1) the ability to carry multiple broods at different developmental stages (superfetation; Turner 1937, 1940b, 1940c), which tends to cause females to produce fewer offspring per brood and to produce broods more frequently, and (2) the provisioning of eggs and developing embryos by the mother, which may occur prior to (lecithotrophy) or after (matrotrophy) fertilization. Their review documented the distribution of these two reproductive traits among poeciliids and their correlates with other life-history traits, including mean and minimum reproductive size, reproductive allocation, brood size, offspring dry weight at birth, interbrood interval, and mean adult size.

Their study yielded several new insights. One was that superfetation may have evolved multiple times within the family and appeared to be largely confined to the closely allied genera *Heterandria*, *Neoheterandria*, and *Poeciliopsis*. The authors cautioned, however, that their data set was limited by uneven representation of genera. Further knowledge of other, hitherto-uninvestigated genera was required before reaching a conclusion that superfetation is confined to these three genera. A second insight was that most species with superfetation were matrotrophic; the strong as-sociation between these two traits suggests that one of the two traits might be more likely to evolve when the other trait is already present (the latter facilitating the evolution of the former). However, the existence of a notable exception in the literature (the lecithotrophic, superfetating *Poeciliopsis monacha*, the only known exception at the time) showed that superfetation and matrotrophy were not strictly linked, indicating that these two traits can evolve independently of each other.

Reznick and Miles (1989a) also proposed a framework for future research that was aimed at evaluating possible causes and mechanisms for the evolution of superfetation and matrotrophy by (1) gathering detailed life-history descriptions of a greater number of poeciliid species, either through common garden studies or from field-collected individuals, (2) comparing superfetating species with their closest nonsuperfetating relatives (and matrotrophic species with their closest lecithotrophic relatives) to test hypotheses about the ecological conditions under which superfetation (or matrotrophy) should be favored, (3) developing a highly resolved, family-wide phylogenetic tree, and (4) applying phylogenetic comparative methods (e.g., phylogenetically independent contrasts; Felsenstein 1985) to test for correlations among different facets of reproductive modes (e.g., superfetation, matrotrophy), on the one hand, and to interpret either the descriptive life-history traits or the outcome of paired experiments, on the other.

Over the last 20 years, much action has been taken on these recommendations. Some of the published work will be synthesized in this chapter, and occasionally we will draw upon our own unpublished data. We concentrate on the evolution of superfetation and postfertilization maternal provisioning (matrotrophy) and address these specific questions:

1. What are the ranges of variation in some traits that are associated with superfetation and postfertilization maternal provisioning (i.e., standard length, offspring size, brood size, and reproductive allocation) within the family Poeciliidae (section 3.2)?
2. To what extent do poeciliids vary in their ability to carry multiple broods (degree of superfetation) and in the amount of postfertilization maternal provisioning (degree of matrotrophy), and are these traits coupled (section 3.3)?
3. What are the patterns of life-history variation within the family, and what do they reveal about evolutionary trends of trait evolution in poeciliids (section 3.4)?
4. What are some of the available hypotheses for the evolution of postfertilization maternal provisioning and superfetation, and to what extent are they supported (section 3.5)?

3.2 Life-history variation in the family Poeciliidae

Poeciliids display wide variation in all life-history traits. Here, we briefly outline the ranges of variation of several life-history traits within the family Poeciliidae known (or suspected) to be associated with postfertilization maternal provisioning. The genera *Cnesterodon*, *Pamphorichthys*, *Phalloptychus*, and *Poecilia* (subgenus *Acanthophacelus*) contain some of the smallest known species in the family (e.g., mean standard length [SL] for *Cnesterodon iguape*, male = 19.7 mm, female = 21.4 mm; *Pamphorichthys pertapeh*, male = 15.9 mm, female = 18.04 mm; *Phalloptychus januarius*, male = 17.3 mm, female = 23.7 mm; *Poecilia picta*, male = 18.3 mm, female = 19.8 mm; Reznick et al. 1992; Rosa & Costa 1993; Lucinda 2005b, 2005a; Figueiredo 2008; Pires et al. 2010). On the other hand, the genera *Poeciliopsis*, *Poecilia*, and *Belonesox* have some of the largest poeciliid species (e.g., mean male/female SL of 72.4/102.9, 63.8/85.7, 36.0/83.5, and 35.0/66.0 mm for *Belonesox belizanus*, *Poecilia catemaconis*, *Poeciliopsis elongata*, and *Poeciliopsis catemaco*, respectively; Miller 1975; Turner & Snelson 1984; Reznick, unpublished data). Poeciliids typically show strong size dimorphism, with males being considerably smaller than females (Bisazza 1993a). Despite this size dimorphism, males display a variation in mean SL similar to that of females (approximately fivefold). Mature males within a population often display polymodal or highly skewed size distributions (Kallman 1989; Kolluru & Reznick 1996; Arias & Reznick 2000). In some of these

species, it has been shown that the different size modes correspond to genes linked to the Y chromosome that control the age and size at maturity (Kallman 1989).

Offspring size (like many other life-history traits) is typically related to female body size (Reznick & Miles 1989a); some of the smallest offspring are produced by the smallest species within the family. Most studies express offspring size in terms of embryo dry weight, which exhibits a 14-fold range of variation (mean neonate mass ranging from 0.58 mg in *Poeciliopsis prolifica* to 8.3 mg in *P. elongata*; Pires et al. 2007; Reznick, unpublished data). An increase in female body size is generally also associated with longer interbrood intervals in both superfetating and nonsuperfetating species (Reznick & Miles 1989a). In nonsuperfetating species, for instance, interbrood interval ranges from 21.9 days in the small-bodied *Poecilia reticulata* (Thibault & Schultz 1978) to 63.8 days in the larger *Poecilia latipinna* (Hubbs & Dries 2002). In superfetating species, there appears to be a similar trend: larger species, such as *Poeciliopsis gracilis* and *Poeciliopsis turneri*, which on average carry two to three broods simultaneously, have an interbrood interval of 11–23 days (Thibault & Schultz 1978; Snelson et al. 1986), while the somewhat smaller-bodied *Heterandria formosa* (Scrimshaw 1944b) and *Phalloptychus januarius* (Pollux, unpublished data) can give birth every 1–3 days to as few as one to five offspring at a time. Brood size (the number of offspring per brood) is related to female body size in nonsuperfetating species, with larger species (as well as larger individuals within species) on average producing larger broods than their smaller counterparts (e.g., Turner & Snelson 1984; Reznick & Miles 1989a; Reznick et al. 1992; Reznick et al. 1993). In superfetating species, however, brood size appears to be independent of female size, both within (e.g., *Xenodexia ctenolepis*; Reznick et al. 2007a) and among (Reznick & Miles 1989a) species. In these species, brood size seems to depend more on the degree of superfetation, with highly superfetating species expected to have smaller broods than similar-sized species that carry fewer simultaneous broods at a time (Reznick & Miles 1989a). Taking all poeciliids into account, there is a remarkable, nearly 100- to 200-fold, variation in mean brood size, ranging from 1 offspring per brood (e.g., *H. formosa*, *Phalloptychus januarius*, *P. prolifica*, *Gambusia hubbsi*; Scrimshaw 1944a; Downhower et al. 2000; Pires et al. 2007; Pollux, unpublished data) to over 99 young per brood in *B. belizanus* (Turner & Snelson 1984) and over 200 per brood in *Gambusia affinis* (Krumholz 1948).

Reproductive allocation (RA), defined as the percent of female dry weight that consists of developing embryos, shows a nearly ninefold variation in poeciliids, being as low as 4.1% in the matrotrophic *Phalloceros caudimaculatus*

(Arias & Reznick 2000) and over 35% in the lecithotrophic *G. affinis* (Stearns 1983c; Reznick & Braun 1987).

Finally, it is noteworthy that life-history traits can vary substantially within species, as a result of either maternally mediated environmental influences (phenotypic plasticity) or genetic differences among spatially isolated populations (local adaptation) (Stearns 1983b, 1983a; Trexler 1989; Johnson & Bagley, **chapter 4**).

3.3 Variations on the theme of livebearing

3.3.1 Maternal provisioning

All poeciliids exhibit internal fertilization, and all but one species, *Tomeurus gracilis*, are viviparous. Viviparity, or "livebearing," is a reproductive mode in which eggs are fertilized internally and then retained in the maternal reproductive system throughout embryonic development until parturition, resulting in a free-living offspring at birth (Wourms et al. 1988).

The lengthy maternal-embryonic interaction resulting from viviparity creates the need for respiratory, osmoregulatory, and endocrinological interactions between the mother and embryos. These intimate interactions in turn create the potential for the evolution of more complex trophic relationships (Wourms et al. 1988; Korsgaard & Weber 1989; Crespi & Semeniuk 2004). Maternal-offspring trophic relationships within poeciliids range from strict lecithotrophy, in which nutrients are provisioned to the embryo solely via yolk allocated to the egg prior to fertilization, to extensive matrotrophy, in which the developing embryo depends largely or completely on a continuous supply of nutrients obtained directly from the mother during gestation (Wourms et al. 1988).

Variation in maternal-offspring trophic relationships in poeciliids can be interpreted within the context of the capital versus income breeding paradigm for reproductive strategy variation, which distinguishes between species that use stored energy to nourish developing offspring (analogous to a lecithotrophic strategy) and species that provision offspring with current energy income (analogous to a matrotrophic strategy; Drent & Daan 1980). Rather than being a true dichotomy (Houston et al. 2007), these trophic modes represent extremes of a continuum in which the embryo relies on prefertilization (indirect) and postfertilization (direct) maternal provisioning in different degrees. Thus, in many poeciliids, lecithotrophy occurs in conjunction with matrotrophy: developing embryos obtain nutrients both from the yolk (prefertilization maternal provisioning) and directly from the mother (postfertilization maternal provisioning). For simplicity, however, species are referred to as "matrotrophic" when postfertilization maternal provision-

ing exists and is large enough to be reflected in some embryonic weight gain during development, while the term "lecithotrophic" is applied to species in which all or most nutrients are provided to the developing embryo before the egg is fertilized (as is the case in an oviparous species). The term "true viviparity" has been used to refer to matrotrophic viviparity (e.g., Scrimshaw 1944a; Trexler 1985), but it should be avoided because it confounds trophic relationships and parity mode (Blackburn 1992).

Matrotrophy is widespread in metazoans, and the mechanism of resource transfer varies greatly across different taxa (e.g., Wourms et al. 1988; Blackburn 1992; Meier et al. 1999; Williford et al. 2004). In poeciliid fishes, matrotrophy is accomplished through a specialized follicular placenta (the "follicular pseudoplacenta" described by Turner 1940b)—the close apposition of the follicle wall (the maternal tissue) to vascularized embryonic tissues (Wourms et al. 1988). The yolk sac is the main embryonic component of the placenta of lecithotrophic poeciliids and facilitates gas exchange and/or transfer of inorganic and a few organic molecules between the mother and the embryos (Wourms et al. 1988; Constantz 1989). In matrotrophic species, however, the yolk sac is greatly reduced, and a modified, highly vascularized pericardial sac allows substantial nutrient transfer between the mother and the embryo (Wourms et al. 1988). The term "placentotrophy" (Wourms et al. 1988; Blackburn 1992) will be used throughout the chapter to refer to this specific form of poeciliid matrotrophy. "Matrotrophy" will henceforth be used to refer to a more general form of postfertilization maternal provisioning.

Because many species have both pre- and postfertilization maternal provisioning, there can be intra- and interspecific variation in the degree of prefertilization yolk allocation (i.e., egg size and yolk composition) and of postfertilization placentotrophic allocation. Strictly lecithotrophic poeciliid species are characterized by a decrease of ca. 35% (ranging from 25% to 55%) in embryonic dry weight over development, which is similar to the change in embryonic dry weight during development observed in oviparous fishes (Wourms et al. 1988). Embryos from placentotrophic poeciliids, on the other hand, may obtain just enough postfertilization resources from the mother to compensate for biomass loss during development due to metabolic costs—in which case there will be no net change (or even a slight loss) in dry weight between the egg at fertilization and the embryo at birth. In the case of extensive placentotrophy, the gain in embryonic dry weight over development can exceed 10,000% (Reznick et al. 2002b).

Placentotrophy has been identified and quantified in poeciliids with the matrotrophy index (MI; Reznick et al. 2002b). The MI is the estimated dry weight of the offspring at birth divided by the estimated dry weight of the egg at fertilization and thus represents the change in embryonic dry weight over development (Wourms et al. 1988; Reznick et al. 2002b; Stewart & Thompson 2003; Thompson & Speake 2006). An MI that is significantly greater than 0.7 or 0.8 (the upper threshold in oviparous species; Wourms et al. 1988) indicates that a placenta is transferring resources to embryos during development and thus characterizes functional placentotrophy. Placentotrophy may be present even if there is a slight loss in dry mass (i.e., MI slightly less than 1 but still significantly greater than 0.7 or 0.8). An MI of 1 thus represents no net change in dry weight over embryonic development.

Trophic patterns of all species in the genus *Poeciliopsis* have been characterized and reflect the diversity of trophic modes that can be found among closely related poeciliid species. There have been three independent origins of extensive placentotrophy in this genus alone (Reznick et al. 2002b). Lecithotrophic species, or those that show a decrease in embryonic dry weight over development of 20%–40%, include the following *Poeciliopsis* species: *P. fasciata, P. monacha, P. infans, P. gracilis, P. hnilickai, P. catemaco, P. turrubarensis,* and *P. scarlii.* Low to moderate levels of placentotrophy range from a decrease in embryonic dry weight over development of ca. 15% to an increase of ca. 60% and are exhibited by the following *Poeciliopsis* species: *P. latidens, P. baenschi, P. lucida, P. occidentalis,* and *P. viriosa.* Finally, species with extensive placentotrophy have an increase in dry weight during development of ca. 500%–11,000% and include the following *Poeciliopsis* species: *P. prolifica, P. paucimaculata, P. elongata, P. presidionis, P. turneri,* and *P. retropinna* (Reznick et al. 2002b). Other independent cases of extensive placentotrophy are found in *H. formosa* (Turner 1940b; Scrimshaw 1944a; Schrader & Travis 2005; increase of ca. 4500%), *Poecilia branneri,* and *Poecilia bifurca* (Pires et al. 2010; increase of ca. 60,000%). Cases of less extensive placentotrophy are found in *Phalloceros caudimaculatus* (Arias & Reznick 2000; increase of ca. 100%) and *X. ctenolepis* (Reznick et al. 2007a; increase of ca. 400%).

Marsh-Matthews et al. (2001, 2005; Marsh-Matthews, chapter 2) and Marsh-Matthews and Deaton (2006) used radioactively labeled amino acids to show that *Gambusia geiseri* and *G. affinis* transfer some resources from the mother to the young during development, but both species are considered lecithotrophic based on the percent of weight lost by embryos during development. Trexler (1997) has shown that *Poecilia latipinna* can be either lecithotrophic or weakly placentotrophic, possibly in response to food availability. These observations suggest that species categorized as lecithotrophic based on the pattern of embryonic dry-weight change over development may still have some capacity to transfer nutrients from the mother to developing young.

3.3.2 Superfetation

Superfetation (also "superfoetation" or "superembryonation"; Veith 1979) is "the occurrence of more than one stage of developing embryos in the same animal at the same time" (Scrimshaw 1944b, 180); that is, it is the occurrence of fertilization and development of a new brood before the former brood is born (Turner 1937; see also Turner 1940c; Thibault 1974; Thibault & Schultz 1978). "Sequential brooding," used to describe multiple, ontogenetically staggered broods coexisting within some bivalves (Cooley & Foighil 2000), is synonymous with superfetation. The term "clutch overlap" (Burley 1980; Hill 1986; Travis et al. 1987), used to describe reproductive characteristics in some birds, is functionally similar to superfetation, but it encompasses simultaneous provisioning of different clutches before and after birth and does not apply to organisms in which maternal provisioning ends at parturition, such as poeciliids. "Litter overlap" was used by Downhower et al. (2002) to describe the simultaneous presence of yolking (or fully yolked) eggs and developing embryos within a female. They considered such overlap to be a form of superfetation. This is a common and interesting phenomenon in poeciliids but is not superfetation, because all developing embryos are part of a single brood and will be born at the same time, before the next clutch of eggs is fertilized.

There is pronounced variation in the number of broods present in the ovary among those species that have superfetation. The degree of superfetation (or a "superfetation index") can be expressed as either the average or the maximum number of developmentally distinct broods found simultaneously within females, depending on whether the goal is to characterize average or maximum reproductive output. All species in the genus *Poeciliopsis* are capable of superfetation (Turner 1937, 1940c; Scrimshaw 1944b; Thibault 1974; Thibault & Schultz 1978; Reznick & Pires, unpublished data), but the maximum number of simultaneous broods per female in a given species ranges from two (e.g., *P. monacha*) to five (e.g., *P. prolifica*). Females of *X. ctenolepis* have been found to carry up to six simultaneous developing broods (Reznick et al. 2007a), and *H. formosa* and *Poecilia branneri* may carry up to five simultaneous developing broods (Turner 1937, 1940b, 1940c; Scrimshaw 1944b; Travis et al. 1987; Pires et al. 2010). Intraspecific variation in superfetation can be largely due to variation in female size and food availability (Travis et al. 1987; Pires et al. 2007; Banet & Reznick 2008). One study also documents differences among populations within a species that appear to be a function of habitat (Zúñiga-Vega et al. 2007; see also section 3.5.1 below).

Reports from aquarists indicate the possible presence of superfetation in *Priapella bonita*, *Phalloptychus januarius*, *Poecilia* (*Micropoecilia*) *branneri* (Stoye 1935, in Turner 1937 and Scrimshaw 1944b), and *Priapichthys fria* (*Pseudopoecilia fria*; Turner 1940c; Henn 1916, in Turner 1937). Turner (1940c) listed additional, unconfirmed reports of superfetation in *Phallichthys fairweatheri* (*Dextripenis evides*), *Gambusia vittata* (*Flexipenis vittata*), and *Priapichthys chocoensis* (*Diphyacantha chocoensis*). Scrimshaw (1944b) described unconfirmed cases of superfetation in *Neoheterandria tridentiger* (*Allogambusia tridentiger*), *Brachyrhaphis cascajalensis*, *Brachyrhaphis episcopi*, *Gambusia nicaraguensis* (*Gambusia dovii*), *Gambusia holbrooki*, *Gambusia nobilis*, *Poecilia reticulata* (*Lebistes reticulatus*), and *Poecilia sphenops* (*Mollienesia sphenops*). He hypothesized that some individuals in all poeciliid species may express superfetation. Our personal observations on *Priapella bonita*, *Phallichthys fairweatheri*, *G. vittata*, *G. holbrooki*, *Brachyrhaphis episcopi*, *Poecilia reticulata*, and *Poecilia sphenops* have failed to confirm these earlier reports. From the species mentioned in these reports, we have been able to confirm superfetation only in *Neoheterandria tridentiger* (described in Stearns 1978), *Poecilia branneri*, *Phalloptychus januarius*, *Priapichthys fria*, and *Priapichthys chocoensis*.

It is clear that an exhaustive survey for the presence and level of true superfetation within poeciliids is still needed. In this process, it is critical that researchers demonstrate the ability to identify distinct developmental stages of embryos (including within-brood variation in embryo development) and abnormal embryos; runts and aborted or deformed embryos can be mistaken for ones that are in early stages of development. In addition, variation in the degree of egg yolking (i.e., prefertilization maternal provisioning) should be excluded from analysis of superfetation, for superfetation refers only to multiple broods of developing embryos. Ignoring any of these factors may lead to misleading reports of true superfetation, as for *Pamphorichthys hollandi* (Casatti et al. 2006a) and *G. hubbsi* (Downhower et al. 2002).

Our numerous observations of dissected reproductive individuals show that a hallmark of superfetation is the simultaneous presence of similar-sized broods in noncontiguous stages of development within a female. As the degree of superfetation increases, however, the researcher's ability to discern simultaneous broods in clearly noncontiguous stages decreases. Superfetation can still be unambiguously differentiated from large within-brood variation if distinct broods have similar numbers of embryos and if there is a stepped distribution of embryo size, which occurs in highly

placentotrophic species. Large broods followed by contiguous broods with a significantly smaller number of embryos (as described in *Poecilia formosa*; Monaco et al. 1983) may represent aborted, regressing embryos rather than true superfetation (R. Riesch, pers. comm.). They may also represent within-brood variation in the stage of development. Such observations are worth recording, for they may reflect important physiological differences among species in their reproductive cycles and hence may represent the necessary variation for the evolution of true superfetation. The underlying biological causes of such variation as well as the extent to which this variation is regulated by environmental and genetic factors are all important questions still to be addressed.

3.3.3 The association between superfetation and placentotrophy

The work of Turner (1937, 1940c), Scrimshaw (1944a, 1944b), and Thibault and Schultz (1978) clearly implied a "parallel development" (Scrimshaw 1944b) between placentotrophy and superfetation. Thibault and Schultz (1978) then hypothesized that there is an adaptive value to the joined presence of both traits in stable environments and proposed that poeciliids could be grouped as either having placentotrophy and superfetation or lecithotrophy and the absence of superfetation. Reznick and Miles (1989a) later described a "nearly perfect" association between superfetation and placentotrophy, with only one species, *Poeciliopsis monacha*, known to be lecithotrophic with superfetation. Since then, a more thorough survey of the genus *Poeciliopsis* revealed that all lecithotrophic ($n = 8$) or incipient placentotrophic ($n = 5$) species are capable of having up to two or three simultaneous developing broods (Reznick et al. 2002b; Reznick and Pires, unpublished data). In addition, Arias and Reznick (2000) and Pires (2007) reported the presence of placentotrophy without superfetation in *Phalloceros caudimaculatus* and in the genus *Pamphorichthys*, respectively. It is thus now clear that superfetation and placentotrophy can evolve independently. However, the frequency with which both traits are found together and the joint expression of extensive superfetation and extensive placentotrophy in *H. formosa*, *Poeciliopsis* spp., and *X. ctenolepis* suggest that these traits are indeed correlated, but imperfectly so. The frequent association between superfetation and placentotrophy suggests that the evolution of one of these traits may facilitate the subsequent evolution of the other. The investigation of such a relationship, and of its significance, will contribute greatly to understanding the evolution of reproductive adaptations in poeciliids, as discussed in the remainder of this chapter.

3.4 Evolutionary transitions in the Poeciliidae

Hrbek et al. (2007) present a well-resolved, DNA sequence–based phylogeny for the Poeciliidae that gives us some basis for making inferences about the evolution of livebearing, and variations on the theme of livebearing, in this clade. First, we will suggest an expected sequence of events, and then we will compare our suggestion with what we can infer from the combination of the phylogeny for these taxa and the distribution of life histories throughout the phylogeny.

Livebearing requires internal fertilization, but many egg layers have this capacity, so we assume that an egg-laying ancestor of the Poeciliidae first evolved internal fertilization, then viviparity; that is, the gonopodium and associated internal fertilization are shared, derived traits of the Poeciliidae. Lecithotrophy without superfetation represents the simplest form of viviparity. Turner (1940c) observed that the eggs of lecithotrophic poeciliids were no different in structure from egg layers in other taxa in the order Cyprinodontiformes, in which the family Poeciliidae is included, so it appears that little or no structural modifications in the egg were associated with this transition. We assume that superfetation and placentotrophy demand subsequent adaptations, such as those that allow for increased flexibility in the yolking and in the timing of egg fertilization and birth of young, as well as for a decrease in the amount of yolk that is provisioned before fertilization and for an increased ability to transfer nutrients after fertilization. Logic thus suggests that the sequence of events was, first, to evolve internal fertilization, then egg retention, then simple viviparity (lecithotrophy without superfetation), then either superfetation or placentotrophy. The distribution of these presumably more derived traits, as discussed in the previous section, suggests that either superfetation or placentotrophy can evolve by itself (Pollux et al. 2009). The question now is whether or not the distribution of these traits in the family provides evidence for such logical transitions.

Tomeurus gracilis is an egg-laying species with internal fertilization, thus exhibiting life-history characteristics that we might expect of a basal species in the family. Hrbek et al. (2007) instead found that *Tomeurus* is not the sister taxon to the remainder of the family; *X. ctenolepis* is. We (Reznick et al. 2007a) have recently confirmed the earlier observations of Hubbs (1950) that *Xenodexia* has both superfetation and placentotrophy. This unexpected distribution of life histories in the basal branches of the family tree suggests alternative hypotheses for the way life histories have evolved in this clade. One is that the common ancestor of the family had a life history that was most similar to *Tomeurus*. For this to be true, there must also have been

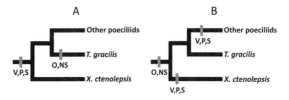

Figure 3.1 Competing hypotheses for the evolution of life-history traits in the basal poeciliid species (based on phylogeny from Hrbek et al. 2007). Hypothesis A requires the common ancestor of poeciliids to have viviparity (V), placentotrophy (P), and superfetation (S), which were lost in the ancestor of *Tomeurus gracilis*. In hypothesis B, the common ancestor of poeciliids has oviparity (O) and nonsuperfetation (NS), thus requiring the origin of viviparity, placentotrophy, and superfetation in *Xenodexia* to be independent of the origin of the same traits in other poeciliids. Under both scenarios, placentotrophy and superfetation (but not viviparity) must have been lost independently many times by other poeciliids. *Phalloptychus januarius* and *Phalloceros caudimaculatus*, the two closest sister taxa to *Tomeurus* in the "other poeciliids" branch in the phylogeny from Hrbek et al. 2007, are both viviparous. *Phalloptychus januarius* exhibits superfetation; *Phalloceros caudimaculatus* does not.

the independent evolution of viviparity, superfetation, and placentotrophy in the ancestor of *Xenodexia*, then again throughout the remainder of the phylogeny. Alternatively, the common ancestor of the family may have been viviparous and may even have had superfetation and/or placentotrophy. For this to be true, these traits must have been lost in the ancestor of *Tomeurus* (fig. 3.1).

It is difficult to distinguish between these alternatives by looking at the poeciliids alone, but this task may be more feasible if we enlarge the analysis to include species from throughout the order Cyprinodontiformes. The sister clade to the family Poeciliidae includes *Oxyzygonectes*, *Jennynsia*, and *Anableps*. The latter two genera have internal fertilization, are viviparous, and are matrotrophic but lack superfetation. *Anableps* has a follicular placenta that is similar to that of some species in the genus *Poeciliopsis* (Turner 1938, 1940b). Development in *Jennynsia* is unlike that of *Anableps* or of any of the poeciliids. The eggs of *Jennynsia* are fertilized while retained in the follicle, but ovulation takes place after the initial development of segments; development is completed in the ovarian lumen (Turner 1940a). *Oxyzygonectes* is an egg layer with external fertilization. It appears that *Oxyzygonectes* is basal to this clade (Hrbek & Meyer 2003), so we are left with alternatives that are similar to those presented by the Poeciliidae: there may have been a common ancestor to both clades that was viviparous, with *Oxyzygonectes* representing a loss of viviparity, or viviparity and placentotrophy may have evolved independently in both clades. We are currently developing a molecular phylogeny that includes representatives from throughout the cyprinodont order with the goal of using it to develop a more highly resolved hypothesis for the evolution of life histories in these taxa. For now, the most likely life history of the common ancestor of the Poeciliidae and the pattern of evolution within the family remain unresolved.

3.5 Ecological hypotheses for the evolution of placentotrophy and superfetation

Despite the uncertainties related to the distribution of reproductive traits within Cyprinodontiformes and, more specifically, poeciliids, all results so far strongly suggest multiple, independent origins of superfetation and placentotrophy within the family. The adaptive significance of these traits, however, remains largely unknown. Some of the hypotheses that have been introduced in the literature to explain the evolution of matrotrophy and superfetation focus on the ecological conditions that may select for these traits. These generally fall into two categories: locomotor performance hypotheses and resource availability hypotheses (Pollux et al. 2009).

3.5.1 Locomotor performance hypotheses

The evolutionary transition from oviparity to viviparity implies an increase in the length of time that a female is physically bound to her developing offspring. Although the female is emancipated from a nest, she must still carry the developing offspring with her during daily activities, including foraging and predator avoidance. Studies in a variety of taxa have shown that egg retention and viviparity reduce locomotor performance (Shine 1980; Bauwens & Thoen 1981; van Damme et al. 1989; Plaut 2002; Ghalambor et al. 2004; Wu et al. 2004). Since matrotrophic species have a smaller initial egg size, the physical burden they carry for a given number of offspring, particularly at the early stages of pregnancy, is smaller than that of lecithotrophic species. Miller (1975) and Thibault and Schultz (1978) suggested that this resulted in a "streamlining" of matrotrophic species, thus reducing the locomotor costs of internal development. Further, they suggested that matrotrophy in concert with superfetation staggers the larger physical burden of later stages of development, amplifying the streamlining effect (fig. 3.2). Several lines of evidence add support to their hypothesis.

The size of the reproductive package a female carries has been linked to locomotor performance in the guppy, *Poecilia reticulata*. Guppies are lecithotrophic; their embryos lose dry mass over the course of development. However, guppy embryos have nearly a fourfold increase in wet mass between fertilization and birth because of an increase in the water content of developing embryos (Ghalambor et al. 2004). Ghalambor et al. (2004) examined different components of the escape response, or C-start (Weihs 1993),

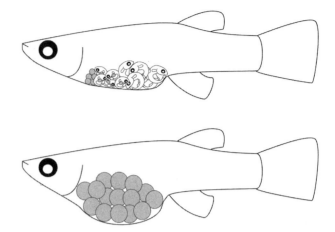

Figure 3.2 Illustration of hypothesized effects of superfetation and placentotrophy on body streamlining of poeciliid species. A lecithotrophic species without superfetation (below) must allocate all resources to offspring prior to fertilization. With superfetation, females are able to have an equivalent fecundity while allocating resources to embryos at different times. When superfetation is coupled with placentotrophy (above), not only is total resource allocation to offspring more spread out over time, but initial allocation is also considerably smaller. A possible functional consequence of this pattern of resource allocation is a more streamlined body, which may reduce the locomotor cost of viviparity.

in pregnant female guppies. They found that stage of pregnancy (and thus the size of the reproductive package) correlated well with maximum velocity, distance traveled, turning angle, and mean rotational velocity during an escape, with females at later stages of pregnancy, and thus with larger reproductive packages, showing impairment. They also found that guppies from high-predation localities, which have a higher reproductive allocation on average (e.g., Reznick et al. 1990; Reznick et al. 1997), performed better in many aspects of the escape response than low-predation guppies; however, they only did so when they were carrying embryos that were in earlier stages of development. High-predation guppies experienced a more rapid decline in velocity and distance traveled during the escape as the pregnancy progressed, suggesting a greater locomotor cost of reproduction for high-predation females due to the larger embryo size at later stages of reproduction.

Walker et al. (2005) verified the assumption that a faster escape response decreases the likelihood of predation during predator-prey interactions. They examined likelihood of predator evasion using the guppy and a natural predator, the pike cichlid *Crenicichla alta*, and found that predator evasion was positively correlated with two locomotor performance variables: rapid rotational velocity in the early stage of the escape response and "rapid tangential acceleration," a composite variable including net distance traveled, maximum velocity, and maximum acceleration. This result thus confirms that the reduced performance associated with pregnancy will increase susceptibility to predation.

Locomotor costs of pregnancy are not limited to predator-prey interactions, however. Routine swimming, such as that used to maintain position against water flow velocity, can also be deleteriously affected during pregnancy. Studies of pregnant *G. affinis* revealed a significant decrease in critical swimming speed (Brett 1964) over the course of pregnancy (Plaut 2002). Body mass and cross-sectional area of the female increased as the pregnancy progressed. Surprisingly, the study did not find any change in tail beat amplitude or frequency when swimming at a fixed speed as the pregnancy progressed. Tail beat frequency and/or amplitude would be expected to increase if the decrease in critical swimming speed was due to the physical burden of pregnancy, which would result in increased drag and reduced flexibility, indicating that a female was working harder to swim at a fixed speed. Plaut (2002) suggests that the locomotor cost is thus caused by a physiological impairment rather than a physical burden during pregnancy.

Zúñiga-Vega et al. (2007) examined the relationship between superfetation and stream velocity in six inland and six coastal populations of the lecithotrophic *Poeciliopsis turrubarensis*. Because superfetation allows a female to carry multiple broods at different stages of development, the proportion of developing embryos that are in the largest stages of development at a given time can be decreased without reducing fecundity (fig. 3.2). Inland populations of *P. turrubarensis* live in areas with higher water velocity than coastal populations and thus are expected to have a higher cost of locomotion. The authors therefore predicted that inland populations would exhibit a higher degree of superfetation. Inland populations did have higher levels of superfetation, as well as larger brood sizes and a higher reproductive investment overall. The authors pointed out that superfetation allowed an increase in the rate of offspring production while bypassing the associated cost in increased physical burden.

Finally, a study of life-history traits in another fish family, the Zenarchopteridae, revealed that, in the genus *Dermongenys*, matrotrophic species produced fewer, larger offspring than lecithotrophs, while the trend was reversed in the genus *Nomorhampus*: matrotrophic species produced more, smaller offspring (Reznick et al. 2007b). The only consistent life-history correlate with matrotrophy was reproductive allocation, which was lower in matrotrophic species in both genera. The cost of locomotion was not directly addressed in this study, but a reduced cost of locomotion is implied in matrotrophic zenarchopterids if decreased reproductive allocation is correlated with increased locomotor performance.

The studies discussed above point to both matrotrophy and superfetation being consistently correlated with reduced reproductive allocation and, consequently, with a reduced burden of pregnancy via improved locomotor performance. It is noteworthy, however, that all evidence thus far is circumstantial; no study conducted to date has specifically addressed differences in locomotor abilities between matrotrophic and lecithotrophic species or between species with and without superfetation.

3.5.2 Resource availability hypotheses

Another subset of ecological hypotheses focuses on resource availability. Matrotrophic females start reproduction with small eggs and provide additional nutrient investment to offspring throughout development, allowing these females to spread investment into offspring over a greater period of time than strictly or predominantly lecithotrophic females can. Such differences in allocation patterns can be studied under the framework of life-history adaptations, as differences between capital and income breeders (Drent & Daan 1980; Houston et al. 2007), and this approach may hold the key to identifying a possible advantage of matrotrophy.

Trexler and DeAngelis (2003) developed a combination of analytical and simulation models to investigate what resource conditions would favor the matrotrophic reproductive mode over the lecithotrophic mode. The analytic model examines the reproductive success of each reproductive mode during a single reproductive event, assuming a size-number trade-off in initial egg number. For simplicity, it also assumes that terminal offspring size is the same for both reproductive strategies. From a set amount of starting resources, a lecithotroph produces fewer, fully yolked eggs. Since the matrotroph starts with a smaller egg, she can produce more eggs initially, which will then need additional resource investment throughout gestation. If resources are consistently abundant during the gestation period, the matrotroph thus has the potential to produce a greater brood size than the lecithotroph. However, if food levels are low or unpredictable, producing a large number of eggs may be counterproductive for the matrotrophic female because she risks having insufficient resources to nourish all offspring and may thus lose the entire brood. This leads to an important assumption in the model: when resources become scarce, the matrotroph is assumed to have the ability to abort and resorb energy from some offspring within the brood. The simulation model expands on the analytical model by examining lifetime reproductive success of each reproductive mode across a range of resource levels and embryo resorption capabilities. It takes growth, storage, and schedule of reproduction into account. Overall, the model suggests that matrotrophy is most likely to evolve in habitats where abundant resources are consistently available. The ability to abort and resorb offspring expands the conditions under which a matrotroph would have higher lifetime reproductive success: without the ability to abort, the predicted habitat range for species with a matrotrophic reproductive mode is narrowed to only areas with the highest, most predictable resource levels.

Recent empirical studies do not support the assumption that matrotrophic species abort offspring in low food conditions. Marsh-Matthews and Deaton (2006; Marsh-Matthews, **chapter 2**) examined the effect of food level on the reproduction of *G. geiseri*. This species is primarily lecithotrophic, but it has been found to allocate small amounts of nutrients to embryos after fertilization, particularly under high resource levels (Marsh-Matthews & Deaton 2006). Although the authors did find indications of abortion, it was independent of food level: both high- and low-food treatments showed similar frequencies of aborted embryos. Banet and Reznick (2008), using closely related placentotrophic and lecithotrophic species from the genus *Poeciliopsis*, found no evidence that placentotrophic species abort offspring in low food conditions. Instead, they found that when food level was reduced after fertilization, placentotrophic females produced smaller offspring and sacrificed body condition to maintain all embryos in a developing brood.

Similarly to placentotrophy, the other facet of reproductive mode variation in poeciliids, superfetation, also allows a female to stagger periods of heightened investment over time without reducing fecundity. Like the evolution of matrotrophy, then, the evolution of superfetation may also be studied in the context of varying resource allocation strategies. For a set rate of offspring production in a lecithotroph, superfetation decreases resource investment in each brood due to a reduction in brood size but increases the frequency at which broods are initiated. In a matrotroph with superfetation, a female spreads resource investment over the course of gestation. In species with MI > 1, the dry mass of embryos increases nonlinearly during development (e.g., Reznick et al. 2007a, for *X. ctenolepis*) in such a way that maternal investment progressively increases as development occurs. Superfetation allows the matrotroph to space out these periods of increased investment, so that fewer offspring are at the growth phase of development at a given time. The Trexler-DeAngelis model is currently being modified to take into account these studies and to include superfetation (J. C. Trexler, pers. comm.).

3.6 Summary and future research

The adaptive significance of matrotrophic reproduction remains one of the least studied aspects of life-history evolu-

tion. The extensive variation on the theme of livebearing exhibited by poeciliid fishes makes them excellent model organisms for studies aiming to address different aspects of the evolution of matrotrophy and, more specifically, placentotrophy. As described throughout the chapter, the past 20 years have seen some progress in the first three study directions proposed by Reznick and Miles (1989a) for such investigations. In summary:

1. Life-history descriptions of key species have revealed the independent origins of placentotrophy and superfetation (e.g., *X. ctenolepis*, Reznick et al. 2007a; the subgenus *Micropoecilia*, Pires et al. 2010; all species in the genus *Poeciliopsis*, Reznick et al. 2002b; Reznick and Pires, unpublished data). In addition, interpopulation comparisons have started to identify the environmental and genetic components of within-species variation in reproductive mode (e.g., Schrader & Travis 2005; Pires et al. 2007), thus providing raw material for investigations of life-history evolution (Reznick & Travis 1996).
2. The descriptive work of life histories within the genus *Poeciliopsis*, coupled with its phylogenetic study (Reznick et al. 2002b) and recent theoretical advances (Trexler & DeAngelis 2003), has made it possible to compare pairs of sister species with contrasting modes of reproduction in common garden conditions within the context of a hypothetico-deductive framework (Banet & Reznick 2008).
3. Phylogenetic studies (e.g., Breden et al. 1999; Ghedotti 2000; Mateos et al. 2002; Lucinda 2005a, 2005b; Hrbek et al. 2007; Meredith et al. 2010) have focused on different taxonomic levels and have started to provide a much needed basis upon which inferences of evolutionary trends within poeciliids can be drawn.

The current knowledge concerning variation and patterns of evolution in life-history traits and reproductive mode in the family Poeciliidae provides a good foundation for understanding selective factors for the evolution of these traits. However, there is still much work to be done; the adaptive significance of placentotrophy, in particular, is still largely unknown. We suggest that the research agenda described above be further explored; in addition, this agenda should be expanded to address the following two general areas:

1. The examination of the relationship among life-history traits. Are there sets of integrated traits that give us clues to understand the observed patterns of evolution?

If so, what are the functional and physiological consequences of the combination of such traits?

2. Experimental studies examining current hypotheses for the evolution of the placenta.

The development of more comprehensive phylogenetic reconstruction of relationships within poeciliids and between the family and its sister groups will allow us to test specific predictions derived from these hypotheses within a phylogenetic, comparative context. Multiple independent origins of placentation within the family make it an excellent group to conduct comparative studies to test current models of life-history evolution. Common garden studies focusing on life-history variation within the family and large-scale comparative analyses will certainly contribute insights into the questions of how and why placentotrophy, superfetation, and their correlated life-history traits evolved in poeciliids.

Acknowledgments

We thank Jon Evans, Ingo Schlupp, Eric Schultz, Stephen Stearns, and an anonymous reviewer for providing several comments and suggestions that improved this chapter. We also thank C. R. Moreira for preparing the illustration for fig. 3.2. Our research has been supported in part by National Science Foundation grants DEB0416085 and EF0623632 to D.N.R.

Chapter 4 Ecological drivers of life-history divergence

Jerald B. Johnson and Justin C. Bagley

4.1 Introduction

FISHES IN THE FAMILY Poeciliidae (Rosen & Bailey 1963) show a wide diversity of life-history traits (e.g., Thibault & Schultz 1978; reviewed in Pires et al., **chapter 3**). This is somewhat surprising, given the complex set of adaptations associated with viviparity that characterize these fishes. For example, all males internally inseminate females using a modified anal-fin gonopodium, and all females (with the single exception of the egg-laying species *Tomeurus gracilis*) give birth to precocial, free-swimming young. Such reproductive adaptations rely upon a complex, coordinated set of anatomical, behavioral, and physiological traits (Constantz 1989). Yet within this highly specialized mode of reproduction, we still find a remarkable amount of phenotypic variation in life-history traits both among and within species.

Several lines of evidence suggest that diversity in poeciliid life histories is the result of natural selection. Poeciliids occupy a wide range of habitat types (Meffe & Snelson 1989a), and associations between ecological conditions and life histories provide clues as to what factors might be driving evolutionary divergence. The sheer number of poeciliid species and life-history differences among them provide an excellent opportunity to employ a comparative framework to infer links between life-history evolution and agents of natural selection. In fact, in their seminal review of poeciliid life-history patterns two decades ago, Reznick and Miles (1989a) called for expanded comparative research, both within and among species. They also demonstrated the value of this approach, first by focusing specifically on

superfetation and its link to the matrotrophy-lecithotrophy continuum of maternal provisioning of embryos (see Pires et al., **chapter 3**, for a follow-up to this work), and second by exploring patterns of covariation in life-history traits within and among species (Reznick & Miles 1989a). However, their review also revealed that much of our understanding about poeciliid life histories at the time was restricted to a few well-studied species (e.g., *Poecilia reticulata*, *Poecilia latipinna*, and *Gambusia affinis*), and that even the most basic description of life histories for most species was absent or incomplete. Even less common were descriptions of ecological correlates associated with life-history traits that might be used to infer the causes of life-history divergence, or experimental studies to confirm causal relationships. This lack of taxonomic breadth and of experimental depth was a fundamental limitation 20 years ago. Here, we consider what progress has been made toward understanding the role that ecological interactions play in driving life-history evolution. Our primary focus in this chapter is on studies that document life-history diversification *within* species. An advantage of considering intraspecific comparisons is that populations are likely to occupy the selective environments actually responsible for evolutionary divergence, thus pointing to probable causes.

We address four topics. First, we highlight the range of life-history variation found within poeciliid species. This exercise exposes those taxonomic groups where basic life-history descriptions are available to make comparisons among populations. Second, we describe the kinds of ecological interactions thought to be important in driving the evolution of poeciliid life histories. We present these hypotheses in the context of life-history theory, pointing to specific predictions of trait evolution where possible. Third, we review several empirical studies of phenotypic life-history divergence in poeciliids. Our objective is to describe the link between theoretical models and actual patterns of phenotypic divergence in select species where sufficient work has been done to infer causal agents of divergence. We find that in some cases, similar agents of natural selection have led to almost identical patterns of intraspecific life-history divergence across multiple species. This suggests that parallel life-history evolution might be common in the family. Finally, we conclude our review by highlighting areas where future research is needed.

4.2 Studying poeciliid life histories: taxonomic scope and depth

Studies of life-history evolution typically begin by focusing on phenotypic variation among populations or among species. The term "life-history phenotype" is used to describe the suite of traits that define the reproductive strategy for an "average" wild-caught individual within a population or species. It is important to recognize that such averaging ignores information on variation among individuals; the issue of individual variation in life-history traits deserves attention but to date has still received little consideration. Instead, most life histories are reported and evaluated as population (or species) averages. One trait, superfetation, is often scored as either "present" or "absent," and when variation is considered, it is usually species that are compared, not populations (Turner 1937; Reznick & Miles 1989a; Pires et al. 2007; but see Zúñiga-Vega et al. 2007).

For females, the life-history phenotype is composed of up to seven interrelated reproductive traits and for males typically just two reproductive traits are measured (box 4.1). Differences in life-history phenotypes observed among populations can be the result of heritable variation, phenotypic plasticity, or a combination of the two. To argue that life-history evolution has occurred, observed differences must be at least partly heritable. This can be demonstrated either by a common-garden experiment or a reciprocal-transplant experiment. If differences observed in the field persist under common environmental conditions after two generations in the laboratory, then we can be sure that evolutionary divergence has occurred, that maternal effects are not responsible for differences, and that phenotypic variation has a genetic component. Yet most available descriptions of poeciliid life-history traits come from wild-caught individuals. In these cases, any differences discussed in an evolutionary context are assumed to be heritable, an assumption that,

---

Box 4.1 Nine life-history traits typically measured in poeciliids fishes

To appreciate the range of life-history strategies employed among poeciliid species, it is helpful to examine the specific traits that make up these strategies. Note that each of these life-history traits is known to vary among species, and several also vary among populations within species.

Females

Most life-history descriptions focus on females and include measures of seven distinct traits: (1) presence or absence of superfetation and, if present, the number of broods carried simultaneously; (2) interbrood interval, defined as the number of days between successive parturition events; (3) degree of maternal provisioning to embryos following fertilization, varying from lecithotrophy to matrotrophy; (4) the amount of energy invested in each brood, known as the reproductive allotment, which is typically measured as the dry mass of all developing embryos in a single brood (sometimes presented as a percentage of total maternal dry mass or with female mass as a covariate); (5) number of offspring per brood, a simple count; (6) size of offspring per brood, measured by mass; and finally (7) age and size at first reproduction, typically measured as the smallest size class in which greater than half of the females are carrying fertilized embryos.

Males

Much less effort has focused on male life histories. Those studies that have examined males typically quantify just two life-history traits: (1) the amount of energy invested in sperm production (male reproductive allotment), measured by testes mass; and (2) size at maturity, taken as the average size of all mature males in a population, which is justified by the fact that upon maturation males almost completely cease growth.

---

when tested, is frequently confirmed (Reznick & Endler 1982; Johnson 2001a; but see Grether et al. 2001b).

Most of what we know about life-history evolution in the poeciliids comes from a small subset of the total number of species in the family. Despite limited sampling of species, the range of interspecific variation in poeciliid life-

Table 4.1 Range of reported phenotypic variation in life-history traits for species considered in this review

Life-history trait[a]	Species	Divergence level (range of trait value)	Number of populations	Reference[b]
Females				
Superfetation	*Poeciliopsis turrubarensis*	1.9-fold (1.4–2.6 broods)	12	1
	Poeciliopsis prolifica	1.1-fold (3.2–3.4 broods)	2	2
Interbrood interval	*Poecilia reticulata*	1.4-fold (21.9–31.1 days)	3	3, 4, 5
	Heterandria formosa	1.4-fold (7.5–10.6 days)	2	5, 6, 7
	Gambusia affinis	1.3-fold (18.0–24.0 days)	6	8
Matrotrophy (index)	*Poeciliopsis turrubarensis*	1.9-fold (0.8–1.5)	12	1
	Xenodexia ctenolepis	1.2-fold (3.4–4.2)	3	9
Reproductive allotment	*Gambusia affinis*	4.6-fold (9.6%–44.0%)	24	10
	Poecilia reticulata	3.0-fold (7.1%–21.1%)	4	11, 33
	Brachyrhaphis episcopi	2.8-fold (3.7%–10.4%)	12	12
	Brachyrhaphis rhabdophora	2.1-fold (6.7%–14.2%)	5	13
	Gambusia holbrooki	2.0-fold (7.7%–14.3%)	2	14
	Phalloceros caudimaculatus	1.8-fold (4.1%–7.5%)	3	15
	Poecilia picta	1.4-fold (15.7%–22.5%)	4	16
	Poeciliopsis prolifica	1.0-fold (8.2%–8.3%)	2	2
Number of offspring	*Gambusia affinis*	19.9-fold (5.5–109.2 ind.)	31	10, 17, 18, 19
	Poecilia reticulata	8.8-fold (1.3–11.4 ind.)	20	11, 20
	Poeciliopsis occidentalis	8.0-fold (2.4–19.22 ind.)	≥6	21, 22, 23, 24
	Brachyrhaphis rhabdophora	6.3-fold (3.2–20.3 ind.)	27	25
	Poecilia latipinna	3.7-fold (10.1–37.3 ind.)	7	26
	Brachyrhaphis episcopi	3.5-fold (2.4–8.4 ind.)	12	12, 27
	Xenodexia ctenolepis	1.7-fold (13.4–22.1 ind.)	3	9
	Poeciliopsis prolifica	1.0-fold (4.1–4.2 ind.)	2	2
Size of offspring	*Brachyrhaphis rhabdophora*	2.8-fold (1.3–3.6 mg)	27	25
	Poecilia reticulata	2.4-fold (0.7–1.7 mg)	28	20, 33
	Neoheterandria tridentiger	1.9-fold (0.7–1.3 mg)	2	28, 29
	Heterandria formosa	1.8-fold (0.4–0.7 mg)	4	30
	Phalloceros caudimaculatus	1.8-fold (0.3–0.6 mg)	4	15
	Gambusia affinis	1.7-fold (1.3–2.2 mg)	24	10

history traits is still quite vast (Reznick & Miles 1989a; Pires et al., **chapter 3**). What is impressive is that variation in life-history phenotypes observed *within* some species actually rivals that found among species (table 4.1), suggesting that much of the diversity observed in the family could have evolved initially at the intraspecific level.

A brief overview of intraspecific life-history variation in poeciliids reveals some interesting patterns. First, some life-history traits show a greater range of phenotypic variation among populations than others (fig. 4.1); and second, there are clear differences among species in the extent of among-population trait variation (table 4.1). Overall, mean trait values among populations within species range from 1- to 20-fold. Two measures of reproductive investment—

number of offspring and reproductive allotment—show the widest range of interpopulation variability across species. In contrast, other components of the life-history phenotype appear far more constrained: interbrood interval tends to show little variability among populations within species, based on current sampling. Here, we consider each of these patterns in turn.

Documentation of intraspecific variation remains uncommon. This may be due, in part, to difficulty associated with collecting the data; for example, interbrood interval measurements must be accomplished using live fish, and usually in a controlled setting. Regardless, the relative paucity of comparative data among populations has implications for the patterns we are presenting. That is, sampling

Table 4.1 (continued)

Life-history trait[a]	Species	Divergence level (range of trait value)	Number of populations	Reference[b]
Females				
Size of offspring	*Brachyrhaphis episcopi*	1.7-fold (2.3–3.8 mg)	12	12, 27
	Gambusia marshi	1.4-fold (2.1–2.9 mg)	7	26, 31
	Poecilia latipinna	1.2-fold (3.0–3.7 mg)	4	26
	Poecilia picta	1.1-fold (0.8–0.9 mg)	4	16
Size at first reproduction	*Brachyrhaphis rhabdophora*	1.9-fold (20.0–38.0 mm SL)[c]	27	25
	Poecilia reticulata	1.7-fold (12.0–20.0 mm SL)	16	20
	Gambusia affinis	1.6-fold (16.0–26.0 mm SL)	4	32
	Brachyrhaphis episcopi	1.5-fold (21.0–31.0 mm SL)	12	12
	Phalloceros caudimaculatus	1.4-fold (19.0–27.0 mm SL)	4	26
	Poecilia latipinna	1.3-fold (27.8–37.1 mm SL)	7	26
	Gambusia marshi	1.3-fold (22.0–28.0 mm SL)	11	26
	Poecilia picta	1.1-fold (18.0–20.0 mm SL)	4	16
Males				
Male reproductive allotment[d]	—	—	—	—
Size at first reproduction	*Poecilia latipinna*	1.8-fold (26.5–48.4 mm SL)	7	26
	Brachyrhaphis rhabdophora	1.7-fold (18.9–31.7 mm SL)	27	25
	Poecilia reticulata	1.4-fold (13.0–18.5 mm SL)	28	20, 33
	Brachyrhaphis episcopi	1.3-fold (17.9-23.9 mm SL)	12	12
	Poecilia picta	1.3-fold (16.0–20.0 mm SL)	4	16
	Phalloceros caudimaculatus	1.2-fold (15.0–18.0 mm SL)	3	15

Note: Reported values are population means. In most cases data come from wild-caught fish and so could include both genetic and environmental effects.
[a]See box 4.1.
[b]1–Zúñiga-Vega et al. 2007; 2–Pires et al. 2007; 3–Rosenthal 1952; 4–Thibault & Schultz 1978; 5–Turner 1937; 6–Cheong et al. 1984; 7–Seal 1911; 8–Stearns 1983b; 9–Reznick et al. 2007a; 10–Stearns 1983a; 11–Rodd & Reznick 1991; 12–Jennions et al. 2006; 13–Reznick et al. 1993; 14–Meffe 1991; 15–Arias & Reznick 2000; 16–Reznick et al. 1992; 17–Scribner et al. 1992; 18–Trendall 1982; 19–Krumholz 1948; 20–Reznick & Endler 1982; 21–Scrimshaw 1944b (cited in Reznick & Miles 1989a); 22–Constantz 1974; 23–Constantz 1979; 24–Schoenherr 1977; 25–Johnson & Belk 2001; 26–Reznick & Miles 1989a, references therein; 27–Jennions & Telford 2002; 28–Miller 1960; 29–Stearns 1978; 30–Leips & Travis 1999; 31–Meffe 1985a; 32–Stockwell & Vinyard 2000; 33–Reznick et al. 1996c.
[c]SL = standard length.
[d]We are not aware of published accounts of variation among populations within species for average male gonad mass.

bias might contribute to the patterns we see, particularly given that certain life-history traits have been studied more extensively than others. Consequently, as life-history data accumulate for other species, variation seen in outliers (e.g., offspring size in *Brachyrhaphis rhabdophora*; fig. 4.1) may not be quite as extreme as it currently seems. However, even in those traits for which little data are currently available, we still observe some degree of variation within species. For example, plasticity in maternal provisioning has been documented in both *Poecilia latipinna* (Trexler 1997) and *Gambusia geiseri* (Marsh-Matthews et al. 2001). Similarly, although relatively little work has been done on variation in matrotrophy index within species, Zúñiga-Vega et al. (2007) showed that even this trait varied among popula-

tions of *Poeciliopsis turrubarensis*, ranging almost twofold from 0.8 to 1.5. Hence, although we recognize that sampling issues present some challenges, the pattern of variability among traits presented here likely reflects overall trends in trait variation present within the family.

Our survey of variation among life-history traits begs an important question: why do some life-history traits vary more within species than others? To date, no hypotheses have been generated to predict such differences. However, several factors could be at play, including physical constraints of design, low levels of heritability, life-history trade-offs, and possible differences in the influence of different selective pressures on specific traits (Stearns 1989; Partridge & Sibly 1991; Roff 1992; Ricklefs & Wikelski

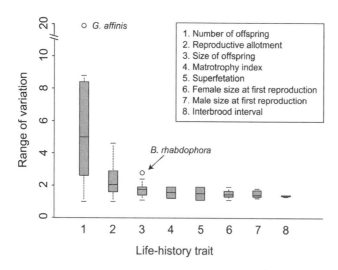

Figure 4.1 Intraspecific variation within species for eight life-history traits. The *y*-axis represents the range of variation among populations within species; for example, a value of 2 indicates a twofold difference between the lowest and highest values reported among populations within a single species (box and whiskers represent median, interquartile, and range; extreme values are marked by open circles). See table 4.1 for species included to generate the boxplots.

2002). Some traits may appear more conserved because there is little genetic variation available at responsible loci. If natural selection has little variation to work with, this would limit evolutionary divergence, even in the face of strongly divergent selective pressures. Differences in environmental variation may also contribute to observed patterns. Some species (e.g., *Poecilia gillii*, which is broadly distributed throughout Central America) occupy a range of habitat types, while others (e.g., *Phallichthys quadripunctatus*, found in small pools along the eastern border of Costa Rica and Panama) are exposed to a more narrow range of selective pressures. Hence, variation in life histories within species is linked to ecological factors that limit or permit species distribution patterns; because of this, the interface of life-history studies with biogeography and phylogeographic inference still holds much promise. The lack of evolutionary divergence seen for some life-history traits might also be linked to speciation itself. That is, intraspecific variation for some traits might be uncommon because variation has been partitioned between species. Understanding how such trait diversity arises within species remains a compelling problem in poeciliid life-history research.

We know comparatively little about variation among populations for male life-history traits relative to our knowledge of female life histories. We found no studies that report variation among populations for average male gonad mass; rather, most work has focused on variation in testes size among individuals in relation to sperm competition (e.g., Constantz 1984). Average male size at maturity has been reported in some species and does vary among populations within species (table 4.1). We note that there is

a reported 1.2-fold difference among populations in *Phalloceros caudimaculatus* (15.0–18.0 mm standard length [SL]; Arias & Reznick 2000) at the low end to a 1.8-fold difference in *Poecilia latipinna* (26.5–48.4 mm SL; see references in Reznick & Miles 1989a). However, the range of variation for male size at maturity is still modest relative to some life-history traits (fig. 4.1). Understanding reproductive timing in male poeciliids is perhaps more complicated than understanding selection on female life-history traits, in part because sexual selection plays a prominent role. For example, in *Brachyrhaphis rhabdophora*, even though differences in average male size at maturity were clearly associated with mortality risk (see below), the range of variation in male body size within populations (as measured by the coefficient of variation) showed equivalent levels of variation, regardless of mortality risk. The likely explanation for this is that males are adopting a range of alternative size-associated mating strategies (Johnson & Belk 2001).

4.3 Ecological hypotheses of life-history evolution

The diversity of life-history traits found within species begs an explanation. Differences in selective pressures among sites may hold the answer. Several selection-based ecological hypotheses have been advanced to explain the wide range of life-history strategies observed in the wild. These hypotheses are underpinned by the fact that poeciliids occupy a vast range of ecological habitats (Meffe & Snelson 1989a; Trexler et al., **chapter 9**; Kelley & Brown, **chapter 16**) and by the expectation that different ecological conditions favor different life-history trade-offs. Most of these hypotheses are also anchored in the framework of life-history theory (Roff 1992; Stearns 1992, 2000). In fact, poeciliid fishes have frequently been used as model organisms for empirical tests of life-history theory (Reznick & Endler 1982; Reznick et al. 1990; Johnson & Belk 2001; Jennions et al. 2006; Marsh-Matthews & Deaton 2006; Zúñiga-Vega et al. 2007). We consider four sets of ecological factors that have been hypothesized to drive the diversification of life histories in poeciliids by natural selection—these factors are (1) differential mortality, typically ascribed to predators, (2) density, (3) resource availability, and (4) abiotic factors such as streamflow, water temperature, and water chemistry. We consider these factors individually but recognize that in nature, multiple selective agents often contribute to evolutionary divergence (Wilbur et al. 1974; Johnson 2002; Reznick et al. 2002a).

4.3.1 Differential mortality

Birds, piscivorous fishes, and aquatic invertebrates are all known to prey upon poeciliids (Britton & Moser 1982;

Rodd & Reznick 1991; Mattingly & Butler IV 1994; Reznick et al. 1996b; Rosenthal et al. 2001; Johansson et al. 2004; Tobler et al. 2007b). Yet we know remarkably little about mortality schedules for most species in the wild. Life-history theory, however, makes several explicit predictions about how traits should evolve if mortality is age or stage specific or if there are overall differences in the level of mortality between species or among populations within species.

Differential mortality among age classes or size classes is predicted to shape the timing and size of individuals at first reproduction, the amount of energy invested in reproduction, the distribution of the number and size of offspring in each brood, and the length of reproductive life span. Both males and females should be affected. Collectively, this suite of theoretical expectations has come to be known as the age-specific mortality hypothesis. This hypothesis makes the following predictions. If mortality is high in adult size classes relative to juvenile size classes, or in large individuals relative to small individuals (assuming size and age are related as they are in females with indeterminate growth), then the population is predicted to evolve (1) decreased age and size at maturity, (2) increased reproductive effort, (3) more offspring per brood, (4) smaller offspring, and (5) decreased reproductive life span (Gadgil & Bossert 1970; Schaffer 1974; Brockelman 1975; Law 1979; Michod 1979; Charlesworth 1980). In species capable of superfetation, an increase in reproductive effort could be manifested by increased energy invested in each brood, decreased interbrood interval (i.e., increased number of broods carried), or a combination of both. The rationale for these predictions is that if the risk of mortality increases as individuals grow older and larger (as is frequently the case with predators that seek the largest possible meal per capture), then the adaptive strategy is to shift reproduction to early in life at smaller body sizes and to give birth to many, small offspring as rapidly as possible. Whereas the classic application of the age-specific mortality hypothesis requires differences in age-specific mortality rates among populations to achieve evolutionary divergence in life histories, more recent theoretical work predicts that overall differences in mortality rates among populations, even if the rates are uniform across all or most age classes, might achieve the same evolutionary outcome (Kozlowski & Uchmanski 1987; Abrams & Rowe 1996; Reznick et al. 1996b).

4.3.2 Density

In nature, poeciliid population densities can vary dramatically among species, and even among populations within species (Schoenherr 1977; Chapman & Kramer 1991a; Schaefer et al. 1994; Leips & Travis 1999; Soucy & Travis 2003). There is also ample evidence that in some species densities change over time, typically in response to seasonal environmental changes. For example, variation in rainfall and its associated effects on stream and river size may be an important factor in mediating temporal changes in tropical systems, whereas changes in temperature are probably more important in temperate systems. Other extrinsic ecological factors—such as predation, parasitism, or competition—might also regulate populations such that populations persist at density levels well below what they might otherwise achieve in the absence of such effects.

Life-history theory makes several predictions about how traits should evolve in response to different density levels. These hypotheses in their simplest form make assumptions about the potential of populations to grow and what kinds of life-history traits would be favored in populations at carrying capacity versus those that are well below it (Boyce 1984; Mueller et al. 1991; Mueller 1997). Although it is perhaps an oversimplification (Nichols et al. 1976; Pianka 1979; Boyce 1984) to classify species as r-selected and K-selected (terms used to reference the intrinsic rate of population increase and carrying capacity, respectively, in the Euler-Lotka equation; Mertz 1970; Reznick et al. 2002a), there are general predictions about how traits might evolve under differing density conditions. Early hypotheses predicted that when density is low and has little effect on survival rates, then the production of small offspring and many offspring will be favored; when density is high, fewer but larger offspring should be favored (MacArthur & Wilson 1967; Pianka 1970; MacArthur 1972). This same body of theory predicts that high density will favor delayed maturity, larger body sizes, and an extended reproductive life span relative to fish in lower-density environments.

Classic models of density-dependent life-history evolution have given way to demographic models that incorporate density with age or stage structure (Stearns 2000; Caswell 2001; Reznick et al. 2002a). Some researchers are now generating estimates of age- or stage-specific fecundity and mortality rates and then using a population matrix modeling approach to estimate the population growth rate (a surrogate for fitness). Density effects are incorporated by asking how each vital rate in the matrix changes as a function of density (Neubert & Caswell 2000; Caswell 2001; Metcalf & Pavard 2007). Sensitivity analysis of the matrix can reveal which life-history traits and which life stages are most important in their overall contribution to fitness, which also reveals which traits are expected to evolve, how rapidly, and in what direction. Another advantage of this approach is that the predicted effects of density on life-history evolution are specific to the poeciliid system under investigation. Thus, unlike the classic r versus K paradigm, this demographic approach allows the researcher to generate specific predictions concerning the nature of life-history evolution in any given population.

4.3.3 Resource availability

As a group, poeciliid fishes eat a variety of foods (Mansfield & Mcardle 1998; Garcia-Berthou 1999; Fares Alkahem et al. 2007). Some species are resource specialists, with extremely restricted diets (e.g., the piscivorous *Belonesox belizanus*). However, most species are classified as omnivores and are quite catholic in food consumption, with diet items ranging from algae and detritus to terrestrial insects that fall into the water or are swept in during rain events. Measuring resource availability in poeciliid populations is not trivial (see Grether & Kolluru, **chapter 6**), and relatively few studies have made direct measures of resource abundance. Most often, surrogates of resource availability (e.g., canopy cover, photosynthetic active radiation) are used to quantify the amount of food available. These studies reveal that there are differences in resource type and quantity among habitats and, not surprisingly, that resource availability fluctuates over time.

Life-history theory makes predictions about how traits should evolve as a function of resource abundance. Heterogeneity in resource availability should exert strong selective pressures on offspring size (Brockelman 1975; Sibly & Calow 1983; Parker & Begon 1986). However, different theoretical models have conflicting predictions about the directions that specific traits (especially offspring size and size at first reproduction) should evolve (Stearns 1992; Mylius & Diekmann 1995; Abrams & Rowe 1996). Some optimality models predict cogradient variation: when resources are scarce, populations evolve low growth rates, resulting in smaller size at first reproduction if the timing of maturity is fixed (reviewed in Arendt 1997). This prediction runs counter to countergradient variation models (reviewed in Arendt 1997) and classic *r*- and *K*-selection reasoning (Pianka 1970). The latter posits that low food availability that leads to higher competition will favor the evolution of fewer but larger offspring that are more competitive and experience later onset of maturity. This, of course, assumes that juvenile survival is size dependent, as would be expected during scramble competition for limited resources (Arendt 1997) or size-selective predation (Brooks & Dodson 1965; Mattingly & Butler IV 1994). Brockelman's (1975) fitness-resource theory, by contrast, suggests that in the face of fluctuating resources natural selection on parents will regulate population dynamics via adjustments in clutch size rather than in parental investment. Consistent and abundant resources may favor the evolution of superfetation or an increase in the degree of superfetation, as this would reduce peak demand for food required under a non-

superfetating strategy (Pires et al., **chapter 3**). Several other life-history traits are known to show a plastic response to decreased food availability, including an increase in interbrood interval, a decrease in reproductive allotment, and decrease in brood size. In several poeciliid systems, resource availability covaries with predation regime and stream location, with fishes from low-predation sites often having lower resource availability than their counterparts positioned downstream in highly productive, open-canopy sites experiencing higher predation pressure (Johnson 2002; Arendt & Reznick 2005). Thus, experiments are required to tease apart the contribution of these major drivers of life-history divergence in natural populations.

Resource availability has also been implicated as a key factor in the evolution of matrotrophy (but see Crespi & Semeniuk 2004 for a nonecological explanation based on parent-offspring conflict). Trexler and DeAngelis (2003) generated a mathematical model and a simulation experiment to predict the resource conditions under which matrotrophy should evolve from a lecithotrophic ancestor. This model assumes that a set amount of resources is available to create an initial brood of offspring, and that final offspring size is the same under both trophic strategies. Given these conditions, matrotrophic species should be able to produce more offspring than lecithotrophic species, because the initial offspring size in the matrotroph is smaller. However, the matrotroph requires a constant food source throughout embryo development, and failure to secure resources could result in the loss of the brood. In contrast, the lecithotroph is not at risk of losing the brood because all nutrients are provisioned to the brood at the beginning of development. Hence, matrotrophy should evolve when resources are abundant and constant, and the opposite conditions should favor lecithotrophy. If matrotrophs are not able to reabsorb embryos, then the resource availability conditions under which matrotrophy can evolve are contracted (Trexler & DeAngelis 2003).

4.3.4 Abiotic factors

Abiotic factors might also shape life-history evolution in poeciliid fishes. Factors such as water temperature and water chemistry are known to induce a plastic life-history response in some poeciliids (Snelson et al. 1986; Meffe 1992; McManus & Travis 1998; Karayucel et al. 2008), yet no specific body of theory predicts adaptive life-history evolution as a function of these factors. In contrast, water velocity has been implicated as a potential driver of poeciliid life-history evolution through its effect on the evolution of body shape. Pregnancy in poeciliids results in a change in body shape that increases the cost of locomotion (Plaut 2002). Hence, there is strong selective pressure that favors streamlining. This pressure is amplified in habitats where rapid swimming or endurance swimming is required. Hence, fish that occupy strong currents should evolve more streamlined body types, which in turn places a constraint on reproduction. Under such conditions we might expect to see selection on body shape resulting in decreased reproductive allotment relative to fish that are not experiencing high-velocity currents. High water velocity is also expected to favor increased superfetation, which reduces the amount of physical space required to produce a set number of offspring over time (see fig. 3.2 in Pires et al., **chapter 3**). In fact, swimming constraints have been suggested as one possible cause for the origin of superfetation in poeciliids (Thibault & Schultz 1978).

4.4 Empirical studies of life-history evolution

Interspecific studies using the comparative method have been important for inferring major evolutionary transitions within the Poeciliidae, including the origins of matrotrophy and superfetation (Pires et al., **chapter 3**). However, much of what we know concerning the adaptive significance of life-history variation in the poeciliids comes from comparisons within species. Here, we focus on empirical studies of life-history evolution that can be used specifically to address the role of ecological interactions in driving life-history divergence. Our intent is to present representative studies that purport to explain observed patterns of life-history divergence within the poeciliids. We offer examples for each of the four major ecological hypotheses described above.

4.4.1 Predator-mediated life-history divergence

We know more about the effects of predation on poeciliid life-history evolution than of any other ecological agent of natural selection. This is due largely to the extensive work on predator-mediated life-history evolution in Trinidadian guppies. Haskins et al. (1961) first noted that natural populations of guppies from the northern slope of the Northern Range mountains of Trinidad occurred in habitats with different suites of predators. In high-predation environments, guppies occur with large fish predators, including the pike cichlid *Crenicichla alta*, which feeds extensively on guppies. Low-predation environments can be found in the same drainages, but above barrier waterfalls that exclude these large predators; at these sites guppies co-occur with a killifish (*Rivulus hartii*) capable of feeding on only small guppies (Liley & Seghers 1975). Mark-recapture experiments show that overall mortality rates are higher in high-predation environments than in low-predation envi-

ronments (Reznick et al. 1996b; Reznick & Bryant 2007), and that *Crenicichla* predators in controlled experiments prefer large guppies to small ones (Johansson et al. 2004). Hence, this "natural experiment" is well suited to test predictions of the age-specific mortality hypothesis.

Several lines of evidence suggest that guppies from high- and low-predation environments have evolved divergent life histories as a result of predator-mediated natural selection. Comparative studies of life-history phenotypes show that guppies from high-predation environments mature at smaller body sizes, have higher reproductive allotment, and produce more but smaller offspring than their low-predation counterparts (Reznick & Endler 1982). This pattern of life-history phenotype divergence is repeated on the northern slope of the Northern Range, where guppies encounter high- and low-predation environments, but with a different suite of predators (most notably *Eleotris* and *Gobiomorus*; Reznick et al. 1996b). Common-garden experiments show that observed differences in life-history phenotypes are heritable (Reznick & Endler 1982). Perhaps most compelling, introduction experiments—where guppies are transplanted from one predation environment to another—show that a change in mortality rate in the field actually leads to rapid life-history evolution in the predicted direction (Reznick et al. 1990). Combined, the results of these studies are completely consistent with expectations from the age-specific mortality hypothesis. Although other ecological factors likely contribute to guppy life-history evolution (Reznick et al. 2002a; discussed below), predator-induced mortality appears to be a dominant agent of selection.

How general are the guppy findings? To answer this question, researchers have examined the effects of predation on life-history evolution in other species. Two species in the genus *Brachyrhaphis* show the same pattern of predator-mediated life-history divergence as guppies. *Brachyrhaphis rhabdophora* is distributed throughout northwestern Costa Rica. Populations of this species also occupy two kinds of predation environments: one is characterized by the presence of the predatory fish *Parachromis dovii*; the second is marked by the absence of piscivorous fishes (reviewed in Johnson 2002). Mark-recapture studies show that fish from predator environments experience high adult mortality relative to juvenile mortality; in contrast, fish from predator-free environments experience lower overall mortality rates that are invariant across ontogenetic stages (Johnson & Zúñiga-Vega 2009). Comparisons of life-history phenotypes showed that reproductive allotment, size at maturity for males and females, and number and size of offspring all differ between habitat types in the direction identical to that found in guppies and as predicted by the age-specific mortality hypothesis (table 4.2; Johnson & Belk 2001). Moreover, these differences are genetically

Table 4.2 Life-history phenotypes from contrasting predation environments for three poeciliid fish species

Species	Predation category	Life-history trait				
		Male size at maturity (mm)	Female size at maturity (mm)	Reproductive allotment (%)	Number of offspring	Offspring size (mg)
Poecilia reticulata[a]	*Crenichthys* (H)	14.9	14.6	16.0	6.6	0.9
	Rivulus (L)	16.4	17.4	12.5	2.9	1.5
Brachyrhaphis rhabdophora[b]	Predator (H)	23.5	27.3	—	11.9	1.6
	Predator-free (L)	27.5	32.2	—	6.1	2.5
Brachyrhaphis episcopi[c]	Characin (H)	20.1	23.6	8.3	5.1	2.5
	Rivulus (L)	22.6	27.8	7.0	4.3	2.9

Note: Values presented are for fish collected during the dry season. Predation categories listed here follow the terminology presented by the original authors (see references below): for each species, the upper term denotes the habitat with large predators capable of preying on adult size classes; the lower term denotes the habitat without fish predators or with gape-limited predators capable of taking only juvenile fish. H = high predation. L = low predation.
[a]Reznick & Endler 1982. [b]Johnson & Belk 2001. [c]Jennions et al. 2006.

based (Johnson 2001a) and have evolved independently in at least five river drainage systems (Johnson 2001b). Life-history divergence in the Panamanian fish *Brachyrhaphis episcopi* follows the same pattern. Fish from populations that co-occur with large predatory fish (Characin sites) have larger reproductive allotment, mature at smaller body sizes, and have more and smaller offspring than their low-predation counterparts (*Rivulus* sites) (table 4.2; Jennions & Telford 2002; Jennions et al. 2006). In *B. episcopi* we still do not know whether the observed differences are heritable, and we do not have estimates of mortality rates in the wild. However, in both *B. rhabdophora* and *B. episcopi*, phenotypic differences persist in both dry and wet seasons (Johnson & Belk 2001; Jennions et al. 2006). Hence, it appears that differential mortality has contributed to life-history divergence in each of these species. This is impressive because guppies, *B. rhabdophora*, and *B. episcopi* appear to have evolved the same pattern of parallel life-history divergence independently and in response to different suites of predators (table 4.2). Moreover, similar patterns of life-history divergence (based on sparser data sets) suggest a link between piscivory and predictions of the age-specific mortality hypothesis in *Gambusia hubbsi* from the Bahamas (Downhower et al. 2000), *Xiphophorus hellerii* from Belize (Basolo & Wagner 2004), and *Xenophallus umbratilis* from Costa Rica (J. B. Johnson, unpublished data).

Piscivorous fishes are likely not the only agents of mortality driving life-history evolution in poeciliids. Cave mollies (*Poecilia mexicana*) experience size-selective mortality by giant water bugs that preferentially eat larger individuals; this in turn could contribute to life-history divergence between populations that occupy different cave habitats (To-

bler et al. 2007b). In some systems, freshwater prawns can also prey upon poeciliids, with mortality risk apparently spread evenly among size classes (e.g., Rodd & Reznick 1991; Reznick & Bryga 1996; Millar et al. 2006). Finally, we know almost nothing about the role that parasites play in shaping poeciliid life histories. However, experimental work on guppies shows that some populations vary in their susceptibility to infection by *Gyrodactylus* pathogens (van Oosterhout et al. 2003a), and that these parasites can inflict high mortality rates, especially in the wild (Scott & Anderson 1984). Hence, the link between parasitism and poeciliid life-history evolution deserves further attention.

4.4.2 Effects of density on life-history evolution

Poeciliid researchers have long known that density could be an important agent of natural selection (MacArthur 1962; MacArthur & Wilson 1967). Yet most early optimality models ignored density-dependent population regulation (Reznick et al. 2002a). Consequently, empirical studies of density-driven life-history evolution in poeciliids have lagged behind other focal research areas (most notably studies of predation and resource availability). Still, growing evidence over the past decade suggests that density can be very important in shaping evolutionary diversification in poeciliids.

Least killifish (*Heterandria formosa*) are small, superfetating poeciliids distributed throughout the fresh and brackish waters of the southeastern United States. Leips and Travis (1999) compared life histories of four populations of least killifish that naturally occur at different density levels. They found that individuals from lake populations occurred at lower densities than their counterparts from spring-fed river populations. However, the ratio of predators to least killifish was also higher in lakes than rivers; such covariance between density level and predation is not unexpected given that predators remove prey from populations, but it does present a challenge for inferring causation in comparative studies. Based on monthly survey samples over a two-year period, the authors found differences among populations in the cumulative variance in density (a statistical measure used to infer density-dependent population regulation). Further, they found that over time, density was negatively correlated with overall female body size in two of the four populations, and that the degree of superfetation and number of offspring were both lowest in the population with the highest density. In fact, offspring size was 45% larger in the highest-density population than in the other three, even after controlling for the effect of phenotypic plasticity. This phenotypic shift in offspring size is consistent with theoretical expectations when juvenile competition is important to fitness. Interestingly, the observed shift in life-history traits was evident only in the highest-density population, suggesting that the effect of density as a selective agent may not be apparent except at very high levels.

Recent work on guppies also suggests that life-history diversification is partly driven by population density. Bronikowski and colleagues (2002) examined the effect of divergent life-history evolution on guppy population dynamics. They conducted a set of controlled experiments in streams in Trinidad to determine what effect changes in density would have on life-history traits. Guppies were removed from natural streams and restocked in pools at three different density levels (one-half natural density, at natural density, and twice natural density). Interestingly, this study was restricted to low-predation sites because preliminary studies revealed that density-dependent population regulation was not occurring at high-predation sites, presumably because predators kept guppy abundance in check. The high-density treatment resulted in a decrease in juvenile survival. Once these data were incorporated into population matrix models, sensitivity analysis showed that high density increases the selective pressure on juvenile growth rates and on the survival of large adult females; low density favors increased adult fecundity. Density also helps explain the evolution of guppy life-history traits in replicated transplant experiments. Guppies that are moved from high-predation to low-predation habitats rapidly evolve low-predation life-history phenotypes (Reznick & Bryga 1987; Reznick et al. 1990), despite predictions from simulation experiments that show that high-predation phenotypes should have a demographic advantage regardless of habitat type (Reznick et al. 2002a). The resolution of this apparent paradox comes only when the effects of density (detailed above) are considered in demographic modeling, in which case high- and low-predation life-history phenotypes are found to have equal fitness in their own environments (Bronikowski et al. 2002).

4.4.3 Resource-driven life-history evolution

Resource availability can have a profound effect on the expression of poeciliid life-history phenotypes (see Grether & Kolluru, chapter 6). Several studies have documented phenotypic plasticity in response to changes in food availability. For example, guppies reared under different food levels in the laboratory show abrupt plastic responses in reproductive allocation, offspring number, and offspring size, even to the extent that changes in these traits can be induced among sequential broods (Reznick & Yang 1993). When food is limited, guppies grow more slowly and have fewer and larger offspring (Reznick & Yang 1993; Reznick et al. 2001). Similarly, seasonal variation in resource availability in the wild predicts life-history phenotypes. Winemiller (1993) showed that when food availability changed between the tropical dry season and the tropical wet season in Costa Rica, the poeciliids *Alfaro cultratus*, *Phallichthys amates*, and *Poecilia gillii* all showed shifts in reproductive allotment. Johnson and Belk (2001) showed that *Brachyrhaphis rhabdophora* maintain life-history differences between predator and no-predator habitats during the wet and dry season, but that reproductive allotment and number of offspring both increase during the wet season, presumably because more insects are available to consume during the wet season. Jennions et al. (2006) showed similar shifts in *Brachyrhaphis episcopi*, where the wet season induced increases in reproductive allotment, offspring size, and average adult male size, all relative to life-history phenotypes observed during the dry season. However, it remains unclear in each of these studies whether phenotypic shifts have any adaptive value or are simply a by-product of having more food available.

In contrast, work on guppies suggests that consistent differences in resource availability can lead to adaptive life-history divergence among populations. Bashey (2006) showed in the laboratory that guppies derived from competitive environments have evolved larger egg sizes than those from resource-abundant environments, and that larger egg size confers a fitness advantage in offspring both via juvenile survival and through the timing of maturity (Bashey 2008). Grether et al. (2001b) identified a clever approach to separating the effects of resource availability from those of predation to determine whether resource availability affects guppy life-history expression in the wild. By focusing on resource variation within six low-predation habitats, they showed that food availability for guppies decreases in river systems as canopy cover increases. Guppies eat algae, so more sunlight results in a larger standing crop of food available for consumption. Mark-recapture studies revealed that increased food availability was associated with increased growth rates for juvenile and female guppies and with a larger size at maturity for males; however, a common-garden experiment revealed that these differences were primarily environmental. In contrast, differences in offspring size and litter size appear to be genetically based, with increased food leading to more offspring and smaller offspring. These results are important because they suggest that increased food availability might be complementary to predation in its directional effect on life-history evolution in guppies (Grether et al. 2001b). In other words, the classic life-history contrast in guppies between high- and low-predation habitats may be due in part to the effect of resources on life-history evolution. Interestingly, in *Brachyrhaphis rhabdophora*, the ecological factors "canopy cover" (a surrogate for resource availability) and "predation" were indistinguishable in their ability to predict life-history divergence (Johnson 2002). Hence, there may be a tight and complementary link between mortality and resource availability as agents of natural selection.

Theory predicts that resource availability explains the evolution of maternal provisioning strategies (Trexler & DeAngelis 2003). Two empirical studies of poeciliids have evaluated components of this model. The first study focused on plasticity in maternal provisioning in *Gambusia geiseri* under different levels of food availability (Marsh-Matthews & Deaton 2006). Females were divided into two treatment groups fed either every day (high-food treatment) or every third day (low-food treatment). This species employs a dual-provisioning strategy in which embryos receive nutrients from yolk placed in the egg prior to fertilization and from maternal nutrient transfer during embryo development. Nutrient transfer was assayed by tracking the transfer of radioactive amino acids from the maternal diet into the embryo tissue. Consistent with the Trexler-DeAngelis model, females fed daily showed higher levels of maternal nutrient transfer than those fed every three days. Daily-fed females also produced more offspring and larger offspring than those fed at three-day intervals. However, contrary to the Trexler-DeAngelis model, there was no evidence that the low-food treatment led to higher levels of embryo abortion than found in the high-food treatment. Differences in food level also led to differences in terminal embryo size, contrary to the assumption of fixed terminal embryo size explicit in the model. The second empirical study evaluated the Trexler-DeAngelis model by comparing two closely related *Poeciliopsis* species from western Mexico that employ different maternal provisioning strategies (Banet & Reznick 2008). *Poeciliopsis prolifica* is a superfetating matrotroph,

and *Poeciliopsis monacha* is a superfetating lecithotroph. Adult females of each species were reared for 30 days at high- and low-food availability. Because gestation period is approximately 30 days for each species, offspring born during this time would have started development prior to the food treatment. According to the Trexler-DeAngelis model, a reduction in resource availability was expected to result in a decrease in number of offspring in the matrotroph but to have no effect on lecithotrophic embryos because the offspring born during the 30-day period would have been fully provisioned prior to the application of the food treatment. The authors found no evidence for embryo abortion as a result of low food in either species: the number of offspring produced by females in high- and low-food treatments was the same. *Poeciliopsis prolifica* responded to low food by reducing offspring size, not offspring number; the authors speculate that this is a maladaptive plastic response to low food. As expected, *Poeciliopsis monacha* in the low-food treatment did not produce smaller offspring. Combined, these two studies suggest that although elements of the Trexler-DeAngelis model are supported by empirical data, several of the model's assumptions should be modified: embryo abortion is not evident in either of the matrotrophic species examined, and terminal offspring size appears to vary considerably. How these modifications to the model affect the resource conditions under which matrotrophy will evolve is an important area for future research.

4.4.4 Effect of water velocity on life-history evolution

Water velocity, through its effect on swimming performance, may be responsible for the evolution of some poeciliid life-history traits. Ghalambor et al. (2004) showed that there is a functional trade-off between locomotor performance and reproductive investment in guppies. They compared swimming efficiency of females from high-predation habitats that had evolved a high reproductive investment strategy (see above) to females from low-predation habitats that had evolved a low–reproductive investment strategy. They found that early in pregnancy females from high-predation sites were better swimmers (measured by acceleration rate and velocity and by the distance traveled when startled) than females from low-predation sites. However, at later stages of pregnancy, when females are carrying a larger reproductive load, the difference in locomotor performance between habitat types was reduced. This difference is apparently due to the greater reproductive investment by high-predation females. Hence, strong selection on swimming performance appears to have a potentially constraining effect on the evolution of reproductive investment in guppies (Ghalambor et al. 2004). How might females maintain the benefits of large reproductive investment while reducing the cost of locomotor performance?

Superfetation may have evolved as a response to this problem. Zúñiga-Vega et al. (2007) examined the link between superfetation and water velocity by comparing six inland and six coastal populations of *Poeciliopsis turrubarensis* from Costa Rica. Low-gradient streams are characteristic of coastal habitats, whereas high-gradient streams are characteristic of inland habitats. The authors predicted that females from inland populations would have a more streamlined body shape and higher levels of superfetation than their coastal counterparts. The advantage of superfetation under high water velocity is that the total volume needed to house developing embryos is reduced because embryos occur at different developmental stages (see fig. 3.2 in Pires et al., **chapter 3**). As predicted, *P. turrubarensis* from inland sites were more fusiform and on average carried more broods per female than fish from coastal populations. Interestingly, not only did inland females have more broods at a given time, but they also had a *greater* reproductive investment per brood than fish from coastal sites (Zúñiga-Vega et al. 2007). Hence, increased superfetation appears to be an adaptation that reduces the locomotive costs of pregnancy, and in this case sufficiently so to allow greater reproductive investment than occurs in habitats with low water velocity.

4.5 Future work

The future holds exciting prospects for poeciliid life-history research. One clear message that has emerged over the past two decades is that continued efforts to expand the taxonomic breadth of life-history studies will be generously rewarded. We presented several examples in this chapter that demonstrate the utility of studying intraspecific life-history divergence. Yet much of this work is still confined to the same select set of "model" species. Some exceptions exist—for example, a growing number of taxa have been used to independently evaluate the role of predator-mediated life-history evolution in poeciliids (see above and table 4.2). Such comparisons are important because they demonstrate the extent to which common selective agents yield similar evolutionary responses. Broadening the taxonomic scope of poeciliid life-history research will also help reveal which life-history traits are most likely to evolve prior to speciation and why. Tests of association among life-history traits suggest that some traits are tightly linked while others appear to evolve independently (Strauss 1990; Johnson & Belk 2001). Comparative studies will allow us to probe for repeated patterns of trait covariance, enhancing our understanding of evolutionary trade-offs and trait coevolution. Moreover, recent advances in phylogenetic comparative methods provide a mechanism by which the effects of common ancestry can be considered, leaving a clearer understanding of the role of selection in shaping life-history traits (Blomberg et al. 2003; Ives et al. 2007). Finally, increasing taxonomic breadth of life-history research will allow researchers to take advantage of emerging poeciliid genome projects. With the use of genomic data, it will be exciting to see whether repeated patterns of life-history divergence within species, and even among species, are in fact controlled by the same genes and developmental pathways. Indeed, our knowledge of life-history diversification in the poeciliids may prove to be ideal for testing the repeatability of evolution in nature.

Acknowledgments

We appreciate the generosity of Brigham Young University in supporting the Evolutionary Ecology Laboratories, a source of many good discussions on poeciliids. We appreciate the helpful suggestions on an earlier version of this chapter from Farrah Bashey, Edie Marsh-Matthews, and one anonymous reviewer.

Chapter 5 Evolution of unisexual reproduction

Ingo Schlupp and Rüdiger Riesch

5.1 General background

THE FOCUS OF THIS CHAPTER is on the ecology and evolution of several highly unusual fishes. In the Poeciliidae, modes of reproduction that are different from the typical sexuality found in almost all vertebrates have evolved at least twice. These fishes are often called asexual, because they reproduce without recombination.

In the scientific literature, the terms "asexuality" and "asexual reproduction" are most commonly used as relatively broad terms encompassing all modes of reproduction that do not pay the full evolutionary cost of meiosis (Maynard Smith 1978). However, in its strictest sense, asexuality should refer to specific modes of reproduction without germ cells and therefore would not include unisexual vertebrates and their reproductive modes of parthenogenesis, gynogenesis, and hybridogenesis. Hence, to avoid confusion, we chose to use the term "asexual" only where we specifically cite literature that uses the term according to its broader definition. We followed the established literature and used the term "sexual" on several occasions to contrast organisms that pay the full evolutionary cost of meiosis (as in "sexuals sometimes copy other sexuals").

Less than a hundred years ago reproduction without recombination was thought to be impossible in vertebrates. A groundbreaking publication by the famous ichthyologists Laura and Carl Hubbs (Hubbs & Hubbs 1932), describing the Amazon molly, *Poecilia formosa*, as the first clonal, yet sperm-dependent, vertebrate opened up a whole new field of research. The description of diploid *Poeciliopsis* as hybridogenetic by Miller and Schultz (1959) established this

mode of reproduction for vertebrates, and the identification of triploid *Poeciliopsis* as gynogenetic (Schultz 1967) was equally important (see below). Several comprehensive reviews of unisexual fishes and the underlying, unusual reproductive mechanisms have been published (Vrijenhoek 1994), including chapters on unisexual vertebrates in two 1989 landmark books (Balsano et al. 1989; Monaco et al. 1989; Rasch & Balsano 1989; Schenck & Vrijenhoek 1989; Schultz 1989; Schultz & Fielding 1989; Turner & Steeves 1989; Vrijenhoek 1989; Wetherington et al. 1989a). More general accounts of unisexuality were published elsewhere (Suomalainen et al. 1987; Beukeboom & Vrijenhoek 1998; Schlupp 2005; 2009). In the present chapter we will therefore focus on advances made between the late 1980s and today. Because this is a book about fishes, we give preference to relevant literature on fishes, not because we are trying to ignore the extremely important findings made using other taxonomic groups. For purely practical reasons we ignore the important question whether unisexual taxa are "species" (according to the biological species concept) or not and how they should be properly named (Echelle 1990), and we refer to them as species, using the nomenclature established by earlier work. In the following pages we will first provide a short introduction to the two genera of unisexual poeciliids and the resulting bisexual/unisexual mating complexes. We will then try to explain why unisexual poeciliids are of great scientific interest, which directly leads us to discuss the various models that have been proposed to explain stability of bisexual/unisexual coexistence and maintenance of sex. Finally, we will discuss the role of ploidy and will end with a short outlook on the role of unisexual poeciliids in cancer research (which is covered in detail by Schartl & Meierjohann, **chapter 26**).

The evolution and maintenance of recombination are major puzzles in evolutionary biology (West et al. 1999). What still remains difficult to understand is how sexual reproduction (meiosis) can overcome the short-term advantages of asexuality (Williams 1975; Maynard Smith 1978; Bell 1982; Barton & Charlesworth 1998). Assuming that all things are equal, asexual females produce twice as many daughters in each generation, and due to meiosis, sexual females pass on only half of their genome (Williams 1975). Furthermore, adaptive genetic combinations are reshuffled in meiosis, and sexual species may pay the "cost of mating" in various forms like efforts to find a suitable mate, exposure to infectious diseases, and increased predation. Obviously, the cost of mating also applies to sperm-dependent unisexual fishes. In contrast to the short-term benefits, clonal organisms pay long-term costs, in that clonal lineages accumulate deleterious mutations irreversibly and rare beneficial mutations cannot spread via recombination (Muller 1964; Kondrashov 1988). However, Mandegar and Otto (2007) argued that mitotic recombination might aid in spreading beneficial mutations in asexual populations, thus effectively negating the previously described long-term cost. Hence, the long-term survival and the evolutionary fate of unisexuals have long been questions of great interest. The argument of a long-term cost to asexuals and a long-term benefit to sexuals rests on the assumption that it is beneficial to adapt to a changing environment (Bell 1982; Salathe et al. 2008). Despite this cost, given the assumed short-term disadvantages of sexuality, asexuals should be able to outcompete sexual females in natural populations—as long as they are not carrying a big genetic load. Asexual populations are likely to face declining fitness over time, almost as the inverse of accumulated deleterious mutations (fig. 5.1). However, while this idea is intuitive, logically compelling, and supported by theory, this outcome seems to be rather rare in nature.

Unisexual vertebrates, all of which are of hybrid origin, have long been used as model systems to study the maintenance of recombination (Vrijenhoek et al. 1989; Vrijenhoek 1994). Surprisingly, unisexuality evolved at least twice in poeciliids, leading to two distinct model systems for studying the evolution and ecology of sexual reproduction. The species complexes in which unisexuality evolved are only distantly related within poeciliids (Hrbek et al. 2007). In both cases females rely on sperm from males of other species for successful reproduction. The two modes of sperm use are gynogenesis, which is found in *Poecilia* and triploid *Poeciliopsis*, and hybridogenesis, which is known from diploid *Poeciliopsis* (Schultz 1967). During gynogenesis females typically produce unreduced eggs, are

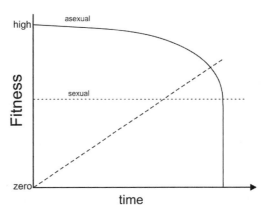

Figure 5.1 Hypothetical fitness decline for a nonrecombining population. Solid line: unisexuals start out with maximum fitness. Initially, fitness declines slowly, but as the cumulative effects of the accumulation of deleterious mutations increases, the decline becomes steeper and rapidly reaches zero. At the same time, mutational load (dashed line) increases. This dynamic can be influenced by any mechanism that slows down genetic decay, such as adding fresh genetic material. Note that maximum asexual fitness can be higher than sexual fitness (finely dashed line) because of hybrid vigor. The area between the lines for sexual and unisexual fitness represents the area where coexistence is difficult to understand, as the unisexuals should outcompete the sexuals.

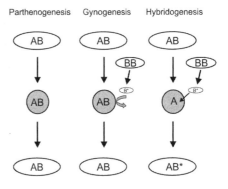

Figure 5.2 Reproductive modes of unisexual vertebrates.

fully clonal, and use sperm only to trigger embryogenesis (fig. 5.2). Therefore, gynogenesis can be viewed as sperm-dependent parthenogenesis. Hybridogenesis, however, is hemiclonal; females produce reduced eggs by excluding the male chromosomes; male sperm then fertilizes these eggs (fig. 5.2). The evolution of these two reproductive modes is especially hard to understand in species with internal fertilization and suggests an important role for male mate choice as males can—in theory—withhold sperm from the unisexuals (see below).

5.2 Unisexual *Poecilia*

The Amazon molly, *Poecilia formosa*, was the first unisexual vertebrate to be detected. Hubbs and Hubbs (1932) noticed that *P. formosa* always had only female offspring and—based on morphological characters—concluded that *P. formosa* must be a hybrid between *Poecilia mexicana* (then *Mollienesia sphenops*) and *Poecilia latipinna*. The hybrid origin was later confirmed using molecular methods (Avise et al. 1991; M. Schartl et al. 1995b; Tiedemann et al. 2005), and currently *P. mexicana limantouri* (for justification of the subspecies, see Menzel & Darnell 1973) is considered to be the maternal ancestor; and *P. latipinna*, the paternal ancestor. This situation can be viewed as a hybridization event (see Rosenthal & García de León, **chapter** 10), leading to instantaneous speciation frozen in time (see section 5.5.2). Attempts to re-create "Amazon-like" de novo hybrids were fruitless and typically resulted in a bisexual F_1 with meiotic recombination. However, these can be used for studies comparing the ancient hybrid with newly created hybrids (Schartl et al. 1991; Schlupp et al. 1992; Dries 2003). There is currently no evidence for a more complicated mechanism of origin (like a backcross), but such mechanisms are possible (Schlupp 2005). MtDNA evidence suggests a single origin for Amazon mollies (Möller 2001; Stöck et al. 2010) approximately 120,000 generations ago (M. Schartl et al. 1995b) in the area around Tampico, Tamaulipas, Mexico. Amazon mollies subsequently dispersed north to reach the current limit of their distribution in the lower Rio Grande valley and the Nueces River in Texas, USA (Schlupp et al. 2002; Costa & Schlupp 2010), and south along the Gulf Coast of Mexico to the Río Tuxpan. Although the age estimate provided is based on a molecular-clock model and has large confidence intervals, it seems as if enough time has elapsed for deleterious mutations to accumulate and negatively affect *P. formosa*'s fitness (Gabriel et al. 1993; Gabriel & Burger 2000; see Loewe & Lamatsch 2008 for a quantification of the extinction risk). The paradox of long-term survival of this species is far from solved (see below). One indication of a fitness decline could be the loss of populations in nature. Amazon mollies are not considered threatened, but McNeely and Wade (2003) noticed a decline of populations in the Brownsville (Rio Grande) area. Such declines, however, could also be explained by metapopulation dynamics (Kokko et al. 2008), by host-parasite cycles affecting male mate choice (Heubel & Schlupp 2008), or by anthropogenic disturbances (e.g., habitat degradation, habitat loss, and urbanization).

The vernacular name of the species, Amazon molly, chosen by Hubbs and Hubbs (1932) refers to the mythical tribe of all-female warriors (Schlupp et al. 2007). Amazon mollies reach a maximum size of about 70 mm. Onset of maturity can be reached when females are around 30 mm. Females of the bisexual hosts are about the same size, and interbrood intervals vary with population (Hubbs & Dries 2002). In Mexico, Amazon mollies use both the Atlantic molly, *P. mexicana* (both subspecies *P. mexicana mexicana* and *P. mexicana limantouri*), and the sailfin molly, *P. latipinna*, as hosts (even though most populations are syntopic with only one parental species in Mexico), while the sailfin molly exclusively provides sperm in southern Texas. Based on this difference in host species, two mating complexes have long been recognized. Schlupp et al. (2002) provided a much more detailed view based on biogeography and availability of hosts, highlighting that in the Río Mante and Río Guayalejo systems a third species, *Poecilia latipunctata* is thought to provide sperm for Amazon mollies (Niemeitz et al. 2002). Both *P. latipinna* and *P. mexicana* occupy large areas in which they do not coexist with Amazon mollies, allowing for comparative studies (Gabor & Ryan 2001; Gabor et al. 2005). Furthermore, both sailfin mollies and Amazon mollies have been introduced into a small area in central Texas (reviewed by Schlupp et al. 2002). With the dates of the introduction fairly well established, these introduced populations have been viewed as a natural experiment and are being used in numerous studies, especially on behavior (Schlupp et al. 1994; Schlupp & Ryan 1997; Schlüter et al. 1998; Schlupp et al. 1999; Gumm & Gabor 2005; Heubel & Schlupp 2006; reviewed by Schlupp 2005). Interestingly, the source for the introduced Amazon mollies was the Brownsville area in southern Texas, whereas the introduced sailfin mollies are thought to have originated in Louisiana (Möller 2001; reviewed in Schlupp et al. 2002). This is important, because these sailfin mollies were "evolutionarily naive" with respect to Amazon mollies.

Hence, the Amazon molly is a natural hybrid that uses only a small fraction of the available host populations: the paternal ancestor, the sailfin molly, *P. latipinna*, ranges from North Carolina along the Atlantic and Gulf coasts south to the Río Tuxpan in northern Mexico (Miller 2005). The maternal ancestor, the Atlantic molly, *P. mexicana*, inhabits the area from the Río San Fernando (Tamaulipas, Mexico) south to at least Honduras (Miller 2005). A recent study implicates mainly abiotic factors (essentially, temperatures) as defining the present range limits of these species (Costa & Schlupp 2010).

Amazon mollies have been classified as ameiotic by Rasch et al. (1982), but a more recent report also found automixis (Lampert et al. 2007; box 5.1). A recent overview of this mating system for nonspecialists was provided by Schlupp et al. (2007).

5.3 Unisexual *Poeciliopsis*

The other taxonomic group in which unisexuality has independently evolved is a species complex of the genus *Poeciliopsis*, which provides an extremely intricate model

Box 5.1   Sources of genetic variation

All unisexual vertebrates are hybrids. At the time of hybridization they have very high variability within the individual genome. Due to their clonal inheritance, however, the population can nonetheless be genetically very uniform. Several mechanisms generate genetic variability:

1. Mutation

Unisexuals seem to experience typical mutation rates. Most mutations are thought to be deleterious (Kondrashov 1984, 1985), but some can be compensatory (Howe & Denver 2008; Loewe & Lamatsch 2008).

2. Introgression (adding a genome)

Because unisexual fishes and amphibians are sperm dependent, unreduced eggs and sperm interact tightly. In rare cases a complete sperm genome can be added, thereby elevating the ploidy of the organisms (Lampert et al. 2005; Lampert et al. 2008). The evolutionary effects of such ploidy elevations are poorly understood.

3. Introgression (adding subgenomic amounts of DNA)

Introgression can also involve incomplete sperm genomes, often with the formation of so-called supernumerary microchromosomes (M. Schartl et al. 1995a). In Amazon mollies these can be passed on for many generations (Nanda et al. 2007), but again, the evolutionary consequences of this phenomenon are not well understood.

4. Gene conversion (mitotic recombination)

This process can lead to exchange between homologous DNA sequences.

5. Incomplete recombination (automictic parthenogenesis)

Amazon mollies are thought to be ameiotic and to form their unreduced eggs via mitosis. There are, however, many more ways that can lead to unreduced eggs, many of which involve subsexual mechanisms, like the fusion of an egg nucleus with a polar body. Limited shuffling of genes can occur (Suomalainen et al. 1987).

system for understanding the maintenance of sex. Six different unisexual *Poeciliopsis* biotypes have been reported, and in all cases the maternal ancestor is the headwater livebearer *Poeciliopsis monacha* (Quattro et al. 1991). Unisexual females are typically between 44 and 50 mm long. Four different species are known to be the paternal ancestor: the lowland livebearer (*Poeciliopsis latidens*), the clearfin livebearer (*Poeciliopsis lucida*), the Gila topminnow (*Poeciliopsis occidentalis*), and the chubby livebearer (*Poeciliopsis viriosa*; Vrijenhoek 1998; Miller 2005). Unlike *P. formosa*, unisexual *Poeciliopsis* of the same biotypes derive from several independent natural hybridization events (Vrijenhoek 1979; Quattro et al. 1991, 1992b; Mateos & Vrijenhoek 2005), and several successful attempts have led to a number of laboratory-created clonal lineages (Vrijenhoek 1994), providing a great tool for understanding the evolution and ecology of this system. This further means that some natural clones may be evolutionarily young, while several old mtDNA lineages were also detected (Quattro et al. 1992a). Similar to the Amazon molly, the mechanisms for long-term survival of unisexual *Poeciliopsis* lineages are still uncertain; however, due to the hemiclonal reproductive mode, deleterious mutations can be masked by male genetic information. Furthermore, recent extant clones can originate as trihybrid hybridogens, via a very complicated pathway (Vrijenhoek & Schultz 1974; Mateos & Vrijenhoek 2002), or as trihybrid gynogenetic triploids (Mateos & Vrijenhoek 2005), involving three different parental species.

Accordingly, unisexual *Poeciliopsis* are syntopic with at least one of the four parental species in the Pacific slope of Mexico. The maternal ancestor, *Poeciliopsis monacha*, occurs in the headwaters of a few tributaries to the Río Fuerte, Río Mayo, and Río Sinaloa (Sonora and Sinaloa, Mexico; Miller 2005). *Poeciliopsis latidens* range from the Río Fuerte basin in the north (Sonora) to near San Blas and Tepic in the south (Nayarit). *Poeciliopsis lucida* can be found from the Río Fuerte basin in the north (Sonora) to the Río Mocorito basin in the south (Sinaloa; Miller 2005). The third paternal species, *Poeciliopsis occidentalis*, occurs from the Gila River in Arizona and New Mexico south to the Río Mayo in Sonora, Mexico. Finally, *Poeciliopsis viriosa* range from the Río Sinaloa basin (Sinaloa) to the Río Ameca (Jalisco and Nayarit; Miller 2005). The biogeography of this group has been studied by Mateos et al. (2002), and because they live in highly fluctuating desert habitats, the conservation status of *Poeciliopsis* has been of concern for a long time (Douglas & Vrijenhoek 1983; Galat & Robertson 1992; Vrijenhoek 1994). With increasing human demands on formerly reclusive habitats (e.g., agriculturalization, industrialization, and urbanization), this situation is likely to deteriorate. Protecting small populations of unusual fishes seems to be especially complex (see Stockwell & Henkanaththegedara, **chapter 12**).

5.4 Why study unisexuals?

Among unisexual vertebrates (Suomalainen et al. 1987; Dawley 1989; Vrijenhoek et al. 1989), unisexual fishes are of central interest mainly because of their peculiar reproductive mechanisms. Commonly three big conceptual areas are addressed using unisexual fishes. A first set of questions revolves around the evolutionary persistence of unisexuals, often also asking if and how long-term survival is possible given the inevitable accumulation of deleterious mutations. This area is addressed using not only unisexual fishes but also other unisexual vertebrates and invertebrates. A second area addresses the mechanisms of coexistence of sexual and asexual organisms. Again, unisexual fishes are among the successful model systems for this. The third area addresses the peculiar sperm-dependent reproductive modes of gynogenesis and hybridogenesis, where sperm is required to initiate embryogenesis, but inheritance is maternal (Dawley 1989; Vrijenhoek et al. 1989; Schlupp et al. 1998; Schlupp 2005). As argued above, both of these reproductive modes are not really asexual, as germ cells are always involved. Also, hybridogenesis clearly contains elements of both clonal reproduction and sexuality. Curiously, there is an unexplained divide along taxonomic lines here: all fishes and amphibians require sperm to trigger embryogenesis, while reptiles do not. Independence of sperm coincides with the evolution of the amnion, but for unknown reasons.

In sperm-dependent systems we find two competing types of females, which rely on the same resource, sperm, but have fundamentally different population dynamics. According to theoretical models (e.g., Jokela et al. 1997), asexuals should quickly outcompete sexuals, thereby driving them to extinction, which would secondarily result in the demise of the asexuals. Thus, the evolution of stable coexistence of asexuals and their sexual hosts and the mechanisms underlying this apparent stability are of great interest in evolutionary ecology (Schlupp 2005).

5.5 Stability of coexistence and maintenance of sex

How can these sexual/asexual (in our case: bisexual/unisexual) mating complexes be stable? Several ideas have been suggested to explain this puzzling fact, all of them basically trying to identify how the twofold, short-term advantage of asexuals is diminished. The underlying reasoning is fairly simple: if all else is equal, sexual females will produce only half the number of daughters as asexual females will. The critical assumption, of course, is that all else is equal. This needs to be tested for each case, as in the Amazon molly. Here size-corrected fecundity is about the same, indicating that the unisexuals actually have a twofold advantage (Schlupp et al. 2010). Essentially, what should be done is compare species that differ only in their reproductive strategy and not much else. In reality, of course, this is difficult to test (Schlupp 2005; Tobler & Schlupp 2008a). A number of models have been developed to analyze the situation, both for unisexual fishes and for other asexuals (Hellriegel & Reyer 2000). Probably the most widely considered and best-supported model is the frozen niche variation hypothesis (see section 5.5.2). It is based on the assumption that clonal diversity results from multiple origins, a condition met for unisexual *Poeciliopsis* but not for the Amazon molly (Möller 2001; Stöck et al. 2010). Another model, by Kokko et al. (2008), stresses the importance of spatial dynamics in these mating systems and predicts that a metapopulation of bisexual hosts and Amazon mollies can persist if a balance exists between local extinctions and recolonizations. However, this newer model remains to be parameterized with field data. An older approach focused on frequency-dependent male mate choice (McKay 1971; Moore & McKay 1971). The reasoning here is elegant and intuitive: if male choosiness increases with increasing numbers of unisexuals, the latter will have less access to sperm and will decrease in the population. At low frequencies, however, unisexuals must benefit from some unspecified advantage. A comparable approach was used in a population-dynamic model by Heubel et al. (2009), which found that male choice can be critical by limiting the reproductive success of the unisexuals.

The ideas that have been empirically tested include versions of the Red Queen hypothesis, regulation via male behavior, and differences in life histories. Interestingly, the Red Queen hypothesis, which posits that recombination produces genetically diverse offspring that are harder to target by pathogens than cloned asexuals (Van Valen 1973), does not easily explain the maintenance of sex in these two systems. Females of *P. latipinna* and *P. formosa* do not differ significantly in parasite loads (Tiedemann et al. 2005; Tobler & Schlupp 2005). Amazon mollies have a reduced population-wide genotypic diversity compared to sailfin mollies from the same habitats. For example, many fewer haplotypes and genotypes based on microsatellite markers are found in populations of Amazon mollies than in sailfin mollies (Schaschl et al. 2008). Because of the hybrid origin of Amazon mollies, on the other hand, diversity within the genome can be higher than in sexuals. This diversity within the genome was comparable to that found in sailfin mollies at major histocompatibility complex (MHC) class I loci but lower than that in sailfins at class IIB loci (Schaschl et al. 2008; see also McMullan & van Oosterhout, **chapter 25**). Also, a comparison of fecundity, size, and parasite load in *Poeciliopsis* does not support the Red Queen model (Weeks 1996b), but an earlier report on black spot disease in nature did (Lively et al. 1990): wild-caught unisexual *P. 2monacha-lucida* triploids were more heavily infected than bisexual *P. monacha* from the same pools. Long-term studies of predicted coevolutionary cycles are clearly needed. The Red Queen hypothesis found good support in other systems when comparing sexuals and asexuals (West et al. 1999; Peters & Lively 2007), but its generality has also been criticized on both practical and theoretical grounds (Salathe et al. 2008; Tobler & Schlupp 2008a). Hence, several other hypotheses have been suggested to explain stability in bisexual/unisexual complexes: (1) the behavioral regulation hypothesis, (2) the frozen niche variation hypothesis, and (3) the life-history regulation hypothesis (Vrijenhoek 1994; Schlupp 2005).

5.5.1 The behavioral regulation hypothesis

Under the behavioral regulation hypothesis, coexistence is regulated by male mate choice. Males do not directly benefit from heterospecific matings, since they are not related to the resulting offspring. This situation has been dubbed "sperm parasitism" by Hubbs (1964). Accordingly, males of the sperm donor species—for both *P. formosa* and the several lineages in *Poeciliopsis*—should prefer conspecific females. This pattern has been documented in a number of studies on both *Poecilia* and *Poeciliopsis*: males have been shown to (a) discriminate between heterospecific and conspecific females and (b) to prefer to mate with conspecific females under many circumstances (Hubbs 1964; McKay 1971; Keegan-Rogers 1984; Keegan-Rogers & Schultz 1988; Schlupp et al. 1991; Ryan et al. 1996; Gabor & Ryan 2001). More detailed studies, however, found that male preferences are not as strong and uniform as predicted. Under certain conditions, male preferences for conspecific females can be diminished (Gabor & Aspbury 2008; Aspbury et al. 2010a; Aspbury et al. 2010b). One example is that male sailfin mollies living syntopically with Amazon mollies have a stronger preference for conspecific females than males from allopatric populations (Hubbs 1964; Ryan et al. 1996; Gabor & Ryan 2001; Gabor et al. 2010). Furthermore, males prefer sexually receptive gynogens over nonreceptive conspecifics (Schlupp et al. 1991), and male preferences vary throughout the mating season, so that Amazon mollies may have easier access to males during certain parts of the year (Heubel & Schlupp 2008). All these studies on differential male mate choice seem to lend support to the hypothesis that male choice can be frequency dependent in these bisexual/unisexual systems (McKay 1971): males selectively favor the rare phenotype (Keegan-Rogers 1983, 1984; Keegan-Rogers & Schultz 1984, 1988) and discriminate against the most common one if conspecific females are rare but not vice versa. Even

though this has been directly addressed and confirmed in *Poeciliopsis* (Moore & McKay 1971) and in an insect (Stenseth et al. 1985), to date no studies have investigated this in Amazon mollies.

Another mechanism by which unisexuals could obtain matings is by visual mimicry: in *Poeciliopsis* unisexual females appear to visually mimic heterospecific females of the host species (Lima et al. 1996), probably making discrimination more costly for males. Finally, unisexuals can compete aggressively with heterospecific females (Schlupp et al. 1992; Marler et al. 1997; Heubel & Plath 2008). Dynamics like this have been viewed as evolutionary arms races (Dawkins & Krebs 1979; Schlupp et al. 1991), which of course assumes that unisexuals are not just F_1's frozen in time (Dries 2003).

Yet another important assumption is that selection should act against males that mate with unisexuals if this reduces mating frequency with conspecifics. Selection should be weaker as the cost of mating decreases for males. A recent study found that costs of interacting with unisexual and sexual females do not differ for males (Schlupp, Reiker, Plante, & Chapman, unpublished data). But even if the cost is low, selection should act because the benefit to males is thought to be zero, as the males make no genetic contribution to the offspring. This assumption was challenged by Schlupp et al. (1994), who showed that mating with heterospecific females made *P. latipinna* males more attractive to observing conspecific females, thus increasing their chances for a subsequent mating with conspecifics (mate copying; see Druen & Dugatkin, **chapter 20**). Hence, under certain circumstances, males may increase their mating success with conspecifics through mating with heterospecific females, potentially relaxing selection on male discrimination. Conspecific mate copying was later also found for *P. latipinna* under field conditions (Witte & Ryan 2002). The first study (Schlupp et al. 1994) was later expanded to look at all possible interactions in this system, including Amazons copying Amazons and sexuals copying sexuals (Heubel et al. 2008). Obviously, males benefit most from conspecific mate copying, and somewhat from conspecifics copying unisexuals, whereas unisexuals copying heterospecifics would be detrimental for the males. The key finding here is that mate copying can be beneficial to males even if unisexuals also copy, because the strength of mate copying varies (Heubel et al. 2008). Interestingly, *P. latipinna* females pay more attention to conspecific females than to *P. formosa* females (Hill & Ryan 2006). Furthermore, studies of mate copying highlight the importance of the social conditions under which mating decisions are made, including sexual harassment, audience effects, and deception (Schlupp et al. 1994; Schlupp et al. 2001; Plath et al. 2007b; Plath et al. 2008a; Plath et al. 2008c).

Amazon mollies have been especially useful in studying communication networks (McGregor & Peake 2000; Matos & Schlupp 2005; see also Druen & Dugatkin, **chapter 20**), because strong predictions can be generated concerning male mate choice. In addition, males might further reduce the cost associated with such matings by providing less sperm to Amazon mollies: Hubbs (1964) found that Amazon and sailfin females produce approximately the same number of eggs, but pregnant sailfin mollies were found to be more fecund than pregnant Amazon mollies. Furthermore, sailfin males from both syntopic and allopatric populations primed more sperm when in the presence of a conspecific female than when in the presence of a heterospecific female (Aspbury & Gabor 2004a, 2004b), and Schlupp & Plath (2005) found that *P. mexicana* males preferred to mate with, and transferred more sperm to, conspecific females. Finally, in a field study, Riesch et al. (2008) found that proportionally more sailfin mollies than Amazon mollies had sperm in their genital tract, and among those females that had sperm, sailfin mollies had more sperm. Amazon mollies in natural populations therefore seem to be sperm limited. Unisexual females and females of the host species also engage in aggressive interactions, which could reflect competition over males or other resources (Schlupp et al. 1992; Marler et al. 1997; Heubel & Plath 2008).

Altogether, male mate choice is clearly a key factor in the stability and maintenance of hybridogenetic and gynogenetic mating systems. Stability based on behavioral regulation can probably work even if unisexuals actually have a twofold demographic advantage over bisexuals.

5.5.2 The frozen niche variation hypothesis

Mating systems with sperm-dependent unisexuals may also be stable if the sexuals and unisexuals differ in niche usage. This idea has been elaborated in the frozen niche variation hypothesis (Vrijenhoek 1994). This hypothesis posits that unisexuals of hybrid origin represent only a fraction of the parental species' genetic variation and consequently should utilize only a fraction of the original niches utilized by the parental species (Vrijenhoek 1979). The frozen niche variation hypothesis is well supported by a large number of studies using both natural and artificial clones (Vrijenhoek 1984; Vrijenhoek & Pfeiler 1997; N. Lima 1998; Weeks 1990, 1995; Wetherington et al. 1989b). Several more recent studies add further support. Gray and Weeks (2001), for example, found that bisexual *Poeciliopsis* populations had broader diets than unisexual *Poeciliopsis* biotypes. Furthermore, a model by Pound et al. (2004) predicts that a combination of both the frozen niche variation hypothesis and the accumulation of deleterious mutations (Kondrashov 1988) will do a better job of explaining why

sexuals are not going extinct in mixed bisexual/unisexual populations than each mechanism alone.

5.5.3 The life-history regulation hypothesis

The life-history regulation hypothesis (Schlupp 2005; for poeciliid life histories, see also Pires et al., **chapter 3**; Marsh-Matthews, **chapter 2**) is based on potential life-history differences between the two types of females. For example, it is possible that the accumulation of deleterious mutations in unisexual poeciliids prevents them from making a full investment in their offspring. Hence, they may need to invest more energy in immune defense or in maintaining homeostasis and growth, leaving less energy available for reproduction. If these trade-offs exist, life-history traits such as age of maturity, interbrood interval, or maternal investment would be likely candidates to be negatively affected. Contrary to the behavioral regulation hypothesis, the life-history regulation hypothesis has so far received only limited attention in *P. formosa*, mainly in a classic paper by Hubbs (1964), while numerous studies on the frozen niche variation hypothesis document differences between unisexual and bisexual *Poeciliopsis*. Basically, stable conditions could arise if the overall success of the unisexuals is around 50% of that of the bisexuals. A reduction in fitness could happen at any stage of the life history, and reductions may accumulate over the different phases of the life cycle. However, basic life-history aspects of these mating complexes—for example, brood sizes, reproductive effort, age at first reproduction, and interbrood intervals (Hubbs & Dries 2002)—are not well documented. Weeks (1995) found that unisexual *Poeciliopsis monacha-lucida* had smaller eggs, had fewer eggs per unit body weight, and also matured later than the maternal *Poeciliopsis monacha* under common garden conditions. On the other hand, wild-caught Amazon mollies were originally described as having approximately the same number of eggs as sailfin mollies, but with fewer eggs actually developing into embryos than sailfin mollies (Hubbs 1964). But preliminary analysis of life-history parameters of syntopic Amazon molly and sailfin molly populations from Texas show lower fecundity rates and lower reproductive allocation in Amazon mollies (Riesch, unpublished data). Juvenile survival under benign conditions was the same for Amazon mollies and sailfin mollies (Hubbs & Schlupp 2008), but under food stress Amazon molly neonates experienced higher mortality (Tobler & Schlupp 2010).

5.6 The role of ploidy

One of the most exciting topics to be studied using unisexual fishes is how ploidy and evolutionary longevity may be connected. This is also intimately connected to the question of whether rare forms of recombination or any kind of gene transfer contribute to the longevity of these mating systems. If, indeed, rare mitotic recombination occurs and is relevant, this raises the question of why most (so-called) higher organisms show full meiosis every generation (Green & Noakes 1995) while other forms of reproduction and propagation (including vegetative propagation such as in many plants) are more common in "lower" organisms. In unisexuals genetic variation is mainly introduced by mutations or by changes in ploidy (see box 5.1). Ploidy changes could be the motor of evolutionary change in these organisms, but are they? The basis for this line of research is that over time deleterious mutations will accumulate in the unisexual genome (for hybridogenetic *Poeciliopsis*, the maternal part of the genome), ultimately leading to extinction. This phenomenon is the subject of many theoretical studies (Muller 1964; Kondrashov 1988; Barton 1995; Charlesworth & Charlesworth 1998; Wilke 2004), but in unisexual vertebrates, including unisexual Poeciliids, partial or complete ploidy elevations have been suggested as potential mechanism for escaping from Muller's ratchet (M. Schartl et al. 1995a), thereby providing longevity for unisexual lineages. These ploidy elevations occur when parts or the whole paternal genome are incorporated into the clonal germ line. Interestingly, in a completely different system, bdelloid rotifers, sequestering genetic material from unrelated species has been implicated in the long-term survival and diversification of that group (Gladyshev et al. 2008).

Most unisexual *Poecilia* and *Poeciliopsis* are diploid, probably still reflecting the original state of their hybridization. In both complexes, ploidy elevations to triploid (and tetraploid in *P. formosa*; Lampert et al. 2008) have been reported. The role of triploids and triploidy in Amazon mollies has been comprehensively summarized (Balsano et al. 1989). Since then several new approaches have been used, all taking advantage of new molecular methods. In most cases these elevations are the result of adding a whole sperm genome to the clonal germ line (Nanda et al. 1995; Lamatsch et al. 2000b; Lamatsch et al. 2004; Nanda et al. 2007). In Amazon mollies, paternal genetic material can also be added in the form of microchromosomes, which are very small, centromere-containing fragments of chromosomes (M. Schartl et al. 1995a; Lamatsch et al. 2004). A recent study reported higher fitness in diploid than in triploid Amazon mollies in the laboratory, but clearly many facts are still missing. In nature, triploids appear to be distributed mainly in two river systems, the Río Soto la Marina and Río Guayalejo drainages, each time with monophyletic but independent origin (Schories et al. 2007). In the laboratory, deviant individuals are sometimes detected because they show gene expression of paternal genes in the form of a mottled phenotype. The black blotches they show indicate introgression of male pigmentation genes because the males used to propagate many stocks are ornamental black mollies. Almost none of these laboratory-bred hybrids produce offspring (Nanda et al. 1995). One of the most intriguing findings is that in rare cases triploids spontaneously turn into males (Lamatsch et al. 2000a). These males were able to trigger embryogenesis in Amazon mollies but did not sire offspring with heterospecific females.

In *Poeciliopsis*, triploids have a different reproductive mode from diploids. Instead of hybridogenesis they use gynogenesis. Why this is the case is unknown, but some differences between gynogenetic and hybridogenetic lineages have been described (Schultz 1967; Dawley et al. 1997; Vrijenhoek & Pfeiler 1997). Another as yet not understood pattern is that all known *Poeciliopsis* triploids combine at least one copy of the *Poeciliopsis monacha* genome and one copy of the *Poeciliopsis lucida* genome. Despite the broad distribution of *Poeciliopsis occidentalis* and *Poeciliopsis latidens*, no triploids carry their genomes (Vrijenhoek 1998).

Clearly, in almost all cases inheritance is clonal or hemiclonal, but in rare cases these sperm-dependent systems turn out to be somewhat leaky. Given that all additional genetic material seems to be of paternal origin, this points to a certain instability of the process of excluding the paternal genetic material. It is unclear, however, whether these are inconsequential "accidents," or whether these incorporations are in any way evolutionarily advantageous for unisexual fishes and/or other unisexuals (Bi & Bogart 2006).

5.7 Cancer Research

The fact that unisexual fishes are essentially clonal has been an attractive feature for scientists trying to understand cancer. Early on, clonality made tissue grafts possible, a technique that was used to study induced cancers (Woodhead et al. 1977; Woodhead et al. 1984). These early studies were also among the first to look at aging (Woodhead 1978, 1984), which again is especially interesting in clonal organisms, and—on the side—provided some early work on the behavior of Amazon mollies (Woodhead & Armstrong 1985). *Poeciliopsis* was also used to understand the effects of carcinogens (Schultz & Schultz 1988). A later study used a microchromosome-bearing strain of *P. formosa* that expressed macromelanophores as black skin pigmentation. Macromelanophores and cancer are well understood in *Xiphophorus* (Schartl 1995; see Schartl & Meierjohann, **chapter 26**) but are normally absent in *Poecilia*. The microchromosomes causing black pigmentation also caused melanoma and could be used to study the formation and induction of skin cancer (Schartl et al. 1997).

5.8 Open questions

Many key questions relative to unisexual fishes remain open, despite the growth in our knowledge of these systems. We still do not fully understand the dynamics of stability of the bisexual/unisexual mating complexes. The picture that emerges points to a complex mix of factors, maybe with male behavior being most relevant. Another unsolved problem is the question of evolutionary longevity of certain lineages. Here major advances can be expected from genetic and genomic approaches and from comparative studies. One of the most exciting developments has to do with the potential of genes wandering from sexual genomes to other sexual genomes using asexuals as stopovers; this mechanism is theoretically possible in hybridogenetic *Poeciliopsis*, which can revert to sexuality via hybridization with males from their maternal ancestor, *Poeciliopsis monacha*, or from their sister species *Poeciliopsis viriosa* (Vrijenhoek & Schultz 1974).

The question of the evolutionary potential of asexuals is very important but hard to tackle. Mostly it is noted that asexuals occur in terminal branches of phylogenies and almost never experience further adaptive radiation. The big exceptions are bdelloid rotifers (Fontaneto et al. 2007) and some ostracods (Butlin et al. 1998; Butlin 2002), but not unisexual vertebrates. Defining and detecting diversification in asexuals is an interesting field for future research.

Looking at variation in ploidy will be very instrumental. Unfortunately, unisexual fishes are not well understood in at least two major areas of biology: physiology and developmental biology. Furthermore, the mechanisms by which asexuality arises following hybridization are fascinating but unknown. Ultimately, they may shed light on the evolution of postzygotic reproductive incompatibility. Finally, if trying to understand sex is the Queen of Questions (Bell 1982), trying to understand why unisexual fishes and amphibians are sperm dependent may be one of the Princesses of Questions.

Acknowledgments

We are grateful to our longtime collaborators Michael J. Ryan, Manfred Schartl, Jakob Parzefall, Katja Heubel, Martin Plath, Michael Tobler, Francisco García de León, and Ralph Tiedemann. A large number of students, both graduate and undergraduate (and far too many to be listed here), worked with us, and we would not be able to report anything without their enormous contributions. We would further like to thank the Mexican government, Texas Parks and Wildlife, and the local communities for allowing us to conduct our research. Funding was provided mainly by the University of Oklahoma, the German Academic Exchange Service, and the Deutsche Forschungsgemeinschaft.

Part II
Evolutionary ecology

Chapter 6 Evolutionary and plastic responses to resource availability

Gregory F. Grether and Gita R. Kolluru

6.1 Introduction

OUTSIDE THE POECILIID LITERATURE, plastic and evolutionary responses to the environment are often treated quite separately, with some researchers focusing on the former and others on the latter. In studies of geographic variation, in which patterns of trait variation are examined in relation to putative agents of selection, phenotypic plasticity is often treated merely as an inconvenient possibility, if it is considered at all. This is symptomatic of the traditional disconnect between developmental biology and evolutionary biology (see box 6.1), but part of the problem is simply that the sort of experiments that are required to distinguish between genetic and environment-induced trait shifts are difficult to achieve with most taxa. The relative ease with which some poeciliids can be raised in captivity is undoubtedly one of the reasons that we know so much about the relationship between phenotypic plasticity and evolution in this group. Guppies (*Poecilia reticulata*), in particular, have proven to be remarkably good subjects for sorting out genetic and environmental sources of variation within and among populations, as well as for studying the evolutionary effects of environmental heterogeneity. Most of the research effort in this area has focused on the influence of predation, but other environmental factors have also received some attention (e.g., see Trexler et al., **chapter 9**; Kelley & Brown, **chapter 16**; Tobler & Plath, **chapter 11**; Pires et al., **chapter 3**; Johnson & Bagley, **chapter 4**). In this chapter, we focus on the effects of resource availability.

The basic idea behind this chapter is to ask: what have we learned about the effects of resource availability on poeciliid biology, and how has this advanced our knowledge of evolutionary processes? Interest in resource availability as an agent of natural selection in poeciliids is relatively new. The vast majority of empirical papers were published in the past 20 years, and most were published in the past 10 years. This is the first general review of the effects of resource availability on poeciliids.

Few studies have been carried out just to develop a better understanding of poeciliids. Usually the goal is to tackle a problem of broader interest. The more is already known about a species, the more attractive it becomes as a model for addressing new questions. One unfortunate consequence of this is that we know a great deal about a few species and very little about the rest. We hope that this chapter stimulates new research on the effects of resource availability on poeciliids, and particularly on species that are poorly represented here.

6.2 General terminology and empirical issues

Resource availability refers to the amount of a resource accessible to an organism per individual or per mass unit of that organism. A "resource" can be any component of the environment that an organism utilizes (Lincoln et al. 1998). In our review of the literature, the resources in poeciliid studies can all be classified as dietary components, and all such studies have been carried out on herbivores or

The role of phenotypic plasticity in evolution is central to proposals for an expanded evolutionary synthesis and the subject of several recent books and reviews (e.g., Pigliucci 2001; Price et al. 2003; West-Eberhard 2003; Grether 2005; Ghalambor et al. 2007; Pigliucci 2007; Duckworth 2009). Here we provide only a brief synopsis of some key concepts and the associated terminology.

The least glamorous view of the role of plasticity in evolution is that it is just environmental noise that reduces the heritability of the affected traits and shields inferior genotypes from selection, thereby slowing the rate of adaptive evolution. A similar conclusion can be reached by considering that an adaptive reaction norm that enables an organism to perform well in multiple environments may preclude divergent selection, which otherwise would cause populations in different environments to diverge genetically. Other theorists have focused on the potential for adaptive plasticity to enable organisms to invade new environments and become exposed to new selection regimes, which in turn may cause rapid genetic divergence from the ancestral population. Environmentally induced phenotypes that are favored by selection may pave the way for genetic changes that enable the same phenotypes to be expressed without environmental induction (genetic assimilation). Conversely, when the environment changes for an entire population (e.g., because of climate change, range expansion, or habitat degradation) and perturbs the development of a trait away from its fitness optimum, selection may restore the ancestral phenotype in the new environment by changing the genetic architecture of the trait (genetic compensation). Each of these hypotheses is supported by empirical research but their relative importance in nature is open to debate.

omnivores. Omnivorous poeciliids typically eat some mixture of terrestrial and aquatic invertebrates, detritus, algae, and vascular plants parts (Meffe & Snelson 1989a).

As with any stochastic population parameter, resource availability can only be estimated, not measured precisely. Some estimates are better than others, of course, and we think poeciliid researchers would be wise to invest some effort in developing reliable estimates. For species that rely heavily on terrestrial inputs (e.g., insects, fruits), there may be no reliable substitute for gut content analysis. For species that primarily consume algae, the most direct method is to estimate the standing crop of algae (periphyton) and divide this by the biomass of the consumers (e.g., Grether et al. 1999; Grether et al. 2001b). Measuring primary productivity may be a suitable shortcut, if the biomass of consumers does not covary with primary production (e.g., Reznick et al. 2001), but it is important to consider that periodic disturbances, such as floods, can decouple resource availability from primary production (Chapman & Kramer 1991a; Grether et al. 2001b). It is not wise to assume that resource availability increases with photosynthetically active radiation or canopy openness, because primary production is not always light limited (Hill 1996).

We would not expect geographic variation in resource availability to have evolutionary effects unless the resource is "limiting," at least at some sites. A resource can be considered limiting if variation within the natural range has measurable phenotypic effects (on growth, reproduction, coloration, etc.). Experiments in which resource levels are manipulated in the field may provide the most definitive tests (Allan 1995), but field observations combined with laboratory experiments can also yield strong evidence for resource limitation (e.g., Grether et al. 1999; Grether et al. 2001b).

Resource availability is often correlated with and may therefore be confounded with predation. Such correlations can arise for multiple reasons. For example, the positive correlation between predation intensity and algae availability for guppies appears to be a product both of algal productivity being higher and of guppies and their competitors being less abundant at downstream sites, where large predatory fish are present, than at upstream sites, where such predators are absent (Gilliam et al. 1993; Grether et al. 2001b; Reznick et al. 2001). With sufficient sampling of sites, it might be possible to separate the effects of predation and resource availability using multivariate statistics (e.g., path analysis; Johnson 2002), but this approach requires multiple assumptions and there is no guarantee that the results will be interpretable. A more straightforward approach, when feasible, is to sample sites that enable comparisons of resource levels within predation levels and vice versa (Grether et al. 2001b; Reznick et al. 2001; Arendt & Reznick 2005).

In principle, geographic variation in phenotypic traits can be partitioned into (1) genetic variation, (2) environmental variation caused by phenotypic plasticity, and (3) the interaction between genetic and environmental variation (fig. 6.1). The second component can be described by a population mean reaction norm, and the third component represents genetic divergence in the mean reaction norm.

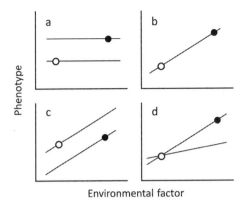

Figure 6.1 Schematic representation of some possible relationships between phenotypic plasticity and genetic divergence. Circles represent population means, and lines represent the corresponding population mean reaction norms. In (a), populations differ genetically in the mean phenotype, and the trait is not plastic with respect to the environmental factor. In (b), populations do not differ genetically in the mean phenotype and occupy different positions along the same mean reaction norm. In (c), population mean reaction norms differ in intercept but not in slope, and the mean phenotypes are more similar than would be the case if representatives of the two populations were raised in a common environment (countergradient variation). In (d), population mean reaction norms differ in slope, and the mean phenotypes are less similar than would be the case if representatives of the two populations were raised in a common environment (cogradient variation). Note that in both (b) and (d), phenotypic differences between populations could disappear in a common environment, illustrating the value of varying environmental factors experimentally.

6.3 Effects of resource availability on poeciliids

We have organized our review of the empirical literature around five kinds of traits: (1) mate preferences, (2) coloration, (3) time allocation and mating tactics, (4) life history, and (5) morphology. These categories are not mutually exclusive but simply serve to help organize the presentation of ideas. We begin each section by briefly presenting the general hypothesis or set of hypotheses that corresponds most closely to the empirical examples that follow. This is intended to be, not an exhaustive survey of all relevant hypotheses and empirical studies, but instead a review of key hypotheses that have been tested using poeciliids and the corresponding empirical studies.

6.3.1 Mate preferences

Mate preferences are expected to evolve to exploit "indicator traits" that correlate with mate quality (see Rios-Cardenas & Morris, **chapter 17**). If the relationship between an indicator trait and mate quality changes, the strength of the corresponding mate preference should evolve to a new optimum (box 6.2). The relationship between indicator traits and mate quality (henceforth "indicator value") is likely to change over time because, by nature, such traits

are sensitive to environmental perturbations, such as shifts in resource availability. There are at least two ways for mate preferences to evolve to track changes in the indicator value of traits. First, mean preference strength could shift genetically, such that populations exposed to different average environments diverge in preference strength. Second, a reaction norm could evolve such that preference strength depends on the same environmental factors that affect the indicator value of the trait (or closely correlated factors). Studies on poeciliid fishes have provided the most direct empirical tests of these predictions. Here we review three

---

Box 6.2 Indicator model prediction

In Iwasa and Pomiankowski's (1999a) model of the indicator process, male ornament size s is a linear function of quality v,

$$s = t + t'v$$

where t is the condition-independent component of ornament size, and t' is the degree of condition dependence, or indicator value, of ornament size. The strength of the female preference is represented by p. Females with positive (negative) values of p prefer to mate with males with larger (smaller) than average ornaments. Females are assumed to benefit, directly or indirectly, by mating with higher-quality males, but the preference also carries a cost b, which increases with p and exponent γ. These costs and benefits balance at the equilibrium female preference,

$$\bar{p} = \left(\frac{\phi \bar{t}'}{\gamma b} \right)^{1/(\gamma - 1)}$$

where ϕ includes the effects of male quality on the direct and indirect components of female fitness (see equation 14 in Iwasa & Pomiankowski 1999a). Thus, the female preference evolves to a level determined by the indicator value of the male trait $\bar{t}'$. The indicator value $\bar{t}'$ is also allowed to evolve in this model and is shown, at equilibrium, to increase in direct proportion to $\bar{p}$. Thus, $\bar{t}'$ and $\bar{p}$ are mutually reinforcing.

Iwasa and Pomiankowski do not model environmentally induced changes in t', but it is clear that a change in t' should cause p to evolve in the same direction.

examples in which resource availability has been linked to mate preference evolution.

6.3.1.1 Guppies. In guppies, a female preference has been shown to respond plastically to the availability of a resource that directly influences the indicator value of male coloration. The polymorphic coloration of male guppies usually includes orange spots with high concentrations of carotenoid pigments, which animals cannot synthesize (Goodwin 1984). Female guppies prefer males with high orange-spot carotenoid concentrations (Kodric-Brown 1989; Grether 2000). In the lab, the carotenoid concentration and thus the chroma (color saturation) of the orange spots increase asymptotically with carotenoid intake (fig. 6.2; Kodric-Brown 1989; Grether 2000; Karino & Haijima 2004) and are positively correlated with algal foraging ability (Karino et al. 2007). Algal foraging ability is heritable (Karino et al. 2005) and correlates positively with size at maturity (Karino & Shinjo 2007). These results suggest that females benefit indirectly from choosing males on the basis of carotenoid coloration, as originally proposed by Endler (1980). Given the asymptotic (diminishing returns) relationship between carotenoid intake and orange-spot chroma, the value of the orange spots as indicators of a male's foraging ability decreases as carotenoid availability increases (fig. 6.2). Thus, the strength of the female preference for carotenoid coloration is predicted to track variation in carotenoid availability in the wild.

In the rainforest streams of Trinidad, West Indies, guppies obtain most of their carotenoids from unicellular algae, which grows on submerged rocks. In the upper reaches of a river drainage, where the streams are narrow and make relatively small gaps in the forest canopy, algae production is strongly influenced by the amount of light available for photosynthesis (Grether et al. 2001b). Small increases in forest canopy cover are associated with significant decreases in algae availability for guppies (estimated as algal biomass divided by guppy biomass), and this is reflected in the growth rates, sizes at maturity, and orange-spot carotenoid concentrations of the fish (Grether et al. 1999; Grether et al. 2001b). By selecting sites above waterfalls that exclude large predatory fish, it is possible to make comparisons between guppy populations exposed to the same predator community but different resource availability levels.

To determine whether the strength of the female preference for carotenoid coloration tracks variation in carotenoid availability, Grether and colleagues reared guppies from 10 low-predation sites under common garden conditions. High- and low-carotenoid-availability sites were paired within drainages to control for phylogenetic effects (five pairs). The strength of the female preference was found to vary genetically among sites but at random with respect to carotenoid availability (Grether 2000; Grether et al. 2005b). Next the researchers examined the effects of diet on the expression of the female preference. Individual females were raised on one of two food levels and one of two food carotenoid concentrations. Females in the low-food, trace-carotenoid diet group showed stronger preferences for carotenoid coloration than females in the other three diet groups (fig. 6.3A). Presumably this reaction norm causes females to base mating decisions more on the color of a male's orange spots and less on other factors where carotenoids are a limiting resource (Grether et al. 2005b).

The sensitivity of the mate preference to carotenoid intake can also be explained as a foraging adaptation. Orange-colored fruits occasionally fall into Trinidadian streams, and guppies treat them as a preferred food source. When presented with small disks of various colors, guppies of both sexes tend to approach and peck the orange disks (Rodd et al. 2002). In lab-reared fish, the degree of attraction to orange disks explains 86% of the variation among populations in the strength of the female preference for carotenoid coloration (Grether et al. 2005b). Like the mate preference, attraction to orange is enhanced by raising females on a diet low in carotenoids (fig. 6.3b; Grether et al. 2005b). A plausible explanation is that guppies especially benefit from consuming carotenoid-rich orange fruits when other dietary sources of carotenoids are scarce. This raises the possibility that the mate preference for carotenoid coloration might merely be an unmodified foraging preference that male coloration evolved to exploit. A problem for this hypothesis is that carotenoid intake does not influence or-

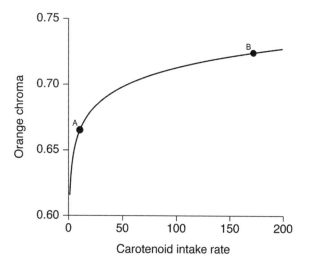

Figure 6.2 Mean carotenoid reaction norm for the orange spots of male guppies. Points A and B represent the mean rates of carotenoid intake at two hypothetical sites. Variation among males in carotenoid intake would have a larger effect on orange-spot chroma at site A than at site B. Modified from Grether 2000.

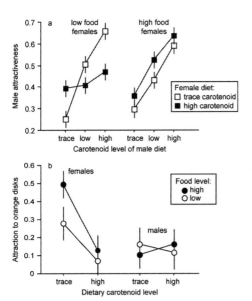

Figure 6.3 Effects of diet on the strength of the female preference for (a) carotenoid coloration and (b) attraction to orange-colored disks in guppies. Plotted points represent least-squares means (± SE) following ANOVA. Modified from Grether et al. 2005b.

ange attraction in males, despite evidence that males benefit more than females do from consuming carotenoids (Grether et al. 2004; Grether et al. 2008). Rather, the sex specificity of the reaction norm suggests that it evolved as a by-product of selection on the mate preference. The existence of such a reaction norm explains why the mate preference has not diverged genetically between populations experiencing different mean levels of carotenoid availability.

6.3.1.2 Cave mollies. Plath and colleagues have documented genetic differences in mate preferences between cave and surface forms of the Atlantic molly (*Poecilia mexicana*) that may be attributable to differences in the energy budgets of fish at cave and surface sites (see Tobler & Plath, **chapter 11**). Cave mollies are generally in poor condition (as measured by abdomen distention), while surface mollies from a nearby river are in uniformly good condition (Plath et al. 2005a). The poor condition of the cave mollies is probably related to the high H_2S content of the water, which interferes with respiration. Cave mollies spend a relatively large proportion of time respiring near the water's surface, where oxygen levels are highest, and this has been shown to be critical for their survival (Plath et al. 2007d). Energetically demanding behaviors, such as aggression and sexual harassment, are reduced or absent in the cave mollies (Parzefall 2001; Plath et al. 2003; Plath et al. 2005b). In mate-choice tests on lab-reared fish, female cave mollies prefer well-fed males with distended abdomens over food-deprived males. This preference is not seen in female surface mollies, which makes sense because a male's condition can

be an indicator of mate quality only where condition is a variable trait. Lab-reared males from both sites prefer large females, but the strength of this preference is greater in cave mollies than in surface mollies (Plath et al. 2006). One of several possible explanations for the greater choosiness of cave males is that copulating is more energetically demanding in the H_2S-rich cave waters than in the river. Whether the sexual behavior of these fish also responds plastically to variation in food availability or H_2S, and whether the cave and surface forms differ in this regard, remain to be studied.

6.3.1.3 Swordtails. Female swordtails (*Xiphophorus birchmanni*) also prefer well-fed males, and this chemically mediated mate preference is affected by a female's own food intake. Compared to well-fed females, food-deprived females are quicker to respond to conspecific male odors and show a stronger preference for (i.e., spend more time near) the odors of well-fed males than for those of food-deprived males (Fisher & Rosenthal 2006b). Female swordtails do not prefer the odor cues of well-fed females over those of food-deprived females, which suggests that the preference for well-fed males is truly a sexual preference, as opposed to a general attraction to food odors (Fisher & Rosenthal 2006a). These results are consistent with the hypothesis that female swordtails, in effect, use their own nutritional state to "decide" how much to weight the nutritional status of prospective mates relative to other factors. This result directly parallels the finding that carotenoid deprivation increases the strength of the female preference for high-carotenoid males in guppies (see section 6.3.1.1).

6.3.2 Coloration

When changes in the environment cause plastic traits to develop sub-optimally, selection favors genetic changes that act to restore the optimal phenotype in the new environment. When this process of genetic compensation occurs along a spatial environmental gradient, it produces a pattern of countergradient variation. Most documented examples of countergradient variation involve the effects of altitude, latitude, or temperature gradients on growth rates (reviewed in Conover & Schultz 1995). Guppies provide one of the few known examples of countergradient variation in coloration (for nonpoeciliid examples, see Grether 2005).

In addition to diet-derived carotenoids, the orange spots of male guppies contain drosopterins, which the fish synthesize de novo from carbohydrates and amino acids (Takeuchi 1975; Grether et al. 2001a; Hudon et al. 2003). In Trinidad, the carotenoid content of the orange spots is constrained by algae availability, which depends on forest canopy cover (see section 6.3.1.1). Geographic variation in the carotenoid content of the orange spots is mirrored by variation in drosopterin content, such that the ratio of the two types of pigments is roughly conserved across streams (fig. 6.4c; Grether et al. 2001a). This pattern is the opposite of what would be expected if males used drosopterins to compensate for the effects of variation in carotenoid availability on orange-spot chroma. The pattern is consistent, however, with the hypothesis that males are under selection to maximize chroma subject to the constraint of maintaining a particular pigment ratio. Carotenoids and drosopterins have different spectral properties, and thus the ratio of the two types of pigments affects the shape, or "hue," of the orange-spot reflectance spectrum (fig. 6.4a and b; Grether et al. 2001a; Hudon et al. 2003). The positive correlation between drosopterin production and carotenoid intake reduces geographic variation in orange-spot hue, relative to the alternative of drosopterin production not correlating positively with carotenoid intake (fig. 6.4d; Grether et al. 2005a). This is a countergradient pattern because genetic differences between populations in drosopterin production

mask the effect of carotenoid availability on the hue of the orange spots. Whether female guppies actually prefer males with a particular pigment ratio remains to be determined.

How resource availability affects other aspects of poeciliid coloration is not clear. Millar and colleagues (2006) carried out a survey of color variation in Trinidad guppies across 29 sites in two river drainages, with the goal of statistically disentangling the effects of predation from the effects of other environmental factors that covary with predation. In one drainage (Marianne), canopy openness (a surrogate for resource availability) correlated negatively with the extent (spot number or area) of black and orange coloration and positively with the extent of green/bronze coloration, after controlling for predation. In the other drainage (Paria), however, the correlations between canopy openness and color were nonsignificant or in the opposite direction.

6.3.3 Time allocation and mating tactics

Animals should invest limiting resources in alternative behaviors, such as foraging, courtship, and aggression, in ways that maximize lifetime reproductive success (Wong & Candolin 2005; Kemp et al. 2006; Fischer et al. 2008). The optimal allocation of time to foraging versus reproductive activities, and the pattern of investment in alternative reproductive tactics, may depend not only on an individual's immediate energy reserves but also on its lifetime food intake and the population average food availability (Kolluru & Grether 2005; Kolluru et al. 2007). Surprisingly few studies have made use of natural gradients in resource availability to test these ideas.

6.3.3.1 Guppies. Guppies are well suited for studying the effects of resource availability on behavior because they occur along a replicated gradient in the availability of their primary food source (algae; see section 6.3.1.1) and are easy to observe in both the field and the lab. Males perform a variety of mating tactics, including the sigmoid courtship display, sneak copulations, and aggressive interference competition (Rios-Cardenas & Morris, **chapter 17**). Aggression in a mating context is distinct from aggression in a foraging context (Magurran & Seghers 1991) and includes displaying, chasing, and biting (Kolluru & Grether 2005; Hibler & Houde 2006). Males regularly jockey for position near receptive females (Houde 1997), which may interfere with mate choice because it interrupts the female response to courtship (Hibler & Houde 2006; Price & Rodd 2006; Kolluru et al. 2009). Male-male aggression also occurs away from the immediate vicinity of females, perhaps for establishing or maintaining dominance relationships (Kolluru & Grether 2005). Males essentially stop

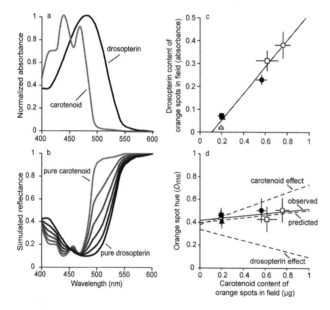

Figure 6.4 Countergradient variation in the sexual coloration of guppies. (a) Absorbance spectra of carotenoids and drosopterins extracted from the orange spots of male guppies. (b) Simulated reflectance spectra for different carotenoid:drosopterin ratios. (c) Geographic covariation between drosopterin and carotenoid content of the orange spots across six populations (field data). (d) Countergradient variation diagram showing that a photoreceptor-based measure of orange-spot hue (D_{ms}) is conserved across populations in the field because of the counterbalancing effects of drosopterins and carotenoids. The solid line in (d) represents the least-squares regression through the observed population means; the upper and lower dashed lines represent the effect of each pigment separately on D_{ms}; the middle dashed line represents the predicted values of D_{ms}, given the observed orange-spot pigment content means (see Grether et al. 2005a). Plotted points represent population means (± SE). Panels a–c were modified from Grether et al. 2001a, and panel d was modified from Grether et al. 2005a.

growing after maturing (Snelson 1989), so postmaturation foraging is largely an investment in future reproduction. Males have been shown to switch predictably between foraging and courtship based on their immediate hunger levels (Abrahams 1993; Griffiths 1996), but until recently nothing was known about how mating tactics are affected by food availability in the wild.

In a behavioral survey of 10 sites in Trinidad with the same predators but different levels of food availability, Kolluru and colleagues (2007) found evidence for genetically variable reaction norms involving three statistically independent axes of variation in male behavior: (1) courtship versus foraging, (2) interference competition versus mate searching, and (3) dominance interactions. The first two axes represent trade-offs between alternative activities. The trends in the field were for males to invest relatively more in courtship and interference competition, and less in foraging and mate searching, at sites with higher food availability. When lab-born males from the same sites were raised on two different food levels, there were no main effects of site food availability but significant interactions were found between lab food level and site food availability. Males from low-food-availability sites showed greater plasticity than males from high-food-availability sites with respect to the first two axes of variation and less plasticity with respect to the third axis. Specifically, when males from low-food-availability sites were raised on the high food level, they courted more but foraged less, and interfered with each other's courtship more but searched for mates less, compared with males from the same sites raised on the low food level. Males from high-food-availability sites showed neither of these responses to food intake in the lab. Food intake in the lab had a positive effect on the rate of dominance interactions in males from both types of sites. The positive effects of food intake on courtship and interference competition support the hypothesized energetic trade-off between investment in current versus future reproduction. The effects of site food availability on these reaction norms support the hypothesis that plasticity in behavior is more likely to evolve in populations with more variable and limited resource availability (Komers 1997).

Although the guppy story is already complex, future studies will undoubtedly add complexity. Mating-tactic variation should be addressed in conjunction with competitive ability in a foraging context to test for synergistic effects. Males vary in their ability to compete for access to food (Magurran & Seghers 1991), and access to food, in turn, affects body size (Grether et al. 2001b), carotenoid intake (Grether 2000), parasite resistance (Kolluru et al. 2006), and mating tactics (Kolluru et al. 2007). A correlation in aggression across foraging and mating contexts would constitute a behavioral syndrome, a consistency in

individual behavior across contexts (reviewed in Kelley & Brown, **chapter 16**; Sih et al. 2004; see also Magellan & Magurran 2007a; Scotti & Foster 2007). An "aggression syndrome" could itself be influenced by food intake if, for example, aggression across contexts is correlated in high-food males but not in low-food males, because the latter are more energetically limited.

Other environmental stressors may interact with food limitation. In guppies, the prevalence of the monogenean ectoparasite *Gyrodactylus* varies among populations and seasons (van Oosterhout et al. 2006b; van Oosterhout et al. 2007a; Martin & Johnsen 2007; see also Cable, **chapter 8**). Infected males are less attractive to females (Houde & Torio 1992; López 1998), display to females less frequently (Houde & Torio 1992; Kolluru et al. 2009), and have reduced success in intermale contests (Kolluru et al. 2009). High food availability partly ameliorates these negative effects, which suggests that parasitized males are less successful, in part, because they are energetically challenged (Kolluru et al. 2009). Males at low-resource sites where *Gyrodactylus* occurs experience the potentially synergistic effects of limited food availability and low carotenoid availability (which may reduce parasite resistance; Kolluru et al. 2006).

6.3.3.2 Other species. There is a paucity of studies of resource availability and mating tactics in other poeciliids, although they inhabit a wide range of habitats and exhibit striking intraspecific and interspecific variation in male mating tactics (reviewed in Farr 1989; Rios-Cardenas & Morris, **chapter 17**). Some species lack courtship altogether (e.g., *Phallichthys quadripunctatus*; Kolluru & Joyner 1997), while males of other species switch among courtship, sneak copulation, and male-male competition (Farr 1989; Houde 1997; Kolluru & Joyner 1997; Plath et al. 2007b; Rios-Cardenas & Morris, **chapter 17**). Even in species without courtship display, females exert some control over mating (Bisazza et al. 2001b), such that males must decide when and how frequently to attempt copulations. Females suffer a cost of sexual harassment by males in the form of reduced feeding time (Magurran & Seghers 1994c; Plath 2008; see also Magurran, **chapter 19**), and this cost can be greater with smaller males, because they attempt sneak copulations more frequently (Schlupp et al. 2001); reduced resource availability could exacerbate these effects by reducing female condition and male size. Promising species in which to investigate the relationship between resource availability and mating behavior include *Brachyrhaphis* spp. and *Alfaro cultratus*, which experience seasonal variation in food availability (Winemiller 1993; Johnson & Belk 2001; Jennions et al. 2006), and extremophile species such as cave mollies, which experience low oxygen levels that may limit the frequency of energetically costly behaviors (see section 6.3.1.2; Tobler & Plath, **chapter 11**).

6.3.4 Life history

Life-history evolution inherently involves trade-offs, such as that between current and future reproduction or between offspring number and quality. Environmental conditions, and especially resource availability, can mediate such trade-offs by determining how much energy is available to devote to different traits and how investment in different traits varies temporally and spatially (see Pires et al., **chapter 3**; Johnson & Bagley, **chapter 4**). Poeciliids afford some of the clearest examples of the effects of variation in resources on life-history traits. For example, offspring size, interbrood interval, and litter size all exhibit plastic responses to maternal food intake in guppies (Reznick & Yang 1993). Interbrood intervals are longer under low food availability, suggesting that females delay brood production to increase egg provisioning. Food levels experienced during the interval between the first and second litters influence the third litter, so that there are long-term effects of food availability on offspring production.

Research on poeciliids has also advanced our understanding of how temporal and spatial variation in resource availability in the wild shapes life-history trade-offs. Seasonal variation in food availability has important consequences for reproductive output (Winemiller 1993; Johnson & Belk 2001; Johnson 2002; Jennions et al. 2006). Winemiller (1993) studied reproductive seasonality of three species of poeciliids (*Alfaro cultratus*, *Poecilia gillii*, and *Phallichthys amates*) in a Costa Rican rainforest. All species showed increases in reproductive output during the wet season, primarily because large females had much larger litters during the wet season. The wet season brings more abundant food (e.g., arthropods) for *A. cultratus* adults and for juveniles of all three species, but food (algae) availability for adults of the other two species declines during the wet season. Thus, the period of peak reproductive output appears to be timed to coincide with the conditions most favorable for juvenile growth (predation on juveniles may also be reduced during the wet season). Jennions et al. (2006) and Johnson and Belk (2001) also describe wet-season increases in reproductive output in insect-eating *Brachyrhaphis* species. Guppy reproductive output is reduced in the wet season, potentially because increased silt deposits reduce algae availability for all age classes (Reznick 1989). These studies emphasize the need to examine life-history variation during wet and dry seasons and in conjunction with environmental variables that may mitigate resource abundance.

The relationship between life-history variables and resource availability has been most intensively studied in guppies, and these studies have emphasized that the effects of predation intensity must either be experimentally separated from the effects of resource availability (Reznick et al. 2002a; Bashey 2006) or be examined independently using populations that vary in resource availability independent of predation intensity (Grether et al. 2001b; Hendry et al. 2006). In some cases, however, the effects of predation and resource availability can be distinguished on theoretical grounds. Arendt and Reznick (2005) raised guppies from multiple sites through two generations under common garden conditions in the lab and found that fish from high-predation, high-resource-availability sites grew faster than fish from low-predation, low-resource-availability sites. The authors argue that only resource availability can explain this pattern because predation should select for more rapid growth at low-predation sites where the predators of guppies are gape limited. The direction of genetic divergence is consistent with optimal–growth rate theory, which holds that conditions of chronic low food availability select for slow growth (reviewed in Arendt & Reznick 2005).

When should we expect local adaptation in life-history traits versus species-level reaction norms? Bashey (2006) tackled this question by studying maternal effects on offspring size in two populations of guppies that experience different levels of resource limitation in nature (see also Bashey 2008). Females from both populations produced larger, more competitive offspring under food-limited conditions, but the low-resource population was less plastic in this regard than the high-resource population. By comparing families with different degrees of plasticity for offspring size, Bashey (2006) found evidence for trade-offs between plasticity and other fitness-related traits. The nature of these genetic correlations suggests that selection favors reduced plasticity in offspring size under conditions of chronically low food availability.

The unique biology of some poeciliids makes them particularly suitable for studies of life history and resource availability. For example, intraspecific genetic variation in the response to resource availability can add "noise" to studies of plasticity in life-history traits; this can be avoided by examining clonal strains, such as *Poeciliopsis monacha-lucida*, in which all individuals are genetically identical and any plasticity in life-history traits can be attributed to the environment. Weeks (1993) and Weeks and Quattro (1991) examined plastic responses of life-history traits in response to density and food availability in females of sexual and asexual *Poeciliopsis* strains. Under high density, females of all strains exhibited delayed maturation, were smaller, and produced fewer eggs than under low density, but there was no effect of density on egg size or egg production per unit of female mass. The lack of response of some life-history traits to variation in resource availability may result from the inability of females to gauge environmental conditions, constraints imposed by the costs of plasticity, or selection favoring genetically fixed alternatives (Weeks & Quattro 1991; Weeks 1993). Fixed differences among strains may result from ecological differences: the sexually reproducing strain may invest more in early reproduction to compensate for reduced expected life span because it experiences lower resource availability in the wild (Weeks 1993).

Another unique feature of poeciliids that makes them amenable to life-history studies involves their reproductive mode, which ranges from lecithotrophy (embryos are nourished by the yolk alone) to varying degrees of matrotrophy (females continue to provision offspring after fertilization; reviewed in Marsh-Matthews, **chapter 2**). The evolution of matrotrophy from lecithotrophic ancestry, and the degree of matrotrophy, are intimately connected with food availability (Trexler & DeAngelis 2003; Banet & Reznick 2008). Recent studies have focused on contrasting patterns of offspring provisioning in lecithotrophic species such as the guppy, with matrotrophic species such as *Gambusia geiseri*, *Poecilia latipinna*, *Heterandria formosa*, and *Poeciliopsis prolifica* and on population-level comparisons within these taxa (Marsh-Matthews & Deaton 2006; Pires et al. 2007; Banet & Reznick 2008; Leips et al. 2009; Pires et al., **chapter 3**). Interestingly, whereas lecithotrophs produce larger offspring under food limitation (reviewed in Bashey 2006), matrotrophs either show no relationship or produce smaller offspring under food limitation (reviewed in Reznick et al. 1996a; Bashey 2006; Pires et al. 2007). The lack of change (or reduction) in offspring size of matrotrophic species under food limitation may reflect constraints inherent to this mode of maternal provisioning (reviewed in Banet & Reznick 2008). Alternatively, matrotrophs may inhabit environments where parental resource availability does not accurately predict offspring resource availability, such that selection does not favor plasticity in offspring size (Reznick et al. 1996a; Trexler 1997; Bashey 2006; Pires et al. 2007). Evidence that matrotrophs are able to plastically alter reproductive output in response to environmental conditions comes from Leips and Travis (1999), who showed that larger *H. formosa* offspring size is associated with higher densities in the field, and Leips et al. (2009), who performed laboratory experiments in which per capita food availability was held constant and showed that females increase offspring size in response to high density.

6.3.5 Morphology

Poeciliids exhibit marked interspecific and intraspecific morphological variation (including sexual dimorphism) in coloration, body size, and body and fin shape. Such

variation can reveal the underlying selective forces shaping phenotypic divergence, especially when viewed in the context of corresponding environmental gradients (Norton et al. 1995; Hankison et al. 2006; Langerhans et al. 2007). Resource availability may influence morphological traits in multiple ways, from selecting for body shapes that favor efficient swimming, to favoring structures that facilitate capture of specific prey types (e.g., benthic versus open-water prey), to indirectly affecting morphology via growth rates (Meffe & Snelson 1989a; Arendt & Reznick 2005; Ruehl & DeWitt 2005). Morphological types that result from the exploitation of different resource types may be associated with behavioral variation, sometimes in contexts other than foraging (e.g., Scotti & Foster 2007). Because they occur along a variety of environmental gradients, poeciliids offer some of the most compelling examples of the link between form and function.

Body shape can have important consequences for maneuvering when escaping from predators or competing for food and mates (Neves & Monteiro 2003; Gomes & Monteiro 2008). Geometric morphometrics enables statistical assessment of shape variation and has been applied to ecological questions in a variety of poeciliids (Neves & Monteiro 2003; Ruehl & DeWitt 2005; Hankison et al. 2006; Hendry et al. 2006; Langerhans et al. 2007). Only rarely have such studies explicitly addressed resource availability. Hendry and colleagues (2006) investigated body shape variation among guppy populations differing in canopy openness (a surrogate for resource availability). Females in open-canopy sites had smaller heads and more distended abdomens than females in closed-canopy sites, potentially reflecting stronger selection on reproductive output in open-canopy sites. Male body shape variation was primarily related to predation intensity rather than canopy openness, and some aspects of morphology in both sexes varied with current speed (Hendry et al. 2006). It is not clear, however, whether the variation in shape represents plasticity or genetic divergence. Body shape variation along ecological gradients has also been examined in *Poecilia vivipara* (Neves & Monteiro 2003; Gomes & Monteiro 2008). These fish inhabit sites in Brazil that fall along a gradient ranging from highly saline, sparsely vegetated, low-predation areas nearer the ocean to freshwater regions with greater macrophytic vegetation and an assemblage of predators further inland. Gomes and Monteiro (2008) found body shape variation among sites consistent with differences in predation intensity. However, there is some evidence that resource availability, which is thought to be higher in the freshwater areas, may also influence morphology (see below). Common garden experiments are needed to determine whether the variation in morphology represents genetic divergence or plasticity.

Resource availability may most directly influence morphology involved with competition for food, and several studies of trophic polymorphisms in poeciliids suggest that the group should be explored further to determine the effects of resource availability in this context. Male guppies from the Turure River in Trinidad exhibited apparently adaptive morphological plasticity in response to the location of the food on which they were raised: males raised on food floating at the surface had a more fusiform body shape and longer, thinner paired fins than males raised on food located on the bottom or at various other locations, corresponding to differences between benthic and open-water morphotypes (Robinson & Wilson 1995). In contrast, females did not exhibit the same morphological plasticity, perhaps because of their livebearing mode of reproduction and associated morphological adaptations. Males and females are more likely to forage in open water and on the stream bottom, respectively, which may be a consequence of differences in morphology and may constrain the evolution of phenotypic plasticity in females (Magurran 2005, 67).

Ruehl and Dewitt (2005) examined morphological plasticity in juvenile *Gambusia affinis* from two populations subject to different levels of predation intensity. Fish fed in the open water developed elongated heads and narrow pelvic fins, which facilitates feeding on free prey in open water. In contrast, fish fed on attached food items acquired wide heads and broad pectoral fins, enabling scraping of food attached to the substrate. Fish from the high-predation population were larger and differed in shape from fish from the low-predation site, but there was no difference in the degree to which fish from the two populations responded to food treatments (i.e., no genetic difference in the mean norm of reaction). Predatory fishes constrain guppies and *G. affinis* to stream margins (away from open water in the center of the stream; reviewed in Reznick et al. 2001), and Ruehl and Dewitt (2005) suggest that this may influence the trophic morphology of *G. affinis*. Indeed, fish in the high-predation site had relatively short fin insertions and arched bodies, consistent with feeding in stream margins (Ruehl & DeWitt 2005). Gomes and Monteiro's (2008) results on shape variation in *Poecilia vivipara* (described above) are consistent with Ruehl and Dewitt's (2005) induced-morphology studies. Body shape differences, including position of pectoral fins and mouth angle, correlate with resource availability: fish in the less vegetated (i.e., more open water), high-salinity environments have more upward-directed mouths, which may aid in feeding on open-water prey. These results offer observational field corroboration of the experimental effects induced by Ruehl and Dewitt (2005) and Robinson and Wilson (1995) and further emphasize that resource availability may influence morphology in conjunction with other environmental factors.

6.4 Summary and future research

Resource availability clearly influences the biology of poeciliids in numerous ways. In guppies alone, resource availability affects mate choice, coloration, mating tactics, foraging behavior, body shape, and several life-history traits. Indeed, the effects of resource availability may prove to be as pervasive as those of predation (Magurran 2005). Most of these effects involve a combination of phenotypic plasticity and genetic divergence between populations. Studying the effects of resource availability on poeciliids has advanced our knowledge of evolutionary processes in several ways. Research on three different species has shown that mate preferences respond plastically to variation in resources that influence the indicator value of preferred traits (section 6.3.1). Geographic variation in guppy coloration provides one of the few known examples of countergradient variation in a secondary sexual character (section 6.3.2). Research on guppies also shows that the optimal allocation of time to alternative behaviors may depend on an individual's lifetime food intake and the population average food availability, as well as on immediate energy reserves (section 6.3.3). Work on several species of poeciliids shows how variation in resources can influence the evolution of life-history traits (section 6.3.4) and morphology (section 6.3.5).

An important challenge for future research will be to disentangle the potentially complex interactions between resource availability and other environmental variables, such as predation, current speed, and parasite prevalence. Multivariate analyses of field data cannot provide the necessary resolution, especially because plastic and genetic responses to the same environmental factors can go in opposite directions. Based on the few studies that have already begun to tackle this problem, it is clear that common garden experiments and manipulations of environmental variables are essential. For example, it was necessary to raise guppies from multiple sites in a common environment to show that resource availability overshadows the effects of predation on growth rate evolution (Arendt & Reznick 2005). Likewise, experimental manipulations of parasites and food levels were necessary to show that food availability ameliorates the effects of parasites on courtship (Kolluru et al. 2009). The beauty of working with poeciliids is that such experiments are often tractable.

Predicting how populations will respond to rapid environmental change is a pressing issue in conservation (see Stockwell & Henkanaththegedara, **chapter 12**), and poeciliids are well suited to serve as model systems for developing such predictions. Research on guppies has shown how plastic and evolved responses of prey to the absence (or removal) of predators may make prey populations vulnerable to extinction when predators are suddenly introduced (or restored to their former range) (Reznick et al. 2008). Whether rapid changes in resource availability could have a similarly destabilizing influence remains to be studied.

Acknowledgments

We thank Jerry Johnson, Ingo Schlupp, and one anonymous reviewer for suggestions and comments that greatly improved the chapter. Our research on guppies was supported by fellowships and grants from the National Science Foundation. We also wish to thank our friends in Trinidad and Tobago and their government for permission to work in protected watersheds and to export guppies.

Chapter 7 Sensory ecology

Seth William Coleman

7.1 Introduction

POECILIIDS ARE FOUND in highly diverse habitats, each presenting a unique sensory challenge for navigation, finding food and mates, and avoiding predators. While a rich literature reveals that many poeciliids rely on multiple sensory modalities in such behavioral contexts, there have been relatively few attempts to rigorously characterize poeciliid sensory systems, and even fewer that quantitatively link sensory systems to sensory environments. This latter deficiency is noteworthy, as poeciliids are, in my opinion, ideal to test Endler's (1993a) elegant model of sensory drive (box 7.1), which proposes that physical properties of the environment direct the evolution of locally adapted sensory systems, which in turn favor the evolution of locally adapted signals for use in communication.

In this chapter I begin (section 7.2) with a brief review of studies that have investigated the use of specific sensory modalities in poeciliid behavior, primarily in the context of mate choice, as this is the context in which much of the poeciliid sensory ecology research has been done (see Rios-Cardenas & Morris, **chapter 17**). Then, in section 7.3, I review what is known about the genetics and physiology of poeciliid sensory receptors. This section focuses primarily on the visual modality, as comparatively little work has been done on characterizing poeciliid chemoreceptors or mechanoreceptors. Finally, I conclude with a section (7.4) suggesting areas for fruitful future research, identifying the need for significant neurophysiological research in poeciliids, and emphasizing the need for cross-disciplinary investigations into the concomitant mechanisms affecting signal production and detection. Such physiological and molecular investigations, combined with quantitative characterization of the sensory environment, can then be coupled to the behavioral assays that have established the Poeciliidae as a model group in behavioral and evolutionary biology (Meffe & Snelson 1989b; Houde 1997; Ryan & Rosenthal 2001; this volume). Applied to other model systems, this cross-disciplinary approach has repeatedly and significantly advanced our understanding of sensory ecology and the evolution of animal communication systems (e.g., Ryan 1985; Ryan & Rand 2003).

7.2 The evolutionary ecology of poeciliid senses

The research on sensory ecology in poeciliids is dominated by studies investigating female mate-choice preferences for visual or chemical cues that males use to attract mates (see Rios-Cardenas & Morris, **chapter 17**). A general tendency seems to be for females to prefer the look and smell of conspecific males that are large, well fed, and brightly colored. While the authors of most mate-choice studies hypothesize how such preferences may have evolved, few conduct tests outside the context of mate choice to test patterns of preference and trait evolution, and fewer still characterize the molecular and physiological mechanisms affecting the evolution of preferences and preferred traits. Such studies are essential for elucidating evolutionary covariances between sensory environments, sensory preferences, and signal evo-

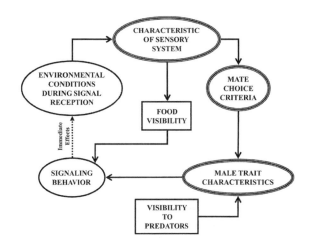

Figure 7.1 A schematic view of the sensory-drive process. The arrows indicate evolutionary effects except for the broken one labeled "Immediate Effects." Double-lined circles are equivalent to Ryan's sensory-exploitation model (Ryan et al. 1990). Modified from Endler 1992.

Box 7.1 The sensory-drive theory

The theory of sensory drive (Endler 1992) posits the causal links between natural selection on sensory systems and sexual selection on signals used in mate choice. The theory has two primary components. First, local ecological conditions, such as the distribution of light in habitat, will favor individuals with visual systems that allow them to most effectively find food and avoid predators (natural selection). As I discuss in the text of this chapter, there are several ways in which a visual system can be changed so as to become "tuned" to local photic conditions. Second, once a locally adapted visual system evolves, then sexual selection should favor the evolution of signals that maximally stimulate these visual systems (fig. 7.1). An analogous evolutionary sequence can be envisioned for chemosensory and auditory systems and subsequent evolution of chemical and acoustic signals. Sensory drive may contribute to speciation (reviewed in Boughman 2002): if different populations occupy different sensory habitats (with concomitant local selective pressures), then given sufficient time, the evolution of locally adapted sexual signals may preclude interpopulation mating and thus facilitate the formation of species. (For a rigorous empirical example of how sensory drive promotes reproductive isolation and how anthropogenic fouling of sensory environments can result in species breakdown, see Seehausen et al.'s [2008] study of African cichlids.)

lution (Endler 1993a; fig. 7.1). In this section, I highlight some exceptional studies investigating the selective forces shaping guppy (*Poecilia reticulata*) visual preferences and highland swordtail (*Xiphophorus malinche*) chemical preferences. Later (section 7.3), I discuss several studies of poeciliids that have characterized sensory systems at the molecular and physiological levels.

7.2.1 Visual preferences for orange coloration in guppies

The sensory-bias hypothesis (Ryan & Keddy-Hector 1992; Ryan 1998) posits that mate-choice preferences are derived from preexisting biases in the sensory system (see Rios-Cardenas & Morris, **chapter 17**); such biases may be the result of natural selection shaping sensory systems to maximize the detection of food or predators. While there is mounting evidence across taxa supporting the importance

of preexisting sensory biases in mate preferences, including in some poeciliid fishes (Basolo 1990; Rodd et al. 2002), the generality of sensory biases in explaining the evolution of female preferences is unclear (for reviews, see Endler & Basolo 1998; Fuller et al. 2005).

In guppies, male sexual traits, such as orange carotenoid coloration, and female preferences for male traits vary among populations in the streams where they are found in Trinidad (Endler 1980, 1995; Endler & Houde 1995; Houde 1997). In an effort to explain the evolutionary origins of female guppy preferences for males with high concentrations of carotenoid pigments in the orange spots on their bodies (Kodric-Brown 1989; Grether 2000), Rodd et al. (2002) characterized the color preferences of male and female guppies outside the context of mate choice. In a series of experiments using colored disks, Rodd et al. (2002) found that guppies of both sexes preferred to peck at an orange-colored disk to disks of other colors. Remarkably, the strength of female preferences for orange disks explained 94% of the variation among populations ($n = 6$) in the strength of female preferences for orange coloration on males. The authors hypothesized that the preference for orange objects may help guppies find highly nutritious orange-colored fruit and thus is favored by natural selection. While this study suggests that the degree of attraction to orange objects and the strength of female preferences for orange coloration share a common genetic basis, phenotypically manifested as a sensory bias affecting behavioral preferences, the authors do not identify the genetic mechanisms involved. In a later section (7.3.1.1), I review research that has empirically characterized the visual system of the guppy.

7.2.2 Chemical preferences for locally tuned sexual pheromones in swordtails

Despite anecdotal evidence that chemical signals are nearly ubiquitous in sexual communication across taxa (Wyatt 2003), studies of chemical signals are rare compared with studies of visual communication (Coleman 2009). Perhaps not surprisingly then, to date only a single study has tested the prediction of the sensory-bias model that male chemical signals are "tuned" to be most effective in local chemical environments.

As with all signals, chemosignals must interact with the medium through which they are transmitted. Water chemistry can be affected by many factors, including soil type, temperature, and anthropogenic inputs. Thus, the chemical environments in which different populations are found may differ in ways critical for chemosignal efficacy. To address how naturally varying chemical environments shape the attractiveness of male chemical signals to females in mate choice, researchers (G. G. Rosenthal, C. C. Wat, Z. Cress, Z. Culumber, & S. W. Coleman, unpublished data) tested the chemical preferences of female highland swordtails (*X. malinche*) from near the headwaters of three separate, nearby drainages characterized by distinct soil types. During preference trials, females strongly preferred male pheromones that were mixed with the drainage-appropriate water. That is, females were more attracted to male signals when the male signal was correctly matched with the male's source water than when the source water and male signal were mismatched.

From the standpoint of sensory ecology, the results are the first, to my knowledge, to show local adaptation in chemical signals produced by males and used by females in mate choice; signal-to-environment matching in visual and acoustic communication has been reported for a wide range of species (Bradbury & Vehrencamp 1998). Whether such local tuning is a result of divergence of female chemosensors due to location-specific natural selection and subsequent evolution of complementary male chemosignals (i.e. "sensory bias"), or more simply reflects changes in the molecular structure of male signals to resist degradation in their specific chemical environments, remains to be determined. In a later section, I discuss research on poeciliid chemosensors and identify potential targets of selection in the chemosensory system.

7.3 How poeciliids perceive the world: perception and processing in the poeciliid sensory periphery

In this chapter, I will focus on the stimulation of the sensory periphery, that is, at the interface between the environment and the sensory receptor, and the morphological, molecular, and physiological characteristics of poeciliid sensory receptors. The analysis, processing, and retention of sensory information collected by the sensory periphery are the subjects of another chapter (see Bisazza, **chapter 15**); for readers interested in the ontogeny of poeciliid sensory systems, Kunz (2007) provides a fascinating review on the development of the guppy retina.

To understand the evolution of communication systems in poeciliids—indeed, in any organism—one must first understand how the physical landscape of the sensory periphery affects perception. It will become evident in this chapter that the poeciliid visual sensory periphery is well characterized relative to the chemosensory and mechanosensory peripheries. Thus, I will focus this initial overview on the aspects of visual perception that are affected by the sensory periphery and concentrate on components of the sensory periphery that have been investigated in poeciliids (for an excellent general review of sensory perception and signal evolution, see Rosenthal 2007).

Photoreceptors (specialized cells densely packed on the retina) are the primary filter for information about the visual environment. Although other aspects of the visual system (such as the transmission and focusing properties of the lens and cornea and the neuronal integration of information at the level of the retina and brain) are important, the ability to transduce light into neuronal information defines the visual capabilities of an animal (Lythgoe 1979; Lythgoe & Partridge 1989). In vertebrates, two types of photoreceptors, rods and cones, are densely packed in a mosaic on the retina (Lythgoe 1979). Rods function primarily in detecting contrast between light and dark surfaces, while cones function in high-resolution color vision. The outer segment of each photoreceptor is lined with light-sensitive visual pigment molecules. Each visual pigment molecule consists of a chromophore, usually 11-*cis* retinal, covalently bound to the protein, opsin. The primary function of the visual pigments is to transduce optical signals in the environment into electrochemical signals in the brain (Yokoyama & Yokoyama 1996). When the pigment in a photoreceptor cell absorbs photons of light, the chromophore is promoted to an excited electrical state, ultimately producing a change in the electrical activity of the cell. The resulting electrical signal enters a neural network, where it is processed and analyzed to re-create an image of the external environment (Archer 1999). Thus, visual pigments represent a direct interface between an organism and its environment at the molecular level (box 7.2).

Theory predicts that natural selection favors the evolution of sensory systems that are locally adapted to ecological conditions (Endler 1992, 1993a; Boughman 2002). This hypothesis has broad support from studies of visual

Box 7.2 Opsin structure and "spectral sensitivity"

The "spectral sensitivity" of a visual pigment—effectively, the probability it will absorb a photon of a given wavelength—is determined by the structure of the pigment's opsin molecule (Yokoyama & Yokoyama 1996; Yokoyama 2002). The three-dimensional structure of opsin is well characterized. Opsins are seven-pass transmembrane proteins (Archer 1999). Changes in a single amino acid residue within the transmembrane region or near the chromophore—the result of a mutation in the corresponding opsin gene—may result in dramatic shifts (up to 35 nm) in the spectral sensitivity of the protein (Archer 1999; Bowmaker & Hunt 1999; Takahashi & Ebrey 2003). In this way, small changes in an opsin gene's sequence can have major effects on the opsin protein's spectral sensitivity, affecting an organism's visual detection phenotype. These changes may have important fitness consequences if they affect an individual's ability to detect predators, resources, and mates.

ecology in fishes (Levine & MacNichol 1979; Lythgoe et al. 1993; Bowmaker et al. 1994; Cummings & Partridge 2001; Cummings 2007). The probability that a photon of light produces a response in a photoreceptor is dependent on the absorption spectrum, optical density, the transmittance properties of the photoreceptor, and the spectral distribution of the light striking the retina of the animal (Lythgoe 1979; Lythgoe & Partridge 1989). The visual system may be adapted to be maximally sensitive to local lighting conditions at any of these loci; here I will focus on factors affecting the spectral sensitivities of photoreceptors.

Locally adapted visual systems can increase an individual's sensitivity to available wavelengths of light, thus affecting an individual's ability to detect predators, resources, and mates. At the molecular level, spectral tuning may be achieved in two ways: (1) changes in the amino acid sequence of opsin proteins (the result of mutations in the opsin genes) can tune spectral sensitivities of specific photoreceptors to the local photic environment (Yokoyama & Yokoyama 1996), and/or (2) the opsin genes may be differentially expressed to maximize an individual's sensitivity to particular wavelengths of light (Carleton & Kocher 2001; Fuller et al. 2004, 2005). In turn, locally adapted sensory systems result in selection for locally adapted communication signals, such as male sexual signals (e.g., Endler 1992; Ryan & Keddy-Hector 1992; Endler 1993a). Thus, in species where visual cues are used in mate choice, the tendency for the visual periphery to be adapted to local photic

environments may select for locally adapted male sexual ornaments, contributing to reproductive isolation between geographically separate populations (Endler 1992, 1993a; Endler & Houde 1995; Boughman 2002).

In addition to affecting an individual's sensitivity to particular wavelengths of light (i.e., color), the sensory periphery also affects spatial acuity (i.e., the spatial resolving power of the eye). To be colloquial: the higher the spatial acuity, the better the eye is at detecting fine detail in a scene. Research in closely related African cichlid fishes has revealed fascinating connections between spatial acuity, environmental complexity, and mating system: polygamous species have higher spatial acuity than do monogamous, pair-bonding species (Dobberfuhl et al. 2005). This stimulating work should motivate similar research among scientists in the poeciliid community. Spatial acuity increases with (1) eye size, with larger eyes having higher spatial acuity than smaller eyes due to differences in eye focal length; (2) photoreceptor density on the retina; (3) the ratio of cones to rods (rods have lower spatial acuity than cones). I suggest that there are many poeciliids that would be excellent for investigating spatial acuity. The Atlantic molly (*Poecilia mexicana*) would seem an ideal study species, with populations inhabiting extreme photic environments: surface, dim-light cave chambers, and perpetually dark cave chambers (see Tobler & Plath, **chapter 11**). Retinal-mapping studies in systems such as this one may offer powerful insight into the relationship between local sensory environments and the strength and direction of selection on sensory systems (for an example in the closely related bluefin killifish, *Lucania goodei*, see Fuller et al. 2003).

The final aspect of visual perception affected by the sensory periphery is temporal resolution. For animals living in water, targets of visual attention are rarely stationary. Thus, the ability of the eye to detect subtle and rapid changes in the state of the target may be important in predator detection and mate choice. Temporal resolution is a function of the membranes of the photoreceptor cells on the retina and must be measured in a living eye. Perhaps it is for these reasons that, to my knowledge, only a single study has investigated temporal resolution in a poeciliid, the green swordtail, *Xiphophorus hellerii* (Crozier & Wolf 1939). The implications of this research for poeciliid behavioral ecologists are discussed below.

Clearly then, the sensory periphery has a powerful influence on the types of signal that may evolve for communication. Returning to the sensory-bias model of signal evolution, it is precisely these elements of the sensory periphery—which determine an individual's visual detection phenotype—that are proposed to be shaped by natural selection. Thus, it is variation within the components of the sensory periphery that may help explain variation in female sexual preferences (see Rios-Cardenas & Morris, **chapter 17**; Druen & Dugatkin, **chapter 20**) and elucidate variation in the signals evolved for communication. In the next several sections, I discuss research that should improve our understanding of how the sensory periphery affects poeciliid behavior.

7.3.1 Molecular and physiological characterization of the poeciliid visual system

The retina is a projection of the brain originating from the optic vesicles of the diencephalon. In most teleosts, the retina is densely packed with a mosaic of photoreceptors, each of which contains a particular visual pigment (Yokoyama 2002). There are four major classes of vertebrate visual pigments (Yokoyama 2002): (1) rod opsin (Rh1), or rod pigments; (2) short-wavelength-sensitive (SWS) opsin; (3) medium-wavelength-sensitive opsin (MWS, typically called Rh2 in fishes); and (4) long-wavelength-sensitive (LWS) opsin (fig. 7.2). As a result of rampant gene duplication events (Yokoyama 2002; Horth 2007), there are multiple subclasses of opsin proteins within these major classes (reviewed in Horth 2007). I will focus here on the genetics and spectral properties of proteins in the four primary opsin subclasses found in fishes: (1) SWS_1 is a blue-violet- and/or ultraviolet-sensitive opsin, with a light wavelength absorbance maximum (λ_{max}) of 360–430 nm (van Hazel et al. 2006); (2) in fishes, the SWS2 opsin ($\lambda_{max} = 440–460$ nm) has undergone a duplication event and is represented by the blue- and violet-sensitive opsins SWS2A and SWS2B, respectively (Chinen et al. 2003); (3) Rh2 is a green-sensitive ($\lambda_{max} = 470–510$ nm) opsin (Sugawara et al. 2000; Hoff-

mann et al. 2007); and (4) LWS is a red-sensitive ($\lambda_{max} = 510–560$ nm) opsin (Trezise & Collin 2005).

While these major subclasses of opsins seem to be present in most fishes that occupy light-rich environments, recent work in African cichlids (Spady et al. 2005; Spady et al. 2006), zebrafish (Chinen et al. 2003), and guppies (Hoffmann et al. 2007; Weadick & Chang 2007; M. Ward et al. 2008) has revealed an even more bifurcating picture of opsin evolution. Each of these species has multiple cone classes, resulting from gene duplication, especially of LWS opsin. This work is discussed in some detail below and highlights an important question: does opsin gene evolution necessarily reflect adaptive evolution? That is, do opsin gene duplication events enhance wavelength discrimination (the central argument to explain opsin gene duplication), and if so, does this affect an individual's fitness? This seems to be the case in the rapidly speciating African cichlids that have eight opsins, six of which are under positive selection depending on the photic environment—clear or turbid water—in which a species is found (Sugawara et al. 2000; Spady et al. 2005). But is it adaptive for a guppy to have six distinct LWS opsins (Weadick & Chang 2007) in addition to a full complement ($n = 5$) of shorter-wavelength opsins (Hoffmann et al. 2007)? Hoffmann et al. (2007) suggest that while the LWS opsin duplication events in guppies may not be the result of adaptive evolution (but see White et al. 2005), they may help to explain female preference variation within (Brooks & Endler 2001b) and among (Breden & Stoner 1987; Houde & Endler 1990; Endler & Houde 1995) guppy populations. Such preference variation may explain the extreme male color polymorphism that characterizes the species (Endler 1980; Houde 1997; Grether et al. 1999). These suggestions are consistent with the sensory-drive and sensory-bias models of male display trait evolution (see Rios-Cardenas & Morris, **chapter 17**).

The variation in the spectral sensitivities of photoreceptors is a function of opsin protein and amino acid sequence, which is itself the product of opsin gene translation. Thus, the sequence of the opsin genes is critical in affecting an individual's sensitivity to particular wavelengths of light (Archer 1999; Bowmaker & Hunt 1999; Takahashi & Ebrey 2003). Opsin genes have been sequenced for only a handful of poeciliids (table 7.1), with most of the research investigating LWS opsin variation in the guppy (discussed in detail below).

To my knowledge, photoreceptor spectral sensitivities have been quantified for only four poeciliids from a single genus: *P. reticulata*, *P. mexicana*, *P. latipinna*, and *P. formosa* (Körner et al. 2006; Hoffmann et al. 2007; Weadick & Chang 2007; M. Ward et al. 2008). None of these studies mapped variation in photoreceptor spectral sensitivity

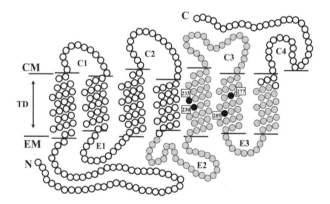

Figure 7.2 Schematic diagram of an LWS opsin. Each circle represents an amino acid residue. The sites known to be variable in guppies are shaded gray; those shaded black are involved in spectral tuning. The numbering of amino acids follows the human LWS opsins. CM, cytoplasmatic membrane; EM, extracellular membrane; C1–C4, cytoplasmatic loops; TD, transmembrane domain; E1–E3, extracellular loops. Modified from Weadick and Chang 2007.

Table 7.1 Opsin genes that have been sequenced in poeciliids

Species	Rh1	SWS1	SWS2A	SWS2B	Rh2	LWS	Reference
Gambusia holbrooki	1						Archer & Hirano 1997
Poecilia bifurca						4	M. Ward et al. 2008
P. parae						3	M. Ward et al. 2008
P. picta						3	M. Ward et al. 2008
P. reticulata	1						Archer & Hirano 1997
P. reticulata	1	1		1	2	2	Hoffmann et al. 2007
P. reticulata						4	M. Ward et al. 2008
P. reticulata						6[a]	Weadick & Chang 2007
Xiphophorus birchmanni		1	NA	1	1	1	Unpublished data[b]
X. hellerii						1	Hoffmann et al. 2007
X. malinche		1	NA	1	1	1	Unpublished data[b]
X. pygmaeus						3	M. Ward et al. 2008
X. malinche/X. birchmanni hybrids[c]		1	NA	1	1	1	Unpublished data[b]

Note: Numbers in each cell indicate the number of distinct genes sequenced in each of the major gene classes. Empty cells indicate that a study did not attempt to sequence a particular gene. NA (no amplification) indicates that the researchers attempted to sequence a gene but observed no PCR amplification. Such failure to amplify is likely due to the absence of the gene in the genome. Sequences for all studies, including those listed as "unpublished data," can be found in Genbank.
[a]Short (390 bp) amplicons from a single individual.
[b]S. W. Coleman, B. Perkins, & G. G. Rosenthal.
[c]Naturally occurring hybrids (see Rosenthal & García de León, **chapter 10**).

on to variation in light environment or on to variation in female preferences for male coloration. Clearly, there is much work to be done on characterizing photoreceptor sensitivities in other poeciliid genera and on integrating such studies with those investigating behavioral and morphological variation.

7.3.1.1 Long-wavelength-sensitive opsin gene duplication in the guppy.

White et al. (2005) showed that selective removal of ultraviolet and short-wavelength light did not affect guppy foraging efficiency on zooplankton, but that guppy foraging efficiency was significantly reduced when long-wavelength light was removed from the foraging environment. These results indicate the importance of long-wavelength vision in guppy foraging (e.g., Rodd et al. 2002; but see Grether et al. 2005b). At the genetic level, guppy sensitivity to long wavelengths may be a function of duplications in the LWS opsin genes. Several studies have revealed seemingly rampant duplications of LWS opsins in the guppy (Hoffmann et al. 2007; Weadick & Chang 2007; M. Ward et al. 2008). The primary advantage of opsin gene duplication may be that it results in diversification in opsin protein spectral sensitivities and hence in wavelength discrimination. If this hypothesis is true, it seems that the guppy should have the best long-wavelength discrimination

of any fish studied to date. Comparative analyses among guppy populations that differ in the number of duplicated opsin genes, and among guppies and other poeciliids that differ in the number of opsin genes, are warranted.

7.3.1.2 Opsin gene expression and sensory dysfunction in *Xiphophorus* hybrids.

It has long been appreciated that crosses between interfertile species frequently result in hybrid offspring that are dysfunctional compared with their parents (Coyne & Orr 2004). To better understand the functional and evolutionary significance of hybrid dysfunction, researchers (S. W. Coleman, B. Perkins, & G. G. Rosenthal, unpublished data) investigated how differences in opsin gene sequences and opsin gene expression affected optomotor performance in two swordtail fishes, *Xiphophorus birchmanni* and *X. malinche*, and their natural hybrids.

To test for opsin gene variation between *X. malinche* and *X. birchmanni*, these researchers (S. W. Coleman, B. Perkins, & G. G. Rosenthal, unpublished data) sequenced four cone opsin genes, SWS1, SWS2B, Rh2, LWS, which in other fishes code for ultraviolet-, blue-, green-, and red-sensitive opsin proteins, respectively (Yokoyama & Yokoyama 1996). They found only a handful of nucleotide differences in the four cone opsins between *X. malinche* and *X. birchmanni*, none of which were located at sites known to have functional effects on the spectral sensitivity of the opsin protein. In contrast to the apparent absence of biologically meaningful opsin gene sequence differences, quantitative PCR (qPCR) revealed significant differences in opsin gene expression among these same populations of *X. malinche* and *X. birchmanni* and among three populations of naturally occurring *X. malinche–X. birchmanni* hybrids. Somewhat surprisingly, the differences in opsin gene expression were not among species (there was no difference in opsin gene expression between *X. birchmanni* and *X. malinche*) but instead were between parentals and hybrids; overall, hybrids had significantly lower opsin gene expression than parentals. These results suggested the tantalizing possibility that hybrids would have poorer color sensitivity than parentals.

In striking contrast to the expression of genes that code for cone opsin proteins (i.e., those associated with color vision) the researchers found a different pattern of rod opsin gene expression. Instead of a reduction in expression in hybrids relative to parentals, rod opsin was expressed as a function of light intensity in the environment. That is, fish found in high-intensity light environments had relatively low levels of rod opsin expression compared with fish found in low-light environments. This result is consistent with other studies showing opsin gene expression changes with local photic conditions (Carleton & Kocher 2001; Fuller et al. 2004, 2005). Altogether, the differences in cone versus rod opsin gene expression between parentals and hybrids suggest that hybrids may have poorer photopic (i.e., color) vision than parentals, while scotopic (contrast sensitivity) vision may reflect local lighting environments. Supporting this hypothesis, in a series of optomotor (OPM) experiments, hybrids performed less well than parentals in OPM trials involving color discrimination. In OPM trials involving sensitivity to contrast (i.e., black and white) performance covaried with local lighting conditions: fish from low-light environments were highly sensitive to contrast at low light levels, while fish from high-light environments ceased to detect contrast between light and dark surfaces at relatively high light levels. These differences in scotopic optomotor performance were correlated with rod opsin gene expression. Whether these results reflect ontogenetic differences or more plastic responses to changing light environments remains to be tested.

These results have several interesting implications. First, the reduction in cone opsin gene expression and in number of cone photoreceptors in hybrids suggests dysfunction at the transcription factor level, either in the transcription factors responsible for building cone photoreceptors or in opsin gene transcription factors, or both. Second, this dysfunction has important functional effects. Compared with parentals, *X. birchmanni–X. malinche* hybrids were relatively insensitive to color contrast, rendering them effectively color-blind at low light levels. Researchers (S. W. Coleman, B. Perkins, & G. G. Rosenthal, unpublished data) suggest that this reduced sensitivity to color may help to explain the broad array of male phenotypes in the *X. birchmanni–X. malinche* hybrid zone (Rosenthal et al. 2003). That is, insensitivity to color may weaken female preferences for male sexual coloration. It is well established that the specificity of receiver sensory mechanisms can drive the evolution of exaggerated male traits (Endler 1992; Ryan 1998). Permissiveness, whereby females respond positively to a broad range of stimulus inputs, can also help explain the evolution of complex displays (Kime et al. 1998; Rosenthal & Evans 1998).

Interestingly, hybrids exhibited neither reduced rod opsin gene expression nor reductions in optomotor performance in trials with black-and-white stripes. These results sharply contrast with those for cone opsin expression and color optomotor trials and suggest that genetic misexpression in these *Xiphophorus* hybrids may be limited to genes associated with cone photoreceptors. These differences may reflect differences in the strength of natural selection on photopic versus scotopic sensitivity: perhaps individuals are under strong selection to maintain highly sensitive scotopic visual systems, as these systems aid in finding food and avoiding predators, especially in low-light conditions; this hypothesis is supported by the relationship between rod opsin expression and natural photic environments. High-resolution color vision may be less important for behavioral tasks related to survival, and therefore, hybrids with misexpression of genes associated with photopic vision are not at a selective disadvantage with respect to individuals with parental-like color vision. This hypothesis should be tested across behavioral contexts—foraging, predator detection, mate choice—and suggests that investigations of sensory dysfunction may yield powerful insights into the performance and evolutionary fate of interspecies hybrids.

7.3.1.3 Variation in opsin gene expression among surface- and cave-dwelling populations of the Atlantic molly. Regressive evolution—the loss or simplification of derived traits—is a widespread phenomenon, the understanding of which provides important insights into the strength of natural selection (reviewed in Porter & Crandall 2003; Wilkens 2007). Most of the research on regressive evolution has focused on embryonic eye development and behavioral differences in the Mexican cave tetra, *Astyanax mexicanus* (Jeffery 2001). While this system provides a powerful model for elucidating the developmental mechanisms associated with regressive eye evolution (reviewed in Jeffery 2001), elucidating the initial mechanisms underlying regressive evolution may be facilitated by studying phylogenetically young systems, in which cave colonization occurred recently or is still in progress.

This is the case in the Cueva del Azufre system in southern Mexico, where the Atlantic molly (*P. mexicana*) has colonized at least two independent caves (Plath et al. 2007a; Tobler et al. 2008a). Cave-dwelling *P. mexicana* occur parapatrically with their epigean ancestors, but despite spatial proximity and the lack of major physical barriers preventing migration, gene flow between surface and cave populations, as well as between different cave chambers, is extremely low (Plath et al. 2007a). Cave populations are morphologically distinct and exhibit reductions in pigmentation as well as in eye size (Tobler et al. 2008a). Unlike in many other cave organisms (see Langecker 2000; Romero & Green 2005), the eyes of cave-dwelling *P. mexicana* are fully functional, and fish readily respond to visual stimuli when given an opportunity (Plath et al. 2004).

Using qPCR, researchers (S. W. Coleman, M. Tobler, & G. G. Rosenthal, unpublished data) compared the expression of opsin genes in surface-dwelling *P. mexicana* and in cave-dwelling conspecifics from two photically distinct cave environments: front cave chambers where natural skylights provide dim light and deep cave chambers with perpetual darkness. Working in this unique system, Coleman et al. investigated differences in spectral tuning in three opsin genes in both natural and lab-reared populations from the surface (full sun conditions), from a front cave chamber (dim light conditions), and from a deep cave chamber (perpetual darkness).

As predicted, the highest opsin expression levels were observed in the surface-dwelling population, with similarly high levels of SWS1 and Rh2 expression in the population living in the low-light front cave chamber. LWS expression, however, was significantly lower in the latter population than in the surface population; this result likely reflects the paucity of long-wavelength light in low-light conditions (Lythgoe 1979). In fish from the deep cave chamber, the expression of all opsin genes was reduced. Coleman et al. found the same gene expression patterns in lab-reared fish from the same three populations, suggesting that the reduction in opsin expression in wild-caught cave-dwelling mollies is not entirely ontogenetic. Instead, the results suggest heritable differentiation among populations, leading to population-specific ontogenetic trajectories, and that the cave-dwelling molly retina has undergone evolutionary reductions not rescued by exposure to light during retinal ontogeny.

It would be interesting to know whether additional changes have taken place in the cave molly visual system.

For example, among surface- and cave-dwelling mollies, are there differences in the number of rods that connect to rod bipolar cells? Rod bipolar cells conduct the nerve response—the result of a rod absorbing a photon of light—into the retina. The effect of connecting multiple rods to a single bipolar cell is additive, and the consequence is increased light sensitivity at low levels. One might predict that mollies occupying the cave vent habitat have rods that are highly bundled, while surface mollies and those that live in perpetual darkness have little selective pressure for visual systems adapted to low-light environments. Additionally, retinal mapping may reveal differences in the ratio of rods to cones, with cave mollies having relatively rod-rich retinas compared with surface-dwelling mollies.

7.3.2 Temporal resolution in poeciliids: flicker fusion, male display, and the use of video in playback experiments

To my knowledge, only a single study has empirically investigated temporal resolution in a poeciliid. Crozier and Wolf (1939) reported that the green swordtail, *X. hellerii*, has a critical flicker-fusion frequency (CFFF) of 43.1 Hz. In practical terms for the *X. hellerii* behavioral ecologist, this means that when using a video playback technique to assay preferences (reviewed in Rosenthal 2000) the image refresh rate (images/s) of the videorecording must be higher than the CFFF of the fish. The television standard of the U.S. National Television Standards Committee displays 60 images/s, appreciably higher than the CFFF of *X. hellerii*. Females in several species of poeciliids have been shown to respond equally to video playbacks of males and of live conspecifics (Rosenthal et al. 1996; Kodric-Brown & Nicoletto 1997; Rosenthal & Evans 1998), suggesting that the refresh rate of the video sequences was not producing perceived deficiencies in the stimuli. Nonetheless, CFFF should be a consideration for poeciliid researchers interested in using video playback techniques.

CFFF may also have important implications for the evolution of male sexual displays. For example, males in many species of swordtails (genus *Xiphophorus*) have highly vigorous courtship displays, which involve back-and-forth lateral movement in close proximity to the female (Ryan & Rosenthal 2001); males in some of the species have vertical bars along their sides that are most conspicuous during courtship. Rosenthal (2007) suggested that display speed may be higher in species lacking bars, as the signal value of the bars would be lost if they were presented at a rate exceeding the female's CFFF. This presents a testable hypothesis for the evolution of complex male display in many species.

7.3.3 Chemical ecology and characterizing poeciliid chemoreceptors

Many studies have shown that poeciliids attend to chemical cues in mate choice (see Rios-Cardenas & Morris, **chapter 17**), yet to my knowledge only three studies have attempted to describe the physical characteristics of poeciliid chemosensors. Calling into question the role of chemical cues in guppy behavior, in a recent electron microscopy study Lazzari et al. (2007) found that the peripheral olfactory organ of the guppy was relatively small and simple compared with those of many other teleosts. The authors go so far as to write that the simple organization of the guppy peripheral olfactory organ "reminds [us of] what is found in early posthatching stages of fish which at the adult state have a well developed olfactory organ" (Lazzari et al. 2007, 782). Such a statement warrants both rigorous research into the use of chemical cues in the guppy and microscopy studies of the olfactory receptorial surfaces in other poeciliids known to use olfactory cues. This latter line of research would provide a powerful comparative approach to elucidate poeciliid chemosensory evolution. It is interesting to speculate that guppies may have high-performance visual discrimination—perhaps due to multiple opsin gene duplications—while being relatively deficient in olfactory sensory performance.

While we tend to think of chemical communication in the context of olfaction, it is important to note that chemosensors are also found in the mouth and are used to "taste information." In the only study to characterize gustatory receptors in poeciliids, Reutter et al. (1974) found that *X. hellerii* have taste buds located between their teeth and on their lips. This observation is tantalizing, as males in many species of poeciliid, including *X. hellerii*, appear to nip at females during courtship (G. G. Rosenthal, pers. comm.). By doing so, perhaps males glean chemical information from the female by tasting her chemical secretions. Clearly, more work on the morphological, physiological, and sensory characteristics of poeciliid taste buds is warranted.

7.3.4 Near-field communication in poeciliids

Among poeciliids, the Atlantic molly (*P. mexicana*) is the most well studied of the cave-dwelling species (see Tobler & Plath, **chapter 11**). From the standpoint of sensory ecology, life in near or total darkness has important implications for communication. Indeed, unlike surface-dwelling females, cave molly females are able to discriminate among males that differ in size (Plath et al. 2004) and nutritional state (Plath et al. 2005a) in total darkness. This discrimination is likely accomplished using both the lateral line,

which in cave mollies is modified with widened pores in the head canal system (Walters & Walters 1965), and chemosensors. It stands to reason that the chemosensory system in cave mollies may have evolved enhanced sensitivity compared with the chemosensory system of surface fish. Early work by Parzefall (1970) revealed that cave molly females have evolved enlarged genital pads with numerous secretory glands (reviewed in Parzefall 2001) and that male cave mollies have an enhanced gustatory system, which may facilitate chemical communication. While studies of cave molly sexual behavior have since shown that males often nip at the female genital pad (Schlupp et al. 1991), the predicted coevolutionary relationship between the male chemosensors and the chemicals secreted by the highly specialized pad has not been investigated.

7.3.4.1 Tactile communication in mollies. While the use of visual and chemical communication in poeciliids is widely appreciated, the role of mechanosensory cues in poeciliid communication is unclear. Recently, Schlupp et al. (unpublished data) used electron microscopy and behavioral assays to elucidate the potential function of mustache-like structures on the upper maxilla of males in some populations of Mexican mollies (*Poecilia sphenops*). They found that the whiskers of the mustache did not contain sensory receptors and thus concluded that the mustaches were not likely used in signal detection. However, in visual preference trials, females preferred males with mustaches to males without mustaches, suggesting that mustaches may have evolved via sexual selection. Moreover, courtship in *P. sphenops* involves intense contact between the male upper maxilla and the female genital region (I. Schlupp, pers. comm.), allowing for the provocative possibility that male mustaches affect female mating behavior, perhaps by enhancing female receptivity. This hypothesis has also been proposed to explain the slight protuberances on the heads of male cave mollies (*P. mexicana*), which likely come into contact with females' enlarged genital pads during courtship (Parzefall 2001; Plath et al. 2004).

7.3.5 A note about nociceptors

As an aside, to my knowledge no studies have investigated poeciliid nociceptors—specialized cells that detect potentially harmful, noxious stimuli and that, when stimulated, invoke behavioral responses in some teleost fishes (Sneddon et al. 2003a, 2003b). This line of investigation may be particularly timely given the intensity of anthropogenic fouling occurring in the habitats where many poeciliids are found (see Stockwell & Henkanaththegedara, **chapter 12**).

7.4 Conclusions and future directions

Theory predicts an evolutionary correlation between sensory environments, sensory systems, and signals (Endler 1992, 1993a). In my estimation, poeciliids seem ideal for rigorous empirical tests of this hypothesis. Genetic and physiological characterizations of poeciliid visual systems are revealing local adaptation (Coleman et al., unpublished data), sensory dysfunction in hybrids (Coleman et al., unpublished data), and unexpected levels of variation (Hoffmann et al. 2007; Weadick & Chang 2007; M. Ward et al. 2008). The hypotheses generated by these studies—such as (1) hybrid female *Xiphophorus* have permissive color preferences; (2) reductions in cave molly visual systems covary with enhanced chemo- and mechanosensory systems; and (3) guppies have high-performance long-wavelength discrimination—should be tested.

Compared with the attention paid to visual preferences, relatively few studies have investigated the importance of chemical information in poeciliid decision-making processes. Chemical preference studies have shown that chemical information is important in mate choice in guppies, mollies, and swordtails (see Rios-Cardenas & Morris, chapter 17) and in shoaling behavior in swordtails (Wong et al. 2005; Coleman & Rosenthal 2006). Despite this growing recognition that chemical communication is widespread in poeciliids—and critically affected by environmental contamination (Fisher et al. 2006)—it has been 35 years since the last study of poeciliid chemosensors was conducted (Reutter et al. 1974). Moreover, recent work in the guppy suggests that this species has quite unsophisticated olfactory senses (Lazzari et al. 2007), questioning the importance of chemical signals in guppy communication. Comparative studies characterizing the morphology of chemosensors, combined with assays of chemical communication, among species and populations of poeciliids would be highly rewarding.

In conclusion, although the literature on poeciliid sensory ecology is growing, it seems somewhat disjointed. Some studies investigate the underlying genetics or physiology of poeciliid sensory receptors (almost exclusively photoreceptors), other studies investigate variation in sensory environments that may affect the evolution of sensory receptors, and still other studies investigate the behavioral preferences associated with particular sensory stimuli. Developing a full understanding of poeciliid sensory ecology will require a truly integrative approach, simultaneously investigating the molecular, physiological, behavioral, and environmental factors affecting the production, reception, and processing of environmental and communicative information. This integrative approach should be the goal of the contemporary behavioral ecologist, who should have intense interest in developing career-spanning collaborations with laboratories in complementary areas of research.

Chapter 8 Poeciliid parasites

Joanne Cable

8.1 Introduction

OVER HALF OF ALL ANIMALS are parasites, and yet, despite their overwhelming prominence, their impact has been neglected in some of the best-studied ecological and evolutionary models. This includes the poeciliids, several species of which have been widely used in laboratory and field research, with the guppy (*Poecilia reticulata*) upheld as an icon of microevolution. In their reviews of guppy biology, Houde (1997) and Magurran (2005) could only briefly mention the dominant pathogens of poeciliids because, until a few years ago, parasites had been overlooked. However, several researchers are now bridging the gap between host ecology, evolution, and epidemiology, with parasites shown to negatively affect poeciliid fitness in the wild. Furthermore, it is increasingly recognized that poeciliids have great potential to introduce novel parasites into native recipient host populations. Poeciliids have been translocated worldwide (1) via the aquarium trade, (2) as biocontrol agents against mosquitoes, and (3) as model organisms for research. Guppies, platies, and swordtails are listed in the top 10 most commonly imported ornamental-fish species, but it is the guppy that is generally considered the most popular tropical fish species, representing, for example, 26% of the total number of tropical fish imported into the United States (Chapman et al. 1997). Through introductions, mosquitofish, particularly *Gambusia holbrooki* and *Gambusia affinis*, have become the most widely distributed freshwater fish in the world (Pyke 2008). These species are detrimental to native fauna, both as competitors and as vectors of disease, and ironically it is argued that they have been no more effective than native fish in reducing mosquito populations (Pyke 2008). Partly due to these wide geographical distributions, there are an increasing number of species descriptions for poeciliid parasites, but the literature on the ecological impact of such parasites is still limited, and this has largely dictated the focus of this review on a particular group of well-studied ectoparasites, the gyrodactylids.

8.2 Diversity of parasites

Like all fish, poeciliids are infected with a variety of phylogenetically diverse parasites (fig. 8.1). Appendix 8.1, which is available online at http://www.press.uchicago.edu/books/Evans, provides a checklist of macroparasites that have been previously recorded from poeciliids. Many of these records have been previously summarized (see, e.g., Pineda-López et al. 2005; Salgado-Maldonado 2006), but this checklist also includes records of experimental infections that would not necessarily occur in the wild. Protozoans are also included, but although other microparasite infections are common (bacteria, fungi, etc.), identification of such pathogens is problematic (requiring microbiological and/or molecular approaches), and thus they are not included in this review. Unfortunately, microparasites are a significant cause of mortality among poeciliids (personal observations) and are often associated with macroparasitic infections (personal observations), but without a full assessment of their diversity it is impossible to predict the relative importance of micro- versus macroparasites.

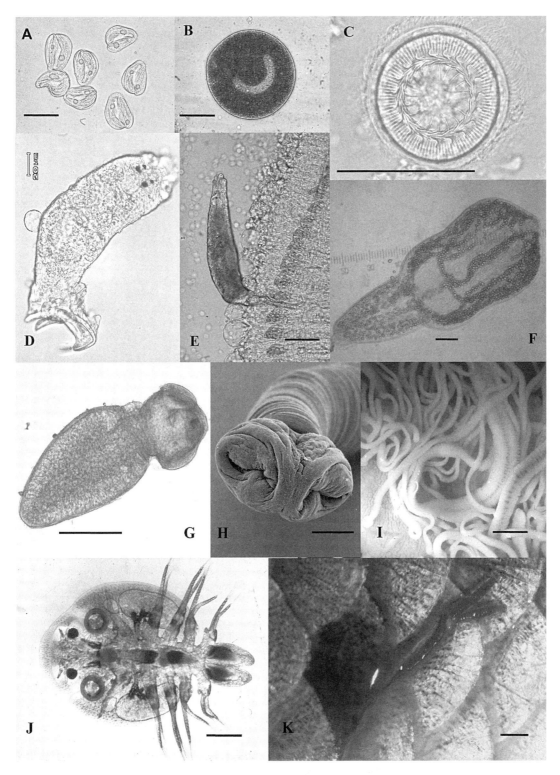

Figure 8.1 Diversity of poeciliid parasites. (A–C) Protozoa: (A) *Chilodonella* sp., (B) *Ichthyophthirius multifiliis*, and (C) *Trichodina* sp. (D–E) Monogenea: (D) *Haplocleidus* sp. and (E) *Dactylogyrus* sp. (F) Unidentified digenean released from its cyst within body cavity of *Poecilia reticulata*. (G–I) Cestodes: (G) *Valipora campylancristrota*, (H–I) Scanning electron micrograph and light microscope image of *Bothriocephalus* sp. (J–K) Crustacea: (J) *Argulus* and (K) *Lernaea*. Scale bars: A, C, E, F = 50 μm; B, G, H, J = 300 μm; D = 20 μm; I = 5 mm; and K = 1 mm. With the exception of F, all images were kindly provided by Dr. Chris F. Williams (Environment Agency, U.K.). These are all generalist parasites and most were recovered from nonpoeciliid hosts.

In total, records of 143 parasite species from 40 host species were found in the primary literature (table 8.1). It would be inappropriate to try and estimate parasite richness or diversity of parasite communities here, because the information required from each individual study (at the very least sample size and sample effort, and for more in depth analyses, host age, diet, sex, etc.) is just not available. Furthermore, it can be no coincidence that the five fish species recorded as having the highest parasite diversity are those that have been the best studied, having gained notoriety as invasive species (particularly *G. affinis*, with 39 parasite species) and/or are of economic importance to the aquarium trade (*Poecilia mexicana*, *Poecilia reticulata*, *Poecilia sphenops*, and *Xiphophorus hellerii*, with either 31 or 33 parasite species each). However, even these figures are underestimates, as illustrated by Overstreet (1997), who recorded 70 parasite species from *G. affinis* from Texas, but the identity of these species has not yet been published. The absence of other poeciliid hosts from the online appendix (equating to 87.5% of the Poeciliidae family listed in *FishBase*; Froese & Pauly 2008) also reflects a lack of data—either these fish have never been examined for parasites or they were not included because their parasites have not been systematically identified. For example, Chandler et al. (1995) assessed the spatial distribution of metacercariae from *Poecilia gilli*, but as these digeneans were not identified to genus level, for simplicity they were not included in the appendix. The first examination of the parasite fauna of *Poecilia picta* revealed a new monogenean species (Cable et al. 2005), and subsequent investigations revealed that this fish can serve as a surrogate host for another well-known ectoparasite, *G. turnbulli* (see King & Cable 2007), as well as *G. bullatarudis* (see King et al. 2009). Only with further work will we be able to assess the full ecto- and endoparasitic fauna of *P. picta* and other poeciliids. Thus, it is impossible to compare parasite diversity of different poeciliids at the current time, but clear trends are apparent from table 8.1.

Digenean metacercariae are the most diverse poeciliid parasites, constituting 35% (50 out of 143) of recorded species (table 8.1). Interestingly, this is not the case for other tropical, nonpoeciliid fishes, in which parasite species composition is extremely variable and no one group is dominant (Choudhury & Dick 2000). In contrast to the diverse larval digeneans of poeciliids, only ten adult digenean species have been recorded, but this is in keeping with other fish communities, as the larval stages in the second intermediate (fish) host are much less host specific than the adults. Also not surprisingly, the prevalence of metacercariae in poeciliids is significantly higher where the density of potential intermediate hosts is highest (Cable & van Oosterhout 2007b). Typically, molluscs are the first intermediate hosts for digeneans, small fishes are the second in-

Table 8.1 Number of parasite species on 40 poeciliid hosts

Parasite group	No. of species	%
Protozoa	20	14.0
Myxozoa	2	1.4
Monogenea	18	12.6
Cestode (adult)	4	2.8
Cestode (larval)	9	6.3
Digenean (adult)	10	7.0
Digenean (larval)	50	35.0
Pentastomida	1	0.7
Nematode (adult)	10	7.0
Nematode (larval)	8	5.6
Acanthocephala (adult)	6	4.2
Acanthocephala (larval)	1	0.7
Annelida: Hirudinea	1	0.7
Arthropoda: Crustacea	1	0.7
Arthropoda: Copepoda	2	1.4
Total	143	100.0

Source: Data derived from the primary literature listed in the online appendix at http://www.press.uchicago.edu/books/Evans.
Note: If two or more records of the same parasite genus from the same host were found but only one case was identified to species level, then this was recorded as a single species.

termediate hosts, and piscivorous fishes and mammals are the definitive hosts. Cercariae actively penetrate through the skin of their fish host and encyst as metacercariae in the body (fig. 8.1F). Some species encyst directly in the gills, and others enter the efferent blood vessels supplying the gills and from here are transported to other organs, such as the heart or brain, before encysting. Yet other digeneans are ingested and encyst in the stomach or intestine or penetrate the gut wall to encyst in other organs. The majority of metacercariae remain dormant until ingested by the definitive host, but it is assumed that they are still energetically costly to the host. However, if metacercariae encyst in vital organs, such as *Ascocotyle (A) felippei* and *Ascocotyle (A) leighi* in the heart, they can cause significant morbidity (Williams & Jones 1994). Relatively little is known about the specific effects of metacercariae on the host, with the exception of those species that actively manipulate the behavior of the intermediate host to facilitate transmission (see section 8.5). Most adult digeneans of poeciliids occur in the intestine and have relatively little effect on the host, with the exception of *Phyllodistomum* sp., which occurs in the urinary bladder, where it can cause osmotic problems (Williams & Jones 1994).

Table 8.1 lists protozoans as the second most diverse group of poeciliid parasites, which are characterized by low host specificity and highly variable pathology. Arguably the most economically damaging protozoan to the aquarium industry and the species recorded from the widest range of poeciliid hosts (see appendix 8.1, online) is *Ichthyophthirius multifiliis* (fig. 8.1B), the causative agent of white spot, which can cause mass mortality in fish stocks (e.g., Traxler et al. 1998). *Tetrahymena* spp. also have a wide host range, causing mortality in both tropical and temperate species, but among poeciliids they have been recorded so far only from *P. reticulata* (see Imai et al. 2000; Hatai et al. 2001) and *Xiphophorus maculatus* (see Pimenta-Leibowitz et al. 2005). However, in Korean fish farms, *T. corlissi* is considered a major threat to ornamental guppy stocks (Kim et al. 2002). Other protozoa, such as *Ichthyobodo* and *Trichodina* (fig. 8.1C), are mostly commensals and become pathogenic only in synergy with other host stressors, such as poor water quality or elevated temperatures. Quite often microparasitic infections are held at bay in the aquarium trade by constant application of water treatments, but once the fish are transferred to clean freshwater on purchase, epidemic outbreaks frequently occur (personal observation). Similarly, low natural-parasite burdens on wild-caught fish tend to flourish if these fish are brought into captivity, partly due to changing water conditions and increased host density influencing transmission but also related to host stress negatively affecting immunocompetence (see references in the online appendix).

The Monogenea (fig. 8.1D–E) are the third most diverse group of poeciliid parasites listed in table 8.1 (second if just including the helminths), with 18 taxa described. Half of these species are gyrodactylids, and as these are one of the most intensively studied and common groups of poeciliid parasites (at least for *P. reticulata*), they are discussed further in section 8.3. The remaining monogeneans are more typical of this group of ectoparasites. They have a direct (single) host life cycle and are oviparous. Eggs released from adults attach to the gills, skin, or fins, embryonate in the environment, and have a short-lived but free-living trans-

mission stage that is responsible for locating and infecting new hosts. There are more than 900 species of dactylogyrids (Gibson et al. 1996), but as yet no specimens have been described to species level in poeciliids. In the popular literature, dactylogyrids are often grouped together with gyrodactylids as "flukes," largely for the purpose of disease control, but nothing is known about the pathology of the former on poeciliids.

Adult and larval nematodes constitute 7% and 5.6%, respectively, of the total poeciliid macroparasites listed in the appendix. Whereas adults occur most commonly in the intestine, larval nematodes have been recorded from the musculature, abdominal cavity, mesenteries, stomach, intestine, and intestinal wall (see references cited in appendix 8.1). Typical of tropical freshwater fish, camallanids appear to be a common component of the enteric helminth community (Choudhury & Dick 2000). *Camallanus cotti* (fig. 8.2) is naturally distributed in tropical Asia but has been disseminated widely to Europe, North America, and Australia (Levsen & Berland 2002). Fish are usually infected via the ingestion of parasitized copepods, but the success of this parasite relates to its wide host range and flexible life cycle that can bypass the normal indirect life cycle (via an intermediate host) and directly infect its definitive fish host (e.g., McMinn 1990; Levsen 2001; Levsen & Jakobsen 2002). This direct development has been questioned by Moravec and Justine (2006), but whatever the reason, this parasite is now common worldwide among aquarium cultures of poeciliids. Low numbers of *C. cotti* reportedly have no noticeable effect on individual guppies, but large numbers are often fatal (Stumpp 1975). In cultured guppies, co-infections of *C. cotti* and *Tetrahymena corlissi* may represent a significant source of host mortality (Kim et al. 2002).

Adult and larval cestodes (fig. 8.1G–I) are relatively rare in poeciliids, representing only 2.8% and 6.3%, respectively, of the total poeciliid parasite fauna. These endo-

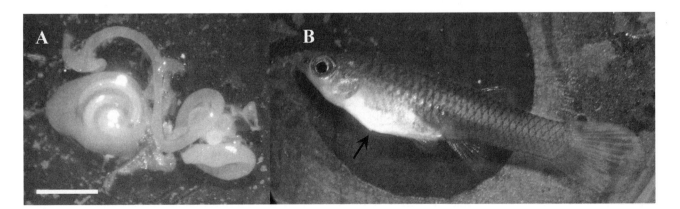

Figure 8.2 Guppy parasites. (A) *Camallanus* sp. removed from the intestine of *Poecilia reticulata*. Scale bar = 3 mm. (B) Nematode-infected host with distended abdomen (arrow). Photographs by Ryan Mohammed (University of West Indies).

parasites show a high degree of host specificity and require at least two hosts to complete their life cycle. Adults inhabit the intestine and larvae occur in visceral organs and musculature. Similarly, acanthocephalans are rare, representing less than 5% of poeciliid fauna. These helminths have complex life cycles and frequently use fish to bridge the gap between the intermediate and definitive host. These so-called transport or parentenic hosts acquire infections accidentally by predation on an infected fish with immature worms, which will continue their development only when ingested by the usual definitive host. DeMont and Corkum (1982) suggested that *Octospiniferoides chandleri* is transferred between *G. affinis* through cannibalism. The remaining four macroparasites of poeciliids consist of the leech *Myzobdella lugubris*, which has been reported on feral poeciliids from Australia (Font 2003), and the three crustaceans *Argulus*, *Ergasilus*, and *Lernaea* (fig. 8.1J–K). All four parasites are generalists that infect many freshwater fish but, due to their relatively large size, could represent a significant cause of pathology to poeciliids.

Overall, poeciliid parasites are dominated by larval, rather than adult, helminths, reflecting the role of fish in the food chain (Choudhury & Dick 2000). For poeciliids, the 1:2.3 ratio of adults:larvae (for cestodes, digeneans, acanthocephalans, and nematodes) is dominated by the presence of many generalist metacercariae. However, the fact that many poeciliids are omnivores and even opportunistic cannibals provides the opportunity for both direct and indirect parasite life cycles.

8.3 Gyrodactylids: the most abundant poeciliid parasites?

Lyles (1990) surveyed the natural parasite fauna of the guppy and found gyrodactylids to be the most widespread and abundant parasites from 807 guppies collected at different sample sites in the Northern Range mountains of Trinidad. Digenean metacercariae were also common, and small gut helminths were mentioned but not identified (Lyles 1990). Similarly, Martin and Johnsen (2007) recorded nongyrodactylid taxa in only 5 out of 568 male Trinidadian guppies. Our ongoing research (with field samples collected from Trinidad and Tobago and mainland South America from 2003 to 2008; Cable & van Oosterhout, unpublished data) has revealed a similar pattern, with the general conclusion that the parasite fauna of guppies is depauperate and dominated by gyrodactylids (see box 8.1; fig. 8.3).

Two *Gyrodactylus* species are commonly reported from both wild and ornamental guppies, *G. turnbulli* and *G. bullatarudis*. The former has been studied extensively, but relatively few studies have focused on the latter (Richards

& Chubb 1996, 1998; Cable & van Oosterhout 2007a). Despite infecting the same host, the parasites differ greatly in a range of life-history traits. Natural (see appendix 8.1) and experimental infections show that *G. bullatarudis* has a wider host range (King et al. 2009) than *G. turnbulli* (see King & Cable 2007). The two species also display different site preferences on the host (Harris 1988; Cable et al., unpublished data), levels of virulence (Cable & van Oosterhout 2007a), and transmission behaviors (Cable et al., unpublished data). Transmission occurs through direct host-to-host contact and can be achieved in milliseconds when one fish brushes past another, with the parasites moving in a rapid leechlike fashion. Nevertheless, such transfer is potentially costly to the parasite. Scott and Anderson (1984) estimated that only 40% of *G. turnbulli* individuals successfully transfer during live host-to-host contact. However, transfer may also occur through contact with a dead host or with a dislodged parasite on the substrate, water column, or water film (reviewed in Bakke et al. 2007). In fact, transmission of at least one guppy gyrodactylid species is more efficient from dead to live fish than between live hosts (Scott & Anderson 1984). At high parasite burdens, *G. turnbulli* migrates into the water film (fig. 8.3D) following the death of its host, a possible adaptation to facilitate transmission to its surface-feeding host (Cable et al. 2002). In contrast, *G. bullatarudis* remains with its dead host, possibly maximizing the chances for transfer when the cadaver is cannibalized (Cable et al., unpublished data).

8.3.1 Epidemiological models

Although gyrodactylids have been studied since the midnineteenth century (reviewed in Cable & Harris 2002), Anderson and colleagues (e.g., Anderson & Gordon 1982) were the first to highlight their potential for studying host-parasite population dynamics. A suite of elegant experimental studies during the 1980s identified key aspects of *G. turnbulli*'s life cycle. The parasite has an average fecundity of 1.68 offspring during its 4.2-day life span at 25°C (Scott 1982); these and other traits are strongly temperature dependant (Scott & Nokes 1984). A typical infection trajectory involves rapid in situ reproduction, with exponential population growth over the first 4–6 days, followed by a population decline, but with peak parasite burdens on individual fish usually attained at 7–9 days. Many fish lose their infections 10–12 days postinfection, but on highly susceptible fish parasite burdens increase until host death (Scott & Anderson 1984; Scott & Robinson 1984; Scott 1985). These experiments by Scott and colleagues were all conducted using ornamental guppies, but we have since found similar results using wild-caught fish (e.g., van Oosterhout et al. 2003a), leading to the sug-

---

Box 8.1 Gyrodactylids

Gyrodactylus species are small ectoparasitic worms that are viviparous hermaphrodites (reviewed in Cable & Harris 2002; see also fig. 8.3). Adult worms give birth to fully grown offspring that attach to the host alongside their mother and are pregnant when born. They employ a variety of different reproductive modes during their life cycle and are renowned for their "Russian-doll" boxing of generations, with both daughter and granddaughter developing within the uterus of their mother at the same time (reviewed in Bakke et al. 2007). They have a short generation time (<24 hours at 25°C for the firstborn daughter in some species), which can lead to rapid population growth on the host. Heavy infections of more than 1000 worms (e.g., Lyles 1990) are not uncommon on laboratory-maintained guppies despite the small size of these fish, resulting in high mortality (e.g., Cable & van Oosterhout 2007a). Scott and Anderson (1984) estimated a daily mortality rate of 7% for infected guppies, compared with 0.4% for uninfected fish. The actual cause of host death is unknown but may be related to secondary infection or osmotic problems caused by the feeding and attachment wounds inflicted by the parasites. In poeciliids, gyrodactylids have been associated with *Aeromonas* infections (Langson 1990), but other hyperparasites have been linked with a range of nonpoeciliid *Gyrodactylus* spp. (Bakke et al. 2006).

Most gyrodactylids are less than 1 mm in length, translucent, and able to flatten themselves against the host (fig. 8.3B and D), and so they can be quite cryptic at low burdens. However, with practice, these worms can easily be detected and accurately counted on anesthetized hosts. Therefore, unlike many parasite infections, the entire trajectory of gyrodactylid infection can be monitored noninvasively. Furthermore, by infecting isolated hosts with a single worm, host-specific virulence data can be gathered on a range of life-history traits (e.g., Gheorghiu et al. 2007). Therefore, in combination with

the guppy (an easily maintained host with a short life span that has well-characterized ecology, behavior, and immunogenetics), gyrodactylids are ideal for addressing epidemiological questions.

Historically, the identification of gyrodactylids has not been straightforward, as all 409 described species (Harris et al. 2008) show remarkably conserved morphology, and some of the earlier epidemiological studies predated the discovery of new species. Bakke et al. (2002) estimated that fewer than 2% of *Gyrodactylus* species have been described, and this includes diversity on poeciliids. However, in the last few years, the number of described poeciliid gyrodactylids has almost doubled, standing now at *G. cytophagus*, three *G. turnbulli*-like species (*G. turnbulli, G. milleri,* and *G. pictae*), and four to five *G. bullatarudis*-like species (*G. bullatarudis, G. gambusiae, G. poeciliae, G. costaricensis,* and *G. rasini*; although Richards et al. [2000] have suggested that the latter should be synonymized with *G. bullatarudis*). P. Faria and I have also recently identified at least two new cryptic species on wild *Poecilia reticulata* samples (Faria & Cable, unpublished data). Due to the advent of molecular markers and the availability of comparative data, identification of specimens is now easier (see Harris et al. 2008). However, unless using an isogenic strain, isolation of parasites for experimental studies should still be treated with caution. When a subsample of specimens is identified at the end of a study (e.g., Kolluru et al. 2006), there can be no guarantee that such experiments did not start with a mixed-species infection. As we know that different *Gyrodactylus* species and even strains vary in their virulence (Cable & van Oosterhout 2007a), mixed infections could significantly affect the outcome of an experiment. Similar caution should be exercised when assessing field samples if the gyrodactylids are not identified to species level. Apparent increases in parasite load in some habitats (see Martin & Johnsen 2007) could just reflect the presence of a different or additional species.

---

gestion that these parasites may represent a significant selection force in the wild (e.g., Cable & van Oosterhout 2007a). Unlike many other helminth systems where parasite-induced host mortality regulates both host and parasite populations, some gyrodactylids can survive host death to reinfect new hosts. Therefore, studying the biology of the parasites when they are detached from the host and assessing how host social behavior mediates gyrodactylid transmission are important for understanding the epide-

miology of this host-parasite system. Attempts to model gyrodactylid infection have so far focused on infrapopulation (single host) dynamics based on our knowledge of the parasite's life-history traits and the localized nature of the host's immune response (van Oosterhout et al. 2008), but the next step will be to model population dynamics across hosts. Such modeling can highlight variables that might help predict virulence and epidemic outbreaks in poeciliids and other taxa, including a major pathogen of salmonids,

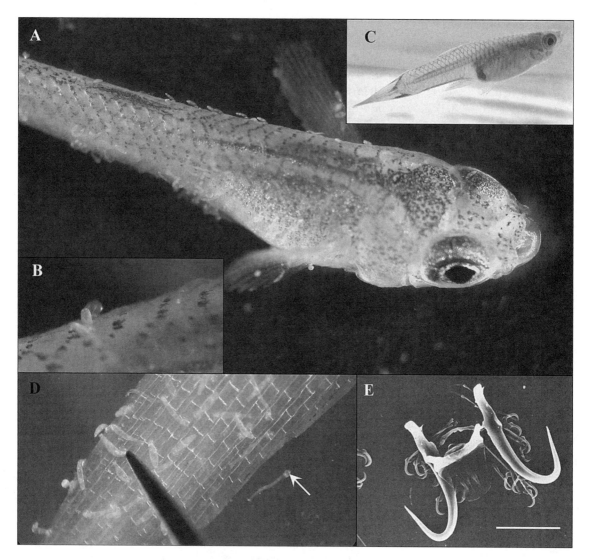

Figure 8.3 *Gyrodactylus. Gyrodactylus turnbulli* infecting *Poecilia reticulata*. (A) Juvenile fish (12 mm standard length) with insert (B) of higher magnification of *G. turnbulli*. (C) Characteristic host pathology with clamped caudal fin. (D) Removing gyrodactylid from host fin on the tip of a pin prior to experimental infection. Arrow indicates a parasite that has naturally left the host and is hanging in the water film like a mosquito larva. The posterior attachment organ (immediately adjacent to the arrowhead) is shown in greater detail in (E). Scanning electron micrograph of the posterior attachment hooks. The size and shape of these hooks are used for morphological identification of the species. Scale bar = 20 µm.

Gyrodactylus salaris (see Cable & van Oosterhout 2007a; van Oosterhout et al. 2008).

The mechanism of the immune response against non-poeciliid gyrodactylids has been reviewed extensively by Buchmann et al. (2003), but both innate (Madhavi & Anderson 1985) and acquired resistance (Scott & Robinson 1984; Richards & Chubb 1996) play an important role in controlling gyrodactylid infections on guppies. There is considerable variation in susceptibility within host stocks, with resistant, intermediate (responder), and susceptible individuals always being identified (reviewed in Bakke et al. 2002), but it is also possible to identify consistent differences between fish stocks. For example, we have demonstrated higher susceptibility of fish from the

Upper Aripo (UA) River in Trinidad than in guppies from the Lower Aripo (LA) during a primary infection on both wild-caught (van Oosterhout et al. 2003a; see also fig. 8.4) and laboratory-bred naive hosts (Cable & van Oosterhout 2007a). Interestingly though, acquired resistance was not consistent across host populations. During a secondary infection with the same strain of parasite, laboratory-bred UA fish significantly improved their performance, whereas the LA fish did not (Cable & van Oosterhout 2007b). This study concluded that the immunocompetence conferred by the adaptive immune response is similar in both populations. Consistent with this interpretation, the genetic variation of genes of the adaptive immune response (major histocompatibility complex, MHC) is similar for both LA

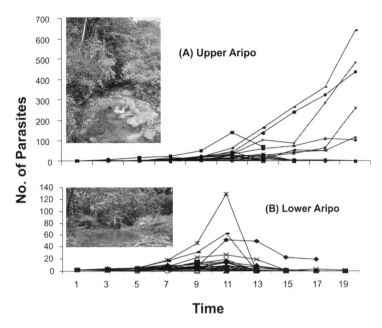

Figure 8.4 *Gyrodactylus* infection. *Gyrodactylus turnbulli* infection trajectories on guppies from the (A) Upper and (B) Lower Aripo River, with number of parasites plotted against day of experimental infection (data from van Oosterhout et al. 2003a). Each line represents a single isolated host.

and UA guppy populations (van Oosterhout et al. 2006a). Earlier studies indicated that the majority of guppies regain full susceptibility to infection 4–6 weeks after a primary infection with *G. turnbulli* (Scott 1985), but we now suspect that this refractory period is much longer (Cable & van Oosterhout 2007b). However, in the wild, most adult guppies will be reexposed to different strains and species of *Gyrodactylus* throughout their lives. Therefore, not only does parasite transmission determine the extent of parasite dispersal throughout the host population but the rate of trickle infection probably plays a key role in continuous stimulation of the host's immune response (see McMullan & van Oosterhout, **chapter 25**).

8.3.2 Impact of gyrodactylids in the wild

Like all parasites, gyrodactylids reportedly have an overdispersed distribution, with a large number of hosts in any population carrying few or no parasites and a small number of hosts carrying the majority of parasites. According to Anderson and Gordon (1982), gyrodactylid hyperviviparity and in situ reproduction contribute further to an overdispersed distribution in this parasite group. However, there are just a few, incomplete, published accounts of gyrodactylid diversity from Trinidadian guppies, and none are ideal for assessing the distribution of gyrodactylids. Harris and Lyles (1992) examined just 19 guppies from Lyles's (1990) field samples and found that most fish were infected with between 1 and 11 gyrodactylid individuals (the maximum

was 38 worms/fish), with both *G. turnbulli* and *G. bullatarudis* co-occurring at three out of nine sites examined from northern Trinidad. Lyles (1990) examined one population in detail from the Paria River, Trinidad, where 54% of guppies were infected, with a mean intensity of 3.9 (maximum, 25) worms/host but with 35.7% of infected hosts harboring only a single worm. Males had a higher prevalence than females in the wild, but mean parasite intensity did not vary significantly between the sexes (Lyles 1990). Adults were more likely to be infected than juveniles (Lyles 1990), probably because parasite loads are largely a function of host size (Cable & van Oosterhout 2007a). A similar level of infection was recorded by Martin and Johnsen (2007), who found an average of 0.68 ± 0.08 gyrodactylids/male fish (range 1–10 with just a single outlier, which had 28 worms). Potentially, both studies underestimated parasitism, as fish were netted for collection. It is important to retain fish in water at all times to obtain an accurate assessment of ectoparasite fauna, particularly as different gyrodactylid species have slightly different mechanisms of host attachment. By ensuring no contact with any substrate during fish collection, we found that 7% of guppies from one population were infected with 10 or more worms (maximum was 100 parasites/fish) (van Oosterhout et al. 2006b). However, based on the observation of much higher infections among laboratory fish, Lyles (1990) suggested that parasite-induced mortality removes the most susceptible fish—those with the heaviest parasite load—from host populations. This suggestion was indeed conclusively corroborated in a mark-recapture experiment in the wild (van Oosterhout et al. 2007a).

To assess the potential impact of gyrodactylids on wild guppies, we tested the hypothesis that more heavily infected fish would be less likely to be recaptured than uninfected or lightly infected fish (van Oosterhout et al. 2007a). All guppies from a section of the Upper Aripo River, Trinidad, were collected at the end of the dry season and screened for parasites. The fish were then individually marked and released at their original sites. For males (but not females) the recapture rate decreased with increasing parasite load; in fact, the chances of a fish being recaught declined by 19% with each additional parasite. Interestingly, parasite load also influenced the position of the fish in the stream on recapture relative to their original site of capture . Males with higher parasite loads tended to be flushed downstream, leading to the suggestion that the more heavily infected fish in a population are lost downstream during spate conditions (van Oosterhout et al. 2007a). Thus, the vulnerability of more highly infected hosts and the fitness costs of these parasites, in addition to spatial and temporal variation, might also explain why Martin and Johnsen (2007) failed to find any heavily infected fish in their study.

Flushing of susceptible fish downstream might also contribute to differential infection rates of upland and lowland guppy populations. Intrariverine differences in host populations (upland guppies tend to be larger, more colorful, and more dispersed than lowland fish, which are smaller, less colorful, and shoal more) are related to predation pressure and, to a lesser extent, availability of resources (Pires et al., **chapter 3**) but could also be related to parasitism. Not surprisingly, lowland populations also show the greatest neutral genetic diversity, largely due to unidirectional gene flow (Breden & Lindholm, **chapter 22**). In contrast, we speculated that the similarity in MHC genes between guppies from the Upper and Lower Aripo (van Oosterhout et al. 2006b) might be related to maintenance of higher immunogenetic diversity among Upper Aripo guppies in response to higher infection loads of these fish compared with their downstream counterparts (McMullan & van Oosterhout, **chapter 25**). We also found differential abundance of the two common guppy gyrodactylids, at least in this river sampled in 2001: although *G. bullatarudis* was more common in both habitats, *G. turnbulli* was relatively more abundant downstream (van Oosterhout et al. 2006b). More in-depth analyses of wider-scale patterns of parasite distribution and host genetic diversity are now required to assess the effects of spatial and temporal variation in the wild. Overall, lowland populations (from high-predation sites) appear to have higher gyrodactylid loads than those from upland (low-predation) sites (Lyles 1990; Martin &

Johnsen 2007). Parasites might even be absent from some small, high-elevation tributaries where guppies are isolated by waterfalls (Lyles 1990; Cable, van Oosterhout, et al., unpublished data).

8.4 Host-parasite speciation and testing the Red Queen hypothesis

Parasites may speciate via coevolution and/or host switching. Parasites that coevolve have a long evolutionary history with their hosts and are more likely to display a high degree of host specificity. In contrast, parasites that speciate following a host-switching event tend to be less host specific (Poulin 2006). Examples of the former are difficult to detect in nature, but there might be some evidence for coevolution among poeciliid gyrodactylids even though host switching is considered the major force for speciation among the group as a whole (reviewed in Bakke et al. 2007). For example, *P. reticulata* and its sister species *P. picta* diverged two million years BP (Breden et al. 1999), whereas the genetic variants of *P. reticulata* in the Northern Range of Trinidad diverged 100,000–200,000 years BP (Fajen & Breden 1992), well before Trinidad separated from mainland South America 10,000 years ago (Carvalho et al. 1991). Sequence divergence (rDNA Internal Transcribed Spacer region) of *G. turnbulli* and *G. pictae*, the gyrodactylids from *P. reticulata* and *P. picta*, suggests that these species diverged 10,000–100,000 years BP (Cable et al. 2005). Further information is now needed about the diversity of gyrodactylids on related hosts in the area.

At the species level, coevolution provides the basis for the Red Queen hypothesis (after Bell 1982), whereby sexual reproduction is an adaptation against parasites (see box 8.2). As in other taxa (see Combes 2001), there is conflicting evidence in support of this hypothesis among poeciliids. In line with predictions of the hypothesis, Lively et al. (1990) found that common clonal strains of *Poeciliopsis* spp. had higher parasite loads (*Uvulifer* sp.) than sexual strains, with the exception of one population, which was suffering from inbreeding due to a founder effect. However, Leberg and Vrijenhoek (1994) found no such correlation between asexual and sexual lineages of *Poeciliopsis* spp. infected with gyrodactylids. Similarly, when assessing multiple-parasite loads, Tobler and Schlupp (2005) found no differences in parasite load between four populations of the sexual fish species *Poecilia latipinna* and its asexual relative *Poecilia formosa*. This finding was not influenced if males were excluded (Tobler & Schlupp 2005) or included (Tobler et al. 2005) in the analysis. Similarly, the authors found no correlation between parasite load and genetic diversity. *Poecilia*

> **Box 8.2**   The Red Queen hypothesis
>
> Van Valen (1973) first coined the term "Red Queen hypothesis" in relation to extinction probabilities in the fossil record, but it has since been widely used to describe the "arms race" between hosts and parasites. This potentially endless battle of adaptation and counteradaptation is otherwise known as coevolution. The name of the hypothesis was taken from Lewis Carroll's book *Alice through the Looking-Glass*, in which the Red Queen explains to Alice that here "it takes all the running you can do, to keep in the same place." The Red Queen hypothesis has subsequently been used to explain a range of fundamental biological observations, such as the evolution of virulence and maintenance of genetic diversity, and even the existence of sex (Bell 1982). The paradox of sex (in which half of the population is male and does not directly produce any offspring, whereas in asexual populations all individuals contribute) has been explained in terms of parasite-escape theory. Through the production of rare genotypes, sexually derived offspring are predicted to have fewer parasites than asexual offspring (e.g., Hamilton 1980; Lively et al. 1990).

latipinna reportedly had higher genetic diversity than *P. formosa*, but the latter did not have more parasites (Tobler & Schlupp 2005 and references therein). In a further study assessing the response of asexual and sexual females to infected (*Uvulifer* sp.) males, only the asexual species avoided infected stimuli (Tobler et al. 2006a; see Schlupp & Riesch, **chapter 5**). However, the authors argue that behavioral adaptations of the infected fish may be masking any potential Red Queen effect, so these studies are inconclusive.

8.5 Impact of parasites on poeciliid behavior and ornamentation

The impact of parasites on fish behavior has been extensively reviewed (e.g., Barber et al. 2000), and various studies have shown how bacteria, monogeneans, digeneans, and nematodes all affect poeciliid behavior. Avoidance of infected fish is one of the most common behavioral traits observed in response to parasites (Barber et al. 2000), but how fish detect infected conspecifics is unknown. Indirect effects of infection, such as secondary pathology (clamped fins, altered coloring) or abnormal behavior, may be the signals that identify infected hosts, or perhaps fish rely on direct visual or chemical detection of parasites on infected conspecifics.

Plath (2004) found that cave mollies (*P. mexicana*), which occupy habitats that preclude the use of visual cues, avoided infected conspecifics under daylight conditions. The fish were infected with *Mycobacterium* sp., which causes large blisters around the eyes. Interestingly, this behavioral trait disappeared if visual cues were hindered or completely removed. Thus, this visually mediate preference has been maintained despite the colonization of a lightless habitat by this host. Plath (2004) speculated that the lack of selection pressure (few parasites) in the cave environment is the reason that nonvisual cues to detect infected individuals have not been maintained (Tobler and Plath, **chapter 11**).

Like bacteria, gyrodactylids are directly transmitted, and it seems that there would be clear costs of shoaling with an infected conspecific. Certainly, fish that are in most frequent contact with infected conspecifics acquire more parasites in aquaria (Richards, van Oosterhout, & Cable 2010), and this is currently being investigated in wild populations using social networks (Croft, Edenbrow, van Oosterhout, Darden, & Cable, unpublished data). In addition to the increased host mortality caused by gyrodactylids, these parasites inflict various indirect effects on their hosts. In guppies, they can reduce feeding rates (van Oosterhout et al. 2003a) and cause fin clamping (fig. 8.3C), which appears to affect swimming (López 1998; Cable et al. 2002), and it is assumed that these fitness costs are greatest in high-predation environments. However, to date they have been studied only in terms of their effects on reproductive behavior. Male showiness (secondary sexual traits) in guppies is known to be negatively correlated with parasite burden (Kennedy et al. 1987; Houde & Torio 1992), as well as positively associated with resistance to these parasites (López 1998). Furthermore, when given a choice of attractive or unattractive male guppies, infected females are less discriminatory than nonparasitized controls (López 1999).

Apparently, infection with *G. turnbulli* reduces the expression of carotenoid-dependent colors in guppies (e.g., Kolluru et al. 2006), and this explains why females prefer colorful males and by doing so also avoid parasitized individuals (Houde & Torio 1992). These and other similar results on nonpoeciliids (Milinski & Bakker 1990; Zuk et al. 1990) have been most commonly explained in terms of Hamilton and Zuk's (1982) parasite–sexual selection hypothesis, which assumes heritable variance in parasite resistance. However, in light of the nutritional benefits of carotenoids and their effects on the immune system (reviewed by Lozano 1994), an alternative explanation is that females choose the healthiest males based on these environmentally induced, not genetic, differences among males (Lozano 1994). However, in contrast to the earlier, laboratory-based studies, Martin and Johnsen (2007) found no correlation between male coloration (orange-spot ornamentation) and gyrodactylid load on wild guppies.

Among digeneans, it is, not surprisingly, the larval stages that modify the behavior of their poeciliid hosts to facilitate transmission to the definitive host. *Diplostomum spathaceum* is a classic example of a digenean that modifies the behavior of its host. This generalist species occurs as metacercariae in the eyes of fish, often causing parasitic cataracts (e.g., Shariff et al. 1980) and leading to various degrees of blindness, which impairs the hosts' ability to evade predation. Guppies experimentally infected with *D. spathaceum* are more prone to predation than uninfected conspecifics, possibly due to decreased activity of infected animals (Brassard et al. 1982). In the wild, other poeciliids are also infected by *Diplostomum* spp., but it is unknown whether their behavior is also affected by this digenean.

An exceptionally novel maternal mode of transmission occurs in the digenean *Apatemon graciliformis* (see Combes & Nassi 1997). Cercariae larvae penetrate female guppies and enter the vitellaria. Just prior to parturition, metacercariae encyst in the embryos and young fry are born with encysted metacercariae. Providing the parasite load is low, the offspring survive but swimming is impaired, presumably facilitating transmission to the definitive host. If a nongravid female becomes infected, the larvae develop as intracellular parasites in the ova, which demonstrates the flexibility of this parasite. On the other hand, male fish are rarely infected, but if cercariae do infect, due to lack of female hosts or high parasite densities, the parasites degenerate (Combes & Nassi 1997).

Metacercariae of several digeneans (including *Uvulifer* spp.) that encyst in the skin, fins, and gills of fish cause a condition known as Black Spot Disease (BSD). Host melanin from surrounding melanophores is concentrated around the parasite and causes the characteristic black spots that give the disease its name. The parasite can influence host reproductive behavior (Tobler et al. 2006a; section 8.4). Furthermore, Tobler and Schlupp (2008b) revealed that both infected and noninfected individuals of *Gambusia affinis* prefer to shoal with uninfected conspecifics. As this parasite is not directly transmitted, there can be no epidemiological reason for avoiding infected fish. However, if BSD infection signals low host immunocompetence and the presence of other parasites, there might still be direct costs of shoaling with infected fish, but Tobler (unpublished, cited in Tobler et al. 2006a) found no correlation between BSD and other infections. An alternative explanation is that BSD is likely to increase conspicuousness and therefore negate the anti-predator benefits of shoaling. Tobler and Schlupp (2008b) speculated that since fish would be unaware of their own infection status, they would not seek out other fish with the same disfigurement. On the other hand, shoaling with infected shoal mates may significantly reduce the benefits of shoaling for all individuals, explaining this avoidance behavior. Tobler and Schlupp (2008b) also noted that infected fish shoaled less, possibly to counteract increased energy demands caused by infection, or it could be active behavioral modification of the host by the parasite to increase transmission to the bird host.

So far only one study has examined the effect of a nematode infection on poeciliid behavior (Kennedy et al. 1987). These authors showed that male guppy display rate (particularly sigmoid displays) was reduced in individuals infected with *Camallanus cotti* and also that females preferred males with fewer camallanids.

8.6 Economic importance and control of fish diseases

The popularity of poeciliids in the aquarium trade is largely due to their ease of maintenance and tolerance of a range of water conditions, although treatment of parasites is costly and problematic (Schelkle et al. 2009). In the wild, some host species also flourish in a range of different environments. Guppies occur in pristine mountain streams and in polluted, industrial sites and even occur in an asphalt lake (Pitch Lake) in Trinidad (Burgess et al. 2005). Tobler et al. (2007a) suggested that hosts inhabiting such extreme environments might do so as a refuge against parasites. They tested this hypothesis by comparing the parasite fauna of *P. mexicana* from natural habitats with normal freshwater with *P. mexicana* from natural habitats containing toxic levels of hydrogen sulfide. As predicted, hosts in extreme environments harbored significantly fewer parasites (*Uvulifer* sp.) (Tobler et al. 2007a). Harvesting natural compounds from extreme environments might yield useful anthelmintics. Aluminum ions are currently used to control gyrodactylid infections in Norwegian salmonid rivers (reviewed in Bakke et al. 2007), but the efficiency of this control measure remains to be seen. Traditionally, livebearers are considered robust in terms of their tolerance to antiparasitic treatments; nevertheless, common infections such as white spot, fin rot, and gyrodactylosis continue to decimate fish stocks in the aquarium trade, frequently causing host mortalities of over 90%. Thilakaratne et al. (2003) suggested that the trend toward ever more exotic strains, obtained through selective breeding and resulting loss of genetic variation, might partially explain the high incidence of infection among ornamental poeciliids. For example, ornamental guppy strains have lost most of their immuno-genetic variation, possessing only 2–3 MHC alleles, compared with 15–16 MHC alleles in wild populations (van Oosterhout et al. 2006a; McMullan & van Oosterhout, chapter 25).

Intraspecific variation in poeciliid resistance to parasitic infection (Madhavi & Anderson 1985; Clayton & Price 1992) and increased susceptibility of inbred fish (Lyles 1990; Clayton & Price 1994; Hedrick et al. 2001a; van Oosterhout et al. 2007b) are well documented. Furthermore, Clayton and Price (1994) found evidence of heterosis in terms of disease resistance, but surprisingly resistance has not yet been examined in F_2 generations or backcrosses. The role of the MHC in poeciliid resistance is reviewed by McMullan and van Oosterhout (chapter 25), who highlight the importance of maintaining immunocompetence in endangered species.

8.7 Conservation threat posed by poeciliid parasites

Introduced pests are a major threat to biodiversity, ranking second only to habitat loss (Vitousek 1990; Wilcove et al. 1998). Through their global distribution, poeciliids have been responsible for the introduction of many diseases, which is particularly well documented in Hawaii (e.g., Font & Tate 1994; Font 2003) and Australia (e.g., Dove 1998). *Bothriocephalus acheilognathi* and *C. cotti* were both introduced into native fishes in Hawaii along with their poeciliid hosts (Font & Tate 1994; Font 1997a, Font 1997b, 1998, 2003; Vincent & Font 2003a, 2003b). In contrast, these same two pathogens were absent from poeciliids released into Australia even though, according to Dove (2000), the Poeciliidae family is the most successful exotic fish family in Australia, with *G. holbrooki*, *X. hellerii*, *X. maculatus*, *P. reticulata*, and *Phalloceros caudimaculatus* accounting for almost 25% of the country's exotic fish fauna (McDowall 1996 cited in Dove 2000). However, many introduced species have less diverse parasite communities than the same host species in their native habitat. This is explained by the "enemy release" hypothesis, in which relatively smaller founder host populations do not carry all the parasites that naturally occur in native populations, purely by chance and/or due to the unsuitability of the new habitat for initial parasite establishment (Kennedy & Gray 1994; Torchin et al. 2002). This lack of parasites theoretically puts the exotics at a competitor advantage, although with time they are likely to pick up infections from native host fauna in their new habitat (Dove 2000). Indeed, this has occurred among the introduced poeciliids in Australia, which have, for example, subsequently acquired *B. acheilognathi* infections from local fishes (Dove 2000).

B. acheilognathi and *C. cotti* are generalist pathogens, and it is by no means just poeciliids that have been responsible for their global translocation. However, specialist parasites are also of conservation concern. In particular, nonnative guppy gyrodactylids have been shown to be a threat to an endangered host, the Gila topminnow (Hedrick et al. 2001a). As we learn more about the host specificity of gyrodactylids (King & Cable 2007; King et al. 2009), it becomes apparent that nothing can be assumed about their host range or pathogenicity unless tested experimentally.

Clearly, poeciliids as exotics are of direct conservation concern themselves, but they have also been used as model organisms to assess the impacts of parasites on conservation programs. Lyles (1990) was the first to show that inbred guppies are more susceptible to gyrodactylid infection than wild stocks but that within each stock there was always considerable variation in susceptibility (Madhavi & Anderson 1985; section 8.3.3). Van Oosterhout et al. (2008) took this one step further by simulating a captive breeding program. Wild-caught guppies from the Upper and Lower Aripo River were transported to the United Kingdom and maintained under three different breeding regimes. In this simulated captive-breeding program, the fish were maintained under parasite-free conditions with no stimulation of the acquired immune response. After four generations, the fish were flown back to Trinidad and housed at the University of West Indies for four weeks in quarantine conditions before release into a seminatural mesocosm, which consisted of a series of large pools that had been seeded with natural plants, invertebrates, and naturally infected guppies (1–2 worms/host). Survival of the fish stocks throughout the captive breeding, quarantine, and release stages of the experiment was consistent with predictions that fish from the genetically more diverse founder populations (Lower Aripo) would perform better than those from small, genetically isolated populations (Upper Aripo). Similarly, fish allowed to mate randomly or maintained under conditions to prevent sib matings performed better than those subject to inbreeding. More interesting was the impact of parasite-induced mortality on these fish, as survival of captive-bred fish was only 58%, compared with 96% for wild fish (van Oosterhout et al. 2007b). Although this study highlighted the impact of disease epidemics on captive-breeding and reintroduction programs, it offered no practical solution to the problem. However, we have just completed a lengthy laboratory experiment (Faria et al. 2010) and shown that the impact of these pathogens can be reduced if the captive-bred fish are exposed to parasites before release and if they are released gradually into the wild populations rather than en masse. Such studies show again what a useful model system the guppy is, here demonstrating for the first time how this vertebrate-pathogen model can be used to optimize conservation strategies.

8.8 Conclusions

Understanding the biology of poeciliid pathogens is crucial not only for reducing the impact of these parasites on the aquarium industry and native fishes but also for assessing the threat they pose to endangered poeciliids. Unlike ornamental guppies, Trinidadian *P. reticulata* has a relatively depauperate eukaryotic parasite fauna, which is dominated by gyrodactylids that have marked effects on host fitness. However, to assess the full ecological impact of parasites on poeciliid hosts, an essential prerequisite is a full taxonomic survey of their diversity, and this includes an assessment of the most contagious microparasite diseases (bacteria, viruses, etc.). I hypothesize that there is considerable temporal variation in parasite abundance, with parasite persistence and levels of parasite-induced mortality being dependent on the immigration of new susceptible hosts, host density, and the level of herd immunity. Testing this and assessing the interaction with other biotic (presence of other hosts and parasites) and abiotic (water quality) factors are our current challenges.

Acknowledgments

Current work on gyrodactylids is funded by the Natural Environment Research Council (NER/J/S/2002/00706) and the European Community Framework Programme 6 (MTKD-CT-2005-030018). Thank you to the editors of this book and to two anonymous reviewers for their constructive comments on an earlier draft of this chapter.

Chapter 9 Community assembly and mode of reproduction: predicting the distribution of livebearing fishes

Joel C. Trexler, Donald L. DeAngelis, and Jiang Jiang

9.1 Introduction

CAN REPRODUCTIVE MODE be used to identify functional species and predict which communities they can invade? We asked this question in order to develop a theory of community assembly for poeciliid fishes that could be used to make testable predictions about their community membership and for their study as a model system in ecology. A useful method to conceptualize community assembly starts with the notion that local communities are constructed of members from a regional species pool filtered by biotic and abiotic factors based on functional traits that permit their persistence under a set of conditions (the trait-environment paradigm; Keddy & Weiher 1999). Functional traits are assumed to trade off as part of competing life functions constrained by energy income, resource allocation budgets, and developmental pathways (Chase & Leibold 2003). For example, it is commonly assumed that dispersal ability is inversely related to competitive ability, or that reproductive output is inversely related to the quality of offspring (generally equated to their size at independence from the parent). Following Clutton-Brock's (1991) broad definition of parental care, we proposed that poeciliid fishes are distinguished ecologically by their mode of parental care: viviparity. Furthermore, the family Poeciliidae includes members representing a remarkable diversity of reproductive adaptations that are presumably tied to the level and timing of energy allocation during development of offspring (Trexler & DeAngelis 2003; part I of this volume) and geared to production of precocious offspring at parturition. In this chapter, we discuss a model developed to make initial steps toward a theory of community assembly for a pool of fish species exhibiting either a viviparous or an oviparous mode of reproduction.

Recent theoretical work in community assembly has included a focus on niche-based models that delineate the role and impact of species in their local communities (Chase & Leibold 2003; Grimm et al. 2006; McGill et al. 2006). In considering the role of poeciliid fishes in communities, we noted that viviparity provides extreme parental care throughout embryonic development and permits production of precocious offspring. In effect, livebearing dramatically increases daily survivorship rates until offspring are released from maternal control. We believe it is reasonable to assume that this early survival boost exceeds any survivorship benefit of parental care provided at the nests of oviparous species. All parental care must come at some cost, and we have focused on energetic ones. Survivorship costs to pregnant females have been documented in the form of decreased mobility and increased susceptibility to predation (Plaut 2002; Ghalambor et al. 2004; see also Pires et al., **chapter 3**, and Kelley and Brown, **chapter 16**), but we have ignored these in our initial model development. Instead, we have considered only costs in the form of lost reproductive opportunities by tracking energy income, storage, and allocation that limit contemporary and future reproductive output. Thus, energy invested in current reproduction may come at an opportunity cost to later reproduction, depending on the rate of energy acquisition. Alternatively, energy invested in growth or storage may come at a cost to current reproductive output but benefit future productivity. We used such rules to define reproductive niches for viviparous and oviparous species and evaluate the environmental conditions in which either mode should yield higher fitness and, thus, persist without requiring recurrent immigration.

Poeciliid fishes serve as an important group for research into the role of interspecific interactions in shaping population biology. Notably, and well documented in this volume, research on the effects of predation in shaping life histories of poeciliid fishes is well developed (see **part I**). Furthermore, there is a well-articulated discussion of the interaction of resource ecology and predation in life-history evolution (Grether et al. 2001b; Bronikowski et al. 2002; Day et al. 2002; Reznick et al. 2002a; Grether & Kolluru, **chapter 6**). In this chapter, we propose that this thinking be expanded to community assembly, building upon the life-history basis established for poeciliid fishes. In particular, we believe that energy budgets are a natural basis for defining a species' niche, as well as for making predictions about community assembly based on those niches. Life-history ecology has focused on three primary drivers: intraspecific competition and resource limitation in population dynamics (r- and K-selection theory), patterns in age-specific survivorship, and environmental predictability (bet hedging). In past work, these drivers were typically treated as mutually exclusive post hoc explanations for observed patterns (Stearns 1992), though increasingly they are treated as predictors for experimental study and even as acting in concert.

In community and ecosystem ecology, the role of interacting environmental gradients in shaping dynamics is well established. The interaction of resource and predation gradients, for example, is a topic of ongoing interest, and theories such as those of "predator permanence" (Wellborn et al. 1996) and environmental grain (Levins 1968; Caswell & Cohen 1995; Brassil 2006; Fox 2007) seek to account for temporal variation in these factors. We have focused on four gradients (resource availability, temporal variation in resource availability, predation rate, and size selectivity of predation) to evaluate how energy costs incurred by alternative reproductive modes could sort species into local communities.

9.2 Model description

Our model envisions a regional pool of species that is filtered by local environmental conditions to yield a local assemblage. We focus on patterns of energy and time investment in reproduction to generate an 800-member pool of species with all equally present at the start of a simulation run. The pool is sorted over a number of generations by exposure to a combination of four environmental parameters

related to resource level and juvenile mortality factors. For each computer run, we assigned one of three levels of food availability (food level = FL), three levels of periodicity in temporal variability of food level (periodicity = PER), three levels of juvenile mortality (juvenile mortality rate = JMR), and three levels of strength of size-selective juvenile mortality (size-selective mortality = SSM). This yields 81 environments by all possible combinations of these four factors ($3^4 = 81$). Our species were characterized by three traits: size of eggs or neonates, maternal care, and energy investment in a growth versus reproduction trade-off. The two reproductive traits, size of eggs or neonates and maternal care, permitted separation of viviparous and oviparous life histories. The distinction between viviparous and oviparous females was made based on the timing of expo-

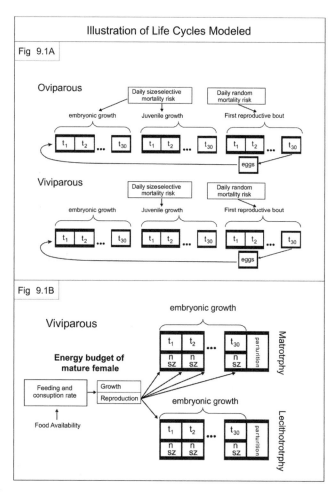

Figure 9.1 Conceptual models illustrating our simulations. (A) Life cycles contrasting oviparous and viviparous life histories. Note that oviparous species experience size-selective mortality while embryos, but viviparous fishes do not. (B) Life cycles of two viviparous strategies, lecithotrophy and matrotrophy. Lecithotrophy provides full investment in embryos prior to fertilization, in contrast to matrotrophy, in which embryos are nourished throughout gestation.

sure of offspring to juvenile mortality, where the offspring of oviparous species were exposed to juvenile mortality starting at the time of fertilization (which we equate with laying by an egg-scattering species), and the offspring of viviparous species were protected from juvenile mortality until parturition. In practice, this meant that cohorts of egg layers began incurring demographic costs through mortality 30 days prior to those of livebearers (fig. 9.1A).

9.2.1 Entities, state variables, and scales

9.2.1.1 Entities. We modeled individual female fish that start as juveniles at the beginning of a simulated year, reach maturity, and begin reproducing in that year. The offspring of fish during the year are represented, not as individuals, but as numbers. Each fish is characterized by a set of hereditary traits:

- For viviparous fish there are 20 strategies along a scale from pure matrotrophy to pure lecithotrophy. In pure lecithotrophy, all the investment in embryos occurs "up front," whereas in pure matrotrophy, there is an equal amount of investment in embryos each day until maturation of the embryos. The 20 strategies are characterized by 2.5%, 7.5%, 12.5%, 17.5%, . . . , or 97.5% up-front investment. Along this scale, neonate size also decreases, from 2 units to 0.3 units of energy invested per egg, either all at once (lecithotrophs) or gradually over the gestation period (matrotrophs). There are an analogous 20 strategies for oviparous fish, but these differ only in egg size, and there is no maternal care (fig 9.1B) (Marsh-Matthews, **chapter 2**).
- There are also different options for how the female invests her surplus energy (the energy beyond that immediately needed to nourish a brood). She can put some fraction of that energy into storage for later use in reproduction, and the rest into growth in biomass. There are 20 strategies, with either 5%, 7.5%, 10%, 15%, 20%, . . . , 90%, or 95% investment in growth, with the remainder going to storage for later use in reproduction. Since all combinations of the two strategies are possible, this adds up to 400 "species" so far.
- Because fish may be either livebearers or egg layers, there are 800 total strategies. In the case of livebearers, the embryos are held for 30 days before neonates are born, while egg layers immediately discharge eggs.
- *Variables*: Each fish also has several variables: age, weight, length, amount of energy currently invested in offspring within the fish, and total number of offspring produced. Age and weight are independent, though

correlated. In addition, the number of surviving offspring produced by each fish during the reproductive season is a variable.

9.2.1.2 Scales. The model is not spatial; all fish are assumed to be exposed to the same environmental conditions. The model has two temporal scales. The first temporal scale is the reproductive season, in which the fish compete to produce offspring. The second scale is that of many generations, over which selection acts to sort the species based on reproductive productivity.

9.2.1.3 Environment. The environment is homogeneous. It consists of a supply of food that has both a mean and variation about the mean (described in detail later). The fish are not assumed to have an impact on the food supply, which is entirely externally regulated. The environment also imposes density- and size-independent mortality on the adult fish and size-dependent mortality on the offspring. The environment is assumed to impose a carrying capacity on fish that allows only 10,000 to come to maturity in a given year.

9.2.1.4 Collectives. The fish are classed into 800 fish "species" (these can equally well be thought of as asexually propagating genotypes). At the beginning of the first year of simulation, there are 10,000 reproductive fish with approximately 12 individuals in each species. For each year after that, the number of individuals of each species is chosen in proportion to its relative representation in the surviving neonates produced by the fish community in the preceding year. Thus, in later years of the simulation, these proportions change, as natural selection acts differentially on the species.

9.2.2 Process overview and scheduling

The model follows individual female fish through a segment of their lifetimes from late juvenile to reproducing adult (assumed to be within one year) in daily time steps and tabulates the daily energetics of each fish, starting with intake of food (see the Energetics of Growth submodel in box 9.1). It describes the allocation of this energy through the reproductive season and tracks the number of offspring produced in order to calculate the reproductive fitness for each fish.

At the start of a given year the fish are assumed to have the same size and to grow at the same maximum rate, governed by allometric relationships between weight and daily intake, respiration, and growth, as each fish has the same available food supply. Depending on their individual strategies, however, fish soon diverge in size, because they begin

Box 9.1 Design Concepts and Submodels

Model Characteristics

Emergence. A stable poeciliid community is expected to emerge from the action of natural selection on the species over many generations. Mutations are not assumed to occur, so the final community will emerge as a single species or a mixture of species from the initially inserted 800 species.

Adaptation. Each fish has three basic traits: (1) whether it is a livebearer or an egg layer, (2) its position along the matrotrophy-lecithotrophy axis (if it is a livebearer), as well as differences in egg or neonate size, and (3) its position along the reproduction-growth axis. Livebearers and egg layers are distinguished in that the eggs of the latter will be vulnerable to mortality in the environment for a longer period of time.

Fitness. Fitness among the species is measured by differential reproduction.

Prediction. Individuals have no predictive ability. However, different species will have different reproductive success depending on the particular environmental conditions that occur.

Sensing. The individuals do not sense their environment. They cannot change strategies.

Interaction. Individual fish compete indirectly in a form of scramble competition. The fish do not have an exhaustive effect on food resources. However, since there is a limit of 10,000 fish that survive to reproduce, species that have greater reproductive success have a negative effect on survivorship of species that have less.

Stochasticity. Mortality of adults is stochastic. Availability of food on a particular day is also stochastic (see Available Food submodel), but the daily intake of food by fish, given its availability, is not.

Collectives. The genotype/species is the main collective. Each of the 800 species consists of a number of individual fish.

Observation. The primary observation of data is the number of fish in each of the genotype/species classes at the end of a 200-generation simulation. Numerous statistics can be derived from this model output.

Initialization

The model is initialized with 10,000 fish divided roughly equally into 800 species on the first day of the growing season of year 1. This is the same for all simulations. All fish have the same weight and length at the start; all are in the juvenile state but can grow and move to the mature state within several days to a few weeks.

External input

Mean values of available food and the temporal variation around the mean values are determined externally (see Available Food submodel).

Submodels

Available Food The mean concentration of available food on a given day is $Food_{level}$ (FL). The amount of food available fluctuates around the mean. Both the magnitude of the fluctuation, $Food_{var}$, and the duration of a fluctuation in one direction or the other in days, called $Food_{fluctuation\_periodicity}$ (PER), can be varied, but both parameters remain fixed throughout a given simulation. Thus, available food concentration is given by

$$Food_+ = Food_{level} + Food_{var} \text{ or}$$
$$Food_- = Food_{level} - Food_{var}$$

The year is divided into period of time, $Food_{fluctuation\_periodicity}$, and in each time period, whether available food is $Food_+$ or $Food_-$ is randomly generated.

Energetics of growth Food assimilated per day by an individual fish is

$$Intake_{food} = Food_{level} Weight^{0.667}$$

so that net energy intake is this minus respiration:

$$NetEnergyGain = Food_{level} Weight^{0.667} - Resp * Weight$$

If $NetEnergyGain > 0$, this surplus energy can be invested in either growth or reproduction or a combination of the two. This growth in weight is calculated by the difference equation

$$Weight_{t+\Delta t} = Weight_t + Alloc_{growth} NetEnergyGain$$

where $Alloc_{growth}$ is the fraction allocated to growth and varies with the fish genotype from 0.05 to 0.95. If $Weight$ for an individual fish exceeds the upper limit on size, $Weight_{limit}$, then $Alloc_{growth} = 0$ and $Alloc_{reproduction} = 1$. It is assumed here that fluctuations in available food, even if they cause low levels of intake, do not affect the ability of a fish to meet its basic respiration demands but will affect growth and reproduction through a trade-off.

investing in reproduction at different sizes, and they invest different fractions of their energy intake.

Reproduction is triggered in an individual fish when it has a combination of stored up-front energy and an energy intake level that allows it to both start and maintain a brood of fish through egg yolking and laying or birth (if viviparous). A fish "checks" its daily intake of energy. As soon as the daily energy intake rate is high enough to sustain what it will need to provide its clutch on a daily basis, the fish starts diverting some fraction (depending on its allocation strategy) of its energy either directly or indirectly into reproduction. "Directly" means that the energy immediately goes to embryos. "Indirectly" means that the energy first goes into storage and then goes into reproduction when needed.

The extreme matrotroph can already produce eggs at the time of switching and start provisioning them. So she starts putting energy *directly* into brood provisioning immediately. What is left over goes partly to growth and partly to storage for later use in reproduction (i.e., the energy is used *indirectly* for reproduction). The other strategies along the matrotroph-lecithotroph spectrum involve storing some energy before embryos can be produced. As soon as the fish has enough stored energy for the up-front investment in eggs, she produces a clutch. At that point she also has enough energy for daily provisioning of the eggs (although the extreme lecithotrophs do not need to provide any more provisioning).

One important aspect of the fish reproductive strategy is what happens to embryos when, because of energy fluctuations, the fish is unable to meet the energy demands of the embryos it is carrying. Livebearing matrotrophs are assumed to have to sacrifice some or all embryos to maintain themselves. Egg layers and livebearing lecithotrophs do not have to sacrifice embryos, as the energy has been put into the embryos up front.

During the year, offspring are subjected to size-dependent mortality. Those offspring that are born or hatched and survive the imposed mortality constitute the reproductive contribution to the following year's reproducers. At the end of each reproductive year, the number of surviving offspring of each of the 800 types is summed up, and the proportions of each type are used to determine the proportions of next year's reproducing fish. This process is carried on over 200 generations, producing what is termed the final stable community (see box 9.1 for a summary of the design concepts of this model).

9.2.3 Reproduction

There are two general types of reproduction, egg laying and livebearing. It is assumed here that egg layers may lay eggs as soon as they have stored enough energy to produce a batch of 100. Livebearers require a gestation period of 30 days before neonates are born. Thus, they offer protection to their offspring longer than egg layers but must wait for at least 30 days to start a new brood.

There are two extremes along the matrotrophy–lecithotrophy axis. At the lecithotrophy extreme, all the energy that an embryo will need to develop is put into the embryo up front. In the matrotrophy extreme, the energy input from the mother to the embryo is spread out evenly over the 30 days. There are many intermediates, in which energy allocation is divided in different ways between up-front energy and later energy input. The model considers 20 gradations along the axis.

For the lecithotrophs in the model, the criteria for reproduction are that at least 30 days have passed since the last reproductive bout, and enough energy has been stored to produce 100 embryos to the neonate stage. If the lecithotroph mother faces low energy intake after initiation of the embryos, it is assumed to have no effect on the embryos. For the pure matrotroph in the model, the criteria are that the matrotroph have enough energy through daily intake and storage to sustain itself and supply its 100 embryos. Intermediates between pure lecithotrophy and matrotrophy must be able to provide some up-front energy plus a continuing supply during gestation.

During the maintenance of a brood, the matrotroph supplies energy for each of its eggs. If the stored energy and energy intake on a particular day are not enough to cover the cost of nourishing all the embryos, enough must be aborted such that the energy intake of the female is still adequate to meet its respiratory demands. Energy from the aborted embryos is not recyclable to the female, because empirical support remains elusive, as does evidence of adaptive patterns of abortion (see Trexler 1997; Banet & Reznick 2008).

To simplify the model, it is assumed that the sizes of viviparous neonates decrease along the axis from matrotrophy (largest) to lecithotrophy (smallest). There is some empirical support for this choice, at least in a relative sense (Thibault & Schultz 1978; Reznick et al. 1996a; Pires et al. 2007; Banet & Reznick 2008). We assume that if the size of eggs of pure matrotrophs is 2 (in arbitrary units), the size of the neonates at the lecithotroph extreme is 0.3 in those units; neonate size varies proportionally with level of lecithotrophy. The energy input needed for complete development of the eggs or neonates is correspondingly scaled with egg size. To make the egg-laying community similar in structure to the livebearers, we assume that the same variation in egg size occurs among the egg layers as among the neonate size of livebearers.

9.2.4 Mortality of offspring

It is assumed that offspring undergo mortality that is inversely dependent on the size of the neonates or eggs. The daily rate of mortality is given by a function

$$M_{offspring\_mortality} = s_M e^{10(JMR/(Egg_{size}+1)-SSM/3)}$$

This is an arbitrary function, but it gives a juvenile mortality rate that can be varied, through parameter JMR, in its strength and, through SSM, in the sharpness of its increase with decreasing egg or neonate size. This rate of mortality is integrated over 2 months for the livebearers and 3 months for the egg layers to calculate surviving offspring for each species.

9.3 Simulations

We conducted two sets of simulations starting with different regional pools of species, one with all having equal fecundity (100 eggs for oviparous females and 100 neonates for viviparous ones) and a second with oviparous species having twice the fecundity of the viviparous ones (viviparous lecithotrophs and matrotrophs had equal fecundity, though they differed in neonate size). All simulations were run for 200 generations, at which point there were generally only a few similar species remaining; a few exceptions were noted where a mixture of viviparous and oviparous species were both present, but we observed no evidence of stable equilibria indicative of a balanced polymorphism and expect that one or the other would have been driven extinct had the models been run longer.

At the end of each simulation, we recorded the number of individuals of each species remaining, as well as information about the age at maturation, number of broods produced in the final generation, size of eggs or neonates produced, reproductive mode (viviparous [coded 1] and oviparous [coded 2]), relative allocation of energy to growth versus reproduction (high value invests more in growth, low invests more in reproduction), and terminal mass. This permitted us to tally two population parameters for each simulation, in addition to the six life-history traits. Because there was generally little interspecific variation remaining, we took the average of these traits and noted the few instances where both oviparous and viviparous species remained and rendered the means poor measures. The two population parameters were population size (number of individuals) and biomass (g). Biomass was calculated as the product of population size and average female terminal mass. We considered biomass to be an index of productivity; if the species were treated as genotypes, final population size could be considered a measure of total fitness (in

contrast to relative fitness of the specific starting genotypes in each combination of environmental conditions).

We characterized the results of this effort by identifying patterns of traits in the survivors of our 81 different environments and by identifying functional groups of species. For the purpose of this model, we defined a functional group as species with similar fitness in an environment. We then examined how these functional groups were distributed across environmental gradients. To identify suites of traits and characterize functional groups, we used a principal-components analysis (PCA) of the average values of the six life-history traits of survivors that we tracked. PCA was conducted on the correlation matrix, and we used a varimax rotation to clarify the results. Both sets of simulations (equal and unequal fecundity) yielded three factors that together explained over 80% of the variation among environmental combinations, so we saved the resulting factor scores for further analysis.

We used analysis of variance to partition relative effect sizes of the independent variables from each set of simulations by calculating the ratio of the sums of squares for each effect relative to the total. This was not a hypothesis-testing exercise, as there was no replication of each treatment combination, but it provides a ranking of the impact of each factor and interactions among factors on our three dependent variables (three sets of factor scores). Each model had four main effects, six two-way interactions, four three-way interactions, and one four-way interaction, underscoring the potential complexity of the environmental conditions that we considered. We considered the effect of food level as consistent with environmental gradients envisioned for r- and K-selection theory, periodicity in random fluctuation of food level as consistent with gradients envisioned for some bet-hedging scenarios, and the level and strength of size-selective juvenile mortality as consistent with gradients in models of age-selective mortality.

9.4 Results

9.4.1 Suites of traits

Suites of correlated traits emerged in the populations we simulated. When fecundity was equal for oviparous and viviparous fishes (100 per brood), reproductive mode and brood number covaried inversely with investment of energy in growth versus reproduction (table 9.1A, factor 1). Viviparous fishes tended to invest relatively more energy in growth than oviparous ones and produced fewer broods of young over their lifetime (fig. 9.2). The terminal mass of females was inversely correlated with their age at maturity (table 9.1A, factor 2), and offspring size (eggs or neonates) was inversely correlated with the number of broods

Table 9.1 Factor scores produced from principal-components analysis of life histories of fish remaining after 200 generations in each model run

A. Equal fecundity run

Variables	Factors		
	1	2	3
Age at maturity	−0.02	0.76	0.252
Female weight	0.041	−0.82	0.253
Number of broods	−0.511	−0.164	−0.747
Relative allocation to growth or reproduction	0.968	0.075	−0.009
Neonate/egg/size	0.029	−0.072	0.956
Reproductive mode	−0.801	0.277	−0.378
Variance explained by rotated components	1.842	1.364	1.742
Total variance explained (%)	30.7	22.7	29.0

B. Unequal fecundity run

Variables	Factors		
	1	2	3
Age at maturity	−0.623	−0.409	0.392
Female weight	0.074	0.927	0.063
Number of broods	−0.225	0.198	−0.866
Relative allocation to growth or reproduction	0.839	−0.072	0.295
Neonate/egg/size	0.82	0.161	0.386
Reproductive mode	0.187	0.302	0.824
Variance explained by rotated components	1.855	1.188	1.823
Total variance explained (%)	30.9	19.8	30.4

Note: Factor scores exceeding 0.4 are highlighted and discussed in the text.

produced in a lifetime (table 9.1A, factor 3). The former relationship may have resulted because age at maturation tended to be inversely related to the number of broods (early maturing females had longer reproductive lives and, thus, more time to add mass) and adult mortality was unrelated to female attributes; a more realistic pattern of adult mortality than we simulated may eliminate this. The latter relationship emerged from our energy allocation and is commonly observed in natural populations.

When fecundity of oviparous fishes was twice that of viviparous ones, offspring size and relative energy investment in growth covaried inversely with female age at maturity (table 9.1B, factor 1). Female age at maturity was inversely correlated with their final mass (table 9.1B, factor 2). There was an inverse relationship between neonate or egg size and

the number of broods a female produced in her lifetime (table 9.1B, factor 3).

We use these three factors for simple examination of environmental gradients, so we label them as follows: factor 1 indicates patterns of energy allocation; factor 2 indicates patterns of maturation; and factor 3 indicates patterns of reproductive output.

9.4.2 Environmental gradients

The partitioning of variance among the three factors that emerged from PCA was surprisingly similar between the two sets of simulations (table 9.2). In both cases, variation in factor 1 was primarily driven by food level, and variation in factor 3 was primarily driven by the severity of size-selective

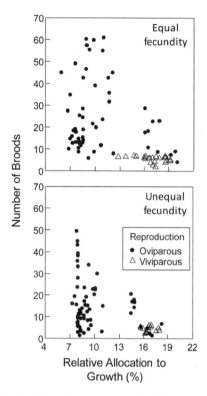

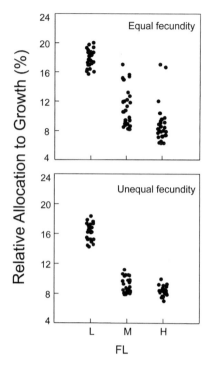

Figure 9.2 Number of broods versus energy allocation to growth (%), plotted separately for models run with equal and unequal fecundity. "Relative allocation to growth" indicates the percentage of energy acquired above maintenance costs that is allocated to growth (100 minus that number is the percentage allocated to reproduction).

Figure 9.3 Energy allocation to growth (%) versus food level (FL) plotted separately for models run with equal and unequal fecundity. L, M, and H indicate low-, medium-, and high-food environments, respectively. Relative allocation to growth indicates the percentage of energy acquired above maintenance costs that is allocated to growth (100 minus that number is the percentage allocated to reproduction).

juvenile mortality. Variation in factor 2 was more complex than in the other two factors, but temporal periodicity in food level was important in both sets; when fecundity was unequal, mean food level and periodicity interacted.

The origins of these patterns can be seen in patterns of the traits correlated with the PCA factors. For example, without regard to reproductive mode, fish in low-food environments allocated relatively more energy to growth than those in medium- or high-food environments (factor 1; fig. 9.3). The minimum terminal mass of females was inversely correlated with variation in food level when fecundity was equal for all females (factor 2; fig. 9.4), yielding a triangular pattern indicative of an effect that interacts with other factors (Thomson et al. 1996); terminal mass was uncorrelated with periodicity when fecundities were unequal. The lifetime number of broods was decreased as the severity of size-selective mortality increased, most markedly in high-food environments (factor 3; fig. 9.5). Interactive effects were pervasive; for example, in both sets of simulations egg or neonate size generally increased with higher food levels when size-selective mortality was high or medium but not when it was low (fig. 9.6). Because the data plotted in fig. 9.6 result from essentially deterministic processes, the wide range of egg or neonate sizes at

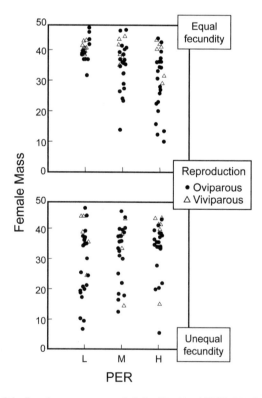

Figure 9.4 Female mass versus periodicity of food level (PER) plotted separately for models run with equal and unequal fecundity.

Table 9.2 Partitioning of variance among effects of six dependent variables

Source	Biomass	Factor 1	Factor 2	Factor 3
Unequal fecundity				
Food Level (FL)	17	65	10	9
Food Level Periodicity (PER)	1	1	2	0
Size-specific Juvenile Mortality (SSM)	22	1	2	57
Juvenile Mortality Rate (JMR)	9	1	2	2
PER*FL	2	1	13	1
SSM*FL	7	8	2	9
PER*SSM	2	1	2	0
FL*JMR	4	1	13	5
PER*JMR	3	1	7	1
SSM*JMR	10	2	7	3
SSM*PER*FL	3	2	5	1
PER*FL*JMR	4	1	9	1
SSM*FL*JMR	3	9	5	3
PER*JMR*SSM	5	3	6	2
PER*SSM*FL*JMR	8	3	16	5
Equal fecundity				
Food Level (FL)	5	68	4	0
Food Level Periodicity (PER)	3	3	21	1
Size-specific Juvenile Mortality (SSM)	11	0	1	71
Juvenile Mortality Rate (JMR)	5	5	2	8
PER*FL	2	1	1	4
SSM*FL	11	0	5	3
PER*SSM	2	3	3	1
FL*JMR	1	1	3	2
PER*JMR	7	4	3	1
SSM*JMR	9	0	4	4
SSM*PER*FL	5	1	2	1
PER*FL*JMR	5	5	4	1
SSM*FL*JMR	9	2	10	2
PER*JMR*SSM	10	0	20	1
PER*SSM*FL*JMR	17	5	16	2

Note: Biomass refers to the mass of all fish at the end of the 200th generation. Factors 1, 2, and 3 refer to PCA results from table 9.1. All results are percentages of total variation.

the low-food treatment with equal fertility is particularly significant. A full range of lecithotrophic, matrotrophic, and oviparous (small and large eggs) life histories "won" in some combination of size-selective juvenile mortality, absolute juvenile mortality, and food-level periodicity at low average food level. Medium and high food levels failed to support lecithotrophic reproduction in our simulations,

possibly because viviparous embryo nourishment yielded relatively small offspring.

Final biomass of fishes is an index of both the overall productivity possible in each environment and variation in life histories because it results from both the production of offspring and the diversion of energy into accumulating somatic mass. Origins of variation in this parameter

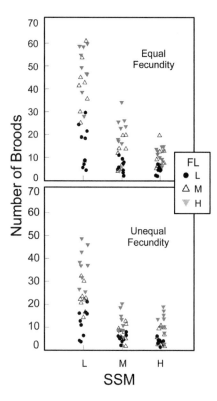

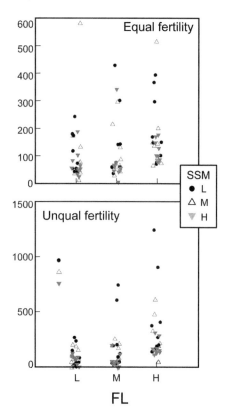

Figure 9.5 Lifetime number of broods produced versus strength of size-selective juvenile mortality (SSM) plotted separately for models run with equal and unequal fecundity.

Figure 9.7 Community biomass versus food level (FL), with strength of size-selective juvenile mortality (SSM) indicated for each point.

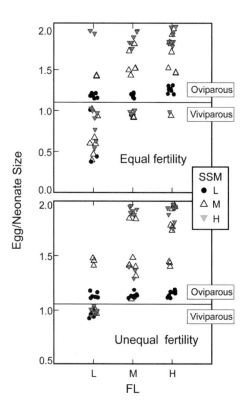

Figure 9.6 Egg or neonate size versus food level (FL) plotted separately for models run with equal and unequal fecundity. The strength of size-selective juvenile mortality (SSM) is indicated for each point. The horizontal line separates oviparous (top) and viviparous (bottom) reproductive modes. Egg or neonate size increases with height on the graph and is scaled in arbitrary units. Decimal portions of numbers indicate egg or neonate size relative to maximum; for example, 1.95 is an oviparous fish laying eggs that are 95% of the maximum size.

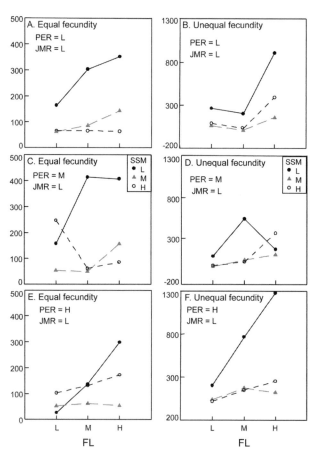

Figure 9.8 Community biomass versus food level (FL), plotted separately for combinations of periodicity; the level of juvenile mortality (JMR) is low for all graphs. Model runs with equal and unequal fecundity are plotted separately. Lines connect points with the same strength of size-selective juvenile mortality (SSM).

depended on the fecundity patterns in our regional species pool. When oviparous fishes had twice the fecundity of viviparous ones, variation in biomass was primarily shaped by both food level and the strength of size-selective juvenile mortality (table 9.2). However, when both reproductive modes had equal fecundity, the four-way interaction explained the most variation. When oviparous fish had high fecundity (unequal fecundity set), the total biomass resulting from low size-selective juvenile mortality and medium or high food level was 1.5–2 times greater than in all other environments (fig. 9.7). The multiway interactions of the equal-fecundity simulations can be seen by drilling down into the results (fig. 9.8). Holding the level of juvenile mortality constant and low, we compared the effects of periodicity in food level on biomass. At low periodicity, low size-selective mortality yields double or higher biomass than medium or high size-selective mortality for all levels of food (fig. 9.8A). However, as the periodicity of food level increases, biomass at low size-selective mortality drops to less than found at medium periodicity, medium size-selective mortality, and low food (fig. 9.8C), and to even less than found at both medium and high size-selective mortality at high periodicity and low food (fig. 9.8E). Since these are deterministic results, these complex dynamics are not trivial noise but result from limitations emerging from competing energy allocation and timing requirements for maintenance of homeostasis in our model fish. When fecundity was equal among all members of the regional pool, the level of complexity revealed in our results was similar to the unequal pool, though the patterns changed (fig. 9.8B, D, F). Notably, the scale of variation was dramatically reduced because the overall productivity was diminished.

9.4.3 Reproductive modes

Our simulations indicated that fish exhibiting viviparous reproduction as we modeled it were superior competitors in environments with low food availability relative to demand and either high juvenile mortality or moderate and high levels of size-selective juvenile mortality (table 9.3). When fecundity was unequal, favoring oviparous reproduction, we found no conditions where lecithotrophic reproduction persisted for 200 generations. In this model, lecithotrophic embryo nourishment was confounded with small offspring size, so this is not a "pure" test of lecithotrophy; these results may simply indicate that viviparity succeeds when it serves to protect relatively large young from mortality during the early days of life. When fecundity was equal for all reproductive modes, gradients from small to large young were present both from left to right (low food level to high) and top to bottom (low size-selective juvenile mortality to high) in table 9.3. Reproductive mode in the middle and bottom of the table (medium to high size-selective juvenile mortality) shifted from lecithotrophic at low food level and low to medium periodicity to matrotrophic at low food and high periodicity and low or medium food and low or medium periodicity to oviparous with medium or large eggs at medium food with high periodicity or high food level. Three- and four-way interactions were manifested in these effects, for example, by the general pattern of small to large eggs diagonally from the top left to bottom right in both halves (unequal fecundity and equal fecundity) of table 9.3.

9.5 Discussion

Our model suggests that viviparous species will be most common in communities of fishes experiencing food conditions that are low relative to demand and where small offspring are at a disadvantage to larger ones. To the extent that lecithotrophic fishes produce smaller offspring than matrotrophic ones, they will be replaced by matrotrophic invaders if fecundity is equal. Of course, in real fish communities the diversity of life-history types in regional pools is less than we simulated. Also, coexisting fishes may experience different realized "food levels" because diet overlap is seldom complete, and food levels of different diet items may not be tightly coupled. For example, in southern Florida least killifish (*Heterandria formosa*), eastern mosquitofish (*Gambusia holbrooki*), and sailfin molly (*Poecilia latipinna*) all coexist and can at times be collected in the same location (Trexler et al. 2001). However, eastern mosquitofish are primarily predatory on invertebrates (amphipods, midge larvae, ants, and spiders from the water surface) and larval or juvenile vertebrates and their eggs, while least killifish eat a mixture of green algae and smaller invertebrates, and sailfin mollies are primarily algivores (they appear to be less selective than least killifish) (Loftus 2000; J. Trexler, unpublished data). The correlation of spatial and temporal variation in density of small invertebrates and algae, particularly when algae are divided into edibility classes, is not strong (J. Trexler, unpublished data). Thus, a low-food environment for a sailfin molly is often not a low-food environment for a least killifish or eastern mosquitofish. This difference possibly explains variation in the spatial distribution of these taxa. Superimposing information on the covariance of resource and predation gradients could be a step toward building a more realistic model of community assembly. In this chapter, we have attempted to make a case for such multifactorial analysis of community composition and dynamics.

Resource level was most important in determining the types of species that would be favored in local communi-

Table 9.3 Distribution of reproductive modes persisting after 200 generations exposed to each combination of environmental conditions

		FL	L			M			H		
		PER	L	M	H	L	M	H	L	M	H
SSM	JMR										
Unequal Fecundity												
L	L		1.15	1.15	1.15	1.15	1.15	1.15	1.15	1.15	1.15
	M		1.15	1.15	1.15	1.15	1.15	1.15	1.15	1.15	1.15
	H		0.96	0.95	0.96	1.16	1.15	1.15	1.16	1.16	1.16
M	L		1.45	1.43	1.44	1.43	1.44	1.44	1.44	1.43	1.43
	M		0.97	0.98	0.98	1.84	1.88	1.81	1.75	1.76	1.76
	H		0.98	0.97	0.98	1.39	1.09	1.48	1.96	1.94	1.94
H	L		0.97	0.97	0.97	1.95	1.88	1.91	1.79	1.80	1.84
	M		0.99	0.99	0.98	1.96	1.96	1.96	1.96	1.94	1.96
	H		1.00	1.00	0.99	1.88	1.38	1.37	1.96	1.96	1.96
Equal Fecundity												
L	L		1.19	1.19	1.19	1.19	1.19	1.19	1.24	1.25	1.25
	M		1.19	1.19	1.19	1.19	1.19	1.19	1.24	1.25	1.28
	H		0.46	0.35	0.96	0.90	1.20	1.21	1.27	1.30	1.29
M	L		0.52	1.46	1.48	1.48	1.48	1.48	1.50	1.50	1.51
	M		0.56	0.42	0.97	0.92	1.76	1.75	1.82	1.80	1.76
	H		0.57	0.64	0.98	0.95	0.96	1.53	0.99	1.94	1.96
H	L		0.61	1.96	0.98	1.82	1.83	1.80	1.83	1.88	1.83
	M		0.58	1.93	0.99	0.96	1.19	1.99	1.96	1.99	1.97
	H		0.75	0.91	0.99	0.99	0.97	2.00	0.99	1.88	1.99

Legend:
- viviparous matrotroph
- oviparous small eggs
- oviparous medium eggs
- oviparous large eggs
- coexist, matrotroph and oviparous with large eggs
- lecithotroph

Note: Food level (FL) and periodicity (PER) are in columns, and size-selective juvenile mortality (SSM) and juvenile mortality rate (JMR) are in rows. Numbers greater than 1 indicate oviparous reproductive modes, and numbers less than or equal to 1 indicate viviparous reproductive modes. Decimal portions of numbers indicate egg or neonate size relative to maximum; for example, 1.95 is an oviparous fish laying eggs that are 95% of the maximum size. For viviparous fish, neonate size is also proportional to matrotrophic nourishment; for example, 0.56 is a relatively lecithotrophic fish with neonates that are 56% of maximum size for viviparous fish, while 0.98 is a matrotrophic fish with neonates that are 98% of maximum size for viviparous fish.

ties with respect to energy allocated to growth and storage or reproduction, while size-selective juvenile mortality was more linked to reproductive output traits. Variation in food availability primarily sorted species by traits related to female maturation. Thus, community biomass was determined by a mix of these mechanisms that was particularly complex when fecundities did not vary among the different reproductive modes. These results are consistent with emerging views that focusing on single drivers of life-history evolution is overly simplistic (Reznick & Travis 1996; Reznick et al. 2002a) and that better models can be generated by considering life-history patterns within a community context (Day et al. 2002; Abrams 2003). Combining such models with carefully selected research may permit an innovative merger of community and life-history ecology (Chase & Knight 2003).

It is common for life-history ecologists to pick a small number of traits, such as fecundity and offspring size, to study because of their assumed linkage to fitness (termed fitness components). However, isolating components of fitness can be misleading because interactive effects of different components of the life history can yield fitness patterns that are not simply predictable from individual traits. In our simulations, final population size was more correlated with r, the generally preferred measure of fitness, than were any of our trait values. Ecosystem ecologists also study productivity (unit-mass/unit-area/unit-time) as indicative of emergent properties of complex systems driving dynamics. Biomass served that role for our model because both area and time were equal to 1. Further consideration of measuring such composite parameters, or their surrogates, may be productive grounds as we seek to reveal causal factors shaping communities (and populations), in addition to studying their components (species or traits).

Unlike in our model, environmental gradients do not vary independently in nature. Thus, some combinations of our environmental gradients may be uncommon or absent from most or all natural systems. For example, food level may tend to be correlated with the level of mortality from predation, particularly if food level for the trophic level simulated is correlated with food level throughout the local food web. Low-resource sites may have fewer predators of large individuals and, by way of relaxing a trophic cascade, more small predators feeding on small offspring or eggs. In aquatic systems, there is speculation that invertebrate predators tend to display less size selectivity than vertebrate ones (e.g., Brendonck et al. 2002). If the mix of vertebrate and invertebrate predators covaries with resource gradients, some combinations of treatment we explored may be absent or rare in a landscape. Thus, sampling studies will lack these cases, and descriptive statistical analyses (correlations or regressions) will suggest patterns that may not be representative of underlying causal linkages.

Covariance of environmental drivers leading to rarity or absence of some combinations is particularly problematic if only one environmental gradient is considered in the analysis. For example, in the case of a species pool with equal fecundity for all fish and a positive correlation between food level and level of size-selective juvenile mortality (holding periodicity of food level and absolute juvenile mortality at our medium level), a regression of egg or neonate size on food level would indicate a positive relationship and only oviparous fishes, whereas if food level and size-selective juvenile mortality covaried inversely, a study would conclude that food level was inversely correlated with offspring size and that there was a gradient with matrotrophs at low food and oviparous reproduction at high. Neither result provides a complete picture of causality. This poses a particular challenge to field-based experimental ecology that seeks general results, because it suggests that experimental manipulations must be considered at multiple, perhaps many, sites to trace out a general picture of causality. Experimental manipulations are typically challenging and thus spatial replication is limited by logistics. These results provide the unfortunate picture that large groups of researchers, probably requiring significant financial and logistic support, may be needed to uncover causal patterns in field conditions. Furthermore, since some combinations of environmental drivers may not be found in natural landscapes, we join the list of those pointing out the absolute necessity of controlled (possibly mesocosm or pond-scale) experiments (e.g., Hairston 1991), their limitations notwithstanding. Contrary to conventional recommendations, however, we point out that it may be necessary to consider combinations of environments not found in nature to disentangle covarying factors. Human impacts on natural environments are already doing just that by perturbing physical factors beyond "natural" limits and introducing species promiscuously, often with surprising results (Holling 1986).

Our model focused on parental care as the basis for defining functional species, but nonreproductive parameters are often the focus of niche-based models. For example, a trade-off between competitive ability and dispersal ability is often assumed in order to evaluate community assembly in a dynamic landscape. We constructed such a model to ask how eastern mosquitofish can coexist with other small fishes in the Everglades when they appear to be both superior competitors and superior dispersers (DeAngelis et al. 2005). That model indicated that an assemblage of inferior competitors can act collectively to swamp the one-on-one superiority of eastern mosquitofish and decrease their abundance to permit coexistence in a hydrologically dynamic landscape. The current model is not spatial and accounts for no aspects of life history beyond reproduction. Thus, it is most relevant to environments where dispersal ability does not limit colonization of local sites: for example, if a propagule rain from some mainland source is preventing stochastic local effects. This may be more generally relevant for long time periods and larger spatial scales.

Perhaps the most apparent contribution of our model is for predicting locations where a nonnative poeciliid fish could become invasive if it were introduced. Making such predictions remains a major challenge for the field of invasion ecology (Murphy et al. 2006). In such a case, either the species must be a good disperser or humans provide the dispersal and recurrent introductions (Rehage & Sih 2004). In southern Florida where we work, the pike killifish (*Belonesox belizanus*) has been introduced and it is well established regionally. It is lecithotrophic and produces relatively large neonates; female terminal size is also the largest of the family, and the fish is uniquely adapted in the family as a sit-and-wait piscivore. South Florida aquatic systems are naturally oligotrophic, but it is not clear whether that means that for this species food level is low relative to demand. A better use of our model results would be to consider more widely the range of environments where poeciliid fishes have been released and either failed or succeeded to expand and thrive. Thus, our local pike killifish example would be one point in a larger study. Ours is a deterministic model that can be used for generating null hypotheses for trait-based mapping of species on environments. Alternative models (neutral models) could be constructed based on random assignment of species to habitats (Temperton & Hobbs 2004). Also, nonadaptive models of evolution of parental care driven by parent-offspring conflicts (Crespi & Semeniuk 2004; O'Neill et al. 2007) could serve as the basis for predicting the distribution of functionally defined species. It seems reasonable that environmental conditions, such as resource availability, could tip the balance in a parent-offspring conflict and be used to drive a trait-environment matching assembly model.

9.6 Conclusions

All models are abstractions, and for all of this one's complexities, it is still a gross simplification of the diversity of fish reproductive modes and of the environments experienced by natural populations. Furthermore, functional traits such as dispersal ability are not considered here but may be equally important as reproductive biology in shaping community assembly under some circumstances (Chase & Leibold 2003; DeAngelis et al. 2005). However, we believe that this model provides provocative predictions worthy of future research. Foremost may be the clear illustration of interactive effects of diverse environmental gradients in sorting among functional species. This work supports other discussions that emphasize the necessity of multifactorial analysis of communities (and populations) along environmental gradients. Though this greatly increases the complexity and sampling/replication demands on ecological research, it is a key to analysis of causation of ecological patterns.

Because of their small size and amenity to laboratory research, poeciliid fishes have proven to be important models for the study of life-history ecology (see the chapters in part I). These characteristics and the battery of knowledge resulting from these studies make them an excellent model group for future studies of niche-based community assembly focusing on groups defined by functional traits. We have started such research by asking what fish communities could be invaded by a species that produces precocious offspring or, more precisely, greatly increases postfertilization survivorship at some cost for reproductive output. This effort made clear that more information is needed on early life survivorship curves. Although it is generally accepted that larger exogenously feeding offspring have lower early-life mortality risk, exactly how the size-number trade-off would be balanced by a competing oviparous and viviparous species remains unclear.

Acknowledgments

We wish to thank the editors of this book for inviting us to participate and causing us to consider the role of poeciliid fishes in community ecology. Also, thanks to Charles Goss for providing helpful comments on a penultimate draft of this chapter. This material was developed in collaboration with the Florida Coastal Everglades Long-Term Ecological Research program under Grant DEB-9910514. DeAngelis was supported by the U.S. Geological Survey's Southern Ecological Science Center. This paper is number 188 from the Tropical Biology Program at Florida International University and publication number 493 from the Southeast Environmental Research Center.

Chapter 10 Speciation and hybridization

Gil G. Rosenthal and Francisco J. García de León

10.1 Introduction

Ποίκιλος, n. ον, many-coloured, spotted, pied, dappled . . .
2. Changeful, varying,—intricate, riddling, ambiguous . . .
(Berry 1962)

In two separate monographs, Eigenmann (1894, 1909) described two new poeciliid species from northeastern South America. Based largely on striking differences in male color and morphology, these were assigned to different genera—*Acanthophacelus melanzonus* (Eigenmann 1909; fig. 10.1a–c) and *Poecilia vivipera* [*sic*] *parae* (Eigenmann 1894; fig. 10.1d). The males were later recognized as distinct morphs of a single species, *Poecilia parae* (Rosen & Bailey 1963), whose divergent appearance was largely driven by a single locus on the Y chromosome (Schartl et al., **chapter 24**; Lindholm et al. 2004): "*Acanthophacelus*" and "*Poecilia*" could have been half brothers.

The sheer exuberance of poeciliids—the diversity of form, color, ornamentation, and behavior—leads us inevitably to think of these fishes as speciose. But much of the variation is maintained as intraspecific—even intrapopulation—polymorphisms, and reproductive barriers, particularly postzygotic barriers, are often weak, with hybridization even occurring among distantly related species (reviewed in Schartl 2008; Coleman et al. 2009). Consider guppies (*Poecilia reticulata*), which have not speciated across their range on several Caribbean islands and mainland South America, despite a million years of geographic

isolation, strong sexual selection, divergent natural selection, and sexual conflict (Alexander et al. 2006).

Viviparity (Zeh & Zeh 2000), geographic isolation (Coyne & Orr 2004), ecological divergence (McKinnon et al. 2004), and sexual selection (Ritchie 2007) have all been cited as important factors in speciation. Poeciliids are viviparous and live on "islands" of freshwater surrounded by land, and they are a canonical system for studying sexual selection (Rios-Cardenas & Morris, **chapter 17**) and ecological divergence (Johnson & Bagley, **chapter 4**). Yet from the northern United States to Argentina, there are fewer than 300 known poeciliid species, compared with more than 500 species of cichlids in Lake Victoria and over 400 species in another group of American vertebrates, the lizard genus *Anolis* (Thorpe et al. 2008). Similarly, there are 66 darter (Percidae) species in Kentucky, according to that state's Department of Fish and Game, compared with just 34 poeciliids in Guatemala, according to *FishBase* (Froese & Pauly 2008). While the most speciose groups may represent extreme examples, it is noteworthy that there is not a single instance of an explosive radiation of livebearing fishes, despite the apparent convergence of several factors that would appear to promote high speciation rates. To paraphrase Felsenstein (1981), why are there so few kinds of poeciliids?

In this chapter, we focus on the mechanistic and ecological factors that contribute to, or hamper, speciation, reproductive isolation, and hybridization. Specific patterns of allopatric speciation (see box 10.1; and also, e.g., Lydeard et al. 1995; Mateos et al. 2002; Mateos 2005; Hrbek et al. 2007) are dependent more on the geological history of the

Figure 10.1 Variation in male morphs of *Poecilia parae*. (a) Red melanzona, (b) yellow melanzona, (c) blue melanzona, (d) parae, (e) immaculate, and (f) female. From Lindholm et al. 2004.

Americas than on the distinctive features of poeciliid fishes and are therefore beyond the scope of this book. Unisexual species of hybrid origin are the focus of Schlupp and Riesch's chapter in this volume (**chapter 5**).

In the following section, we examine the factors that influence genetic divergence among and within poeciliid populations. Poeciliids represent a convergence of multiple factors that should result in high speciation rates: they are highly polygynous, livebearing animals with strong sexual selection and chromosomal sex determination; and they are restricted to fresh and brackish waters. As we detail below, there is support for each of these as drivers of genetic incompatibility among populations. In section 10.3, we discuss premating barriers to gene flow: in poeciliids, these primarily take the form of robust behavioral biases toward conspecific sexual partners. We argue that postmating barriers (section 10.4), by contrast, are surprisingly weak in poeciliids. Sections 10.5 and 10.6 focus on hybrid zones, hybrids, and introgression. Here, we argue that these zones tend to arise due to a breakdown of behavioral barriers, and that hybridization is an important diversifying force. We conclude by returning to the original question: what does our current understanding of speciation and hybridization in poeciliids tell us about the relative paucity of poeciliid species?

10.2 Factors influencing genetic divergence

10.2.1 Geographic barriers to gene flow

Physical isolation is the primary mechanism for speciation (Dobzhansky 1937; Mayr 1942) and is often a prerequisite for at least initial genetic divergence in response to ecological or sexual selection (Ritchie 2007). Distributions of poeciliid species vary widely, and this is usually associated with barriers to dispersal. For example, the sailfin molly, *Poecilia latipinna*, is euryhaline (Gonzalez et al. 2005) and ranges from North Carolina, USA, through Veracruz,

---

Box 10.1   Modes of speciation

The origin of species (Darwin 1859) remains a central focus of evolutionary biology (Coyne & Orr 2004). Allopatric speciation, caused by the elimination or reduction of gene flow due to geographic barriers between populations, is relatively easy to demonstrate when barriers are evident. Since freshwater habitats are "islands" in terrestrial "oceans," allopatric speciation has often been invoked as a primary cause of diversity among freshwater fishes (Hugueny 1989). Indeed, major poeciliid clades are congruent with the geographic history of Mesoamerica and the Caribbean (Hrbek et al. 2007). Divergence in allopatry can occur without invoking selection, purely as a result of chance fixation of alleles in separate populations (Coyne & Orr 2004). Conditions for parapatric speciation, which involves restricted but ongoing gene flow among spatially connected species, and sympatric speciation among members of the same population are substantially more restrictive, since mating and recombination prevent the formation of genetically distinct subpopulations (reviewed in Bolnick & Fitzpatrick 2007). Parapatric and sympatric speciation require both ecological selection against hybrids and divergence of sexual communication systems (Arnegard & Kondrashov 2004).

Mexico. Schlupp et al. (2002) have argued that coastal species' ranges might be limited by nearshore ocean currents. At the other extreme, species at stream headwaters often inhabit an "island" of cool, piscivore-free water surrounded by land and unsuitable downstream habitats. The northern swordtail, *Xiphophorus continens*, for example, is restricted to the headwaters of the Río Ojo Frio in San Luis Potosí (Rauchenberger et al. 1990).

Interestingly, a single species can often have a disjunct distribution, with populations separated by long-standing physical barriers, yet still retain the capacity to produce viable offspring. This is the case with guppies, mentioned above, where neither Trinidadian mountain ranges nor the Paria Gulf have led to speciation among disjunct populations (but see section 10.4). Populations of *Xiphophorus malinche*, restricted to headwaters of the Sierra Madre Oriental of Hidalgo, Mexico, share identical polymorphisms at mitochondrial control region and several nuclear intron loci (Culumber et al. 2011). There is therefore little genetic divergence among these populations despite the presence of apparently formidable barriers to dispersal, including the main ridge of the Sierra Madre Oriental and unsuitably warm habitat downstream. Recent climate fluctuations, recent uplift, low-lying catchment boundaries, or bird-borne dispersal could account for connectivity across apparent physical barriers, although most of these alternatives have yet to be thoroughly investigated.

10.2.2 Ecological selection

Poeciliids are a canonical model for the evolution of life histories (see the chapters in **part I** of this volume), sexual ornaments (Brooks & Postma, **chapter 23**), and other traits under divergent ecological conditions. Variation in the physical environment and in predator regimes selects for sharply different suites of phenotypes. Surprisingly, only a handful of studies have addressed ecological speciation (Schluter 2009) in poeciliids. Two studies have shown partial reproductive isolation associated with among-population differences in predation risk. In guppies (Schwartz & Hendry 2007) and in *Gambusia hubbsi* (Langerhans et al. 2007), intraspecific ecological divergence is associated with divergent mate preferences, leading the authors of these studies to suggest a role for ecological speciation. Poeser (1998) suggested that ecological character displacement might account for the pattern of geographic variation in jaw tooth morphology among six widely distributed *Poecilia* species. Evidence for ecological divergence of cave- and surface-adapted mollies is provided in Tobler and Plath (**chapter 11**).

It should be noted that there is abundant evidence for predictable phenotypic divergence in response to ecological conditions (*Gambusia* spp.: Langerhans & DeWitt 2004; Langerhans et al. 2004; guppies: Hendry et al. 2006; *Brachyrhaphis rhabdophora*: Johnson 2001a). In at least a few stickleback populations, comparable ecological divergence leads to sexual isolation and ultimately speciation (Boughman et al. 2005); the effects in poeciliids are much more modest. Possible reasons for limited ecological speciation are discussed in section 10.7.

10.2.3 Sexual selection

Poeciliids are a classic model system in the study of sexual selection (see Brooks & Postma, **chapter 23**, and **part IV** on sexual selection in this volume). Sexual selection plays two critical roles in speciation. First, it involves a mechanism—mate choice—that plays an integral role in the formation and maintenance of reproductive isolation via assortative mating. Second, divergent selection on secondary sexual traits can enhance genetic divergence. The vast literature on conspecific mating preferences in poeciliids will be discussed in section 10.3. Alexander and Breden (2004) argued that guppies from Cumaná represent a case of incipient speciation due to sexual selection: females show highly divergent mating preferences corresponding to strikingly different male phenotypes among populations.

There was no evidence for postzygotic incompatibilities between Cumaná and other populations (Alexander & Breden 2004). The interaction of sexual selection with conspecific mate recognition is discussed in section 10.3.2; sexual selection in natural hybrid zones is treated in section 10.5.2.

Heritable variation among females in mating preferences can provide a mechanism for *sympatric speciation* (box 10.1; Ritchie 2007). Female *Xiphophorus cortezi* show polymorphic preferences for vertical bars, which are in turn polymorphic in males (Morris et al. 2003). However, these preferences may be a function of female size, as in the closely related *X. malinche* (Morris et al. 2006), and therefore are likely ontogenetic in nature. Guppies show little additive genetic variance in female mating preferences (Brooks & Endler 2001b; Hall et al. 2004), so there may be little opportunity for divergent trait-preference associations within a population.

10.2.4 Sex determination

Poeciliids show considerable variation in genetic sex determination mechanisms (Schartl et al., **chapter 24**; Volff & Schartl 2001), ranging from XX/XY (male heterogametic) to ZZ/ZW (female heterogametic) to complex polygenic systems. Chromosomal divergence can be a barrier to gene flow. Eastern mosquitofish, for example, are ZZ/ZW, while the parapatric western mosquitofish are XX/XY; the presence of the heterologous chromosome pair causes hybrid inviability (Black & Howell 1979). A subsequent study by Scribner (1993), however, found no evidence of hybrid inviability in crosses between the two species.

Haldane's rule indicates that hybrids of the heterogametic sex should suffer greater inviability or sterility, and there are large effects of the homogametic chromosome on reduced hybrid fitness (reviewed in Presgraves 2008). Theoretical models predict that the mode of sex determination should influence the evolution of elaborate traits (Kirkpatrick & Hall 2004) and, by extension, the likelihood of speciation via reinforcement of mating preferences (Hall & Kirkpatrick 2006). A survey of sex determination mechanisms across actinopterygian fishes (including poeciliids), however, revealed no effect of sex determination mode on male trait elaboration (Mank et al. 2006). In poeciliids, ZZ/ZW fish, including mosquitofish and mollies, are, if anything, less elaborate than XX/XY guppies, platies, and swordtails.

10.2.5 Viviparity-driven conflict

Zeh and Zeh's (2000) "viviparity-driven conflict hypothesis" proposed that viviparous species should be particularly prone to postzygotic isolation. Viviparity provides an important theater for sexual conflict (see Magurran, **chap-**

ter 19). This is because mothers, fathers, and offspring have different fitness optima with regard to maternal provisioning of a given embryo. Roughly, selection favors fathers and offspring that maximize onetime provisioning and mothers that optimize lifetime allocation. Intersexual and intergenerational conflict over maternal allocation therefore results in antagonistic coevolution (Rice & Holland 1997), which in turn leads to interpopulation incompatibilities in fetomaternal interactions. Furthermore, if females practice multiple mating, this should favor an aggressive paternal genome—that is, postzygotic competition among embryos with different fathers (Zeh & Zeh 2000).

Poeciliids provide an opportunity to test the effects of both multiple mating and maternal provisioning on postzygotic isolation. There is substantial variation within and among species in the degree of multiple mating: for *P. reticulata*, see Kelly et al. 1999; Hain & Neff 2007; for *P. latipinna*, see Trexler et al. 1997; for *Gambusia holbrooki*, see Zane et al. 1999; for *Heterandria formosa*, see Soucy & Travis 2003; for *Xiphophorus multilineatus*, see Luo et al. 2005 (see also table 18.1 in Evans & Pilastro, **chapter 18**). Maternal provisioning is similarly variable. Pseudoplacentas, associated with high levels of maternal provisioning, have independently evolved three times within the genus *Poeciliopsis* (Reznick et al. 2002b) and once in the lineage leading to *Xenodexia* (Reznick et al. 2007a). Pires et al. (2007) suggested that antagonistic coevolution could explain genetic variation in maternal provisioning in the placental *Poeciliopsis prolifica*.

To date, there is limited evidence to support the viviparity-driven conflict hypothesis as a widespread mechanism of speciation in poeciliids. Intriguingly, two of the three progenitors of the trihybrid, clonal *Poeciliopsis* (Mateos & Vrijenhoek 2002; see also Schlupp & Riesch, **chapter 5**) are highly matrotrophic, placental species, suggesting that high levels of maternal provisioning do not automatically yield unviable progeny. It would be straightforward to test levels of postzygotic isolation (Mendelson et al. 2007) in species varying in maternal provisioning and multiple mating. In section 10.4, we discuss the puzzling dearth of postzygotic barriers in poeciliids.

10.3 Premating isolating mechanisms

The range of many *Xiphophorus* forms overlaps; in some waters several species may live alongside each other. In spite of this, the many thousands of specimens collected in the wild have yielded not a single hybrid. Apparently, therefore, the isolating mechanisms preventing hybridization are fully efficient in the wild, where the species live in shoals. This is all the more astonishing because, under favourable conditions, the different species and subspecies interbreed very readily in

the laboratory and produce viable, fertile offspring. (Jacobs 1971, 392)

Perhaps the greatest contribution of poeciliid biology to the Modern Evolutionary Synthesis (Mayr 1982) was the identification of robust prezygotic isolating mechanisms (Dobzhansky 1937; Mayr 1942) in the form of sexual behavior (Clark et al. 1954). In one of the first experimental studies outside *Drosophila* on the problem of sexual isolation, Eugenie Clark and colleagues (1954) showed that mating behavior was the primary barrier to interspecies hybridization between the southern platyfish, *Xiphophorus maculatus*, and the distantly related, sympatric green swordtail, *Xiphophorus hellerii*. In controlled observations, individuals strongly preferred to interact with opposite-sex conspecifics over heterospecifics. When heterospecific individuals were paired in isolation, however, production of viable hybrids often ensued. Indeed, platyfish-swordtail crosses were already routine in the aquarium hobby and had been used to establish long-standing strains in genetic research (Gordon 1937; Kazianis 2006; Rosenthal & García de León 2006). Behavioral mechanisms were thus likely to play a determining role in maintaining reproductive isolation between these sympatric species.

Since Clark et al.'s (1954) study, many others have demonstrated behavioral preferences for conspecifics in a host of poeciliid species (Liley 1966; Crapon de Caprona & Ryan 1990; Hurt & Hedrick 2003; Kozak et al. 2008). Reviewing them all is beyond the scope of this chapter. We focus instead on four aspects of conspecific mate preferences critical to both speciation and hybridization.

10.3.1 Reproductive character displacement and reinforcement

If hybrids have reduced fitness, selection should favor mating preferences that minimize the risk of mating with heterospecifics (reproductive character displacement) or conspecifics from a genetically divergent subpopulation (reinforcement); as a consequence, selection will also favor mating signals that differentiate the two groups.

A critical test of both reproductive character displacement and reinforcement is whether mating preferences and mating signals are more divergent in sympatry than in allopatry. Given the amount of work that has been done on conspecific mate preferences and on geographic variation in poeciliids, it is surprising how few studies have performed this critical comparison. Male sailfin mollies (*P. latipinna*) from sympatric populations discriminate more strongly against unisexual *Poecilia formosa* (Schlupp & Riesch, **chapter 5**; Ryan et al. 1996; Gabor & Ryan 2001; Gabor et al. 2005).

If mate choice depends on experience with a particular phenotypic distribution, this should influence the propensity to mate with heterospecifics. For example, Walling et al. (2008) recently showed that preferences for sexually dimorphic "sword" ornaments in female green swordtails can be reversed by early experience with swordless males. Similarly, female northern swordtails, *Xiphophorus birchmanni*, exposed to *X. malinche* males for eight days shift their mating preferences toward heterospecifics (M. Verzijden, Z. W. Culumber, and G. G. Rosenthal, unpublished data). Learning, however, can evolve. Male guppies from allopatric populations must learn to avoid attempting to mate with *Poecilia picta*, but this behavior is innate in sympatric males (Magurran & Ramnarine 2005).

The opportunity to study reinforcement and reproductive character displacement arises from the fact that poeciliids are often distributed along altitudinal gradients within drainages. Gene flow is likely to be biased in an upstream-to-downstream direction, as has been shown in guppies (Crispo et al. 2006). Fish from downstream populations should thus experience stronger selection to avoid hybridization, and conspecific mate preferences should be correspondingly stronger. Female *X. birchmanni* strongly prefer conspecific visual cues over those of the highland *X. malinche* (Fisher et al. 2006; Wong & Rosenthal 2006). *X. malinche* females, meanwhile, fail to show visual preferences for conspecifics (G. G. Rosenthal, unpublished data). Within a species, however, female guppies from downstream, high-predation populations show behavioral preferences for more-ornamented males from upstream locations, although low-predation males garner more matings through sneak copulations (Magellan & Magurran 2007b). Further attention needs to focus on places where congeners co-occur; for example, the platyfish *X. maculatus* and *Xiphophorus variatus* are each sympatric with several swordtail species, setting the stage for replicated studies of preference evolution in the presence of heterospecifics. A particularly intriguing case concerns the closely related *Xiphophorus montezumae* and *X. continens*. Males of the latter species are small and have secondarily lost ornamentation and courtship behavior, yet females retain (albeit weaker) preferences for large size and vertical bars (Morris et al. 2005). Furthermore, female *X. continens* prefer the olfactory cues of *X. montezumae* (McLennan & Ryan 2008). It remains to be seen which prezygotic mechanisms, if any, apparently prevent hybridization between these sympatric species.

10.3.2 Asymmetric mating preferences

Although seldom studied explicitly in the context of prezygotic isolation, mating-preference asymmetries—where females of one species are more likely to mate with males of the other species than the reverse—appear to be widespread in poeciliids. Mate choice is covered by Rios-Cardenas and Morris (**chapter 17**). Female poeciliids often have directional biases for exaggerated traits (Ryan & Keddy-Hector 1992), although several studies have shown that preferences for extreme traits in poeciliids can be secondarily weakened (Basolo 1998), lost (Rosenthal et al. 2002), or reversed (Wong & Rosenthal 2006; Fisher & Rosenthal 2007). In *Xiphophorus*, the male sword ornament evolved in response to a preexisting female bias (Basolo 1990, 1995a) for males of large apparent size (Rosenthal & Evans 1998); a similar mechanism may explain the attractiveness of the large dorsal fin in some mollies (MacLaren & Rowland 2006). Given this, it is not surprising that females often prefer heterospecific cues, even when conspecific males have never borne the trait (*Priapella olmecae*, Basolo 1995a; *X. maculatus*, Basolo 1990; *Poecilia mexicana*, MacLaren & Rowland 2006) or when the trait has been secondarily lost in conspecific males (*X. continens*, Morris et al. 2005; *Xiphophorus pygmaeus*, Ryan & Wagner 1987). As we discuss in section 10.5, such preference asymmetries can have important consequences for the directionality and occurrence of hybridization.

10.3.3 Conflicts between mate choice and species recognition

Given that females often have strong directional preferences for male traits, can these interfere with avoidance of heterospecifics? In several species of molly (genus *Poecilia*) a directional preference for large overall size (lateral projected area) overrides, or cancels out, any visual preference for conspecific males (Ptacek 1998; Kozak et al. 2008). In other systems, preferences along a univariate axis force a compromise between selecting among conspecifics and avoiding heterospecifics (pierid butterflies, Wiernasz & Kingsolver 1992; spadefoot toads, Pfennig 2000). In the pygmy swordtail, *X. pygmaeus*, large size, ornamentation (swords and vertical bars), and courtship have all been secondarily lost in males. In some populations, female *X. pygmaeus* prefer the visual cues of male *Xiphophorus nigrensis* (Ryan & Wagner 1987) and *X. cortezi* (Hankison & Morris 2002) over conspecifics. Preference for large size in *X. pygmaeus* varies among populations (Morris et al. 1996). Females from populations where the preference had been lost showed stronger visual preferences for conspecific males over *X. nigrensis* and *X. malinche*; further, there was a strong negative correlation, among individuals, between the strength of species recognition and the strength of the preference for large size. In other words, females who strongly preferred the largest males were more likely to prefer heterospecifics (Rosenthal 2000).

The trade-off between avoiding heterospecifics and attending to attractive conspecifics can be avoided by paying attention to additional cues. Indeed, female *X. pygmaeus* prefer olfactory cues of conspecifics and prefer males that lack vertical bars. When multiple cues are presented in tandem, females prefer conspecifics (Crapon de Caprona & Ryan 1990; Hankison & Morris 2003).

10.3.4 Sensory modality and species recognition: chemical signals as prezygotic isolating mechanisms

As discussed in the preceding two sections, preferences for *visual* traits are often asymmetric, with females preferring trait values outside the range of variation for conspecifics. With *chemical* traits, females almost always prefer conspecific cues (*P. mexicana*, Plath et al. 2007c; *X. pygmaeus*, Hankison & Morris 2003; *X. cortezi*, McLennan & Ryan 1997; *X. nigrensis*, McLennan & Ryan 1999; *X. birchmanni*, Fisher et al. 2006; *X. malinche*, G. G. Rosenthal & H. S. Fisher, unpublished data); the exception is *X. continens*, mentioned above. Sensory mechanisms provide a clue as to why olfactory traits should generally be more species specific. Increased size, brightness, or motion will all increase conspicuousness across species, simply by increasing overall stimulation. Olfactory traits, however, may be more sensitive to the properties of the local sensory environment, since these traits are likely to be directly influenced by water chemistry. In poeciliids (Fisher et al. 2006) and other systems (Brown 2002; Heuschele & Candolin 2007) the chemistry of the aquatic environment affects the efficacy of chemical communication. Female *X. malinche* respond preferentially to male olfactory cues when these are mixed with water from males' native stream, suggesting that olfactory systems and olfactory signals may be coevolving in response to environmental constraints (see Coleman, **chapter 7**).

However, neither open-ended visual preferences nor tightly coupled chemical preferences appear to be universal. Females can prefer less conspicuous visual traits (Morris 1998; Wong & Rosenthal 2006), suggesting that sensory stimulation and stimulus evaluation are separate processes. Conversely, male *X. birchmanni* prefer the olfactory cues of *X. malinche* over those of conspecific females (Wong et al. 2005), although this may reflect an open-ended preference for steroid metabolites.

10.4 Postmating isolating mechanisms

Postzygotic isolating mechanisms seem to be relatively modest in poeciliids. The X-*mrk* oncogene in *Xiphophorus* (Schartl & Meierjohann, **chapter 26**) has been proposed as one of the few documented examples of a "speciation gene" arising from Dobzhansky-Müller incompatibilities (Noor

2003; Orr et al. 2004). In a recent review, Schartl (2008) argued on several counts that X-*mrk* is unlikely to play an important role in postzygotic isolation. This is because malignant oncogenetic effects are present in only a fraction of laboratory F_2 hybrids; F_2 hybrids show heterosis with respect to other traits; and X-*mrk* incompatibilities are found in only a subset of sympatric species pairs. Indeed, platyfish-swordtail hybrids are fertile, despite divergence tests estimated as greater than 10 million years (Coleman et al. 2009). Bolnick and Near (2005) reviewed divergence times in a range of taxa. While the centrarchid fishes, on which they focused, retained viability for an unusually long 25 million years, other taxa (birds, *Drosophila*, butterflies, anurans) reached total hybrid inviability in 5.5 million years or less. At least some poeciliids, therefore, retain hybrid viability substantially longer than other well-studied groups.

Although total inviability is slow to evolve, some poeciliid taxa show more limited viability effects over shorter time scales. In guppies, male F_1 hybrids between populations from different drainages show reduced fertility, and F_2 hybrids show reduced embryo viability, brood size, and sperm counts relative to parentals (Russell & Magurran 2006). With their XX/XY sex determination, guppies thus follow Haldane's rule (1922) in that fitness effects are greater on males (Russell & Magurran 2006). Postcopulatory but prezygotic sexual selection may also serve as an isolating mechanism; following artificial insemination with sperm from males from two drainages, heteropopulation sperm were at a disadvantage in sperm competition (Ludlow & Magurran 2006; see also Evans & Pilastro, **chapter 18**, section 18.4.4). Given the prevalence of forced copulations by multiple males in guppies, divergence with respect to fertilization success may arise from antagonistic coevolution between sperm and female reproductive mechanisms (e.g., Holland & Rice 1999). Russell and Magurran (2006) argued that postcopulatory isolating mechanisms were likely to be stronger among these populations than prezygotic isolation via mate choice.

In contrast, two closely related, allopatric *Poeciliopsis*, *P. occidentalis* and *P. sonoriensis*, show strong assortative mating in addition to hybrid inviability (Hurt et al. 2004). Hurt and Hedrick (2003) demonstrated that hybrids of these two species showed reduced fitness. Furthermore, sex ratios were male biased. In the absence of information about sex-determining mechanisms, the authors suggested that if females are the heterogametic sex, the male-biased sex ratio is consistent with Haldane's rule. Alternatively, if males are the heterogametic sex, the authors proposed a role for "incompatibilities between the paternal X chromosome and the maternal cytoplasm" (Hurt & Hedrick 2003, 2839). The *P. occidentalis*–*P. sonoriensis* system is also the only instance, in poeciliids, where pre- and postzygotic isolation have been explicitly compared. Hurt et al. (2005) went on to examine cytonuclear signatures in experimental populations. They found that most of the deviation from the null expectation of random mating could be explained by assortative mating rather than by postzygotic factors. O'Neill et al. (2007), however, recently showed that a gene involved in *Poeciliopsis* placental function was undergoing strong positive selection as a result of viviparity-driven conflict. Postzygotic isolation in *Poeciliopsis* clearly deserves further attention.

It is perhaps not a coincidence that the strongest evidence for postzygotic isolation comes from a system with low levels of sexual dimorphism, the sympatric eastern and western mosquitofish. As discussed in section 10.2.4, some F_1 hybrids may show chromosomal incompatibilities. In addition, F_2 and backcross progeny show hybrid breakdown with respect to viability (Reznick 1981).

Overall, postzygotic isolating mechanisms in poeciliids appear to be relatively weak. Closely related species kept in aquaria often interbreed with little encouragement (Kazianis 2006), and as in cichlids (Seehausen et al. 1997), human disturbance to the environment is sufficient to cause rampant hybridization in at least one species pair (Fisher et al. 2006). As recognized by Clark et al. (1954) and predecessors, the primary barrier to hybridization in poeciliids is often behavioral. In the next section, we will discuss how the breakdown of premating isolating mechanisms can lead to hybridization in the wild.

10.5 Mate choice and evolutionary genetics in natural hybrid zones

Given the ubiquity of interspecific crosses in the laboratory and in the aquarium hobby, hybrids are surprisingly rare in the wild. The strikingly transgressive expression of sexually dimorphic traits in artificial hybrid males makes it unlikely, for example, that natural hybridization of any frequency between *X. maculatus* and *X. hellerii* would have gone undetected by Clark et al. (1954). Some "natural" hybrids arise from introductions of exotic species. The Monterrey platyfish *Xiphophorus couchianus* has effectively been lost due to hybridization with *X. hellerii–X. maculatus* hybrids from the aquarium trade (Kallman & Kazianis 2006). The threatened Clear Creek *Gambusia heterochir* hybridizes with the introduced *Gambusia affinis* (Hubbs 1971), but there is little evidence for introgression into *G. heterochir* (Davis et al. 2006).

Hybridization between native species has been documented in only a handful of cases. Hybridization has been documented in two separate instances within mollies. Kittell et al. (2005) documented four individual male mollies from

the Yucatán Peninsula that were found to be sailfin-shortfin molly hybrids. A single male *Poecilia butleri–Poecilia sphenops* hybrid was found in the Río Papagayo, Guerrero, Mexico (Schultz & Miller 1971). Since females of closely related species are often difficult to distinguish, it is not surprising that male hybrids are more frequently documented; however, if male hybrids are less likely to be viable (e.g., due to Haldane's rule), this may give a misleading estimate of the rarity of hybridization.

There are few known cases of naturally sympatric or parapatric poeciliids forming populations with high frequencies of hybrids. In the upper Río Lacantún and Río Salinas drainages of Guatemala, Rosen (1979) found morphological intergrades both between *X. alvarezi* and *X. hellerii*, and between *Heterandria obliqua* and *H. bimaculata*. Meyer (1983) argued, again from morphological evidence alone, that *X. kosszanderi* populations were in fact natural hybrids between northern platyfish *X. variatus* and *X. xiphidium*, and *X. roseni* were hybrids between *X. variatus* and *X. couchianus*. Both molecular and morphological evidence show that northern platyfish and *X. nezahualcoyotl* hybridize (Kallman & Kazianis 2006), but as Meyer (1983) pointed out, *X. variatus* coexist with several other congeners (*X. birchmanni, X. cortezi,* and *X. pygmaeus*) without hybridizing. *Xiphophorus birchmanni* and *X. cortezi* also hybridize in the wild (Schartl 2008). With the exception of parthenogens (Schlupp & Riesch, **chapter 5**), two natural hybrid systems have been the focus of more intensive study: *G. affinis–G. holbrooki* and *X. birchmanni–X. malinche*.

10.5.1 Life-history evolution and *Gambusia affinis–Gambusia holbrooki* hybrid zones

The eastern mosquitofish, *G. holbrooki*, hybridizes with the western mosquitofish, *G. affinis*, throughout much of the southeastern United States (Wooten et al. 1988). The grandchildren of *affinis* females and *holbrooki* males suffer hybrid breakdown (Reznick 1981). The species show marked differences in life-history traits, with parental and hybrid offspring of *holbrooki* females exhibiting higher embryo weight, faster growth rates, and shorter time to maturity (Scribner 1993). Within natural hybrid zones, therefore, selection appears to favor *holbrooki* genotypes in most environments, and *holbrooki* genes are introgressing into *affinis* populations (Scribner 1993; Scribner & Avise 1993). Intriguingly, Scribner and Avise (1994) similarly found directional selection favoring *holbrooki* genotypes in seminatural populations established in Biosphere 2. It is surprising that there has been little recent work on this potentially fascinating system in the evolutionary genetics of life-history traits.

10.5.2 Mate choice and *Xiphophorus birchmanni–Xiphophorus malinche* hybrid zones

Mate choice plays a determining role in the origin and subsequent dynamics of hybrid zones among *X. birchmanni* and *X. malinche* (fig. 10.2). The two species are members of the nine-species monophyletic northern swordtail clade. Phylogenetic hypotheses differ on their relationship: trees based on phenotypic data place them as sister species, while sequence data place them more distant (Marcus & McCune 1999; Gutierrez-Rodriguez et al. 2007). *Xiphophorus birchmanni* is broadly distributed over lowland areas (elevation 161–300 m) of the southern Río Pánuco drainage of the Atlantic slope of central Mexico, in the states of Hidalgo and Veracruz (Rauchenberger et al. 1990), while pure *X. malinche* are known only from six highland sites in Hidalgo (658–1499 m). Hybrid populations are found at intermediate elevations (272–1188 m).

Xiphophorus birchmanni and *X. malinche* are highly divergent in sexually dimorphic visual traits (Rauchenberger et al. 1990; Rosenthal et al. 2003; fig. 10.3). *Xiphophorus malinche* is an archetypical swordtail (fig. 10.2), with a long caudal extension, a relatively short dorsal fin, and irregular vertical bars. *Xiphophorus birchmanni* has a prominent cephalic hump and a high dorsal fin, lacks pigmented swords altogether, and expresses a parallel series of vertical bars.

Breakdown of prezygotic isolating mechanisms appears to be responsible for the recent (since the 1980s; Rauchenberger et al. 1990; Fisher et al. 2006) origin of these hybrid zones. Hybridization appears to be the result of recent human interference via organic pollution (see Stockwell & Henkanaththegedara, chapter 12). Wild-caught female *X. birchmanni* showed a strong preference for conspecific chemical cues when tested in clean highland water but failed

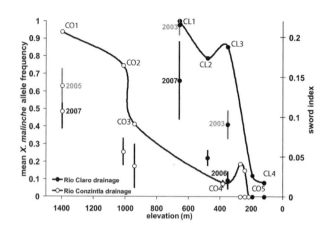

Figure 10.3 Clinal variation in genotype and a sexually selected trait in two *Xiphophorus birchmanni–Xiphophorus malinche* hybrid zones. Solid lines show mean frequency of *X. malinche* alleles at one mitochondrial and four unlinked nuclear loci; error bars show mean ± SE of sword index (sword extension length/standard length). Decreases in sword index with corresponding years are shown for three populations. Filled circles: Río Claro drainage; open circles: Río Conzintla drainage.

to discriminate when tested in stream water subject to sewage effluent and agricultural runoff (Fisher et al. 2006).

10.5.2.1 Phylogeography and hybrid-zone structure. Phylogeographic analysis using sequence data from one mitochondrial gene and four intron loci on separate linkage groups of the *Xiphophorus* genome (Walter et al. 2004) suggests that hybridization has occurred independently in six drainages constituting separate upstream-to-downstream gradients. *Xiphophorus malinche* genotypes are shared among highland populations in disjunct drainages separated by mountain ranges. Despite these geographic barriers, *X. malinche* and *X. birchmanni* are reciprocally monophyletic, suggesting independent hybridization events between the two species (Culumber et al. 2011).

As is typical for hybrid zones across an environmental gradient (Barton & Hewitt 1985), each hybrid zone shows clinal variation, in this case from *X. malinche* alleles upstream to *X. birchmanni* downstream (fig. 10.3). In general, wild-caught *malinche-birchmanni* natural hybrids are as well adapted to their local environment as parentals. For several ecologically important traits, the performance of both parentals and hybrids covaries with habitat properties—for example, thermal tolerance and heat-shock protein gene expression with temperature variation (Z. W. Culumber, M. Tobler, S. W. Coleman, & G. G. Rosenthal, unpublished data), low-light sensitivity with light regime (S. W. Coleman, B. Perkins, & G. G. Rosenthal, unpublished data), and boldness (J. B. Johnson & G. G. Rosenthal, unpublished data) and swimming performance (K. Kruesi, G. Alcaraz, & G. G. Rosenthal, unpublished data) with habitat structure.

With respect to natural selection, *Xiphophorus* hybrid

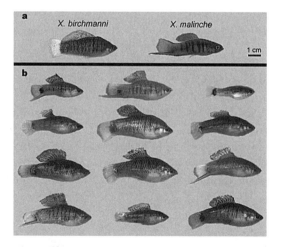

Figure 10.2 *Xiphophorus birchmanni–Xiphophorus malinche* hybrids. (a) Males from parental species, *X. birchmanni* and *X. malinche*. (b) Natural hybrids from Chahuaco, Río Calnali, Hidalgo, Mexico. From Fisher et al. 2006.

zones thus appear to conform to a model of "bounded hybrid superiority" (Moore 1977; Barton & Hewitt 1985). Hybrids appear to outperform or at least equal parentals at intermediate elevations. The major differences between parentals and hybrids involve traits directly relevant to the process of mate choice. Hybrid males show transgressive expression of sexual phenotypes, with both individual trait values and trait combinations outside the range of parentals (fig. 10.2; Rosenthal et al. 2003). In turn, hybrid visual systems show reduced expression of cone-opsin genes and, correspondingly, reduced discrimination of color contrast (S. W. Coleman, B. Perkins, & G. G. Rosenthal, unpublished data). Both signal diversity and receiver permissiveness are therefore increased in hybrids.

10.5.2.2 Mate-choice mechanisms and hybridization. As discussed in section 10.3.2, X. birchmanni and X. malinche exhibit strong preference asymmetries. Female X. birchmanni prefer visual cues of conspecific males, including a disdain for the sword ornament (Wong & Rosenthal 2006). Female X. malinche are indifferent with respect to visual cues (G. G. Rosenthal, unpublished data). Meanwhile, male X. birchmanni prefer olfactory cues of X. malinche females (Wong et al. 2005). Thus, two predictions regarding hybrid populations can be made: first, X. birchmanni traits should introgress into X. malinche populations; and second, there should be a signature of X. malinche female–X. birchmanni male matings among hybrids. The first of these predictions is supported; the second is not. Swords have been decreasing in frequency since 1997 and introgressing against X. malinche genetic backgrounds (fig. 10.3; Z. W. Culumber and G. G. Rosenthal, unpublished data), which is consistent with female mate choice in X. birchmanni selecting against swords. However, cytonuclear signatures and backcross analyses reveal no directionality to hybridization. Intriguingly, patterns of assortative mating vary widely among populations: in some populations, the two parental species and hybrids are reproductively isolated, while in others, mating is indistinguishable from random expectations (Z. W. Culumber and G. G. Rosenthal, unpublished data).

The effects of hybridization on sexual signals can have far-reaching evolutionary consequences. Some hybrid males are more attractive to parental females than males of their own species; this may be due to novel phenotypic combinations in hybrids (Rosenthal et al. 2003) breaking up antagonistic correlations with respect to mate preference (Fisher et al. 2009) or to a more general preference for novel stimuli (Burley & Symanski 1998). While F_1 hybrids are extremely rare in the hybrid zones, consistent with the notion that conspecific mate recognition is only episodically interrupted by anthropogenic disturbance, putative backcrosses to parentals are common. Attractive male trait

combinations might therefore compensate for deleterious fitness effects, as with Y-linked ornamentation in guppies (Brooks 2000), resulting in rampant hybridization.

10.6 Hybrid speciation and introgression

The gynogenetic Poeciliopsis trihybrid, which carries three haploid genomes from sexual species, and the dihybrid Amazon molly are discussed by Schlupp and Riesch (chapter 5). Xiphophorus clemenciae, by contrast, is a sexual species of hybrid platyfish-swordtail origin. Xiphophorus clemenciae is nested within the platyfish clade in mitochondrial DNA–based phylogenies, and within the southern swordtails when nuclear genes are used; and X. clemenciae females prefer males with the long swords characteristic of green swordtails. Meyer et al. (2006) argued that this indicated hybrid origin, possibly arising via mate choice favoring conspicuous, heterospecific males. Hybrids of X. birchmanni and X. malinche, which are reproductively isolated from parentals in some populations, may represent another case of hybrid speciation (section 10.5.2.2).

Introgressive hybridization, whereby a small portion of a heterospecific genome is incorporated as a result of an initial hybridization event, followed by recurrent backcrossing, is likely to be a far more widespread phenomenon in poeciliids. Indeed, given the fragility of barriers to hybridization, reticulate evolution (Arnold 2006) may be widespread in poeciliids and may provide a more general explanation for incongruent topologies among phylogenies.

10.7 Conclusions and future directions

10.7.1 Postzygotic versus prezygotic barriers

In general, many poeciliid species pairs are capable of producing viable lines of F_2 and later offspring, yet documented cases of natural hybridization are rare. This appears to be due to the strength of premating isolating mechanisms relative to postmating mechanisms, although more studies along the lines of Mendelson et al.'s (2007) study of darters, which directly compared the two mechanisms, are highly desirable in poeciliids. Russell (2003) used a comparative analysis to argue that hybrid fitness in fishes suffers less than in other taxa, although his analysis considered only F_1 crosses. It is remarkable that there is no evidence of strong hybrid breakdown in hybrids of such divergent crosses as swordtails and platyfishes (Meyer et al. 2006), particularly given the opportunity for strong divergence of mechanisms involved in embryonic provisioning (Zeh & Zeh 2000) and the variation in maternal provisioning among and within taxa (Pires et al. 2007; Marsh-Matthews, **chapter 2**). It would be fruitful to investigate the evolutionary genetics of parental care in poeciliids. It may be that functional constraints on critical aspects of provisioning limit the possibility for genetic divergence (but see O'Neill et al. 2007). This may also shed light on the fact that hybridization in two poeciliid lineages has created unisexual hybrid species (Schlupp & Riesch, **chapter 5**).

The importance of prezygotic barriers to current isolation between sympatric species does not imply that these barriers were important in the process of speciation itself. Biogeography suggests that most poeciliid species diverge in allopatry. Sexual signaling systems then diverge, either arbitrarily (Pomiankowski & Iwasa 1998) or in response to environmental differences (Endler 1992). Indeed, experimental work with swordtails suggests that the local chemical environment may drive the sensory tuning of chemical signals and receptors (G. G. Rosenthal et al., unpublished data). Behavioral barriers between species may thus be an effect of signal-receiver divergence in allopatry.

The evolution of hybrid zones depends on isolating mechanisms in two fundamental ways. First, with prezygotic barriers to hybridization—specifically, male signals and female preferences—the phenotypes of hybrids may in fact reinforce the process of hybridization. For example, female *X. birchmanni* show antagonistic preferences for correlated male traits, preferring males with large bodies but small dorsal fins; the net direction of preference is therefore orthogonal to that of male trait variation (Fisher & Rosenthal 2007). The allometry of dorsal fin size in hybrids is much weaker, and some hybrid males are more attractive to parental females (Fisher et al. 2009). Novel trait and preference phenotypes in hybrids may thus increase the probability of hybridization.

The second critical aspect of prezygotic barriers is their susceptibility to disturbance. Individuals do mate when only heterospecifics are available (Clark et al. 1954), distantly related species produce fertile offspring (Schartl 2008), and disruptions to the sensory environment cause females to stop discriminating against heterospecific mates (Fisher et al. 2006). Given the presence of visually apparent genetic markers, the lack of documented poeciliid hybrids prior to 1971 is unlikely to be due to lack of effort on the part of earlier generations of aquarists and ichthyologists.

10.7.2 Sympatric and parapatric speciation: life, dinner, and sex

As chapters throughout this volume attest, poeciliids offer striking examples of divergence in both ecological traits and sexual signaling. In other systems, notably western Canadian sticklebacks (Boughman 2001) and African Rift Lake cichlids (Seehausen et al. 2008), ecological divergence coupled with divergence in sexual communication has resulted in rapid sympatric speciation. Why has this not been the case in poeciliids?

First, sympatric speciation is unlikely to occur solely via sexual selection; rather, sexual isolating mechanisms evolve in response to disruptive natural selection (Arnegard & Kondrashov 2004). Second, it is perhaps noteworthy that the best-known examples of ecological divergence in poeciliids involve predator-mediated selection on life-history traits (see **chapters 1–4**). By contrast, the stickleback and cichlid systems feature divergence both in sensory biology and in trophic ecology. There are at least three reasons why disruptive selection via predation may have less of an impact on genetic divergence in sympatry than trophic and sensory divergence:

1. By contrast with divergence driven by predation, divergence in sensory systems can automatically cause divergence in sexual communication and therefore reproductive isolation. Poeciliids tend to inhabit shallow freshwater environments, perhaps limiting the number of sensory microhabitats and reducing the opportunity for sensory divergence.

2. Divergent selection via predation is likely to be less consistent over space and time than selection via trophic pressures or sensory microhabitats. From the perspective of poeciliid populations, predators will always be patchier in space and time than prey; predator-free populations can thrive, but food-free populations cannot. The patchiness of predation-driven selection means that predator-mediated divergence may be counteracted by migration.

3. With a handful of exceptions, poeciliids feed on a relatively general suite of microinvertebrates, algae, and detritus, possibly limiting the opportunity for divergence based on diet. Divergent feeding systems, such as those seen in sticklebacks and cichlids, may be more likely than divergent responses to predation to produce dysfunctional hybrids. Exploitation of a novel food resource requires adaptations involving perception, behavior, morphology, and physiology, and hybrids are likely to express maladaptive phenotypes (Schluter 1996). By contrast, responses to predation risk are labile both evolutionarily (Reznick et al. 1990) and ontogenetically (Kelley & Magurran 2003).

Intriguingly, one of the few examples of poeciliids diverging in close geographic proximity involves a steep ecological gradient that imposes divergent selection on both sensory biology and trophic biology (Tobler & Plath, **chapter 11**). Understanding the role of ecology in poeciliid speciation will benefit from comparisons with the genetics underlying ecological divergence in sticklebacks and cichlids (Streelman et al. 2007).

10.7.3 Genomic architecture, speciation, and hybridization

Reproductive isolation depends on the nature and distribution of genetic incompatibilities among taxa (Noor & Feder 2006). Sex chromosomes (Schartl et al., **chapter 24**) may be a critical influence on the tempo and mode of speciation in poeciliids. Male traits in many species are linked with the Y chromosome (guppies: Brooks 2000; Brooks & Postma, **chapter 23**), and in several cases these Y-linked traits show discrete polymorphisms (e.g., Ryan & Wagner 1987; Lindholm et al. 2004). One possibility, readily testable, is that divergent Y chromosomes have introgressed from different species. This is predicted to be rare due to Haldane's rule (Coyne & Orr 2004); however, it remains a possibility in poeciliids given that sexual selection on Y-linked sexual traits can overcome natural selection against Y-linked traits deleterious to fitness (Brooks 2000), and given that hybrids are often viable. The intraspecific diversity that confounded Eigenmann (1894, 1909) may thus be a signature of reticulate evolution (Arnold 2006).

10.7.4 Why are there so few kinds of poeciliids?

A tale from the aquarium literature, in the days before the Internet, has it that the "Reverend" Guppy, awed by the remarkable variability of color patterns in the fish that would bear his name, wanted to describe each male as a distinct species. Mr. Guppy wasn't a reverend, and the story is apocryphal, but scores of reef-fish species have been described on the basis of more subtle distinctions than those of the blue and gold morphs of male *X. pygmaeus*. Poeciliids are some of the most diverse creatures on earth, but much of that diversity, particularly with regard to communication signals, occurs within a population. Poeciliids rose to prominence as a classical genetic model both because of the ease of generating interspecific crosses among divergent species and because of the ubiquity of Mendelian polymorphisms associated with striking differences among visual traits (reviewed in Rosenthal & García de León 2006). Could the paucity of among-species phenotypic diversity be causally related to intraspecific variation? The coastal sloughs and mountain creeks haunted by poeciliids are the perfect arena for diversification in allopatry or parapatry, followed by secondary contact and hybridization. Genetic exchange in the wild occurs between taxa as behaviorally, morphologically, and genetically divergent as swordtails and platyfishes (Meyer et al. 2006; Schartl 2008). Within a population, genetic variation could reflect a mosaic of genetic contributions from incompletely isolated relatives (Arnold 2006).

The maintenance of intrapopulation genetic diversity is, along with speciation, a fundamental problem in evolutionary biology. In poeciliids, behavioral decisions play a central role in maintaining both genetic variation within species and reproductive barriers between species. More so than other mechanisms (such as regulatory incompatibilities or balancing selection at the molecular level), behavior is contingent on a host of ecological forces, from predation risk (see Kelley & Brown, **chapter 16**) to disturbance of sensory channels, that can modify or impair its expression. Such behavioral variation in turn determines the strength and direction of selection on hybrids, introgressed traits, or polymorphic alleles. As the preeminent vertebrate system in the evolution of behavior, poeciliids are poised to make an important contribution to our understanding of how genetic divergence and genetic exchange shape organic diversity.

Acknowledgments

We are grateful to the Lara, Hernández, and Zapata families for their continued help over the years; to the Movimiento Indígena Unión Sierra y Huasteca; and to the Mexican federal government for permission to collect fish. We are especially grateful to the editors for putting together this volume and for providing valuable comments. We thank M. Tobler, D. Hughes-Phillips, R. Serva, M. Verzijden, J. B. Johnson, A. Jones, Z. Culumber, C. Carlson, J. Endler, and two anonymous reviewers for critiques and discussion. We are indebted to M. Schartl for setting us straight on ecclesiastical matters. Funding was provided by Texas A&M University, National Science Foundation grant IOB-0636712, and the American Livebearer Association to G. G. Rosenthal.

Chapter 11 Living in extreme environments

Michael Tobler and Martin Plath

11.1 Introduction

POECILIID FISHES have broad environmental tolerances. Many species are able to sustain extremes in salinity (Chervinski 1984; Nordlie et al. 1992), high temperatures (Carveth et al. 2006; Nordlie 2006), and hypoxia (Timmerman & Chapman 2003; Carter & Wilson 2006), as well as high concentrations of natural toxicants (Tobler et al. 2008d) or human-induced pollution and habitat alterations (Klerks & Lentz 1998; Casatti et al. 2006b; Franssen 2008). These broad physiochemical tolerances not only make poeciliids successful invasive species (Lindholm et al. 2005) and allow populations to sustain seasonal extremes in abiotic conditions (Nordlie 2006) but also allow some species to inhabit continuously harsh environments (Casatti et al. 2006b).

Continuously harsh environments can be considered extreme if at least one physiochemical stressor lies outside the range normally experienced by a population, and the successful maintenance of homeostasis requires costly morphological, physiological, or behavioral adaptations absent in closely related taxa (Townsend et al. 2003). Hence, extreme environmental conditions lead to a sharp fitness reduction in individuals lacking specific adaptations, which ultimately can lead to local population extinction (or a failure to colonize a particular habitat). Even for adapted individuals, withstanding a physiochemical stressor incurs costs. Such costs may come in the form of energy allotment for expressing or maintaining certain traits or in the form of trade-offs, where traits beneficial in the presence of a stressor are disadvantageous in its absence (which can lead to local adaptation and speciation).

Overall, organisms residing in extreme environments (extremophiles) can serve as model systems to study not only how organisms cope with the direct effects of particular physiochemical stressors but also how physiochemical stressors have a profound impact on ecological communities and ecosystem functions. Hence, these systems may be a platform to study variation of ecological functions and evolutionary trajectories along complex selective gradients. Here, we introduce poeciliids living under extreme conditions, discuss ecological and evolutionary consequences of living in extreme habitats, and identify areas for future research.

11.1.1 Poeciliids in extreme habitats

A number of poeciliids have been reported from extreme habitats, including populations in hypoxic (McKinsey & Chapman 1998) and high-temperature springs (Winkler 1985). However, the most striking examples of extreme habitats inhabited by poeciliids are hydrogen sulfide–rich and cave habitats. Hydrogen sulfide (H_2S) is a potent respiratory toxicant lethal for most metazoans even in micromolar amounts (Evans 1967; Bagarinao & Vetter 1992; Grieshaber & Völkel 1998). H_2S can be of biogenic origin and periodically occurs in hypoxic habitats in which organic material is decaying due to microbial activity (such as marshes or hypertrophic lakes: Jørgensen & Fenchel 1974; Jørgensen 1982, 1984). However, H_2S can also be

Table 11.1 Poeciliids in cave and sulfidic habitats

Species	Location	References
Cave habitats		
Gambusia holbrooki[a]	Spunnulate Dolinas system, Torre Castiglione, Italy	Camassa 2001
Poecilia mexicana	Cueva del Azufre and Cueva Luna Azufre, Río Tacotalpa drainage, Mexico	Gordon & Rosen 1962; Tobler et al. 2008c
Sulfidic habitats		
Gambusia affinis	Platt National Park, Oklahoma, USA	Covich 1981; Tobler, unpublished data
Gambusia eurystoma[b]	Baños del Azufre, Chiapas, Mexico	Miller 1975; Tobler et al. 2008d
Limia sulphurophila[b]	Enriquillo National Park, Dominican Republic	Rivas 1980; Hamilton 2001
Poecilia mexicana[c]	Multiple locations in the Río Tacotalpa and Puyacatengo drainages, Mexico	Tobler et al. 2006b; Tobler et al. 2008a; Tobler & Plath, unpublished data
Poecilia reticulata	Poza Azufre, Río San Juan drainage, Venezuela	Winemiller et al. 1990
Poecilia sulphuraria[b]	Baños del Azufre, Chiapas, Mexico	Alvarez del Villar 1948; Tobler et al. 2008d

[a]Introduced population; referred to as *G. affinis* by Camassa (2001).
[b]Endemic in sulfidic springs.
[c]Potentially includes nondescribed species.

of volcanic origin in some freshwater systems (Rosales Lagarde et al. 2006). In this case, high and sustained levels of H_2S are often found (Rosales Lagarde et al. 2006; Tobler et al. 2006b), rendering such systems extreme habitats for any aquatic animals residing in them. The absence of light in caves inhibits the use of visual senses, and cave dwellers must cope with perpetual darkness, especially if they evolved from diurnal surface-dwelling forms as the poeciliids did. Consequently, living in caves entails complex adaptations characteristic of extreme habitats (Howarth 1993; Langecker 2000; Plath et al. 2004).

To date, we know of only two caves that are naturally inhabited by poeciliids; a third cavernicolous population of a poeciliid fish (*Gambusia holbrooki*) has been introduced to an Italian cave (table 11.1). Poeciliids from different genera, however, have independently colonized a number of sulfidic springs (table 11.1). Some species described are even endemic to such spring habitats (Tobler et al. 2008d). Although reports of poeciliids in sulfidic habitats are mounting, few systems have been thoroughly investigated. Most of the research has been conducted on *Poecilia mexicana* occurring in and around a sulfur cave (the Cueva del Azufre) near the village of Tapijulapa in the southern Mexican state of Tabasco, where two physiochemical stressors (H_2S and darkness) occur in all possible combinations within a perimeter of a few kilometers (Tobler et al. 2008a). *Poecilia mexicana*, which has colonized all habitat types in this system, will serve as the predominant example in the subse-

quent sections of this chapter. We will concentrate on darkness and toxic H_2S as extreme environmental factors and examine how their occurrence affects various aspects of the ecology and evolution of extremophile *P. mexicana*.

11.2 Consequences of life under extreme conditions

Physiochemical stressors can affect the ecology of individuals as well as the evolutionary trajectory of populations in two ways: directly by interfering with the maintenance of homeostasis in individuals, and indirectly by altering other aspects of the abiotic and biotic environment (fig. 11.1; see Congdon et al. 2001 for a review).

11.2.1 Direct effects

Extreme stressors directly affect physiological functions. In the absence of specific adaptations, this can lead to homeostatic imbalance, which increases the risk of mortality and reduces the chance of successful reproduction. In contrast, through physiological adaptations, extremophiles have evolved the ability to maintain homeostatic balance even in the presence of an extreme stressor. For example, H_2S is an inhibitor of the cytochrome *c* oxidase, blocking electron transport in aerobic respiration and thereby hampering the function of mitochondria and the production of ATP (Evans 1967; Nicholls 1975; National Research Council 1979). It

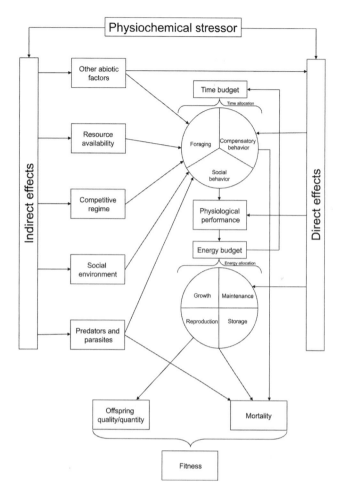

Figure 11.1 Direct and indirect effects of physiochemical stressors on fitness-relevant factors in extremophile poeciliids. Physiochemical stressors interfere directly with bodily functions (i.e., the maintenance of homeostasis) and indirectly affect selective regimes by altering other aspects of the abiotic and biotic environment. Modified after Congdon et al. 2001.

that allow for a more efficient energy acquisition (e.g., different foraging strategies) and energy use (e.g., reduction of energetically costly traits, shifts in life-history strategies; see section 11.3).

11.2.2 Indirect effects

Physiochemical stressors can also have profound indirect effects on organisms inhabiting extreme habitats in that they dramatically alter abiotic and biotic environmental characteristics. Hence, not only do extreme and nonextreme habitats differ in the presence or absence of a physiochemical stressor, but in fact they are connected by a complex environmental gradient usually involving multiple abiotic environmental factors as well as major differences in the biotic communities. Due to the correlation with other environmental factors, extremophiles are exposed to additional selective agents compared with closely related forms in nonextreme habitats, but at the same time some selective agents may be relaxed (table 11.2).

The presence of H_2S in a habitat coincides with a number of other differences among habitat types (table 11.2). H_2S is highly reactive at room temperature and spontaneously oxidizes in water, which leads to and aggravates hypoxia in aquatic systems (Cline & Richards 1969; Chen & Morris 1972). Also, the temperature in sulfidic habitats is usually higher and less variable compared with adjacent habitats.

As most species cannot cope with the toxicity of H_2S, species richness appears to be reduced in sulfidic habitats (Tobler et al. 2006b; Tobler et al. 2008d). This has consequences for species adapted to the environmental stressor(s). First, predation by piscivorous fishes is relaxed due to the absence of other fish species in sulfidic habitats. Predation by piscivorous birds, however, appears to be higher in sulfidic systems, because spring pools are shallow and provide little shelter, and poeciliids under hypoxia spend a considerable portion of their time performing aquatic surface respiration (see below), which increases their risk of aerial predation (Kramer 1983; Riesch et al., forthcoming; box 11.1).

Second, the absence of similar-sized species with comparable ecological (especially trophic) niches leads to a reduction of interspecific competition in extreme habitats (i.e., competitive release). For example, more than half a dozen poeciliid species are regularly found sympatrically in nonsulfidic waters in southern Mexico (depending on the site, these may include species of the genera *Belonesox*, *Carlhubbsia*, *Gambusia*, *Heterandria*, *Heterophallus*, *Phallichthys*, *Poecilia*, *Priapella*, and *Xiphophorus*), but maximally two species—*Poecilia sulphuraria* and *Gambusia eurystoma* in the Baños del Azufre—are simultaneously found in water systems with high levels of H_2S (Tobler et al. 2006b; Tobler et al. 2008d; Plath et al. 2010b). H_2S also leads to funda-

can also modify oxygen transport proteins (Carrico et al. 1978; Park et al. 1986). Exposed but nonadapted individuals succumb to the toxicant within short periods of time as cell respiration is blocked (Bagarinao & Vetter 1992; Tobler et al. 2008d). However, H_2S-tolerant organisms can cope with the toxic effects by sulfide oxidation and subsequent excretion from the body (Curtis et al. 1972; Bagarinao & Vetter 1992).

Consequently, both nonadapted and adapted individuals face costs when exposed to extreme physiochemical stressors. But while the cost for nonadapted individuals is—by definition—fatal, the cost for adapted individuals constitutes the energy investment necessary for maintaining the coping mechanism. Assuming that energy is a limited resource, we predict a shift in energy allocation toward a higher investment in somatic maintenance (Sibly & Calow 1989), and we expect extremophiles to have not only adaptations to maintain homeostasis but potentially also traits

Table 11.2 Indirect effects of physiochemical stressors caused by changes in abiotic and biotic environmental characteristics

Category	Environmental factor	In the presence of H₂S	In the absence of light
Other abiotic factors			
	Temperature	+, less variation	+, less variation
	pH	−	∅
	Dissolved oxygen	−	∅
	Dissolved minerals	+	∅
	Down-dwelling light	+ (through reduced vegetation cover)	
	Transmitted light	− (due to presence of colloidal sulfur)	
Biotic factors			
Plant growth	Contact vegetation	−	−
	Submersed vegetation	−	−
Resource availability	Species diversity	−	−
	Resource type	shift from photo- to chemoautotrophic	shift from auto- to heterotrophy
	Abundance	?	?
Competitive regime	Number of competing species	−	−
	Type of competitors	shift from inter- to intraspecific	shift from inter- to intraspecific
Social environment	Sex ratios	more balanced	more balanced
	Population density	+	+
	Male sexual harassment	−	−
Predators	Number of predators	likely −	−
	Type of predators	reduced diversity, predominantly avian predation	only *Belostoma* sp.
Parasites	Number of parasites	−	−
	Type of parasites	reduced diversity	reduced diversity

Note: Observed indirect effects were estimated in southern Mexican sulfidic springs and the Cueva del Azufre system (Plath & Tobler 2010). Indirect effects of lightlessness may differ in other cave systems for various geological and biological reasons. + indicates increase, − indicates decrease, ∅ indicates no change.

mental changes in the resource dynamics of a habitat, because H₂S is toxic for most phototrophic primary producers as well (Bagarinao & Vetter 1992). Furthermore, it allows for chemoautotrophic primary production by sulfide-oxidizing microbes (Nelson & Jannasch 1983). The reduction of interspecific competition and changes in the availability of food sources are accompanied by shifts in resource use. While *P. mexicana* is predominantly algivorous/detrivorous in typical (nonsulfidic) habitats, the algal part of the diet is replaced almost entirely by sulfide bacteria and invertebrate prey in sulfidic habitats (Tobler 2008 and unpublished data). Furthermore, fish generally have wider dietary niches in extreme habitat types (Tobler 2008). The shift in resource use is also paralleled by heritable differences in viscerocranial and intestinal-tract morphology (Tobler 2008); fish in extreme habitats have thicker and wider jaws (which may be advantageous in handling larger and more evasive prey; Hulsey & García de León 2005; Higham et al.

2007) and shorter intestinal tracts (which reflects the lower proportion of plant matter in the diet; Kramer & Bryant 1995). Like mollies, *Cyprinodon* species that colonized a habitat low in species diversity (the Laguna Chichancanab in Mexico) underwent comparable shifts in resource use (Horstkotte & Strecker 2005; Horstkotte & Plath 2008).

Third, parasites may occur at lower abundance or may be completely missing in sulfidic habitats, either because the stressor affects free-living stages of parasites or because other host species in the parasite's life cycle are locally absent (Tobler et al. 2007a). Metacercarian cysts of the digenean trematode *Uvulifer* sp., which are common in *P. mexicana* populations from nonsulfidic habitats, are reduced in sulfidic surface habitats and are lacking in the sulfur cave (Cueva del Azufre). A similar pattern has been described for trematode infections in White Sands pupfish (*Cyprinodon tularosa*, Cyprinodontidae) from a high-salinity desert creek in New Mexico (Rogowski & Stockwell 2006b).

Box 11.1 Predator-prey interactions between an insect predator and a cave fish

Most cave habitats are essentially predator-free habitats for cave fishes, so cave fishes typically rank in the uppermost portion in the trophic chain of subterranean ecosystems (Poulson & Lavoie 2000). However, an aquatic heteropteran (*Belostoma* sp.) occurs at high densities in the Cueva del Azufre. Using mark-recapture analysis, the population size of this sit-and-wait predator was estimated as 336 ± 130 (mean ± SE) individuals in cave chamber V of the Cueva del Azufre, which results in a density of over 1 *Belostoma*/m² (Tobler et al. 2007b). *Belostoma* are voracious predators of cave mollies. Hence, in the Cueva del Azufre an aquatic insect acts as the top predator, leading to a unique predator-prey relationship between an insect and a fish.

The susceptibility of cave mollies to *Belostoma* predation is at least in part driven by the adverse environmental conditions in the sulfidic Cueva del Azufre. During aquatic surface respiration, which mediates survival in the hypoxic and toxic water, fish remain just underneath the water surface (Plath et al. 2007d). Here, *Poecilia mexicana* may encounter *Belostoma*, which catch mollies with their raptorial forelegs while sitting on a rock and holding their abdomen in the air for gas exchange (Tobler et al. 2007b). Male *P. mexicana* have a higher susceptibility to the adverse abiotic conditions

than females and spend more time performing aquatic surface respiration (Plath et al. 2007d). Consequently, male cave fish were found to be preyed upon by *Belostoma* more often than equally sized females (Tobler et al. 2008b). This is unique in that differential susceptibility to adverse environmental conditions rather than morphological and behavioral differences among the sexes (e.g., Endler 1980; Trexler et al. 1994; Macías-Garcia et al. 1998) seems to be driving the differences in susceptibility to predation.

Two independent experiments in different years also demonstrated that *Belostoma* exhibit a prey preference for larger cave mollies over smaller fish (Plath et al. 2003; Tobler et al. 2007b). This prey preference of the heteropterans may have implications for some aspects of life-history trait evolution of cave mollies—namely, the persistence of male body size polymorphism despite strong selection for large male body size by female choice (Plath et al. 2004). Water bugs also preferentially prey upon immigrants from ecologically divergent habitat types, thus contributing to the low levels of gene flow between cave and adjacent surface habitats (Tobler 2009). Recent investigations also revealed that some large-bodied orthognath and opistognath spiders act as molly predators in the Cueva del Azufre (Horstkotte et al. 2010), but their impact on the evolution of cave mollies is probably much lower.

The complexity of environmental gradients not only causes organisms inhabiting extreme habitats to diverge in traits that can be related to the direct effects of a physiochemical stressor, but multiple-trait divergence occurs as a response to a set of selective factors. The fact that inhabiting a seemingly uninhabitable place results in the relaxation of certain selective pressures (such as predation and interspecific competition) further highlights that—given some coping mechanism—the colonization of extreme habitats also provides benefits to colonizers, thereby compensating for some of the costs associated with the presence of the stressors.

11.3 Trait divergence along environmental gradients

11.3.1 Hydrogen sulfide

There is some evidence that *P. mexicana* from sulfidic habitats have unique physiological adaptations that are absent in conspecifics from nonsulfidic habitats. For example, *P. mexicana* from sulfidic habitats can sustain motion in sulfidic water, while those from nonsulfidic habitats lose

motion control within seconds after exposure to H₂S (Tobler et al. 2008d). Sulfide tolerance appears to be at least partially heritable, as laboratory-reared animals from sulfidic habitats retain the higher tolerance even when reared for several generations in a nonsulfidic environment (Peters et al. 1973; reanalyzed in Plath & Tobler 2010). However, the physiological pathways that allow for sulfide detoxification in poeciliids have yet to be identified. Sulfide detoxification has predominantly been studied in benthic and deep-sea hydrothermal vent invertebrates, which can oxidize sulfide either themselves or through symbiotic bacteria (Grieshaber & Völkel 1998; Bright & Giere 2005). Physiological pathways have also been studied in fish from temporarily sulfidic habitats (e.g., salt marshes; Bagarinao & Vetter 1989; Ip et al. 2004). In *Fundulus parvipinnis* (like poeciliids, a member of the order Cyprinodontiformes), sulfide oxidation has been reported by mitochondria and ferric hemoglobin (Bagarinao & Vetter 1992, 1993).

Due to the hypoxic conditions in sulfidic habitats, oxygen available for respiration is generally limited, but at the same time oxygen is required for detoxification processes to cope with the adverse effects of H₂S. Organisms living in sulfidic habitats should therefore have respiratory adaptations facilitating efficient oxygen acquisition. *Poecilia mexicana* from nonsulfidic and sulfidic habitats diverged primarily in total gill filament length and head size; larger heads seem to allow for an increased gill size (Tobler et al. 2008a). When exposed to H₂S, *P. mexicana* also rely on compensatory behavior—aquatic surface respiration (ASR)—in which the fish exploit the more oxygen-rich air-water interface using their gills (Plath et al. 2007d; Tobler et al. 2009a). The ability to perform ASR (i.e., the ability for efficient oxygen acquisition) is the best predictor for short-term survival in fish exposed to sulfidic water (Plath et al. 2007d). *Poecilia sulphuraria* further exhibit conspicuous lip appendages (fig. 11.2a), which are thought to increase the efficiency of ASR (Winemiller 1989; Tobler et al. 2008d). Changes in gill morphology as well as the performance of ASR are not unique to sulfide-spring inhabitants but can also be found in fishes exposed to nonsulfidic but hypoxic environments (Lewis 1970; Kramer & Mehegan 1981; Kramer & McClure 1982; Chapman et al. 1999; Chapman et al. 2000; Chapman & Hulen 2001; Timmerman & Chapman 2004).

Although direct physiological measurements of the costs involved with sulfide detoxification have yet to be performed, there is indirect evidence that living in sulfidic environments (i.e., coping with the stressors) imposes direct energetic costs. In the sulfidic Cueva del Azufre, energy availability (i.e., experimental feeding with energy-rich food) significantly increases short-term survival when fish are placed in mesocosms in the natural environment (Plath et al. 2007d). Furthermore, sulfide-spring residents exhibit a low body condition factor (Tobler et al. 2006b) and have only about a third of storage lipids in body tissues compared with fish from nonsulfidic surface habitats (Tobler 2008). Besides energetic costs for physiological detoxification mechanisms (Ip et al. 2004), ASR also imposes opportunity costs because it detracts from the time available for foraging (Kramer 1983; Weber & Kramer 1983; Chapman & Chapman 1993). A comparison of time budgets among different *P. mexicana* and *P. sulphuraria* populations from sulfidic and nonsulfidic habitats revealed a significant trade-off between the amount of time spent feeding and time spent performing ASR (Tobler et al. 2009a). Besides these costs that directly affect the energy budget, evidence is also emerging that traits beneficial in sulfidic habitats come at a cost under nonsulfidic conditions. Reciprocal transplant experiments indicated that *P. mexicana* from sulfidic habitats have a significantly lower survival in nonsulfidic water than the residents (Tobler et al. 2009b).

Low energy availability and/or high energetic costs of living in a stressful environment seem to be strong selective factors acting on extremophile *P. mexicana*. A multitude of traits may have evolved as a response to energy limitation, ranging from shifts in life-history patterns consistent with energy economy (Franssen et al. 2008; Riesch et al. 2009a; Riesch et al. 2010) to reductions of costly behavioral patterns. For example, reductions in aggression (Parzefall 1974, 1979), male sexual activity (Plath 2008), and shoaling behavior (Plath & Schlupp 2008) have all been shown to have a genetic basis in *P. mexicana* inhabiting sulfidic habitats. Changes in some behavioral traits (e.g., shoaling behavior) may have been facilitated by the relaxation of selection, as piscivorous fish are absent.

11.3.2 Darkness

Trait divergence has also been documented along the surface/cave environmental gradient. Cave populations of *P. mexicana* have elaborated nonvisual senses, and they have more taste buds and wider pores of the head canals of the mechanosensory lateral-line system than surface fish do (Walters & Walters 1965; Parzefall 1970). Female cave mollies also develop a "genital pad" (fig. 11.2b), which is a large, gland-rich, folded protuberance around the gonopore (Parzefall 1970). This structure presumably facilitates communicating receptivity to males, which nip at the female gonopore prior to copulation (Zeiske 1968). This elaboration of nonvisual sensory structures is congruent with the evolutionary trends of other organisms living in cave and other continuously dark habitats (Langecker 2000; Poulson 2001).

Divergence in nonvisual sensory structures is accompanied by a sensory shift and the ability to exercise mate choice in darkness, at least in the Cueva del Azufre

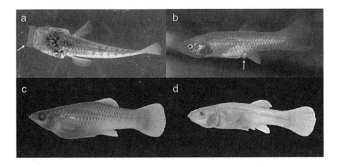

Figure 11.2 Poeciliids from extreme habitats are characterized by divergent morphological traits. (a) *Poecilia sulphuraria* female viewed from the top. The arrow indicates the characteristic lip appendages that appear to increase the efficiency of aquatic surface respiration in hypoxic water. (b) *Poecilia mexicana* female from the Cueva del Azufre. The arrow indicates the female genital pad expressed in this population. (c) *Poecilia mexicana* female from a nonsulfidic surface habitat. (d) *Poecilia mexicana* female from a sulfidic cave habitat. Note the reduction in pigmentation and eye size as well as the relative increase in head size in the specimen from the sulfur cave.

Box 11.2 Evolution of nonvisual mate choice in cave-dwelling poeciliids

Animals colonizing lightless caves provide a unique opportunity to examine how mate-choice behavior evolves in relation to changed sensory environments. The cave molly, *Poecilia mexicana*, from the Cueva del Azufre is one of very few cave fishes in which the evolution of mating preferences has been investigated in detail. One male trait that tends to be universally preferred by females is large body size (Ryan & Keddy-Hector 1992; Andersson 1994). Surface-dwelling *P. mexicana* use visual cues to differentiate between differently sized males (Plath et al. 2004). How does this female mating preference evolve once visual communication breaks down after the colonization of a subterranean habitat?

To study this question, individual molly females from different populations were allowed to choose between a large and a small stimulus fish in simultaneous choice tests, whereby the females could perceive either multiple cues (visual plus nonvisual) from the stimulus fish (i.e., the stimulus males were presented behind a wire-mesh grid under normal light conditions), solely nonvisual cues (infrared observations using the same mode of presentation of males), or solely visual cues (presenting the males behind Plexiglas in light). Whereas surface fish were unable to exhibit a preference in darkness, cave fish had evolved the ability to discriminate between large and small males in darkness (Plath et al. 2004). Even light-reared cave molly females showed this behavior, so nonvisual mate choice seems to be a heritable trait, not just a plastic response to life in darkness (Plath et al. 2004). Subsequently, nonvisual mate choice has also been documented for male cave mollies (Plath et al. 2006) and traits other than mate size (e.g., body condition: Plath et al. 2005a). Cave mollies of the Cueva del Azufre population thus not only have undergone a sensory shift, where an ancestral mating preference is maintained by employing a new set of sensory organs, but—at least in the case of condition-dependent mate choice—have also evolved divergent mating preferences absent in surface populations. Both the mechanosensory lateral line and the sense of smell appear to play a role in nonvisual mate choice in cave mollies (Plath et al. 2007c).

population of *P. mexicana* (box 11.2). Surface-dwelling *P. mexicana* exhibit only a visual preference for large mating partners, but cave fish also exhibit this preference in darkness. Both chemical and mechanosensory cues seem to be involved (Plath et al. 2007c). Incidentally, the second cave population (Cueva Luna Azufre) of *P. mexicana* has not evolved the ability to discriminate between differently sized mating partners in darkness, although the ancestral visual preference has been retained in both cave populations (Tobler et al. 2008e).

As stabilizing selection from light is lacking in subterranean habitats, traits like eye size (Tobler et al. 2008a), opsin gene expression (Tobler et al. 2010), and pigmentation (Peters et al. 1973) are reduced in cave mollies (see also fig. 11.2c and d). Unlike in other cave organisms (Jeffery et al. 2003; Porter & Crandall 2003), and despite some regressive trends in eye development, eyes in both cave molly populations are still functional (Körner et al. 2006), and cave mollies are even positively phototactic (Parzefall et al. 2007).

11.4 Phenotypic plasticity and genetic assimilation

To cope with physiochemical stressors, all organisms have evolved phenotypically plastic traits that can be expressed facultatively upon exposure to different environmental conditions (defining their reaction norm). Among the best-known cellular responses are heat-shock proteins (HSP) and other molecular chaperones, the expression of which is transcriptionally upregulated when encountering high temperatures or stressful conditions to prevent the stress-induced denaturation of other proteins (Schlesinger 1990; Feder & Hofmann 1999; Sørensen et al. 2003). HSP expression in nature may vary among species and is correlated with resistance to stress and the levels of stress encountered in natural habitats (for HSP expression in tropical and desert poeciliids, see Norris et al. 1995; DiIorio et al. 1996). Most divergent traits examined so far in extremophile poeciliids (from physiology to morphology and behavior), however, are not based on phenotypic plasticity but appear to have a heritable basis (Peters & Peters 1968; Peters et al. 1973; Plath et al. 2004; Plath et al. 2006; Plath 2008; Tobler et al. 2008a). For example, fish from cave and sulfidic habitats retain their distinct morphological features over multiple generations when raised in the laboratory in the absence of H_2S and in the presence of light (Tobler et al. 2008a). Similarly, the ability to perform nonvisual mate choice in cave and surface fish is not dependent on whether individuals were raised in light or darkness (Plath et al. 2004; Plath et al. 2006).

Many adaptations observed in extremophile poeciliids are facultatively expressed (as phenotypically plastic traits) in related forms from nonextreme habitats, even though trait expression is typically weaker than in extremophile populations. For example, *P. mexicana* females from surface habitats reduce their fecundity when reared in darkness, whereas cave molly females exhibit a low fecundity that does not vary in response to different rearing conditions (Riesch et al. 2009b). Future studies will need to test the hypothesis that many of the divergent traits in extremophiles evolved through genetic assimilation (Pigliucci 2001) of previously plastic traits, followed by an increase in trait expression due to strong directional selection.

11.5 Speciation in extreme habitats?

If the evolutionary response to the presence of a persistent physiochemical stressor is heritable, local adaptation can proceed to speciation (Schluter 2000; Streelman & Danley 2003). Ecological speciation occurs when divergent selection leads to the correlated evolution of reproductive isolation. In traditional models of ecological speciation, reproductive isolation evolves incidentally as a by-product (Schluter 2000, 2001; Dieckmann et al. 2004; Rundle & Nosil 2005), but whenever divergent natural selection occurs among populations, there may be direct selection for premating isolation mechanisms to evolve (Schluter 2001; Rodriguez et al. 2004). Direct evidence for ecological speciation in extreme habitats is thus far lacking. However, phenotypic differentiation between extreme and nonextreme habitats does coincide with extremely low rates of gene flow among habitats and strong genetic differentiation despite the lack of major physical barriers that would prevent migration (Plath et al. 2007a; Plath et al. 2010a; Tobler et al. 2008a). Field experiments indicated that natural and sexual selection against immigrants seems to play a key role in mediating reproductive isolation among habitat types (Tobler 2009; Tobler et al. 2009b; Plath et al. 2010b). Furthermore, some poeciliids from sulfidic habitats were described as distinct species based on morphological divergence (see table 11.1).

11.6 Perspectives

Poeciliids in extreme environments provide ideal model systems to study the ecological and evolutionary conse-

quences of physiochemical stressors, evolution along complex environmental gradients, and, potentially, ecological speciation. In addition, poeciliids from sulfidic habitats may also serve as a model system to understand the mechanisms of sulfide toxicity and of detoxification and thus contribute to a better understanding of the long-term effects of human-induced pollution, as naturally occurring H_2S is essentially a "natural pollutant" that has been present for a prolonged time.

Although poeciliids have long been established as model systems in other fields of research, *living in extreme habitats* is a relatively young and emerging field in poeciliid research. Consequently, many basic questions are still unanswered. Especially the physiological and biochemical mechanisms of sulfide tolerance remain a major black box. Interestingly, several extremophile freshwater fishes are found in the order Cyprinodontiformes, such as pupfish (e.g., in a lake with calcium sulfate reaching saturation: Humphries & Miller 1981; in sulfidic springs: Lozano-Vilano & Contreras-Balderas 1999) and poeciliids (this chapter), but it is unclear why cyprinodontiform fishes prevail among extremophile freshwater fishes. A detailed understanding of the mechanisms employed to cope with stressors as well as their genetic basis will be crucial to the study of both ecophysiological and evolutionary aspects of life in extreme habitats. Furthermore, the patterns of divergence uncovered in *P. mexicana* of the Cueva del Azufre system also need to be investigated in other extremophile poeciliids to obtain an estimate of the generalizability of the current findings. The increasing number of known extremophile populations in various phylogenetic lineages will allow for such comparisons in the future.

Acknowledgments

C. Bleidorn (Potsdam, Germany), C. M. Tobler (Stillwater, Oklahoma, USA), and G. Rosenthal (College Station, Texas, USA), as well as S. Würtz (Porto, Portugal), kindly provided comments on earlier stages of the manuscript. We are deeply indebted to the people in the community of Tapijulapa, the Universidad Intercultural del Estado de Tabasco in Oxolotán, and the Municipio de Tacotalpa for their continuous support and for their hospitality during our visits. Financial support came from the Swiss National Science Foundation to M.T. (PBZHA-121016) and from the Deutsche Forschungsgemeinschaft to M.P. (PL 470/1-1, 2).

Chapter 12 Evolutionary conservation biology

Craig A. Stockwell and Sujan M. Henkanaththegedara

12.1 Introduction

POECILIIDS ARE OF PARTICULAR interest to conservation biologists for a number of reasons. First, they have become excellent models for conservation biologists due to their small size and rapid generation time (Quattro & Vrijenhoek 1989; Leberg 1990, 1993). Second, many poeciliid studies have provided important insights on the rate of evolutionary diversification (e.g., Endler 1980; Reznick et al. 1990; Reznick et al. 1997; see also Pires et al., **chapter 3**; Breden & Lindholm, **chapter 22**; Grether & Kolluru, **chapter 6**; Johnson & Bagley, **chapter 4**; Schlupp & Riesch, **chapter 5**), the ultimate source of biodiversity. Third, poeciliids are vulnerable to the same anthropogenic factors driving the current extinction crisis (e.g., habitat loss, invasive species). Paradoxically, one of the greatest threats to poeciliids is the spread of heterospecific poeciliids (Minckley & Deacon 1968; Meffe 1985b; Minckley & Jensen 1985; Courtenay & Meffe 1989; Belk & Lydeard 1994). Thus, poeciliids have been extensively evaluated in the context of conservation biology (Johnson & Hubbs 1989; Leberg 1990, 1993; Stockwell et al. 1996; Stockwell & Weeks 1999).

Johnson and Hubbs (1989) provided an overview of the conservation status of the poeciliids in the United States. Since that time, poeciliids have mirrored the decline of other freshwater fishes (see Jelks et al. 2008). Furthermore, our understanding of relationships among taxa has increased, allowing reevaluation of the systematics and distribution of poeciliid biodiversity (Rosen & Bailey 1963; Parenti 1981; Parenti & Rauchenberger 1989; Hrbek et al. 2007).

Much applied research has focused on endangered poeciliids (e.g., *Poeciliopsis* spp.) as well as common and invasive poeciliids (e.g., *Gambusia* spp., *Poecilia reticulata*). In fact, many poeciliids have been used as model systems to examine questions central to the field of conservation biology. Here, we provide an overview of the poeciliids from the perspective of conservation biology. First, we review the geographic distribution of poeciliid biodiversity as well as associated threats. We then consider the impact of nonnative poeciliids on native species and ecosystems. We conclude by summarizing the role of selected poeciliid species as models for studies in the emerging field of evolutionary conservation biology (as conceptualized by Ferrière et al. 2004).

12.2 Diversity, conservation, and threats

Poeciliid conservation status has changed considerably during the last few decades (Jelks et al. 2008). Here, we provide a brief overview of poeciliid biodiversity with special focus on the conservation status of these fishes. We also review the primary threats faced by poeciliids.

12.2.1 Poeciliid biodiversity and distribution

The *FishBase* data base (Froese & Pauly 2009) and recent reviews (Parenti 1981; Parenti & Rauchenberger 1989; Nelson et al. 2004; Hrbek et al. 2007; Eschmeyer & Fong 2008; Jelks et al. 2008; Scharpf 2008) recognize more than 250 species of poeciliids, belonging to 22–28 genera. The most species-rich genera are *Gambusia* (43 species),

Poecilia (34 species), *Xiphophorus* (28 species), *Poeciliopsis* (23 species), *Phalloceros* (22 species), and *Limia* (21 species) (table 12.1). The Poeciliidae family is widely distributed, occurring as native species from the southeastern United States to northeastern Argentina (Rosen & Bailey 1963; Nelson 2006). This clade of fishes originated in South America and dispersed to Central America and North America (Hrbek et al. 2007).

Our understanding of poeciliids in South America and parts of Central America has increased markedly in the last few decades (Meyer & Etzel 2001; Poeser 2002; Meyer et al. 2004; Lucinda 2005b, 2008; Lucinda et al. 2005;

Table 12.1 Number of species and distribution of poeciliid genera in the world

Genus	Number of species	Distribution
Alfaro	2	CA, SA
Belonesox	1	NA, CA
Brachyrhaphis	12	NA, CA
Carlhubbsia	2	CA
Cnesterodon	9	SA
Gambusia	43	NA, CA, SA
Girardinus	7	CA
Heterandria	10	NA, CA
Heterophallus	2	NA, CA
Limia	21	CA, SA
Micropoecilia	5	CA, SA
Neoheterandria	3	CA, SA
Pamphorichthys	6	SA
Phallichthys	4	CA
Phalloceros	22	SA
Phalloptychus	2	SA
Phallotorynus	6	SA
Poecilia	34	NA, CA, SA
Poeciliopsis	23	NA, CA, SA
Priapella	5	NA, CA
Priapichthys	7	CA, SA
Pseudopoecilia	3	SA
Quintana	1	CA
Scolichthys	2	CA
Tomeurus	1	SA
Xenodexia	1	CA
Xenophallus	1	CA
Xiphophorus	28	NA, CA

Source: FishBase (Froese & Pauly 2009).

Note: NA = North America; CA = Central America, and SA =South America.

Poeser et al. 2005; Lucinda et al. 2006). For example, a recent expedition in South America added 21 new species to the genus *Phalloceros*, which previously contained only 1 described species (Lucinda 2008). The known biodiversity of poeciliids (subfamily Poeciliinae = family Poeciliidae in Rosen & Bailey 1963) thus has grown from 194 species in the late 1980s (Rauchenberger 1989) to more than 262 valid species by 2008 (Eschmeyer & Fong 2008). This number is likely to grow as more areas are intensively sampled (Lucinda 2008).

Our knowledge of poeciliid diversity is further complicated by discrepancies among workers in the recognition of particular taxa. For instance, topminnows native to the Gila River and Yaqui River drainages were originally described as two species, *Poeciliopsis occidentalis* and *P. sonoriensis* but were synonymized by Minckley (1969) as *P. o. occidentalis* and *P. o. sonoriensis*, respectively. However, based on a series of studies including a variety of molecular markers, these two taxa are now considered distinct species (Vrijenhoek et al. 1985; Quattro et al. 1996; Minckley 1999; Hedrick et al. 2006).

Here, we follow Parenti (1981) in recognizing Poeciliidae as livebearing fishes excluding South American *Fluviphylax* and Old World relatives. Biodiversity of poeciliids should also consider the recognition of conservation units—evolutionarily significant units (ESUs)—although this concept has not been widely applied within Poeciliidae (but see box 12.1).

12.2.2 Poeciliid conservation status

The conservation status of poeciliid species is best known for populations in the United States; thus, knowledge of the threats faced by poeciliids is based largely on studies conducted for more northern species. For instance, the International Union for Conservation of Nature (IUCN 2008) evaluated the conservation status for a good portion of North American (including Mexico) poeciliids (22 species) but for only 1 species in Central America and 1 species in South America (table 12.2). This is mainly due to poor understanding of the diversity, distribution, and natural history of Central American and South American poeciliids (Lucinda et al. 2005; Lucinda 2008). Furthermore, the Endangered Species Committee of the American Fisheries Society (AFS-ESC) identified 37 imperiled poeciliid taxa from North America, including 32 (33% of total poeciliids) described species and 5 undescribed taxa/subspecies or populations (Jelks et al. 2008; fig. 12.1).

Unfortunately, the conservation status of poeciliids in North America has deteriorated in the last two decades, with the list of imperiled taxa growing from 21 to 37 (Williams et al. 1989; Jelks et al. 2008). Furthermore, the con-

Box 12.1 Conservation units and the management of desert topminnows

The Gila topminnow (*Poeciliopsis occidentalis*) provides an excellent opportunity to consider genetic management of protected species. This species was once widespread but is now restricted to a small number of remnant populations (Minckley 1999). Because gene flow must be accomplished by translocation of fish, delineation of evolutionarily significant units (ESUs) is of particular interest to managers.

Moritz (1994) proposed the operational yet restrictive criterion that ESUs be defined as taxonomic units that are reciprocally monophyletic at mitochondrial DNA (mtDNA) markers. Crandall et al. (2000) suggested that conservation units be considered along a gradient based on genetic and ecological exchangeability. Units that are nonexchangeable are thus unreplaceable, making them evolutionarily significant.

Fortunately, *Poeciliopsis* species have been well characterized with a wide battery of molecular markers. Vrijenhoek et al. (1985) observed low levels of genetic variation (allozymes) in remnant populations, leading them to recommend the experimental mixing of populations within the major groups as a means to increase genetic diversity (Vrijenhoek et al. 1985). Quattro et al. (1996) also observed no genetic diversity (mtDNA, RFLP) within the Gila River populations of topminnows and suggested that differences among Arizona populations of *P. occidentalis* were most likely associated with recent anthropogenic isolation. They nevertheless recommended caution in applying managed gene flow among these populations (Quattro et al. 1996). Significant differences at both microsatellites (Parker et al. 1999) and major histocompatibility complex (MHC; for more on MHC, see McMullan & van Oosterhout, **chapter 25**) loci, along with ecological differences, led Hedrick et al. (2001b, 2006) to argue that *P. occidentalis* was composed of two ESUs (box-fig. 12.1). However, these populations

of *P. occidentalis* exhibit no mtDNA variation (sequence data for 2626 bp) and therefore do not meet the criterion of reciprocal monophyly. Hedrick et al. (2006) argued that the requirement of reciprocal monophyly for ESU designation was too restrictive. In fact, the absence of mtDNA variation was consistent with the theoretical expectations based on the period of known divergence (10,000 years) (Hedrick et al. 2006).

Based on the ESU criteria set forth by Crandall et al. (2000), the two conservation units can be scored as ecologically and genetically nonexchangeable. Thus, a consensus seems to have emerged to recognize two ESUs of *P. occidentalis* (box-fig. 12.1) (Minckley 1999; Parker et al. 1999; Hedrick et al. 2006).

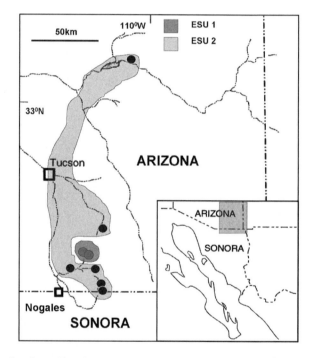

Box-figure 12.1 The distribution of two ESUs of Sonoran topminnow (*Poeciliopsis occidentalis*). Monkey Spring and Cottonwood Spring populations were assigned to ESU 1, while all other populations within the Gila drainage were assigned to ESU 2 (Hedrick et al. 2001b; Hedrick et al. 2006).

servation status has not improved for any of the species listed in 1989, and status has declined for 34% of the listed species (Jelks et al. 2008; table 12.3).

12.2.3 Threats to poeciliids

The primary threats to poeciliids fall into four broad categories: (1) restricted range, (2) habitat destruction/degradation, (3) overexploitation, and (4) impacts associated with nonnative species (Jelks et al. 2008; table 12.3). Many habitats are restricted in part due to natural, as well as anthropogenic, isolation. For instance, eight (73%) *Gambusia* species in the United States are endemic to small spring systems or restricted to portions of streams. Fish species with restricted ranges are more vulnerable to extinction. For instance, the construction of the Amistad Reservoir in 1968 caused the extinction of the Amistad gambusia (*G. amistadensis*) (Johnson & Hubbs 1989).

Table 12.2   Poeciliid fishes evaluated according to IUCN Red List listing criteria

Species	Status	Distribution
Poecilia latipunctata	Critically endangered	Mexico
Poecilia sulphuraria	Critically endangered	Mexico
Poecilia sphenops	Data deficient	Mexico to Colombia
Priapella bonita	Extinct	Mexico
Xiphophorus clemenciae	Data deficient	Mexico
Xiphophorus couchianus	Critically endangered	Mexico
Xiphophorus gordoni	Endangered	Mexico
Xiphophorus meyeri	Endangered	Mexico
Gambusia alvarezi	Vulnerable	Mexico
Gambusia amistadensis	Extinct	United States
Gambusia eurystoma	Critically endangered	Mexico
Gambusia gaigei	Vulnerable	United States
Gambusia georgei	Extinct	United States
Gambusia heterochir	Vulnerable	United States
Gambusia hurtadoi	Vulnerable	Mexico
Gambusia krumholzi	Vulnerable	Mexico
Gambusia longispinis	Vulnerable	Mexico
Gambusia nobilis	Vulnerable	United States
Gambusia senilis	Lower risk/near threatened	Mexico, United States
Gambusia speciosa	Data deficient	Mexico, United States
Gambusia nicaraguensis	Data deficient	Guatemala to Panama
Poeciliopsis monacha	Data deficient	Mexico
Poeciliopsis occidentalis	Lower risk/near threatened	Mexico, United States

Source: IUCN 2008.

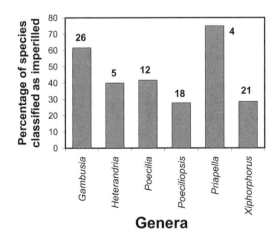

Figure 12.1   The percentage of imperiled taxa for poeciliid genera. The numbers above the bars represent total imperiled taxa for each genus. Adapted from Scharpf 2008 and modified according to Jelks et al. 2008.

Likewise, the type locality of the Big Bend gambusia (*G. gaigei*), Boquillas Spring, dried up in 1957, causing the extirpation of this population (Hubbs & Springer 1957).

Habitat impacts are often associated with water management practices (dewatering, diversions) that have resulted in habitat loss and/or habitat fragmentation (Minckley 1999). For example, topminnows (*P. occidentalis*) probably functioned as large metapopulations that went though periods of isolation and connectedness (Minckley 1999). However, these habitats are now highly fragmented, resulting in widely dispersed, isolated remnant populations. The disruption of migration corridors now calls for managers to evaluate the costs and benefits of human-assisted migration to facilitate gene flow and recolonization (boxes 12.1 and 12.2) (Minckley 1999). Also, the only habitat for Clear Creek gambusia (*G. heterochir*) was dammed prior to the formal description of this species (Johnson & Hubbs

Table 12.3 Imperiled poeciliid fishes of North America and their status

	Species	Status[a]	Status change[b]	Threats[c]
1	Gambusia alvarezi	E	stable	H R
2	G. amistadensis	X	stable	H I R
3	G. clarkhubbsi	E	new	H R
4	G. eurystoma	V	stable	H R
5	G. gaigei	E	stable	H I R
6	G. sp. cf. gaigei	E	new	H R
7	G. georgei	Xp	stable	H R
8	G. heterochir	E	declined	I R
9	G. hurtadoi	E	declined	H R
10	G. sp. cf. hurtadoi	E	declined	H I R
11	G. krumholzi	V	new	H R
12	G. longispinis	E	declined	H R
13	G. nobilis	E	declined	H I
14	G. senilis	T	declined	H I
15	G. sp. cf. senilis	E	declined	H R
16	G. speciosa	T	new	H I
17	Heterandria jonesii	V	new	H R
18	H. sp. cf. jonesii	V	new	H I R
19	Poecilia catemaconis	V	new	H O R
20	P. chica	V	new	H R
21	P. latipunctata	E	declined	H R
22	P. sulphuraria	T	declined	H R
23	P. velifera	V	new	H R
24	Poeciliopsis catemaco	V	new	O I R
25	P. latidens	T	new	H
26	P. occidentalis	E	declined	H I
27	P. sonoriensis	T	stable	H I R
28	P. turneri	V	new	H R
29	Priapella bonita	X	declined	H I R
30	P. compressa	T	new	R
31	P. olmecae	T	new	R
32	Xiphophorus clemenciae	T	declined	H R
33	X. couchianus	E	stable	H I R
34	X. gordoni	E	stable	H I R
35	X. kallmani	V	new	I R
36	X. meyeri	E	stable	H I R
37	X. milleri	E	new	H I R

Source: Jelks et al. 2008.
[a]V = vulnerable; T = threatened; E = endangered; X = extinct; Xp = possibly extinct .
[b]Compared with Williams et al. 1989 .
[c]H = habitat modification/destruction; O = overexploitation; I = impact of nonnatives; R = restricted range.

1989). Deforestation and associated habitat fragmentation remain an important threat to poeciliids in Central America and South America (Bussing 2008). These impacts are hard to quantify, as new poeciliid species have been recently discovered in South American Atlantic forests (Lucinda 2005b, 2008).

Habitat degradation is also associated with land use practices such as overgrazing, as well as with the introduction of various pollutants. Degradation of water quality associated with land use practices has contributed to the decline of some poeciliid populations (Minckley 1999). Pollution effects are often sublethal but are nevertheless important. Pollution associated with paper mill effluent has been shown to affect sexual development of eastern mosquitofish (*Gambusia holbrooki*) (Howell et al. 1980; Toft et al. 2004; Orlando et al. 2007). Females exposed to paper mill effluent were masculinized, developing a gonopodium-like anal fin (Howell et al. 1980). Furthermore, reduced pregnancy rates (Toft et al. 2004) and decreased embryo production (Orlando et al. 2007) were reported for paper-mill-exposed mosquitofish.

Pollution can also disrupt chemical communication and thereby facilitate hybridization (see Rosenthal & García de León, **chapter** 10). Fisher et al. (2006) provided experimental evidence that *Xiphophorus birchmanni* females preferred conspecific males in clean water but mated indiscriminately (*X. birchmanni* females and *X. malinche* males) in water polluted with agricultural runoff. This appeared to be mediated by humic acid, which results from degradation of organic matter. *Xiphophorus birchmanni* females exposed to elevated humic acid did not discriminate between conspecific and heterospecific males (Fisher et al. 2006).

Selected poeciliids have been threatened by overexploitation. The aquarium trade is the major cause for overexploitation, with 90% of freshwater aquarium fishes coming directly from the wild (Olivier 2001). For instance, both *Poecilia catemaconis* and *Poeciliopsis catemaco* are imperiled due to commercial exploitation (Miller 2005). The other threat of the aquarium trade is the release of ornamental fishes outside their native range (section 12.3).

The introduction of nonnative species has co-occurred with the rapid decline of many poeciliids, especially the topminnows. Minckley and Deacon (1968) suggested that the rapid decline of *P. occidentalis* is mainly due to the introduction of aggressive *Gambusia affinis*. Furthermore, Meffe (1985b) hypothesized that niche overlap combined with the predaceous nature and high reproductive rate of *Gambusia* explained the decline of *Poeciliopsis*. Collectively, it is important to note that the decline of many poeciliid species has been due to various combinations of the above-listed threats (table 12.3; Johnson & Hubbs 1989; Minckley 1999).

12.3 Impacts of poeciliids on native biota

Poeciliids have been introduced worldwide primarily through the ornamental-fish trade and their presumed ability to control mosquitoes. For example, poeciliids have been introduced to Australia (Arthington & Lloyd 1989; Morgan et al. 2004), Europe (Almaca 1995; Garcia-Berthou et al. 2005), the Mediterranean basin (Crivelli 1995; Goren & Ortal 1999), Africa (Welcomme 1988), and various countries in Asia (Ng et al. 1993; Pethiyagoda 2006). Two poeciliid species were introduced to Sri Lanka for malaria control, and two additional species were introduced via undocumented ornamental-fish releases (Pethiyagoda 1991). In Australia, five poeciliids escaped due to the ornamental-fish trade, while one species was introduced for biological control of mosquitoes (Lintermans 2004). The freshwater fish fauna in Singapore is dominated by four poeciliids and one cichlid introduced through the ornamental-fish trade (Ng et al. 1993). Moyle (1976, 2002) reported six nonnative poeciliids from California, five of which were introduced as ornamental-fish releases and one as a mosquito control agent.

Mosquitofish (*G. affinis* and *G. holbrooki*) have received much attention for their presumed value in controlling mosquito-borne diseases. These two species are native to the southeastern United States but now occur on every continent except Antarctica due to aggressive introduction programs which began in the early twentieth century (Van Dine 1907; Krumholz 1948; Welcomme 1988; Pyke 2008). For example, the western mosquitofish (*G. affinis*) was introduced from Texas to Hawaii in 1904 (Van Dine 1907, 1908) and to California in 1922 (Moyle 1976). Mosquitofish were subsequently introduced from Hawaii to the Philippines (Seale 1917) and New Zealand (McDowall 1990). Likewise, mosquitofish populations were rapidly established throughout the state of California, and by 1934 mosquitofish had been introduced to Nevada (Stockwell et al. 1996). Finally, mosquitofish introductions have often been facilitated by ordinary citizens attempting to control local mosquito populations (Stockwell et al. 1996).

Nonnative poeciliids have been very successful at invading new habitats. Garcia-Berthou et al. (2005) reported a 96.8% establishment rate of *Gambusia* species in Europe, the highest among the top 10 most frequently introduced aquatic species in the world. Establishment success of mosquitofish is associated with a number of key characteristics, including broad diet, broad physiological tolerance, rapid population growth rates, high genetic variability, high levels of aggression, and high dispersal tendencies (Arthington & Mitchell 1986; Ehrlich 1986; Arthington 1989; Courtenay & Meffe 1989; Leberg 1990;

Grether & Kolluru, **chapter 6**). Recent work has shown that dispersal tendencies are higher for invasive species of mosquitofish (*G. affinis* and *G. holbrooki*) than for noninvasive species of mosquitofish (*G. geiseri* and *G. hispaniolae*) (Rehage & Sih 2004). Furthermore, Alemadi and Jenkins (2008) reported that *G. holbrooki* readily dispersed in shallow water, increasing the likelihood that populations could establish themselves by traveling through networks of drainage ditches.

The efficacy of mosquitofish in controlling mosquitoes is controversial (Bence 1988; Pyke 2008), but their negative impacts on native biota have been well documented (Deacon et al. 1964; Minckley & Deacon 1968; Schoenherr 1981; Courtenay & Meffe 1989; Pyke 2005, 2008). Here, we briefly review the wide array of poeciliid impacts on native biota, focusing on invertebrates, fish, and amphibians.

12.3.1 Poeciliid impacts on invertebrates

Poeciliids are generally omnivorous, feeding on floating insects, chironomid larvae, zooplankton, odonate nymphs, mites, molluscs, crustaceans, ephemeropterans, and oligochaetes, as well as fish eggs and larvae (Hurlbert et al. 1972; Farley & Younce 1977; Rees 1979; Dussault & Kramer 1981; Hurlbert & Mulla 1981; Bence 1988). Hurlbert et al. (1972) provided one of the earliest experimental evaluations of mosquitofish impacts on ecosystems. Their controlled mesocosm experiments demonstrated that mosquitofish caused reductions of zooplankton (rotifers and crustaceans) and aquatic-insect populations and a subsequent increase in phytoplankton populations (Hurlbert et al. 1972). Thus, mosquitofish apparently can act in a top-down manner (Carpenter & Kitchell 1988) by causing a reduction in zooplankton densities, which in turn release phytoplankton from zooplankton grazing.

Mosquitofish can also affect the community composition and biodiversity of a system. For instance, *G. affinis* eliminated *Daphnia pulex* and *Ceriodaphnia* sp. populations and significantly reduced many other zooplankton and macroinvertebrate taxa in experimental ponds (Hurlbert & Mulla 1981). In another controlled study, back swimmer, damselfly, and dragonfly populations were significantly reduced in the presence of mosquitofish (Farley & Younce 1977). Also, *Megalagrion* damselflies native to Oahu, Hawaii, were absent from all lowland habitats occupied by introduced poeciliids: *G. affinis*, *Poecilia mexicana*, *Poecilia reticulata*, and *Xiphophorus hellerii* (Englund 1999). A Hawaiian native atyid shrimp was eliminated from its anchialine pool habitats within six months of the introduction of guppies (Brock & Kam 1997).

12.3.2 Poeciliid impacts on fish

The impacts of mosquitofish and other poeciliids on native fish have received much attention. In some cases the impacts have been well documented, but in many others impacts have been inferred from the simultaneous establishment of mosquitofish and decline of native fish(es). In fact, many authors have attributed the decline and extinction of native fishes to the concurrent introduction of nonnative mosquitofish (Miller 1961; Hubbs & Brodrick 1963; Deacon et al. 1964; Minckley & Deacon 1968; Pister 1974; Soltz & Naiman 1978).

Earlier work was largely correlative, yet the repetition of species replacement across a variety of systems provided convincing evidence for the negative impacts of nonnative poeciliids on native species. For example, endangerment of White River springfish (*Crenichthys baileyi*) and Moapa dace (*Moapa coriacea*) followed the establishment of introduced guppies (*P. reticulata*) and shortfin mollies (*P. mexicana*), respectively (Deacon et al. 1964; Scoppettone 1993). Likewise, Hubbs and Brodrick (1963) reported the loss of many populations of the endangered Big Bend gambusia (*G. gaigei*) following the establishment of western mosquitofish (*G. affinis*).

There has been considerable experimental work evaluating the effects of poeciliids on native species (table 12.4). Experimental work and field surveys implicated heavy predation by mosquitofish (*G. affinis*) on Sonoran topminnow (*P. sonoriensis*) juveniles as the primary cause for the rapid replacement of topminnow populations by introduced mosquitofish (Meffe 1985b; Galat & Robertson 1992). The results of controlled mesocosm experiments showed that *G. holbrooki* had negative impacts via size-selective predation on experimental populations of the least killifish (*Heterandria formosa*) (Lydeard & Belk 1993; Belk & Lydeard 1994). More recently, Mills et al. (2004) reported that *G. affinis* caused reduced body growth and severely reduced survival of young of the year in least chub (*Iotichthys phlegethontis*). *Gambusia affinis* also caused reduced population growth of experimental populations of the White Sands pupfish (*Cyprinodon tularosa*) (Rogowski & Stockwell 2006a).

Mechanistically, impacts by mosquitofish are largely due to hybridization, resource competition, and predation on fish eggs, larvae, young, or even adults (Myers 1965; Moyle 1976; Shakunthala & Reddy 1977; Meffe 1985b; Courtenay & Meffe 1989; Kelley & Brown, **chapter 16**). In a few circumstances, nonnative poeciliids hybridized with native congeners, which increased risk of extinction for the native species. For example, Contreras and Escalante (1984) reported hybridization of introduced *X. hellerii* and *X. varia-*

tus with endangered *X. couchianus*. Furthermore, western mosquitofish (*G. affinis*) hybridized with endangered Clear Creek gambusia (*G. heterochir*) (Johnson & Hubbs 1989), although recent work has shown that this hybridization has been limited (Davis et al. 2006).

Predation by nonnative poeciliids on the eggs and larvae of native species has been shown to be important in many systems. For instance, Barrier and Hicks (1994) provided experimental evidence for *G. affinis* preying on fry of Australian black mudfish (*Neochanna diversus*). Rincon et al. (2002) reported evidence for heavy predation by *G. holbrooki* on juveniles of two endangered Spanish toothcarps (*Aphanius iberus* and *Valencia hispanica*).

Nonnative poeciliids have also been shown to be aggressive toward adult fish. For instance, Gill et al. (1999) conducted tank experiments to evaluate the impact of *G. holbrooki* on an endemic Australian fish. This work showed that the degree of caudal fin damage and mortality of western pygmy perch (*Edelia vittata*) was directly correlated with *G. holbrooki* density (Gill et al. 1999). Likewise, reduced growth of adults was reported for another Australian fish, *Pseudomugil signifer*, in the presence of *G. holbrooki* (Howe et al. 1997).

Another likely aspect of poeciliid impact on native fish is transmission of exotic parasites and diseases (Arthington & Lloyd 1989; Eldredge 2000). Native freshwater fishes in Hawaiian streams had helminth parasites but only in sympatry with exotic poeciliids such as guppies (*Poecilia reticulata*) and green swordtails (*X. hellerii*) (Font & Tate 1994). Furthermore, Font (1997b, 2003) showed that 4 out of 11 helminth parasites of native Hawaiian freshwater fishes originated from exotic poeciliids. Similarly, the Asian tapeworm (*Bothriocephalus acheilognathi*) was presumably co introduced with mosquitofish (*G. affinis*) and now infects endangered Mohave tui chub (*Siphateles bicolor mohavensis*) (Archdeacon 2007).

Although most case studies reported negative impacts of introduced poeciliids on native species, neutral interactions and coexistence of nonnative and native fishes have been observed occasionally (Barrier & Hicks 1994; Maddern 2003; Ling 2004; S. Henkanaththegedara and C. Stockwell, unpublished data). It appears that coexistence can occur under special circumstances such as reciprocal predation and/or minimum niche overlap (Barrier & Hicks 1994; Ling 2004; S. Henkanaththegedara and C. Stockwell, unpublished data). In an unpublished thesis, Maddern (2003) reported no significant impacts of nonnative *Phalloceros caudimaculatus* on *Edelia vittata*, an Australian endemic species.

We recently found that nonnative mosquitofish (*G. affinis*) may not impact native populations of the endangered

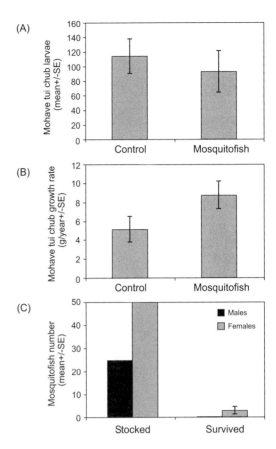

Figure 12.2 Mohave tui chub–mosquitofish interactions as measured during a controlled mesocosm experiment. (A) Mosquitofish presence had no effect on Mohave tui chub (*Siphateles bicolor mohavensis*) larval survival ($t = 0.567$; $P < 0.05$). (B) Mohave tui chub had higher growth rates in the presence of mosquitofish ($t = 0.567$; $P < 0.05$). (C) The number of surviving mosquitofish is depicted relative to the initial number of mosquitofish stocked (25 male and 50 female mosquitofish in each mesocosm). S. Henkanaththegedara and C. A. Stockwell, unpublished data.

Mohave tui chub (*Siphateles bicolor mohavensis*). These two species have coexisted for at least eight years at one site (S. Parmenter, California Department of Fish and Game, pers. comm.). Furthermore, a controlled mesocosm experiment revealed that mosquitofish presence had no effect on Mohave tui chub larval survival (fig. 12.2A). In fact, Mohave tui chub gained higher body mass in the presence of mosquitofish (fig. 12.2B). Interestingly, mosquitofish survival in experimental mesocosms was low, probably because of predation by Mohave tui chub (fig. 12.2C). Only large females that exceeded the gape size of Mohave tui chub survived the mesocosm experiment (S. Henkanaththegedara & C. Stockwell, unpublished data).

12.3.3 Poeciliid impacts on amphibians

Nonnative poeciliids also negatively impact native amphibians by preying on amphibian eggs and larval stages (Grub

Table 12.4 Experimental evidence of negative impacts of mosquitofish (*Gambusia spp.*)

Taxa	Impact	Overall impact	Reference
Impacts on invertebrates			
G. affinis	Reduced zooplankton and insect populations; high algal densities	Negative	Hurlbert et al. 1972
G. affinis	Reduced aquatic macroinvertebrate abundance in rice fields	Negative	Farley & Younce 1977
G. affinis	Reduced pelagic aquatic invertebrates; increased algae and some benthic invertebrates	Negative	Hurlbert & Mulla 1981
G. affinis	Reduction of aquatic macroinvertebrates in rice fields	Negative	Bence 1988
Impacts on fish			
G. affinis	Replacement of *Poeciliopsis occidentalis* by predation	Negative	Meffe 1985b
G. holbrooki	Reduced population growth of *Heterandria formosa*	Negative	Lydeard & Belk 1993
G. holbrooki	Size-selective predation on small *Heterandria formosa* in experimental mesocosms	Negative	Belk & Lydeard 1994
G. holbrooki	Reduced growth and lack of egg survival of *Pseudomugil signifer*	Negative	Howe et al. 1997
G. holbrooki	Caudal fin damage and mortality of *Edelia vittata*	Negative	Gill et al. 1999
G. holbrooki	Heavy predation on *Aphanius iberus* and *Valencia hispanica* juveniles	Negative	Rincon et al. 2002
G. affinis	Reduced growth and survival of *Iotichthys phlegethontis* young of year	Negative	Mills et al. 2004
G. affinis	Reduced population size and biomass of *Cyprinodon tularosa*	Negative	Rogowski & Stockwell 2006a
G. affinis	No impact on larval survival; increased body growth of *Siphateles bicolor mohavensis*	Neutral	Henkanaththegedara & Stockwell, unpublished data
Impacts on amphibians			
G. affinis	Elimination of *Hyla regilla* tadpoles	Negative	Hurlbert & Mulla 1981
G. affinis	Predation on *Taricha torosa* larvae	Negative	Gamradt & Kats 1996
G. affinis	Predation on *Hyla regilla* tadpoles	Negative	Goodsell & Kats 1999
G. affinis	Delayed metamorphosis and reduced growth rates of *Rana aurora draytonii*	Negative	Lawler et al. 1999
G. holbrooki	Reduced survival of endangered *Litoria aurea* tadpoles	Negative	Hamer et al. 2002

1972; Bradford et al. 1993; Brönmark & Edenhamn 1994). The decline of many amphibian populations has been correlated with the introduction of nonnative poeciliids to previously fishless water bodies (Gamradt & Kats 1996; Goodsell & Kats 1999; Hamer et al. 2002). Controlled laboratory experiments and field surveys have revealed neg-ative impacts of mosquitofish on amphibians, as measured by the survival of tadpoles and larvae, delayed metamorphosis, and reduced larval growth rates (Gamradt & Kats 1996; Goodsell & Kats 1999; Lawler et al. 1999; Hamer et al. 2002). For example, Lawler et al. (1999) showed that *G. affinis* caused injury to the tadpoles of the endangered

California red-legged frog (*Rana aurora draytonii*) and caused reduced growth rates. Likewise, *G. holbrooki* reduced the survival of endangered green and golden bell frog (*Litoria aurea*) tadpoles in southeastern Australia (Hamer et al. 2002).

12.4 Poeciliids as model organisms in conservation biology

As in evolutionary and ecological studies (e.g., see Schlupp & Riesch, **chapter 5**; Grether & Kolluru, **chapter 6**; Langerhans, **chapter 21**), poeciliids have been used extensively to test many important questions in the field of conservation biology. This is largely because their short generation time and small size make poeciliids excellent model organisms for controlled experimental studies. Furthermore, many poeciliid species have been extensively translocated (Hendrickson & Brooks 1991), allowing retrospective analyses of the factors associated with population persistence (Sheller et al. 2006), as well as the genetic and evolutionary consequences of historic translocations (Stearns 1983b; Stockwell et al. 1996; Stockwell & Weeks 1999).

Much of the work on the conservation biology of poeciliids has focused on population genetics. For example, understanding population structure is central to identifying appropriate conservation units (box 12.1). This information is particularly useful for managing metapopulations that have experienced recent habitat fragmentation, such as the case with desert topminnows (*Poeciliopsis* spp.) (Minckley 1999). The population genetic structure of poeciliids is influenced by the interactive effects of genetic drift, gene flow, and divergent selection (sexual and/or natural) (see Breden and Lindholm, **chapter 22**). Management practices that influence gene flow among population segments could potentially erode genetic structure, some of which may be adaptive (box 12.2). Thus, the benefits and costs of gene flow should be considered before establishing an artificial gene flow regime (boxes 12.1 and 12.2).

Poeciliids have also been used to examine questions central to the management of endangered species concerning the effects of inbreeding depression, outbreeding depression, and the loss of genetic variation (Leberg 1990, 1993; Stockwell et al. 1996; van Oosterhout et al. 2007b). For instance, Leberg (1990) reported that experimental populations of *G. holbrooki* with reduced genetic variation had lower population growth rates. More recently, van Oosterhout et al. (2007b) examined the effects of inbreeding during captivity on reintroduction success using captive and wild populations of Trinidadian guppies (*Poecilia reticulata*). Captive-reared and wild fish were released to a streamside mesocosm to evaluate population responses.

They found that captive-reared guppies were more vulnerable to gyrodactylid parasites (58% survival) than wild guppies were (96%). They attributed this effect to reduced variation at immune function genes due to inbreeding and lack of previous parasite exposure during captivity (van Oosterhout et al. 2007b).

Many workers have considered the importance of genetics to the conservation and management of poeciliids (Vrijenhoek et al. 1985; Quattro & Vrijenhoek 1989; Leberg 1990; Stockwell et al. 1996; Sheffer et al. 1997). For instance, Vrijenhoek et al. (1985) reported that the Monkey Spring population (located in Arizona) of the Gila topminnow (*P. occidentalis*) had no genetic variabiliity as measured with 25 allozyme loci. These findings were striking because this population had been used as brood stock for extensive restoration efforts (Vrijenhoek et al. 1985; Simons et al. 1989). Theory indicates that populations with little to no genetic variability are not likely to survive over the long term when introduced to a variety of habitat types. Quattro & Vrijenhoek (1989) reported that the lower genetic variation of the Monkey Spring population was associated with lower fitness compared with other populations of *P. occidentalis* with greater genetic variation. This work was used to argue for changing the restoration brood stock from the Monkey Spring population to a population with greater genetic variation (Sharp Spring) (Quattro & Vrijenhoek 1989).

Subsequent work by Sheffer et al. (1997) compared these same populations and found no relationship between population genetic diversity and fitness. These contrasting results may in part be explained by different experimental environments. Sheffer et al. (1997) pointed out that mortality rates were substantially lower for their study than for Vrijenhoek et al.'s study (Vrijenhoek et al. 1985). Thus, it is possible that a correlation between genetic diversity and fitness can be observed only under harsh environmental conditions (Stockwell & Leberg 2002).

These two studies emphasize that factors other than genetic variation should also be considered in selecting brood stock for restoration efforts (Sheffer et al. 1997). For instance, Sheffer et al. (1997) recommended that managers use geographically proximate populations for restoration. Managers should consider the evolutionary history of the species when designing restoration plans. Thus, it is critical to have sound data concerning the evolutionary distinctiveness of subspecies and population segments to guide conservation and restoration efforts (see box 12.1).

The persistence of translocated poeciliid populations has broad implications for conservation biology due to the extensive use of translocation as a tool for managing and restoring various protected species (Hendrickson & Brooks 1991; Sheller et al. 2006). For instance, the Gila topminnow (*P. occidentalis*) has been extensively translocated in

Box 12.2 Evolutionary considerations in the replacement of native poeciliids by nonnative poeciliids

The rapid replacement of many native desert fishes by nonnative species has in part been explained by evolutionary constraints. Mosquitofish (*Gambusia affinis*) have replaced native Gila topminnows (*Poeciliopsis occidentalis*) throughout much of their range. However, some topminnow populations persist in the presence of mosquitofish (Meffe 1984), raising questions about the ecological and evolutionary factors that promote coexistence.

Meffe (1984) pointed out that topminnow populations often persisted in habitats vulnerable to flash flooding. Following such events, topminnows typically dominated the fish community, and mosquitofish increased in number until the next flood (Minckley 1999). Minckley and Meffe (1987) hypothesized that the evolutionary history of these two species likely selected for different responses to flooding. Whereas topminnows had evolved in an environment with periodic flash flooding, this was not the case for western mosquitofish (Minckley & Meffe 1987). Indeed, Meffe (1984) found that topminnows outperformed mosquitofish in swimming trials.

Evolutionary history may also have played a role in the response of native species to nonnative species (Miller 1961). Because topminnows evolved in species-depauperate habitats, the introduction of *G. affinis* represented a novel selective threat in terms of both predation and competition. Indeed, nonnative poeciliids seemed to have their greatest harmful effects as novel predators (section 12.3.2; Miller 1961; Courtenay & Deacon 1983). Thus, earlier workers have recognized the importance that evolutionary histories can have in determining the outcome of native and nonnative species interactions. The previous work essentially treated these species as evolutionarily static, although it is plausible that contemporary evolution plays an important

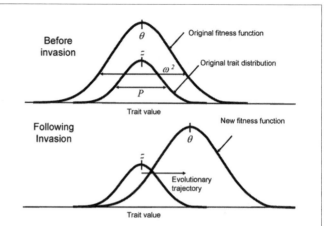

Box-figure 12.2 A representation of trait variation and selection before and after the invasion of a poeciliid likely to prey on small larvae. The upper curve in each panel represents the fitness function acting on a trait (the higher the curve, the higher the fitness of a given trait value). The lower curve represents the frequency distribution of trait values in the population. *P* is the phenotypic variance for the trait, θ is the optimal trait value, ω^2 is the strength of stabilizing selection around the optimum, and *z* is the mean trait value. Evolution for faster larval growth rates would be expected under this scenario. Figure adapted from Stockwell et al. 2003.

role in the persistence of native poeciliids in the face of novel selection pressures (Stearns 1983b; Reznick et al. 1997; Stockwell & Weeks 1999).

Here, we briefly describe how contemporary evolution can play a role in the demise or persistence of populations. The evolutionary response of a population Δz is the product of the selection gradient β and the additive genetic variance G: $\Delta z = G\beta$ (box-fig. 12.2). Thus, we may ask if a native population of topminnows has sufficient genetic variation to evolve in response to novel selection pressure associated with nonnative mosquitofish. For instance, mosquitofish are highly piscivorous on fish larvae (Courtenay & Meffe 1989) and thus may exert strong selection pressure on early life-history traits. From previous poeciliid studies (Stearns 1983b; Reznick et al. 1990; Stockwell & Weeks 1999), we hypothesize moderate levels of additive genetic variation for poeciliid

an attempt to restore this species to its former range (Simons et al. 1989; Hendrickson & Brooks 1991; Minckley 1999). Sheller et al. (2006) found that a number of factors were important for the persistence of translocated Gila topminnow populations. For instance, populations founded late in the year (July–December) persisted up to five times longer than populations established in May or June. Most translocations occur early in the year, so persistence time may increase simply by translocating populations later in

the season. They also found that topminnows persisted longer when introduced to ponds/lakes than when introduced to streams, wells, or tanks. Finally, they found that populations established from Monkey Spring did not persist as long as populations exhibiting higher levels of heterozygosity (Sheller et al. 2006). By contrast, they did not find any evidence that initial population size had any effect on population persistence (Sheller et al. 2006). Collectively, they recommended that translocations should be conducted

life-history traits, suggesting that they should exhibit a response to selection due to predation (see Reznick et al. 1990).

Thus, we can envision an undisturbed topmminnow population as being centered on an optimum value and the introduction of moquitofish changing this optimum to a new location (i.e., larger embryo size). This creates a mismatch between current phenotype and optimal phenotypes (box-fig. 12.2), leaving the topminnow population subject to directional selection, which can be represented as

$$\beta = \frac{-(z - \theta)}{\omega^2 + P}$$

where z is the mean trait value, θ is the optimal trait value, P is the phenotypic variance for the trait, and ω^2 is the strength of stabilizing selection around the optimum (the original optimum before disturbance, the new one after disturbance) (Arnold et al. 2001). An evolutionary response is expected if there is sufficient genetic variation.

Even evolving populations can go extinct if selection is too strong and/or there is insufficient genetic variation (Lynch 1996; Boulding & Hay 2001). Population size declines during the early phases of evolution. Thus, if selection is too strong or if the population is too small, population extinction is likely. In fact, a highly adapted population can go extinct due to demographic stochasticity (Gomulkiewicz & Holt 1995).

Gene flow can also have important effects on evolutionary potential. Gene flow has been characterized as the Jekyll and Hyde of conservation due to its potential benefits and costs (Stockwell et al. 2003). Gene flow can facilitate adaptive divergence by enhancing genetic variation and thus evolutionary potential (Frankham 1995). Alternatively, gene flow may limit adaptive divergence if the immigrants are not locally adapted (Lenormand 2002; but see also Räsänen & Hendry 2008). For example, gene flow imposed across a predator selection gradient should be carefully considered (section 12.4). This example is especially relevant, as predators have repeatedly been shown to have important evolutionary effects on poeciliids (Reznick et al. 1990; Reznick et al. 1997; Johnson 2001a; Langerhans et al. 2005)

Therefore, from an evolutionary perspective, one may examine the role of these factors in topminnow population persistence in the presence of nonnative mosquitofish. For instance, population sizes and associated genetic variation vary among topminnow populations (Minckley 1999; Hedrick et al. 2001b). Thus, evolutionary potential is also likely to vary among populations. Furthermore, contemporary evolution can be facilitated or constrained by factors such as gene flow (see Räsänen & Hendry 2008).

It is also important to further examine the evolutionary response of invading species. Meffe (1984) pointed out that topminnows were evolutionarily adapted to periodic flooding, whereas mosquitofish had evolved in the absence of such selection pressure. We suggest that an important unresolved question is whether this relationship has been maintained. It seems plausible that nonnative mosquitofish populations may well have evolved in response to novel selection pressures associated with flooding (see Collyer et al. 2005). Thus, it seems prudent to assess whether mosquitofish swimming performance has evolved in response to the novel flow regime associated with desert aquatic systems: periodic severe flash floods.

We conclude that an evaluation of the evolutionary dynamics of both native and invasive poeciliids can provide important insights regarding species replacement and coexistence. Poeciliids are ideal for such studies, as a number of important life-history and morphological traits have been well studied (Stearns 1983b; Reznick et al. 1990; Reznick et al. 1997; Stockwell & Weeks 1999; Langerhans et al. 2005).

---

in late summer or later and utilize fish from sources other than Monkey Spring and target ponds and lakes as introduction sites (Sheller et al. 2006; see box 12.1). These data, combined with genetic data on conservation units (Hedrick et al. 2001b; Hedrick et al. 2006), have allowed for a rich scientific foundation for managing desert topminnows (see also box 12.1).

Poeciliids can also serve as useful models of the evolutionary consequences of translocating populations to new habitats (Stockwell et al. 1996; Stockwell & Weeks 1999). Mosquitofish provide a best-case scenario for the retention of genetic diversity because females have multiply sired broods (Constantz 1989; see table 18.1 in Evans & Pilastro, **chapter 18**). Thus, translocated populations are likely to include the genes of males that were not actually relocated (Stockwell et al. 1996). Mosquitofish also have rapid population growth rates (Leberg 1990), greatly reducing the likelihood of prolonged genetic bottlenecks.

Many workers have conducted retrospective genetic surveys to determine the loss/retention of genetic variation of nonnative poeciliid populations (Scribner et al. 1992; Stockwell et al. 1996; Lindholm et al. 2005; Shoji et al. 2007). Scribner et al. (1992) examined genetic diversity for populations of G. affinis translocated from three sites in Texas to Hawaii in 1905. They found no evidence for a loss of genetic diversity. In another study, Stockwell et al. (1996) studied populations in California and western Nevada derived from the translocation of 900 fish from two populations in Texas in 1922. Despite predicting that the California and Nevada populations would retain high levels of genetic diversity, they found that allelic diversity was severely reduced, with the elimination of most rare alleles. These data combined with theoretical expectations (Allendorf 1986) suggested a severe initial bottleneck of fewer than 10 individuals (Stockwell et al. 1996). These findings were surprising and suggested that even under ideal conditions, genetic diversity of introduced populations can be compromised.

Lower levels of genetic diversity are expected to limit the evolutionary potential of managed populations (Frankham 1995). However, two poeciliid case studies have reported results contrary to this expectation. Stockwell & Weeks (1999) reported rapid life-history evolution for recently established populations of mosquitofish, despite the fact that these populations recently descended from a population that had experienced a severe bottleneck (Stockwell et al. 1996). Likewise, guppies introduced to Australia had low genetic variation as measured by microsatellites (Lindholm et al. 2005), despite the fact that these populations exhibit high levels of additive genetic variation for morphological traits (Brooks & Endler 2001a).

These case studies illustrate the importance of obtaining data sets that include both neutral markers and phenotypic variation. In fact, as Lynch (1996) pointed out, conservation of biodiversity is based on the preservation of phenotypic variation. These case studies also illustrate the importance of considering evolution on a contemporary time scale. In fact, it now appears that evolution on ecological time scales is rather common, requiring managers to take an evolutionary approach to the conservation of biodiversity (Stockwell et al. 2003; Kinnison et al. 2007). Studies of poeciliids have shown that the same factors associated with the current extinction crisis are also important selective factors that promote contemporary evolution: habitat destruction/degradation, habitat fragmentation, exotic species, and altered harvest schedules (Reznick & Ghalambor 2001; Stockwell et al. 2003; box 12.2).

Many cases of contemporary evolution have been documented for introduced poeciliid populations (Stearns 1983b; Reznick et al. 1990; Stockwell & Weeks 1999).

Thus, evolution may play an important role in the invasion dynamics of nonnative species (Stockwell & Weeks 1999; see also box 12.2).

Evolutionary dynamics should be also considered in the conservation and management of predators (Reznick et al. 2008). Following predator removal, certain key traits under natural selection are likely to evolve, such as size at maturity and escape performance (Reznick et al. 1990; Ghalambor et al. 2004). For instance, rapid life-history evolution toward later age of maturity has been observed for guppies released from high predation pressure (Reznick et al. 1990). By contrast, guppy populations exposed to increased predation pressure fared poorly. Simply stated, in these circumstances fish are often consumed before they reach maturity. These effects apparently increase the risk of extinction for naive prey species (Reznick et al. 2008). In fact, Reznick et al. (2004) found that guppy (Poecilia reticulata) populations adapted to predator absence were likely to go extinct when exposed to predators. These findings are consistent with observations of nonnative fishes severely impacting native fishes in the American Southwest deserts. In these regions, fishes that have evolved in the absence of predation have been extremely vulnerable to the introduction of nonnative fish predators (Miller 1961; see also box 12.2).

Environmental novelty is also of concern for the expected environmental changes associated with global climate change. For instance, increased sea levels are likely to have important consequences for coastal ecosystems. Purcell et al. (2008) examined evidence for local adaptation to salinity among coastal populations of G. affinis. They found variation among populations in salinity tolerance, suggesting that populations may vary in their response to rising sea levels (Purcell et al. 2008).

12.5 Conclusions and future directions

Poeciliids have received considerable attention from conservation biologists and evolutionary ecologists. This work has examined threats to protected species as well as the threat of nonnative poeciliids to native species and ecosystems. Poeciliids have been excellent model/surrogate species for addressing basic questions in conservation biology and evolutionary ecology.

Despite this attention, many poeciliid species are at risk of extinction, and little progress has been made in securing poeciliids in the last 20 years. Furthermore, the conservation status of poeciliids outside the United States is largely unknown, mainly due to poor understanding of species diversity, distribution, and natural history. Thus, one critical direction for further research is to obtain more data on the taxonomy, distribution, and conservation status of poeciliids in Mexico, Central America, and South America.

Evolutionary theory and ecology should be integrated to address the issues concerning the invasion biology of poeciliids. Recently, workers have examined ecological factors such as niche breadth to determine habitat features that may be modified to promote coexistence between native species and invasive mosquitofish (Laha & Mattingly 2006; Ayala et al. 2007). A similar approach that incorporates evolutionary theory could prove profitable. For instance, additive genetic variation should be estimated for key traits (i.e., swimming performance, temperature tolerance, and early life-history traits) in both invasive and protected fish species to better predict the evolutionary prospects for species coexistence. Such evolutionarily enlightened management may offer important insights regarding factors that will promote the persistence of poeciliid populations under the stress of novel anthropogenic induced selection.

Acknowledgments

We thank R. Rader, B. Kowalski, J. Fisher, J. Evans, A. Pilastro, I. Schlupp, and two anonymous reviewers for reviewing an earlier version of this chapter. This work was partially supported by a National Park Service Grant administered through the Mojave National Preserve (Dr. Debra Hughson) to C.A.S. We dedicate this chapter to the late Dr. Clark Hubbs for his lifelong dedication to the study and conservation of native fishes and especially the many species of *Gambusia*.

Part III

Behavior and cognition

Chapter 13 Group living

Jens Krause, Richard James, and Darren P. Croft

13.1 Introduction

POECILIID FISHES in general and the Trinidadian guppy, *Poecilia reticulata*, in particular have been important study systems for the investigation of group living over the last 50 years or so. Most poeciliids live in groups at some stage of their lives, and the results from numerous studies have provided important insights into the costs and benefits associated with living in groups (Godin & Davis 1995; Krause & Godin 1995; Dugatkin & Wilson 2000), the importance of familiarity and kinship in grouping (Griffiths & Magurran 1997a), the role of social learning (Laland & Williams 1997; Webster & Laland, chapter 14), and the evolution of group living and cooperative behavior (Magurran et al. 1995; Dugatkin 1997). Poeciliids are such good study species in this context because many species are relatively easily observed in the wild and at the same time they are easy to keep and breed in captivity. They are also well suited for experimental laboratory research due to their small size and hardiness. Another aspect that has long fascinated evolutionary biologists is that different populations of a given poeciliid species can often be found in close geographic proximity but under very different ecological conditions (Magurran 2005). This exposure to different ecological factors makes them excellent study organisms for evolutionary processes. We can look for correlations between ecological factors and phenotypic or genotypic differences between populations and thereby learn how natural selection acts on animal populations—a method known as the comparative approach (Harvey & Pagel 1991).

Group living is a widespread behavior found in many different taxonomic groups (Krause & Ruxton 2002). A number of adaptive benefits associated with living in groups have been identified. For example, in a classic experiment, Neill and Cullen (1974) demonstrated that the success of various predators (including both cephalopod and fish predators) was reduced when attacking large shoals. Such antipredator benefits are thought to result from a combination of factors, including early predator warning (the many-eyes effect), predator confusion, and the encounter dilution effect (Krause & Ruxton 2002; Magurran 2005; Kelley & Brown, chapter 16; see box 13.1 for mechanisms of antipredator defenses). As we will see in the latter sections of this chapter, the role of predation in the evolution of grouping behavior has been highlighted in many studies on poeciliids. In addition to the antipredator benefits of group living, there may also be foraging benefits for shoaling fish. For example, fish in larger shoals may locate food with greater speed (Pitcher et al. 1982). Another proposed benefit of group living includes a reduction in the energetic costs of locomotion (Krause & Ruxton 2002). An interesting aspect of social behavior in a number of poeciliid species is that grouping by females may actually reduce sexual harassment (Pilastro et al. 2003). As the benefits of grouping increase with an increase in group size, so too do many of the associated costs, particularly competition for limited resources (Pitcher & Parrish 1993). For example, fish in larger shoals may allocate more time to feeding to compensate for the increased competition for food (Grand & Dill 1999; Johnsson 2003). The decision to join a social group thus represents the outcome of a dynamic

Box 13.1 Brief overview of antipredator defenses of grouping

The predator confusion effect manifests itself in a reduced ability of predators to single out and successfully attack prey (Neill & Cullen 1974; Milinski 1977; Ioannou et al. 2008). The processing of multiple prey targets is assumed to result in a lower strike accuracy by the predator (Krakauer 1995). Therefore, information degradation in neural networks has been suggested as the underlying mechanism (Tosh et al. 2006; Tosh & Ruxton 2006). To date, little work has been done integrating neural network modeling with experimental work (but see Ioannou et al. 2008 for an exception).

The encounter dilution effect was first proposed by Turner & Pitcher (1986) and comprises two components: the probability of predators encountering prey groups and the probability of an individual being attacked when a group has been encountered. The encounter dilution effect can provide a lower per capita predation risk if the probability of predators encountering groups is not proportional to group size.

Predator swamping occurs when mass emergence or synchronized reproduction reduces per capita predation risk. Given that prey handling time cannot be reduced to zero, there is an upper limit to the number of prey that a predator can capture and consume per time unit. Synchronized emergence of prey can exploit this constraint on predators (Krause & Ruxton 2002).

Colony breeding can potentially reduce offspring mortality, particularly in centrally located nests, because of group defense. The predator is likely to be attacked by multiple colony members (bluegill sunfish, *Lepomis macrochirus*: Dominey 1981).

The probability of detecting a predator is generally believed to increase with increasing group size through the many-eyes effect (Godin et al. 1988; Cresswell 1994). The underlying idea is that a larger number of group mates will result in more individuals being vigilant. There is good empirical support for this theory despite the fact that some of the underlying details of vigilance models have been challenged (Krause & Ruxton 2002).

social behavior between populations both within and between species.

In this chapter we give an overview of what has been learned about the grouping behavior of poeciliids in the field and in the lab. We begin by considering the literature on shoal choice preferences of poeciliids and follow this by reviewing data on population comparisons to draw wider conclusions on the evolutionary causes of variation in preferences for shoal size and composition. Finally, we introduce a social-networks approach (which is increasingly recognized as a potentially very powerful approach in the behavioral sciences) to bring all the available information together (Krause et al. 2007; Croft et al. 2008).

13.2 Shoal choice preferences

13.2.1 Shoal size

Detailed data on group size distributions in the field are surprisingly hard to find in the animal behavior literature in general (Krause & Ruxton 2002), and poeciliids are no exception despite the fact that they have been studied extensively in the wild. Trinidadian guppies are usually found in small- to medium-sized shoals of under 10 fish per shoal, with larger shoals of 20 or more fish being rare (Magurran & Seghers 1991; Croft et al. 2003; fig. 13.1). For example, Magurran and Seghers (1991) found that the percentage of solitary individuals ranged from 0.4% to 84% in eight different rivers. Another striking observation was that individuals encountered another fish or a shoal on average every 14 seconds, which indicates that the social behavior of these fish is likely to be highly dynamic—a topic we will return to in section 13.4, on social networks (Croft et al. 2003).

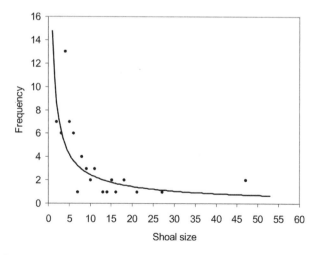

Figure 13.1 Group shoal size distribution of a guppy population in the Arima River, Trinidad. Power trend line was fitted as $Y = 14.8x^{-0.78}$; $r^2 = 54$, $p < 0.001$. Modified from Croft et al. 2003.

cost-benefit trade-off, which may be influenced by a number of factors such as predation risk, parasite load, body size, nutritional state, the availability and distribution of resources, and reproductive strategies (Krause & Ruxton 2002). As a result, there is great variability in the degree of

Under laboratory conditions it has been shown that poeciliids generally prefer the larger of two stimulus shoals when introduced to a novel environment (*P. reticulata*: Lindstrom & Ranta 1993; Krause & Godin 1995; Lachlan et al. 1998; Ledesma & McRobert 2008; *Poecilia formosa* and *Poecilia latipinna*: Schlupp & Ryan 1996; *Xiphophorus hellerii*: Buckingham et al. 2007; *Gambusia holbrooki*: Agrillo et al. 2007; Agrillo et al. 2008b). The preference for the larger of two shoals has been attributed to the larger amount of activity due to the greater number of moving individuals in species such as guppies (Krause & Godin 1995) and zebrafish (Pritchard et al. 2001). It is possible to decouple fish number and activity by keeping larger shoals in colder water (reducing activity) and smaller shoals in warmer water (increasing activity). By contrast, Agrillo et al. (2008b) reported an activity-independent preference for the larger of two groups in mosquitofish (*G. holbrooki*). The preference of fish for larger group sizes in novel environments in captivity is probably explained by the fact that per capita risk of predation is lower in larger groups, which has been demonstrated experimentally in several fish species, including poeciliids (guppy: Krause & Godin 1995; guppy and other species: Neill & Cullen 1974; minnow, *Hybognathus nuchalis*: Landeau & Terborgh 1986; anchovy, *Stolephorus purpureus*: Major 1978). Therefore, the ability or willingness to discriminate between groups of different sizes might strongly depend on the selection pressure experienced by individuals in different populations. For instance, Landeau and Terborgh (1986) reported that predation risk in shoals of 8 minnows (*H. nuchalis*) or larger was very low because antipredator defenses were highly effective for such group sizes. This would suggest that at least in the context of a predator threat (and the predator species tested) these fish would not need to be able to discriminate on the basis of group size for groups larger than 8 fish.

Experiments in which the environmental context was varied have demonstrated that shoal size changed according to context, with large shoals forming in the presence of olfactory cues of predators and fish dispersing when they sense food cues (Hoare et al. 2004). To do this they do not need to assess shoal size as such but simply need to modify their local interaction rules (Hoare et al. 2004). To join larger groups, for instance, individuals simply need to extend their attraction zone—the distance within which they respond to conspecifics with an approach. Moreover, when it comes to the formation of larger shoals, it seems unlikely that fish assess actual numbers when simple rules of self-organization are sufficient to explain not only the formation of groups and the observed group size distributions but also the spatial position of individuals within groups (Couzin et al. 2002; Couzin & Krause 2003). Some groups of pelagic fish can be extremely large. A study by DeBlois and Rose (1996) reported a school of Atlantic cod

(*Gadus morhua*) that was more than 10 kilometers across. It seems highly unlikely that individual fish can accurately assess the size of such groups or monitor where exactly they are within such a school. Nevertheless, the modification of local interaction rules (attraction, repulsion, polarization) provides a simple means by which individuals can take up particular positions within a group (front, middle, or back) that are adaptive in terms of foraging rates and predation risk (Couzin et al. 2002). Another important but often overlooked aspect is that fish shoal size is not simply a result of individual decisions but instead can be influenced on a more global scale by fish density (Hensor et al. 2005). Hensor et al. (2005) successfully predicted shoal size as a function of fish density in captivity using an individual-based model. However, when the same model was used to predict shoal size under field conditions for the littoral zone of a lake, it turned out that the model greatly underestimated observed shoal size. The reason for this underestimate was that in addition to the social responses (of fish to each other), nonsocial aggregation factors such as environmental heterogeneity (e.g., variation in water temperature or food abundance) can play an important role for group size (fish: Hensor et al. 2005; locusts: Bouaïchi et al. 1996).

Recent work by Chapman et al. (2008a) showed that, in guppies, fish density during early development can also play an important role for shoaling behavior in later life. Juveniles that were reared at low densities showed a stronger shoaling tendency than individuals reared under high densities.

13.2.2 Shoal composition

Group size is not the only factor influencing shoal choice. Another factor is shoal composition. Over the years a number of important factors have been identified that play a role, such as body size, body color, sex, familiarity, kin, species, and parasite load (Krause et al. 2000). In most cases individuals assort by phenotype to form uniform shoals, in which predation risk is lower than in heterogeneous ones due, for instance, to the predator confusion effect (Krause et al. 2000). Several studies have demonstrated that phenotypically "odd" fish that differ in size or color from the rest of the shoal are preferentially targeted by predators (Landeau & Terborgh 1986; Theodorakis 1989). This so-called oddity effect reduces the predator confusion effect because it allows the predator to visually lock on to a particular prey individual (see box 13.1). Interestingly, this results not only in a higher predation risk for the individual that is being targeted but also in a slightly raised per capita predation risk for all shoal members (Landeau & Terborgh 1986). This phenomenon should result in a tendency to associate with phenotypically matched con- or heterospecifics and might explain why individuals in many fish shoals look so homogeneous.

Most of the information that we have on shoal mate preference in poeciliids comes from laboratory studies, and relatively little is known about the composition of free-ranging poeciliid shoals. We review these studies in the subsequent sections.

13.2.2.1 Body size. Most of the available data come from investigations on guppies and reveal that body size assortment is an important factor influencing shoal composition (field: Croft et al. 2003; lab: Lachlan et al. 1998), as in other shoaling freshwater teleosts (Krause et al. 2000). There is evidence that shoal assortment in guppies is driven by direct social choice of shoal membership based on body size (Lachlan et al. 1998; Croft et al. 2003). However, indirect social sorting by habitat preferences based on water depth and shoal fission is also a likely candidate for maintaining body size assortment in guppies (Croft et al. 2003). Chapman et al. (2008b) demonstrated the interactions between juvenile and adult guppies can also partly depend on early experiences. Juveniles reared with adults spent less time shoaling with adults in later life than juveniles that had not experienced adult fish before.

13.2.2.2 Body color. A preference for specific color morphs has been reported from domesticated black mollies and white mollies, *P. latipinna* (McRobert & Bradner 1998). However, color preference may have been confounded with familiarity because the two color morphs were kept in separate holding tanks. This would mean that black mollies become familiarized with black ones, and white ones with white ones. Therefore, an alternative interpretation of McRobert and Bradner's result is that mollies simply preferred fish with whom they were familiar, regardless of color. The same authors also reported that background coloration had an influence on color segregation in mollies, suggesting that these preferences could have adaptive benefits in terms of reducing predation risk (Bradner & McRobert 2001). However, given that black mollies and white mollies are pet-shop fish that have been bred in captivity for color, it is not clear what selective forces would have produced these preferences in captivity in the absence of any predators. Experiments with different strains of zebrafish have shown that body patterns are learned during an early period in life and individuals that show these body patterns are preferred as shoal mates in later life (Engeszer et al. 2004).

13.2.2.3 Sex. Male guppies are known to switch shoals more often than females (Croft et al. 2003), which is presumably triggered by their search for new mating oppor-

tunities given that males prefer novel females to familiar ones (Kelley et al. 1999). In contrast, adult female guppies often form long-lasting associations with other females in connection with high site fidelity (Croft et al. 2005). Shoal choice experiments by Lindström and Ranta (1993) on guppies and by Agrillo et al. (2008a) on mosquitofish (*G. holbrooki*) demonstrated that males prefer to shoal with females rather than with males. Females, on the other hand, are unlikely to prefer shoals of males given that males engage in sexual harassment, which can reduce females' foraging success (guppy: Magurran & Seghers 1994a; mosquitofish: Pilastro et al. 2003; see also Kelley and Brown, **chapter 16**). Given the differences in size and pattern between the sexes in many poeciliid species, we might also expect that females would prefer to shoal with other females (and not males) to keep the shoal homogeneous in appearance, thus reducing predation risk (Krause & Ruxton 2002).

In some species of poeciliids females may actually experience reduced sexual harassment in social groups, which is thought to be an important factor in determining shoal choice behavior. For example, Dadda et al. (2005) reported that female mosquitofish form more cohesive shoals in the presence of males, while Agrillo et al. (2006) reported that females prefer to shoal with the larger of two groups in the presence of a male. In mosquitofish, when several males compete for the same female, mating attempts tend to be monopolized by the dominant male, which in turn reduces the level of sexual harassment endured by females (Bisazza & Marin 1995). Accordingly, female mosquitofish have been found to vary their shoaling behavior when harassed by a male, approaching relatively larger males to promote male-male interactions and reduce sexual harassment (Dadda et al. 2005). Taken together, the results of these studies suggest that levels of male sexual harassment and male-male competition may be important determinants of social organization and shoal structure in some poeciliids.

In most cases shoaling preferences are tested in the laboratory first and then investigated under field conditions (Krause et al. 2000). An interesting exception in this regard is the phenomenon of sex segregation by guppies, which has so far been documented only in the field (Croft et al. 2006a). Females prefer the deeper parts of the river where they can avoid sexual harassment from males, but this comes at the cost of potentially higher risk of predation. Given that Godin and McDonough (2003) showed that cryptic coloration reduces predation risk, this type of segregation probably occurs because the larger body size of females and their more cryptic coloration allow them to inhabit deeper habitats at lower risk than males. A recent study (Darden & Croft 2008) showed that females turn to, and swim into, the deeper part of the river when attempting to evade harassment from males, providing the potential mechanism behind the pattern of segregation described by Croft et al. (2006a).

13.2.2.4 Familiarity. A preference for familiar individuals over unfamiliar ones has been found in guppies (Griffiths & Magurran1997a, 1997b) and persists over at least 5 weeks in isolated fish (Bhat & Magurran 2006). Preferences based on familiarity have also been reported in many other fish species, and even between different species (Ward et al. 2003; Ward & Hart 2003). In some cases it has been demonstrated that the fish prefer individuals that smell of a familiar microhabitat, whereas in others genuine individual recognition has been invoked (Ward et al. 2004; Griffiths & Ward 2006). The former can be established relatively quickly (e.g., within hours in the three-spined stickleback, *Gasterosteus aculeatus*: Ward et al. 2007), whereas the latter seems to take considerably longer (e.g., 12 days in the guppy: Griffiths & Magurran 1997a). Familiarity preferences have also been demonstrated in the context of diet odors in guppies (Morrell et al. 2007) and also seem to be mediated by predation pressure (Godin et al. 2003).

The benefits of shoaling with familiar individuals have received relatively little attention in poeciliid fishes, despite the fact that such information is essential if we are to understand the adaptive basis of the behavior. Studies of the shoaling behavior in species such as fathead minnows, *Pimephales promelas*, and three-spined sticklebacks, *Gasterosteus aculeatus*, suggest that familiar individuals show greater shoal cohesion (than unfamiliar ones) (Chivers et al. 1995; Webster et al. 2007), thereby possibly promoting the predator confusion effect and thus reducing predation risk. Previous work on guppies also suggests that shoals of familiar individuals may outperform nonfamiliar shoals in foraging tasks (Morrell et al. 2008), which is similar to findings reported for three-spine sticklebacks (Ward & Hart 2005). Social familiarity in fish is also known to be important for mediating aggression (Utne-Palm & Hart 2000) and stabilizing group hierarchy (Hojesjo et al. 1998). Clearly there is a need to collect more information on the potential benefits of familiarity in poeciliids (and other fish species).

13.2.2.5 Kin. There is considerable speculation on why shoaling with kin might be beneficial to individuals but no direct experimental support that demonstrates clear fitness benefits. It has been suggested that grouping with kin could promote cooperative behavior, which in turn could reduce predation risk during behaviors such as predator inspection because siblings might be less willing to desert each other in such risky situations (Dugatkin 1997). Whether guppies assort with kin in the wild is unclear. Russell et al. (2004) found no kin assortment in adult guppies of the Quaré and lower Tacarigua rivers in Trinidad's Northern Range mountains (which are both high-predation sites). However, this could have been due to the lack of the more powerful genetic and statistical tools now available. Hain and Neff (2007) reported no kin structure for low-predation adult fish populations from the Paria River but found that juveniles used phenotype matching to identify kin. In a recent laboratory study Evans and Kelley (2008) showed that juvenile guppies preferred to shoal with full siblings rather than half siblings, indicating that relatedness plays a role in shoaling decisions, at least in juvenile fish. Recent work on juveniles indicates that high-predation populations are assorted by kin, whereas low-predation populations are not (Piyapong et al., in press). However, similar studies on a different population of juvenile guppies from the Tacarigua River showed that fish had no preference to shoal with unfamiliar kin but strongly preferred familiar non-kin (Griffiths & Magurran 1999; see also Warburton & Lees 1996). Hain and Neff (2007) explained the discrepancy between their study and that of Griffiths and Magurran (1999) by differences in the degree of multiple paternity (intrabrood multiple paternity) and predation risk in the Paria and Tacarigua rivers.

In summary, new and more powerful methods of detecting relatedness are continually being developed and may shed more light on kin-structured interactions. It is also important to carefully distinguish between the juvenile and adult stages and between different populations. We might expect to find kin preferences to be more strongly expressed in high-predation sites where movement is restricted and cooperation with kin in antipredator contexts is more highly rewarded. Similarly, it is more likely that we should find kin assortment in juveniles than in adults because the fish are most vulnerable to predation early in life, and the association with cooperative kin could raise survivorship. This means that juveniles in high-predation areas are probably the most likely candidates for kin assortment. However, Hain and Neff (2007) pointed out that the degree of multiple paternity could be an additional factor to which we should pay attention because in populations where multiple paternity is low, kinship preferences could be acquired through familiarity, whereas in populations with high multiple paternity, familiarity is not necessarily a good indicator of close relatedness.

13.2.2.6 Species. Examples of species preferences are found in different species of mollies and in guppies. Amazon mollies (*P. formosa*) and sailfin mollies (*P. latipinna*) both prefer to shoal with conspecifics but, if presented with a larger shoal of heterospecifics, may eventually join the latter if the shoal is large enough to outweigh the benefits of conspecific shoaling (Schlupp & Ryan 1996). Warburton and Lees

(1996) reported that guppies more readily associated with swordtails (*X. hellerii*) if they had been raised with them, demonstrating that species preference has an underlying developmental component. Evidently, early rearing experience is important in assessing assortment preferences.

13.2.2.7 Parasites. Parasites are known to have profound effects on the behavior of poeciliid fishes (see Cable, **chapter 8**), including social interactions. For example, parasites have the potential to alter the costs and benefits of group living. From the perspective of an uninfected individual, association with infected fish potentially increases the risk of becoming infected if the parasite can be transmitted directly between individuals (Barber et al. 2000). Furthermore, the presence of an infected fish in a group may increase the risk of predation for other group members because parasitic infections often change the behavior and physical appearance of infected individuals, which in turn may reduce levels of shoal coordination and make groups more heterogeneous looking, thus increasing predation risk (see box 13.1). From the perspective of an infected individual, parasitism may decrease competitive ability and/or increase energy demands, which could make group living disadvantageous (Krause & Godin 1996a; Barber et al. 2000). The effect of parasitism on social behavior in poeciliids has been investigated in a recent study by Tobler and Schlupp (2008b). In their study, female western mosquitofish (*Gambusia affinis*) were infected with the trematode *Uvulifer* sp. (Diplostomatidae), a parasite that alters the phenotypic appearance of infected individuals by causing the formation of black spots on the body surface. Tobler and Schlupp's (2008b) work suggests that the shoaling tendency of infected fish is significantly reduced because of the greater energy demands on them due to the parasite. Their work further suggests that associating with infected conspecifics may be costly. When given a choice between an infected and a healthy stimulus fish, both infected and healthy focal fish preferred to associate with the healthy stimulus fish. This latter result may be because associating with a phenotypically odd individual (i.e., an individual that has black spots due to parasitism) could increase the risk of predation due to the oddity effect. Interestingly, Tobler and Schlupp (2008b) speculated that the reduced shoaling tendency of infected fish may be due to manipulation of the host by the parasite. The final host of the parasite is a piscivorous bird, and a reduction in shoaling tendency is expected to increase predation risk, which could be beneficial to the parasite.

13.3 Population comparisons

One of the major strengths of the guppy system in Trinidad, as well as other poeciliid systems, is that comparisons between multiple populations are possible (Seghers 1973; Magurran & Seghers 1991; Reznick et al. 1997; Brown & Braithwaite 2004; Magurran et al. 1995; see also Kelley and Brown, **chapter 16**). This means that we can look beyond the behavior of individuals in a particular population and examine general evolutionary trends. This approach can address long-standing questions such as whether grouping tendency and/or phenotypic assortment increase with predation pressures. Seghers (1973, 1974) was probably the first to document the variation in shoaling behavior among guppy populations in Trinidad. His work suggested that differences in predation risk were responsible for the observed variation in shoaling behavior—a topic that has attracted major attention in this system and in other comparable poeciliid systems. Seghers's approach led to the designation of high-predation and low-predation sites, defined mainly by the presence or absence of a major guppy predator, the pike cichlid, *Crenicichla frenata* (Magurran 2005). Magurran and Seghers (1991) measured shoaling tendency in eight guppy populations and found considerable variation in this behavior among populations (0.4%–84% of solitary fish as a percentage of the total number of individuals observed). This variation in shoaling behavior appeared to be related to the predation pressures experienced in the different rivers, with populations experiencing high predation having a higher tendency to shoal than those inhabiting low-predation sites (Magurran & Seghers 1991).

One of the strengths of the comparative approach is that it can identify broad evolutionary trends by means of population comparisons, but one of its weaknesses is that it uses correlations, which do not necessarily indicate causal relationships. More direct evidence for the link between predation pressures and shoaling behavior came from transplantation experiments. In 1957 Caryl Haskins and Edna Haskins moved about 200 guppies from a high-predation river to a low-predation river (that contained only *Rivulus hartii*, a minor guppy predator). An analysis of the shoaling behavior of the descendants of these fish showed that the levels of shoaling behavior had decreased over an approximately 30-year period (Magurran et al. 1992). To deal with potentially confounding environmental factors that may influence grouping decisions, shoaling behavior was tested in fish that had been raised under standardized conditions in the lab and that were the first-generation descendents of wild-caught individuals (Magurran et al. 1992). Further transplantations of guppies by Endler in 1976, in which fish were also moved from a high-predation to a low-predation river, paved the way for studies that investigated a greater range of life-history traits. This work demonstrated that descendants of the introduced fish had changed relative to the ancestral population in phenotypic traits such as offspring size, body size at sexual maturity, and brood size (Reznick et al. 1997; for a review, see Magurran 2005).

One of the challenges to the comparative approach is to accurately measure the ecological variables of interest. Recent studies that presented small guppy groups in glass bottles that can be dangled at different water depths in the field made it possible to assess the relative predation risk experienced by guppies in terms of the number of attacks and the predator species attacking in different rivers (Croft et al. 2006a; Botham et al. 2008). This way of quantifying predation risk allows a refinement in the study of predator-prey interactions because we can now assess the predation risk on a continuous scale rather than a binary one (i.e., presence or absence of *Crenicichla* spp.). The use of a continuous scale for predation risk means that quantitative predictions for how social behavior should change with predation risk can be made. The potential of this approach is illustrated by a recent study by Botham et al. (2008) that showed that the strength of antipredator behavior across seven guppy populations was affected by both the type of predator to which they were exposed and the level of predation risk they experienced in the wild. Interestingly, guppies from high-risk populations showed a heightened response to predators compared with those from lower-risk populations, but only when exposed to the predator species that posed the greatest risk.

Previous work on the effect of predation risk on shoaling tendency, foraging behavior, mating strategies, and sex segregation has been largely restricted to diurnal predators of guppies. However, there are important nocturnal predators as well (such as *Hoplias malabaricus* and *Rhamdia* catfish) that could have important effects on guppy behavior. Yet little is known about the relationship between day- and nighttime risks to guppies. The only study to date is the one by Fraser et al. (2004), which showed that guppies (in Ramdeen Stream and Guanaopo River, Trinidad) will engage in night feeding if predation pressure is low and that night feeding can be as profitable as daytime feeding. Further studies on nighttime predation risk and the consequences for guppy behavior are needed.

So far population comparisons of social behavior in guppies have focused on population differences in predation risk. However, for guppy populations inhabiting streams in the Northern Range mountains of Trinidad, habitat differences in predation risk are often correlated with productivity (Grether et al. 2001b; Reznick et al. 2001). High-predation rivers tend to have higher levels of productivity than low-predation rivers, where the forest canopy is usually less dense (Grether et al. 2001b; Reznick et al. 2001). This raises the question of whether differences in behavior and life-history strategies such as growth rate are primarily driven by predation or by resource availability. Arendt and Reznick (2005) argued that differences in growth rate are mainly driven by differences in resource availability rather than predation. The interaction between habitat productiv-

ity and group living in guppies remains to be investigated and poses an exciting area for future research.

13.4 Social networks

A promising way to study the social organization of a population is to construct a social network (fig. 13.2; Krause et al. 2007; Croft et al. 2008). Social networks are basically a reflection of the interaction patterns in a population and can be used to capture information on any type of interaction, be it sexual, aggressive, cooperative, associative (as in shoaling), predatory, etc. Most of the above information from the sections on shoal size and shoal composition in poeciliids can be related to the network approach, which means that we can put all this individual-based or group-based information into the context of the population and apply network metrics to it. For instance, if shoaling tendency is high, we might expect to find a more interconnected social network, which in turn may have consequences for how socially learned information is transmitted or for how fast ectoparasites might be transmitted in a population. This means that based on the architecture of the social network, we can make predictions for information transfer, disease transmission, sexual selection, and gene flow in populations—to name but a few possibilities (for reviews, see Krause et al. 2007; Croft et al. 2008).

The methods for studying social networks in poeciliid fishes are well developed (see box 13.2), and the application of social-network analysis to wild populations of guppies has provided new insights into their social organization (fig. 13.3). Despite the fact that guppy shoals meet and

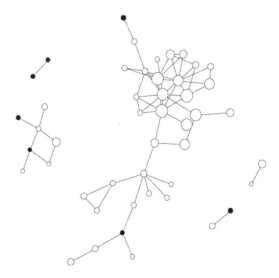

Figure 13.2 A social network of a guppy population where nodes symbolize individuals and lines symbolize connections between them. Node size is an indication of body size, with filled nodes being males and unfilled ones females. Modified from Croft et al. 2008.

Box 13.2 Quantifying social networks in poeciliid fishes

A common (but by no means the only) way to build a network is to use the "gambit of the group" (Croft et al. 2004a; Croft et al. 2008). The approach works best with populations that have low rates of emigration and immigration so that we can study them in relative isolation. Once a promising location (an isolated pool or an area with restricted access) has been identified, all fish are removed from it and taken to the lab, where each one is given a different mark (e.g., using fluorescent elastomer). After marking, they are immediately returned to their natural habitat and left at least for a few hours (or overnight) to recover. With the help of a beach seine it then is possible to remove entire shoals of fish from some areas of the pool on repeated visits to the study site. On each visit multiple shoals are removed from the study site, and each shoal is kept in a separate bucket until no more shoals are found in the sampling area, and we can now identify which individuals are found with which other ones together in shoals. On the basis of shoal membership we assign connections in the network (fig. 13.3). The use of the "gambit of the group" (see Croft et al. 2008 for further details) for constructing a network is based on the assumption that associations in small groups allow opportunities for social learning (see Webster & Laland, **chapter 14**), transmission of diseases, mating opportunities, etc. This assumption is probably justified in the case of the guppy, where shoals are usually relatively small. Once all fish have been identified, each bucket is emptied in the location where the fish shoal was collected. Depending on how often shoals meet and exchange members, the above procedure is carried out once a day or repeated two to three times a day (Croft et al. 2008). Useful tools for visualizing the network (which is generated from observation of shoal memberships over multiple sampling days) and for carrying out basic calculations are NetDraw and UCINET. A visual inspection of the network is usually a good starting point for all further work. It is easy to see at a glance whether the population is largely fragmented into many substructures or dominated by one large network. Using the above methodology, it was found that guppy populations quickly form one large network, which provides a useful basis for many descriptive statistics in network analysis (see review in Croft et al. 2008).

a)
Day 1
2 shoals: (1, 2, 3); (4, 5, 6)

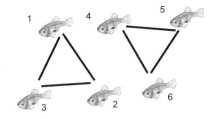

b)
Day 2
1 shoal: (1, 2, 4)

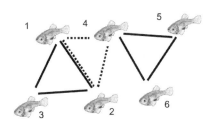

Figure 13.3 Illustration of how a network is constructed from the "gambit of the group." (a) On day 1 two separate shoals are caught, and fish that belong to the same shoal are interconnected. (b) On day 2 another shoal is caught, leading to a single network, where the new connections are indicated by dashed lines.

interact very frequently (Croft et al. 2003), which creates a great potential for social mixing, we found strong social connections between pairs of females in most pools that we studied (Croft et al. 2005). If given a choice between a control fish (of the same sex, size, and pool) and their usual associate, females prefer the latter, indicating that these associations are based on a genuine association preference (Croft et al. 2006a). This observation could have important implications for the evolution of cooperation because repeated interactions with known individuals are important prerequisites.

Further work on the inspection behavior of pairs of females comprising either natural or manufactured pairs (fish that we assigned to each other) strongly supported the idea that females that naturally occur together behave more cooperatively toward each other (Croft et al. 2006a). Furthermore, the network approach also allows the detection of social substructures, so-called communities, which could form important building blocks of the social organization of guppies and partially segregate populations along the lines of sex and size (fig. 13.4) or microhabitat/diet preferences. Given that networks are built from the gambit of the group over extended time periods, communities are not simply the equivalent of shoal composition but represent

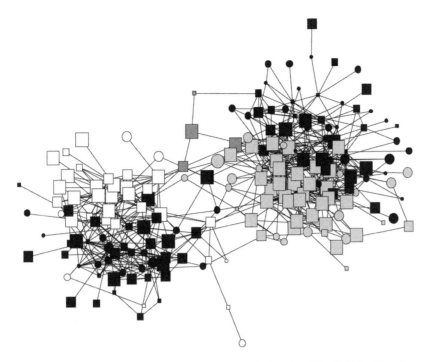

Figure 13.4 Guppy network where five different communities have been identified. The different shades of gray correspond to the different communities. The white and the dark gray communities (top left) inhabit the lower pool, and the light gray, medium gray, and black ones inhabit the upper pool of two interconnected pools. Modified from Krause et al. 2007.

a layer of social organization between the group and the population. The identification of a community structure can help identify phenotypic or genotypic characteristics according to which the population is structured. Furthermore, we could test whether transmission processes (e.g., information, disease) are linked to community structure. For example, a socially learned behavior may be initially restricted to within a community before it starts spreading elsewhere (see also Webster & Laland, chapter 14).

13.5 Future directions

Fish shoaling behavior has been important in helping us understand the functional benefits of living in groups because small freshwater species such as poeciliids have facilitated the study of predator-prey interactions (Krause & Ruxton 2002). As a result we have a good understanding of some of the shoal-size-related benefits of shoaling fish.

Modeling work of shoaling behavior has given us a good appreciation of the mechanisms underlying the fission-fusion processes of shoal formation (Couzin et al. 2002; Couzin & Krause 2003). It is fairly well understood how interactions between shoal members produce a self-organized group, and how individual fish can position themselves within such a group (Couzin & Krause 2003). However, scale changes remain a challenge when model-

ing not only the social interactions between individuals but also interactions of individuals with the nonsocial environment (Hensor et al. 2005). Poeciliids should make for good study systems in this context because their relatively small size and relatively low dispersal rate make it easier to study entire populations and to achieve replication at the population level.

Shoal composition has received less attention than shoal size, probably in part because it is not as easy to study. The oddity effect attracted considerable attention in the past, and some elegant studies (Landeau & Terborgh 1986; Theodorakis 1989) have provided insights into the functional benefits of phenotypic homogeneity of fish shoals. The mechanism underlying phenotypic oddity has been much speculated on, and it is only due to the recent advances in neural-network modeling that we have obtained insights into this complex problem (Tosh et al. 2006). Poeciliid fishes may be particularly suitable for investigating the role of the oddity effect in driving phenotypic assortment in the field. For example, a comparative study of levels of shoal assortment under different predation risks would elucidate the role of predation in selecting for phenotypic assortment.

Relatively little is known about the mechanisms and functions of familiarity preferences and kin preferences. It is clear that olfactory cues (rather than visual ones) play an important role in familiarity and kin preferences in most fish species. Microhabitat-related cues have been identified to be important in some studies on familiarity (Ward et al. 2004), and the major histocompatibility complex (see McMullan & van Oosterhout, **chapter 25**) is believed to play a role in kin recognition (Olsen et al. 1998). However, in experimental studies, the actual cues used for social recognition have not been clearly identified. This should be a major growth area for research on fish studies and work on poeciliids in particular.

Network theory is an important tool for studying cooperation. Theoretical papers have shown that the network structure of a population has far-reaching implications for the evolution of behavioral strategies, including cooperation. The predictions made by these theoretical studies provide a fertile field for experimental biologists (Lieberman et al. 2005; Ohtsuki et al. 2006; Santos et al. 2006; Taylor et al. 2007). Previous work on the evolution of cooperation was entirely focused on interindividual interactions (in the context of the Tit-for-Tit and Prisoner's Dilemma problems; Dugatkin 1997). However, the more recent literature on the evolution of cooperation goes beyond pairwise interactions and takes the social structure of entire populations into account. Explaining the details of this highly theoretical literature would lead us too far in the context of this chapter. However, it is clear that small freshwater fishes such as poeciliids continue to provide a good test bed for theoretical predictions regarding the evolution of cooperation. One of the main reasons is that they are small enough that individuals in entire populations can be marked and observed in the field, which makes population-level predictions testable.

A particular strength of the guppy system in Trinidad, and indeed of several poeciliid systems, is that they often inhabit extremely varied habitats (Magurran 2005). This makes within-species population comparisons (and interspecies comparisons) possible, which aids in determining how different ecological conditions influence social organization. The network approach is an ideal tool for quantifying such social interactions, providing us with a multitude of descriptive statistics (for details, see Krause et al. 2007; Croft et al. 2008). We anticipate much progress in this area over the next few years.

The network approach is not restricted to the gambit of the group. Studies on mating networks could provide insights into sexual selection and the level of gene flow within populations. The fact that guppies have internal fertilization makes them good study subjects in this regard because the network approach is well suited to situations where females mate with multiple males. With the help of networks we can quantify the number of different males that females are, on average, sexually connected to in a mating network. The network approach also allows us to understand who is connected via whom, which means we can quantify the exclusivity of sexual relationship. The latter is particularly important in the context of sperm competition. Furthermore, we can investigate whether individuals with many sexual partners are connected to individuals that also have many partners, which is known as a positive degree correlation in network statistics (Croft et al. 2008). In conclusion, the application of the network approach to the sexual behavior of poeciliids might enhance our understanding of sperm competition and sexual selection (see also Evans & Pilastro, **chapter 18**).

Acknowledgments

We are grateful to Tommaso (the Swan) Pizzari, Jon Evans, and Jenny Kelley for many stimulating discussions. Critical comments by the editors, Culum Brown, Ashley Ward, and one anonymous referee improved an earlier version of this chapter. Funding was provided by the Natural Environment Research Council (J.K.: NE/D011035/1; D.P.C.: NE/E001181/1).

Chapter 14 Innovation and social learning

Mike M. Webster and Kevin N. Laland

14.1 Introduction

THE DAY-TO-DAY LIVES of all animals are complicated by a tangle of competing needs, threats, and options. Individuals can effectively address these challenges only if they possess appropriate knowledge, and animals that live in complex, changeable environments must therefore regularly gather accurate and up-to-date information about their surroundings if they are to operate efficiently (Dall et al. 2005). In simple terms, animals can learn about their environment in two different ways: they can gather private information by personally exploring and sampling the environment, or they can obtain social information by observing or interacting with other animals. The first route may be reliable, but it can also be costly in terms of personal risk or energy expenditure. Given the fact that others too, particularly conspecifics, will face the same challenges, and may already have invested effort in gathering this information, the second route to adaptive behavior, which involves copying others, offers many potential benefits. However, social learning can be error prone too—if, for instance, the information that is acquired is out-of-date, inappropriate, or suboptimal. Learning from others therefore offers rapid and cheap, if sometimes unreliable, solutions to many of life's problems, and such socially acquired information will directly affect an individual's fitness. Accordingly, recent decades have seen an explosion of interest in animal social learning and in the evolution of learning strategies (Heyes & Galef 1996; Galef & Giraldeau 2001; Fragaszy & Perry 2003; Laland 2004; Galef & Laland 2005; Kendal et al. 2005). A great deal of research effort, both theoretical and experimental, has focused on identifying the environmental conditions that promote reliance on social sources of information, determining the trade-offs associated with copying others versus learning from personal experience, and on understanding the psychological processes that underpin social learning (Danchin et al. 2004a; Kendal et al. 2005).

In this chapter we review the research on animal innovation and social learning. Poeciliid fishes gather and exploit information in much the same manner as other animals. They are subject to the same challenges, in that they must feed, avoid predators, and reproduce, and to these ends they adaptively employ a balance of information that they have gathered themselves and information that they have acquired from others. Innovations, which lead to the emergence of new behaviors, and social learning, which facilitates their diffusion through the population, are widespread processes in the animal kingdom and are important processes in virtually all vertebrate, and many invertebrate, taxa. Because of the practical advantages of working with small, cheap, readily available laboratory fishes, poeciliids have proved to be useful models for experimental biologists working in this expanding field. Many poeciliids occur across a diverse range of environments, resulting in population-level variation in a number of life-history traits and leading to their status as established model organisms in numerous disciplines in ecology and evolution (this volume). This is especially true of the guppy, *Poecilia reticulata*, which has been extensively studied within its natural habitat in Trinidad. In this system, research into the various selective pressures acting upon different populations, and their influence upon these populations' adaptive trajectories, spans several decades. This "natural experiment" (Magurran 2005), along with the depth of knowledge about the guppy's ecology and life history and the ease with which it can be maintained and reared under laboratory conditions, has led the guppy to become very widely used in behavioral research, including social learning and innovation. Our review therefore reflects this bias.

We begin by defining "innovation" and "social learning" and describing a number of social-learning processes. We consider the theoretical basis of innovation and social learning, the trade-offs associated with different types of learning, and the conditions under which they are most likely to be deployed. We then review studies of innovation and social learning. While only a handful of studies (exclusively on guppies) have investigated innovation, this work has enabled us to make significant inroads into elucidating the characteristics and environmental conditions under which individuals might innovate. By contrast, a far larger body of work, incorporating a range of poeciliid species, has focused on social learning. Finally, we outline some future challenges facing researchers working in the field and suggest ways in which the poeciliids could be employed to address these.

14.2 Defining innovation and social learning

14.2.1 What is innovation?

Innovation refers to novel learned behavior exhibited by animals: for instance, exploiting new opportunities or overcoming problems brought about by unfamiliar situations or changes in their environments (Reader & Laland 2003). Innovations include existing behavior exhibited in a novel context or new behavior performed in a familiar context (Kummer & Goodall 1985). However, the term "innovation" can also describe the introduction of new solutions into the repertoire of a population (Wyles et al. 1983). Thus, "innovation" can be applied both to the *product* of the creative process (the new or modified behavior pattern itself) and to the creative *process* itself (i.e., the means by which the new or modified behavior pattern is acquired). By this definition, learning is a key component of innovation. Further discussion of the concept of animal innovation can be found in Reader and Laland 2003.

14.2.2 What is social learning?

Social learning is any process through which one or more individuals ("demonstrators") influence the behavior of another individual (the "observer") in a manner that leads the observer to learn something (Heyes 1994; Hoppitt & Laland 2008). Research over the last few decades has identified several social-learning processes that, directly or indirectly, can increase the likelihood of learning in others (Whiten & Ham 1992; Heyes 1994; Galef 1998; Hoppitt & Laland 2008). In the online Glossary on the book's Web site we present definitions for several recognized social-learning processes that have been demonstrated to apply to, or which might plausibly occur among, poeciliid fishes (Hoppitt & Laland 2008). These processes may operate alone or in concert. For fuller discussion of social-learning processes, we refer the reader to comprehensive reviews on the topic (Whiten & Ham 1992; Heyes 1994; Hoppitt & Laland 2008).

14.2.3 Social-learning strategies

Historically, social-learning researchers have tended to assume that learning from others is adaptive (Rogers 1988). Individuals are assumed to benefit by copying because by doing so they take a shortcut to acquiring adaptive information, saving themselves the costs of asocial learning. However, it is now known that this reasoning is flawed (Boyd & Richerson 1985; Rogers 1988; Giraldeau et al. 2002). Despite its cheapness, reliance upon social information alone is not inevitably adaptive. Successive transmission of information between individuals can lead to the accumulation of errors and the formation of cascades of misinformation (Giraldeau et al. 2002). In temporally variable environments social information also has the potential to become outdated or obsolete over time, while in spatially variable environments differences in local optima mean that the relative usefulness of social information will vary among sites. Evolutionary game theory and population genetic models therefore lead to the prediction that animals ought to be highly selective with respect to both the circumstances under which they rely on social learning and the individuals from whom they learn (Boyd & Richerson 1985; Giraldeau et al. 2002). The use of private versus social sources of information by animals can be viewed as a trade-off between accuracy and cost. Boyd and Richerson (1985) proposed a *costly-information hypothesis*, predicting that as the costs of gathering accurate but expensive personal information rise, individuals should come to rely more heavily upon the relatively cheap but potentially inaccurate social information provided by others.

Indiscriminate social learning is therefore not an evolutionarily stable strategy (Boyd & Richerson 1985; Rogers 1988; Giraldeau et al. 2002). Rather, selection should favor a mixed strategy of reliance on both asocial and social sources. Hence, we might expect the evolution of learning strategies that specify when animals should innovate, gather private information, or learn socially, from whom they should learn, and the circumstances under which they should exploit this information. Laland (2004) discusses a number of social-learning strategies that animals can potentially use, some of which are summarized in box 14.1. "When" strategies refer to the circumstances under which individuals should exploit social information, while "who" strategies determine from which individuals they should seek to learn. Social-learning strategies are defined in functional terms and in principle may operate through any of the social-learning processes described in the Glossary.

Despite extensive research on social-learning processes, comparatively little is known about the learning strategies employed by poeciliids, or indeed by nonhuman animals from any taxa. Recent work on nonpoeciliid fish (Cyprinidae, Webster & Laland 2008; and Gasterostidae, Kendal et al. 2009) suggests that fishes have evolved social-learning strategies—in these cases "copy when asocial learning is costly" and "copy if better," respectively (see box 14.1). Many poeciliids represent excellent study systems for the explicit investigation of the evolution of social-learning strategies. In fact, it may be possible for the findings of some previous studies, such as those focusing on directed learning mediated by preferences for shoaling with familiar

---

Box 14.1 Social-learning strategies

Social-learning strategies are evolved rules that specify when individuals will learn from others (when strategies) and from whom they will learn (who strategies).

When strategies refer to the circumstances under which individuals should gather and employ socially learned information. Laland (2004) identifies a number of when strategies, including the following:

The *copy when asocial learning is costly* strategy is derived from Boyd and Richerson's (1985) costly-information hypothesis, which predicts that greater reliance on social learning should be favored as the costs of asocial learning increase.

A *copy when uncertain* strategy may be adaptive to animals that range between areas that differ in risk. Habitat-specific differences in risk require that animals adopt habitat-specific behavior patterns, which are potentially maladaptive when performed in the wrong habitat. If an animal is unable to accurately gauge the risk posed by the habitat that it is occupying, it might be adaptive for it to copy the behavior of others rather than to risk following the wrong course of action and paying the costs of doing so.

Who strategies specify from which individuals observers should seek to learn (Laland 2004). Examples include the following:

Copy the majority strategies may be commonplace. Theory suggests that where selection favors the evolution of social learning, it often promotes conformity, with the probability of acquiring a trait increasing disproportionately with the number of demonstrators performing it (Boyd & Richerson 1985; Henrich & Boyd 1998). Work in guppies provides some evidence consistent with this idea, with observers increasingly adopting behaviors in proportion to the number of demonstrators

(Sugita 1980; Laland & Williams 1997; Lachlan et al. 1998).

Copy if better and *copy the most successful* strategies require individuals to evaluate the payoffs associated with their own and a demonstrator's behavior and to adopt the demonstrator's behavior if its returns are greater. Public information use by nine-spined sticklebacks (*Pungitius pungitius*) provides evidence consistent with the deployment of such strategies by fish (Coolen et al. 2003; van Bergen et al. 2004; Coolen et al. 2005; Kendal et al. 2009).

A *copy familiar individuals* strategy may stem from preferences for associating with familiars, which gives rise to enhanced opportunity to learn from them, both because observers are more likely at any given time to be closer to familiars than to unfamiliar would-be demonstrators and because observers are more attentive to their actions (Swaney et al. 2001). Many fish species are capable of recognizing other individuals. Guppies, for example, recognize conspecifics specifically, by learning their identities (Griffiths & Magurran 1997a), and generally, using self-referent matching of habitat- or diet-derived chemical labels (Ward et al. 2009). Copying familiars is potentially adaptive, since observers are more likely to be sharing the same environment with familiars than with nonfamiliars.

Copying kin may be adaptive since kin are more likely to share genes than are distantly related individuals, making them more likely to perform the same behaviors, react similarly to certain stimuli, or exploit similar resources (CoussiKorbel & Fragaszy 1995). Guppies have been shown to be capable of kin discrimination (Hain & Neff 2007; Evans & Kelley 2008), but it has yet to be shown whether any poeciliid species preferentially learns from kin.

individuals (e.g., Lachlan et al. 1998; Swaney et al. 2001) or on mate-choice copying (reviewed by Witte 2006; Druen & Dugatkin, **chapter 20**), to be interpreted as forms of social-learning strategy.

14.3 Poeciliids and the study of innovation

14.3.1 Experimental studies of innovation

Among the most famous examples of animal innovations are milk-bottle opening by various birds (Fisher & Hinde 1949) and sweet potato washing by macaques (*Macaca fuscata*, Kawai 1965), with much of the subsequent work on this topic also focusing on birds and primates. More recently, Laland and co-workers have used guppies to address fundamental questions about the characteristics of innovators, such as the following: To what extent is the likelihood of innovation determined by internal state? Are some individuals intrinsically more likely to innovate than others? We summarize this work in this section.

In a series of experiments, Laland and Reader (1999a, 1999b) studied foraging innovations by guppies, focusing on individual latency to solve a maze to locate food in relation to other characteristics such as sex, body size, competitive ability, and nutritional state. Being the first individual within a population to solve a maze is not an indication of innovation per se, since solving a maze does not imply that the route has been learned, and learning is central to the definition of innovation adopted here (Reader & Laland 2003). However, these authors investigated both the latency of guppies to complete a given maze upon first exposure and the total numbers of exposures required to learn the maze (i.e., to reach a predetermined criterion latency) and found them to be highly correlated within individuals. In subsequent studies the latency to complete the maze upon first exposure was therefore adopted as a valid proxy measure of innovation.

This work revealed substantial intraspecific variation in the propensity to innovate, variation that was tied to intersexual differences in motivation in adults (but not in sexually immature juveniles), as well as differences in competitive ability, body size, and internal state. In one experiment Laland and Reader (1999a) found that within mixed-sex groups, female guppies were significantly more likely to be the first to solve a maze than males. Hunger level also played a significant role, with food-deprived individuals being more likely to solve the maze before their recently fed group mates. Gross differences in body size also exerted an influence, with smaller individuals solving the maze significantly sooner than larger fish. This latter finding was found to be independent of sex, despite the marked sexual dimorphism in body size exhibited by guppies.

Laland and Reader (1999b) investigated the relationship between competitive ability and innovation, finding an association in male, but not female, guppies. Focusing upon scramble competition, Laland and Reader (1999b) considered two measures of competitive ability: the individual's share of the prey resource during a single bout of competition (a short-term measure of competitive ability) and weight change over a two-week period (a longer-term measure of competitive ability). Scramble competition is a specific form of, usually nonaggressive, competition in which each forager can detect its competitors and attempts to reach and consume prey items before they do. It is distinct from exploitation or depletion competition, where individuals consume resources that are patchily distributed, thus rendering them unavailable to other foragers that arrive later on the scene at a given patch, or contest completion, in which individuals physically and often aggressively compete for ownership of prey (Ward et al. 2006). After monitoring competitive ability, Laland and Reader (1999b) subjected male guppies to maze tasks and reported an inverse relationship between competitive ability and the ability to solve maze tasks.

Finally, individual-level variation in the tendency to innovate, a potential facet of the bold-shy axis (Wilson et al. 1994), is apparently also important in determining the likelihood of a given individual innovating. In an experiment that controlled for the effects of sex, hunger, body size, and other state-dependent factors, Laland and Reader (1999a) showed that individuals that had innovated on two previous occasions were more likely to innovate a third time than those that had failed to innovate when twice presented with the opportunity to do so.

In summarizing the contribution of these experiments to our understanding of animal innovation, Laland and van Bergen (2003) argue that for guppies, and most likely other species, the propensity to innovate when foraging is largely a product of necessity. It was seen that females, small fish, hungry individuals, and poor competitors tend to innovate more readily than males, large fish, satiated individuals, and good competitors. These findings are consistent with the interpretation that hunger, driven by a lack of foraging success or by the energetic costs of growth and pregnancy, leads individuals to be driven to access resources in novel ways. For instance, female guppies can store sperm, and as a result, once they reach adulthood they are virtually always gravid; it is the availability of food, rather than mates, that constrains female fitness. Female guppies exhibit indeterminate growth (while males stop growing when they reach maturity), and in females there appears to be a direct relationship between how much food a female gathers and her fecundity (Reznick 1983). Given the absence of maternal care in guppies, the cost-benefit trade-off appears to favor female foraging innovation. Conversely, male fitness is constrained by the availability of receptive mates, rather than food, and they will seemingly switch priority to search for novel food sources only when very hungry (Griffiths 1996).

14.3.2 Trade-offs between conformity and innovation

Innovation and social learning could be regarded as opposing tendencies, since the first introduces novelty and the latter encourages homogeneity in behavior. Social learning in guppies has been shown to increase with the frequency of the prevailing behavior (Laland & Williams 1997; Lachlan et al. 1998). Furthermore, the costs of nonconformity to common behavior may be severe, especially if nonconformity results in differences in space use, because smaller groups are known to forage less efficiently than larger groups (Day et al. 2001) and because individuals will forfeit some of the antipredator benefits of shoaling (Krause & Ruxton 2002). Given this, the transmission of social information provides the opportunity to conform, while the disadvantages of not conforming impose fitness costs on innovation. This gives rise to a paradox: if individuals have a strong tendency to copy others, how can new behavior patterns arise? The *social-release hypothesis* (Brown & Laland 2002) was proposed to resolve this dilemma and predicts that individuals will increase their rate of innovation in the absence of knowledgeable demonstrators. Indeed, using an antipredator task, Brown and Laland (2002) found that innovation became more frequent when the availability of social information was reduced, relaxing the pressure to conform.

14.4 Poeciliids and the study of social learning

A substantial body of research has focused on identifying social learning and the processes that lead to it, as well as detecting individual differences in social learning. Little is known about the social-learning strategies that are used by poeciliids, although the findings of some studies suggest that specific strategies are being deployed. In the following sections we review this research, focusing on social foraging, antipredator behavior, and mate-choice copying.

14.4.1 Foraging

Foraging as part of a group enables individuals to learn about the characteristics or location of food by monitoring others. The majority of research on this topic has focused on how individuals use visual cues to learn the location of food patches. Less is known about how (or if) poeciliids use this information to learn about patch or prey quality, or whether they use information from other sensory modes, such as food-derived chemical cues, to learn about prey availability or distribution.

14.4.1.1 Group size and the transmission of foraging information. The opportunity to learn from others is likely to be directly influenced by sociality, but the relationships between group size, density, and social learning remain unclear. Conceivably, fishes that shoal to a greater extent, or that form larger or more cohesive groups, may be exposed to social information more frequently. At one extreme, fish that rarely interact with others will have few opportunities for social learning. However, at the other extreme, very high densities may not promote reliance on social learning either, since individuals would be overwhelmed by social information.

Recent work has revealed that the degree to which guppies shoal and are capable of learning routes from others is dependent upon their early social environment. Chapman et al. (2008a) demonstrated that juveniles reared at low densities had significantly higher shoaling tendencies and were more proficient social learners than juveniles reared under high-density conditions. Laland and Williams (1997) found that simply shoaling with other fish was sufficient to generate social learning in female guppies. Focal fish, which were allowed to interact with trained demonstrator fish that were familiar with the layout of an experimental arena, were able to locate food patches by following their demonstrators. Later, when tested alone, the focal fish preferentially used the route demonstrated to them, revealing that they had learned how to navigate to a food patch through shoaling. The probability of social learning was found to increase with the number of trained demonstrators. Lachlan et al. (1998) showed that individual guppies preferred to follow larger groups of conspecifics over smaller ones and that by following such a group on just three occasions they were able to learn the route to a food patch.

It is well established that larger groups tend to find food faster than smaller groups (Pitcher et al. 1982). In an experiment where food was hidden in one of six boxes, with open water between them, Day et al. (2001) found that in guppies the latency to detect the prey patch was lower for the first and focal fish in larger groups than in smaller ones. However, in a subsequent manipulation, in which fish were required to pass through a small gap in an opaque partition to locate food, they found that the opposite pattern occurred; the latency for the first and focal fish to reach the food patch was now lower in the smaller than in the larger groups. This finding may be a consequence of conformist behavior, favoring the rapid transmission of information between group members when visual contact could be maintained between group members but militating against it when the partition was in place, and visual contact was prevented. At least early on in the diffusion, the tendency to remain with the shoal and do what they were doing discouraged individuals from leaving the shoal to locate a novel food source, with large shoals exerting more pull than small shoals. In another experiment, replicating the second but using a transparent partition, larger shoals were again advantaged, confirming this interpretation.

In a different experiment, Laland and Williams (1997) added naive fish to small groups of demonstrators and allowed them to follow the demonstrators to the food patch via one of two routes. Over successive days the demonstrators were removed and replaced with new, naive fish, simulating the recruitment of new generations, until none of the original founders remained. Upon testing it was found that these untrained fish preferentially used the route used by the original founders and that route preferences remained in the population in spite of a complete replacement of its members. This suggests the potential for cross-generational transmission of information, as well as lateral, between-individual transmission. Laland and Williams (1997) suggest that such a mechanism might allow arbitrary behavioral preferences to be maintained within fish populations. Though not demonstrated in the wild for poeciliids, this simple mechanism may underpin the foraging-route and schooling-site traditions observed in reef fishes (Helfman & Schultz 1984; Warner 1988).

Foraging techniques, as well as routes, can be maintained as traditions in laboratory populations of poeciliids. Stanley et al. (2008) presented guppies and platies (*Xiphophorus maculatus*) with a feeding task that they were unlikely to solve by themselves. Demonstrators were initially trained to feed by swimming into horizontal tubes to collect food held at the far end. Over a period of several exposures these tubes were rotated to vertical. While trained fish reliably fed from these tubes, no naive fish presented with a vertical tube learned to feed from it on its own. However, when placed in groups with experienced demonstrators, naive fish readily learned to feed from the vertical tubes, demonstrating the social transmission of a novel feeding behavior. When the experienced demonstrators were gradually removed and replaced with further naive fish, it was seen that all group members continued to exploit the feeding tubes. Larger groups of fish showed more stable transmission of this behavior than smaller groups, although this was found to be related to their slower rate of turnover rather than a direct effect of group size. Taken together, the findings of Stanley et al. (2008) and Laland and Williams (1997) illustrate how social transmission might operate in nature to give rise to stable traditions in behavior, such as learning foraging routes.

14.4.1.2 Social organization and social learning of foraging information. Female guppies are known to recognize, and to preferentially shoal with, individuals with which they have previously had prolonged interaction (Griffiths & Magurran 1997a). Swaney et al. (2001) found that guppies learned the route to a food patch more readily from familiar group mates than they did from unfamiliar ones. This suggests that familiarity may generate a form of directed social learning and is consistent with the idea that guppies might have evolved a "copy familiar individuals" social-learning strategy (see box 14.1).

Morrell et al. (2008) investigated the structure of natural and artificial guppy shoals, consisting respectively of individuals that were captured while shoaling together in the wild or of individuals that were collected separately and placed together in the laboratory. Social-network analyses (see Krause et al., **chapter 13**) revealed that individuals within natural shoals tended to form smaller subgroups than those in artificial shoals. While the mean latency to begin feeding was lower in natural shoals, and the number of instances of fish feeding simultaneously was greater in natural shoals than in artificial shoals, there was no relationship between patterns of pairwise associations and information transmission. This may be due to the fact that visual information, such as when, where, or what a demonstrator is eating, can be transmitted and detected over distances that exceed the interindividual spacing seen within shoaling networks. In other words, an observer need not spend a disproportionate amount of time in the immediate proximity of a given shoal mate in order to be able to receive the information that it produces as it forages.

14.4.1.3 Social learning of maladaptive foraging behavior. Laland and Williams (1998) also demonstrated that shoaling could lead to the transmission of maladaptive information in guppies. They found that guppies following trained demonstrators could socially learn to follow long and energetically costly paths to feeding grounds, ignoring shorter, more direct routes. As in previous studies, they found that observer fish persisted in the use of the longer routes even in the absence of the trained founders, and they took longer to learn the faster routes than control guppies that had never been exposed to the social demonstration. These findings can be interpreted as individual guppies pursuing a '*copy when asocial learning would be costly*' strategy. In the real world, moving away from the group to take a shorter route could be costly, since it would leave an individual vulnerable to predation, and here such costs militate against individual exploration and innovation.

14.4.1.4 Foraging and social-learning summary. Collectively, these studies reveal that the social learning of foraging routes and sites can occur passively as a result of shoaling behavior and that selection, likely shaped by predation pressure, has strengthened shoaling tendencies, resulting in social-learning tendencies that promote learning and information pooling. The observed preference for shoaling with familiar, rather than unfamiliar, fish adds a directional bias, such that individuals become more likely to learn from certain members of their population than others. This body of work has focused on the use of visual cues by observers, with no explicit emphasis on the roles of diet or habitat-derived chemical cues. Research focusing on other fish families (Gasterostidae, Ward et al. 2004; Ward et al. 2005; Ward et al. 2007; Webster et al. 2007, 2008b; Cyprinidae, Webster et al. 2008a) has revealed that some species are able to recognize patterns of prey and habitat use by others using derived chemical cues and that they preferentially shoal with these individuals. Morrell et al. (2007) and Ward et al. (2009) reported evidence for diet cue matching in guppies, suggesting that this mechanism influences social organization in this species, and potentially in other poeciliids too. Self-referent shoaling behavior such as this might serve to promote directed social learning and potentially even social enhancement of food preferences (Galef & Wigmore 1983) and warrants further investigation.

14.4.2 Social learning and predation risk

One of the key advantages of group living is protection from predators (Krause et al. chapter 13). Individuals benefit not only from attack dilution and confusion effects but also from the opportunity to gather information from their group mates about the nature of the predation threat and the best means to avoid it (Krause & Ruxton 2002). This information can be direct, such as predator location or direction of attack, or it can be indirect, such as learning escape routes from a particular location.

14.4.2.1 Shoaling with predator-experienced group mates. Kelley et al. (2003) found that naive guppies could improve their antipredator responses by associating with predator-experienced individuals. Guppies from low-predation sites exhibit only a limited range of antipredator behaviors compared with those from high-predation sites. By adding conspecifics from high-predation sites to groups of fish from low-predation rivers, in the presence of a model predator, they were able to show that over time the low-predation fish

significantly increased the rate at which they shoaled and the distance from which they inspected the model predator, compared with control group fish that were paired with other low-predation site demonstrators. These findings suggest that individuals might also learn antipredatory behavior from more experienced group mates.

Using a similar protocol to those employed by Laland and colleagues to investigate social learning of prey-patch locations (e.g., Laland & Williams 1997), Reader et al. (2003) found that wild guppies were able to evade a simulated predator by following trained demonstrators through one of two possible escape routes. Just as in the prey-patch learning experiments, observer fish continued to use the routes learned from following the demonstrators even after the demonstrators had been removed. This suggests that the same directed social-learning processes can serve to impart fitness-enhancing information to naive individuals in markedly different contexts.

14.4.2.2 Predator inspection. Predator inspection is a behavior in which prey animals, having detected a nearby predator, approach and investigate it (see Kelley & Brown, chapter 16). This seemingly counterintuitive behavior may be adaptive if it yields information about the status of the predator (Fishman 1999) or reduces the likelihood of attack or capture (Magurran 1990; Godin & Davis 1995). The costs, in terms of energy expenditure and lost feeding or reproductive opportunities, of leaving an area to avoid a predator might be significant. Predator inspection may serve to inform potential prey whether the predator is hungry or satiated or whether it is hunting or resting. If the predator is not actively foraging, it might pay the prey to continue current activities rather than undertaking an expensive relocation to another area (Fishman 1999). For many predators, successful attacks depend on low prey vigilance (e.g., Krause & Godin 1996b). Another function of predator inspection may therefore be to signal to predators that they have been detected, and there is evidence that predators are more likely to attack noninspectors rather than inspecting fish (Magurran 1990). Predator inspection in fishes is usually carried out only by small subsets of the shoal or individuals, and fish can inspect predators both visually and chemically by moving up gradients of predator-derived chemical cues (even when the predator itself is obscured from view; Brown & Godin 1999b). Research on other fishes has shown that predator inspection can lead to social learning in at least two different ways, both of which may plausibly operate in poeciliids.

First, even though predator inspection is usually carried out only by a small subset of the shoal, inspecting fish can transmit information to observing group mates, even when these observers cannot see the predator, causing the observers to alter their behavior appropriately. Working with minnows, *Phoxinus phoxinus* (a cyprinid), Magurran and Higham (1988) reported that observers decreased their activity (an antipredatory response to reduce conspicuousness) after observing the startle responses of predator-inspecting group mates. Thus, predator inspection can lead to social information about predator status being passed from inspectors to observers. Many poeciliids are known to engage in predator inspection behavior, including guppies (Botham et al. 2006) and the mosquitofishes *Gambusia holbrooki* (De Santi et al. 2001) and *Gambusia affinis* (Smith & Belk 2001). It is possible that information is transmitted from inspectors to observers in the same manner in these species too.

Second, studies reveal that inspectors can learn about predator status via chemical alarm cues released from the skin of ingested prey, probably via the urine or feces of the predator (Brown et al. 1995). A number of nonpoeciliid fishes are known to respond to alarm cues transmitted via predators, causing them to inspect at a greater range (Brown & Godin 1999c; Brown et al. 2000; Brown et al. 2001). Little is known about alarm cues in poeciliid fish relative to other families, although there is some evidence that guppies respond to cues in conspecific skin extract (Nordell 1998; Brown & Godin 1999c). Accordingly, Brown and Godin (1999c) report that free-ranging wild guppies were less likely to approach a model predator paired with conspecific skin extract than they were to approach an identical model paired with distilled water. These studies imply that the inspecting fish associate the presence of prey alarm cues in the predator's odor profile with increased predation risk, suggesting that such cues indicate that the predator is actively hunting.

14.4.3 Mate-choice copying

Mate choice, which can be adaptive for both sexes (Kokko & Johnstone 2002), is underpinned by genetic predispositions but can also be influenced by experience, including socially learned behavior and the social environment (Witte 2006; Rios-Cardenas & Morris, **chapter 17**; Druen & Dugatkin, **chapter 20**). Mate-choice copying is a specific form of social learning in which the partner preference of an observer is influenced by having seen that partner mating or sexually interacting with an opposite-sex conspecific. Mate-choice copying might have evolved in response to the costs of courtship and mate-quality assessment. Such costs could include heightened conspicuousness and reduced attentiveness in one or both parties brought about by courtship displays, leading to enhanced predation risk (Andersson 1994). From a female perspective, unsolicited matings,

which can lead to reduced foraging efficiency (Magurran & Seghers 1994a; Matthews & Magurran 2000), might also select for mate-choice copying. By eavesdropping on potential mates, females can gauge mate quality without paying the cost of receiving unsolicited mating attempts.

Mate-choice copying has been widely reported in poeciliids, including laboratory studies of female guppies (Dugatkin 1992a), female and male sailfin mollies (Schlupp et al. 1994; Schlupp & Ryan 1997), and female humpback limia (*Limia nigrofasciata*, Munger et al. 2004). Witte & Ryan (2002) also demonstrated mate-choice copying by both sexes in free-ranging (wild) sailfin mollies. However, several studies (such as Brooks 1996 and Lafleur et al. 1997, both in guppies) have failed to find evidence of mate-choice copying, suggesting that this behavior may vary among populations, perhaps due to differences in the costs of direct mate assessment. There is also evidence for mate-rejection copying (Dugatkin & Godin 1992b) and mate-choice reversal (Witte & Ueding 2003) in poeciliid fishes, suggesting that the social influence on mate preferences can operate in both directions.

Mate-choice copying has implications for the diffusion of behaviors within populations—for example, by initiating chains of mating preferences, such that preferred males amassed significantly more matings than males who mated in the absence of female observers (Dugatkin et al. 2002) or by generating female preferences for males of similar phenotype to the initially observed mating male (Godin et al. 2005). Such mechanisms could lead to directed social transmission of mate-trait preferences, generating traditions of preferred male characteristics (Kirkpatrick & Dugatkin 1994; Laland 1994).

Several studies on female guppies have focused on intrinsic and extrinsic factors affecting the propensity to engage in mate-choice copying. Dugatkin and Godin (1993) found that young females were more likely to copy the mate choices of older (and presumably more experienced) demonstrators, whereas old guppies were not influenced by the mate choices of young demonstrators. This finding is consistent with the deployment of a fairly specific "who" social-learning strategy (see box 14.1).

Further work by this group found no evidence that hunger (Dugatkin & Godin 1998) or heightened predation risk (Briggs et al. 1996) affected the degree of mate-choice copying. These findings are surprising since increased mate-choice copying could serve respectively to maximize foraging time and minimize exposure to predation risk. It may be that the benefits of mate-choice copying are independent of hunger or predation risk or that selection had primed the fish to always behave as though predation risk were high. Further work is necessary to determine the factors that might promote variation in the tendency to engage in mate-choice copying, at both the population and the individual level.

14.4.4 Aggression

Aggression, usually associated with conflict over resources such as territory, food, or mates, is potentially costly, since it can lead to injury, lost time and energy, and greater conspicuousness to predators (Slotow & Paxinos 1997). Consequently, there is good reason to expect selection favoring the adoption of strategies that minimize the frequency or intensity of aggressive encounters. Socially learned information about the competitive ability of a potential opponent can benefit an individual by promoting rapid concession in a contest against a superior fighter or escalation of an encounter with a poor one, in order to minimize time spent in contests.

When two individuals are fighting, others in the vicinity might observe the outcome of the contest and in doing so learn the relative competitive abilities of the two contestants, an example of a process termed eavesdropping (McGregor 1993). Eavesdropping allows individuals to tailor their own response should they subsequently engage either opponent in an aggressive encounter. Eavesdropping presumably evolved because the costs of personal sampling (i.e., in this context engaging in a potentially damaging fight) can be severe. This phenomenon has been described in a number of fish species (e.g., Johnsson & Akerman 1998; Oliveira et al. 1998; Doutrelant et al. 2001), including the green swordtail (*Xiphophorus hellerii*, Earley & Dugatkin 2002). In this study, male swordtails were found to be less likely to initiate, escalate, or win fights against others that they had previously seen winning a contest against another opponent.

14.5 Summary and future directions

The studies reviewed here provide clear evidence that poeciliid fishes employ flexible behavioral repertoires that allow them to meet the demands of living in changeable and unpredictable environments. These studies illustrate the conditions under which fish are likely to innovate and to some extent also allow us to predict the characteristics of individuals that might predispose them to innovate. The large body of work on social learning has revealed that acquired information is readily transmitted among individuals and that rates of transmission are mediated and directed by the social environment. This work has established that social learning by poeciliids is an important process in many, if

not all, behavioral contexts but that the use of social information when unchecked by personal sampling is not per se an adaptive strategy. Rather, social learning is adaptive only as part of a broader complex of learning strategies that draws information from multiple sources and deploys it flexibly in response to prevailing environmental pressures. Elsewhere in this review we have identified specific gaps in our knowledge about social learning in poeciliids. In this final section we outline four broad areas that warrant investigation and suggest ways in which poeciliids might be useful study species.

First, the concept of social-learning strategies (Laland 2004) is one area in which there is much scope for further experimental research. To date, much of the research into social learning in poeciliids (and indeed in other animals) has focused on social-learning processes, that is, psychological mechanisms that can directly or indirectly lead to social learning. By contrast, as we highlight throughout this review, comparatively little research has focused on exploring the evolution of social-learning strategies that determine when animals should gather private or social information, from whom they should learn, and the conditions under which they should exploit this information. Theory predicts that learning strategies should differ among populations as a function of variation in the social environment, predation pressure, resource distributions, and other factors. The "natural experiment" of discrete high– and low– predation pressure populations found in the guppy system of northern Trinidad potentially represents an excellent study system for answering questions about the evolution of social-learning strategies.

Second, little is known about how the social organization of a group might affect the way in which information diffuses through it. Social-network analyses have revealed that guppies can live in complexly ordered groups (Krause et al., **chapter 13**). While this potentially has profound implications for the diffusion of information through populations, work to date has failed to link network structure to diffusion pathways either in guppies (Morrell et al. 2008) or in other animals (Boogert et al. 2008). We need to determine whether this reflects methodological limitations whereby information is transmitted over greater distances than those used to define "associations" when constructing networks or whether current methods of analysis lack power to detect weak or subtle transmission patterns.

Third, recent work has revealed that leadership and collective decision making affect how nonpoeciliid fish shoals use their environment. Reebs (2000, 2001), working with golden shiners (*Notemigonus crysoleucas*), and A. Ward et al. (2008), studying three-spined sticklebacks (*Gasterosteus aculeatus*), found that a minority of fish were able to lead naive shoal mates toward food patches or shelter. They showed that larger proportions of leaders were more effective at entraining the shoal, that demonstrator phenotype influences following behavior, and that more leaders were required to initiate more movements as absolute group size increased. A. Ward et al. (2008) suggested that all group members are likely to adopt a behavior only when the proportion of the group engaging in it reaches some threshold level. Such quorum responses can facilitate rapid and accurate decision making compared with choices made individually, allowing the group to behave adaptively even though not all of its members possess pertinent information. Shoaling with informed individuals has already been shown to facilitate social learning in poeciliids (Laland & Williams 1997). Further research should integrate collective behavior with work on directed social learning to determine how group composition affects group decision making and the diffusion of information.

Finally, future work should focus on social learning and the diffusion of information in free-ranging populations. While several research groups have taken their experimental paradigms into the wild and confirmed that social learning does indeed occur in free-ranging fishes (Witte & Ryan 2002; Reader et al. 2003), the majority of the experiments have been performed in the laboratory. Relatively little is known about the diffusion of information through natural populations, its longer-term persistence though time, or its influence on the fitness of the animals that receive and exploit it. Several species of poeciliid fish could be employed as model systems with which to study these behavioral and evolutionary questions, questions that traditionally have been addressed using larger-brained vertebrates. Doing so could yield valuable insights not only into the biology of poeciliids but also into the biology of social animals in general.

Acknowledgments

Our research was supported in part by a NERC grant to K.N.L. We thank Jon Evans and three anonymous referees for helpful comments.

Chapter 15 Cognition

Angelo Bisazza

15.1 Introduction

POECILIIDS HAVE COLONIZED a wide range of environments from the equator to temperate regions of both hemispheres, in some cases occupying extreme habitats such as caves, hot springs, or brackish waters (e.g., see Tobler & Plath, **chapter 11**). While some species are adapted to a particular habitat, many others are able to live in extremely different conditions, and when introduced into different continents they often rapidly adapt to the new conditions to the point of displacing native species (Courtenay & Meffe 1989; Stockwell & Henkanaththegedara, **chapter 12**). Although the key to the poeciliids' success undoubtedly lies in their physiological adaptations (above all, viviparity), a further contribution to their adaptability is that they are equipped with a complex behavioral and cognitive repertoire—attributes that have made the family excellent subjects for cognitive research over the last fifteen years.

Poeciliids have only recently become the focus of cognition research, and most studies have primarily focused on only a few selected topics such as social learning and information transfer, predator-prey interactions, social cognition (individual recognition, cooperation, etc.), numerical cognition, and the lateralization of cognitive functions. The first three of these topics are covered elsewhere in this volume (Krause et al., **chapter 13**; Kelley & Brown, **chapter 16**; Webster & Laland, **chapter 14**; Druen & Dugatkin, **chapter 20**). In this chapter I summarize the available information on learning, memory, and spatial orientation and navigation. Then I focus on two very specialized topics, numerical abilities and lateralization of cognitive function, subjects that

in recent years have been the main focus of research in my laboratory. For a more general treatment of cognition in fishes, I refer the reader to Brown et al. 2006a.

15.2 Learning, memory, and spatial cognition

Learning influences almost every aspect of the life of poeciliids. Newborns are equipped with a behavioral repertoire that allows them to find food, avoid predators, and interact with conspecifics (Magurran & Seghers 1990a; Magurran 2005). Nonetheless, they possess the ability to modify their behavior in relation to both individual experience and information obtained from conspecifics. For example, guppies (*Poecilia reticulata*) can rapidly learn an antipredator response when exposed to a novel, simulated predation hazard in the presence of experienced conspecifics (Kelley et al. 2003; see also Kelley & Brown, **chapter 16**) and learn routes of escape from predators and new foraging sites from conspecifics (Reader et al. 2003; see also Webster & Laland, **chapter 14**). They also learn to recognize individuals on the basis of previous experience. In guppies this familiarity develops gradually over a period of 12 days (Griffiths & Magurran 1997a) and has important consequences for social dynamics (Croft et al. 2006a; see also Krause et al., **chapter 13**).

Poeciliids can gain information passively by observing social interactions among conspecifics. For example, in the green swordtail, *Xiphophorus hellerii*, male bystanders consistently modify their behavior toward individuals that they have witnessed winning or losing a fight (Earley & Dugatkin 2002). This behavior seems to represent eavesdropping, since bystanders do not modify their behavior toward naive individuals (Earley et al. 2005; see also Druen & Dugatkin, **chapter 20**). In at least two species, the guppy and the sailfin molly (*Poecilia latipinna*), females alter their individual mate preferences after witnessing a male being chosen by another female (Dugatkin 1992a; Schlupp et al. 1994; see also Rios-Cardenas & Morris, **chapter 17**). A number of studies have shown that early social experience can influence male and female mate preferences (Ferno & Sjolander 1973; Breden et al. 1995). Sexual experience also plays an important role in species recognition. Inexperienced male guppies offered either hetero- or conspecific females mate indiscriminately, but males allowed prior interactions with females of both species learned to distinguish conspecific females in about a week (Haskins & Haskins 1949; Magurran & Ramnarine 2004).

15.2.1 Orientation and spatial cognition

Poeciliid habitats typically exhibit variability in the spatial and temporal distribution of resources and hazards. Numerous laboratory and field studies reveal that poeciliids are equipped with multiple, highly flexible orientation mechanisms that allow them to exploit information from their environment. For example, many poeciliids preferably inhabit shallow waters along the shorelines of rivers and lakes or areas of dense vegetation where they are less exposed to predators (Barney & Anson 1921; Mattingly & Butler 1994). Nevertheless, fish may move temporarily into open water, either individually or in schools, to escape aerial predators, avoid sexual harassment, find more favorable temperatures, or reduce competition for food (Maglio & Rosen 1969; Goodyear 1973; Darden & Croft 2008) and thus need to find the route back to favorable locations.

Goodyear and Ferguson (1969) found that when mosquitofish (*Gambusia affinis*) were displaced to unfamiliar locations, they use a sun compass to reach the shore from which they were captured. This movement toward shallow water seems to function principally as a mechanism of predator avoidance since it was absent in mosquitofish captured from environments lacking predators. Goodyear (1973) found that mosquitofish could also orient to shore using local landmarks when available. However, when both sources of information were available, most individuals oriented by sun compass (Goodyear 1973). Bisazza and Vallortigara (1996) tested wild-caught eastern mosquitofish (*Gambusia holbrooki*) in a circular tank with a light placed in the middle. Females (but not males) swam in different directions with respect to the light depending on whether they were tested in the morning or the afternoon. This pattern was not observed under diffuse light or among females

raised in captivity—a result suggesting that swimming direction in the laboratory is primarily determined by a sun compass direction learned prior to capture.

Many poeciliids can learn to orient themselves using local features of the environment (*Girardinus falcatus*: Sovrano et al. 2005; *P. reticulata*: Burns & Rodd 2008; *Brachyrhaphis episcopi*: Brown & Braithwaite 2005; *P. latipinna*: Creson et al. 2003). For example, Sovrano et al. (2005) have shown that goldenbelly topminnows, *G. falcatus*, are able to combine geometric with nongeometric information from the environment to locate a goal. Previously, this capacity was shown to be present in human adults but absent in preverbal children and in rats, leading to speculation that the ability to combine different types of spatial information was somehow related to development of human language or other higher functions (Cheng 1986; Hermer & Spelke 1994). However, the ability of poeciliids and other fishes (Sovrano et al. 2002; Vargas et al. 2004) to use information in the same way as humans suggests that interspecific differences in spatial abilities probably reflect specific ecological adaptations rather than different levels of complexity of the nervous system (Sovrano et al. 2002; Vallortigara 2004).

15.2.2 Other learning studies in poeciliids

A few studies have employed operant conditioning to train poeciliids to discriminate between stimuli (Rensch 1956; Agrillo et al. 2009b; Dadda et al. 2009). Cantalupo et al. (1995) provide an anecdotal example of place avoidance learning in poeciliids. They examined the direction of escape by young *G. falcatus* in response to a simulated approaching predator. The stimulus was presented to individual fish when they were swimming in the center of the tank and their bodies were aligned with the stimulus predator. After repeated testing, the subjects tended to avoid the center of the tank and, if they moved through the center, tended to avoid alignment with the predator. Furthermore, guppies that witnessed the behavior of adults under attack during the first 48 hours of life were more likely to escape attacks by cichlids 10 weeks later than those that lacked this early experience (Goodey & Liley 1986). This appears to be an example of procedural memory, since fish were apparently learning how to escape from a predator rather than learning the characteristics of the predator itself.

15.2.3 Memory

Memory has rarely been investigated in poeciliids, and our current knowledge of memory processes and their neural bases in fishes comes from other species (for a review, see Rodriguez et al. 2006). Learning studies show that poeciliids can retain memories for varying lengths of time, depending on the task and the information required. In guppies and sailfin molly females, observation of a sexual interaction between a male and a female can influence subsequent mate-choice decisions after 24 hours (Witte & Massmann 2003; Godin et al. 2005). Guppies trained in a spatial-memory task continued to perform well when retested two weeks later (Burns & Rodd 2008). In a visual-discrimination learning task (two vs. four dots) Rensch (1956) found that the majority of guppies and swordtails (*X. hellerii*) retained the discrimination after 30 days. Goodyear and Ferguson (1969) reported that western mosquitofish that learned the direction of the shore when young retained this knowledge when tested two weeks later, while fish that learned this as adults lost their shoreward orientation more quickly. Feral eastern mosquitofish females apparently retain a sun compass direction more than two months after capture (Bisazza & Vallortigara 1996).

Research on numerical abilities (see section 15.3.3) suggests that poeciliids can keep at least short-term track of a small number of objects. Griffiths and Magurran (1997b) found that female guppies living in schools with more than 50 individuals do not develop schooling preferences, suggesting that this is approximately the upper limit of the number of different individuals a female can memorize.

15.2.4 Intraspecific variation

Most studies on learning and cognition involve subjects of one sex only, and differences between the sexes in these abilities are rarely considered. However, Laland and Reader (1999a) found that female guppies were more likely to be innovative foragers than males, and in a subsequent study they observed that novel foraging information was transmitted at a faster rate among females than males (Reader & Laland 2000). As mentioned above, female but not male mosquitofish showed evidence of a learned sun compass route when tested in the laboratory (Bisazza & Vallortigara 1996). To date, however, there is no clear evidence that male and female poeciliids differ in learning and memory, and all the reported variation could be explained by other factors such as sex differences in risk sensitivity (Magurran et al. 1992), feeding strategies (Dussault & Kramer 1981), or schooling tendencies (Griffiths & Magurran 1998).

Two studies have compared cognitive abilities among populations that differ in the level of predation. Burns and Rodd (2008) found no effect of predation intensity on the time necessary to learn a spatial task in guppies. However, distinct strategies were used by high- and low-predation guppies, the latter being more inclined to make quick but inaccurate decisions than their high-predation counterparts. Brown and Braithwaite (2005) investigated

differences in a spatial task in *B. episcopi* from populations with different predation regimes (see also Kelley & Brown, **chapter 16**). Low-predation fish learned spatial tasks about twice as quickly as high-predation fish. The causes of these differences are not known, but high- and low-predation populations also differ in temperament and in the pattern of cerebral lateralization, two factors that are known to affect learning (Sneddon 2003; Rogers et al. 2004). Direct evidence of the influence of cerebral lateralization on learning was reported in a study on *G. falcatus* in which fish artificially selected for a high degree of lateralization (see section 15.4.1) performed significantly better than fish selected for low lateralization in two tasks (Sovrano et al. 2005).

15.3 Numerical cognition

Abilities such as recording the number of events, enumerating items in a set, or comparing two different sets of objects can be adaptive in many ecological contexts. For example, wild chimpanzees enter intergroup contests only if they outnumber the opposing side by a factor of 1.5 (Wilson et al. 2001), while lions prefer to hunt smaller, rather than larger, prey groups, as smaller groups are more vulnerable (Scheel 1993). In recent years research on numerical cognition has broadened to include other mammals and a few bird species, leading to the conclusion that rudimentary numerical abilities are widespread among these two vertebrate classes (Hauser et al. 2003; Kilian et al. 2003; Rugani et al. 2008; Irie-Sugimoto et al. 2009). Despite this progress, we have yet to determine whether animals have a mental representation of numbers, whether one or several different mechanisms underlie these abilities, and whether numerical abilities are innate or learned through experience. Another question that has remained unresolved until recently is whether numerical abilities are confined to mammals and birds or are evolutionarily ancient adaptations shared by most animals.

15.3.1 Evidence for two distinct numerical systems in poeciliid fish

Some authors have contended that in primates there are two distinct nonverbal systems for representing numerosity: an object-tracking system, which is precise but has an upper limit of 3–4 units, and an analog-magnitude system, which allows approximate discrimination of large quantities (Feigenson et al. 2004; Revkin et al. 2008). Agrillo et al. (2008b) systematically investigated the limits of quantity discrimination in mosquitofish (*G. holbrooki*) using the widespread tendency of fish to choose the largest social group when placed in a new, potentially dangerous environment (Hager & Helfman 1991; Pritchard et al. 2001). In this study, fish discriminated between two shoals that differed by one element (i.e., individual) when the choices of shoal size were 1 versus 2, 2 versus 3, and 3 versus 4, but not when the choice was 4 versus 5 elements, a limit that coincides with that observed in nonhuman primates in comparable tasks (reviewed in Feigenson et al. 2004). However, fish were able to discriminate between larger shoals provided that the number ratio was at least 1:2 (e.g., 4 vs. 8 or 8 vs. 16 fish but not 4 vs. 6 or 8 vs. 12 fish), in accordance with Weber's law. Buckingham et al. (2007) reported identical results for female green swordtails.

The study by Agrillo and co-workers (2008b) provides strong evidence for two distinct systems: one for representing exact small numbers of objects or events and one for representing large approximate numerical quantities. In accordance with Weber's law, when stimulus shoals of mosquitofish contained more than 4 fish, the comparison became more difficult as the ratio became smaller, the typical signature of the large-number system in mammals. No such relationship was found with 1 versus 2, 2 versus 3, and 3 versus 4 comparisons, where discrimination was independent of the numerical difference between groups but showed a set-size limit, the typical signature of the object-tracking system. In addition, a single mechanism obeying Weber's law can hardly explain all the results, since mosquitofish were unable to discriminate when presented with ratios of 2:3 or 3:4 in the range of large numbers, while such discrimination was possible with small numbers.

15.3.2 Estimating the number of conspecifics in other contexts

The ability to enumerate could be used in other contexts besides antipredator responses. When pursued by an active male, pregnant female mosquitofish show a tendency to join the largest shoal to dilute sexual harassment (Agrillo et al. 2006). In this situation the females' capacity to discriminate different numbers of fish is identical to that observed in the previous study, suggesting that the same cognitive systems operate in both contexts (Agrillo et al. 2007).

In many poeciliids females form shoals of varying sizes and sex ratios, and males seeking mating opportunities must often decide which group to join. Male guppies are primarily influenced in their choice by group size, whereas shoal sex ratio plays a marginal role (Lindstrom & Ranta 1993). Similar results have been found in a series of experiments conducted with male mosquitofish (Agrillo et al. 2008a). In tests with all-female shoals, males exhibited a preference for shoals rather than a single female and preferred larger shoals to smaller ones. They also preferred an all-female shoal to a mixed-sex one, even if the latter

was larger. Both preferences appear sensible, since larger groups of females offer more protection and more mating opportunities. Yet Agrillo et al. (2008a) reported that male mosquitofish did not choose shoals with more favorable sex ratios, leading the authors to suggest that there may be a cognitive limit in the capacity of males to simultaneously take into account the number of males and females in the different groups.

15.3.3 Can fish use numerical information?

Animals can judge numerosity using cues other than the number of the items. This is because numerosity normally covaries with several other physical attributes, for example, the sum of areas or contours of the objects and the total area occupied by the set, and an individual can use the relative magnitude of these continuous variables to estimate which group is larger (Feigenson et al. 2002b). Agrillo et al. (2008b) investigated whether focal mosquitofish were still able to use numerical information when nonnumerical variables were controlled experimentally. In one experiment, the total areas occupied by stimulus (i.e., nonfocal) fish in a small and a large group were held experimentally constant by using slightly larger fish as subjects for the smaller shoal and slightly smaller fish for the larger shoal. In this way, stimulus fish occupied the same total area in both groups. Under these conditions, focal female mosquitofish did not choose the larger shoal when presented in either small (2 vs. 3) or large numbers (4 vs. 8). The total activity of fish within a shoal was also manipulated by keeping the two stimulus groups at different water temperatures. Once the total number of movements within shoals were approximately equivalent, focal fish also exhibited no preference for the larger shoal in the 4 versus 8 comparison, but interestingly they chose the larger shoal in the 2 versus 3 comparison. These results do not necessarily imply that mosquitofish are unable to discriminate two groups solely on the basis of their numerosity. Using perceptual cues of the stimuli may simply be the easiest way to obtain a quick numerosity judgment. Indeed, there is compelling evidence that species such as humans, apes, and dolphins, which have the capacity to count, typically base their quantity judgments primarily on stimulus properties such as area, contour, or density and use number only when no other cues are available (Durgin 1995; Kilian et al. 2003).

A recent study suggests that fish are capable of true numerical representation. Female mosquitofish were trained to discriminate between two quantities (2 vs. 3 small geometric figures) by reinforcing the correct choice with the possibility of rejoining their social group (Agrillo et al. 2009b; fig. 15.1). In the first experiment, during the training the figures varied in shape, size, and distance but there

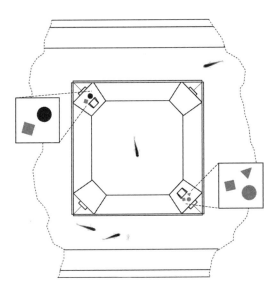

Figure 15.1 Apparatus used to train female *Gambusia holbrooki* to discriminate between sets containing different numbers of geometric figures. Subjects were placed individually in a test chamber provided with two doors (one associated with three and the other associated with two elements) and could pass through the reinforced door to rejoin shoal mates in the outer tank.

was no control for nonnumerical variables. After subjects had learned the discrimination, they were tested without reward while controlling for one nonnumerical variable at a time. A good numerical discrimination performance was maintained once the total luminance of the stimuli and the sum of perimeters of the figures were matched. Conversely, mosquitofish did not select the previously reinforced numerosity when stimuli were matched for the sum of the areas of the figures or for the total area occupied by the sets, suggesting that these two cues had been spontaneously used during the learning process. In a second experiment fish had to learn a 2 versus 3 discrimination while these nonnumerical variables were simultaneously controlled for. All fish were able to learn the task and they made approximately the same number of errors, as did the fish in the first experiment when nonnumerical cues were available. It therefore seems that fish, like mammals and birds, are able to compare two quantities using only numerical representations, although, as in the other species, this capacity is utilized only when nonnumerical cues are unavailable.

15.3.4 Development of numerical skills

As in other poeciliids, guppies shoal soon after birth (Magurran & Seghers 1990a), which makes it possible to study how counting ability develops during ontogeny. In a recent study (A. Bisazza, G. Serena, L. Piffer, & C. Agrillo, unpublished data) newborns were tested in a miniature version of the apparatus used for assessing shoal choice in adults. When offered a choice between shoals composed of "small"

numbers of fish (2 vs. 3), they chose the larger shoal significantly more often, but they exhibited no preferences when "large" numbers (4 vs. 8 or 4 vs. 12 fish) were used. The latter result was not due to newborns being motivated to avoid very large shoals, since a shoal of 8 fish was significantly preferred over shoals comprising 3 fish. Instead, this study suggests that newborn guppies have the ability to discriminate among groups comprising small numbers, while the ability to discriminate among larger shoals occurs later. To assess the relative role of maturation and experience, guppies were tested at 20 and 40 days of age after they had been raised without the possibility to count fish (i.e., with just one companion) or with normal experience (raised in a large group with some adults present). At 40 days fish from both treatments were able to discriminate 4 from 8 fish, while at 20 days this was observed only in fish that had had normal social experience. These results again suggest the existence of two separate numerical systems in poeciliids. The precise small-number system is innate and displayed immediately at birth, while the approximate large-number system emerges later as a consequence of both maturation and social experience.

15.3.5 Cognitive limitations and decision rules

The above studies suggest that poeciliids exhibit numerical skills that closely match those found in animals such as apes and dolphins, which possess much larger and more complex brains. Indeed, in the domain of small numbers, the skills displayed by poeciliids appear to exceed those of many other vertebrates. Mosquitofish and guppies (including newborns) easily distinguish 3 from 4 items, a capacity not observed in salamanders, toads, chicks, and preverbal children (Feigenson et al. 2002a; Uller et al. 2003; Rugani et al. 2008).

Research over the last two decades has revealed that human and nonhuman primates share an object-tracking system, a mechanism that allows individuals to track up to four objects in parallel even if these are moving in space, provided they remain in view or undergo only brief periods of occlusion (Trick & Pylyshyn 1994; Scholl & Pylyshyn 1999). A considerable body of evidence suggests that the precise small-number system is based on this object-tracking system (reviewed in Feigenson et al. 2004). Recent studies indicate that similar systems may operate in other mammalian taxa and in birds (Kilian et al. 2003; Rugani et al. 2008), and the experiments with guppies and mosquitofish suggest that this system may be phylogenetically very ancient and shared by most extant vertebrates. Investigation in this area is highly desirable since cognitive limitations are likely to affect many fitness-related traits such as mate selection, foraging, antipredator responses, and so on. Although the possibility that all vertebrates encounter similar cognitive limitations is speculative, it is intriguing to note that in the majority of species that have been studied extensively, females sample on average three to four males before making a mating decision (reviewed in Gibson & Langen 1996) and that similar effects have been observed in foraging contexts (Langen 1999). The argument that cognitive limitations can constrain comparative evaluation mechanisms can, however, be reversed. In both mate sampling and food patch selection, searching costs are normally proportional to the number of alternatives sampled, while the benefits follow the law of diminishing marginal returns (Stephens & Krebs 1986; Real 1990). Put simply, after visiting four males, sampling one or two additional males would add little to the selectivity of the female while considerably increasing her costs (risk of predation, travel time, energy consumed, etc.; see, e.g., Byers et al. 2005; Vitousek et al. 2007). It is therefore possible that selection on optimization of search rules has shaped cognition instead of the other way round.

15.4 Lateralization of cognitive functions and its ecological consequences

Hemispheric specialization refers to sensory, motor, and cognitive abilities that are specialized to either the left or right cerebral hemisphere. Lateralization of cognitive functions has traditionally been investigated in humans and in a few mammalian and avian models (Andrew & Rogers 2002). Research on fish started very recently when Cantalupo et al. (1995) documented a significant population bias to turn right during fast escape responses in young *G. falcatus* when presented with a simulated predator attack. Since this initial study, many other instances of behavioral lateralization in fishes have been reported, regarding either motor (e.g., sound production, fin use, gonopodial thrusts) or sensory biases (visual, olfactory, lateral line); for a review, see Vallortigara and Bisazza 2002. Poeciliids have quickly become a prominent group in the study of animal lateralization. Early studies, conducted mainly on two species, the goldbelly topminnow (*G. falcatus*) and the eastern mosquitofish (*G. holbrooki*), revealed a pattern of lateralization of cognitive functions similar to that observed in mammals and birds, where each hemisphere is specialized for different functions and the resulting left-right motor and sensory differences affect everyday behavior (Bisazza et al. 1997a; Bisazza et al. 1998a). Mosquitofish, for example, have a significant preference for keeping a shoal mate on their left, while they tend to inspect a potential predator with their right eye (De Santi et al. 2001). Bisazza et al. (1999) have studied the laterality of cooperative predator inspection using a procedure introduced by Milinski (1987) in which a mirror is placed parallel to the tank during inspection so that the image appears to swim along with the fish, simulating a cooperative partner. In this experiment mosquitofish performed significantly closer inspections when the mirror was on their left side so that each of the two stimuli was seen with the preferred eye. These apparently surprising results occur because in vertebrates with laterally placed eyes, such as fish, the optic nerve fibers completely cross at the optic chiasm, and thus each eye projects almost exclusively onto the contralateral hemisphere. Since fish lack the efficient interhemispheric communication through the corpus callosum that characterizes mammals, information is primarily processed by the hemisphere that receives it.

15.4.1 Individual variation in lateralization

Population level analyses often mask any underlying individual variability. For example, in mosquitofish, a minority of individuals show a reverse laterality pattern to that described above, while some other individuals show little or no eye bias (De Santi et al. 2001). The significance of such individual variation in laterality has become the focus of more recent research. In two fish species, *Danio rerio* and *G. falcatus*, the degree and direction of cerebral asymmetries were found to be partly under genetic control (Bisazza et al. 2000b; Barth et al. 2005; Facchin et al. 2009), and a selection experiment was undertaken in the latter species to obtain lines of fish that either preferentially turned to the right (right detour = RD), to the left (LD), or had no turning preferences (nonlateralized = NL) when encountering a dummy predator that was visible behind a barrier (Bisazza et al. 2007). RD and LD lines significantly diverged after a few generations (fig. 15.2), and the fish from these two lines also showed an opposite direction of lateralization in many other cognitive tasks, including motor activities and visual functions as well as other sensory modalities (Facchin et al. 1999; Bisazza et al. 2001a; Bisazza et al. 2005). This suggests that LD and RD fish may be similar but with complete mirror-reversed organization of cerebral functions, while NL fish have a bilateral representation of most cognitive functions.

The discovery that lateralization is widespread and that individual differences have a genetic basis has raised new and important questions. Is there a selective advantage of lateralization of cognitive functions? Are there costs associated with having motor and perceptual asymmetries? What mechanisms maintain variability in the strength and direction of laterality? Why do population biases exist, and why is the direction of bias sometimes consistent across different species?

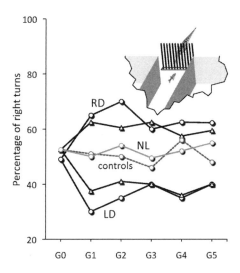

Figure 15.2 Selection experiment on *Girardinus falcatus* for turning direction when facing a dummy predator beyond a barrier. The frequency of right turns is reported over five generations for two replicated lines selected for right turning (right detour = RD), two for left turning (LD), one line selected for no turning bias (nonlateralized = NL), and one unselected control line (controls). Redrawn from Bisazza et al. 2007.

15.4.2 Selective advantages of cerebral lateralization

Animals are often constrained by how much attention they can simultaneously focus on different activities (Dukas 2004). Rogers (2000, 2002) suggested the intriguing hypothesis that hemispheric specialization evolved because it allows individuals to cope with limited attention. In her view, cerebral lateralization is one way to increase the brain's capacity to carry out simultaneous processing, by channeling different types of information into the two separate halves of the brain and by enabling separate and parallel processing to take place in the two hemispheres. To test this idea, Rogers et al. (2004) compared normally and weakly lateralized chicks (the latter obtained by incubating eggs in the dark during the final days before hatching). Chicks had to learn to discriminate between food and nonfood while a model of an avian predator was moved overhead. Lateralized chicks learned faster and were more responsive to the model predator than weakly lateralized chicks, whereas no difference in learning ability was found in the control experiment without the predator.

Dadda and Bisazza (2006a, 2006b) performed similar studies comparing lines of *G. falcatus* selected for high and low degrees of laterality (see section 15.4.1) in situations requiring the sharing of attention between two simultaneous tasks. In one study (Dadda & Bisazza 2006a) hungry individuals entered a compartment adjacent to the home tank to capture live brine shrimps in either the presence or the absence of a live predator situated at some distance. With the predator visible, fish of both lateralized lines (LD and RD, collectively called LAT) were twice as fast at catch-

ing shrimps than nonlateralized fish of the NL line, while no difference in capture rate was recorded when the predator was absent and subjects were not required to share attention between vigilance and prey capture. A more detailed analysis of fish movements in this test revealed that LAT fish tended to monitor the predator with one eye (the right eye in LD and the left in RD fish) and to use the other eye for catching prey, whereas NL fish swapped between tasks, using each eye for both functions. In the second study (Dadda & Bisazza 2006b), lateralized females proved to be better foragers than NL females when they had to share attention between retrieving food items scattered on the surface and avoiding unsolicited male mating attempts, while no difference between LAT and NL fish was evident in control tests where females could forage undisturbed.

A superior performance of LAT individuals was also shown in situations that did not explicitly involve multitasking, suggesting other possible advantages of cerebral lateralization. LAT lines of topminnows proved to be better than NL fish at using physical features or geometric cues to reorient themselves in a small environment (Sovrano et al. 2005; see section 15.2). In another study (Bisazza & Dadda 2005), schools of LAT fish showed significantly more cohesion and coordination than schools of NL fish. Moreover, in schools composed of both LAT and NL fish, the latter were more often at the periphery of the school, while lateralized fish occupied the center—a position normally safer and energetically less expensive (Bumann et al. 1997; Svendsen et al. 2003). In these experiments, fish were observed in a novel environment, so it is possible that during the test they were actively scanning the surroundings for predator presence, an additional attention task that would favor fish with specialized hemispheres.

15.4.3 Costs of cerebral lateralization

In organisms such as fish with laterally placed eyes, the complementary specialization of hemispheres often translates into differential responsiveness to sensory input on the left and right side of the body (Deckel 1995; Vallortigara et al. 1998). The appearance within the visual field of biologically relevant objects such as a predator or a prey is often unpredictable, and it is easy to see the potential disadvantages arising from having side biases in the promptness or effectiveness of response to a particular class of objects, as well as the possibility for competitors, predators, or prey to exploit such asymmetries. Whether these left-right differences in the way an animal analyzes and responds to environmental stimuli also translate into a disadvantage for more highly lateralized individuals remains largely unexplored. An attempt to test this possibility involved comparing the selected lines of *G. falcatus* for their promptness in reacting to a predatory stimulus appearing on the right or the left visual hemifield (Agrillo et al. 2009a). No difference between LAT and NL fish or between the two eyes was detected, but the setting used in this study was very simplified and the measurement technique was rather crude, so a more thorough investigation is needed before drawing further conclusions.

A recent study utilized the selection lines described above (Dadda et al. 2009) to explore the possibility that marked hemispheric functional asymmetries may interfere with a cognitive task when it relies on hemispheric communication and cooperation to be accomplished. Two situations were devised in which visual inputs were divided between the visual fields so that each eye (and contralateral hemisphere) had access to only one half of the information necessary to accomplish the task. The first experiment was an adaptation of a widely used neuropsychological test, the line-bisecting task. When right-handed human subjects are required to mark the middle points in a straight line, they tend to transect slightly to the left, an effect that is commonly ascribed to the right-hemisphere dominance for spatial tasks (reviewed in Jewell & McCourt 2000). Topminnows were required to use the middle door in a row of nine in order to join their social group (fig. 15.3). In the majority of trials NL fish correctly chose the central door, while fish from LD and RD lines made systematic errors to the left or the right of the central door, respectively.

In the second experiment, isolated individuals emerged into an unfamiliar area where they could choose between two shoals differing in quality (number and size of fish). The apparatus allowed subjects to see each shoal with a different eye prior to the choice. NL fish chose the high-quality shoal significantly more often than the lateralized fish, which in most cases chose the option seen with the eye that in their line was dominant for analyzing social stimuli, irrespective of its relative quality. Interestingly, *B. episcopi*

from high-predation populations are, on average, more lateralized than their low-predation counterparts (Brown & Braithwaite 2005) and surprisingly take longer to complete a maze task since their laterality interfered with an efficient exploratory behavior (C. Brown et al. 2004).

It is not easy to estimate the extent to which these constraints affect an individual's fitness. These experiments suggest that lateralized fish may make frequent suboptimal decisions about mates, prey, shoals, or refuges any time they need to make a rapid decision and the alternatives are placed at the opposite sides of the body. Perhaps this situation is not so uncommon in their natural habitat. Like most fish, poeciliids posses only a small overlap between the two visual hemifields, and importantly they lack the kind of mobile neck that allows them to rapidly scan relevant stimuli sequentially with the two eyes and send information to both hemispheres (Clayton & Krebs 1994).

The picture emerging from studies of *G. falcatus* is that the advantages associated with having an asymmetric brain may balance the ecological disadvantages associated with left-right differences in the response to stimuli. The relative costs and benefits of lateralization are likely to vary with ecological conditions such as predation pressure. Intuitively, better schooling performance and efficiency in multitasking (i.e., strong lateralization) should be favored under high-predation regimes, a hypothesis that recently received support from a field study of *B. episcopi* (C. Brown et al. 2004; see Kelley & Brown, **chapter 16**). However, high predation risk should also favor fish that make quick and accurate decisions about escape trajectories. Perhaps the often-used dichotomy of high versus low predation is too simplistic, and a finer distinction is required (Botham et al. 2006; Croft et al. 2006b). For example, the capacity to mate or feed while monitoring predators should be most effective where predators are always in sight or come from predictable directions (e.g., from deep water), while ambush predators should favor fish with equal capacity to react on the left or the right side.

15.4.4 Evolutionary significance of population biases in laterality

Directional asymmetries in laterality are by far the most frequent condition in all vertebrate taxa (reviews in Rogers 1996; Bisazza et al. 1998b). Advantages in brain efficiency outlined in section 15.4.2 may explain the existence of individual lateralization but cannot account for the alignment in the direction of lateralization in most individual of a species. Indeed, consistent population biases in laterality are potentially disadvantageous since they make individual behavior more predictable to other organisms, for example, allowing predators to learn the most frequent direction of escape in their prey. Rogers (1989) has proposed

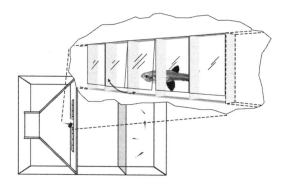

Figure 15.3 Apparatus used to study line-bisecting performance in *Girardinus falcatus*. Subjects from lines selected for high and low degrees of cerebral lateralization were trained to use the middle door in a row of nine in order to join their social group. Fish from LD and RD lines made systematic errors to the left or the right of the central door, respectively, whereas NL fish chose the correct door more often.

that the direction of lateralization in a population occurs as a consequence of the need to coordinate social behavior (see also Brown 2005). A game-theoretical analysis showed that population-level lateralization can arise as an evolutionarily stable strategy when the benefit to an asymmetrical individual of coordinating with others with the same laterality equals the costs arising from predators having more success with the more common prey type (Ghirlanda & Vallortigara 2004). Empirical evidence in support of this hypothesis is equivocal. Rogers and Workman (1989) found that the social hierarchy was more stable in groups of light-exposed (lateralized) chicks than in groups of dark-incubated (nonlateralized) chicks. Bisazza et al. (2000a) examined 16 fish species and found some evidence that population biases were more frequent in species with a strong shoaling tendency than in solitary ones, but this result needs to be confirmed using a larger sample size and correcting for phylogeny. Nevertheless, Bisazza and Dadda (2005) found no evidence that schools composed of female *G. falcatus* with the same laterality were more coordinated than schools composed of females of mixed laterality.

15.5 Conclusions and future directions

Although classical studies on learning and memory have benefited research on other fish groups, in the last fifteen years poeciliids have emerged as important model organisms for investigating complex cognitive behaviors such as mate-choice copying, eavesdropping, recognizing individuals, social learning, and numerical abilities (see **chapters 13, 14, 16,** and **20**). Studies on poeciliids continue to challenge the once-dominant view that a complex flexible behavior is a prerogative of mammalian and avian species, suggesting instead that these abilities are exhibited much more widely among animal taxa (see Bshary et al. 2002 for a recent discussion). As clearly highlighted here and elsewhere in this volume, in many cases complexity of behavior, learning abilities, and memory capacities of poeciliids are comparable to, and in some cases exceed, those of mammals and birds. Poeciliids can recognize up to 40 familiar individuals, remember the outcome of past cooperative predator inspections and bias behavior accordingly, exploit information obtained from observing the outcome of a fight or the mate choice of another individual, and learn new foraging and antipredator habits from an expert conspecific.

The recent rapid expansion of animal cognition studies has prompted considerable interest in understanding the selective factors that promote the evolution of specific cognitive capacities. Comparative studies have proved a useful tool to identify the ecological factors associated with variation in encephalization, brain organization, and cognition (for recent discussions, see Byrne & Bates 2007; Cunning-

ham & Janson 2007; Healy & Rowe 2007). Poeciliid fishes appear to be an ideal group for investigating such topics using both intra- and interspecific comparative approaches. The habitats occupied by poeciliids show extreme variability in structure and complexity, and poeciliids exhibit considerable variation in diet and predation regimes (Meffe & Snelson 1989a). Pioneering studies (see section 15.2.4) have reported learning differences among populations with different predation regimes in two poeciliid species, suggesting promising directions for future studies.

Another topic of increasing interest concerns differences in cognition that are expected to arise from the differential action of sexual selection on the two sexes and as a consequence of niche divergence that might derive from sexual dimorphism. The latter effect is expected to be particularly relevant in poeciliids, as they exhibit some of the most extreme examples of sexual dimorphism of all vertebrates (Bisazza 1993b). Yet no study has directly addressed this topic in poeciliids, although anecdotal observations (see section 15.2.4) suggest that sex differences in cognition may be present in some species.

Individual differences in cognition represent an emerging area of study that to date has been extensively examined only in humans (e.g., O'Boyle et al. 2005; Posner et al. 2007). Lateralization studies in *G. falcatus* have revealed the existence of individual differences in the organization of brain functions that appear to have a genetic basis and that exert their effects on multiple aspects of behavior and cognition (see sections 15.4.1–15.4.3). One promising area for future investigation is the potential for covariation of personality and cognitive abilities (Dugatkin & Alfieri 2003; Brown & Braithwaite 2005; see also Kelley & Brown, **chapter 16**). Another virtually unexplored potential source of variation in cognition is represented by differences among males in mating tactics. Since the two mating tactics of the poeciliids, sneak mating and courtship, require quite different skills (Bisazza 1993b; see also Rios-Cardenas & Morris, **chapter 17**, and Magurran, **chapter 19**), it will be interesting to see if these tactics select for different suites of cognitive abilities in poeciliids with genetically determined alternative male mating tactics (Zimmerer & Kallman 1989; Travis 1994).

Acknowledgments

I would like to thank Jenny Kelley, Sue Healy, Jon Evans, Culum Brown, and Ingo Schlupp for their helpful comments on the manuscript. I am also indebted to John Endler for an enlightening discussion about the possible relationship between cognitive limitations and searching rules. I am indebted to Marco Dadda and Christian Agrillo for the productive discussion of the issues addressed in this chapter and for useful suggestions during the preparation of the manuscript.

Chapter 16 Predation risk and decision making in poeciliid prey

Jennifer L. Kelley and Culum Brown

16.1 Introduction

VARIATION IN PREDATOR abundance is one of the strongest selective pressures operating in natural ecosystems and influences a wide range of physiological, morphological, and behavioral traits in prey species. Several influential reviews on this topic (Lima & Dill 1990; Endler 1995; S. Lima 1998) have demonstrated that predation risk exerts a considerable influence on life-history patterns (see Pires et al., **chapter 3**), genital morphology (Langerhans, **chapter 21**), male courtship and female mate choice (Rios-Cardenas & Morris, **chapter 17**), dominance and aggressive interactions (Magurran, **chapter 19**), foraging behavior and habitat use (Grether & Kolluru, **chapter 6**), and, of course, the expression of antipredator strategies such as schooling and predator inspection (Krause et al., **chapter 13**; this chapter). The risk sensitivity hypothesis posits that prey optimize the trade-off between predation risk and other important behavioral activities in order to maximize fitness (Sih 1980; Lima & Dill 1990; S. Lima 1998). At one extreme, the cost of not responding to an increase in predation risk might result in the loss of life. On the other hand, overreacting to changes in risk will cause a reduction in other activities (e.g., foraging, courtship), which can also severely affect an individual's fitness.

Optimizing the balance between predation risk and behavioral activities requires two key processes. First, prey must acquire accurate and reliable information about predation risk (Kats & Dill 1998). Second, they must act on this information and implement corresponding adjustments to their behavior. Both of these processes—information gathering and decision making—require a considerable level of cognitive processing. The aim of this chapter is not to review the literature on behavioral responses of prey to changes in predation risk, as this subject has already been covered in some detail for a number of taxa (see reviews by Lima & Dill 1990; S. Lima 1998). Rather, we extend the discussion of behavioral decisions under predation risk by considering the risk-related cues that drive these decisions and also the cognitive requirements that underpin the decision-making process. Specifically, we will argue that decision making under the risk of predation is fundamental to a prey animal's survival and fitness and, consequently, that predation risk may be partly responsible for driving the evolution of cognition.

Research on the evolution of cognition has tended to focus on the importance of social interactions (the social-intelligence hypothesis: Humphrey 1976; Byrne & Whiten 1988) or complex foraging tasks (Clutton-Brock & Harvey 1980) among animals that form hierarchical social groups (e.g., primates) rather than the role of predation risk per se (Zuberbuhler & Byrne 2006). However, responding to predators has a strong social component, particularly since grouping is one of the most common responses to increased predation risk (Krause & Ruxton 2002). Prey that form groups not only benefit through safety in numbers (e.g., risk dilution) but may also increase their chances of survival as a result of their social interactions with other group members (Krause & Ruxton 2002; Krause et al., **chapter 13**). For example, the use of predator-specific alarm calls (e.g., Marler 1957), cooperation among individuals during predator inspection (e.g., Milinski 1987), and the ability to learn by observing/listening to the responses of others (social learning: Heyes & Galef 1996; Brown & Laland 2003; Webster & Laland, **chapter 14**) require a considerable level of cognition. Although other tasks (e.g., foraging, competitive interactions) are likely to have facilitated the evolution of social complexity, we expect strong selection on any social behaviors (and their underlying cognitive processes) that increase a prey individual's chance of surviving an encounter with a predator.

Predator-prey interactions are often likened to an evolutionary arms race; as the predator gets the upper hand over its prey, the prey responds in some manner to attempt to stay one step ahead. Much of the literature regarding the predator-prey arms race has tended to revolve around morphological or behavioral traits, such as attack speed and escape velocities of predators and prey, respectively (Dawkins & Krebs 1979; Abrams 2000). However, such traits are underpinned by physiological or cognitive mechanisms that have received far less attention in this context. Considering the arms race from a cognitive perspective, we expect the increased cognitive capacity of prey living in high-risk environments to be counteracted by heightened levels of cognition underlying predator attack strategies. Thus, we hypothesize that predator and prey are trying to "outsmart" one another (Brown & Braithwaite 2005). If this argument is correct, we expect to observe a relationship between cognitive ability and variation in predation risk.

Fishes, and poeciliids in particular, are an ideal group for evaluating the experimental support for these ideas. First, fishes display surprisingly complex cognitive abilities (e.g., cooperation, tool use, spatial memory) that are traditionally assigned to high-intelligence organisms, such as primates (Bshary et al. 2002; Brown et al. 2006b; Bisazza, **chapter 15**). Second, members of the same species often occupy a large variety of habitats, making it possible to establish correlative links between ecological variables (e.g., predation risk) and cognitive ability. Here, we draw on evidence from several species of poeciliid to evaluate the evidence that predation risk is an important factor driving the evolution of cognitive ability in prey (Zuberbuhler & Byrne 2006). We first examine how the interaction between brain morphology and the detection of sensory cues allows poeciliid prey to gain information about their risk of predation (section 16.2). We then evaluate how the decision-making process—for example, deciding when and how to respond to a predator-related cue—is influenced by both predation risk and an individual's personality (section 16.3). Finally, we assess the relationship between predation risk and cognitive ability, focusing on how living with risk affects brain development and the evolution of cerebral lateralization (section 16.4).

16.2 Information gathering: prey assessment of predation risk

Poeciliids, like other prey fishes, have a large number of sensory modalities available to them to assess changes in their risk of predation (Coleman, **chapter 7**). Specializations in each of these sensory systems are revealed by variations in brain morphology that tend to correlate with the behavior and ecology of the species (Kotrschal et al. 1998; Gonzalez-Voyer et al. 2009). For example, the optic tectum and telencephalon (part of the forebrain) are well represented in the brains of shallow-water fishes that rely heavily on visual cues (e.g., for foraging and navigation). In contrast, fishes living in deep waters (>1000 m) tend to rely more on mechanoreception and olfaction, and the size of the optic tectum and telencephalon is much reduced (Kotrschal et al. 1998). Environmental characteristics, therefore, exert considerable influence on brain morphology and the way in which prey can detect and manage their level of risk.

The sensory cues that are available to prey fishes to assess their level of predation risk depend on the physical properties of the water (e.g., depth, turbidity) as well as habitat structure (e.g., open water versus complex environments). In addition, environmental variation, such as habitat heterogeneity and changing light levels that are related to diurnal patterns of predator activity, may serve as important cues regarding changes in risk. We discuss each of these in turn and highlight the manner in which poeciliid studies have contributed to our understanding of how prey animals assess their risk of predation.

16.2.1 Chemical cues

Chemical cues provide an important source of information to prey regarding local predation risk (Kats & Dill 1998). These cues may originate from a predator in the form of predator odors (sometimes referred to as "kairomones") that are passively released through normal metabolic processes or odors associated with the recent feeding habits of a predator (e.g., urine, feces) (Wisenden & Chivers 2006). Other chemical cues that can provide important information to both predators and prey are damage-released chemical alarm cues (hereafter referred to as "alarm cues") and disturbance cues. The detection of such cues may cause prey to increase their level of vigilance, adopt risk-sensitive behavioral strategies, or directly engage in predator avoidance tactics, depending on the severity of the threat posed (Chivers & Smith 1998).

Alarm cues are chemicals that occur in the epidermis of the skin and are released when a fish is captured or injured by a predator (Pfeiffer 1977; Smith 1977, 1982, 1992; Pfeiffer 1982; Chivers & Smith 1998). Alarm cues occur in many fish species, including poeciliids (Chivers & Smith 1998; reviewed in Brown & Chivers 2006), and the detection of these cues elicits a generalized behavioral response in fishes known as the "fright response" or "fright reaction" (hereafter referred to as "fright response"), which is characterized by avoidance of the cue source, an increase in shoal cohesion, a reduction in activity, movement toward the substrate, and erratic swimming such as dashing and skittering (Chivers & Smith 1998).

A response to alarm cues has been reported in only a few species of poeciliid: mosquitofish (Reed 1969; Garcia et al. 1992), swordtails (Mirza et al. 2001), and guppies (Nordell 1998; Brown & Godin 1999b; Brown et al. 2009). For example, Brown and Godin (1999b) used a combination of laboratory and field-based studies to show that guppies (*Poecilia reticulata*) respond to alarm cues from injured conspecifics but not to cues from swordtails (which were previously thought to lack alarm cues, but see below). Interestingly, the nature of the fright response was dependent on the population origin of the alarm cue donor; guppies from a high–predation risk population showed a stronger response (movement toward the substrate) to alarm cues from a high-predation donor than to those from a low-predation donor (Brown & Godin 1999b). The finding that both juveniles and adults avoided alarm cues in these experiments suggests that this response may be largely genetically based or formed at an early developmental stage (Brown & Godin 1999b).

Alarm cues that originate from heterospecifics can also serve as indicators of increased predation risk (Chivers & Smith 1998; Smith 1999), and cross-species responses are particularly beneficial for prey that form mixed-species shoals. Mirza et al. (2001) exposed swordtails (*Xiphophorus hellerii*) to either conspecific skin extract, guppy skin extract, or distilled water. The swordtails displayed a fright response toward both skin extracts (containing alarm cues) but showed a stronger reduction in activity levels in the presence of conspecific extract than heterospecific extract.

Alarm cues not only function as indicators of increased predation risk but also provide an opportunity to acquire a learned response to novel stimuli. Many studies have demonstrated that prey fishes can acquire a fright response by associating novel visual, auditory, and/or chemical cues with alarm cues that originate from either conspecifics or heterospecifics (reviewed in Brown 2003; Kelley & Magurran 2003, 2006; Brown & Chivers 2006). Following a single association of the stimulus and alarm cue, prey can learn a response to novel predators (Magurran 1989; Mathis & Smith 1993; Chivers & Smith 1994), risky habitats (Chivers & Smith 1995), or even nonbiological stimuli such as flashing lights (Hall & Suboski 1995; Yunker et al.

1999) and artificial noises (Wisenden et al. 2008), although learning occurs more rapidly in response to biologically relevant cues (Csanyi 1986). Other mechanisms of learning have been shown to occur in poeciliids (e.g., social learning, Brown & Laland 2003; see also Webster & Laland, **chapter 14**), but poeciliids are noticeably absent from the literature on associative learning using alarm cues.

Disturbance cues are more generalized chemical cues that originate from fish that are distressed but not injured. Disturbance cues are thought to consist of ammonia that is excreted from the gills during periods of increase metabolic activity (Wisenden et al. 1995). The detection of disturbance cues influences the behavior and survival of a variety of fishes (e.g., Iowa darters, *Etheostoma exile*, Wisenden et al. 1995; Jordao & Volpato 2000; Bryer et al. 2001; Mirza & Chivers 2002; Jordao 2004) but has not yet been demonstrated in poeciliids.

16.2.2 Visual cues

Visual cues from predators, such as exposure to live fish predators or to fish or bird predator models, have been used to assess antipredator behaviors in a number of poeciliids (e.g., Magurran et al. 1992; Magurran & Seghers 1994b; Smith & Belk 2001). Commonly observed responses to these cues include a reduction in activity levels, increased shoaling and hiding, movement toward the substrate, and predator inspection. For example, *Brachyrhaphis episcopi* from high- and low-predation populations were exposed to a series of scenes hidden behind a trapdoor. Each shoal was exposed to each scene (blank control, live cichlid, or novel object) for 15 minutes, and their shoaling and inspection behavior assessed. Fish from high-predation areas formed tighter shoals and exhibited higher rates of inspection when exposed to a live cichlid than did those from low-predation populations (fig. 16.1a and b). No population differences were observed in responses to the blank scene (control) or the novel object (fig. 16.1a and b).

In addition to generalized schooling behavior, predator inspection is perhaps the best-studied response to visual predator cues, and studies with guppies have provided some key illustrations of the predator-deterrent function of this apparently paradoxical behavior. Predator inspection occurs when a small group of fish leave the shoal and slowly approach the predator, often swimming along the length of its body, before returning to the shoal (Pitcher et al. 1986). In the guppy, predator inspection is variable across populations, with fish from high-risk populations inspecting in larger group sizes (Magurran & Seghers 1994b). Most studies of predator inspection behavior have focused on the role of visual cues, but prey fishes will also inspect chemical

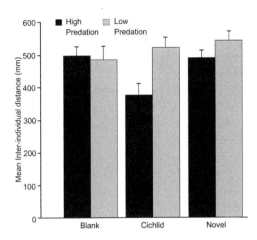

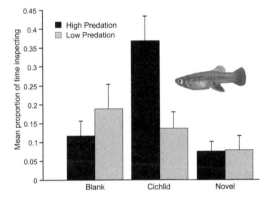

Figure 16.1 Mean (± SE) (a) interindividual distance and (b) proportion of time inspecting the compartment for shoals of *Brachyrhaphis episcopi* exposed to a blank scene (control), a live cichlid predator, and a novel object. There was no overall treatment effect on mean interindividual distance (repeated measures ANOVA: $F_{2,72} = 1.928$, $P < 0.05$), but post hoc analysis revealed that high-predation fish formed tighter shoals in response to the live cichlid than low-predation fish did (Fisher's PLSD, $P = 0.007$). There was a significant overall treatment effect on time spent inspecting the compartment; shoals spent a greater proportion of time inspecting the compartment containing the cichlid than either the blank scene or the novel object ($F_{2,72} = 5.495$, $P = 0.006$). C. Brown, unpublished data.

cues by moving toward the source (Brown & Godin 1999a; Brown & Godin 1999b; Brown & Cowan 2000; Brown et al. 2000).

Several studies have demonstrated that fish tend to show a bias in the eye that faces a predator during inspection (i.e., lateralization of inspection behavior; see also section 16.4.3 and Bisazza, **chapter 15**). For example, mosquitofish (*Gambusia holbrooki*) tended to explore a novel environment using their left eye but closely approached a predator using their right eye (De Santi et al. 2001). Interestingly, this eye bias during predator inspection depends on the predation pressure of the population; *B. episcopi* from high-predation populations preferred to use their right eye when viewing a predator, whereas those from low-predation populations favored use of their left eye (C. Brown et al. 2004). These patterns may be attributable

to differences in the stimuli that are perceived as threatening (C. Brown et al. 2004); thus, prey from high-risk populations may view a predator as a potential threat, while fish from low-risk populations may perceive the same stimulus as a harmless object. Alternatively, the population differences in left-right eye bias could be dependent on the frequency of left-right lateralized individuals in the population (i.e., frequency-dependent selection; see section 16.4.3).

Important information about potential predation threats can also be obtained by individuals that observe the behavior of others. For example, because predator inspection is thought to provide the inspectors with important information regarding the attack motivation of the predator (Pitcher et al. 1986; Murphy & Pitcher 1997), any changes in the behavior of the inspectors arising as a result of this information may be adopted by observing (noninspecting) fish (Pitcher et al. 1986; Magurran & Higham 1988). Observing the behavior of conspecifics (i.e., social learning, Brown & Laland 2001; Webster & Laland, **chapter 14**) may be an important way to acquire information regarding the threat posed by novel predators. Fish can also learn about potential threats by observing the behavior of heterospecifics (Krause 1993b; Mathis et al. 1996), but this has yet to be demonstrated in poeciliids (see Webster & Laland, **chapter 14**).

16.2.3 Spatial and temporal variation in risk

Predation risk is rarely constant, and prey fishes are usually subjected to risk that can vary on either a spatial or a temporal scale. Spatial changes in predation risk are usually associated with habitat heterogeneity and depend on the preferred habitat of the predator(s). In guppies, for example, the deeper parts of streams are often occupied by dangerous predators such as the pike cichlid (*Crenicichla frenata*), whereas riffles at the river's edge may be occupied by less dangerous predators such as Hart's rivulus (*Rivulus hartii*), which predominantly preys on juvenile guppies (Liley & Seghers 1975). Guppies respond to this spatial variation in risk and have been observed occupying the outer edges of stream margins, which may allow them to avoid both species of predator (Fraser et al. 2004).

Temporal variations in predation risk can be predictable if, for example, risk varies diurnally or in accordance with lunar or seasonal changes in predator activity or abundance (Lima & Bednekoff 1999). Alternatively, periods of increased predation risk can be highly irregular and depend on predator-prey encounters and/or detection of their associated cues. In the guppy, diurnal changes in predation risk may arise as a result of the activity patterns of different predators. For example, guppies that occur with the nocturnal wolf fish (*Hoplias malabaricus*) are inhibited from feeding at night and suffer reduced growth and daytime courtship compared with their upstream counterparts that do not occur with this predator (Fraser et al. 2004). Theories examining the temporal nature of predation risk suggest that the level of predation pressure and its temporal distribution have several implications for the evolution of, and the amount of "effort" invested in, antipredator responses (the risk allocation hypothesis, Lima & Bednekoff 1999; Sih et al. 2000). For example, relatively low densities of predators that are highly unpredictable in space or time might correspond to a relatively high risk environment in comparison to those situations where predator density is high but highly predictable. Experiments examining these phenomena have yet to be conducted in poeciliids, but a study using cichlids found a temporal-predictability effect for low-predation levels but not for high-predation levels (Ferrari et al. 2008).

16.3 Decision making under risk of predation

16.3.1 When to respond?

Once a prey individual has detected cues that are associated with increased predation risk (section 16.2), it must decide how to respond to these cues. Since many predator-related cues are continuously present in the environment, the initiation of antipredator behaviors may be contingent on the detection of threshold levels of a particular cue. For example, glowlight tetras (*Hemigrammus erythrozonus*) exposed to low concentrations of a putative alarm cue do not show an overt antipredator response but do increase their vigilance toward other sensory cues involved in risk assessment (G. Brown et al. 2004). Due to the time and energy constraints associated with living in high-risk environments, we expect there to be population differences in the threshold levels of predator-related cues that initiate an overt antipredator response. Specifically, we expect that higher levels of risk-related cues will be required to stimulate antipredator behavior in fish from high-predation populations than in those from low-predation ones (but see Brown et al. 2009) because there are likely to be higher levels of background predator-related cues in high-predation environments. However, threshold responses may also depend on variability in the level of risk-related cues; in line with the risk allocation hypothesis (see above), prey are expected to exhibit stronger responses when cues are unpredictable and weaker responses to cues that are constantly present in the environment (Ferrari et al. 2008). Given that poeciliid populations are often exposed to varying levels of predation risk, this hypothesis could be tested relatively easily.

An extension of the ideas presented above is the prediction that prey from high- and low-risk populations should show differential stress responses to risk-related cues. We predict that prey living in high-risk environments will be better adapted for coping with stress; thus, higher levels of stress-inducing stimuli will be required to induce a stress response in fish from high-risk populations than in fish from low-risk ones. Indeed, it appears that fish from high–predation pressure environments are better able to cope with stress induced by constant harassment by predators. This is evidenced by their reduced opercula beat rates when exposed to mild stresses, suggesting that both the physiology and the psychological response of the fish have altered in some fashion via natural selection or ontogenetic plasticity (Brown et al. 2005a). Ontogenetic plasticity, however, has received very little attention in this context and would be a rewarding avenue for future research.

Laboratory-reared fish from high- and low-predation areas also differ in their response to a mild stressor in a fashion reminiscent of their wild-caught parents, suggesting that the stress response has a heritable component (fig. 16.2). Interestingly, lab-reared fish that showed a small change in opercula beat rate over the 15-minute experimental period had higher levels of blushing (sequestration of blood) in the nasal cavity. However, this relationship was significant only in fish reared from high-predation parents and not from low-predation parents (fig. 16.3). This suggests that the fish sequester blood in this region rather than increasing their opercula beat rate when under stress. It is likely that reduced opercula movement lowers the probability of detection by predators when the fish choose a cryptic antipredator response rather than schooling (Brown et al. 2005a).

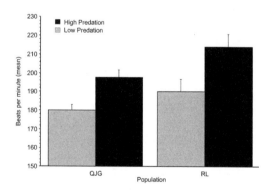

Figure 16.2 Mean (± SE) opercula beat rate for two populations (QJG = Quebrada Juan Grande, RL = Río Limbo) of lab-reared *Brachyrhaphis episcopi* from high- and low-predation areas. (Predation effect ANOVA: $F_{1,30} = 11.814$, $P = 0.002$.) C. Brown, unpublished data.

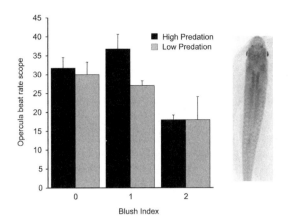

Figure 16.3 Change in opercula beat rate (scope: max. − min. opercula beat rate) over a 15-minute stress trial for lab-reared fish from high- and low-predation areas against nasal blush index. Lab-reared fish had a higher rate of blushing in the nasal cavity ($F_{2,31}$ = 5.258, P = 0.011), but this relationship was significant only in fish reared from high-predation parents ($F_{2,12}$ = 8.528, P = 0.005) and not from low-predation parents ($F_{2,16}$ = 1.184, P = 0.331). C. Brown, unpublished data.

16.3.2 How to respond? Predator-specific defenses and threat sensitivity

Although it seems intuitive that prey should tailor their responses toward particular types of threat, this can be costly if a prey's response to one predator increases its risk to another (Sih et al. 1998). Again, from the cognitive perspective, we anticipate that greater cognitive skills are required to recognize and respond differentially to different types of predator threat, as multiple risk factors represent a greater level of complexity. It may not be the diversity of predators per se that results in selection on prey for increased cognitive ability but, rather, the variation in attack strategies that the multipredator environment presents. In mammals, for example, the different attack modes displayed by avian and ground predators (which require incompatible prey escape responses) are thought to have facilitated the evolution of complex predator-specific alarm calls in vervet monkeys and ground squirrels (Macedonia & Evans 1993).

In fishes, avian predators (e.g., herons, kingfishers, cormorants) present a different type of threat than piscivorous predators do because birds approach from above and perpendicular to the fishes' plane of movement, while fish predators tend to attack either from the same level or below (Katzir & Camhi 1993). Avoiding attacks from these two classes of predators involves strategies that are diametrically opposed. In Panamanian streams, for example, *B. episcopi* is frequently found in the shallows, thereby avoiding predation by large in-stream piscivores (C. Brown, personal observation). This behavior, however, makes them highly vulnerable to avian predators such as the tiger heron (*Tigrisoma lineatum*). Presumably their very presence in shal-

low areas suggests that the in-stream predators represent a greater threat than that posed by avian predators. Thus, the different attack modes utilized by piscivorous and avian predators present an ideal opportunity to examine the relationship between predator attack strategies and prey cognitive complexity in fishes.

Templeton and Shriner (2004) compared the response of guppies from a high- and a low-risk population toward a live piscivorous predator (resembling the pike cichlid) and an aerial predation threat (a three-dimensional model of a green kingfisher). The populations differed in their behavioral responses depending on the type of threat; the aerial-predator stimulus caused high-predation fish to seek cover, whereas low-predation fish tended to freeze. In contrast, the aquatic predator caused high-predation fish to spend more time performing inspection behavior than low-predation fish (Templeton & Shriner 2004). Maintaining vigilance toward aerial threats while also being alert to piscivorous predators is likely to present a significant challenge. Lateralization of the brain hemispheres may be an important mechanism by which prey in high-risk environments can remain vigilant to different types of threat (see section 16.4.3). Given the potential complexity resulting from multiple threat sources, it may be that fish develop a generalized antipredator response that minimizes the overall threat (a kind of predator-community response) rather than specific responses to each predatory species. However, this latter approach may well place them in danger of secondary attacks from other predators.

Prey that have encountered predators or risk-related cues must also decide on the strength of their antipredator response. This is often termed "risk sensitivity" and refers to the observation that prey individuals display different levels of antipredator behavior according to the strength or severity of the threat posed (Helfman 1989). For example, guppies are more likely to avoid hungry predators than satiated ones (Licht 1989). Smith and Belk (2001) examined the theory of threat sensitivity in detail by presenting western mosquitofish with predators (live green sunfish, *Lepomis cyanellus*) that varied in size (large or small), diet (fed mosquitofish or chironomid larvae), and hunger level (hungry or satiated). Shoals maintained a greater distance from mosquitofish-fed predators than chironomid-fed predators and were more likely to occupy the upper half of the aquarium in the presence of hungry predators than satiated ones (Smith & Belk 2001). Inspection behavior was also affected by diet, with mosquitofish maintaining a greater distance from hungry predators than satiated ones. Mosquitofish and other poeciliids are therefore capable of quite complex behavioral decisions depending on the information contained in different types of sensory cue.

16.3.3 Individual variation in responses to risk

Behavioral ecologists have long been interested in explaining population-level differences in the behavior of fishes with reference to the environment that they occupy. More recently, however, it has become apparent that prey often display individual-specific responses to predation. Sex- and age-specific responses to predation risk are perhaps among the best studied of these. For example, female poeciliids must maximize their reproductive fitness via longevity, and they display better-developed antipredator responses than males (Magurran & Seghers 1994c; Magurran & Garcia 2000). In contrast, males are often blasé about the presence of predators and will continue to court females irrespective of the level of threat. Furthermore, young and old (small and big fish) may respond differently to a given predator based on the relative threat that it poses (Werner & Hall 1988; Persson et al. 1996; Magnhagen & Borcherding 2008).

These sources of individual variation in responses to risk are often overlooked because many studies take a single snapshot in time and individuals are only observed once and often in a group context. Making multiple observations of the same individual, however, may reveal individual responses that are stable or consistent over a variety of contexts and that are suggestive of personality traits. Personality traits typically transcend demographic variability and are relatively stable over time. Use of the word "personality" to describe consistent individual differences in behavior has been avoided in the fish literature largely due to fear of anthropomorphism. In its place we are often confronted with alternative terminology such as "coping style" (Clement et al. 2005), "temperament" (Shaklee 1963), or "behavioral profile" (Magellan & Magurran 2007a). Despite a general reluctance to accept that fish have personalities, there is a growing body of literature that suggests that not only are they pervasive (Wilson 1998; Bell 2005; Dingemanse et al. 2007), but personality traits have fitness consequences (Smith & Blumstein 2008) and may therefore be subject to natural selection (Gottlieb 2002; Brown et al. 2007a; see also section 16.3.3.3).

16.3.3.1 The shy-bold continuum. In fishes, two personality traits in particular—boldness and aggression—have been the subject of a number of studies. From a poeciliid perspective much of the attention has focused on the former. The shy-bold continuum refers to the likelihood that an individual is prone to risk taking (Wilson et al. 1993). It is immediately evident, therefore, that there is likely to be a link between predation pressure and the evolution and development of personality traits. Budaev (1997) investigated the social tendencies and exploratory behavior of 29 gup-

pies and found that, in the broader context, individual guppies could be placed on two personality dimensions that he defined as the "approach" and "fear avoidance" continua. The contexts in which the fish were tested included exploration of a novel environment, predator inspection, and schooling. Each of these contexts represents a situation where the subject has to make a decision between opposing tendencies (i.e., a trade-off scenario).

Brown et al. (2005b) utilized replicate high- and low-predation streams in Panama to determine how predation pressure influences an individual's position on the shy-bold continuum. Other studies with this species have concentrated on exploration of a novel environment (open-field paradigm) and the time taken to emerge from cover (C. Brown et al. 2004; Brown et al. 2005b). Later studies revealed correlations between the tendency to explore a novel environment and the tendency to leave a shoal to investigate a novel object (Brown et al. 2007a). Once again, all contexts rely on the fish making relatively dichotomous decisions in a mildly stressful situation. In addition to variation that could be explained by demographic variables (size and sex), fish from high-predation areas tended to be bolder than those from low-predation areas. It seems likely that the constant threat of predators in the environment favors individuals that are more likely to accept a higher degree of risk for a given payoff. In a high-predation context, shy fish would spend a significant proportion of their lives hiding from predators and would likely have reduced growth rates and lower fecundity as a result of missed foraging opportunities. In contrast, bold fish would emerge from cover sooner to gain access to resources. Indeed, there is expected to be a strong link between growth rates, fecundity, and personality, as fast-growing individuals must behave in a risky manner to acquire resources and meet their high metabolic demands (Stamps 2007; Biro & Stamps 2008).

16.3.3.2 The development of personality. Personality traits in humans are known to be influenced both by genetics (i.e., they are heritable) and by individual experience (McGue & Bouchard 1998). Indeed, Budaev (1997) suggests that the term "temperament" is best reserved for the heritable components of personality, which are then shaped by experience to form personality. The fact that personality traits are heritable paves the way for identifying the genes responsible. Quantitative trait loci analysis in zebrafish (*Danio rerio*, Cyprinidae) suggests that regions on chromosome 9 and 16 are significantly associated with boldness, showing signs of dominance and additive effects, respectively (Wright et al. 2006). A similar approach could be adopted in poeciliids when the guppy genome is sequenced. Brown et al. (2007a) investigated the relative contributions of the genetic and experiential components of personality traits by collecting poeciliids (*B. episcopi*) from high- and low-predation populations, breeding them, and rearing their offspring in the lab under two conditions. Half of the lab-reared offspring were left undisturbed (simulating a low-predation experience) while the other half were chased with a net every day for two weeks (simulating a high-predation experience). Prior to initiating the treatment, fish from each population showed boldness scores that were consistent with their wild-caught parents. However, simulated predator exposure increased boldness scores irrespective of the source population, demonstrating the manner in which predation experience during ontogeny shapes personality traits.

16.3.3.3 Personality and fitness. Such population (and individual) variation in response to predation pressure suggests that there are likely to be fitness consequences associated with displaying various personality traits in a given context (Stamps 2007; Biro & Stamps 2008; Smith & Blumstein 2008). The tendency to explore, for example, is likely to open up new resources to bold individuals and thus potentially expose them to novel predators and pathogens (Wilson et al. 1993; Fraser et al. 2001). Recent work with poeciliids shows that the expression of bold behavior has direct influences on fitness measures. In *B. episcopi* there is a strong correlation between a fish's position on the shy-bold continuum and its relative body mass (Brown et al. 2007a). Bold fish tend to be heavier than shy fish, and the rate at which they put on mass per unit length is also greater. Boldness may also enhance male reproductive success via female mate choice. During a simulated predator encounter, female guppies preferred to mate with males that approached a model predator over males that did not (Godin & Dugatkin 1996). This preference was expressed irrespective of male coloration (which is an important cue in mate choice and is correlated with boldness) and body length (Godin & Dugatkin 1996).

16.4 Relating predation risk to cognitive ability

16.4.1 Predators as a form of "environmental complexity"

The presence of predators in the environment may be thought of in terms of habitat complexity. Predators are just another (although very important) variable that prey must keep track of in space and time. Research on rodents has revealed that exposure to an enriched environment stimulates neuronal development and has an important influence on cognitive capacity (Park et al. 1992; Gomez-Pinilla et al. 1998). Similar observations have been made in invertebrates (Lomassese et al. 2000). Environmental enrichment is commonly employed by zoos to reduce stereotypy and to encourage the expression of more "natural" behavior in captive animals. Such manipulations are rare in fishes and totally absent in poeciliids, despite the fact that the brains of fishes are more plastic than those of most vertebrates (Zupanc 2006), making them an ideal model system. In general, the brains of fishes are more highly developed if they have been reared in complex environments (Marchetti & Nevitt 2003). The optic tectum and the telencephalon are most profoundly affected, both being key regions in the analysis of visual and spatial information (Salas et al. 1996; Rodriguez et al. 2006). Moreover, when exposed to enriched environments, cognitive function is also enhanced (Brown et al. 2003). No such manipulative studies have been conducted on poeciliids to date; however, some progress has been made by examining the behavior of fish captured in high- and low-predation environments (Brown & Braithwaite 2005).

16.4.2 Does predation pressure influence brain development?

Brown & Braithwaite (2005) addressed the question of whether predation pressure affects brain development using *B. episcopi*, which inhabits a number of streams each containing high- and low-predation environments. The fish were exposed to a radial maze in which they had to locate a hidden foraging patch. Contrary to expectation, fish from high-predation regions took significantly longer to learn the location of the foraging patch than fish from low-predation areas. When the video footage was analyzed more carefully, it was evident that the manner in which the fish navigated the maze varied tremendously. High-predation fish had a very strong preference to move in a counterclockwise direction, irrespective of their location relative to the rewarded arm of the maze. In comparison, the low-predation fish gradually altered their behavior and began to head directly to the rewarded arm. Despite these unanticipated results, further experimentation examining the cognitive capacity of high- and low-predation fishes is likely to be fruitful. For example, rainbowfish (*Melanotaenia duboulayi*) from high-predation areas are better at solving spatial tasks than those from low-predation areas (Brown & Warburton 1999).

Burns and Rodd (2008) examined differences in the spatial memory of guppies collected from high- and low-predation populations and compared these differences with brain size. Although there were no differences among populations in the time taken to complete the spatial-memory task or the number of errors made, high-predation fish took significantly longer than low-predation fish when deciding which maze chamber to enter. There were no differences in brain morphology between high- and low-predation fish, but within populations, fish that made decisions most rapidly (i.e., "hasty" individuals) had smaller telencephalons (Burns & Rodd 2008). Note that male guppies were used for the spatial-memory task in this experiment, and as males are more likely to take risks than females (see section 16.3.3), population differences in memory and brain anatomy are least likely to arise in males. Future work should examine the brains of females, as females are known to show strong antipredator behavior. Although further research is required to examine the link between predation risk, cognitive performance, and underlying brain structure, it is important to note that variation in predation risk is correlated with other important ecological factors (e.g., light intensity, food availability, water temperature; Endler 1995) that are also likely to affect cognitive processes and brain morphology. Thus, a controlled, manipulative experimental approach will be necessary.

16.4.3 Does predation pressure influence cerebral lateralization?

Cerebral lateralization refers to the selective partitioning of cognitive function to the left or right hemispheres of the brain. Cerebral lateralization is often overtly expressed in fish during navigation in the form of turn preferences (see Bisazza, **chapter 15**, for more on lateralization). Such biases are common in the animal world and reflect the underlying interhemispherical organization of the brain in terms of both sensory processing and motor control. Over the last 10 years studies on fishes, and poeciliids in particular, have greatly contributed to our understanding of cerebral lateralization in vertebrates, and fishes are considered ideal model taxa, especially for distinguishing between lateralization at the population and individual levels. On a species, population, and individual level, fishes show remarkable variation in the strength and direction of lateralization that has yet to be revealed in any other taxa. Even within the family Poeciliidae intriguing variation exists (Bisazza et al. 1997b). In addition to turn biases, cerebral lateralization is also evidenced by eye preferences while viewing particular objects or scenes. For example, while the majority of species view predators with their right eye, some species prefer their left eye or show no preference at all. This variation is even greater when fish belonging to different families are compared (Bisazza et al. 2000a).

For some years biologists have struggled to explain why some species or populations show strongly lateralized behaviors while others do not. One likely explanation is that such variation is linked to predation pressure and the propensity to rely on group-oriented antipredator responses (Brown 2005). The escape behavior of group-living species relies heavily on synchronized responses, the success

of which is reliant on the degree of coordination between group members (chorus line effect, Potts 1984; oddity effect, Landeau & Terborgh 1986; conformity effect, Brown & Laland 2002). Theoretically, coordination would be increased if individuals within a group displayed similar lateralized phenotypes. Thus, group antipredator responses provide examples of traits under positive frequency-dependent selection because the fitness advantages associated with conforming to the behavior of other individuals *increase* with increasing frequency of the expression of that behavior within the population (Brown 2005). Thus, variation in group antipredator behavior may be one of the major selective forces favoring the skewed distribution of lateralized phenotypes observed in a wide variety of animals.

Most studies have examined the potential benefits of lateralization by examining the behavior of individuals in isolation, and there have been few investigations of the fitness consequences of lateralization in the group context. In group-living species, it is apparent that the dual processing associated with lateralization provides a great advantage since each animal can concentrate on its group mates and predators simultaneously (C. Brown et al. 2004; Brown 2005, 2007a). It is conceivable that animals in a group that prefer to observe predators with their right eye would preferentially take up a position on that side of the group, whereas those that use the left eye would be on the opposite side. With the correct ratio of left- and right-eyed individuals in the group, coordinated antipredator responses could be optimized. It would be very interesting to examine whether there is an optimum ratio of left-/right-eyed individuals that confers enhanced coordination and vigilance to the group. Although it is known that fish do take up preferred positions within shoals based on hunger level and mortality risk (Krause 1993a; see also Krause et al., **chapter 13**), research suggests that shoal positions are also based on laterality (Bisazza & Dadda 2005). Schools of the poeciliid *Girardinus falcatus* comprising strongly lateralized fish are more cohesive than those comprising nonlateralized individuals, and strongly lateralized fish tend to assume optimal (central) positions in the shoal (Bisazza & Dadda 2005).

Heuts (1999) hypothesized that differential exposure to predators owing to the occupation of varying habitats (benthic vs. limnetic) might explain interspecific variation in lateralized escape responses in fish. Benthic species that rely on hiding to escape predators do not require coordinated responses at the population level, whereas those in open waters do. Likewise, Bisazza et al. (2000a) explained interspecific variation in eye use by examining species-specific shoaling habits. Brown et al. (2004, 2007b) tested this hypothesis directly by examining the lateralized responses of a single poeciliid species collected from regions of high and low predation pressure and found substantial differences in lateralized behavior. As predicted, fish from high-predation areas are more likely to be strongly lateralized in comparison to those from low-predation environments. Lateralized behavior is not only heritable (Bisazza et al. 2000a; Brown et al. 2007b) but can be induced by simulated predation attacks during ontogeny (Brown et al. 2007b). Although this has been observed in only a single species, it provides the first solid evidence that predation pressure plays an important role in the evolution of lateralization in vertebrates.

16.5 Summary and future directions

Our review has revealed that living under the threat of predation has allowed poeciliids, like many other prey fishes, to evolve sophisticated cognitive mechanisms for information gathering and processing. Poeciliid fishes use a variety of sensory modalities to detect the presence of cues that are associated with changes in predation risk, but it is noticeable that our understanding of chemical cues for risk assessment in this group (e.g., alarm cues and predator odors) lags behind our understanding of other (i.e., visual) sensory cues. The extensive variation in predation risk observed among many populations of poeciliids presents an ideal opportunity to compare population differences in the sensitivity and responsiveness of prey to different levels of risk-related cues. It would also be interesting to further examine population differences in prey stress responses in order to reveal how predation risk can shape the physiological and psychological mechanisms that allow animals to deal with stress.

Poeciliids occupy a huge diversity of habitats and thus present an ideal opportunity to examine the impact of environmental (and other) constraints on sensory perception and brain morphology. The structure of fish brains is highly variable among species and tends to be correlated with factors such as habitat type and social complexity, though empirical studies are lacking (Kotrschal et al. 1998). The next step is to conduct manipulative studies to examine the effect of these (and other factors) on brain complexity and development. For example, fish from different populations could be reared in different environments and their corresponding morphological, physiological, and behavioral responses assessed. This would allow a rigorous investigation of the impact of predation risk on brain development while controlling for other ecological variables. For example, it would be interesting to compare cognitive plasticity (i.e., brain development and behavior) among fish originating from high– and low–predation risk environments and to examine the specific stimuli that are required to induce ontogenetic change.

It is evident from our review of the literature that poeciliids have contributed greatly to our understanding of the function and significance of cerebral lateralization in vertebrates, although there are many avenues for future research. In particular, the link between lateralization and the ability of prey to perform several tasks simultaneously (i.e., multitasking) seems to be an exciting area for future investigation. For example, exactly how does cerebral lateralization allow prey to maintain social contact with group members while remaining vigilant to predation threats, and are there constraints on the types of cognitive activity that can be performed in the separate hemispheres of the brain? What are the disadvantages of using the nonpreferred eye for a particular task (e.g., predator inspection)? Does cerebral lateralization bear some form of cost? If we return to the concept of a cognitive arms race between predators and their prey, then it is important to consider how the attack strategies of predators are shaped by cognitive processes such as cerebral lateralization. For example, do predators display lateralized attack strategies, and how successful are they at counteracting the lateralized escape responses of their prey? Such a line of thought leads to the possibility that predator and prey attacks and defense may fluctuate temporally in a frequency-dependent fashion (Hori 1993).

Viewing predator-prey relationships from a cognitive perspective allows us to make a number of predictions about the cognitive constraints on predators and prey that are imposed by the high energetic cost of maintaining expensive brain tissue. While in high-predation areas it may be worthwhile to increase expenditure on cerebral tissues owing to the potential gains to be made with respect to moving safely through the deadly predatory maze, no such investment should be made in low-predation areas. Furthermore, if prey develop generalized (predator community) responses due to environmental or cognitive constraints, one would predict that although it minimizes risk on a global scale, such a strategy increases the relative vulnerability of prey to each predatory species. Multiple predators have an important impact on community-level dynamics and may act to either enhance (through conflicting prey responses to multiple predators) or reduce (e.g., through predator-predator interactions) the level of predation risk experienced by prey (the "multipredator hypothesis," Sih et al. 1998). Understanding how multiple predators (and their foraging modes) influence the defense strategies and cognitive investment of prey is likely to yield significant insights into the factors (and constraints) driving the evolution of cognition. We feel that these intriguing ideas—relating cognitive investment to individual behavioral strategies—represent an exciting new research direction in poeciliid predator-prey ecology.

Acknowledgments

We would like to thank Angelo Bisazza, Dan Blumstein, Jean-Guy Godin, the editors, and an anonymous reviewer for their valuable thoughts, comments, and discussion of the topics covered in this chapter. J.L.K. gratefully acknowledges support from a University of Western Australia Postdoctoral Research Fellowship. C.B. would like to thank the Australian Research Council for their continued support.

Part IV
Sexual selection

Chapter 17 Precopulatory sexual selection

Oscar Rios-Cardenas and Molly R. Morris

17.1 Introduction

DARWIN (1871) DESCRIBED sexual selection as a struggle of two kinds: between individuals of the same sex to drive away or kill their rivals (intrasexual selection) and between individuals of the same sex to charm the opposite sex (intersexual selection). These two mechanisms of sexual selection can operate both before mating (precopulatory sexual selection) and after mating (postcopulatory sexual selection; Evans & Pilastro, **chapter 18**). Here we highlight the continuing role that poeciliid fishes play in expanding our understanding of precopulatory sexual selection (for reviews see Farr 1989; Bisazza 1993b; Houde 1997).

Two previous reviews of sexual selection within the Poeciliidae focused mainly on color dimorphisms and male aggression as possible indicators of mate choice and mating competition, respectively (Farr 1989; Bisazza 1993b). Based on these criteria and an apparent lack of courtship in most members of the family, these reviews suggested that mate choice may not be an important factor in over half of poeciliid species (Farr 1989; Bisazza 1993b). Although this is unlikely to be true (see Bisazza et al. 2001b), we still know little about the mating behaviors in most of these species, as detailed studies have focused on fewer than a dozen species. In this chapter, we do not attempt to extensively review studies of precopulatory sexual selection in the family but instead examine recent advances in which studies of poeciliid fishes have played an important role in furthering our understanding of the topic. In addition, we attempt to identify areas of research where poeciliid fishes are likely to provide us with new insights into precopulatory sexual selection.

17.2 Mate preference

Two complementary questions can be asked about female mate choice: What role has it played in the evolution of male traits? And what drives the evolution of female mating preferences? In this section we review studies that address both of these questions and offer new insights into the evolutionary implications of female mate choice.

17.2.1 Variation in female mating preferences

The idea that variation in male mating behaviors, both among and within individuals of a species, can be adaptive is well accepted and has generated extensive research on alternative male mating strategies/tactics and the mechanisms that produce and maintain them (see below). Studies of variation in female mating behaviors, however, have lagged far behind (Jennions & Petrie 1997; Alonzo & Warner 2000; Hunt et al. 2005). Not all females prefer the same male or have the same strength of preference, nor do their preferences remain inflexible to environmental influences. Variation in female mate preferences has been detected at both population and individual levels in several species (Godin & Dugatkin 1995; Brooks & Endler 2001a; Morris et al. 2003; Rios-Cardenas et al. 2007). Wagner and Basolo (2008) argue that selection will often favor female reproductive tactics that are conditionally based on the past costs and benefits involved in mate preference, which has particularly interesting implications for models of sexual conflict (see Magurran, **chapter 19**). Studying variation in female preferences is important to better understand the evolution of female preference, as well as the role of female preference in selecting for male traits.

Some of the first studies to report variation in female preferences in poeciliid fishes focused on *Poecilia reticulata* and revealed that these preferences covary with male color patterns among populations (Houde & Endler 1990), while both female preference for the male trait and the trait itself seem to depend on levels of predation (Endler & Houde 1995). Furthermore, in the presence of predators, female guppies from natural populations that experience high predation reduce their preferences for male coloration (Godin & Briggs 1996) and were less likely to choose males exhibiting the most conspicuous displays (Stoner & Breden 1988). These results suggest not only that there are costs associated with female mating preferences, but also that it can be adaptive for a female to be flexible about the circumstances in which some preferences are expressed. In guppies, female mating preferences can be modified based on prior experience with males (e.g., Breden et al. 1995; Dugatkin 1996) and can shift from ornamental traits to behavioral traits that indicate dominance when fighting among males is frequent (Kodric-Brown 1993). Furthermore, mate-choice copying has been shown to occur in guppies (Dugatkin 1992a; Dugatkin & Godin 1993; see Druen & Dugatkin, **chapter 20**, for more on mate-choice copying). There is some evidence to suggest that the willingness of a female to invest in mate choice, as well as the shape of her preference function, may be condition dependent (Brooks & Endler 2001a) or be influenced by the resources she has available to allocate to particular life-history functions (Hunt et al. 2005). And yet, what we have learned about variation in female preference has yet to be incorporated into models for the evolution of female preferences. One way to incorporate both environmental and genetic influences on preferences into an evolutionary model is to consider preferences as context-dependent traits (West-Eberhard 2003; Robinson & Morris, forthcoming).

Identifying the factors that influence variation in female mate preference will provide valuable insights into why females assess particular male traits. In addition, the implications of variation in female mating preference for the evolution of male traits are likely to be important. Here we highlight two evolutionary consequences of variation in female mating preferences: the maintenance of male alternative mating strategies and speciation.

Female *Xiphophorus multilineatus* have an overall significant preference for large courting males (courters) over small males that use sneaky matings (sneakers), and

this preference results in courting males having a higher fertilization success than sneakers (Zimmerer & Kallman 1989). However, the strength of this preference was positively related to female size, with smaller females having a weaker preference for courters (Rios-Cardenas et al. 2007). Since there were significant differences in mean female size among subpopulations and across time, as well as a relationship between mean female size and the relative frequencies of these two genetically distinct male strategies (courters were significantly more common in those samples with the largest females), Rios-Cardenas et al. (2007) suggest that variation in female preferences over space and/or time among subpopulations of *X. multilineatus* could maintain the alternative strategies in the species as a whole (without negative frequency dependence). A more recent study demonstrated that the relationship between female size and strength of preference results in smaller females being more likely to have mated with sneakers in the field, and larger females with courters (Morris et al. 2010). Therefore, the mating success of sneakers will be greater in populations with smaller females.

For variation in female preferences to maintain the different strategies of *X. multilineatus*, either the average size of females must vary across time, or there must be gene flow between these subpopulations. In contrast, genetic isolation between populations that vary in female preferences could lead to speciation. As described above, in guppies both female choice and selection due to predation shape male reproductive strategies, resulting in variation in male mating behavior among populations. Because different populations have consequently diverged genetically (Endler 1995), this variation in preference could provide a mechanism for the divergence in mate recognition leading to speciation (Lande 1981). Brooks (2002) argues that variation in mate choice within guppy populations, as well as enhanced mating success for new immigrants to a pool, could explain why guppy populations with divergent mate recognition have not speciated (see also Magurran 2005; Rosenthal & García de León, **chapter 10**).

17.2.2 Female preference for multiple male traits

Early models of sexual selection considered only one female mating preference for one male trait. We now know that females often prefer several different traits and that many elaborate male traits are actually complex sets of traits, with females having preferences for one or more of these components (Candolin 2003). For example, studies of female preference in guppies have found that males with a higher display rate have higher mating success (Farr 1980b). However, when a male displays, his dorsal fin is erected, allowing females to assess attributes of the dorsal fin. Indeed, females have preferences for the size, shape, and color of the dorsal fin (Bischoff et al. 1985; Houde 1987), suggesting that the increased mating success of males that display more could be due to multiple female mating preferences. Ultimately, the presence of multiple male ornaments may be due to multiple female mating preferences (Brooks 2002). Even though several studies have demonstrated that female guppies exhibit multiple preferences (Kodric-Brown 1993; Endler & Houde 1995; Brooks 1996; Brooks & Couldridge 1999) and that females prefer different traits in different contexts (Kodric-Brown & Nicoletto 2001), how such multiple mating preferences interact to influence male sexual attractiveness is still poorly understood.

Theoretical (Iwasa & Pomiankowski 1994; Johnstone 1995, 1996) and empirical studies (Dale & Slagsvold 1996; Brooks & Couldridge 1999; Künzler & Bakker 2001) suggest that multiple mating preferences can select for multiple male traits. The question of how these preferences interact to select for male traits will require more complex analyses that examine correlations between single variables and sexual selection. Using quadratic regression analyses, along with knowledge about genetic variation and covariation in male traits, Blows et al. (2003) have highlighted the complexities of the selection on male traits. Previous work with guppies (Brooks & Endler 2001a) had identified three linear selection gradients on male sexual ornaments. Further analyses by Blows et al. (2003) suggested that there were complex interactions among traits in determining male attractiveness, including both disruptive and stabilizing selection.

Several hypotheses have emerged to explain why females assess multiple cues in mate choice; some suggest that multiple preferences are adaptive, and others suggest that they are not (see table 17.1). The multiple-message hypothesis (Møller & Pomiankowski 1993; Johnstone 1996) suggests that multiple male cues provide females with information about different male conditions. An implicit but underappreciated outcome of this hypothesis is that preferences may conflict, selecting for different males. The backup-signal hypothesis (Johnstone 1996; similar to the redundant -signal hypothesis; see Møller & Pomiankowski 1993) is also based on benefits females gain from assessing multiple cues. Here the theory suggests that each signal gives a partial indication of a male's condition, and together the signals function to increase the accuracy of assessment. A third hypothesis, the unreliable-signal hypothesis, suggests that female preference is weak for most ornaments of multiple-signal systems because they provide unreliable information (Møller & Pomiankowski 1993).

Evidence to support both the backup- and multiple-message hypotheses comes from studies of mate preferences for conspecific versus heterospecific males in the northern swordtail fishes (*Xiphophorus* spp.). Generally, sexual selection and species recognition reinforce one another; the mechanisms of species recognition are often based on sexually selected traits (for review, see Ryan & Rand 1993; Anderson et al. 2005). However, these two components of mate choice can conflict when females have preferences for traits of conspecifics that overlap with traits of heterospecifics (Pfennig 1998). *Xiphophorus pygmaeus* females have a preference for large male size (Ryan & Wagner 1987; Morris et al. 1996), which, as the only criterion for mate choice, would drive them to mate with larger sympatric heterospecifics (*Xiphophorus cortezi*) (Hankison & Morris 2002).

In addition to body size, *X. pygmaeus* and *X. cortezi* males differ in that the latter, but not the former, have a vertical-bar pigment pattern. This pigment pattern would appear to aid *X. pygmaeus* females in species recognition, as *X. pygmaeus* females preferred barless males to naturally barred males when size was held constant (Morris 1998; Hankison & Morris 2002). And yet the presence of vertical bars did not result in a significant preference for smaller conspecifics when females were presented with a choice between smaller, barless conspecifics and larger, barred *X. cortezi* males (Hankison & Morris 2002). *Xiphophorus pygmaeus* females also prefer chemical cues from conspecific males to those from allopatric *Xiphophorus nigrensis*

Table 17.1 Hypotheses concerning the function and evolution of mate preference for multiple male traits

Proposed hypothesis	Adaptiveness of preferences	Information content of trait	Predicted change in preference strength when females are allowed to assess more cues
Multiple message	Adaptive	Informative	Increases or stays the same (depending on whether or not preferences conflict)
Redundant signal	Adaptive	Informative	Increases
Unreliable signal	Nonadaptive	Uninformative	No change

Source: Modified from Candolin 2003.

males (Crapon de Caprona & Ryan 1990) and sympatric *X. cortezi* males (Hankison & Morris 2003). However, similar to the results found with respect to vertical body bars in the previous study, preference for the chemical cues was not strong enough to reverse the preference for larger heterospecific males (Hankison & Morris 2003). It was only when females were allowed to assess both odor and the visual vertical body bars cue (more than one species-specific cue) that they spent more time with conspecifics. These results provide support for the backup-signal hypothesis because together these two cues of odor and presence or absence of vertical bars provided enough species-specific information for females to associate with conspecifics (Hankison & Morris 2003). These results also provide support for the multiple-message hypothesis because they demonstrate that the preference for body size conflicted with the preferences for the species-specific cues.

Both vertical-bar pigment pattern and odor turn out to be more complex traits than initially realized and are not easily categorized as traits that function solely during sexual selection or species recognition. Recent studies suggest that odor is more than a species-specific cue in swordtails (McLennan & Ryan 2008) and guppies (Shohet & Watt 2004). McLennan and Ryan (2008) found that *Xiphophorus montezumae* females had a clear preference for the odor of conspecific males over that of sympatric *Xiphophorus continens* males, whereas *X. continens* females preferred the odor of *X. montezumae* males. One explanation for these results is that odor provides information not only about species identity but also about male mating strategy (McLennan & Ryan 2008). *Xiphophorus continens* males all use sneak-chase behavior (Morris et al. 2005), while *X. montezumae* males court females. A similar suggestion was made by Shohet and Watt (2004) to explain why female guppies associated with some males more than others based on olfactory cues alone. In this study, female guppies preferred the odor of females over the odor of males, and therefore, preferred males may have smelled less "male-like" (Shohet & Watt 2004). In addition, the preference based on odor conflicted with the preference based on visual cues (Shohet & Watt 2004), providing support for the multiple-message hypothesis.

As the studies on preference for odor cues and vertical bars demonstrate, the division between traits used for sexual selection and species recognition will continue to become blurred as we gain a better understanding of the role of sexual selection in speciation and the processes of speciation in general. Depending on the range of cues available, and the way in which preferences for those cues interact, comparative studies will be able to unravel the evolution of multiple mating preferences and their role in speciation. For further studies on the role of odor as a species-specific

cue for species of northern swordtails that hybridize, see Wong et al. 2005, Fisher et al. 2006, and Rosenthal and García de León, **chapter 10**.

Farr (1989) suggested that many poeciliid fishes inhabit variable environments where heterozygosity, or increased genetic variability, would be advantageous. Fluctuating asymmetry (FA) may be a cue that females use to assess heterozygosity in some cases (Vøllestad et al. 1999), and therefore, symmetry could be an important cue for mate choice in poeciliid fishes. This does not seem to be the case for guppies, since inbreeding depression does not increase FA in the sexual coloration of males (Sheridan & Pomiankowski 1997b), and even though females prefer symmetrical males (Sheridan & Pomiankowski 1997a), female guppies do not appear to discriminate against related males or their sperm (Evans et al. 2008; Pitcher et al. 2008). However, female preference for bar number symmetry has been clearly detected in *X. cortezi* (Morris & Casey 1998). These studies represent only the beginning of our understanding of female preference for symmetry. We suggest that as our understanding of both the interactions among multiple preferences and variation in female preferences increases, we will find that some of the conflicting evidence on the importance of FA is due to our lack of consideration of conflicting preferences, the role of sexual conflict in the evolution of honest signals between males and females, and overly simplified models of what it means to be a high-quality mate in variable environments. Taking into consideration our increased understanding of the interaction between multiple preferences, we believe that preference for symmetry in other poeciliid species warrants further investigation.

17.2.3 Evolution of mating preferences

Several different mechanisms have been proposed for the evolution of mating preferences, and distinguishing between these has proven to be much more difficult than demonstrating that female mating preferences influence the evolution of male traits. Models for the evolution of mate preference fall into two categories: those based on direct selection on the preference and those based on indirect selection of a preference through its genetic correlation with the preferred traits (Kirkpatrick & Ryan 1991). Here we examine some of the evidence in support of both direct- and indirect-selection models from studies of poeciliid fishes.

The most intuitive method by which mate preferences are expected to evolve is by direct selection on females (Andersson & Simmons 2006). The benefits females gain by being choosy can include mating with a male that has a high-quality territory or one that provides females with nutrition, parental care, or protection from harassment.

While most of these benefits are uncommon in livebearing poeciliid fishes, evidence from the mosquitofish (*Gambusia holbrooki*) suggests that females gain direct benefits from proximity preference in the form of reduced sexual harassment (Pilastro et al. 2003; Dadda et al. 2005; Agrillo et al. 2006; Dadda et al. 2008). Females that choose males that provide more sperm or more viable sperm would also be under direct selection, although there is currently no evidence for this in poeciliids (Pilastro et al. 2008). In addition, direct benefits can include a reduction in search costs. Female preference for larger males (*Heterandria formosa*, *P. reticulata*, and *X. nigrensis*) (Ryan et al. 1990; Houde 1997; Aspbury & Basolo 2002; Magellan et al. 2005) and more conspicuously pigmented males (*P. reticulata* and *X. cortezi*) (Endler 1980, 1983; Morris et al. 2001) could have evolved due to direct selection on females to reduce search costs.

Distinguishing between the current function of a trait such as female mate preference and the context in which the trait initially evolved requires the use of comparative studies. The first clear evidence that a female mating preference initially evolved in a context other than preference for the preferred male trait (a preexisting bias) came from a study of *Xiphophorus* fishes (Basolo 1990). This now classic work, in which Basolo attached artificial swords to *Xiphophorus maculatus* males, a species in which males have no swords, and found a significant female preference for the sword, led to numerous studies across diverse taxa demonstrating similar preexisting biases. Further work by Basolo (1996, 1998) has provided additional support for this preexisting bias in poeciliid fishes. The context in which the preference initially evolved, however, is still unclear. There is some evidence to suggest that the preference for swords may have initially evolved due to female preference for large male size (Rosenthal & Evans 1998). The detection of preferences for specific components of the sword that would not increase the male's apparent size suggests that, at the very least, there has been subsequent evolution of the preference after the sword evolved (Basolo & Trainor 2002).

Female mate preferences for males that have good genes could provide females with indirect benefits and could therefore evolve through direct selection on the females' preference (Kokko et al. 2003; but see Kirkpatrick & Ryan 1991 for an example of female preferences for good genes evolving due to indirect selection, described below). Demonstrating that the expression of the preferred male trait is condition dependent has often been considered evidence for "good genes" models (e.g., Nicoletto 1993). However, a study by Nicoletto on guppies (1995) revealed no significant differences between the offspring of preferred and nonpreferred males with respect to constitution (e.g., critical swimming speed), ornamentation, or sexual behaviors. One possible explanation for these results is that the best traits for a female to provide to her offspring may vary depending on her age, the sex ratio of her brood, or even the time of year. This is also true for the "sexy son" hypothesis, which predicts that females can have indirect benefits by producing males that are better at attracting females (but see Kirkpatrick 1985). Here we see that increasing our understanding of why female preferences vary helps increase our understanding of the benefits females gain from their mate preferences. At the same time, a better understanding of selection on male traits is needed before assuming that we understand the benefits females gain from mating with particular males. For example, Reynolds and Gross (1992) suggested that in guppies, female preference for large male size indirectly benefited females due to heritable benefits to their offspring. Larger guppy fathers produce both larger male and female offspring with higher growth rates and higher daughter fecundity due to their larger size. While the benefit of producing larger female offspring seems to some extent inarguable, recent work with the green swordtail *Xiphophorus hellerii* suggests that increased growth rate may not necessarily be beneficial in all situations (Walling et al. 2007). When considering the evolution of female preference for large male size, trade-offs with functions other than mating success are too seldom considered.

A recent example of direct selection on female mating preferences comes from work by Fernandez and Morris (2008) on female preference for males with the spotted caudal pigment pattern (Sc; see fig. 17.1), which is always associated with the oncogene *xmrk* (see also Schartl & Meierjohann, **chapter 26**). They suggest that direct selection against preferring males with Sc could explain why females from a population of *X. cortezi* with high frequencies of *xmrk* prefer males without Sc, compared with other populations of this species where females prefer males with this pigment pattern. When *xmrk* is overexpressed, the macromelanophore pattern Sc can form melanomas in natural populations of *X. cortezi*, reducing an adult's reproductive life span by approximately half (A. Schartl et al.

Figure 17.1 *Xiphophorus cortezi* male with vertical bars and spotted caudal fin. Photo by K. de Queiroz.

1995; Weis & Schartl 1998). In addition, there is evidence to suggest that offspring with two copies of *xmrk* are unviable (Kallman 1971). In the population where females prefer males without Sc, the frequency of females with Sc is much higher, which means that if females were to prefer males with Sc, they would increase the probability of producing offspring with two copies of *xmrk*.

Indirect selection on female mating preferences relies on a genetic correlation between preference and preferred traits (Kirkpatrick & Ryan 1991). These correlations form when mating is nonrandom and there is sufficient genetic variation in preference and trait (Fisher 1958; Lande 1981). Two different types of study have tested the hypothesis that preference and trait are genetically correlated in poeciliid fishes. The first involves selection experiments, in which one determines whether, by selecting on the male trait, female preference evolves as a correlated response. Using this approach, Houde (1994) demonstrated that female guppies from lines in which males were selected for increased orange tended to show stronger preferences for orange than females from lines where males were selected for decreased orange. However, evidence from Houde's study also suggested that such genetic correlations may break down over time in laboratory conditions, possibly due to situations that are not favorable to continued nonrandom mating.

Another way to test for genetic correlations between preference and trait is to examine the congruence between female preferences and male traits (both trait and preference present or absent in a taxon) across populations or species in a phylogenetic context. If genetic correlations can explain the evolution of diverse preferences across taxa, these correlations should result in the loss of preferences when the trait is lost and should persist over speciation events. Morris and Ryan (1992) compared the congruence of a male trait and female preference with congruence of the same male trait and male response to the trait in the sister species *X. nigrensis* and *X. multilineatus*. They found that while male response was congruent (*X. multilineatus* with vertical bars had male response to the bars, *X. nigrensis* without vertical bars had no male response to the bars), female preference was not (females preferred the bars in both species). These results suggest that any genetic correlation between preference and trait was lost when the bars were lost in *X. nigrensis*. Further studies of both male (Moretz & Morris 2006) and female (Morris et al. 2007) responses to vertical bars in the northern swordtail clade confirmed this earlier conclusion. Over the same evolutionary events (loss of vertical bars and speciation), male response remained correlated with the trait in every case (three out of three changes once male response evolved), while female preference was incongruent in three out of six changes in bar state (once female response had evolved; fig. 17.2). Lack of sufficient genetic variation could be responsible for the

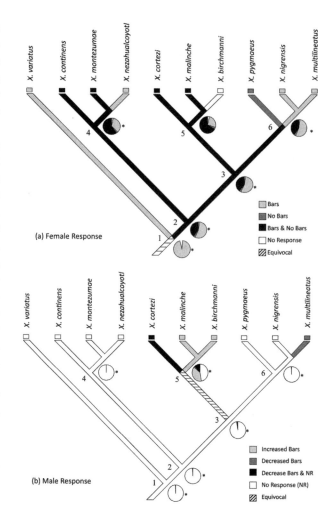

Figure 17.2 Reconstruction of ancestral states of both female and male response to vertical bars using maximum parsimony and maximum likelihood and using the Rauchenberger et al. (1990) phylogeny. Branches are shaded based on maximum-parsimony reconstruction. Area of pies indicates relative support for different ancestor states using maximum likelihood. Likelihood decision threshold was *T* = 2 (* next to pie indicates one or more significant states). (a) Ancestral states for female response to bars; (b) ancestral states for male response to bars. Modified from Morris et al. 2007.

loss of the genetic correlation, especially if speciation resulted from dispersal and involved population bottlenecks. However, if this were the case, one might expect the same pattern of congruence for male response as for female response, and this does not appear to be the case. While the difference in the coevolution of female and male response with the trait has not yet been explained, one possibility that is currently being explored is a potential sexual difference in phenotypic plasticity of the response to the bars.

17.3 Competition for mates

This section will focus on direct competition for mates in the narrow sense (i.e., mechanisms that increase access to mates through contests or fights). Contests between members of the same sex over access to mates is very common in nature, and when fighting and ritualized contests are involved, such contests can be the most dramatic and obvious form of sexual selection. Among poeciliids, it has been suggested that male-male competition is more important in species that defend territories, as well as in species lacking courtship where males inseminate females without cooperation (Bisazza 1993a). Here we review the recent literature on competition for mates in poeciliids.

17.3.1 Traits used by males to determine the outcome of male-male contests

Precopulatory intrasexual selection should favor sexual differences in body size and shape, strength, and traits that can serve the bearer as weapons for defense or to threaten opponents (Darwin 1871). Nevertheless, it has been suggested that male contests over mates can select not only for traits that serve as physical weapons but also for conspicuous signals that serve as either indicators of strength or indicators of aggressiveness (e.g., as badges of status; Fisher 1958). Male traits that indicate strength and aggressiveness may be under selection if they help males win aggressive interactions and also if they make good fighters recognizable and memorable (Andersson 1994). Among poeciliids, dorsal fins have significance in both courtship and male-male competition: examples are the sailfin mollies *Poecilia velifera* (Bildsøe 1988) and *Poecilia mexicana* (MacLaren & Rowland 2006) and the swordtail fish *Xiphophorus birchmanni* (Robinson et al., forthcoming; Fisher & Rosenthal 2007). In the green swordtail *X. hellerii* the sword serves to attract females but is also used during male-male competition. Males with longer swords win more contests (Benson & Basolo 2006) and also do so more quickly (Prenter et al. 2008). These recent studies suggest that in the green swordtail, both male-male competition and female choice (see above) play a role in the evolution of longer swords.

A similar phenomenon occurs with pigment patterns. An example of this situation occurs in the Jalapa population of *X. hellerii*, which has two types of male: those with a black or dark brownish midlateral stripe (black males) and those with a red or brownish midlateral stripe (red males). In addition to a strong female preference for the red males in this population (which are, on average, larger), red males are dominant over black males, even when red males are smaller (Franck et al. 2003).

In some species of the northern swordtail clade (including *X. cortezi*, *X. multilineatus*, *X. nigrensis*, and *X. birchmanni*), the vertical-bar pigment pattern functions as a signal of aggressive intent, as males have the ability to intensify the expression of the bars at the onset of aggression (prior to the first bite), and the subordinate male suppresses expression of the bars at the end of the contest (Moretz &

Morris 2003). The males of *X. cortezi* are polymorphic for vertical bars, and barred males respond to the expression of the bars with reduced aggression. In this species aggression and fighting ability are correlated; barless males are more aggressive and dominant over their barred counterparts, and they seem to have a higher resource-holding potential (RHP) than barred males of the same size (Moretz 2005). Finally, further comparative studies that included the remaining members of the northern swordtails (*X. pygmaeus*, *X. nezahualcoyotl*, *X. montezumae*, and *X. malinche*), as well as a platyfish (*X. variatus*), have suggested that because the ability to vary the expression of the bars evolved before the male response (either reduced or increased aggression toward the expression of the bars), the bar's function as a sexually selected trait seems to have evolved first in the context of female mate preference and later as an honest signal of aggressive intent in male-male competition (Moretz & Morris 2006; Morris et al. 2007).

In the guppy, the contribution of male-male competition to the overall pattern of sexual selection is probably minor relative to other poeciliid species (but see Luyten & Liley 1991; Bruce & White 1995; Houde 1997). However, studies with guppies have shown that aggressive interactions that occur during competition for mates may have important consequences in the maintenance of pigmentation patterns that are not necessarily preferred by females. In this species, despite strong female preference for visually attractive males, visually unattractive but behaviorally dominant males were able to obtain a substantial proportion of matings by restricting mating opportunities and suppressing the courtship of subordinate males (Kodric-Brown 1992, 1993). Similarly, in the mosquitofish *G. holbrooki*, a melanic (black) body coloration is correlated with aggressive mating behavior, and it has been suggested that the persistence of this melanic morph in nature may be related to its advantage in male-male competition (Horth 2003).

Although prior experience contributes to the outcome of fights in *X. hellerii* (Franck & Ribowski 1987, 1989; Beaugrand et al. 1991), as in many animal species body size is usually a good indicator of RHP in poeciliid fishes in general (Farr 1989; Bisazza 1993b) and in *Xiphophorus* fishes in particular (Beaugrand & Zayan 1985; Ribowski & Franck 1993). Male-male competition contributes to the greater reproductive success of large males than small males in *X. nigrensis*; both field and laboratory experiments showed that large males exclude smaller males from access to females (Morris et al. 1992). Further studies showed that in both *X. nigrensis* and *X. multilineatus*, aggressive motivation in addition to body size seems to influence the outcome of contests (Morris et al. 1995). In *X. cortezi* body size is a moderate predictor of RHP (but see above and Moretz 2005 for the effect of the presence or absence of bars); however, contrary to predictions, smaller males are more likely to initiate a conflict than larger males (Moretz 2003).

17.3.2 Alternative male mating strategies

In the past, with the exception of sex and age differences, individuals within populations were traditionally regarded as similar in ecology and ethology. However, a few decades ago a new emphasis on individual selection made clear that many populations had large, discontinuous differences in mating behaviors and morphology among individuals (Gross 1996; Shuster & Wade 2003). We now realize that such within-population variation has important ecological and evolutionary consequences and also that this variation seems to be strongly influenced by sexual selection in the form of competition for mates (Shuster & Wade 2003). Alternative reproductive phenotypes may be categorized as "real" alternative strategies (*sensu* Gross 1996) or conditional strategies, depending on the degree of genetic variation underlying the phenotypic variation (Gross 1996). In this section we review studies where alternative ways of achieving fertilization have evolved; generally, this implies using different strategies or tactics such as sneaky matings or female mimicry. Some of the early descriptions of alternative male reproductive matings in poeciliids included guppies (Liley 1966; see Magurran 2005 for a recent review). Here we focus on the species where most research has been done (table 17.2), but for a comprehensive list we refer the reader to the reviews by Bisazza (1993b) and Taborsky (1994, 2008).

Male guppies have two methods to achieve fertilization: they can either perform a sigmoid display to attract females (courters), or they can use a sneaky mating attempt, also termed a gonopodial thrust (sneakers) (Liley 1966). Males from populations with different predation regimes vary in their use of these tactics (Luyten & Liley 1985, 1991; Endler 1995). Furthermore, these differences among populations relate to the male's ability to switch from conspicuous displays to sneaky mating attempts in the presence of a predator. Fish from predator-safe streams performed sigmoid displays, even when exposed to predators; alternatively, fish from predator-rich sites seem risk sensitive (but see below), as they decrease the rate of sigmoid displays while increasing the rate of sneaky matings when exposed to predators (Endler 1987; Magurran & Seghers 1990b; Magurran & Nowak 1991; Godin 1995). In the case of the guppy, sneaking is not adopted by inferior competitors since this behavior is preferentially adopted by larger males under predation risk (Reynolds et al. 1993). Finally, Evans et al. (2002a) report that males do not switch their mating behavior to avoid risk to themselves but instead switch to exploit changes in female behaviors brought on by the presence of predators.

Alternative strategies have never been rigorously demonstrated because it has proven extremely difficult to simultaneously show that the alternatives are the result of a genetic polymorphism, that the strategies have equal average fitness, and that this equality is maintained through negative frequency-dependent selection. However, some of the best evidence demonstrating the existence of this type of strat-

Table 17.2 Poeciliid species where alternative reproductive matings have been well described

Common name	Species	References
Mosquitofishes	*Gambusia affinis*	Hughes 1985
	G. holbrooki	Pilastro et al. 1997
	Limia perugiae	Erbelding-Denk et al. 1994
Sailfin mollies	*Poecilia latipinna*	Travis & Woodward 1989
	P. velifera	Oliveira, Gonçalves, & Schlupp, unpublished data
Guppy	*Poecilia reticulata*	Liley 1966; Farr 1980b; Bisazza 1993a
Gila topminnow	*Poeciliopsis occidentalis*	Constantz 1975
Platy	*Xiphophorus maculatus*	Halpern-Sebold et al. 1986
Northern swordtails	*Xiphophorus montezumae*	Kallman 1983
	X. nezahualcoyotl	Morris et al. 2008; Rios-Cardenas et al. 2010
	X. multilineatus	Zimmerer & Kallman 1989; Rios-Cardenas et al. 2007
	X. nigrensis	Zimmerer & Kallman 1989; Ryan et al. 1990

Note: In all cases differences between males are size correlated (small males tend to sneak, while large males tend to court). About 40% of poeciliid species practice alternative matings, about 55% only sneak, and about 5% only court (Bisazza 1993b; Bisazza & Pilastro 1997).

egy comes from poeciliids, and in particular the swordtails. In the pygmy swordtail, *X. nigrensis*, and the high-backed pygmy swordtail, *X. multilineatus*, maturation age and hence adult male size are under genetic control, and in the presence of large males, small males adopt sneaky matings (Zimmerer & Kallman 1989; Ryan et al. 1990). In the former species Ryan et al. (1992) presented evidence to suggest that courters and sneakers have equal fitness due to a mating advantage for the courters and a higher probability of reaching sexual maturity for the sneakers. Current studies of *X. multilineatus* (fig. 17.3) are evaluating the same prediction as well as other possible mechanisms (besides negative frequency-dependent selection; see section 17.2.1) that could be maintaining both strategies in natural populations (Bono et al., forthcoming).

Mank and Avise (2006a, 2006b) considered the evolution of alternative male mating strategies of ray-finned fishes (Actinopterygii) in a phylogenetic context and showed that the evolution of alternative mating behaviors was significantly correlated with the presence of sexually selected traits in bourgeois males (see Taborsky 1997). They focused on the effect of male-male competition on producing alternative behaviors (bourgeois, parasitic, and cooperative; Taborsky 1997) but did not consider the role of female mate choice. To determine how and when coercive behaviors evolved in relation to coaxing behaviors or to evaluate the relative roles of male-male competition and female mate choice in driving the evolution of alternative mating tactics, alternative tactics should be considered not only in the context of circumventing male-male competition but also in relation to circumventing female mate choice. For example, adding or removing male competitors did not always affect the mating behaviors used by *X. nezahualcoyotl* males. When

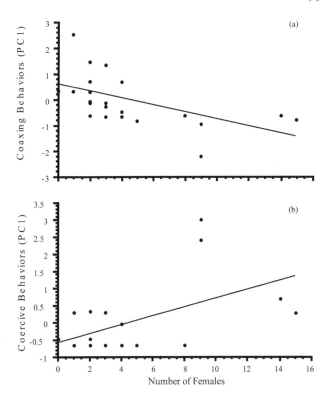

Figure 17.4 Relationship of the number of *Xiphophorus nezahualcoyotl* females and the rate of mating behaviors that males directed toward them. The number of females around focal males was significantly correlated with the mating behaviors that males directed toward those females, with (a) the rate of coaxing behaviors (first principal component shown; PC1) decreasing with the number of females; (b) in contrast, the rate of coercive behaviors (first principal component shown; PC1) increased with the number of females.

the behavior of males with and without larger competitors was examined in the laboratory, the largest males never used fast chase (a coercive behavior) even when a larger competitor was present, and the smallest males never performed headstands (a coaxing behavior) even when alone with a female. However, analysis of field data suggests that the number of females available plays an important role in the mating behaviors employed by *X. nezahualcoyotl* males (fig. 17.4). In this particular case, female mate choice rather than male-male competition appears to determine the tactic adopted by males. Further studies in other systems should also consider this possibility.

17.4 Interaction between mate preference and male-male competition

It was previously thought that female mate choice played little to no role in the mating system in which males do not court (Farr 1989; Bisazza & Marin 1995; Kolluru & Joyner 1997). The assumption was that males either coax females to mate through courtship, in which case females would have the opportunity to choose their mates, or co-

Figure 17.3 *Xiphophorus multilineatus* males. (a) Large courting male photo by K. de Queiroz. (b) Small sneaker male photo by Lisa Bono.

erce females to mate, in which case females would not be able to exercise mate choice. We now know that the situation is not so clear-cut, and that the interactions among the diversity of male and female mating behaviors can be complex. For example, female mate choice is expected to diminish as male-male competition increases, especially when the operational sex ratio (OSR) becomes male biased. However, Jirotkul (1999) demonstrated that both modes of sexual selection (female mate choice and male-male competition) became more intense in guppies when they experimentally shifted the OSR toward males. Finally, although it had initially been suggested that traits that function in both male-male competition and female mate choice evolved in the context of male-male competition and were co-opted by females (Berglund et al. 1996), this does not appear to be the case for all dual-functioning traits. Morris et al. (2007) found that female preference for vertical bars was present before male response to the bars in the context of male-male competition evolved (see fig. 17.2), suggesting that males may co-opt courtship signals for use in male-male competition.

17.5 Future directions

We hope that this chapter has made evident how the study of sexual selection using poeciliids as model systems has greatly advanced our understanding of sexual selection and evolution over the past 20 years. Most studies have focused on just a few representatives of the clade (mainly guppies, swordtails, platies, mollies, and mosquitofish). It is important to continue to investigate these model systems, as

building on what is known will facilitate the next wave of investigations into the interactions among different female and male preferences, between preference and competition, and the interactions between environmental and genetic influences.

Throughout the chapter we have identified several areas where future research on precopulatory sexual selection within these model species could take sexual selection research to the next level, which we view as providing us with a more integrative perspective on the different components of sexual selection, as well as incorporating the interaction between environmental and genetic influences on these traits. In particular, by incorporating multivariate analyses of multiple mating preferences and interactions between female and male mating preferences, we will gain a more integrative perspective on the evolution of mating preferences within each species. Nevertheless, we also suggest that future research should expand to include other members of the family. Information about the mating systems and sexual selection patterns of additional species will provide a more general and comparative perspective on the important factors that have shaped the evolution of the multiple mating systems that exist among the poeciliid fishes. With more information on the mating systems of all the species in this group, comparative phylogenetic studies could be used to determine if plasticity of mating behaviors is ancestral or derived, why courtship behavior evolved in some species and not others, and the factors that promote and maintain alternative mating behaviors, to name a few of the many questions that could be addressed with comparative studies in this group.

Chapter 18 Postcopulatory sexual selection

Jonathan P. Evans and Andrea Pilastro

18.1 Introduction

IN THE FIRST REVIEW of sperm competition in poeciliid fishes, included as part of Smith's *Sperm Competition and the Evolution of Animal Mating Systems*, Constantz (1984) highlighted the burgeoning progress on sperm competition in the family, which was arguably ahead of its time. These auspicious beginnings were fueled by early studies on genetics and reproductive biology (box 18.1), where researchers clearly recognized that poeciliid fishes were predisposed to high levels of sperm competition because of the females' propensity to mate with multiple partners and their capacity to store sperm over prolonged periods. Indeed, as Magurran (2005) notes, the historical literature on sperm competition in poeciliids dates back as early as 1917–1918, when Johannes Schmidt (1920) used Y-linked color markers to establish parentage in twice-mated guppies, *Poecilia reticulata*. Since then, poeciliids have emerged as evolutionary models for investigating postcopulatory sexual selection (hereafter PCSS).

In this chapter we critically review the literature on PCSS. Initially, we review the evidence for polyandry in natural populations, as this information provides the foundations for studies that investigate the phenomenon in the laboratory. We then focus on the patterns of selection that result from postcopulatory selective processes and attempt to identify the traits that are thought to underlie these patterns. We conclude by highlighting several key topics that await further investigation.

18.2 Mechanisms of postcopulatory sexual selection

Postcopulatory sexual selection comprises two broad mechanisms: sperm competition and cryptic female choice. Sperm competition was defined by Parker (1970) as the competition between the sperm from two or more males for the fertilization of a female's ova. Parker (1970) noted that sperm competition will favor the evolution of traits that either enhance the competitiveness of a male's ejaculate (e.g., the production of larger or otherwise competitively superior ejaculates) or reduce the likelihood of sperm competition with other males (such as mate guarding or sperm removal tactics). More recently, biologists have noted that females are far from passive participants in these processes, culminating in the theory of cryptic female choice (Thornhill 1983; Eberhard 1996), which is defined as nonrandom paternity biases resulting from female morphology, physiology, or behavior that occur during or after mating (see Pitnick & Brown 2000). A further source of selection on male and female reproductive traits arises when the evolutionary interests of males and females differ. Such sexual conflict can lead to coevolutionary "arms races," favoring *persistence* traits in one sex (usually males) that are costly to the other and *resistance* traits that mitigate these costs for the other sex (reviewed by Arnqvist & Rowe 2005).

18.3 Polyandry

Polyandry is often studied from an adaptive perspective by those asking why it is advantageous for females to mate

Box 18.1 Early insights into sperm competition

Early insights into sperm competition from poeciliid fishes came several decades before Geoff Parker formally introduced the concept to evolutionary biologists (Parker 1970). It was no surprise, therefore, that when sperm competition was first reviewed on a taxonomic basis by Smith (1984), the single chapter devoted to fishes focused exclusively on poeciliids (Constantz 1984). Then, as now, poeciliids seemed ideal for investigating sperm competition. Females were known to store sperm over several reproductive cycles (Schmidt 1920; Dulzetto 1928; Purser 1937; Winge 1937) and mate with a succession of males both within and among these cycles (Haskins et al. 1961; Borowsky & Kallman 1976). Sperm competition is therefore an inevitable consequence of poeciliid reproductive biology, a fact that did not escape the attention of the famous geneticist Øjvind Winge (1937) when studying sperm storage in the guppy *Poecilia reticulata* (formerly *Lebistes reticulatus*):

Plainly enough there is competition between the spermatozoa, and it is remarkable that when a female has been fertilized by a male of one race of *Lebistes* and has produced some few broods within a few months, and then a new male of another race of *Lebistes* is put into the tank immediately after a birth, the next brood will be from fertilization by the new male. The old spermatozoa cannot compete with the fresh ones. (Winge 1937, 467)

This early insight (probably the first mention of what we now term *last-male sperm precedence*) was corroborated by Hildemann and Wagner (1954) (see section 18.4.1). Meanwhile, Vallowe reported that sperm mixing gave rise to multiple paternity in *Xiphophorus maculatus*, noting that "competition and selection exist among the spermatozoa within the ovary and genital tract of the female" (Vallowe 1953, 246). Around the same time, Clark (1950) had developed a method for artificial insemination in livebearing fishes using *Xiphophorus hellerii*, commenting, again with remarkable foresight, that the "experimental investigation of the problem of sperm competition could be greatly facilitated in these viviparous fishes by a method for controlled inseminations."

Box 18.2 The genetic benefits of polyandry

Several studies ask why females should mate with more than one male when a single ejaculate is capable of fertilizing their eggs. The prevailing view is that where males offer no resources at mating, females mate multiply to obtain genetic benefits (theory and reviews of empirical data in Curtsinger 1991; Keller & Reeve 1995; Yasui 1997; Jennions & Petrie 2000; Simmons 2005). These benefits can be categorized according to whether paternal genes have an additive effect on a female's fitness or nonadditive effects. The former category proposes that PCSS will increase the probability of fertilization by intrinsically high-quality males, who in turn pass on genes that influence the viability or reproductive success of their offspring (Curtsinger 1991; Keller & Reeve 1995; Yasui 1997). In contrast, nonadditive benefits depend on the interacting effects of male *and* female genotypes so that PCSS will favor particular combinations of males and females (Zeh & Zeh 1997).

In poeciliids, the question of why female engage in polyandry is especially pertinent because females can produce a succession of broods following a single copulation. Furthermore, multiple mating is likely to entail significant costs for females, including physical damage (Constantz 1984). Work on guppies confirms the potential for indirect costs to arise through polyandry, because multiple inseminations by different males can lead to greater variability in offspring relatedness, with potentially important reductions in cooperative behaviors among offspring (Evans & Kelley 2008). Tentative evidence for genetic benefits of polyandry comes from guppies, where multiply mated females produce relatively large offspring (Ojanguren et al. 2005) with enhanced antipredator skills (Evans & Magurran 2000). Nevertheless, the mechanisms underlying these findings have yet to be established, and environmental influences (including differential maternal effects) have yet to be excluded before these benefits can be attributed to the genetic quality of sires.

with several males when the sperm from just one male are capable of fertilizing their eggs (e.g., see reviews by Keller & Reeve 1995; Jennions & Petrie 2000; Simmons 2005). This question is especially relevant for species in which males offer nothing but sperm during reproduction. In such species, the prevailing view is that females gain indirect genetic benefits by mating with multiple males (Jennions & Petrie 2000) (box 18.2). Irrespective of the selective pro-

Table 18.1 Estimates of multiple paternity in wild populations of poeciliid fishes

Species	Method	N_{loci}	$N_{alleles}$	N_{pops}	$N_{families}$	% multiply sired	Mean$_{sires}$	Reference
Gambusia affinis	Allozymes	3	2	1	25	56	—	Chesser et al. 1984
	Allozymes	2	2–3	1	200	49–81[a]	—	Greene & Brown 1991
	Allozyme	1	3	1	28	50	—	Robbins et al. 1987
Gambusia holbrooki	Microsatellites	3	4–5	2	9 and 11	84–88	2.2	Zane et al. 1999
Poecilia reticulata	Color patterns	—	—	—	—	—	2[b]	Haskins et al. 1961
	Microsatellites	3	10–13	1	22	95	3	Hain & Neff 2007
	Microsatellites	2	2–17	10	21–30	10–82[c]	—	Kelly et al. 1999
	Microsatellites	3	3–28	10	9–11	70–100	3.5	Neff et al. 2008b
Poecilia latipinna	Allozymes	14	2–5	1	23	52	—	Travis et al. 1990
	Allozymes	3	3–5	4	18–42	9–85	—	Trexler et al. 1997
Xiphophorus maculatus	Color patterns	—	—	2	71d	66	1.9	Borowsky & Kallman 1976
Xiphophorus variatus	Color patterns	—	—	1	43	42	1.4	Borowsky & Khouri 1976
Xiphophorus multilineatus	Microsatellites	7	2–16	1	18	33	1.4	Luo et al. 2005
Xiphophorus hellerii	Microsatellites	5	3–5	1	14	71	1.6–2.3[e]	Simmons et al. 2008
	Microsatellites	9	12–39	2	69	64	1.8	Tatarenkov et al. 2008
Poeciliopsis monacha	Allozymes	1 and 5	2–3	2	12 and 81	23[f]	—	Leslie & Vrijenhoek 1977
Heterandria formosa	Microsatellites	3	5.4–8.0[g]	3	11–13	15–66	1.15–1.75	Soucy & Travis 2003

Note: Where estimates for a given species come from two or more populations, ranges are reported for the number of loci used to assign paternity (N_{loci}), the number of families included in the analysis ($N_{families}$), the percentage of broods that were found to be sired by more than one male, and the mean number of males siring broods (Mean$_{sires}$). Where provided, we report the range for the number of alleles that were present at each genetic marker ($N_{alleles}$).
[a]The range given for the percentage of multiply sired broods came from estimates at four successive sampling dates.
[b]Haskins et al. (1961, 324–325) refer to their prior investigations (of an unstated number of populations) and note that the majority of broods were sired by two, and never more than three, sires.
[c]Values are estimates from fig. 1 in Kelly et al. 1999. These data were reanalyzed by Neff and Pitcher (2002) using a mathematical model to calculate the probability of detecting multiple mating (PrDM).
[d]Total number of families analyzed (the number of families that were used in the genetic analysis was not specified for each population individually).
[e]The range reported for Mean$_{alleles}$ comes from three methods for estimating sire number: allele counting and the software packages GERUD and PARENTAGE.
[f]The percentage of broods that were multiply sired came from just 26 broods that were large enough for the genetic analysis.
[g]Average numbers of alleles per population for the three loci are provided.

cesses that are thought to promote polyandry (which are reviewed in detail elsewhere; see Keller & Reeve 1995; Jennions & Petrie 2000; Zeh & Zeh 2003; Simmons 2005), a fundamental consequence of polyandry is that sperm from two or more males can overlap temporarily at the site of fertilization, resulting in PCSS for male traits that enhance fertilization success. Thus, an immediate question is whether, and to what extent, polyandry occurs in nature.

18.3.1 How important is polyandry in natural populations?

The question of the importance of polyandry has been addressed in several poeciliids using a variety of techniques that estimate mating frequency in natural populations. Such methods, including the analysis of sex-linked offspring phenotypes, allozymes, and microsatellite markers, have confirmed that multiple paternity (within a single brood) occurs in at least 10 species spanning five genera (table 18.1).

According to the evidence presented to date, guppies have the highest frequencies of multiple paternity of any poeciliid fish (table 18.1), and interestingly these estimates are among the highest reported for any vertebrate (Hain & Neff 2007; Neff et al. 2008b). However, it is important to note that while approaches that document patterns of multiple paternity in natural populations are helpful for evaluating the potential for PCSS (and therefore the utility of a given species for investigating such phenomena), they can offer only minimum estimates for female mating frequency in nature (see below). This is because estimates of mating frequency are typically derived by genotyping offspring rather than the sperm that are present at the site of fertilization (note that this limitation can be overcome by directly genotyping the sperm stores of wild-caught females; see Simmons et al. 2007). Selective processes that target particular males in the competition to fertilize eggs (i.e., sperm competition or cryptic female choice) will affect realized rates of polyandry when paternity data are used to estimate multiple mating by females.

18.3.2 Patterns of paternity skew

Paternity skew describes the degree to which paternity is shared equally (low skew) or is monopolized by one or a small subset of mated individuals (high skew). High skew, therefore, can be symptomatic of intense PCSS, since it is indicative of strong selective processes that exclude some mated males from achieving fertilization (or, at least, genetic representation in broods that are genotyped for paternity analysis). This highlights another inherent problem in interpreting studies that document polyandry in the wild: estimates of polyandry, taken in isolation from information of paternity skew, provide limited resolution for estimating the strength of PCSS. Studies that have combined both measures have revealed high levels of paternity skew in natural populations of the swordtails *Xiphophorus hellerii* (Simmons et al. 2008) and *Xiphophorus multilineatus* (Luo et al. 2005), the eastern mosquitofish *Gambusia holbrooki* (Zane et al. 1999), and both wild and captive populations of the guppy *P. reticulata* (Becher & Magurran 2004; Neff et al. 2008b). In a recent and highly powerful study of the green swordtail, *X. hellerii*, Tatarenkov et al. (2008) genotyped individuals from a natural population to estimate both the levels of multiple paternity and reproductive skew. By sampling the population during the dry season, they were able to exhaustively sample discrete pools for all pregnant females and putative sires (males were also collected from adjacent pools to account for the fact that fertilizations may have been due to stored sperm from matings occurring before the pools had become isolated). Tatarenkov et al. (2008) also estimated mating success for

each "successful sire" by counting the number of females that produced at least some of his offspring within their broods (note that this parameter therefore provides a minimum estimate of mating success). Their genetic analysis, encompassing 69 females and 158 candidate sires, revealed pronounced reproductive skew (average proportion of offspring sired by the most successful sire was 80% across 44 multiply sired broods; fig. 18.1a) and a positive correlation between their estimates of mating success and reproductive success (fig. 18.1b).

Together these studies indicate that paternity is typically strongly biased toward just one, or a small proportion, of mated males, a phenomenon known as sperm precedence. Such patterns of paternity skew are usually estimated in the laboratory, where they are expressed as the proportion of offspring sired by the second of two males to mate with a female during a single bout of mating (termed P_2) (Boorman & Parker 1976).

18.4 Patterns of postcopulatory sexual selection

Studies of sperm precedence in poeciliids are confined to the guppy and focus on patterns of selection in relation to male ornamentation, genetic similarity, and population origin. These patterns, along with the mechanisms underlying them (where identified), are reviewed in this section.

18.4.1 Sperm precedence

Hildemann and Wagner (1954) established patterns of sperm precedence using a domestic strain of guppies and homozygous genetic color markers to assign paternity to the offspring from twice-mated females. They mated virgin female guppies that were homozygous for autosomal, recessive alleles first to a male of the same strain and then, after the production of their first brood, to a second male that was homozygous for dominant color markers. They found that freshly inseminated sperm (from males carrying the dominant markers) had a competitive advantage over stored sperm from previous matings, supporting early research showing that stored sperm do not compete efficiently with fresh sperm in guppies (see box 18.1). The precedence of fresh sperm over stored sperm has also been reported in "wild-type" guppies (Matthews 1998) and may arise because spermatozoa that are not used immediately for fertilization are stored within sperm storage pockets lining the oviduct (Potter & Kramer 2000; Kobayashi & Iwamatsu 2002), making them unavailable for fertilization. Alternatively, sperm quality may decline over time, placing fresh sperm at a competitive advantage over stored sperm. Hildemann and Wagner (1954) did not perform reciprocal

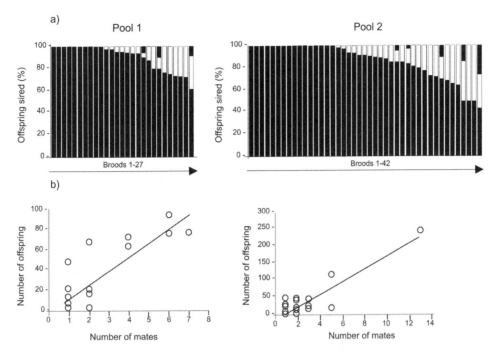

Figure 18.1 Reproductive skew and the relationship between mating success and reproductive success in the green swordtail, *Xiphophorus hellerii*. (a) Percentage of offspring from various sires across 69 broods from two pools within a natural population. Shading type within each bar represents different sires and their relative genetic representation within each brood. (b) The relationship between mating success and reproductive success in males at the same two sites (see text for details). Modified from Tatarenkov et al. 2008.

mating crosses for their different genetic strains and therefore could not rule out the possibility that females were simply more compatible with males carrying the dominant color marker gene or that males carrying the dominant gene had competitively superior sperm.

Other studies on guppies have focused on competitive fertilization success when sperm from two males compete for fertilization during the same reproductive cycle. Lodi (1981) used artificial insemination to control the relative number of sperm from males of divergent genotypes ("normal" and heterozygous "palla" mutants) and found that both sperm types initially exhibited similar competitive abilities, but normal sperm were increasingly successful in fertilizing eggs in subsequent broods, suggesting possible differences between the long-term viability of sperm from different genotypes (Lodi 1981). By contrast, successive copulations by female guppies result in a tendency for last-male sperm precedence, whereby the second of two males to mate with a female tends to sire a higher proportion of the female's brood than predicted by chance (Evans & Magurran 2001; Pitcher et al. 2003). Evans and Magurran (2001) reported that clutches were never sired in equal proportions by the two males; instead, the distribution of P_2 exhibited strong bimodality (see fig. 18.2a) and some degree of last-male sperm precedence (see also Pitcher et al. 2003). This bimodal paternity distribution was not apparent in a

subsequent study of the same population in which sperm from two rival males were artificially inseminated in equal proportions into a female (Evans et al. 2003b). Like Lodi's (1981) experiment, artificial insemination produced a more uniform distribution of paternity, where paternity was shared more equally between competing males (fig. 18.2b). The contrast between the two paternity distributions illustrates an important attribute of artificial insemination: the exclusion of components of PCSS (e.g., mating order, female sperm ejection, differential ejaculate expenditure; see below) and the consequent reduction in the level of reproductive skew (e.g., see fig. 18.1).

Neff and Wahl (2004) used Pitcher et al.'s (2003) paternity estimates to evaluate the mechanistic basis of sperm precedence in guppies. Their mathematical model tested whether sperm competition operates according to a fair raffle (where sperm from competing males have an equal chance of fertilizing eggs) or a loaded raffle (where ejaculates exhibit different fertilizing abilities irrespective of their numerical representation) (see Parker 1990). Their model also incorporated Pitcher et al.'s (2003) estimates of relative ejaculate size to examine the relationship between fertilization success and the number of sperm competing for fertilization. A nonlinear relationship between paternity and relative ejaculate size would be indicative of a loaded-raffle mechanism. Their analysis hinted at last-male sperm

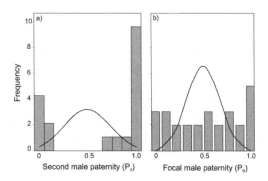

Figure 18.2 Comparing distributions of paternity between natural matings and artificial insemination. Graphs show the proportion of offspring sired by focal sperm competitors when two males copulate (a) naturally with a female or (b) simultaneously through artificial insemination. In both panels, the expected paternity distributions are depicted by lines and are based on the expectation of random sperm use according to a fair raffle. In each experiment these expected distributions were compared with observed patterns (filled bars). P_2 = the proportion of offspring sired by the second of two males to consecutively mate with a female; P_B = the proportion of offspring sired by a focal sperm competitor arbitrarily labeled B within each pair. Modified from Evans and Magurran 2001 and Evans et al. 2003b.

precedence with diminishing reproductive returns for increased sperm production, thereby supporting a loaded-raffle mechanism. This conclusion is supported by studies documenting nonrandom paternity biases when the number of sperm from each male is regulated through artificial insemination (Evans et al. 2003b; see also fig. 18.1). Nevertheless, the relative sperm "loadings" used by Neff and Wahl (2004) came from sperm counts from rested males (Pitcher et al. 2003) and therefore may not have reflected the size of natural ejaculates (e.g., see Pilastro et al. 2002a). Furthermore, females may have manipulated sperm transfer so that second males inseminated higher number of sperm (Pilastro et al. 2004). Such a process would therefore not preclude a fair raffle entirely. Indeed, female responsiveness to subsequent mating partners is an increasing function of their attractiveness (Pitcher et al. 2003), and the number of sperm transferred during solicited copulations is positively associated with the degree of male ornamentation (Pilastro et al. 2002a) and is under female control (Pilastro et al. 2004; Pilastro et al. 2007). We explore such cryptic female preferences in section 18.4.2.2.

18.4.2 Linking pre- and postcopulatory sexual selection

Recent research has examined whether PCSS works in concert with, or antagonistically to, precopulatory sexual selection (e.g., Danielsson 2001; Birkhead & Pizzari 2002; Pizzari et al. 2002; Malo et al. 2005; Hosken et al. 2008). When both processes target the same phenotypic traits, PCSS will augment patterns of precopulatory sexual selection. Evidence for this relationship comes from guppies,

where both sperm competition and cryptic female choice target phenotypically attractive males.

18.4.2.1 *Sperm competition.* Indirect evidence for congruent patterns of pre- and postcopulatory sexual selection comes from studies revealing associations between sperm traits and male secondary sexual traits such as the area of orange pigmentation (Pitcher & Evans 2001; Pilastro et al. 2002a; Locatello et al. 2006; Pitcher et al. 2007), courtship behavior (Matthews et al. 1997; Evans et al. 2002b), and body size (Pilastro & Bisazza 1999). More direct evidence comes from work linking these secondary sexual traits to paternity success. For example, Evans et al. (2003b) used artificial insemination to regulate the relative number of sperm from competing males and found that competitors with large areas of orange spots in their color patterns enjoyed a greater share of paternity than less ornamented males.

The competitive advantage of relatively colorful males in sperm competition could be due to the association between ejaculate quality and color ornamentation. Male guppies with relatively large orange spots produce ejaculates with higher swimming velocities (Locatello et al. 2006; Pitcher et al. 2007; Skinner & Watt 2007) and increased sperm viability (Locatello et al. 2006). This association between sperm performance and color ornamentation may be due to the antioxidant properties of carotenoids in the diet, which are thought to reduce oxidative damage to sperm while concomitantly contributing to the phenotypic attractiveness of males (Blount et al. 2001). This effect may arise if antioxidant-rich foods defend against reactive oxygen species (containing free radicals), which can culminate in high levels of oxidative stress and result in damage to spermatozoa while simultaneously affecting the expression of carotenoid-based color pigments in the male's display traits (these ideas are reviewed and expanded by Blount et al. 2001). Interestingly, male aging does not affect sperm competitiveness in guppies (Gasparini et al. 2010a), which might be expected if sperm are susceptible to oxidative stress. Nevertheless, in humans, antioxidants such as lycopene are reduced in the seminal plasma of subfertile men (Palan & Naz 1996; Lewis et al. 1997), and the dietary supplementation of these antioxidants can improve male fertility (Gupta & Kumar 2002). To test these ideas vigorously it is necessary to show that free radicals can cause concomitant reductions in ornament expression and sperm quality, something that has yet to be tested in guppies or any other species.

18.4.2.2 *Cryptic female choice.* Pilastro et al. (2002a) reported that in guppies the proportion of the male's body covered in orange spots was positively associated with ejaculate size, as estimated by counting sperm extracted from the female's reproductive tract. They also uncovered a positive association between mating speed (the time a female takes to accept a copulation) and ejaculate size. Since mating speed is under the female's control during solicited matings and is known to predict competitive fertilization success when two males mate consecutively with a female (Evans & Magurran 2001), the authors speculated that females may exert control over the number of sperm transferred so that attractive males have an advantage during sperm competition. This suggestion is intuitive because the relationships between ejaculate size and both mating speed and male ornamentation were not evident following forced copulations, where females had no control over matings (Pilastro et al. 2002a).

To test the idea that females cryptically "choose" sperm from different males, Pilastro et al. (2004) performed a subsequent experiment that manipulated the female's perception of male attractiveness independent of any direct manipulation of the focal males themselves. In a paired design, focal males were made to appear either relatively attractive (preferred role) or unattractive (nonpreferred role) by placing them alongside a relatively dull or colorful stimulus male, respectively. Virgin females viewed the focal males in their assigned role and then copulated with them once. The inseminated sperm were then extracted from their oviducts and counted. On average, ejaculates contained 68% more sperm when focal males mated in the preferred role than in the nonpreferred role. Since copulation duration is positively associated with both ejaculate size and the degree of orange coloration (Pilastro et al. 2007), control over the duration of copulations could be the proximate mechanism of cryptic female choice in guppies.

18.4.3 PCSS for genetically compatible or similar mates

So far we have focused on directional sexual selection (where selection favors a single *optimal* phenotype). However, the residual (unexplained) variance in relative fertilization success is typically high in studies reporting relative paternity success. For example, although fertilization success is significantly repeatable when ejaculates from two rival male guppies compete across different females, the unexplained variance is high (Evans & Rutstein 2008). In other taxa, there is an increasing awareness that the outcome of sperm competition can depend on interactions between male and female genotypes, which in turn can be attributable to postcopulatory mechanisms of inbreeding avoidance, or selection against genetically incompatible mates (e.g., Clark et al. 1999; Tregenza & Wedell 2000; Zeh & Zeh 2003; Birkhead et al. 2004; Rosengrave et al. 2008). For example, in birds, reptiles, mammals, and insects the level of genetic similarity between mating partners can influence competitive fertilization success (Olsson et al. 1996; Wilson et al. 1997; Stockley 1999; Kraaijeveld-Smit et al. 2002; Mack et al. 2002; Bretman et al. 2004; Thuman & Griffith 2005; Jehle et al. 2007). Although these effects are not universal (e.g., see Stockley 1997; Jennions et al. 2004; Denk et al. 2005; Lane et al. 2007), the accumulating evidence that paternity success can be skewed toward unrelated or genetically dissimilar males has been taken as evidence that polyandry facilitates postcopulatory mechanisms of inbreeding avoidance (e.g., mediated by sperm-egg interactions or differential zygote mortality). Until now, postcopulatory inbreeding avoidance, and indeed the potential for inbreeding depression on traits involved in PCSS, have received relatively little attention in poeciliids, although recent work failed to uncover evidence for postcopulatory inbreeding avoidance when female guppies mated successively with an unrelated and a sibling male (Pitcher et al. 2008) or when artificial insemination was used to deliver the combined ejaculates from first cousins and unrelated males simultaneously (Evans et al. 2008). However, second-generation inbred males were shown to have lower sperm competitiveness (Zajitschek et al. 2009), suggesting a cross-generational cost of inbreeding.

18.4.4 PCSS and reproductive isolation

Evidence for putative reproductive isolating mechanisms through postcopulatory (prezygotic) processes is limited to Trinidadian guppies, where geographically isolated populations vary in a wide range of ecological, morphological, and behavioral characteristics (Endler 1995; Houde 1997; Magurran 2005). In particular, male mating tactics and female preferences exhibit considerable population divergence (Endler & Houde 1995). Luyten and Liley (1991) exploited these differences by crossing males from different populations with females from either their own or a different population. They then used radioactive labeling to identify sperm from the competing males within the female's reproductive tract. They reported that males from both environments were less successful at inseminating heteropopulation females, but only when competing within their own environmental conditions (i.e., males from headwater streams outcompeted their lowland rivals in clear water and vice versa). A more recent study confirms that PCSS has the potential to reinforce these potential reproductive barriers. Ludlow and Magurran (2006) used artificial insemination to impregnate females descended from fish collected from the geographically (and genetically) disparate Caroni and Oropouche drainages in Trinidad's Northern Range mountains. Artificial insemination enabled the researchers to regulate the relative number of competing sperm from males from each drainage, thus controlling for differential sperm priming (Aspbury & Gabor 2004b, 2004a) and sperm investment (Schlupp & Plath 2005) by males from either population, as well as differences in female preferences that may favor "native" over heteropopulation males (e.g., Luyten & Liley 1991; but see Magellan & Magurran 2007b). They found that native male sperm had precedence over heteropopulation sperm, and that this effect was symmetrical for both drainages and absent in control inseminations involving mixed inseminates from two male populations within the same river drainage. This offers compelling evidence that PCSS can limit gene flow between divergent populations.

Interestingly, the precedence of native sperm over foreign sperm in guppies may be counteracted by opposing female preferences that favor foreign over native males. In a mate choice experiment involving a single focal female population descended from fish collected from a high-predation population in Trinidad, Magellan and Magurran (2007b) reported that females preferred males originating from a low-predation (upstream) site. Males from low-predation populations are typically more colorful than their high-predation counterparts, so instead of representing innate preferences for migrant males, these results may simply reveal female preferences for highly ornamented males. Magellan and Magurran (2007b) did not perform the reciprocal mate-choice trials to exclude this possibility.

18.5 Traits targeted by postcopulatory sexual selection

18.5.1 Ejaculates

Head et al. (2007) used linear selection analyses (see Lande & Arnold 1983) to estimate directional sexual selection on a range of male sexual traits, including sperm production (estimated from the size of stripped ejaculates taken after the study). They reported that selection favored males with *lower* sperm reserves, prompting them to speculate that components of sperm quality may be traded off against sperm production. An alternative explanation, however, is that male sperm reserves after mating reflect higher ejaculate expenditure *during* the mating trials (e.g., Evans et al. 2003a; Aspbury 2007). If this alternative scenario holds, selection imposed through sperm competition would have targeted males that produced larger ejaculates, which in turn would be reflected by lower available sperm loads after the trials.

Several studies indicate that male poeciliids are capable of adjusting their ejaculate expenditure according to cues that signal the value of a particular breeding event (Bozynski & Liley 2003; Aspbury & Gabor 2004a; Schlupp &

Plath 2005; Aspbury 2007). Aspbury and Gabor (2004b) examined sperm production in the sailfin molly, *Poecilia latipinna*, when males were in the presence of either a conspecific female or the heterospecific (gynogenetic) Amazon molly, *Poecilia formosa*. In this species complex, the gynogenetic *P. formosa* is clonal and requires sperm from males of closely related bisexual species (*P. latipinna* or *Poecilia mexicana*) to initiate embryogenesis (see Schlupp & Riesch, **chapter 5**). Their results confirmed that mating preferences for conspecific females (Ryan et al. 1996; Gabor & Ryan 2001; Schlupp & Plath 2005) are likely to be reinforced by differences in sperm production; males from both an allopatric and a sympatric population produced more sperm when in the presence of a conspecific female than a heterospecific female. Schlupp and Plath (2005) subsequently demonstrated that males within this species complex reduce the size of their ejaculates when mating with heterospecific females.

Interestingly, when Robinson et al. (2008) subsequently used a different technique to investigate sperm allocation by *P. latipinna* (estimated by the number of sperm recovered from recently mated males), they reported that males had higher available sperm reserves after mating with conspecific females than after mating with the heterospecific *P. formosa*. This somewhat nonintuitive result suggests either that male *P. latipinna* expend more sperm on heterospecific (parasitic) females or alternatively that spermiation (see Greven, **chapter 1**) is triggered when males encounter conspecific females (Aspbury & Gabor 2004b).

More direct evidence for PCSS on patterns of ejaculate expenditure comes from studies that have addressed theoretical models of sperm competition. Parker et al. (1996, 1997) developed a series of game theory models designed to predict the level of ejaculate expenditure according to either the risk or the intensity of sperm competition. The risk model applies to situations where sperm competition is relatively infrequent, involving at most two rival males (fig. 18.3a). According to this model, when the probability of sperm competition is low, males are expected to allocate relatively few sperm to a given mating because paternity is ensured with the minimal investment (Parker et al. 1997). As the risk of sperm competition increases, males are predicted to increase ejaculate expenditure—a situation analogous to a raffle in which an individual's investment in tickets relative to a rival competitor's investment is directly proportional to the probability of success.

Support for the risk model comes from studies that have manipulated sex ratios to simulate the perceived risk of sperm competition. In *P. latipinna*, for example, Aspbury (2007) reported that focal males increased their sperm expenditure (estimated from the number of sperm left after the mating trials) when a potential sperm competitor was

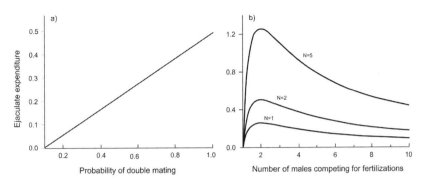

Figure 18.3 Predictions for evolutionarily stable strategy (ESS) ejaculate expenditure under sperm competition risk and intensity. (a) Predictions for ejaculate expenditure in relation to the frequency of double matings in the population (risk model). The line depicts optimal ejaculate expenditure when the outcome of sperm competition follows the principle of a fair raffle. Modified from Parker et al. 1997. (b) Predicted ejaculate expenditure according to the number of males competing for fertilizations (intensity model). The three curves are for populations that differ in the average number of males (N = 1, 2, and 5) competing for fertilizations. Modified from Parker et al. 1996.

in view. By contrast, sperm production was unaffected under high–sperm competition risk (but see below for situations involving more than one competitor). These findings are similar to those for eastern mosquitofish, where sperm production was unaffected by the perceived risk of sperm competition, but sperm expenditure was higher when males had previously been maintained under sex ratios simulating a high risk of sperm competition (Evans et al. 2003a). In the guppy, Evans and Magurran (1999) similarly failed to find any effect of manipulating the perceived level of sperm competition risk on sperm production (estimated through the gonadosomatic index and from sperm counts), but in these trials they did not estimate ejaculate size. More recently, Evans (2009) similarly found no evidence for rapid sperm-priming effects (influencing sperm numbers and sperm quality) in response to fluctuations in the level of sperm competition, despite the fact that males can exhibit very rapid sperm-priming responses (increases in sperm number and velocity) to the presence of females (Bozynski & Liley 2003; Gasparini et al. 2009).

Unlike the risk model, the intensity model applies to situations where sperm competition is relatively frequent, with "intensity" referring to the number of ejaculates competing to fertilize a female's eggs. According to this model, males should invest maximally with a single rival (i.e., the same as the risk model) but allocate progressively fewer sperm to each ejaculate as the number of rivals exceeds one (Parker et al. 1996; see fig. 18.3b). The decline in sperm expenditure is due to the progressively lower reproductive returns as the number of competitors increases beyond one, again following the principle of a lottery. Although support for this model comes from other fish taxa (e.g., bitterling: Candolin & Reynolds 2002; gobies: Pilastro et al. 2002b), the only tests of the hypothesis on poeciliids have failed to offer support for the model (Aspbury 2007; Evans 2009).

However, in the case of Aspbury's (2007) experiment, her sex ratio manipulations involved visual-only exposure to stimulus males, and these stimulus males did not directly interact with the female. Focal males may therefore have underestimated the intensity of sperm competition (see Engqvist & Reinhold 2005). As we will see in the following section, the mere presence of "audience" males may not be sufficient to elicit a response to elevated levels of sperm competition.

18.5.2 Behavior

Recent experiments also suggest that male mating preferences can be targeted by PCSS. For example, Dosen and Montgomerie (2004b) examined male mating preferences in guppies to determine whether males adjust their courtship according to both the presence of rival males and female mating status (mated or nonmated)—situations simulating different levels of sperm competition. In an initial experiment, where focal males observed females either on their own or in close proximity to four males, they found that focal males did not discriminate against females in either group (fig. 18.4a). However, in a second experiment, where the rival males in the sperm competition treatment could mate with the female, focal males spent more time in close association with, and directed more courtship toward, females that they had previously seen alone than toward those that they had previously observed mating (fig. 18.4b). Similarly, Guevara-Fiore et al. (2009) found that male guppies direct more courtship toward virgin than toward nonvirgin females. Finally, in the eastern mosquitofish (*G. holbrooki*) males reduce their preferences for initially preferred females after viewing these females in the vicinity of a potential rival male (Wong & McCarthy 2009). Taken together, these studies suggest that males may be sensitive to

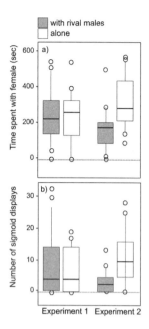

Figure 18.4 Male mating behavior in relation to the risk of sperm competition. In Experiment 1, focal males had previously viewed females in close proximity to four rival males (filled boxes) or on their own (unfilled boxes). In Experiment 2, focal males had previously seen the females directly interacting (and copulating) with the rival males. Panels show (a) the amount of time that focal males spent with each female and (b) the number of sigmoid displays (courtship) that the focal males directed toward each female. Boxplots show the median (line), interquartile range (box), and 10% and 90% values (whiskers). Data points outside this range are also shown. Modified from Dosen and Montgomerie 2004b.

visual and/or olfactory cues that predict the level of sperm competition, but only when they have accurate information about the number of rival males that have actually mated with the females (see Engqvist & Reinhold 2005).

Plath et al. (2008a) reported that in surface-dwelling Atlantic mollies (*P. mexicana*), males spend significantly less time near an initially preferred female when another conspecific male is visible (see also Plath et al. 2008c). Interestingly, this "audience" effect is also observed in the cave-dwelling form of the same species despite the absence of visual communication in its natural habitat (Plath et al. 2008b). These findings prompted the authors to suggest that the reduction in the strength of male sexual preferences for initially preferred females may arise because males conceal their mating preferences when the perceived risk of sperm competition is high (Plath et al. 2008b). According to this view, the concealment of mating preferences from conspecific males functions to reduce the future risk of sperm competition with the focal male's preferred mate (i.e., the female that he is most likely to have mated with), because rival males copy other males' mate preferences (Schlupp & Ryan 1997). However, an alternative explanation is that by dividing their attention more equally among potential (or

past) mating partners, males may reduce the risk of sperm competition in *both* females, not just their preferred mate.

18.5.3 Sperm quality and morphology

It is well known in other animal taxa that ejaculate quality and/or sperm morphology can influence the outcome of sperm competition (reviewed by Snook 2005). In the green swordtail and guppies, recent work employing artificial insemination also reveals effects of sperm traits on sperm competitiveness. When artificial insemination was used to inseminate mixed ejaculates from rival male green swordtails, differences in sperm velocity (average path velocity = VAP) were positively associated with relative paternity success (Gasparini et al. 2010b). In the case of guppies, both sperm swimming velocity (curvilinear velocity = VCL) and relative sperm numbers were significant predictors of paternity, while sperm morphology (sperm head, midpiece, and tail length) was uncorrelated with paternity success (Boschetto et al. 2010). These findings confirm previous indirect evidence for such effects, where male guppies with relatively large orange spots (which enjoy relatively high success in sperm competition; Evans et al. 2003b) produce ejaculates containing more viable sperm (Locatello et al. 2006) and sperm exhibiting greater swimming velocities (Pitcher et al. 2007; but see Skinner & Watt 2007) than their less ornamented counterparts.

18.5.4 Genitalia

The extreme variability in male genital morphology in poeciliid fishes is striking, a fact that did not escape Eberhard (1985) when he described genital morphology in the genus *Gambusia* as part of his review of sexual selection on animal genitalia. Although the potential for sexual selection to drive this divergence has been recognized since Eberhard's review (see Langerhans, **chapter 21**), no studies have drawn an explicit link between these traits and relative fertilization success. Yet there is anecdotal evidence that PCSS may fuel the evolution of gonopodial length in poeciliid fishes. For example, Kelly et al. (2000) reported that males inhabiting streams characterized by relatively high levels of predation (e.g., by *Crenicichla crenulatum* and *Hoplias malabaricus*) had relatively longer gonopodia than those from low-risk populations (where minor predators such as *Rivulus hartii* occur). Since males tend to switch from courtship to forced mating attempts under high predation risk (Magurran & Seghers 1994c; Evans et al. 2002a), Kelly et al. (2000) suggested that selection would favor the evolution of longer gonopodia in high-predation environments, assuming that relatively longer gonopodia are more efficient at transferring sperm during nonsolicited matings and that this increases the level of sperm competition. This assumption seems plausible, given the finding that larger gonopodia are associated with the higher prevalence of forced mating activity, both within species (Reynolds 1993) and among poeciliid genera (Rosen & Tucker 1961).

Further speculative evidence that PCSS may target male genital traits comes from Peden's (1972b) comparative work on *Gambusia*, which revealed that male and female genital shapes exhibit correlated divergence among species (the occurrence of the urogenital papilla in females occurs in species where males have pointed gonopodia) (see Langerhans, **chapter 21**, for further comparative analyses of these traits). This observation led Constantz (1984) to hypothesize that the function of the papillae of female poeciliids was to thwart male mating advances and therefore enhance the female's control of paternity. Accordingly, these correlated patterns of selection on male (persistence) and female (resistance) traits may be the hallmark of sexually antagonistic selection (Arnqvist & Rowe 2005). Poeciliid fishes offer good opportunities for testing these predicted patterns of covariance, both within species (e.g., where genital traits vary among populations) and among species (where divergent mating tactics may select for very different genital morphologies). Indeed, interspecific comparative studies indicate that the relative incidence of forced matings (via "gonopodial thrusts") in the mating systems of various poeciliid species is associated with corresponding variation in gonopodium length (see Rosen & Tucker 1961; Farr 1989).

18.6 Summary and future directions

Our review highlights progress in documenting patterns of PCSS in natural and captive populations of poeciliid fishes, but it also reveals important gaps in our understanding of the mechanisms and selective processes that underlie them. For example, we need to identify specific traits in males (e.g., ejaculate quality, sperm or genital morphology) that influence fertilization success and female traits that mediate the process. Research on poeciliids lags far behind other taxa in this regard, where traits such as ejaculate size, sperm morphometry, sperm quality, behavior, genital shape, and nonadditive genetic effects have all been implicated as important brokers of fertilization success (reviewed by Snook 2005; Evans & Simmons 2008). There is also a need for studies that explore the genetic basis of such traits, as much of the theory proposed to explain polyandry (e.g., so-called good- or sexy-sperm models: Curtsinger 1991; Yasui 1997) depends critically on a detailed understanding of the genetic architecture of traits involved in mediating PCSS (Evans & Simmons 2008). Recent evidence from

guppies confirms that there are very high levels of additive genetic variation underlying ejaculate traits (Evans 2010, 2011), thus fulfilling this basic requirement of good- and sexy-sperm theory. Nevertheless, multivariate quantitative genetic studies documenting patterns of genetic variance and covariance for traits involved in PCSS are badly needed in other poeciliid fishes.

Our review also highlights the need for a better understanding of the physiological mechanisms that underlie variance in fertilization success. In guppies, for example, the evidence linking carotenoid-based sexual ornaments to variation in sperm quality (Locatello et al. 2006; Pitcher et al. 2007) and sperm competitiveness (Evans et al. 2003b) suggests that these traits may be functionally linked, yet the mechanisms accounting for these relationships remain elusive, despite promising verbal arguments proposed to explain such effects (Blount et al. 2001; Dowling & Simmons 2009).

Poeciliids present some of the most striking examples of male alternative mating strategies and intraspecific male polymorphism (e.g., *Poecilia parae*: Lindholm et al. 2004). The maintenance of such extreme behavioral and morphological polymorphisms, including body size (which in several cases has a strong genetic basis), is one of the most fascinating evolutionary problems in poeciliid research. There is indirect evidence that PCSS may contribute to the maintenance of body size polymorphism in poeciliid fishes. For example, in *G. holbrooki* body size is negatively associated with insemination success (Pilastro et al. 1997), while in guppies sperm competition appears to favor relatively small males (Evans et al. 2003b; Becher & Magurran 2004). Clearly, there is enormous scope for investigating the variation of postcopulatory traits in relation to male mating tactics in the family.

Poeciliids also offer good opportunities to study adaptive differential maternal investment, given the extended period that developing offspring are physiologically dependent on their mothers (Heath & Blouw 1998). This is especially so for species exhibiting matrotrophy, where nutrients are transferred to developing embryos (see Marsh-Matthews, **chapter 2**). In the majority of poeciliids the female's contribution to developing offspring is confined to minerals and respiratory gases. The scope for differential maternal investment would appear more limited in these species (but see Marsh-Matthews, **chapter 2**). Nevertheless, it is possible that females still manipulate the timing of fertilization so that the offspring from preferred males benefit by being fertilized from more developed ova. Indeed, in ovoviviparous poeciliids, fertilization of a female's entire clutch typically proceeds over several days (Turner 1937; Thibault & Schultz 1978). By extending the period of fertilization or by adjusting the duration of brood retention by varying the interbrood interval (which can vary within species; see Johnson & Bagley, **chapter 4**), females may increase the opportunities to exert postcopulatory choice (see discussions by Evans & Magurran 2000; Pitcher et al. 2003). Indeed, prolonging the retention of embryos may enable ovoviviparous females to increase their investment in developing offspring (e.g., see Shine & Olsson 2003). In guppies, for example, females are capable of adjusting embryo retention according to extrinsic ecological factors such as predation risk (Evans et al. 2007), so it is at least conceivable that they may also manipulate the timing of parturition in relation to sexually selected cues. Differential maternal investment is clearly a topic that warrants investigation in poeciliids.

Finally, our chapter focuses on just a few "model" species, as do most of the topics covered in this volume (see Endler, **chapter 27**). To illustrate the bias in our own field, a search of the literature using the ISI Web of Science® and the keywords "sperm competition AND poecil*" yielded 107 search results for the period 1984 to February 2010 (the period since George Constantz's 1984 review). Of these, 27 papers included empirical data on sperm competition (or were otherwise directly relevant to our topic of postcopulatory sexual selection). Twenty of these (74%) focused on guppies (*P. reticulata*), while the remaining seven studies were on *P. latipinna*, *P. mexicana*, *G. holbrooki*, *Heterandria formosa*, and the swordtails *X. multilineatus* and *X. hellerii*. There are several reasons for this bias, but

chief among them has to be the shear ease of maintaining and breeding guppies in captivity. These attributes, coupled with the availability of genetic markers (based on color polymorphisms of Y-linked phenotypic traits), enabled early geneticists to establish patterns of parentage from multiply sired broods (Schmidt 1920; Winge 1922a) and would have made guppies very attractive models for the early pioneers of sperm competition. However, we anticipate that the increasing availability of genetic markers in other poeciliids (Breden & Lindholm, **chapter 22**), coupled with ongoing genome-sequencing projects, will make an enormous contribution to this research field, and we expect a broadening of research on PCSS and other topics in the family in the coming years.

Acknowledgments

We are grateful to our collaborators, students, and colleagues who have helped shape our thoughts on these topics. We especially thank Trevor Pitcher, Bryan Neff, Anne Magurran, Iain Matthews, and Clelia Gasparini. We are also grateful to John Fitzpatrick, Bob Montgomerie, and Trevor Pitcher for comments that helped us improve the manuscript. Finally, financial support came from the Australian Research Council (J.P.E.) and the University of Padova and the Italian Research Ministry (A.P.).

Chapter 19 Sexual coercion

Anne E. Magurran

19.1 Sexual coercion in poeciliids

SEXUAL COERCION OCCURS when one partner attempts to undermine or manipulate the mating choices of the other (Clutton-Brock & Parker 1995). It may take the form of a forced, or nonconsensual, mating. The vocabulary used to describe the noncourtship elements of poeciliid mating activity—gonopodial thrusting, sneaky mating, forced copulation—implies that coercion is not only widespread in this taxon but a common occurrence in the daily lives of individual fish. A few minutes' observation of guppies or mosquitofish is usually enough to see multiple instances of females being persistently followed by males, and the fact that these females often try to evade such mating attempts suggests that they are unwelcome. Although the evolutionary ecology of poeciliids has been studied for over a century (Magurran 2005) and the coercive nature of many mating attempts commented upon (Breder & Coates 1935; Farr 1989; Bisazza 1993b), it is only recently that the causes and consequences of coercion have been investigated in detail. Mating attempts that do not lead to insemination may also represent sexual coercion through the imposition of costs on the target sex, which in poeciliids is almost invariably the female. Males may even direct mating attempts at one another in the absence of females, but this is an extremely unlikely occurrence in wild populations (Bruce & White 1995).

This chapter begins by asking whether forced copulations can ever be consensual. This is an intriguing area and one that is—as yet—not fully resolved. I then review the costs of coercion and the attempts by females to mitigate them. The next section explores the ecological setting in which coercion occurs and briefly discusses reasons for individual responses to coercion. The chapter concludes by examining coercion directed toward heterospecific females. Although sexual coercion probably occurs in all poeciliids to a greater or lesser extent (Bisazza [1993b] estimates that over 95% of species engage in gonopodial thrusting), Atlantic mollies (*Poecilia mexicana*), Trinidadian guppies (*Poecilia reticulata*), and mosquitofish (*Gambusia* spp.) have been particularly well studied and therefore contribute many of the examples that follow.

The terms "forced copulation," "sneaky mating attempt," "sexual harassment," "gonopodial thrusting," and "sexual coercion" are often used interchangeably. Sexual coercion is itself a form of sexual conflict. Box 19.1 briefly explains the relationship between sexual conflict and sexual coercion.

19.1.1 Coercion—or cooperation?

The common thread that extends across the poeciliids is the presence of the male intromittent organ, the gonopodium, and a dependence on internal fertilization (see Langerhans, **chapter 21**). However, there is considerable variation in the means by which fertilization is achieved and the extent to which females control the mating process. Males of all species employ gonopodial, or copulatory, thrusts (which are sometimes described as "forced copulations"); fewer than half of these species also engage in courtship (Farr 1989; Bisazza 1993b). Phylogenetic analysis reveals that a dependence on gonopodial thrusts may be the ancestral condition (Ptacek & Travis 1998; but see Hrbek et al. 2007 for a more recent phylogenetic analysis that uncovers a differ-

Box 19.1 Sexual coercion and sexual conflict

Sexual conflict was defined by Parker as "a conflict between the evolutionary interests of individuals of the two sexes" (1979). Because selection in one sex may be opposed by selection in the other sex, another way of describing sexual conflict is "sexually antagonistic selection" (Rowe & Day 2006). Sexual conflict exists because males and females have different reproductive priorities (Darwin 1871; Parker 1979; Smuts & Smuts 1993; Parker & Partridge 1998; Hosken & Snook 2005; Hosken & Stockley 2005). Females are usually limited in the number of offspring they can produce and for this reason try to ensure that their offspring are of as high quality as possible. Males, on the other hand, have the potential to sire large numbers of progeny, but they also run the risk of producing none at all. This asymmetry produces greater variance in reproductive success in males than in females—the so-called Bateman gradient (Bateman 1948). Poeciliids exhibit this pattern of reproductive skew (*Poecilia reticulata*: Becher & Magurran 2004; Neff et al. 2008b; *Xiphophorus hellerii*: Simmons et al. 2008; Tatarenkov et al. 2008; *Xiphophorus multilineatus*: Luo et al. 2005; *Gambusia holbrooki*: Zane et al. 1999; and see Evans & Pilastro, **chapter 18**).

In nature both sexes will strive to maximize their reproductive fitness. This selects for adaptations in one sex that may have detrimental consequences for the other (Rice 1996b). Mate guarding, harassment, forced copulation, and toxic sperm are just some of the sexually antagonistic traits that males use to increase their mating success (Parker 1979). Females counter these through avoidance and resistance and by exerting post-copulation choice. Sexual conflict may also be expressed at the genomic level (Rice & Holland 1997; Parker & Partridge 1998), but this aspect of conflict is outside the scope of this review.

It is not always clear who wins this battle of the sexes. Theory predicts a variety of outcomes and in some cases evolutionary arms races will result, with one sex briefly gaining control only to be outmaneuvered by the other (Arnqvist & Rowe 2005). Sexual coercion, or nonconsensual mating, allows one sex (usually males) to gain an advantage in this conflict. In poeciliid fish persistent sneak mating tactics and matings that occur because resistance is too costly for females are examples of sexual coercion. Frenetic gonopodial thrusting seems to suggest that males are ahead in the poeciliid battle, but as this chapter stresses, females can and do counteract this coercion.

---

ent pattern). It is widely assumed that gonopodial thrusting is synonymous with sexual coercion (see, e.g., Hosken & Stockley 2005), and indeed this is probably correct in many, if not most, species, particularly those in which male courtship is also present. This view is supported by the observation that "consensual copulations"—which require obvious female cooperation—are apparently preferred by females. Males also seem to benefit from consensual copulations, as more sperm will be transferred through this route than as a result of sneaky mating (Pilastro & Bisazza 1999). Nonetheless, the fact that females are not exhibiting clear receptive responses does not necessarily mean that they are unwilling recipients of the sperm delivered via gonopodial thrusts (fig. 19.1). Moreover, in species in which gonopodial thrusting is the only male mating tactic it is unlikely that every sexual encounter is an instance of sexual coercion. Farr (1989) notes that females may cooperate with a mating attempt by not moving away from a pursuing male. Females may also position themselves close to better-quality males.

The evidence that females that are the target of sneaky mating attempts might be actively choosing or at least passively cooperating with these males is slight but tantalizing. For example, there appears to be female cooperation in

Poeciliopsis latidens (Rosen & Gordon 1953), a species in which males engage solely in gonopodial thrusting (Rosen & Tucker 1961). A similar observation has been reported for *Phallichthys amates* (Rosen & Gordon 1953). These and other accounts (described by Farr 1989) suggest that sexual coercion is by no means the rule. However, the absence of comprehensive studies of more than a handful of species makes it difficult to draw general conclusions about sexual coercion in the poeciliids as a whole. In addition, it is probably unwise to make broad generalizations based on easily quantifiable morphological characters such as sexual color dimorphism or gonopodium length or to extrapolate intraspecific trends (Ptacek & Travis 1998). Nonetheless,

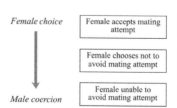

Figure 19.1 The coercive-consensual mating continuum. There is likely to be a continuum from male coercion to consensual mating even in those species in which gonopodial thrusting is the sole mating tactic.

we do know that sexual coercion is almost universally present in the poeciliids (the exception being some cave-dwelling and extremophile populations that are discussed in more detail below) and that there is no reduction in the costs of forced matings in those species in which courtship is present (Plath et al. 2007b).

One major hurdle is the challenge of providing compelling evidence that females are voluntarily participating in or even actively encouraging the copulation. McAlister (1958) argued that because male *Gambusia hurtadoi* approach females from behind and are outside their field of view during a gonopodial thrust, female choice is unlikely. These and similar observations led Farr to comment: "No female response or cooperation has ever been observed in the genus (*Gambusia*)" (1989, 99). However, Hughes (1985) found that female *Gambusia affinis* deprived of males for 30 days spent increased time with males and preferentially associated with larger males, while Bisazza and Marin (1991) discovered that *Gambusia holbrooki* females, although showing little overt propensity to mate, also actively positioned themselves near the largest male—a phenomenon that also occurs with male-deprived *G. holbrooki* females (Bisazza et al. 2001b). Condon and Wilson (2006) similarly argued that these mosquitofish females can influence the outcome of mating attempts. Of course, position need not imply preference, as proximity to larger dominant males may reduce the level of interference that a female receives from other nearby males (Pilastro et al. 2003; Dadda et al. 2005). My point is simply that sexual coercion should not be assumed in all cases and that the female contribution to mating outcomes in situations where gonopodial thrusts are deployed deserves much greater attention.

A parallel can be seen in the guppy, a species now hailed as a textbook case of female choice. Breder and Coates suggested that male guppies slip up behind "seemingly unsuspecting females" (1935), while Haskins et al. commented that they had "not been able to pin down any firm evidence of male selection by the female" (1961, 387), and Liley (1966) was willing to commit himself only to the suggestion that female choice "may operate." Careful experiments by Houde (1987, 1988, 1997) and others were needed to settle the matter.

One intriguing instance where females clearly benefit from gonopodial thrusting occurs in those unisexual *Poeciliopsis* species that depend on subordinate congeneric males for inseminations (Schultz 1969; McKay 1971; Constantz 1975).

19.2 Costs and benefits of coercion

It is hard not to feel sympathy for female poeciliids when they are being relentlessly pursued by males. The intensity and persistence of the behavior suggest that costs are involved. But what is the evidence that sneaky mating is actually detrimental to females? Potential costs fall into three categories. First, females may be prevented from engaging in some other activity, such as foraging or predator avoidance. Next, females may suffer actual harm, such as injury or infection by parasites or sexually transmitted diseases. Finally, persistent courtship may potentially undermine female preferences or lead to a loss of fitness. There is some evidence for each of these costs but also opportunities for further research.

19.2.1 Loss of opportunity

Because food availability is linked to growth and fecundity in female poeciliids (Hester 1964; Reznick 1983; Reznick & Miles 1989a), the loss of opportunities to feed may have significant impacts on fitness. There are now a number of studies showing that sexual harassment leads to a reduction of foraging (examples include *P. reticulata*: Magurran & Seghers 1994c; *Poecilia latipinna*: Schlupp et al. 2001; *G. holbrooki*: Pilastro et al. 2003; *P. mexicana*: Plath et al. 2003). Although most of the focus has been on forced copulations, "conventional" courtship activity might also, at least in some contexts, reduce female foraging efficiency. Interestingly, a comprehensive study of nine poeciliid species (*P. reticulata, P. latipinna, P. mexicana, Poecilia orri, Xiphophorus cortezi, Xiphophorus variatus, G. affinis, Gambusia geiseri,* and *Heterandria formosa*) by Plath et al. (2007) revealed that the type of mating behavior has little impact on foraging reduction.

Females may suffer further costs through being forced to leave a preferred location (Brewster & Houde 2003) or by having to divide their attention between avoiding males and avoiding predators (Magurran & Nowak 1991). Indeed, coercive mating by male guppies increases when females become responsive to predators (Evans et al. 2002a).

19.2.2 Increased risk of injury or disease

Sexual harassment can have direct, as well as indirect, costs. Forced copulations may be preceded by gonopore nipping (Parzefall 1969). Poeciliid gonopodia are often festooned with hooks and claws (Rosen & Gordon 1953; Rosen & Bailey 1963), and there are records of these causing injury to females (Clark et al. 1954; Kadow 1954; Constantz 1984; Greven 2005; see also Langerhans, **chapter 21**).

Sexually transmitted diseases are another risk (Able 1996). Wild populations of poeciliids are exposed to a variety of parasites (Lyles 1990; Harris & Lyles 1992; Cable et al. 2002; van Oosterhout et al. 2003a; see also Cable, **chapter 8**) that have adverse consequences for infected individuals (Kennedy et al. 1987; Houde & Torio 1992;

López 1998, 1999). Since forced-mating attempts involve physical contact and may occur as often as once a minute (Magurran & Seghers 1994c; Condon & Wilson 2006), sexual coercion may increase a female's chances of becoming infected by parasites.

19.2.3 Loss of fitness

The substantial reduction in foraging that females experience as a result of coercion (see, e.g., Magurran & Seghers 1994a; Plath et al. 2007b) suggests that fitness costs are inevitable. However, Head and Brooks (2006) found that elevated rates of sexual coercion had no impact on several female fitness traits (including number of offspring and offspring survival) in guppies. It is possible that fitness reductions are revealed over longer periods than the 21-day treatments used by Head and Brooks. Sexual coercion may also have a greater impact on wild fish that must forage for scarce resources and respond to predators as well as cope with harassment. Evidence that coercion is costly is provided by a study showing that female guppies pursued by several males produce smaller broods than females who encounter the same number of males sequentially (Ojanguren & Magurran 2007). Interestingly, Smith and Sargent (2006) and Smith (2007) found that high densities of females (in the western mosquitofish *G. affinis*) also negatively impact female fitness. Indeed, the presence of conspecific females may mask the costs of male harassment.

It seems probable that persistent attention from males reduces a female's ability to exert choice. Female guppies are sometimes vigorously pursued by a number of males, and it would be interesting to assess the extent to which preexisting preferences are maintained in such circumstances. The presence of predators is an important variable here. Predation risk can cause females to become unreceptive or to change previous preferences (Gong & Gibson 1996; Gong 1997), and it also leads to an increase in coercive male tactics (Endler 1987; Magurran & Seghers 1990b).

19.3 Ameliorating coercion

Females can potentially ameliorate coercion at a number of points during the reproductive sequence. Their first option is to move to a place where fewer males are found. Female guppies position themselves in deeper or faster water to avoid harassment (Croft et al. 2006b; Magellan & Magurran 2006). Females may also flee from approaching males (Brewster & Houde 2003). A second option is to dilute harassment by joining a larger shoal or to redirect harassment by positioning themselves beside more attractive females, such as those that are postpartum or have a

larger body mass (Pilastro et al. 2003). Eastern mosquitofish (*G. holbrooki*) females exhibit preferences for larger shoals and for shoals composed of larger females—but only in the presence of a harassing male (Agrillo et al. 2006). The distance between females decreases when they are harassed. Female mosquitofish will also join groups of males if harassed (Dadda et al. 2005) and, given a choice of males, will prefer to be beside the largest ones, which is advantageous since the success rate of gonopodial thrusting declines with male size (Pilastro et al. 1997). Proximity to males might seem counterproductive, but it can actually reduce rates of harassment in those species in which males compete for copulations (Bisazza & Marin 1995; Pilastro et al. 2003).

Familiarity is another means by which coercion may be mitigated. We know that males show less interest in familiar females (Kelley et al. 1999), so it would be interesting to learn whether females can exploit this tendency, perhaps by preferentially associating with familiar males. Similarly, shoals of familiar females may be more adept at evading unwanted mating attempts. Perhaps surprisingly, given that they are often larger than males, females rarely act aggressively toward their pursuers.

It is possible that in their attempts to mitigate sexual coercion, females avoid the habitats and settings to which they are best adapted and in doing so trigger a cascade of evolutionary consequences. For example, deeper (Croft et al. 2006b; Darden & Croft 2008) and/or faster-flowing (Magellan & Magurran 2006) water may select for improved antipredator responses or swimming ability, stronger schooling tendency, or modified foraging behavior.

19.4 Sexual conflict in poeciliids: the ecological context

Fish in the wild, particularly in the small streams and water bodies in which poeciliids are often found, can experience marked temporal and spatial variation in both the biotic and the abiotic environment. In what follows I draw together examples of extrinsic influences on levels of coercion. These influences may offer an advantage to one or the other sex and in doing so alter the dynamic in the battle of the sexes. Alternatively, the nature of the interaction between the sexes—such as the proportion of sneaky mating attempts successfully avoided by females—may stay roughly constant, but its intensity may be either muted or amplified (see fig. 19.2).

Investigations of species in their natural environment can provide crucial insights into the evolution of sexual conflict and the ecological arena in which it is played out (Magurran & Seghers 1994c; Rowe & Day 2006). There are relatively few examples of sexually antagonistic traits in

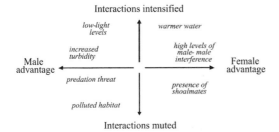

Figure 19.2 External influences on sexual conflict. External influences will cause the interactions between the sexes to be either muted or intensified and can result in either male or female advantage. An individual may move through this space on different time scales, for example, through the diurnal light cycle or during ontogeny, while populations and species will also experience different conditions as the ecological community changes or the environment is modified.

females—an issue that warrants much greater attention—and studies that draw on the natural history of species, in conjunction with carefully designed experiments that manipulate key ecological variables, are an important means of plugging that gap (Rowe & Day 2006). The examples given here are by no means exhaustive but they do illustrate the rich opportunities for exploring the ecological context of conflict and the relationship between coercion and both the abiotic and the biotic environment.

19.4.1 The physical environment

19.4.1.1 Temperature. There can be marked temporal and spatial variation in water temperature, even in tropical environments. For example, guppies in Trinidad are distributed across a wide range of thermal regimes (Magurran 2001; Magurran & Phillip 2001), and in some localities the diurnal and seasonal temperature range can be considerable (Alkins-Koo 2000; Magurran 2005). There have been a number of investigations of thermal acclimation in guppies (e.g., Chung 2001) but rather few attempts to understand the influence of temperature on patterns of mating behavior or levels of sexual coercion. Fortunately, a series of experiments on *Gambusia* species has shed light on the physiological and behavioral responses of both males and females to variation in temperature, as well as on the way in which a changing thermal environment mediates the battle of the sexes.

The eastern mosquitofish remains reproductively active across an unusually broad range of temperatures for an ectotherm. Wilson (2005) found that males pursued copulations across the entire range of temperatures examined (14–38°C), leaving open the possibility that mating activity may occur in yet cooler or warmer water. Interestingly, although the time spent following females and the total number of mating attempts increased as the water became warmer, before reaching a plateau at around 22–26°C, the

total number of copulations (defined as occurring when the gonopodium was inserted into the female's genital tract) did not vary with temperature (Wilson 2005). Female mosquitofish endeavor to avoid mating attempts and manage to do so on about 80% of occasions (Wilson et al. 2007a). However, the thermal regime that females are accustomed to can influence the outcome of these mating attempts, sometimes in rather unexpected ways. Warm-acclimated nonvirgin females receive three times as many copulations as cool-acclimated females when both sets are tested at 30°C (Wilson et al. 2007a). This counterintuitive result, which is also evident with virgin females (Condon & Wilson 2006), appears to be due to a change in female response rather than a result of an increase in mating attempts by males. Although the benefits, to females, of accepting—or at least failing to avoid—about two matings per minute at these temperatures are unclear, it does seem that they are more than passive recipients of male activity. Acclimation also influences male behavior. Males in the environment to which they are acclimated pursue matings at a higher rate than males from other thermal regimes, irrespective of whether acclimation has been to cool or to warm water (Wilson et al. 2007b).

Temperature can also have a profound influence on growth, swimming behavior, and asymptotic body size in males (Liley & Seghers 1975; Johnston & Wilson 2005), traits that will in turn affect the propensity of males to engage in sneaky mating attempts.

19.4.1.2 Seasonality. Temperature is not the only feature of a habitat that exhibits seasonal variation. Water levels and flow can vary (Alkins-Koo 2000), as can food availability (Reznick 1989) and fecundity (Winemiller 1993), not to mention levels of competition and predation (Magurran 2005). The complex patterns of seasonal variation that affect poeciliid populations in the wild (Chapman & Kramer 1991a, 1991b; Winemiller 1993; Magurran 2005) will have interesting, but as yet little investigated, consequences for sexual conflict.

19.4.1.3 Stressed environments. Poeciliids are increasingly found in polluted or other stressed environments. Indeed, in Trinidad, guppies are one of the few species that can survive—and even breed—in severely impacted sites (Magurran & Phillip 2001). One of the earliest reports of their ecology notes that guppies thrive in eutrophic habitats (Regan 1906). Carter and Wilson (2006) found that *G. holbrooki* males that had been acclimated to hypoxic conditions obtained more copulations than males that had not previously experienced such conditions. This distinction vanished when competing males were present. Carter and Wilson's study shows that poeciliids can exhibit a flexible physiological response that enables them to persist with potentially coercive behavior, even in suboptimal water conditions. The observation that there were no overall differences in female-following behavior and copulation rates in hypoxic and normoxic conditions underlines the resilience of male mating activity in this species.

A growing number of studies demonstrate that male secondary characteristics and courtship behavior (Toft & Baatrup 2001; Bayley et al. 2003; Toft & Baatrup 2003), sperm production (Haubruge et al. 2000), and the ability to sire offspring (Kristensen et al. 2005) are reduced in polluted environments. Female poeciliids are also affected. Female guppies can experience masculinization (i.e., they express some coloration in their anal fin, which takes on the form of a male gonopodium) following exposure to effluent (Larsson et al. 2002), while female swordtails *Xiphophorus birchmanni* lose their preference for olfactory cues from conspecific males in sites affected by sewage and agricultural runoff (Fisher et al. 2006), leading to increased levels of hybridization with *Xiphophorus malinche* (see Rosenthal & García de León, **chapter 10**). However, the relationship between environmental contaminants and sexual coercion is probably complex, because it will depend not just on the types and concentrations of pollutants present but also on the opportunities that the population has had to adapt to local conditions.

Insights into this adaptive process can be gleaned from investigations of poeciliids found in habitats with extreme environmental conditions. Plath (2008) observed a reduction in male sexual activity in populations of the Atlantic molly (*P. mexicana*) exposed to toxic hydrogen sulfide. Furthermore, male harassment did not appear to impose costs on females in these extreme habitats. It has been suggested that this reduction in sexual activity is a consequence of poor body condition, which in turn is the result of physiologically costly adaptations to high concentrations of H_2S (Plath et al. 2007d). Low food availability may also play a role (Plath 2008).

19.4.1.4 Turbidity. Although the popular image of poeciliids is one of small, colorful fish inhabiting crystal-clear streams, many of the localities in which these species occur are naturally turbid. Guppies that occur in turbid sites exhibit higher rates of gonopodial thrusting than those found in clear water conditions (Luyten & Liley 1991). One explanation for this is that sigmoid displays are less effective in turbid water. It is also possible that females find it harder to avoid approaching males and that sexual coercion becomes easier in these conditions. However, this is by no means invariably true, as *P. latipinna* males spend less time with any form of female stimulus in turbid environments (Heubel & Schlupp 2006).

19.4.1.5 Microhabitat. Microhabitat choice can provide the opportunity to engage in sexual coercion or shelter from it. Seghers (1973) noted that guppies in predator-free habitats occupied all regions of a stream, whereas those in high-risk sites were concentrated in shallow water and at the stream edges. These observations were supported by Odell (2002). It appears that females move to areas where water velocity is greater in an attempt to avoid mating attempts and in doing so have fewer interactions with males (Magellan & Magurran 2006). Although females can reduce persistent courtship by moving to faster flow, they cannot entirely avoid it, as some gonopodial thrusting occurs even there. These microhabitat choices involve trade-offs due to the energetic costs imposed by the current (Nicoletto 1993, 1996; Kodric-Brown & Nicoletto 2005) and an increased risk of predation in localities where predators congregate offshore (Magurran 2005). Deeper water can also provide a refuge from sexual coercion (Croft et al. 2006b; Darden & Croft 2008), but here too females may be exposed to increased predation risk.

19.4.1.6 Light. Trinidadian guppies reduce conspicuous courtship displays in favor of gonopodial thrusting when light levels are high, which in the wild coincides with the middle of the day—a time when predation risk is elevated (Endler 1987). The observation that the proportion of sneaky mating attempts increases in high light levels even when predators are absent supports the idea that light level has become a proxy for risk (Reynolds et al. 1993). Most wild poeciliids experience fluctuating light levels (Long & Rosenqvist 1998), not just seasonally and diurnally but also during those moments when a cloud obscures the sun. This points to considerable temporal variation in the extent of sexual coercion, even in places where the habitat is stable and the social structure constant. The exception to this general pattern occurs in cave populations that have evolved to cope with perpetual darkness (Walters & Walters 1965). Sexual behaviors appear to be muted in the cave populations of the molly *P. mexicana*, suggesting that sexual conflict and the avoidance of harassment play no more than a minor role in the reproductive repertoires of these troglodyte fish (Plath et al. 2003; Plath et al. 2006; Plath et al. 2007b; Plath 2008).

19.4.2 The biological environment

19.4.2.1 Sex ratio, population density, and the presence of competitors. The observation that there is often an excess of females in poeciliid populations (e.g., Geiser 1924; Haskins et al. 1961; Turner & Snelson 1984) led Snelson to conclude that female-biased adult sex ratios "are characteristic of most species of the family" (1989). However, the sex ratio in natural populations can vary dramatically in space and time. Guppy populations in streams where there is no serious threat from predation show a stronger female bias than those in high-risk localities (Haskins et al. 1961; Seghers 1973; Liley & Seghers 1975). A similar trend can be seen in other poeciliids, including *H. formosa* (Belk & Lydeard 1994) and *Brachyrhaphis episcopi* (Brown et al. 2005b). It has also been suggested that female-biased sex ratios predominate in stressed or ephemeral habitats (Snelson & Wetherington 1980). In addition, the sex ratio at a given locality may show considerable temporal variation, even when predation risk and other habitat characteristics remain roughly constant (Pettersson et al. 2004; Magurran 2005). Fish density can also vary dramatically, both seasonally and spatially (see, e.g., Grether et al. 2001b). Moreover sex ratio, fish density, and food availability interact to shape male mating responses (Kolluru et al. 2007). As a result the social background against which sexual conflict is played out is in a constant state of flux.

Sex ratio has a well-documented influence on sexual conflict. When male guppies predominate, male-male competition tends to increase, as does the rate of gonopodial thrusting experienced by females (Head & Brooks 2006). In contrast, sigmoid displays may become the dominant courtship mode when females are in excess (Farr 1976, 1980b; Evans & Magurran 1999; Jirotkul 1999). The relationship between sexual coercion and sex ratio will depend not just on the densities of individuals and proportions of each sex but also on the reproductive status and attractiveness of females. Pheromones and other chemical cues produced by females during their receptive period shift the balance between courtship and gonopodial thrusting (Crow & Liley 1979; Farr 1980a; Sumner et al. 1994; Hallgren et al. 2006). Larger—and particularly virgin and/or receptive females—are strongly preferred by male guppies (Dosen & Montgomerie 2004a; Herdman et al. 2004; Ojanguren & Magurran 2004). Male guppies will switch from displaying to a preferred female to persistent gonopodial thrusting in the presence of competing males. Male behavior is also socially mediated in species with a single mating tactic. Larger *G. affinis* females, for example, attract more copulation attempts than small ones (Deaton 2008).

19.4.2.2 Familiarity and kinship. It is now well established that fish use familiarity when deciding when and how to interact with other individuals (Magurran et al. 1994; Griffiths & Magurran 1997a; Griffiths 2003). Familiarity also mediates sexual coercion. For example, male guppies direct fewer mating attempts toward familiar females (Kelley et al. 1999). Female guppies also exhibit stronger preferences for unfamiliar males (Hughes et al. 1999), but whether they are more likely to accept forced matings from strangers is unknown (see also Rios-Cardenas & Morris, **chapter 17**).

In many species mating decisions are mediated by relatedness, though as yet there is little evidence that it plays a role in sexual coercion in poeciliids. Neither guppy sex discriminates against closely related mates (Viken et al. 2006; Pitcher et al. 2008), making it unlikely that inbreeding avoidance is an important consideration during frenetic forced encounters.

19.4.2.3 Predation risk. The influence of predation risk on sexual coercion has already been well documented, particularly in the context of the Trinidadian guppy system (Magurran 2005). In brief, males from sites with historically elevated levels of predation engage in higher rates of sneaky mating (Farr 1975; Luyten & Liley 1985; Magurran & Seghers 1994c; Matthews et al. 1997). There is evidence that this tendency is heritable (Rodd & Sokolowski 1995), but males are also adept at adjusting their behavior to local conditions and exhibit higher levels of gonopodial thrusting at times of day when the risk is greatest (Endler 1987) or when females are preoccupied with predator avoidance (Magurran & Nowak 1991; Dill et al. 1999; Evans et al. 2002a). Reproductive skew is higher among males than among females (e.g., Becher & Magurran 2004; Luo et al. 2005; Neff et al. 2008b; Simmons et al. 2008), and females can ensure their continued reproductive success simply by managing to stay alive long enough to produce the next brood. Females switch from avoiding males to avoiding predators and receive more unsolicited mating attempts as a consequence, while males appear to trade off increased risk of predation against an improved chance of successful insemination through gonopodial thrusting (Magurran & Nowak 1991; Kelly & Godin 2001). This asymmetry can lead to rapidly fluctuating rates of sexual conflict.

19.5 "Internal" influences

These extrinsic factors will be modulated by what I call "internal" influences—that is, the combination of factors such as motivation, experience, learning, personality, growth rate, and genes that shape an individual's responses to a set of external conditions. Although internal factors have received less attention than the environmental ones reviewed above, they can lead to marked individual differences in responses to the same set of circumstances.

The evolutionary history of a population or species can act as a dial that either increases or lowers the intensity of sexual coercion, and these differences may persist even when fish are raised in a common garden environment. High-predation risk appears to select for increased levels of coercion (Magurran & Seghers 1994c; Rodd & Sokolowski 1995; Matthews et al. 1997), while cavernicolous and extremophile habitats lead to an overall reduction in sexual activity (Plath 2008). Indeed, there is evidence in some cases, notably cave-dwelling species, that patterns of reproductive activity (Arndt et al. 2004) and the proclivity toward sexual coercion (Plath 2008) are largely innate. Counteradaptations to coercion may also be hardwired. Dadda et al. (2008) found that early experience did not influence eastern mosquitofish female responses to sexual harassment. However, in other instances genetic differences may be modulated by early experience, and Rodd and Sokolowski (1995) stress the importance of the gene × environment interaction in shaping adult reproductive behavior in guppies. The sex ratio that a male encounters during development can have a substantial impact on his reproductive repertoire. For example, male guppies reared in male-biased groups engage in more forced copulations following maturity (Evans & Magurran 1999). It is now abundantly clear that poeciliids have the cognitive ability to learn and to make subtle behavioral decisions (see Bisazza, **chapter 15**; Kelley & Brown, **chapter 16**; Webster & Laland, **chapter 14**). Further investigation of the role of learning in sexual coercion (and its avoidance) would be productive.

Another theme that would repay exploration is the relationship between personality and coercion. Early researchers dismissed individual variation as uninteresting, not realizing that there are marked and consistent differences among individuals. The correlated patterns that emerge have been referred to as personalities (Budaev 1997) or syndromes (Sih et al. 2004). Male guppies have a tendency to engage in either consensual or coercive mating tactics. Interestingly, these individual differences are preserved across contexts, even when the relative advantages of courtship versus sneak mating are reversed (Magellan & Magurran 2007a). Some of these personality differences may be linked to underlying physiological factors. For example, given that size influences a male's tendency to engage in sexual coercion (Houde 1997), as well as the outcome of these encounters (Pilastro et al. 1997), the growth trajectory of an individual (Walling et al. 2007) may predict his mating phenotype.

Equally intriguing, but extremely challenging to measure, is the role that female motivation plays in determining the outcome of mating encounters. As I noted at the outset, there are interactions that might appear to represent instances of coercion when, in fact, the mating is being controlled by the female. This is likely to be the case, at least occasionally, in species where forced copulations are the norm. It is also possible that some sneak matings in species with alternative mating behaviors, such as guppies, are accepted by females. Indeed, it is conceivable that a male's

ability to inseminate a female at a time when they are both in grave danger of being attacked by a predator is not due to his manipulation of her but rather is an indication to the female that the male possesses good genes.

19.5.1 Male phenotype

Male phenotype can have an impact on coercion. In some species, such as the sailfin molly, *P. latipinna*, larger males have a reproductive advantage (Farr & Travis 1986; Travis & Woodward 1989). However, although courtship rates tend to increase with male body size, the relationship between male phenotype and gonopodial thrusting is complex (Ptacek & Travis 1996). Evidence for phenotype-linked mating behavior has been reported in other poeciliids (e.g., the guppy: Jirotkul 2000; Karino & Kobayashi 2005; Magellan et al. 2005), but again there is often no simple pattern, because genetic background and social context also influence behavior (Rodd & Sokolowski 1995). In contrast, in the genus *Xiphophorus*, different male behaviors have been attributed to a single locus (Ryan & Causey 1989; Zimmerer & Kallman 1989).

One interesting hypothesis is that sexual size dimorphism is driven by selection for small male advantage during gonopodial thrusting (Bisazza & Marin 1995; Bisazza & Pilastro 1997; Pilastro et al. 1997). Again, the issue is complex, since other factors such as growth rate, temperature, and time to maturation also influence morphology. A related observation is that males have relatively longer gonopodia in localities where predators are abundant (*P. reticulata*: Kelly et al. 2000; *B. episcopi*: Jennions & Kelly 2002). As these localities are associated with elevated sneaky mating rates, it is plausible that this geographic variation in genitalia is driven by opportunity for coercion (see Langerhans, **chapter 21**). Nonetheless, natural selection linked to varying river conditions could be involved—perhaps even as the primary driver of this pattern.

19.6 Postmating effects

This chapter has focused on the events that occur before and during an insemination. The story does not, of course, end there. Poeciliid females can store sperm from several males and draw on these reserves to father successive broods (Winge 1937). Indeed, nonvirgin females will continue to produce offspring for some months even in the absence of males (see Greven, **chapter 1**). The relationship between the composition of a brood and a female's previous sexual encounters is one of the most interesting topics in poeciliid research today (Magurran 2005). What is clear is that offspring are not the result of a random mix of the available inseminates (Evans et al. 2003b). Rather, a female is able to exert postcopulatory choice (Birkhead 2000; Birkhead & Pizzari 2002) and use it to influence the paternity of her brood (see Evans and Pilastro, **chapter 18**). It will be interesting to discover whether females can use postcopulatory mechanisms to purge sperm from forced matings (Liley [1966] reports that recently inseminated sperm is sometimes extruded by females)—or to utilize sperm from matings that were not solicited but were from preferred males.

19.7 Coercion of heterospecific females

Persistent mating by male poeciliids is well documented, so it is perhaps not surprising that forced matings can be redirected toward heterospecific females. In some circumstances—as with unisexual poeciliids that are dependent on heterospecific sperm (Schultz 1969; Schlupp & Riesch, **chapter 5**)—these matings are necessary, though they may also impose costs (Heubel & Plath 2008). In the majority of cases these interactions have the potential to be harmful. Guppies in Trinidad pursue congeneric *Poecilia picta* females and attempt forced matings (Haskins & Haskins 1949; Magurran & Ramnarine 2004), which can successfully transfer sperm (Russell et al. 2006). These males can learn to discriminate heterospecifics (Magurran & Ramnarine 2004)—a skill that eventually becomes innate under sympatry (Magurran & Ramnarine 2005). However, guppies are now being introduced widely and have encountered many other species. In Mexico, for example, guppies are invading the habitats of endangered goodeid species. Guppies persistently court morphologically similar but phylogenetically distant *Skiffia bilineata* females—even when females of their own species are in excess (Valero et al. 2008). This behavior seems to be less influenced by experience than the encounters with *P. picta* (Valero et al. 2009) mentioned above and may place an additional burden on populations that are already under threat.

19.8 Conclusions

Sexual coercion is a dominant feature of poeciliid societies. Males typically devote considerable time and energy to the pursuit of females, while females experience costs as a result of this activity. However, although persistent mating activity can be detrimental to females, there are probably instances where apparent coercion is accepted by females. An important theme for future research will be the unraveling of the extent of female participation in seemingly coercive mating behavior. Nevertheless, there are robust indications that female time budgets, particularly as they relate to foraging, are being disrupted by males. There have been only a few attempts to link feeding disruption with fitness; doing so would strengthen the case that sexual coercion is genuinely harmful to females. Equally, the rapid advances in the study of postcopulatory mechanisms will shed light on the ability of females to counteract coercion.

Another observation that emerges is that male-female interactions take place against a constantly changing biological and environmental landscape. Larger females or bigger shoals appear, populations increase in size, predators move farther upstream, the canopy closes in, or a river becomes a popular picnic site. Natural habitats are increasingly subject to anthropogenic disturbance and pollution, and the impact of this on poeciliid mating systems in general and sexual coercion in particular is only just beginning to be recognized. Longer changes—over evolutionary time—will also occur. By conducting experiments in relatively uniform laboratory aquaria we may underestimate the dynamic nature of the battle of the sexes. Rapid transitions in behavioral responses of male and female guppies exposed to variable predator risk have already been documented (Magurran & Nowak 1991), but there is considerably more scope for tracking temporal shifts in the interactions between the sexes.

Individual or personality differences also deserve more attention, especially now that there is growing awareness of how physiological, and ultimately genomic, processes underpin behavior. A formal comparative analysis of coercion among the poeciliids would also be illuminating. Indeed, over the longer term I hope that research will be able to shed light on why this group of fishes has evolved mating systems in which coercion plays such a major role.

Acknowledgments

I am grateful to Jon Evans, Andrea Pilastro, and Ingo Schlupp for their invitation to write this chapter and to the editors, Locke Rowe, and an anonymous reviewer for their helpful and thoughtful comments.

Chapter 20 Information societies: communication networks and sexual selection

Matt Druen and Lee A. Dugatkin

20.1 Introduction

WITHIN THE FIELD of sexual selection, poeciliids have emerged as model systems in which to test ideas about how socially acquired information might affect animal mate choice and aggression (Basolo 1990; Houde 1997; Moretz 2003; Earley & Dugatkin 2005; Magurran 2005; Moretz & Morris 2006; Dugatkin 2009). Inclusion of the social context within which these behaviors occur marks the logical progression from a tractable dyadic conceptualization of animal signaling to a more complex information-trafficking network view of communication. In this light, variation in a population's social structure and composition serves as a potentially powerful environment imposing its own set of empirically unexplored selection pressures (McGregor & Peake 2000; Brown & Laland 2003; Bonnie & Earley 2007). As a result, information exchange between two or more individuals may have a far larger sphere of influence than has been historically conceived (Krause et al. 2007; Arnott & Elwood 2008; see also Krause et al., **chapter 13**).

Compelling evidence for different kinds of social-information use by animals has accumulated rapidly over the last 15 years and has been documented in amphibians, fish, birds, and mammals (Laland & Williams 1997; McGregor et al. 2000; Sontag et al. 2006), including poeciliid fishes (e.g., Webster and Laland, **chapter 14**). One of the most complex forms documented so far involves gathering information about a conspecific by observing that individual's social interactions with others, a practice often referred to as social eavesdropping. Eavesdropping and copying/imitation produce communication networks (McGregor 1993; McGregor & Peake 2000; Johnstone 2001).

Within a communication network, animals may assume any of three general behavioral roles: signaler, receiver, or bystander/eavesdropper. Bystanders become eavesdroppers when they intentionally acquire and use socially generated information (Earley & Dugatkin 2005). Therefore, eavesdropping applies uniquely to groups of animals in a way that other forms of information transfer do not (e.g., social learning, public-information use) because it requires at least three individuals—two interactants and one observer (Bonnie & Earley 2007). Social eavesdropping can enhance fitness in two ways. First, it allows animals to augment information acquired through direct experience. Second, directly learning about individuals could be potentially costly, and eavesdropping allows such costs to be minimized.

The primary goal of this chapter is to provide an overview of how social eavesdropping in poeciliid communication networks expands our understanding of mate choice and aggression. First, we discuss empirical work and theoretical models of social eavesdropping in three contexts: aggression in green swordtails (*Xiphophorus hellerii*) and mate choice in guppies (*Poecilia reticulata*) and in sailfin mollies (*Poecilia latipinna*). Next, we provide a summary and critical discussion of existing data. We conclude by considering how animal personalities may help link individual behavior to group-level phenomena.

20.2 Aggression and the role of information in green swordtail contests

Intrasexual competition determines control of resources essential for reproduction, such as mates (Darwin 1871). In wild populations, this often plays out in the form of agonistic interactions: fights between males contesting the same resource. To a large extent, winning or losing a fight is a function of contestants exchanging information about fighting abilities — or resource-holding potential (RHP) (Parker 1974; Peake & McGregor 2004). RHP arises out of a complex interaction between two broad categories of traits that contribute to fighting ability: intrinsic traits such as body size and weaponry, and extrinsic traits such as recent social experience or motivation. Extrinsic effects are also called winner-loser effects in the context of animal aggression (Landau 1951; Dugatkin & Dugatkin 2007). Through the course of a fight, participants exchange visual and possibly physical signals that convey information about each other's RHP (Enquist & Leimar 1983). Among poeciliids, there is arguably no better example of intrasexual competition than that documented in the green swordtail, *X. hellerii*.

Fights between nonterritorial males often progress through four stages: initiation, escalation, resolution, and reinforcement. During the initiation stage, individuals signal hostile intentions to one another using relatively low-cost, noncontact, frontal and lateral sigmoid displays in which the fins are fully spread and opercula flared. In theory, if contestants perceive large enough asymmetries in RHPs, the interaction may terminate without escalation. However, if accumulated information is insufficient, or if contestants make similar estimations of their own RHP relative to a competitor (i.e., they are closely matched), a fight may escalate to physical combat (Enquist et al. 1990; Beaugrand et al. 1991).

Escalated swordtail combat can be quite spectacular to observe, and a fight between well-matched males may last for over an hour in the laboratory. A typical escalated fight consists of numerous bouts of mouth wrestling (gripping the upper or lower jaw and thrashing), incessant circling, chasing, high-speed ramming maneuvers, and repeated bites to the body, including the opercula (gill coverings) and gonopodium (reproductive organ) (Franck & Ribowski 1989; M. Druen, personal observation). As the contest progresses, individuals exchange increasingly reliable information about relative fighting prowess. Information becomes more reliable because escalated aggressive behaviors are more physiologically costly to produce (Hurd 1997). As a result, it becomes more difficult to conceal one's true fighting prowess (Ribowski & Franck 1993). Eventually, a clear winner and loser emerge from the mêlée. In essence, contestants learn which is the winner and which the loser through aggressive reciprocation. Although sustained aggression is abandoned at this point, winners may continue to intermittently inflict attacks against losers, apparently reinforcing their dominance.

Both fighters sustain potentially significant costs by simply engaging in fights. For example, physiological costs may manifest as spikes in stress hormone levels, accumulation of anaerobic metabolites (lactic acid), and depletion of energy reserves (Hannes et al. 1984; Neat et al. 1998; Hsu & Wolf 1999, 2001; Oliveira et al. 2001). In addition, injuries or physical exhaustion might increase the likelihood of infection by pathogens, as well as impede mobility. Moreover, there are a host of "opportunity costs" for behaviors subject to time-budget trade-offs such as resting, foraging, predator evasion, and searching for mates (Dugatkin & Godin 1992a).

While some costs are borne by both contestants, the ultimate payoffs to winners and losers are quite different. The benefit to winners is an increase in social rank (or the maintenance of the status quo if the winner was already the more dominant male). In populations of swordtails, higher social rank translates into more mating opportunities. For example, alpha-ranked males can monopolize an overwhelming proportion of mature females, accounting for up to 75% paternity in the next generation (Luo et al. 2005). Conversely, defeat often leads to a decreased rank in a hierarchy and, presumably, a reduction in mating opportunities.

In addition, the experience of winning and losing may significantly impact the probability of dominating a rival in subsequent bouts. Winner and loser effects are usually defined as an increased probability of winning at time T, based on victories at times T-1, T-2, and so on, and an increased probability of losing at time T, based on losing at time T-1, T-2, and so on, respectively. Although loser effects are more common than winner effects, both have been documented many times (Chase et al. 1994; Dugatkin & Druen 2004).

Given the implications of winning and losing and the shared risks of agonistic behavior, when should male green swordtails initiate aggression or avoid it? In general, when the costs of fighting are high, individuals should be sensitive to when aggression will be more or less profitable to fitness. Therefore, the decision heuristic is "avoid fighting unless you can (probably) win." In such a system, a premium is placed on any behavioral mechanism that can acquire and integrate accurate information about a potential opponent's RHP before actually engaging it. Social eavesdropping is just such a behavior.

For our purposes, a male green swordtail may alternately assume any one of several roles in an aggression network. It may be a signaler, actively fighting with another; a bystander/eavesdropper within detection range of a fight; or a solitary individual (Earley & Dugatkin 2005). Based on this simple arrangement, several intriguing questions emerge. Do eavesdroppers respond differently to observed winners and losers? Are eavesdroppers sensitive to the dynamics of an observed fight? Does watching fights affect aggressive tendencies in a general way or is behavior altered toward specific individuals? Indeed, do swordtails eavesdrop at all?

20.2.1 A case for eavesdropping in swordtail aggression networks

As a practical consideration, how does one actually go about setting up a controlled environment to test whether eavesdropping occurs? In swordtails, and other systems in which vision is a primary sensory (and signal) modality, an effective approach is to design a test apparatus that can manipulate what each member of a group is able to see the others do. In a series of experiments on social eavesdropping in green swordtail aggression networks, Ryan Earley and his colleagues (Earley & Dugatkin 2002; Earley et al. 2005; Earley & Dugatkin 2005) accomplished this by employing multiple-compartment test arenas separated by partitions of varying opacity, including one-way mirrors (fig. 20.1). Using this setup, they could manipulate the flow of information in staged fights between pairs of males and a bystander. Subsequently, eavesdroppers could be pitted against observed winners and losers to test whether information acquired in this manner altered fight dynamics or outcomes.

At the start of a trial, a male was placed into one compartment and exposed to a staged fight between two males. Because large asymmetries in body size strongly influence contest outcomes, all participants in a trial were closely matched according to lateral surface area. Subsequently, bystanders were pitted against the winner or loser of the fight that they had observed. Results indicated that watching fights had a statistically significant impact on swordtail agonistic behavior. The effects were most pronounced when eavesdroppers fought previous winners. When facing winners, eavesdroppers exhibited a decreased propensity to initiate and show aggression compared with individuals that had not viewed the fight. An "avoidance response" was present in eavesdroppers when they fought previous winners: eavesdroppers showed a decreased propensity to initiate aggression against observed winners. An eavesdropper's assessment of a winner's fighting ability was not sensitive to a fight's dynamics (whether it had witnessed an escalated or nonescalated match). Fight intensity, however, did affect eavesdroppers when they interacted with individuals that they saw lose a contest. Eavesdroppers were less likely to initiate fights and win against previous losers that had persisted for long periods or that had initiated escalated behavior in their fight. These findings suggest that eavesdropping allows individuals to decrease their odds of engaging in potentially costly fights by fine-tuning aggressive responses to particular types of individuals.

Next, Earley and his colleagues examined aggressive priming as a possible alternative explanation for changes in the observer's responses after watching fights. Priming occurs when bystanders exhibit a generalized agonistic response that is insensitive to a conspecific's identity (i.e., aggression is shown to all males) (Hollis et al. 1995). To test this, fights were staged in the presence of a bystander. Bystanders then fought naive individuals that had not been observed. Watching and then confronting lone individuals increased bystanders' probability of initiating aggression and winning the fight if their opponents were slightly larger. This finding was opposite that of the first experiment, in which eavesdroppers exhibited an avoidance response when presented with previous winners. Moreover, the directionality of a bystander's aggressive response was not evident when they confronted unfamiliar opponents. In other words, bystander response appeared to be based on familiarity and was not driven by the mere observation of aggression. Overall, these experiments suggest that male green swordtails acquire specific information that allows them to make a distinction between winners and losers of observed fights. Moreover, eavesdroppers were able to accurately differentiate between the fighting ability of "average" losers and highly aggressive ones. That is, eavesdroppers were sensitive not only to the status of each individual as a winner or a loser but also to fighting ability irrespective of status. As a result, eavesdropping in swordtail aggression networks appears to be a useful and surprisingly

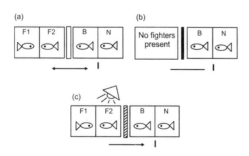

Figure 20.1 Test arena for eavesdropping on aggressive interactions between male green swordtails. (a) Clear glass, (b) opaque partition, and (c) one-way mirror. Arrows indicate which way information can flow between B (bystander) and F1/F2 (fighters). Modified from Earley et al. 2005.

Box 20.1 Game theory and the evolution of aggression

Under what conditions do both strategies coexist in the Hawk-Dove game? Hawks split the value of a resource (V) and pay a cost (C) when they encounter each other. Hawks always win against Doves. Doves do not incur a cost and split the resource's value when they encounter each other. When $C > V$, Hawk is not an evolutionarily stable strategy (ESS). When $V > C$, Dove is not an ESS.

	HAWK	DOVE
HAWK	(V−C)/2	V
DOVE	0	V/2

Box-figure 20.1 Payoff matrix for the classic Hawk-Dove game. In a population let p = frequency of Hawks; $1 - p$ = frequency of Doves. Payoff to Hawks is $p(V - C)/2 + (1 - p)V$; payoffs to Doves is $p(0) + (1 - p)/V$. Strategies coexist when $p(V - C)/2 + (1 - p)/V = p(0) + (1 - p)/V$. Hawk and Dove are ESSs when $p = V/C$.

nuanced strategy for acquiring indirect information about a potential opponent's fighting ability.

20.2.2 Eavesdropping on aggression: a theoretical perspective

The classic Hawk-Dove game (box 20.1) makes predictions about the consequences of aggressive behavior for the fitness of two individuals competing for control of some valued resource. The strategy Hawk is aggressive and always escalates in a fight. The strategy Dove submits to Hawks and divides the resource equally with other Doves. To begin the game, a value is assigned to the resource and a cost to fighting. Next, random pairs of individuals from within a large population interact. When the process is iterated many times (to simulate an indefinite number of "fights" or generations), strategies can be tested over different value-to-cost ratios to see whether Hawks and Doves can coexist or if one drives the other to extinction.

In many instances, there is a negative frequency-dependent effect such that Hawks profit when they are rare but do poorly when numerous (Maynard Smith 1982). For example, Hawk is a profitable strategy when it occurs as a small proportion of a population because it defeats more numerous Doves. Hawks do not fare as well when common, because two aggressive individuals who encounter one another become embroiled in an escalated and costly

fight. If the cost of fighting is great enough, the game yields a mixed equilibrium containing both strategies. While the Hawk-Dove model of aggression incorporates signalers and receivers, it does not consider the role of eavesdropping on the evolution of aggression. This suggests a number of important questions. For example, can eavesdropping exist as an evolutionarily stable strategy in an expanded Hawk-Dove game? Once established, how might it influence individual levels of aggression or the overall frequency of aggression in the population? Finally, how might variation in fighting ability modify the utility of eavesdropping?

In 2001 Rufus Johnstone was the first to incorporate an eavesdropping strategy into the Hawk-Dove model of aggression (Johnstone 2001). In each cycle of the game, pairs of individuals are chosen at random and fight over a resource. Johnstone's model defines Hawk and Dove strategies in the conventional manner but adds a new strategy to the game—eavesdropping, defined as "play hawk when facing an opponent who lost in the previous round and dove when facing an opponent who won." One stipulation is that eavesdroppers' ability to retain information about individuals is limited to the most recent round. In other words, information cannot accumulate about specific individuals over successive rounds, and eavesdroppers cannot remember information about more than one individual at a time. Johnstone evaluated the interactions of these strategies over a range of cost-benefit ratios (fig. 20.2).

Across much of the cost-benefit distribution, eavesdropping reaches evolutionary stability and coexists with Hawks and Doves. At high ratios, eavesdroppers outnumber Doves

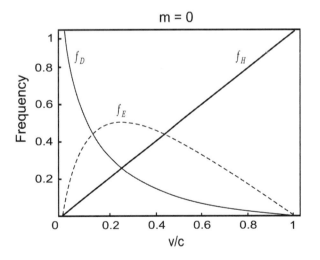

Figure 20.2 Effects of social eavesdropping in the Hawk-Dove game. Equilibrium frequencies of hawks (f_H), doves (f_D), and eavesdroppers (f_E) as a function of the value of victory relative to the cost of losing an escalated fight (V/C), for eavesdropping without mistakes $(m = 0)$. Modified from Johnstone 2001.

but occur less frequently than Hawks. Somewhat surprisingly, eavesdropping does not evolve to be the lone strategy under any of the examined value ratios. This is due to a frequency-dependent effect based on the probability that one eavesdropper encounters another. For example, eavesdroppers proliferate when they represent a small proportion of the population because of their ability to predict the behavior of most other individuals. However, as the population's proportion of eavesdroppers grows, so too does the probability that they encounter other eavesdroppers. This erodes the value of eavesdropping because eavesdroppers cannot predict other eavesdroppers' behavior. In the end, eavesdropping is most advantageous only as long as the probability of encountering either a Hawk or a Dove is higher than the probability of meeting another eavesdropper. Under these conditions, eavesdropping is maintained at higher frequencies than either pure strategy while simultaneously limiting its own spread.

Johnstone's model also makes nonintuitive predictions about how eavesdropping affects overall levels of aggression in the population. For example, the frequency of escalated aggression is higher when eavesdropping is present than it is in the standard Hawk-Dove game. Since individuals use socially acquired information to avoid potentially escalated and costly fights, why does eavesdropping not decrease overall aggression? It turns out that the explanation is due to a kind of winner effect. Because Hawks always beat Doves, and because eavesdroppers always avoid fighting with observed winners, victory imparts a "reputation," which enhances the probability of winning future rounds. This model establishes unambiguous conditions in which eavesdropping coexists with other evolutionarily stable strategies and paves the way for further theoretical experimentation on the evolutionary significance of social eavesdropping. For example, one important assumption of the model is that differences in fighting ability have a discrete, rather than continuous, distribution. In wild populations, there is normally much variation in fighting ability.

Mesterton-Gibbons and Sherrat (2007) examined the effect of social eavesdropping among triads of individuals when variation in fighting ability is continuously distributed and when stronger individuals are more likely to escalate than weaker individuals. The explicit eavesdropping rule is that "a sufficiently weak animal may defer to a prior winner it has observed, but that no individual defers to a loser or to an individual that it has not observed" (2007, 1259). During a round, a fight occurs between two of the three in a triad. After its resolution, the bystander/eavesdropper initiates a fight with the loser (by definition, the weaker of the two initial interactants). As this second fight ensues, the previous winner adopts a bystander/eavesdropper role. If the original loser wins its second fight, the original winner must initiate a fight with the initial eavesdropper. But if the first loser loses a second time, the first winner escalates—if it is strong enough. This model makes quite different predictions from Johnstone's model. For example, eavesdropping reduces the overall frequency of aggression because avoiding the costs of initiating a fight against an observed winner trumps the potential benefits of winning for weaker animals. In addition, continuously distributed fighting abilities cause a strength threshold to emerge that determines decision criteria to fight. As the costs of fighting rise, there is a higher threshold for aggression.

Though somewhat different in their assumptions, each model demonstrates that the broader social dimension in which aggression takes place can strongly alter contestants' behavior. One obvious necessity for future models is to extend the game beyond triads of players. Analysis of triads provides tractability but still falls short of capturing most natural instances, in which aggression occurs in groups of more than three individuals. Though a daunting task, such investigations are necessary before we will fully understand how communication networks affect the flow of information and the consequences of such information flow in wild populations.

20.3 Mating games: the interaction of genetic and cultural transmission

Early models of mate choice assumed that females choose mates independently of the choices made by other females in their social group (Fisher 1958). Recent work has expanded these models to examine the extent to which information acquired from observing others affects mating preferences and therefore variation in male mating success. The issue is important because some forms of social information can be transmitted horizontally (nongenetically) as well as vertically (genetically through offspring) (Cavalli-Sforza & Feldman 1981; Boyd & Richerson 1985). Such cultural transmission may affect behavior at greater speeds—one or a few generations—than genetic changes, which occur over evolutionary time (Boyd & Richerson 2002).

For a number of reasons poeciliids are attractive model species to researchers wishing to address these issues. First, most species are easily kept and cared for, breed readily under captivity, and behave similarly in field and laboratory settings (Beaugrand et al. 1984; Franck & Ribowski 1987; Houde 1997; Earley & Dugatkin 2005; Magurran 2005). Second, a large number of investigations have confirmed that several poeciliid species attend to the behavior of other group members in the contexts of foraging, mate choice, and antipredator behavior (Dugatkin 1992b; Dugatkin & Godin 1992a; Witte & Ryan 1998; Laland & Reader 1999a). Finally, genetically based preferences for certain male traits have been identified in a number of poeciliids, including guppies, mollies, mosquitofish, and swordtails (Basolo 1995b; Bisazza & Pilastro 2000; see also Rios-Cardenas and Morris, **chapter 17**). For example, female guppies from the Paria River in Trinidad exhibit a genetic preference for males with more or brighter orange coloration (Houde & Endler 1990; Houde 1994). Similarly, female mollies possess a genetic preference for larger males (Ptacek & Travis 1997).

Here we will look at mate-choice copying as a form of eavesdropping. Mate-choice copying can be defined in technical language as follows: "the conditional probability of choice of a given male by a female is either greater or less than the absolute probability of choice depending on whether that male mated previously or was avoided, respectively" (Pruett-Jones 1992, 1001). In principle, mate-choice copying can have dramatic effects on the distribution of male traits in a population by altering the distribution of genes passed on to the next generation (Kirkpatrick & Dugatkin 1994; Agrawal 2001). Whereas our examination of eavesdropping in swordtail aggression takes place between three males, the female mate-choice scenario occurs among a courting male/female dyad that is being watched by another female. Guppies led the way in early studies on mate-choice copying. Sailfin mollies have reinforced original findings and extended our perception of how social information affects decision rules used in mate assessment.

20.3.1 Mate-choice copying in guppies

In the absence of mate-copying opportunities, guppies consistently prefer males with more and brighter orange carotenoid pigments (Houde & Endler 1990). Such males are more active, are likely to be less parasitized, and are better foragers than drabber individuals (Houde & Torio 1992). However, the nature of guppy social structure provides numerous opportunities for females to observe the mating decisions of other females and acquire information about male quality (Croft et al. 2004b; Morrell et al. 2008). Dugatkin and his colleagues examined this possibility and its potential consequences in the guppy.

In one set of experiments, a focal female was allowed to make an initial choice between two stimulus males that were behind glass partitions and that had been matched for body size and percent orange color (Dugatkin & Godin 1992a). Preference was assessed based on the proportion of time a female spent with a particular male. After the initial choice, the focal was then allowed to observe the same two

males again, this time with a female "model" placed next to the male that had not been chosen in the first test. Next, the model was removed and the focal female was immediately allowed to choose again between the original stimulus males. Females that had the chance to copy reversed their initial choice significantly more often than females in control treatments. This was the first experimental evidence that observing signaling and social cues exchanged between conspecifics had a large impact on assessment of males made by focal females. The validity of mate-choice copying in guppies was questioned in two studies that did not use populations obtained from Trinidad (Brooks 1996; Lafleur et al. 1997). One hypothesis was that procedural disturbances during the course of a trial were responsible for apparent copying events (Lafleur et al. 1997). However, this view was not supported in a subsequent experiment that tested this hypothesis explicitly (Dugatkin et al. 2003).

The discovery that social information could strongly affect a female's preference for a particular male provided an opportunity to investigate how asymmetries in male phenotypes interacted with social information and genetically based preferences in determining a female's ultimate assessment of a potential mate. To accomplish this, tests were conducted in which pairs of stimulus males were created with predetermined differences in percentage of orange body pigmentation. Three categories of male pairs were created: pairs that differed in the amount of carotenoid pigments by 12%, 24%, and 40%. Focal females preferred males with more orange coloration in the absence of model females. However, in trials in which the model was placed near the drabber male, bystander females often reversed their initial choice and associated with the less colorful male. Specifically, when males differed by 12% or 24%, focal females preferred the previously unattractive male. However, females did not copy a model female's choice when males differed by 40% (fig. 20.3), suggesting the complex nature of the interaction between genetic and culturally based information.

Mate-choice copying also presents us with the opportunity to continue to pursue the phenomenon of cultural transmission. Cultural transmission occurs when changes in behaviors spread throughout a population by nongenetic means (Dawkins 1976). For example, how many females in a group might be affected when one of them copies the mate choice of another? In other words, once a single copying event has occurred, what proportion of a population's females is likely to "copy the copier"? Dugatkin and colleagues (2003) found that an initial act of mate-choice copying affected the mating preferences of significantly more observer females, tested consecutively in a series, than of control females, which were not given any prior

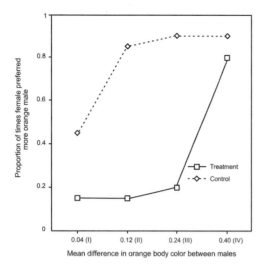

Figure 20.3 Effects of social learning on mating preferences: the proportion of times females preferred the more orange of the two males. "Treatment" refers to the trials in which an observer female saw the less orange male adjacent to the model female, and "Control" refers to the trials in which a female did not observe the mate choice of any other individuals. Modified from Godin and Dugatkin 1996.

opportunity to observe the apparent mate choice of another (model) female. These results suggest that the social effect of mate-choice copying is not restricted to the initial copier but rather can subsequently affect the mating preferences of other females in the population at a frequency greater than expected by chance alone.

For a number of reasons, not all female guppies are equally likely to copy. For example, young female guppies have a higher probability of copying an older female, but not vice versa (Dugatkin & Godin 1993; Amlacher & Dugatkin 2005). This finding suggests that naive females value social information differently than their older group mates do. Future experimental work could test whether naive or young female guppies are more likely to make errors in mate assessment than more experienced or older females of their group. This would help us better understand the ways that individuals value types of social information and demonstrate that decision-making processes can change as a function of experience or developmental stage.

Like all compelling experiments, work in the guppy system suggests more questions than it answers. For example, are preferences for certain males acquired by eavesdropping stable over time? In other words, how long do socially modified preferences last before the resurgence of genetically based assessment strategies? In addition, selecting a mate is only one-half of the mate-choice paradigm. To complete the picture we need to incorporate the possibility that females *reject* some courting males as well. How would females react to a model female's apparent rejection of a male

that was initially preferred? Experiments on sailfin mollies shed light on this question.

20.3.2 Mate-choice copying in sailfin mollies

Klaudia Witte and her colleagues in 1998, 2002, and 2003 performed a series of experiments that examined the relative strength of a molly's genetic preference for larger males against information arising from observed courtship behavior between male and female conspecifics (Witte & Ryan 1998, 2002; Witte & Ueding 2003; Witte & Massmann 2003). Consistent with work in the guppy system, mollies copied the mate choice of another female when males did not differ substantially in size. However, unlike guppies, where social cues sometimes overrode genetic predispositions, bystander mollies failed to copy when moderate to large differences in male size existed.

In an important extension of the mate-choice-copying paradigm, Witte and Massmann (2003) tested the duration of the copying effect. They gave bystander females an opportunity to copy the choice of a model female after three different time intervals: immediately after viewing, one hour after viewing, and one day after viewing. When tested immediately after viewing a model in proximity to the male that was not initially chosen, a majority of molly females reversed their preference. Further trials revealed that the effect of observing an apparent mate choice by a model female persisted for both the one-hour interval and the twenty-four-hour interval.

As mentioned above, copying the apparent mate choice of a model is only one of the ways in which a female might alter its assessment of the suitability of a prospective mate. Another possibility is that a bystander might copy a model's rejection of a courting male. Witte and Ueding (2003) tested this by using video playback to show focal females instances of a model rejecting the courtship of a male. To begin, a female's intrinsic preference for one of two males was documented. Next, females observed a playback video that depicted a female rejecting the male that was initially chosen. Subsequently, the focal female was presented with an opportunity to choose again between the same males as before. Results showed that more than half of the females reversed their initial preference and instead associated with the male that had been avoided in the initial assessment trial. Thus, bystander females appear to devalue a male's attractiveness based on observing a model female avoid that male.

Until recently, no study existed that documented the existence of mate-choice copying in wild populations. Witte and Ryan (2002) conducted a field study to determine if populations of sailfin mollies copied the mate choice of con-

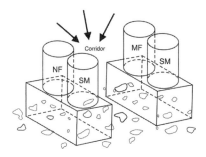

Figure 20.4 Testing mate-choice copying in the field. Two upside-down plastic tanks rest on the streambed. Two jars are placed on each inverted tank. A stimulus male (SM) is placed in each of two jars, the model female (MF) is in one jar beside a stimulus male, and the fourth jar has no fish (NF). Modified from Witte and Ryan 2002.

specifics in a natural setting. Their apparatus was a modified version of laboratory test arenas (fig. 20.4). In female mate-choice trials, a male and female were presented in one of the tanks and a lone male was presented in the other. Members of nearby shoals were then free to approach through the corridor created by the setup and associate with either a "mating" pair or the lone male. Consistent with laboratory experiments, significantly more females chose to associate with the male-female combination rather than with the lone male in 19 out of 20 trials. These results are consistent with the notion that mate-choice copying occurs in wild populations; however, additional studies are needed to replicate these results, and several alternative hypotheses should also to be explored to allow more certitude about mate-choice copying in mollies. For example, females may have preferred a larger shoal or may have been avoiding a lone male. For example, sexually unreceptive females will prefer groups of conspecifics and avoid lone males. Presumably, this strategy dilutes what would otherwise be continuous sexual harassment by the male (Schlupp et al. 2001).

20.4 Audience effects

In the eavesdropping scenario, a bystander alters its own behavior in response to information that it has acquired from watching the social behavior of others. In essence, information flows from social interactants to a bystander. However, the presence of a bystander may affect the very interactions it observes. This phenomenon is commonly referred to as "audience effects" (McGregor & Peake 2000; Doutrelant et al. 2001; Dzieweczynski et al. 2005). Audience effects represent the strategic balance to social eavesdropping and conceptually "closes the circuit" in animal communication networks. Recent findings through work with poeciliid systems are providing important insights

into how individuals respond to audiences within the context of mate choice.

Using Atlantic mollies (*P. mexicana*), Martin Plath and his colleagues (Plath et al. 2008a, 2009) examined whether the presence of an unfamiliar same-sex audience altered affiliative behavior in focal males and females under controlled conditions. Groups of males or females were first given a standard mate-choice test (two stimulus males and no audience). Next, the same scenario was run again but with the addition of a conspecific audience, no audience, or a heterospecific audience (*X. hellerii* for males and *P. formosa* for females) enclosed in a transparent acrylic cylinder within the arena. Results showed that the presence of an audience affected focal males and females in different ways. Overall, audience effects were related to a reduction in the relative amount of time subjects spent with the previously preferred stimulus fish. However, the impact of an audience on the behavior of focal males was significantly stronger than that observed for focal females. Moreover, focal males were sensitive to whether the audience was a conspecific or a heterospecific. A swordtail audience produced a weaker response than when the audience was another Atlantic molly. Further analyses suggested that females' reduction in time spent with the male that was previously preferred did not appear to be due to audience effects per se but rather to time-budget trade-offs associated with a focal female's interaction with the audience. Changes in the behavior of focal females were not statistically different between treatments. Such sex-based differences in response to social cues represent a treasure chest of new research opportunities for undergraduate and graduate students alike.

In another study with the northern swordtail (*Xiphophorus birchmanni*), Fisher and Rosenthal (2007) tested the effect of an audience on the courtship behavior of focal males. The authors had previously observed that males employ a display in which the dorsal fin is fully raised—and do so in the contexts of both courtship and aggressive encounters with other males. In a series of video playback tests, they discovered that females typically preferred males with reduced displays (by virtue of a smaller fin) and that males tended to direct more aggression toward males of this type as well. In other words short-finned males appeared to have lowered competitive ability. Next, a scenario was staged in which the behavior of focal males that had been paired with females was examined in the presence of a male, a female, or no audience placed within small netcages within the arena. Focal males engaged in the erect dorsal-fin display more frequently with a male audience than with a female or no bystander. These results suggest that fin raising during courtship is a signal directed toward nearby males rather than toward the courted female. Since females find such males less attractive, the authors note that the only way such a system could be evolutionarily stable is if males incur a net fitness cost by not using the signal during courtship. One factor driving such signal evolution may be consistently high population densities, which place reproductive events firmly within the broader social milieu. Because males must court in the presence of rival males, the net effect is a balance between mate choice for diminished fin displays and intrasexual selection for its use in that context because of its role as an honest signal of RHP to nearby rivals.

Taken together, audience effects and social eavesdropping are providing powerful tools to examine animal communication networks in the contexts of sexual selection and intrasexual aggression. At last the stage is fully set and we may begin to examine how individuals and groups respond to socially generated information as it is gathered and given across a range of fitness-determining contexts associated with fitness outcomes.

20.5 Summary and discussion

Poeciliids have played a major part in our efforts to understand questions related to aggression, mate choice, and the role of social information in each. We have seen that forms of eavesdropping occur among a number of poeciliid species. In addition, eavesdropping has apparent fitness benefits in both intrasexual competition and mate-choice interactions. We have also discovered that social information can dramatically influence genetically based behaviors for relatively long durations. Nevertheless, there is ample opportunity for additional variations on these themes.

For example, swordtail bystanders were tested against observed winners and losers immediately following a fight. Testing while the information is very recently acquired is different from finding that the effect persists for hours or days (as in the work with mollies). This is an important issue because the rate at which information decays suggests the temporal scale at which it has the highest survival value. Presumably, if the intervening period is moderate—a few minutes to a few hours—an eavesdropper's response would be consistent with findings from other investigations of social-information use in swordtails. If information decays more rapidly, it would suggest that information relevant to assessment of a potential opponent's RHP remains valuable only if the eavesdropper perceives a fight as imminent.

Another question for future work is whether observing additional fights interferes with information acquired about an individual through eavesdropping. For example, imagine that an eavesdropper observed a fight between contestants A and B. At some future time, the eavesdropper

must fight the winner of that match. During the intervening period, the eavesdropper does not engage in any fights but happens to observe fights between males C and D, E and F, etc. The question is, when the eavesdropper finally fights A or B, does the original information that it learned about A and B remain retrievable, unchanged, and relevant? In practice, an eavesdropper could be allowed to observe several fights and then tested against winners or losers from fights at different points in time. Questions such as these will also be better answered through the inclusion of studies on the mechanisms underlying cognitive processes related to numerical abilities (see Bisazza, **chapter 15**).

Of immediate importance is the need to document real fitness effects of social-information use. This approach would involve measuring correlations between instances of social eavesdropping and variables such as lifetime reproductive success and survival. Such experiments would help to confirm the assumptions about the survival value of social eavesdropping and would tell us the extent to which social information determines variance in mating success over time.

20.6 Future directions: behavioral identities and information use

Understanding diversity in nature is one of the primary goals of evolutionary biologists. Since Darwin, there has been a historical progression of examining the consequences of natural selection at increasingly finer scales of biological organization (Wilson 1998). The evolutionary approach has been remarkably successful at revealing the effects of natural and sexual selection on shaping adaptations and readily explains much of the variation between species, subspecies, and populations. Yet, the finest scale of behavioral variation—within and between individuals—remains largely unexamined (Clark & Ehlinger 1987; Sih et al. 2004). The critical observation that suggests an evolutionary approach in this instance is that individuals routinely exhibit consistent behavioral responses across time and ostensibly unrelated contexts (Sih & Bell 2004). In other words, individuals display behavioral strategies—personalities that distinguish them from other members of their group.

For example, one individual might exhibit behavioral inhibition (fear) in the contexts of antipredator behavior and sociability, while another might consistently exhibit boldness in the exact same settings. In addition, different ecological situations or developmental phases can impose conflicting adaptive pressures (Stamps 2007). As a case in point, in guppies enhanced risk-taking behavior may increase fitness, as when competing for scarce forage, but it may reduce fitness when a dangerous predator is encountered. Or, a juvenile male guppy may benefit from reduced risk-taking behavior because that strategy helps it to avoid dangerous situations and increases the odds of surviving to sexual maturity, but timidity may reduce fitness in the same individual during adulthood because boldness is a quality important in mate attraction (Godin & Dugatkin 1996).

Animal personalities may account for variation in the tendency to use social eavesdropping and may also affect the ways it is evaluated. As we have seen, female guppies are not all equally predisposed to copy the mate choice of other females. In addition, none of the studies we have highlighted has reported that every individual responded equally to social information. This kind of variation is traditionally overlooked by most researchers as simply nonadaptive "noise" around an adaptive mean (Wilson 1998). However, several recent reviews suggest that such variation reflects responses to the selection pressures (Reale et al. 2000; Bouchard & Loehlin 2001; Gosling 2001; Dingemanse & Reale 2005; Koolhaas et al. 2007).

One way to examine how communication networks interact with animal personality traits is to test whether individuals that consistently differ with respect to, for instance, sociability are more or less easily influenced by socially generated information. For example, less sociable individuals (e.g., those that prefer smaller groups, interact less often with others) would probably *view* fewer social interactions than more sociable group members. From this, we might ask whether the tendency to be in proximity to social interactions translates into fitness consequences beyond the traditionally studied advantages to sociality (foraging efficiency, dilution effect). To test this idea within the mate-choice paradigm, a number of females could be scored on a scale of sociability. Next, they could be given a standard mate-choice test between two matched males. Finally, females with high and low sociability scores could be given an opportunity to copy the choice of a model female. If females lower on the sociability scale were less likely to copy

than more social females, it would suggest that sociability is positively correlated with the extent to which social cues influence the mating decisions of individual females. From here, one could examine how different proportions of such individuals within a population could affect the way information spreads throughout it.

Consistent patterns of individual behavior affect cultural transmission dynamics because the rate and penetration of information are a function of the number of individuals that will integrate it, as well as their relationship with the rest of that population. If we continue with our sociability and mate-choice copying example, we might predict that a population with a large proportion of highly sociable individuals (more likely to have the opportunity to copy) presents a more fluid medium through which a single copying event might spread than a population with a high proportion of less sociable individuals. In highly social groups, information and its transmission could dramatically decrease the variance in male mating success. Conversely, in groups with less sociable individuals, the influence of a copying event may not spread as readily, thereby potentially increasing variation in male mating success.

Throughout this chapter we have tried to highlight some of the important ways that poeciliids are contributing to our understanding of how communication networks arise from behavioral strategies that sensitize individuals to socially generated information. Although still in its infancy, this work has already expanded how we view the adaptive landscape of group-living animals and provides a strong basis for continued exploration of the role that the social environment plays in the evolution of sexual and aggressive behavior. The practicality of their use in the lab and field, combined with the impressive body of literature that has accumulated around them, will ensure the status of poeciliids as productive animal models for social-information use and sexual selection for decades to come.

Chapter 21 Genital evolution

R. Brian Langerhans

21.1 Introduction

GENITAL MORPHOLOGY, particularly in males, is strikingly variable in animals with internal fertilization (Eberhard 1985; Edwards 1993; Sirot 2003; Hosken & Stockley 2004; Evans & Meisner 2009; Eberhard 2010). Indeed, genitalia may experience more rapid, divergent evolution than any other animal character—but why? Several hypotheses have been proposed to explain this remarkable degree of variation (table 21.1), with those involving postmating sexual selection currently enjoying the strongest empirical support (mostly from insects and spiders). Here, I review and synthesize existing data to assess the possible importance of each hypothesis for genital evolution in poeciliid fishes. Each hypothesis proposes an important role for a distinct process in genital evolution; the hypotheses are conceptually distinct but not mutually exclusive.

Poeciliids display conspicuous variability in genital morphology, with gonopodium diversity being particularly well studied (fig. 21.1; see also fig. 1.5 in Greven, **chapter 1**). As with many internal fertilizers, male genital characters are critically important for distinguishing among close relatives (Eigenmann 1907; Regan 1913; Henn 1916; Rosen & Bailey 1963). Owing to the dramatic diversity of genital morphology, range of mating strategies employed, breadth of habitats occupied, ease of laboratory experimentation, and the existence of some fairly well-resolved phylogenies, poeciliid fishes represent a model system for studies of genital evolution. So far, we have only scratched the surface—much future work is needed to gain a strong understanding of the causes and consequences of genital diversification in poeciliid fishes.

This chapter assumes a basic familiarity with poeciliid reproductive biology (see Greven, **chapter 1**, for details) and focuses on the mechanisms potentially responsible for genital evolution. I particularly focus on male gonopodium morphology and female urogenital aperture (and surrounding integument) morphology. However, nonmorphological traits associated with genitalia, such as sperm properties (mobility, chemistry), accessory fluids, or pheromones might exhibit similar levels of variability and experience similar forms of selection (Aspbury & Gabor 2004b; Mendez & Cordoba-Aguilar 2004; see also Evans and Pilastro, **chapter 18**). In this chapter, I inevitably focus on genitalic structures that possess the most available data.

Many previous tests of hypotheses of genital evolution in poeciliids either lacked statistical analyses, used small sample sizes, ignored phylogenetic relationships, or used indirect evidence rather than direct observations for basing functional conclusions. Thus, I conduct a number of new analyses here to provide at least crude tests of both old and new hypotheses. These tests (presented in online appendices; see online supplementary material [OSM] at http://www.press.uchicago.edu/books/Evans) were all performed within a phylogenetic context and are meant to guide future work. Family-wide analyses used the molecular phylogeny of Hrbek et al. (2007). Because resolution of much of these data is currently available only at the generic level, data were collected and analyzed at this scale. The phylogeny was pruned to yield a genus-level topology (28 genera) with all branch lengths equal to 1 (see appendix 21.1,

Table 21.1 Primary hypotheses for the evolution of genital diversity

Category of explanation	Hypothesis
Premating sexual selection	Male contest competition: intermale competition over access to mates
	Mate choice: selection of mates based on phenotypic values
	Premating sexual conflict: sexually antagonistic selection over control of mating
Postmating sexual selection	Sperm competition: competition between sperm of different males over fertilization
	Cryptic female choice: postmating ability of females to bias fertilization success among males
	Postmating sexual conflict: sexually antagonistic selection over control of fertilization
Natural selection	Lock-and-key: selection against hybridization favors species-specific complementarity in male (key) and female (lock) genitalia
	Nonmating natural selection: selection on genitalia via agents independent of mating
Neutrality	Pleiotropy: genitalia not under direct selection but genetically correlated with such traits

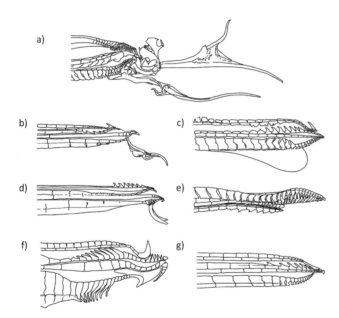

Figure 21.1 Lateral depiction of gonopodial distal tips at rest, with the fish facing left (i.e., tip pointing caudally). (a) *Tomeurus gracilis*, (b) *Cnesterodon decemmaculatus* (ten-spotted livebearer), (c) *Poecilia elegans* (elegant molly), (d) *Girardinus cubensis* (Cuban topminnow), (e) *Carlhubbsia kidderi* (Champoton gambusia; note the asymmetry), (f) *Xiphophorus hellerii* (green swordtail), (g) *Priapichthys nigroventralis*. Adapted from Rosen and Tucker 1961 and Rosen and Bailey 1963.

genetically independent contrasts (PIC; Felsenstein 1985), with correlations forced through the origin. Character data sets are provided in appendices 21.3 and 21.4 (OSM). In all cases where a priori one-sided hypotheses exist, I use one-tailed *P*-values.

21.2 Basics of poeciliid copulation

All poeciliids possess internal fertilization, whereby males use their gonopodium to transfer spermatozeugmata onto or into female genitalia. Prior interpretations of how this process occurs were largely based on deductions from genital morphology, actions of gonopodia forcibly manipulated on preserved or anesthetized fish, and photographic glimpses of extremely rapid, complex copulatory behaviors of a few species (e.g., Rosen & Gordon 1953; Warburton et al. 1957; Rosen & Tucker 1961; Peden 1975). Put simply, we have a very incomplete functional understanding of how spermatozeugmata are released and travel along the gonopodium, how sperm reach the female reproductive tract, and how accessory structures (e.g., pectoral and pelvic fins), gonopodium size (e.g., length, surface area), and gonopodial distal-tip morphology influence this process. To accurately and thoroughly assess alternative hypotheses for genital diversification, we would ideally begin with a strong functional knowledge of the mechanics of copulation—thus, future work on this topic is paramount. In the meantime, I will briefly evaluate what we do know so that we may use this knowledge in our appraisals of the hypotheses discussed throughout the chapter.

During a successful copulation (i.e., one in which insemination occurs), a male locates a female, circumducts the gonopodium, and transfers spermatozeugmata to the female genitalia. It has been suggested that spermatozeug-

OSM). For analyses performed within the genus *Gambusia*, I generated a topological hypothesis based on molecular and morphological data and set all branch lengths equal to 1 (appendix 21.2, OSM). When testing for correlated evolution between two discrete binary traits, I used Pagel's (1994) Discrete method using maximum likelihood (Pagel 2000). To test for correlated evolution between continuous traits and between a continuous trait and a binary trait, I used the PDAP module of the Mesquite package (Maddison & Maddison 2008; Midford et al. 2008) to examine phylo-

mata travel along the gonopodium by unknown means (e.g., cilia, centrifugal force) through either a permanent groove in species with bilaterally asymmetric gonopodia or a transitory groove formed by the folding of anal-fin rays during circumduction in species with bilaterally symmetric gonopodia. This suggestion is primarily based on deductions from morphology of asymmetric gonopodia (Rosen & Bailey 1959; Chambers 1987) and direct manipulations of anesthetized specimens of *Gambusia affinis* (western mosquitofish; Kuntz 1914) and *Xiphophorus hellerii* (green swordtail; Rosen & Gordon 1953). However, recent work has demonstrated that the groove is actually permanent, not temporary, in at least two genera exhibiting bilaterally symmetric gonopodia—*Gambusia* (Rivera-Rivera et al. 2010) and *Belonesox* (R. B. Langerhans, unpublished data)—calling into question the accuracy and generality of these previous descriptions. Prior work suggested that pelvic fins sometimes aid in guiding spermatozeugmata along the gonopodium (Rosen & Gordon 1953; Rosen & Tucker 1961), but detailed descriptions of direct observations are lacking. In *Gambusia*, pectoral fins have long been thought to support the gonopodium during copulation (Hubbs & Reynolds 1957; Warburton et al. 1957; Peden 1972b, 1975; Rosa-Molinar et al. 1994; Rosa-Molinar et al. 1996). Yet, recent work produced what is probably the most detailed observations of copulatory behaviors ever reported for poeciliid fish—simultaneous ventral, lateral, and frontal digital recordings of copulation attempts (1000 Hz, 1024- × 1024-pixel resolution)—and failed to confirm this purported behavior in *G. affinis* (Rivera-Rivera et al. 2010). Finally, functions of the distal-tip elements of gonopodia (e.g., hooks, spines, serrae) are less than obvious (Rosen & Gordon 1953; Clark et al. 1954). Clearly, we need a more detailed knowledge of the biomechanics of poeciliid copulation.

21.3 Sexual selection

To begin evaluating the possible mechanisms underlying genital diversity in Poeciliidae, let us first consider sexual selection. The mating process in poeciliids comprises mate acquisition, copulation, and fertilization—sexual selection can act on genital morphology at any of these stages. First, conspecifics of the opposite sex must locate one another: premating sexual selection on genitalia can occur at this stage (sections 21.3.1–3). Once copulation is initiated, insemination of sperm and fertilization of ova must occur if embryos are to result from mating: postmating sexual selection on genitalia can occur at this stage (sections 21.3.4–6; see also Evans & Pilastro, **chapter 18**).

21.3.1 Male contest competition

Although male poeciliids exhibit considerable variation in the degree of male-male agonistic interactions (Farr 1989; Bisazza 1993a; Earley & Dugatkin 2005; see also Rios-Cardenas & Morris, **chapter 17**), the role of male contest competition in genital evolution is virtually unexplored. Fighting among males for mating rights appears common in several poeciliid species and might influence gonopodium evolution in two ways: (1) gonopodia functioning as male-combat weapons, and (2) gonopodia serving as badges of status. Since gonopodia reach very large sizes in some species (e.g., >70% of standard length in *Cnesterodon*) and sometimes exhibit conspicuous pigmentation, and since males of some species exhibit gonopodial swinging during male-male interactions, it seems that the opportunity for male contest competition to drive gonopodial evolution exists—partially analogous to the use of swords (elaborate, ventral-ray elongations of the caudal fin) in male-male competition in *X. hellerii* (Benson & Basolo 2006; Prenter et al. 2008).

If males used gonopodia either as weapons or as honest status signals, we might expect larger or more colorful gonopodia in species with a greater intensity of intermale competition. Yet this prediction would be naive, as it assumes that costs associated with gonopodium size and color are similar across species, which could obviously be violated if costs in aggressive species were higher than in nonaggressive species (i.e., traits can have costs of both somatic growth and aggression toward the trait bearer). Furthermore, both scenarios (i.e., weapons, badges of status) might result in strong, directional selection favoring larger gonopodia, which should generate hyperallometry of gonopodia (Green 1992), assuming that selection from other sources, like the braking effects of natural selection (section 21.4.2), is negligible (Bonduriansky & Day 2003; Bertin & Fairbairn 2007; Bonduriansky 2007). Gonopodial allometry has been examined in only a few species (Kelly et al. 2000; Jennions & Kelly 2002). Nevertheless, other mechanisms (e.g., female preference, sexually antagonistic selection) can also produce positive allometry of gonopodia (Bonduriansky 2007; Eberhard 2009). Thus, the importance of male contest competition will unlikely be deduced from comparative trends or allometric patterns but instead necessitates experimental approaches. No study has yet demonstrated the use of gonopodia as either male-combat weapons or badges of status.

Male contest competition could also influence the evolution of female genitalia. First, this could occur as a simple by-product of accommodating gonopodium modification resulting from male contest competition (e.g., enlarged gonopodia might favor enlarged urogenital apertures). Alternatively, a female might advertise her fertility to intensify inter-male competition and thus enhance her probability of mating with a more aggressive, socially dominant male (Farr 1989). I will describe here how this might be accomplished via genital evolution. Females often produce chemicals that stimulate male mating behaviors (Constantz 1989; see Greven, **chapter 1**). These chemicals likely derive from the urogenital aperture and could conceivably intensify male contest competition, although such a possibility has not been tested. Additionally, some female poeciliids (especially *Gambusia* species) exhibit contrasting coloration of the genital region (e.g., yellow spots in *G. affinis*) and/or conspicuous anal spots (darkened pigmentation of the urogenital aperture or nearby integument). In some species, these spots exhibit cyclic expression, intensifying during ovulation, and have been suggested to influence male mating behavior (Peden 1973; Kodama et al. 2008). If male contest competition has been important in female genital evolution, we might expect to find genital modifications for fertility advertisement more prevalent in species where male mating success is greatly influenced by male contest competition. This test has not yet been conducted.

21.3.2 Mate choice

Mating preferences based on visual and olfactory cues are known to occur in both male and female poeciliids (see Rios-Cardenas & Morris, **chapter 17**). If genitalia provide mating cues during precopulatory behaviors, mating preferences could influence the evolution of genital diversity. The male gonopodium is often relatively conspicuous, exhibits wide variability among species in size and color, and is sometimes extended or abducted during mating displays (Rosen & Gordon 1953; Rosen & Tucker 1961; Hughes 1985; Basolo 1995b; Langerhans et al. 2005). For these reasons, it seems plausible that female mating preferences might influence gonopodial diversity. Female preference for males with larger gonopodia has been demonstrated using an experimental approach in *G. affinis* (Langerhans et al. 2005) and *Gambusia holbrooki* (Kahn et al. 2010) and with correlational observations in *Poecilia reticulata* (guppy; gonopodium length; Brooks & Caithness 1995). Tests of female preferences for gonopodium size in other species, or for gonopodium color in any species, do not yet exist to my knowledge. If female preferences for larger, more colorful gonopodia have played a major role in gonopodial evolution not just within species but across species, we might predict that species with mating displays would exhibit larger, more colorful gonopodia. For gonopodium size, previous work and new analyses suggest that the opposite pattern actually exists (see section 21.3.3). For gonopodium color, new results cannot rule out the role of female mate choice in influencing its evolution (appendix 21.5, OSM).

Female genitalia are often relatively inconspicuous and therefore at first glance would seem ill-suited for evolution via male mate choice. Nevertheless, there are a couple of possibilities that deserve further attention. First, males might exhibit preferences for particular chemical cues associated with the urogenital aperture, possibly driving the evolution of pheromone diversity (e.g., McLennan & Ryan 1999; Hankison & Morris 2003; Shohet & Watt 2004; Plath et al. 2006). Second, the aforementioned anal spots possessed by some female poeciliids vary considerably among species in size, intensity, location, and shape (Peden 1973). It is possible that male mate choice could influence the evolution of anal-spot diversity, but no study has yet examined this. In sum, the importance of mate choice in the evolution of genital diversity appears variable across hierarchical levels for male gonopodium size, has some cautious support for gonopodium color, and is unclear for females.

21.3.3 Premating sexual conflict

Sexual conflict over control of mating and fertilization is a common phenomenon and can result in numerous forms of sexual selection (Arnqvist & Rowe 2005). The premating sexual conflict hypothesis of genital evolution is explicitly concerned with the role of sexually antagonistic selection on traits that function prior to copulation. Under this scenario, trait values that are beneficial to one sex impose direct costs on the other sex. Sexually antagonistic selection could be due to the fact that males typically benefit from increasing mating frequency, while females often benefit from rejecting some males, which reduces costs associated with superfluous male mating attempts. Female poeciliids are likely to suffer considerable costs due to harassment by males, including lost feeding opportunities, exposure to predators, and physical injury from aggressive interactions (Plath et al. 2007b; see Magurran, **chapter 19**). Because sexually antagonistic selection might involve nongenitalic traits, coevolution of male and female genitalia is not necessarily predicted under this scenario. However, genital morphology certainly might evolve as a response to selection on either side of this conflict of interests between the sexes.

Males might circumvent female mating preference/resistance by evolving a longer gonopodium to achieve mating attempts from longer distances or otherwise enhance mating success against the female's wishes. Whether longer gonopodia function to facilitate mating from longer distances is unknown, but prior work based on indirect evidence has suggested that a longer gonopodium somehow increases mating success during gonopodial thrusting (i.e., copulation attempts not preceded by any form of display) (Rosen & Gordon 1953; Rosen & Tucker 1961; Reynolds et al. 1993; Greven 2005). If longer gonopodia effectively prevent female choice, then we would predict that species with longer gonopodia would not exhibit mating displays. While Ptacek and Travis (1998) found the opposite trend across three *Poecilia* species, a recent analysis conducted within a phylogenetic context across 65 species found confirmatory evidence (Martin et al. 2010). Indeed, examining this question across genera in a phylogenetic context yields confirmatory results, as shorter gonopodia tend to evolve in concert with male display (appendix 21.6, OSM). Although this work grossly categorizes taxa as having either short or long gonopodia, Martin et al. (2010) demonstrated that poeciliids appear to exhibit a bimodal distribution of relative gonopodium length (fig. 21.2), indicating that such a categorization will likely capture the major trends at this scale of analysis. Furthermore, intraspecific studies sometimes report that males with relatively longer gonopodia exhibit higher rates of gonopodial thrusting (Farr et al. 1986; Reynolds et al. 1993; Travis 1994; but see Schröder et al. 1996; Ptacek & Travis 1998), and that males in environments characterized by higher predation risk, where female choice is expected to be less important than in low-risk environments, tend to exhibit longer gonopodia (Kelly et al. 2000; Jennions & Kelly 2002; but see Cheng 2004; Langerhans et al. 2005). While these findings alone do not reveal causation (we need functional approaches that directly address the hypotheses), they do provide supporting evidence for a possible role of premating sexual conflict in the evolution of gonopodium size.

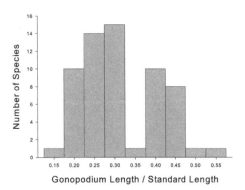

Figure 21.2 Gonopodium lengths in poeciliids. Frequency distribution of poeciliid fish species (*n* = 61) sharing common relative gonopodial lengths. Adapted from Martin et al. 2010.

21.3.4 Sperm competition

Because poeciliid females often mate with multiple males and can store sperm, postmating sexual selection plays an important role in poeciliid evolution, with sperm competition receiving much attention in the group (see Evans & Pilastro, **chapter 18**). However, the role of sperm competition in genital diversification is largely unknown. With respect to genital evolution, sperm competition will generally favor male genital traits that (1) increase insemination success (e.g., number of spermatozeugmata delivered per copulation), (2) increase postinsemination fertilization success, or (3) decrease insemination or postinsemination fertilization success of other males. In poeciliids, male gonopodial morphology may be a target for such selection. Although little evidence so far exists regarding the influence of variation in gonopodial morphology on insemination or postinsemination fertilization success (e.g., Clark & Aronson 1951; Kadow 1954), there are many reasons to believe this is a promising arena for future investigation.

Two obvious ways that males might enhance insemination or postinsemination fertilization success are by exhibiting gonopodial traits that increase the duration of copulation (which is known to increase the number of sperm inseminated in guppies; Pilastro et al. 2007) or that place spermatozeugmata in more favorable locations on or in female genitalia. Four general male reproductive traits that appear to exhibit wide diversity among poeciliids provide intuitive candidates for such traits: gonopodium size (e.g., length, surface area), gonopodial armament (e.g., hooks, serrae, acuteness of tip), accessory mating structures (e.g., gonopodial hood/palp, gonopodial bony extensions, modified paired fins), and the type of gonopodial groove (i.e., permanent asymmetrical folding, permanent dorsal groove, transient folding) (see Greven, **chapter 1**, for details on these structures). It is possible that males with longer gonopodia are capable of achieving deeper penetration or can better position the gonopodium during copulation with the aid of visual cues. Some types of gonopodial armament might function as holdfasts during copulation (Clark & Aronson 1951; Rosen & Gordon 1953; Clark et al. 1954; Cheng 2004) and subsequently increase the duration of copulation. Armament might also facilitate the release of spermatozeugmata in more favorable locations, such as providing deeper penetration or altering trajectories of spermatozeugmata release. Accessory structures might help guide and stabilize the gonopodium during copulation in order to place the gonopodial tip in a position that enhances insemination or fertilization success (e.g., palps might limit depth of insertion; Rosen & Bailey 1963; Greven 2005) or perhaps reduce the loss of spermatozeugmata after copulation (Leo

& Greven 1999). Variation in the gonopodial groove might influence the efficiency of spermatozeugmata transfer from the urogenital pore to the gonopodial tip. No direct tests of any of these hypotheses have yet been performed.

Seven genera of poeciliids exhibit bilaterally asymmetric gonopodia (see appendix 21.3, OSM; and see also fig. 21.1e; Greven, **chapter 1**, fig. 1.5), which possess a permanent groove in either the dextral or the sinistral position. A permanent groove might enhance insemination success relative to a temporary one, but asymmetry is not a prerequisite for permanent groove formation, and it is unclear what selective effects related to insemination/fertilization success bilateral symmetry per se might confer. Previous studies have discussed an interesting pattern in which all asymmetric gonopodia are also relatively long (>35% of standard length; Rosen & Tucker 1961; Rosen & Bailey 1963; Greven 2005), suggesting the correlated evolution of gonopodium symmetry and length due to a functional integration of the two traits where certain combinations improve insemination/fertilization success. A test of this association within a phylogenetic context suggests that it does indeed represent a fairly robust pattern of correlated evolution, at least at the level of genera (appendix 21.7, OSM), although whether this pattern is the result of sperm competition is unknown. Moreover, it is not clear why symmetric gonopodia evolved in the first place. Based on ancestral-state reconstruction using maximum parsimony, it now appears that gonopodial asymmetry was likely the ancestral condition of the family—both the sister lineage to all other poeciliids (*Xenodexia*) and the family Anablepidae contain species with asymmetric gonopodia—and was lost early in poeciliid evolution but regained four or five times. Perhaps gonopodial symmetry evolved in concert with holdfast devices, as the two might be functionally integrated—that is, selection for holdfasts might lead to bilateral symmetry since they might be more effective when symmetric than when asymmetric. Across poeciliid genera, the evolution of symmetry is indeed significantly associated with the evolution of potential holdfast devices (appendix 21.7, OSM). Thus, it is plausible that symmetric gonopodia may sometimes evolve as a means of enhancing the effect of holdfasts, perhaps in response to selection via sperm competition.

Gonopodial morphology might also serve to decrease the insemination or fertilization success of rival males. One previously proposed function of larger gonopodia and increased armament is to cause injuries to female genitalia that tend to keep females chaste (Constantz 1984). Copulations in some species are known to at least occasionally injure females (Clark et al. 1954; Peters & Mäder 1964; Constantz 1984; Horth 2003), and the tearing and subsequent swelling of the urogenital sinus could reduce insemination or fertilization success of later matings. This hypothesis can be tested easily by examining the effects of gonopodial morphology on injuries and the effects of injuries on the success of subsequent matings, but no such tests have been performed. An alternative means of decreasing the fertilization success of rival males is to remove sperm from the urogenital sinus or reproductive tract of females. Perhaps some hooks, spines, or gonopodial extensions (palps, bony processes) sometimes serve this role. Again, the ability of such traits to remove sperm from females has not yet been tested.

Previous studies suggest that some of these male reproductive traits evolve in a correlated fashion, with one group of species exhibiting relatively long gonopodia that have few holdfasts and that are guided visually during copulation without the aid of accessory structures, and another group of species exhibiting short gonopodia that have numerous holdfasts and that are aided by several accessory structures (Rosen & Tucker 1961; Rosen & Bailey 1963; Greven 2005). Sperm competition could have driven this correlated evolution to enhance postmating insemination/fertilization success via functionally integrated traits. New analyses performed across poeciliid genera within a phylogenetic context were only moderately consistent with these previous claims (appendix 21.8, OSM). First, it seems that once adjusted for phylogenetic relatedness, gonopodium length and holdfasts do not exhibit as tight an association as previously suggested. Second, while a suggestive, positive relationship between holdfasts and accessory structures was found, the pattern is not indicative of a major evolutionary association, and holdfasts are also known to exhibit a modest level of variation within several genera. In contrast, it is quite clear that gonopodium length and accessory structures have indeed evolved in a correlated manner among poeciliid genera. This suggests that functional explanations by previous researchers may be correct, in that longer gonopodia that reach to or beyond the eye are effectively guided by visual cues, while shorter gonopodia that cannot be seen by the bearer are generally aided in positioning by accessory structures (Rosen & Tucker 1961; Rosen & Bailey 1963; Chambers 1987). Such correlated evolution suggests that these patterns have resulted from selection via sperm competition as a means of enhancing insemination and fertilization success or perhaps via premating sexual conflict to increase mating frequency.

21.3.5 Cryptic female choice

After copulation is initiated, females might influence the probability of insemination or fertilization by a given male—and this bias in paternity might depend on the male's genital morphology. This female-driven influence on male reproductive success may derive from a number of sources, including female genital morphology or chemistry, and essentially represents a challenge to males to find the optimal means of stimulating a female. Under this scenario, females indirectly benefit from rejecting some males based on their genital morphology via increased offspring quality. Recent work suggests that cryptic female choice is important in poeciliid evolution (see Evans and Pilastro, **chapter 18**), although we have virtually no knowledge so far of its role in genital evolution.

Cryptic female choice has received considerable attention and support in the study of genital evolution in insects and spiders, where the hallmark of cryptic female choice is the evolution of highly unusual male genital morphologies (Eberhard 1985, 1996). The distal tip of poeciliid gonopodia would seem to meet the criterion of peculiarity, as some distal-tip elements are so bizarre (see figs. 21.1 and 21.3) that it would be surprising if cryptic female choice did not play some role in their evolution. To date, however, no direct evidence exists in support of this hypothesis of genital evolution, and well-designed experiments are needed to assess its potential importance.

21.3.6 Postmating sexual conflict

Poeciliid females might suffer a direct cost from copulation, such as copulatory injuries (see section 21.3.4). Thus, selection might favor traits that allow females to gain control of insemination or fertilization, reducing such costs. The postmating sexual conflict hypothesis of genital evolution focuses on sexually antagonistic selection resulting from direct costs to females after the initiation of copulation (rather than indirect costs of offspring quality, as in cryptic female choice).

The postmating sexual conflict hypothesis of genital evolution makes three predictions: (1) the male genital traits that enhance male fitness reduce female fitness, causing females to directly benefit from rejecting some conspecific males by reducing direct costs of unwanted inseminations, (2) female genital morphology directly influences the insemination or fertilization success of males, sometimes conferring the ability to completely exclude some conspecific males, and (3) a tight coevolutionary arms race of male and female genitalia occurs. In support of the first prediction, copulatory injuries have been shown to occur (see section 21.3.4), suggesting that females might directly benefit from rejecting at least some males for this reason. However, we do not know injury frequencies within species, their variability among species, or the extent of their possible negative consequences. Moreover, no studies have yet examined whether such copulatory injuries to females represent male adaptations (direct selection favoring injuries; "adaptive harm") or negative side effects of traits that evolved because of other selective advantages to males (selection favoring other performance attributes incidentally resulting in injuries; "collateral harm") (Lessells 2006). Here, I assume that substantial costs may be incurred from copulation, and I evaluate the possible ways that postmating sexual conflict might drive genital diversification; clearly, this assumption should be tested in the future. Related to the second prediction, some species exhibit urogenital apertures and sinuses that appear to contain defensive structures (Constantz 1984), although no tests have yet been conducted to determine their function. Finally, previous work has suggested that coevolution of male and female genital morphology may occur in some poeciliid groups (Peden 1972a; Constantz 1984), but detailed tests of how these structures are associated and their possible offensive and defensive functions have not yet been conducted. Extending the work of Peden (1972a), I have performed the first quantitative test of genital coevolution among the sexes in poeciliids (for results, see appendix 21.9, OSM).

What genital structures might we expect to evolve by postmating sexually antagonistic selection? We might envision a range of possible solutions for minimizing injury in females, such as placing obstructions in the urogenital aperture or sinus (i.e., genital papillae), covering the aperture with tissue, shifting the location of the sinus or reproductive tract, enlarging the sinus to minimize contact with tissue, and reinforcing the region with strengthened tissue to absorb gonopodial blows. Variation in all these traits exists in poeciliids, although whether they function as defensive structures is unknown (Peden 1972a; Constantz 1984; Greven 2005). It is also possible that these features provide a means of exerting cryptic female choice rather than, or in addition to, serving as defensive structures to reduce injury. For males, a host of possible offensive structures are known in poeciliids, such as large tips, highly acute tips, and large and numerous hooks, spines, and serrae. Yet, a greater diversity of solutions is possible. For instance, rather than attempt to break through a defensive structure, males might circumvent intromission and simply deposit spermatozeugmata on the exterior of the female genitalia, with sperm only later traveling into the reproductive tract (external depositors do occur, e.g., *Tomeurus*; Rosen & Tucker 1961). The occurrence of such a diversity of apparently offensive and defensive genital structures is certainly suggestive of sexual conflict, although functionally oriented studies are needed to elucidate how these structures work, and comparative studies are needed to test whether patterns of genital evolution are consistent with predictions from sexual conflict.

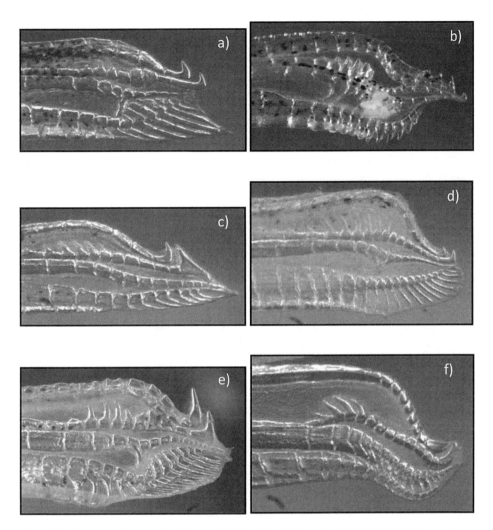

Figure 21.3 Representative examples of gonopodial tips of *Gambusia* species, illustrating the high degree of interspecific variation. (a) *G. atrora*, (b) *G. echeagarayi*, (c) *G. nicaraguensis*, (d) *G. panuco*, (e) *G. punctata*, and (f) *G. vittata*.

Gambusia males exhibit obvious, dramatic variation in the distal-tip morphology of the gonopodium (fig. 21.3), and Peden (1972a) demonstrated a correspondingly high degree of variation in female genitalia among *Gambusia* species, suggesting that male and female genital morphology coevolved in *Gambusia*. Constantz (1984) suggested that some female genital structures represent defensive traits resulting from sexually antagonistic selection. While the functions of these structures have not yet been examined, we can now address more thoroughly the question of whether a tight coevolutionary relationship truly exists among male and female genital morphology in *Gambusia*. Although coevolution of genitalia is a possible outcome of other processes, no other hypothesis of *sexual* selection explicitly predicts strong, genital coevolution.

I found striking evidence for coevolution of external genital morphology among the sexes in *Gambusia* (appendix 21.9, OSM; fig. 21.4). Thus, Peden's suggestion of

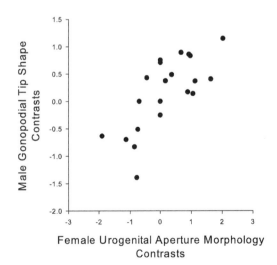

Figure 21.4 Coevolution of external genital morphology among the sexes in *Gambusia*. Each axis depicts phylogenetically independent contrasts.

coevolution is strongly supported after greatly expanding the sample size and placing the analysis within a quantitative, phylogenetic context. This suggests that male and female genitalia evolve in concert—when males evolve blunt/rounded gonopodial tips, females evolve reduced or absent genital papillae within small or enclosed apertures; whereas in species with males exhibiting sharply pointed tips, females tend to exhibit a large papilla within a large aperture. While consistent with the hypothesis of postmating sexual conflict, without corroborating evidence regarding function of the structures we cannot be sure of the underlying process. Moreover, such a pattern is also predicted by the lock-and-key hypothesis, which additionally has further support in this case (section 21.4.1).

21.4 Natural selection

During the past several decades, the role of natural selection has received far less attention and support than sexual selection in the evolution of genital diversity (Eberhard 1985; Arnqvist 1998; Hosken & Stockley 2004; House & Lewis 2007). However, a long-standing hypothesis of taxonomists is that genitalia evolve via natural selection against hybridization (lock-and-key), and previous work in poeciliids suggests it is a viable hypothesis in need of future work (see below). Moreover, unlike genitalia in taxa traditionally studied in this research arena, gonopodia are external, nonretractable, and sometimes quite large relative to body size. Although genitalia are often viewed as doubtful targets of natural selection (Eberhard 1985; 1993; Andersson 1994), gonopodia might often be subjected to various forms of natural selection, such as through their effects on locomotion. While little research has investigated these questions in poeciliids so far, I demonstrate here that numerous lines of evidence suggest that natural selection likely plays an important role in genital evolution of poeciliids.

21.4.1 The lock-and-key hypothesis

The lock-and-key hypothesis has received little convincing support despite a long history of investigation (Shapiro & Porter 1989; Arnqvist 1998). It describes the scenario where selection favoring hybridization avoidance leads to patterns of morphological (or chemical) complementarity of genitalia among the sexes. This hypothesis makes several predictions: (1) female genitalia should reduce mating, insemination, or fertilization success of sympatric, heterospecific males compared with conspecific males, (2) there should be tight coevolution of male and female

genitalia, and (3) reproductive character displacement in genitalia should occur, where genital differences are greater between sympatric populations/species than between allopatric populations/species. Although strong coevolution of genitalia between the sexes might also result from other processes (see above), the other two predictions are unique to the lock-and-key hypothesis. Although it seems possible for genitalic traits to experience such selection in poeciliids, none of these predictions have previously received significant attention. Experimental work could easily test the first prediction, but no such work has yet been conducted. The second prediction now enjoys strong support in *Gambusia* (see above), and some anecdotal evidence indicates that broader trends might exist in the family (see Greven, **chapter 1**, fig. 1.5). The third prediction has never been investigated in detail to my knowledge in poeciliid fishes, and thus in appendix 21.10 (OSM) I provide a first test in the genus *Gambusia*.

In poeciliids, early work suggested that differences in genital morphology provided a poor means of reducing hybridization among species (Sengün 1949; Clark et al. 1954; Liley 1966). However, Peden's (1972a, 1973, 1975) work with *Gambusia* species, which represents the most thorough set of comparative studies of male and female genitalia across poeciliid species to date, resurrected the viability of the lock-and-key hypothesis. Peden suggested that the observed covariance in copulatory behaviors, gonopodial morphology, and female genital morphology observed among species was indicative of their acting together in a lock-and-key fashion, producing "more efficient sperm transfer in conspecific than in heterospecific copulation" (1975, 1296). Although it is true that the lock-and-key hypothesis could have generated the apparent associations described by Peden, other mechanisms could also have generated such patterns (see above). Building from Peden's findings, in appendix 21.10 (OSM) I provide the strongest test to date of the lock-and-key hypothesis in poeciliids, testing for reproductive character displacement in *Gambusia* species.

Support for the lock-and-key hypothesis in *Gambusia* is strong for male gonopodial tip shape and female urogenital aperture morphology and is suggestive for female anal-spot location (fig. 21.5). While this pattern of reproductive character displacement in both male and female genitalia provides strong support for the lock-and-key hypothesis, it does not necessarily indicate the exact process by which the pattern emerged; the trend could reflect either evolutionary adjustments to minimize heterospecific insemination or community-level assortment of preexisting differences between species. Yet in either case, the same underlying mechanism is at work—selection against hybridization—merely

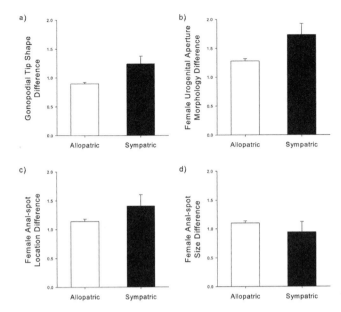

Figure 21.5 Reproductive character displacement in *Gambusia*. Divergence between allopatric and sympatric species pairs in (a) male gonopodial tip shape, (b) female urogenital aperture morphology, (c) female anal-spot location, and (d) female anal-spot size, controlling for phylogenetic relatedness. See appendix 21.10 (OSM) for methodology. Means ± 1 SE presented.

acting at different hierarchical levels. Thus, combining two lines of evidence from *Gambusia* presented in this chapter (tight genitalic coevolution and reproductive character displacement), the lock-and-key hypothesis appears to play a significant role in genital evolution in the genus and deserves future consideration in other poeciliids. Importantly, these findings do not exclude the role of postmating sexual conflict, as both processes might operate simultaneously (or at different times during the evolutionary history of the lineages).

21.4.2 Nonmating natural selection

Natural selection on genital form might also arise from selection independent of mating. There are three obvious sources from which such selection could occur: (1) costs of somatic growth, (2) conspicuousness of genitalia, and (3) effects of gonopodia on locomotor abilities. First, gonopodial development requires resources, and growing a large gonopodium requires more resources than growing a small one. It is unclear exactly how costly gonopodium growth might be relative to other sources of selection on gonopodium size, but it is doubtful that costs of somatic growth have a large influence on selection on female genitalia. Second, both male gonopodia and female genitalia (especially anal spots) are sometimes visually conspicuous features of poeciliid fish. Larger, more colorful, or otherwise more conspicuous genitalia (e.g., potent pheromones could increase

female conspicuousness) could draw the attention of predators or aggressive heterospecifics, similar to the known costs of increased attention by predators for poeciliids with bright coloration, larger size, large fins, or elaborate swords (e.g., Rosenthal et al. 2001; Basolo & Wagner 2004; Johansson et al. 2004; see also Kelley & Brown, **chapter 16**). This hypothesis seems plausible but has not yet been tested. Finally, gonopodium size and stability can affect locomotor performance, and thus any source of selection on locomotion might affect gonopodium size. Because strong selection on swimming abilities appears widespread in poeciliid fish (O'Steen et al. 2002; Ghalambor et al. 2004; Langerhans & DeWitt 2004; Langerhans et al. 2004; Walker et al. 2005; Hendry et al. 2006; Langerhans et al. 2007; Zúñiga-Vega et al. 2007; Langerhans & Reznick 2010), this source of selection seems like a particularly promising area for future research. It is also the only component of nonmating natural selection on gonopodia that has so far received much attention in poeciliids.

Gonopodia are expected to influence locomotion primarily through costs of drag, as it is unlikely they contribute much useful thrust. Poeciliids employ steady swimming (constant-speed cruising) for a variety of important activities, such as searching for food, courtship chases, male-male agonistic interactions, and seeking favorable abiotic conditions. Selection should generally favor various means of reducing the energetic costs of movement. We might thus expect natural selection to favor reduced gonopodium size (relative to body size) and increased gonopodial stability, particularly in environments where steady swimming is of paramount importance. This is because a gonopodium with a larger surface area should incur greater drag, and an unstable gonopodium that freely swings about during steady swimming should also incur greater drag than a sturdy gonopodium (Lighthill 1970; Beamish 1978; Webb & Gerstner 2000). To maintain stability and reduce surface area of the gonopodium, we might expect males to often press the gonopodium against their body during cruising. However, it might actually be possible for larger gonopodia to reduce energetic costs of steady swimming by delaying the separation of the boundary layer (Anderson et al. 2001; Fish & Lauder 2006), especially when males depress the gonopodium to the ventral surface of their body. No previous study has directly examined the link between gonopodium size and steady-swimming performance, although Basolo and Alcaraz (2003) showed that larger swords—which are superficially similar to gonopodia—do incur energetic costs. Of course, swords are not gonopodia, and recent results from a swim-tunnel experiment with *G. affinis* suggest that males with relatively larger gonopodia actually tend to exhibit higher endurance (time before fatigue) during steady swimming (R. B. Langerhans, unpublished data). The relationship between steady-swimming performance and gonopodium size requires further investigation.

In addition to steady swimming, poeciliids also frequently employ unsteady locomotor behaviors, such as rapid acceleration and turning. One of the most important unsteady swimming activities routinely performed by poeciliids is the C-start escape burst used when avoiding a predator strike (Domenici 2010). As expected, relative gonopodium surface area has been shown to exhibit a negative association with burst-swimming speed (*G. affinis*; Langerhans et al. 2005). Based on this locomotor cost of gonopodium size, combined with its possible effects on conspicuousness, we would expect selection via predation to favor smaller gonopodia. I recently tested this hypothesis by measuring selection on gonopodium length in *Gambusia hubbsi* (Bahamas mosquitofish) males in the presence of a predatory fish (*Gobiomorus dormitor*, bigmouth sleeper) within large experimental tanks (400 L) (R. B. Langerhans, unpublished data). Consistent with the prediction, I found strong evidence for selection against gonopodium length (fig. 21.6; logistic regression of survival on relative gonopodium length: one-tailed $P = 0.016$) after allowing overnight predation to occur (~8 hours). Thus, predators can generate selection for smaller gonopodium size, but whether this form of selection exists in the wild is unknown.

If selection via predation generally drives smaller gono-

podia, then we might predict divergence in gonopodium size between environments differing in predation intensity. Indeed, some poeciliids exhibit relatively smaller gonopodia in populations experiencing higher levels of predation from piscivorous fish (*P. reticulata*: Cheng 2004; *G. affinis*, *G. hubbsi*: Langerhans et al. 2005). However, other studies have found the opposite pattern, where males possessed longer gonopodia in environments with higher predation intensity (*P. reticulata*: Kelly et al. 2000; *Brachyrhaphis episcopi*: Jennions & Kelly 2002). Two possible explanations for these latter findings are that either (1) selection for longer gonopodia via increased mating frequency (section 21.3.3) or insemination success (section 21.3.4) is stronger in high-predation environments and outweighs any potential locomotor costs, or (2) confounding effects of increased water velocity (likely leading to stronger selection favoring increased steady-swimming performance) in high-predation localities led to longer gonopodia. These studies, combined with the possibility that gonopodia influence conspicuousness to predators, suggest that further investigations in the context of divergent predator regimes might elucidate some complex interactions between natural selection and sexual selection in poeciliid genital evolution.

21.5 Pleiotropy

Some researchers have contended that genital diversity arises as pleiotropic effects of selection on other traits, with genitalia being selectively neutral (Mayr 1963; Eberhard 1985; Arnqvist 1997; Arnqvist & Thornhill 1998). This hypothesis is contentious, however, as it is unclear how such a process could produce the observed rapid and divergent evolution of genitalia unless genital traits were disproportionately affected by pleiotropic effects of functionally important genes—an assumption for which we have no reason a priori to suspect is true. Yet, if genitalia do not experience direct selection, and instead diversify via pleiotropy, four testable consequences are predicted: (1) genitalia should not experience direct, contemporary selection, (2) genitalia should not exhibit strong signatures of past selection, (3) genes under selection should tend to pleiotropically affect genital organs more frequently than other traits, and (4) genitalia should often exhibit high intraspecific variability.

Based on the results discussed above, considerable evidence exists that at least some features of poeciliid genitalia either experienced strong selection in the past or continue to experience such selection today. These findings directly contradict the first two predictions of the pleiotropy hypothesis. The third prediction could be confirmed either by

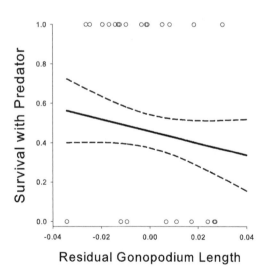

Figure 21.6 Selection in *Gambusia hubbsi* favoring smaller gonopodium length in the presence of a predatory fish (*Gobiomorus dormitor*). Relative gonopodium length measured as residuals from regression of log-transformed gonopodium length on log-transformed standard length. Fitness function estimated using the nonparametric cubic-spline regression technique. The solid line represents mean survival probability, and the dashed lines indicate ± 1 SE of predicted values from 1000 bootstrap replicates of the fitness function. R. B. Langerhans, unpublished data.

demonstrating that genes underlying functionally important traits tend to influence genitalia more often than nongenital traits or by showing stronger genetic correlations between functionally important traits and genital traits relative to other traits. Recent work has demonstrated a genetic association between gonopodial development and the growth of swords in some *Xiphophorus* species (Zauner et al. 2003; Offen et al. 2009). Specifically, gonopodium and sword development are both associated with expression of the *msxC*, *rack1*, *dusp1*, *klf2*, and *tmsβ-like* genes, some of which may influence the growth of long anal-fin rays in general. If so, selection on fin size, such as selection for larger fins via predation or by mate choice, could indirectly affect gonopodium size. However, if gonopodium size has evolved more rapidly than other fin sizes, this finding would be insufficient to explain its diversity. One could test the consequences of this pleiotropic link by examining genetic correlations among gonopodium length and unpaired fin-ray lengths, as well as by testing for correlated evolution of these traits among species. Of course, genetic correlations can arise for other reasons, including correlational selection, and are not exclusive to the pleiotropy hypothesis. In the data I have personally collected, I have never observed phenotypic or genetic correlations involving gonopodium size that are stronger than correlations involving other traits (including unpaired fin lengths)—indeed, correlations involving gonopodium size are typically smaller than other trait correlations (R. B. Langerhans, unpublished data). Finally, the fourth prediction derives from the fact that neutral traits can accumulate considerable amounts of variation within species. Evidence to date does not suggest that variation in gonopodium size (length, surface area) exhibits higher intraspecific variation than other traits (Kelly et al. 2000; Jennions & Kelly 2002; R. B. Langerhans, unpublished data), although other genital traits have not yet been examined. Overall, predictions of the pleiotropy hypothesis have not fared well in the face of empirical evidence in poeciliid fishes, suggesting that pleiotropy is unlikely to play a major role in driving genital diversification.

21.6 Consequences of genital evolution

Has genital diversification played a major role in lineage proliferation of poeciliids? It is certainly plausible that rapid divergence of genitalia might lead to speciation more readily than similar levels of divergence in nonreproductive traits. Considering the numerous sources of natural and sexual selection that might act on genitalia, and the remarkable diversity in genital form, an important role for genital diversity in promoting speciation is a reasonable expectation. Future work could test for a relationship between rates of genital evolution and speciation rates within lineages with well-resolved phylogenies. The finding in *Gambusia* that genital differences are greater between sympatric species than between allopatric species is consistent with a role of genital evolution in speciation but does not rule out postspeciation processes. One means of addressing the role of genital evolution in speciation is to conduct empirical investigations at scales where the process of speciation is more easily observed or inferred: sister species, incipient species, and populations within species.

21.7 Conclusions

This chapter provides the first review of the evidence for all major hypotheses of genital evolution in poeciliids. There are many unanswered questions, yet it is now clear that genital evolution in poeciliids is complex, resulting from multiple processes (table 21.2). Many genital traits are likely to be shaped by multiple selective processes, serve multiple, simultaneous functions, and differ in functionality among species. Moreover, major causes of genital evolution in poeciliids may not match those in other taxa. For instance, the lock-and-key hypothesis has gained little previous support in most taxa and yet appears important in poeciliids. Additionally, an important gap in our understanding of poeciliid genital evolution is that no study has yet directly tested for cryptic female choice, despite the fact that this hypothesis has garnered much empirical support in other taxa (Eberhard 1985, 1996; Cordero & Eberhard 2003). Furthermore, we badly need a stronger, functional

Table 21.2 Summary of the existing evidence for each hypothesis of genital diversification in poeciliid fishes

Hypothesis	Importance for poeciliid genital evolution
Male contest competition	Largely untested
Mate choice	Some role confirmed
Premating sexual conflict	Suggestive evidence
Sperm competition	Suggestive evidence
Cryptic female choice	Largely untested
Postmating sexual conflict	Highly suggestive evidence
Lock-and-key	Some role confirmed
Nonmating natural selection	Some role confirmed
Pleiotropy	Unlikely

understanding of the mechanics of copulation. We also have little knowledge of the importance of genital divergence in the speciation process. Clearly, we have much work ahead of us. Understanding the causes and consequences of genital diversification in poeciliid fishes will require integrative approaches, and I am optimistic about the advancement in this understanding that will be described in the next review of this field.

Acknowledgments

I am grateful to the editors for inviting me to contribute to this book. I thank J. Evans, M. Jennions, B. Mautz, A. Pilastro, and an anonymous reviewer for comments that improved the chapter.

Part V

Genetics

Chapter 22 Genetic variation in natural populations

Felix Breden and Anna K. Lindholm

22.1 Introduction

ONE OF THE EXTRAORDINARY FEATURES of poeciliids is the high degree of phenotypic variation among populations and species, which is presumably driven by adaptation to local environmental conditions. This has motivated many of the varied types of genetic studies of poeciliids, from surveys of genetic polymorphisms using presumably neutral markers to studies that investigate the extent to which phenotypic differences are genetically determined and are the result of selection or other evolutionary processes. Since the surveys of genetic variation presented by Echelle et al. (1989) and Smith et al. (1989) in the first synthesis of research on the evolutionary ecology of poeciliid fishes (Meffe & Snelson 1989b), DNA markers have almost completely supplanted allozyme markers. There has also been an increasing interest in the link between phenotypic and genetic variation, as well as an intense focus on describing genetic variation in one particular species, the guppy (*Poecilia reticulata*). Our purpose in this chapter is to survey what has been learned about the genetic structure of natural populations of poeciliids using molecular genetic markers and foreshadow what is to come (for a review of genetic variation in introduced or endangered poeciliids, see also Stockwell & Henkanaththegedara, **chapter 12**). We start with a survey of within-population genetic variation and ask if there is evidence that sexual selection, natural selection, migration, and/or drift shape within-population genetic variation. We then address the classic question concerning how genetic variance is distributed within and among populations in streams and drainages, and how this compares with variation among drainages. Finally, we connect this information to the evolution of phenotypic diversity in poeciliids by surveying studies that have used information on the distribution of molecular polymorphisms to explicitly test hypotheses concerning which evolutionary processes determine the distribution of genetic and phenotypic variation.

Poeciliids have provided many models for systematics, biogeography, evolutionary and ecological studies, and biomedicine. However, despite this interest in the group as a whole, only a small proportion of species has been studied in terms of the distribution of molecular genetic variation in natural populations. This is shown in table 22.1, in which we list the genera as defined by Rauchenberger (1989), the number of species per genera, and the number of studies within each of the species that address the partitioning of genetic variation within and among populations, rivers, drainages, or larger biogeographic regions within the species. When the first book on poeciliids (Meffe & Snelson 1989b) was written, two groups stood out as having large data sets regarding population genetic structure: *Gambusia* and *Poeciliopsis*. The studies on *Gambusia* concentrated on the two species in the southeastern United States, *G. affinis* and *G. holbrooki* (although at the time these were considered subspecies). These studies were motivated by interest in the taxonomic status of these two groups and what these species could tell us about the biogeographical zones of the southeastern United States. Much of the motivation for studies of *Poeciliopsis* was driven by interest in the distribution of genetic variation in the clonal lineages and in the conservation status of the desert *Poeciliopsis* that were

Table 22.1. Studies with molecular-genetic population structure data available

Genus[a]	Number of species in genus[a]	Species	Number of studies	Approximate proportion of range covered	References
Poecilia	49	P. reticulata	At least 18	Entire range	Selected references: Carvalho et al. 1991; Alexander et al. 2006; Crispo et al. 2006
		P. latipinna	2	Half of species range	Trexler 1988; Gabor et al. 2005
		P. velifera	1	Most of range	Hankison & Ptacek 2008
		P. petenensis	1	Most of range	Hankison & Ptacek 2008
Xiphophorus	17	X. cortezi	1	Entire range	Gutierrez-Rodriguez et al. 2007
		X. birchmanni	1	Entire range	Gutierrez-Rodriguez et al. 2008
Gambusia	45	G. affinis	3	Most of range	Wooten et al. 1988; Echelle et al. 1989; Smith et al. 1989
		G. holbrooki	3	Most of range	Wooten et al. 1988; Echelle et al. 1989; Smith et al. 1989
		G. geiseri	1	Most of known range	Echelle et al. 1989[b]
		G. nobilis	1	Entire range	Echelle et al. 1989[b]
		G. hubbsi	2	One of two or more islands	Schug et al. 1998; Langerhans et al. 2007b
Brachyrhaphis	8	B. rhabdophora	1	Most of range	Johnson 2001b
Poeciliopsis	21	P. occidentalis occidentalis/ P. o. sonoriensis	3	Entire range for at least some markers	Vrijenhoek et al. 1985; Quattro et al. 1996; Hedrick et al. 2006
Heterandria	9	H. formosa	3	Entire range	Baer 1998a, 1998b; Soucy & Travis 2003

[a]From Rauchenberger 1989. No molecular genetic population structure data are available for 16 genera.
[b]Reported in Echelle et al. 1989 from Milstead 1980.

greatly threatened. Of course, most of the data presented in the first poeciliid book consisted of surveys of allozymic variants by electrophoresis.

Since the publication of Meffe and Snelson's book, surveys of genetic variation have relied more on mtDNA and nuclear-sequence variation and variation in highly polymorphic microsatellite loci. As seen in table 22.1, the intense interest in the ecology and evolutionary history of guppy populations has led to a proliferation of studies on the distribution of genetic variation within and among populations of this species. But even in the well-studied genus *Poecilia*, there is information on the distribution of variation within and among populations for only 4 species (out of 49, following Rauchenberger's [1989] classification). One group that stands out in terms of paucity of data is *Xiphophorus*, certainly one of the most important groups within poeciliids in terms of ecology and evolution, systematics, and medical interest, and yet only two species, *X. cortezi* and *X. birchmanni*, have been examined rigorously for genetic variation within and among populations. In summary, of 194 species, only 14 are the focus of studies explicitly examining population structure; in the 6 genera covered (out of 22), the percentage of total species studied ranges from 5% to 12%. Given that many of the

approaches that attempt to factor out the relative contributions of drift, migration, selection, and phylogenetic constraint on phenotypic evolution depend on understanding the distribution of genetic variation (e.g., Althoff & Pellmyr 2002; Räsänen & Hendry 2008), it is critical that we conduct systematic surveys of molecular genetic variation in more of these important model organisms. Ultimately, such knowledge will also help identify conservation priorities in this family.

22.2 Genetic variation within populations

22.2.1 Role of sexual selection

Estimates of genetic variation within populations have often emerged as by-products of studies that attempt to measure the strength of sexual selection in natural populations and to associate these estimates with variation in behavioral or morphological traits or sex ratios. These estimates could potentially be used to test whether the intensity of sexual selection influences neutral genetic variation within populations. When individuals are equally successful in reproduction (and, thus, sexual selection is weak), effective population sizes will be greater, and genetic variation higher, than when reproduction is dominated by a few individuals (Falconer & Mackay 1996). In an early study on *Poecilia latipinna*, Simanek (1978) found a significant negative correlation between observed heterozygosity at 21 allozyme loci and an index of reproductive skew, which he defined as the ratio of breeding females to dominant males (males with full sailfins). Simanek assumed that only dominant males reproduce; more than 30 years later, the actual reproductive share that dominant males achieve is still unknown. With the aim of testing this hypothesis further, we review studies that use genetic markers to make inferences about reproductive skew in poeciliids.

Estimates of reproductive skew based on success of individual males in natural populations (rather than estimates of paternity skew within broods; see Evans & Pilastro, **chapter 18**) have been obtained only in two studies of poeciliids (Neff et al. 2008b; Tatarenkov et al. 2008). The best study is that of Tatarenkov et al. (2008) on *Xiphophorus hellerii*, in which pregnant females and their embryos were sampled together with possible sires. All individuals were genotyped at nine highly variable microsatellite loci; these genotypes were used for paternity analyses. Reproductive skew could then be calculated taking into account males that had no reproductive success. Unfortunately, this study did not provide population-specific heterozygosities; thus, we have analyzed only the results of Neff et al. (2008b). Neff et al. (2008b) genotyped mother and offspring guppies and then reconstructed paternal genotypes using the software COLONY (Wang 2004). Mating success of individual males was scored, and skew subsequently calculated. Males that sired no offspring were not detected in this analysis, which means that skew was underestimated. We used these results to test the hypothesis of a negative association between population reproductive skew and genetic variation. From tables 1 and 2 of Neff et al. 2008b, we calculated for each population the average expected heterozygosity per population, based on three microsatellite loci, and compared this with mean reproductive skew. We found no evidence for a relationship between genetic variation and male reproductive skew in 10 populations of guppies (linear regression, $\beta = 0.19$, $t = 0.95$, $p > 0.36$). Three loci

are a very small sample size for heterozygosity estimates (DeWoody & DeWoody 2005), and with the underestimation of skew, this test represents only a weak test of the hypothesis. Nonetheless, there is no further evidence available to support a role for sexual selection in determining within-population genetic variation.

22.2.2 Role of natural selection, migration, and drift

Genetic variation estimated as expected heterozygosity is lower in upstream, low-predation populations of guppies (Shaw 1991; Shaw et al. 1994; see table 22.2). The same pattern is seen in upstream and downstream populations of *Gambusia holbrooki* (Hernandez-Martich & Smith 1997). If we exclude a role for sexual selection, what then are the relative roles of natural selection, migration, and drift in determining within-population genetic variation?

The relative importance of different evolutionary processes for neutral genetic variation within populations of guppies in Trinidad has been addressed in several studies (Shaw et al. 1994; Crispo et al. 2006; Barson et al. 2009). Shaw et al. (1994) conducted a detailed study of multiple sites along the Aripo and Tacarigua rivers, while Crispo et al. (2006) focused on the Marianne River, sampling at 20 locations. Barson et al. (2009) used a course-grained approach across a larger number of rivers—they sampled one upstream and one downstream site (in one case only downstream) in each of six rivers. These latter two studies employed coalescent models to estimate parameters such as migration rates and effective population sizes. Crispo et al. (2006) and Barson et al. (2009) also used Bayesian clustering models to investigate the natural divisions within these regions, without predefining the population divisions. Shaw et al. (1994) and Crispo et al. (2006) concluded that differences in natural selection between populations, in the form of predation regime, did not influence genetic variation.

All studies viewed constraints on migration as key. Genetic variation was related to the geographic distance between upstream and downstream sites (Shaw et al. 1994; Crispo et al. 2006). Shaw et al. (1994) suggested that founder effects and bottlenecks could explain low genetic

Table 22.2 Estimates of expected heterozygosity, averaged over microsatellite loci, with range, for multiple river surveys of guppies in Trinidad

Reference	Number of loci	HE upstream/low predation	HE downstream/high predation
Suk & Neff 2009	7	0.45 (0.13–0.69)	0.66 (0.58–0.71)
Barson et al. 2009	5–7	0.11 (0.02–0.25)	0.64 (0.59–0.72)
Neff et al. 2008b	3	0.68 (0.59–0.77)	0.84 (0.82–0.88)
Kelly et al. 1999	2	0.55 (0.41–0.72)	0.77 (0.54–0.89)

Note: High-predation upstream populations were excluded.

diversity in far-upstream populations. However, in the analyses of Barson et al. (2009), no evidence was found for recent bottlenecks in (mid-)upstream populations, which had very low genetic diversity. Effective population sizes were, on average, about half as large in upstream as in downstream populations (Barson et al. 2009). Shaw et al. (1994) further proposed that waterfalls, as physical barriers, isolated upstream populations and that gene flow predominantly downstream resulted in higher genetic diversity downstream. The analyses of Crispo et al. (2006) tested this idea and found support for the role of waterfalls in restricting gene flow. Across waterfalls, genetic diversity was higher downstream, and gene flow was predominantly, but not always, downstream. Barson et al. (2009) concurred that gene flow was generally higher downstream than upstream. In studying variation in microsatellites of guppies from two sites in the Aripo River using the coalescent-type model IM, van Oosterhout et al. (2006b) came to the same conclusion. Somewhat surprisingly, Crispo et al. (2006) found, in estimates of contemporary dispersal using genetic admixture analysis in GENECLASS, that guppies often disperse upstream when there are no waterfalls in the way. Barson et al. (2009), using BAPS (Corander et al. 2004) to infer recent gene flow, found mostly downstream-biased contemporary dispersal but made no comparisons at a fine scale. Thus, the difference between the two results could be due to asking the same question at fine versus coarse geographic scales or to the difference between movement of individuals and gene flow, or the difference could be a strong, but not insurmountable, effect of waterfalls in Trinidadian rivers. Whether guppies are unusual among poeciliids in exhibiting high rates of upstream dispersal remains to be seen. More studies would be welcome in evaluating the relative contributions of evolutionary processes to neutral genetic variation within populations. Thus far, we have no evidence for a role of selection, only for migration, and we can only infer an effect of drift from studies of small, isolated populations of poeciliids (e.g., Schug et al. 1998).

22.3 Rates of migration and partitioning of genetic variation within and among rivers and drainages

We now turn our focus to a larger physical scale than genetic variation within populations. We address the following question: how is genetic variation in natural populations partitioned into variation among populations and variation among sets of populations or regions? For stream fish this often equates to questions of variation among populations within rivers and variation among rivers, or perhaps among even more inclusive units, such as drainages consisting of several rivers all draining to the sea together. In general,

reduction in effective population size will decrease variation within populations and increase variation among populations, while effective migration among populations will counteract these effects (Wright 1951). Such questions are of course important for setting the geographical scale over which adaptation occurs.

Description of genetic partitioning within and among populations is also relevant to conservation biology because it helps to measure the genetic effect of habitat fragmentation or of reduction in population size due to the introduction of nonnative species and helps to determine the geographical scale over which reintroductions should be made (e.g., in *Poeciliopsis*; Vrijenhoek et al. 1985; Quattro et al. 1996). Most studies reporting among-population genetic variation employ at least one of two main types of analyses: hierarchical F-statistics (Wright 1951), partitioning standardized genetic variances at multiple levels of biological organization, or the analysis of molecular variance (AMOVA), partitioning measures of genetic distances within and among populations, as developed by Excoffier et al. (1992). Table 22.3 is a summary of these types of studies in poeciliids, emphasizing those studies that report the partitioning of molecular variation among populations within rivers or drainages, and then among these larger units.

When Meffe and Snelson's book was published (1989b), *Gambusia* was the best-studied genus of poeciliids in terms of genetic population structure. These studies are summarized in table 22.3 and show that a large proportion of the genetic variance for the species in *Gambusia* is partitioned both among populations within regions and among regions. In fact, this pattern is repeated for most of the studies reported in table 22.3, which summarizes data on genetic variation among populations and among drainages from 14 species. F_{ST} within rivers varies from 0.02 to 0.6 (average, 0.21), while that among rivers/drainages ranges from 0.038 to 0.471 (average, 0.29). In 10 of 11 comparisons, F_{ST} among drainages is greater than within drainages. This trend agrees with the results from the AMOVAs reported in table 22.3; most often, the proportion of variance among rivers is greater than that among populations within rivers. Two studies reverse this trend: *Gambusia hubbsi* on Andros Island (Langerhans et al. 2007) and certain *P. reticulata* populations in Trinidad (Suk and Neff 2009).

As seen in tables 22.2 and 22.3, since the publication of Meffe & Snelson 1989b, several groups have undertaken studies of the partitioning of genetic variation within and among drainages in the genus *Poecilia*, especially in the guppy, *P. reticulata*. The intense study of guppy population structure is of course motivated by attempts to understand the forces influencing the distribution of adaptive phenotypic variation, especially between upstream and down-

Table 22.3 Partitioning of genetic variation within and among rivers/drainages

Species	Marker[a]	F_{ST} between sites within rivers/drainages	F_{ST} among rivers/drainages	AMOVA partition among sites within rivers/drainages[b]	AMOVA partition among rivers/drainages	$N_e m$[c]	Sample design	References
Poecilia reticulata	4 alloz.	0.18	0.36	na[d]	na	na	6 sites from 5 rivers, Trinidad	Carvalho et al. 1991[e]
P. reticulata	mtDNA control region	na	na	21.5	32.2	na	35 sites, 3 major regions, NE South America	Alexander et al. 2006[f]
P. reticulata	7 micro.	0.30	na	12.5	22.5	0.2–12.3	20 sites Marianne R., Trinidad	Crispo et al. 2006[g]
P. reticulata	8 micro.	0.16	na	na	na	0.12–0.54	2 sites, Aripo R., Trinidad	van Oosterhout et al. 2006b[h]
P. reticulata	5–7 micro.	Mean 0.25	Mean 0.47	na	na	0.04–8.8	11 sites, Trinidad	Barson et al. 2009[i]
P. reticulata	7 micro.	Mean 0.21	Mean 0.31	25.2	4.6	na	15 sites, 4 rivers, Trinidad	Suk & Neff 2009[j]
P. latipinna	30 alloz.	0.07	0.21	na	na	~3.5–5	25 sites, 3 geographical regions, SE USA	Trexler 1988[k]
P. latipinna	16 alloz.	na	na	8.0	1.4	2.4	12 populations, SE USA	Gabor et al. 2005
P. mexicana	10 micro.	0.02–0.05 among sites in cave	>0.29 low-sulfur and other sites	23.9 among all sites	na	1–3	3 cave sites, 2 surface sites, Mexico	Plath et al. 2007a
P. petenensis	9 micro.	Mean 0.16	0.28	11.3	27.0	na	9 sites, 3 major regions, Yucatán Peninsula, Mexico	Hankison & Ptacek 2008
P. velifera	9 micro.	0.14	0.43	5.9	15.0	na		Hankison & Ptacek 2008
Xiphophorus cortezi	mtDNA control region	na	na	42.1	53.7	na	10 sites, 3 drainages, Mexico	Gutierrez-Rodriguez et al. 2007
Xiphophorus birchmanni	mtDNA control region	na	na	3.4	87.7	na	11 sites, 3 drainages, Mexico	Gutierrez-Rodriguez et al. 2008
Gambusia holbrooki and G. affinis combined	5 alloz.	0.11	0.36	na	na	15–20 among sites	76 sites, 19 drainages, SE USA	Table 12.2 in Echelle et al. 1989
G. nobilis	6 alloz.	0.26	0.35	na	na	15–20 among sites	2–6 populations in each of 4 spring systems, New Mexico/Texas, USA	Table 12.3 in Echelle et al. 1989; Smith et al. 1989

(continued)

Table 22.3 *(continued)*

Species	Marker[a]	F_{ST} between sites within rivers/drainages	F_{ST} among rivers/drainages	AMOVA partition among sites within rivers/drainages[b]	AMOVA partition among rivers/drainages	$N_e m$[c]	Sample design	References
G. hubbsi	mtDNA ND2,	0.60	na	56.9	3	na	12 sites, Andros Is., Bahamas	Langerhans et al. 2007[l]
G. hubbsi	21 alloz.	0.36	na	na	na	0.41	16 sites, Andros Is., Bahamas	Schug et al. 1998
Brachyrhaphis rhabdophora	11 alloz., mtDNA control region	0.36	0.17	na	na	0.44	12 sites, 5 drainages, Costa Rica	Johnson 2001b[m]
Heterandria formosa	16 alloz.	na	na	na	na	6.0–14.0	34 sites, 10 geographic regions, SE USA	Baer 1998b
H. formosa	6 alloz.	0.01	0.04	na	na	na	9 sites, St. Johns R.; 7 sites Ocklawaha R., SE USA	Baer 1998a[n]
Poeciliopsis occidentalis	12 alloz.	na	0.53	na	na	na	1–6 sites in each of 6 Sonoran Desert rivers	Echelle et al. 1989[o]

[a]alloz. = allozyme electrophoretic markers (polymorphic only); micro. = microsatellite markers; mtDNA = various regions of the mtDNA molecule.

[b]Partitioning of genetic distances between haplotypes according to analysis of molecular variance (Excoffier et al. 1992).

[c]$N_e m$ = estimate of contemporary dispersal, or effective number of migrants.

[d]na = not available.

[e]G_{ST}, coefficient of gene diversity, not F_{ST}, is reported.

[f]Estimates for within and among major regions: Venezuela/west Trinidad; east Trinidad; east Guyana.

[g]AMOVA results reported for groups of populations separated by waterfalls.

[h]G_{ST} reported.

[i]Estimates do not include 12th population surveyed from the island of Tobago.

[j]Averages reported within and among drainages: Caroni, Oropuche, Paria, and Yarra. Tururue populations not included because they are influenced by introduction from Caroni drainage.

[k]$N_e m$ estimated among sites within separate regions, from private alleles method.

[l]F_{ST} calculated between two sets of populations: all high-predation and all low-predation sites.

[m]Within and among rivers within drainages.

[n]$N_e m$ estimated among neighboring populations assuming a two-dimensional stepping-stone model.

[o]Reported in Echelle et al. 1989 from Vrijenhoek (pers. comm.).

stream populations (e.g., Alexander et al. 2006; Crispo et al. 2006; Barson et al. 2009; Suk & Neff 2009). How this distribution of genetic variation can be used to explicitly test evolutionary hypotheses will be described in section 22.4.

Only two species in the genus *Xiphophorus*, *X. cortezi* (Gutierrez-Rodriguez et al. 2007) and *X. birchmanni* (Gutierrez-Rodriguez et al. 2008), have been described rigorously in terms of genetic variation among sites within drainages and among drainages. These studies observed low levels of variation within populations, but this is common for the marker used in these studies, the mtDNA control region. In general, there was strong differentiation among sites and among drainages.

Table 22.3 also presents migration estimates for most of these species, given as $N_e m$, the per generation effective number of migrants. It is difficult to compare these estimates of effective number of migrants among species and studies, because the geographic range for each study, and the methods used for estimating $N_e m$, vary greatly. More recent studies employ coalescent approaches to estimate migration, such as IM (Hey & Nielsen 2004) and MIGRATE (Beerli 2006). Some of the estimates are high, with 6 of 10 estimates as high as 5 migrants per generation. Given this potential for high rates of migration, the amount of differentiation among populations and drainages is somewhat surprising and may be due to adaptive divergence restricting effective gene flow, despite dispersal (Räsänen & Hendry 2008).

The partitioning of genetic variation has also been studied for several species in the genus *Poeciliopsis* (Vrijenhoek et al. 1985; Quattro et al. 1996; Hedrick et al. 2006), motivated by the endangered status of some of these species due to habitat destruction, including the introduction of nonnative fishes (often another poeciliid, *G. affinis*) (Vrijenhoek et al. 1985). These studies have concentrated on describing the distribution of genetic variation to determine species boundaries, to identify evolutionarily significant units (ESUs) for conservation purposes, and to direct reintroduction programs. These studies on *Poeciliopsis* are not listed in table 22.3 because they do not report standardized *F*-statistics or AMOVAs; one F_{ST} value for among-region differentiation was reported by Echelle et al. (1989) based on unpublished data. Overall, the groups studied exhibited well-differentiated lineages, supporting species status for *Poeciliopsis occidentalis* relative to *Poeciliopsis sonoriensis* and ESU status for the Monkey Spring population of *Poeciliopsis occidentalis* (Hedrick et al. 2006).

In summary, only a few species, and in only a few genera, have been extensively studied in terms of the partitioning of genetic variation. Table 22.3 shows that most species exhibit strong differentiation among populations and

among rivers or drainages. The two studies that reported low among-population differentiation (F_{ST}) were done on temperate, widely distributed species (*Heterandria formosa*, $F_{ST} = 0.01$, Baer 1998a; *P. latipinna* from the southeastern United States, $F_{ST} = 0.07$, Trexler 1988). We excluded the low values observed in the cave molly (*Poecilia mexicana*) because these were calculated from populations separated by only a few hundred meters (Plath et al. 2007a). This general pattern of strong differentiation among populations and drainages for most of the species listed in table 22.3 extends the summary presented by Echelle et al. (1989). The concordance between surveys suggests that this pattern is a general characteristic of poeciliid phylogeography.

22.4 Using data on molecular genetic variation within and among populations to test hypotheses regarding evolutionary processes

In this section we highlight several studies on poeciliids that report variation within and among populations in both molecular markers, presumed to be neutral, and phenotypic characters, presumed to be under selection, in order to explicitly test hypotheses on the forces driving variation in adaptive phenotypes. These studies attempt to evaluate the relative effects of selection, drift, and phylogenetic constraint on phenotypic variation among populations and among rivers and drainages. This approach is now more feasible due to a greater reliance on DNA polymorphism data and the development of coalescent models for estimating population sizes and migration rates.

As described in section 22.3 and in table 22.3, some of the most extensive studies of genetic variation in poeciliids have been conducted on *Gambusia* species, especially in the southeastern United States (summarized in Echelle et al. 1989; Smith et al. 1989). Smith et al. (1989) concentrated on several populations of both *G. affinis* and *G. holbrooki* and described patterns of differentiation in allozyme markers over both space and time. Several types of data pointed to selection as being important in determining the pattern of genetic variation. The authors concluded that sexual selection was indicated by differences in genotypic frequencies between pregnant and nonpregnant females and that selection acting through coadapted gene complexes was indicated by strong correlations among alleles at different loci (although the latter observation could be explained by other processes). However, the authors' overall conclusion was that drift was responsible for much of the frequency variation in these allozyme variants across small time and geographic scales.

More recent studies of *Gambusia* have concentrated on *G. hubbsi* on Andros Island, occurring in "blue holes," vertical caves that have filled with freshwater since the rise of sea levels following the last glacial maximum. Migration is low among these blue holes, and drift seems to be the predominant factor determining the distribution of neutral markers, such as allozymes (Schug et al. 1998). The predation environment differs among the blue holes, primarily due to the presence or absence of the redfin needlefish (*Strongylura notata*). Langerhans et al. (2007) measured variation in mtDNA sequences, mating behavior, and morphological characters and combined these data with allozyme data from Schug et al. (1998). They used a phylogeny based on molecular polymorphisms to reject the hypothesis that all the high-predation populations shared a common phylogenetic history and thus were able to conclude that selection had most likely independently shaped the behavior and morphology of this rapidly evolving species. This use of molecular markers to build phylogenies, ultimately to test for independent, repeated evolution of adaptive traits, is a common approach in poeciliids. Another approach to testing for contemporary selection is to compare the distribution of molecular genetic variation, assumed to be neutral, with that of phenotypic variation.

Hankison and Ptacek (2008) adopted this latter approach, studying two species of mollies, *Poecilia velifera* and *Poecilia petenensis*. The main thrust of their approach was to use the distribution of neutral molecular markers to estimate the overall effects of random genetic drift, gene flow, and geographic barriers. By then comparing the distribution of characters most likely under selection—in this case, morphology and components of male mating behavior—with the genetic structure of the neutral genetic markers, they reasoned that any marked differences would reflect recent selection on these phenotypes. Such tests of course assume that the effect of homoplasy is not strong enough to significantly alter the phylogeographic signal estimated from the microsatellite markers.

The two species studied are in the derived clade of sailfin mollies with respect to the other shortfin mollies (Ptacek & Breden 1998). They occupy a range of drainages and ecological zones across the Yucatán Peninsula and differ in their ecological requirements. The saltwater *P. velifera* is restricted to coastal habitats, while *P. petenensis* occupies freshwater habitats. Thus, populations of these species across these ecological zones, from within and among different drainages, are ideal for examining the effect of contemporary selection, gene flow, and phylogeographical constraint on partitioning of neutral genetic markers and phenotypes under selection.

Hankison and Ptacek (2008) surveyed molecular genetic variation (nine microsatellite loci), linear body measurements in males (on dorsal and caudal fins), and male mating characteristics (rate and duration of courtship displays and gonopodial thrusts) in nine populations of each species. There was strong population structuring in both species, with high F_{ST} values among populations within geographical regions and among geographical regions. Two types of analyses used genetic information to test hypotheses about selection versus other forces. The BAPS algorithm (Corander et al. 2004) uses information on the distribution of variation in any type of character and estimates where virtual barriers might exist that could explain discontinuities in the distributions. The barriers predicted based on the partitioning of genetic variation did not correspond to the barriers predicted from the distribution of the phenotypic variation in the male morphological or behavioral characters. This result was supported by Mantel tests (Smouse et al. 1986), comparing matrices of genetic distances with those calculated for morphological tests. In both species, the Mantel tests overall showed very little concordance between the genetic and morphological or behavioral measures. Hankison and Ptacek (2008) therefore concluded that contemporary selection had a very strong effect on the distribution of male morphology and courtship behavior in both species.

Alexander et al. (2006) described the distribution of mtDNA variation across the entire range of the guppy, *P. reticulata*. Similar to Hankison and Ptacek's (2008) approach, their goal was to determine the partitioning of genetic variation across the species' range and to compare these data with patterns of phenotypic variation in order to test for the effects of contemporary selection. In this way they wanted to test the generality of the patterns of selection observed in the streams of Trinidad.

The phenotypes they measured were male coloration and body length, characters that have often been shown to respond to differences between populations in predation pressure and other environmental factors within streams of the Northern Range of Trinidad (Grether et al. 2001a; Magurran 2005). They surveyed 45 sites for mtDNA variation and 36 sites for morphological variation across the range of the species from Venezuela and Guyana to Suriname and the well-studied populations in Trinidad. Again, similar to the findings of Hankison and Ptacek (2008) and to results obtained from populations in Trinidad, a discordance between morphology and neutral genetic markers across the species' range suggested that many of the patterns of phenotypic variation for both color and male length were driven by contemporary selection. Also, the clades in the phylogenetic analysis of Trinidadian populations included upstream/downstream pairs of populations, suggesting that the adaptive divergence observed between these types of guppy populations had independently evolved several times. However, there was also some association between molecular genetic patterns and patterns of phenotypic variation at this broad level of biogeography, so selection was not the only process determining the partitioning of phenotypic variation among these populations.

Guppy populations in the streams of Trinidad's Northern Range have been extensively studied for adaptive phenotypic variation, and based on this interest, several groups have measured the distribution of molecular variation (e.g., Carvalho et al. 1991; Shaw 1991; Fajen & Breden 1992; Shaw et al. 1994; Alexander & Breden 2004; Becher & Magurran 2004; Alexander et al. 2006; Crispo et al. 2006; van Oosterhout et al. 2006b; Neff et al. 2008b; Barson et al. 2009; Suk & Neff 2009; see box 22.1). From these we highlight one study that uses this molecular variation to explicitly test evolutionary hypotheses.

Andrew Hendry and colleagues have examined whether gene flow restricts adaptive divergence by swamping local populations with genes that are not locally adapted or whether adaptive divergence restricts gene flow by reducing effective migration, and how these effects can be differentiated (Räsänen & Hendry 2008). They have applied these ideas to several fish species, including salmon (e.g., Hendry et al. 2000), sticklebacks (e.g., Moore et al. 2007), and guppies. Their study on guppies (Crispo et al. 2006) addressed this question by examining individuals from a large number of sites, 20, within one drainage, the Marianne River in Trinidad. In this river, different sites are separated by physical distance, physical barriers such as waterfalls, and differences in predation regime; there were enough sites that it was possible to independently estimate the effect of these factors on genetic variation among populations. They found that waterfalls consistently affected gene flow but that differences in predation regime did not. This strongly suggested that adaptive divergence due to differences in level of predation was not limiting gene flow. This could be one explanation for why reproductive isolation has never been observed among populations within these rivers despite the fact that divergent selection has been seen to drive ecological speciation in several other systems (reviewed in Rundle & Nosil 2005).

A study by van Oosterhout et al. (2006b) compared the distribution of polymorphisms in functional genes versus neutral molecular markers to test whether these different types of genes had different effective migration rates. They compared the distribution of major histocompatibility complex (MHC) variation, functional genes of the adaptive immune system (see McMullan & van Oosterhout, **chapter 25**), and variation at eight microsatellite loci in guppies from one upstream and one downstream site in the Aripo River in Trinidad. Effective population sizes and effective

Box 22.1 Reproductive isolation among guppy populations

Guppies from the Caroni and Oropuche drainages within the Northern Range of Trinidad are highly differentiated in terms of molecular variation, including allozyme and mtDNA polymorphisms, although high- and low-predation phenotypes can be found in both drainages. Russell and Magurran (2006) and Ludlow and Magurran (2006) tested for reproductive isolation among populations drawn from these two drainages. Both studies found evidence of reproductive isolation in various forms, including reduced fitness in F_1 individuals due to reduced rates of mating behavior and reductions in embryo viability, brood size, and sperm counts.

This is not the only observation of reproductive isolation among populations of this well-studied species. Alexander and Breden (2004) observed behavioral reproductive isolation between a highly phenotypically differentiated population of guppies from Venezuela, called Endler's livebearer or the Cumaná guppy, and Venezuelan populations showing more typical coloration patterns and shapes. The Cumaná population shows asymmetric reproductive isolation, in that females from Cumaná exhibit a preference for males from their own population, while females from neighboring guppy populations do not show a preference for their own population (Lindholm & Breden 2002; Alexander & Breden 2004). By applying the criteria outlined by Panhuis et al. (2001), Alexander and Breden (2004) concluded that sexual selection most likely has driven this reproductive isolation and possible incipient speciation between the Cumaná guppy and regular guppy populations. One of the criteria from Panhuis et al. (2001) is that populations that exhibit strong differentiation in sexually selected characters should not also show elevated levels of genetic differentiation in neutral molecular markers. The reasoning behind this criterion is that such a result suggests that the sexually selected characters have differentiated rapidly, while most other characters have not differentiated overall. In this case, Alexander and Breden (2004) used mtDNA polymorphisms to show that the Cumaná population nested within other regular guppy populations in northeast Venezuela.

migration rates were estimated by the IM algorithm, employing a maximum-likelihood, coalescent approach (Hey & Nielsen 2004). They observed more variability in the neutral markers in the downstream population, a commonly observed pattern (table 22.2). However, there was little difference between the upper- and lower-stream populations in MHC variability. The authors suggested that this could be due to higher effective migration rates among populations for the MHC polymorphisms than for microsatellite variants, thus making the distributions of the MHC alleles more similar among populations. This could be due to balancing selection at the MHC locus: immigrant alleles into the upstream population, which showed overall lower variability, would more likely be found in a heterozygous state, increasing effective migration rates due to positive selection on those heterozygous individuals. They tested this hypothesis using forward Monte Carlo simulations and showed that strong balancing selection on the MHC is consistent with the observed low level of differentiation (McMullan & van Oosterhout, **chapter 25**).

Two other studies on mollies in the genus *Poecilia* examined genetic variation and phenotypic adaptation. Plath et al. (2007a) studied several cave populations and two surface populations of the cave molly, *P. mexicana*. The habitat of one of the surface populations had extremely high concentrations of hydrogen sulfide. One of the researchers' main goals was to evaluate the relative importance of gene flow versus adaptive divergence in terms of determining the distribution of phenotypic differences. It was clear from the distribution of genetic variation and from estimates of recent migration that adaptation to the high-sulfide environment restricted gene flow between this population and the other sites.

Gabor et al. (2005) measured allozymic and behavioral variation across the range of one of the sailfin mollies, *P. latipinna*. They compared populations that were sympatric and allopatric with *Poecilia formosa*, the Amazon molly. The parthenogenetic females of *P. formosa* must mate with a *Poecilia* male to initiate embryogenesis, and it was observed that *P. latipinna* males that were sympatric with *P. formosa* were more choosey and rejected these heterospecific females (see also Schlupp & Riesch, **chapter 5**). By comparing the distribution of male choosiness with the distribution of the neutral allozymic markers, the authors concluded that contemporary selection shaped the geographical distribution of male mating behavior.

Finally, one species in the genus *Brachyrhaphis*, *B. rhabdophora*, has been studied in great enough detail that data on genetic partitioning can be compared with data on among-population variation in phenotypic traits. In the coastal streams in Costa Rica, upland and lowland populations differ strongly in predation pressure, in a situation similar to that of the Trinidad guppy populations (Johnson 2001b). Johnson studied 12 localities in northwest Costa Rica, 4 of which were reported to have low predation. A phylogenetic tree based on genetic distances calculated from 11 allozyme markers showed that each of the 4 low-predation populations clustered with 1 or more high-predation populations in the same stream or drainage. The most parsimonious explanation for this pattern is that there has been independent evolution of low-predation phenotypes in each of these low-predation populations.

22.5 Future directions

One of the most pressing goals for the future is to expand both the number of poeciliid species that are studied in terms of the distribution of genetic variation and the geographical range across which these species are studied. As table 22.1 indicates, very few groups have been covered by more than one or two studies. Also, for those groups for which there are multiple studies, these often concentrate on a few well-known species or sets of populations. Consequently, some of these species have become important models for ecology and evolution, such as swordtails in Mexico and guppies in the Northern Range in Trinidad. However, we would like to be able to draw general conclusions from the identifying characteristics of poeciliids. One such question concerns how the livebearing life history affects patterns of migration, which in turn could determine the potential for adaptive divergence and subsequent evolution of reproductive isolation. Addressing such general questions for poeciliids awaits studies on a greater number of species and a broader coverage of each species' range.

A second area in which poeciliids can play a major role is the emerging approach to population genetics referred to as population genomics. This approach uses information from a large number of markers throughout the genome to examine classic questions in population genetics, but in much greater detail and with greater precision. Major genomics projects in at least two groups, *Xiphophorus* and *Poecilia*, are developing the knowledge necessary for this approach. Pioneering work by Don Morizot (Morizot et al. 1977) based on allozyme polymorphisms introduced the power of comparative genomics to explore synteny among vertebrate genomes and reconstruct the ancestral vertebrate genome. Further interest in *Xiphophorus* genomics was motivated by the identification of oncogenes and tumor suppressor genes for melanoma in the Gordon-Kosswig cross between *X. maculatus* and *X. hellerii* (Anders 1991) and in other *Xiphophorus* crosses (Kazianis et al. 2004b; Walter et al. 2004). This connection with cancer motivated the production of a high-density, genetic linkage map with over 400 markers for *Xiphophorus*; this information can now be used to map genes contributing to phenotypic adaptation and reproductive isolation. The guppy is a second species for which genomic resources are being developed, in this case directly in order to map and identify genes underlying adaptive phenotypic variation. Detlef Weigel, Christine Dreyer, and colleagues at the Max Planck Institute for Developmental Biology have undertaken this project, and genomic resources include sequences from more than 18,000 expressed-sequence tags (ESTs) (Dreyer et al. 2007), a 35,000 clone BAC library, and a high-density linkage map based on over 1000 single-nucleotide polymorphism (SNP) markers (Tripathi et al. 2009b). In addition, many of the markers for the guppy map are applicable to other species in the genus *Poecilia* (Tripathi and Dreyer, pers. comm.).

One of the major advantages of the population genomics approach is the ability to examine population differentiation using polymorphic markers across the entire genome. In this way it is possible to "scan the genome" and determine whether certain regions of the chromosomes are more "permeable" to reproductive barriers than others ("heterogeneous genomic divergence"; reviewed in Nosil et al. 2009). Polymorphisms in chromosomal regions that freely introgress between phenotypically differentiated groups are presumed to be neutral. On the other hand, chromosomal regions that show strong differentiation between such groups may harbor loci that are under divergent selection (e.g., Rogers & Bernatchez 2007), and this approach can identify candidate genes for the control of important phenotypic variation. Also such regions may harbor candidates for genes that are responsible for reproductive isolation (i.e., "speciation genes"; Ting et al. 2001; Orr et al. 2004). At least one study has indicated that a phylogeny based on a chromosomal region associated with reproductive isolation more closely reflects the most likely species-level phylogeny (Ting et al. 2000). Given that there are high-density linkage maps available for at least two groups of poeciliids, population genomics can be used to identify the most likely phylogenetic relationships of these groups and identify chromosomal regions underlying adaptive phenotypic variation (Willing, Dreyer, and Weigel, pers. comm.).

Finally, one of the major questions in evolutionary genetics is the molecular basis of adaptive phenotypic variation, and again the extensive knowledge of the adaptive significance of phenotypic variation in poeciliids makes them ideal models for such "functional" genetic studies. In addition to the population genomic approaches described above, which scan the whole genome, it is possible to use sequence variation in candidate genes among populations to determine whether variation in these genes is responsible for adaptive phenotypic variation. Such studies have recently been undertaken on opsin and MHC genes. Opsins are the protein components of the visual pigments in cone cells, and variation in such genes partially determines the tuning of color vision in vertebrates (Yokoyama 2000). As such, these genes are good candidates for controlling some of the variability that has been observed in color-based female preferences in poeciliids (e.g., Endler & Houde 1995). Variation in long-wave-sensitive opsin repertoire (number of duplicated genes), sequence variation, and expression differences have all been explored in guppies (Hoffman et al. 2006; Weadick & Chang 2007; M. Ward et al. 2008), although this variation has not yet been tied to differences in preference functions (see also Coleman, **chapter 7**). As discussed in the previous section, sequence variation has also been studied in MHC class IIB genes, which are important components of the adaptive immune system and which contribute to protection against parasites (McMullan & van Oosterhout, **chapter 25**).

In summary, there are a few general patterns that emerge when genetic variation within and among populations is compared across poeciliid species. Typically, upstream populations have lower genetic variation, and most species exhibit strong differentiation among populations within rivers and among rivers and drainages. This may be due to adaptive divergence tending to limit gene flow, but determining the underlying processes causing such patterns of genetic differentiation can be extremely difficult (Räsänen & Hendry 2008). However, surveys of genetic variation of more species and across a broader geographical range of these species, combined with data on adaptive phenotypic variation, should allow explicit tests of important evolutionary hypotheses in many poeciliid species.

Chapter 23 Genetics of male guppy color patterns

Robert C. Brooks and Erik Postma

23.1 Overview

THE GENETICS OF MALE ornaments in guppies (*Poecilia reticulata*) presents the first documented and still one of the clearest cases of Y-linked genetic variation. Guppies were a major genetic model organism, before *Drosophila* or *Saccharomyces*, and they were the first species used to describe, for example, sex limitation and sex linkage in the early 1900s (e.g., Schmidt 1920; Winge 1922b, 1922a, 1927). Indeed, studies of color pattern inheritance in guppies, as well as in close congeners such as *Poecilia parae* (Lindholm et al. 2004) and in more distantly related poeciliid species like *Xiphophorus maculatus* (Basolo 2006), have led to new and important insights into the evolution of male ornamentation and coloration, on the one hand, and sex chromosomes, on the other.

In this chapter we will first review what we know about the inheritance of color pattern genes and about the quantitative genetic inheritance of colors. We will then make the case that many sexual ornaments, and the color patterns of male guppies in particular, are sexually antagonistic traits. That is, the genes involved confer a net evolutionary benefit when expressed in one sex but not when expressed in the other. We will then interpret the genetic evidence, including the pleiotropic effects of these genes in light of the sexually antagonistic status of guppy color patterns. We will also briefly consider nonadditive forms of genetic variance and the intriguing possibilities that they present. To complete the chapter, we will consider the outstanding feature of guppy color patterns—extreme polymorphism—and discuss what we know about the potential processes underlying the maintenance of such variation.

23.2 The color patterns of male guppies

One of the most distinctive features of wild (including feral) guppies is the bright and highly polymorphic color patterns of males (fig. 23.1). Females are larger than males, but with a very few exceptions they are unpigmented. The color patterns of males are expressed only at maturity, making them typical sex-dependent secondary sexual traits.

Most studies distinguish three basic categories of color in male guppies: the oranges, the blacks, and the iridescent structural colors. The oranges include a variety of colors from yellow through to red and contain drosopterin (a pteredine) and carotenoids (Grether et al. 2001a). Black and dark-brown spots are due to the presence of melanins, and many authors distinguish between permanently expressed hard-edged black spots and facultatively expressed "fuzzy black" spots and bars (Baerends et al. 1955; Kodric-Brown 1998), which reach their maximum extent during courtship (Endler 1983). The structurally produced iridescent blues, greens, violets, bronzes, and silvers occur as spots, as background, and sometimes as overlays to orange spots (Kemp et al. 2008). These structural colors are probably due to multilayer guanine crystals (Vukusic & Sambles 2003), and their color and reflectance depend, like all structural iridescent colors, on the angle of incident light and the angle of the receiver relative to the guppy's body.

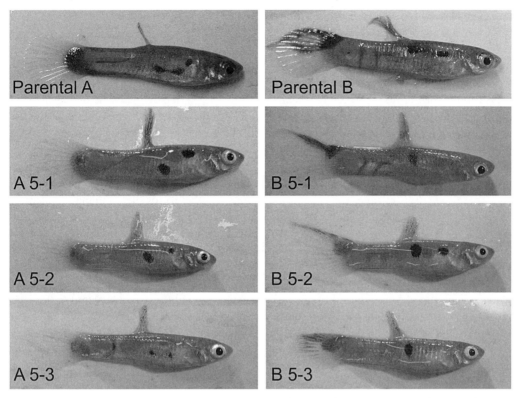

Figure 23.1 The color patterns of male guppies from two patrilines. Two males (top row, A-parental and B-parental) and their fifth-generation descendants (rows 2–4, males in a column are partrilineal descendants of the parental male at the head of that column) from a pedigree inbreeding experiment (Brooks & Postma, unpublished data). The color pattern elements shared by males within a column are probably Y linked, as males are autosomally related to one another by between 3% and 25%. As this figure illustrates, there are also many elements of the color pattern that are not strictly patrilineally inherited. Parental guppies are second-generation lab-reared descendants of animals captured in Alligator Creek, North Queensland. Some photographs have been retouched using the clone tool in Adobe Photoshop where particular spots were partially obscured by reflectance from moisture on the dorsal edge of the fish.

The color patterns of male guppies comprise several spots and some lines that differ dramatically in intensity, shape, size, and position among individuals (Haskins & Haskins 1951; Endler 1980, 1983; Houde 1992; Brooks & Endler 2001a; Hughes et al. 2005) and among lines descended from wild-caught animals (Winge & Ditlevsen 1947; Farr 1981). Furthermore, the frequencies of these color pattern elements, and thus the mean and variance in total coverage of each color, vary strongly among populations (Haskins & Haskins 1951; Endler 1980; Reznick & Endler 1982, 1983). It is this variability in color patterns among males that makes for one of the most extreme polymorphisms in the natural world and, as we show below, one that is to a large degree genetic.

23.3 Guppy sex chromosomes

Although chromosomal sex determination is less common than environmental and autosomal sex determination in fishes (Purdom 1993), the majority of poeciliid fishes have been shown to have heterogametic sex chromosomes (Angus 1989a; Lindholm & Breden 2002). There is, however, extreme variation among and in some cases even within species in the specific form of chromosomal sex determination, including XX/XY, ZZ/WZ, and multifactorial sex determination (Angus 1989a; Volff & Schartl 2001; Lindholm & Breden 2002) (see also Schartl et al., **chapter 24**).

Guppies have a major sex-determining region on the Y chromosome, and recombination is suppressed in and around this region (Winge 1922b, 1927; Winge & Ditlevsen 1947). These Y chromosomes are large, not distinguishable visibly from X, and still possess a large recombining (i.e., pseudoautosomal) region (Angus 1989a; Nanda et al. 1990). Thanks largely to studies of color pattern inheritance in guppies, as well as in other poeciliid species (Lindholm et al. 2004; Basolo 2006), there has been considerable interest in the guppy X and Y chromosomes. Even though it can be inferred from their inheritance that some alleles that influence the presence of particular color spots or background color patterns are X linked or autosomal, these traits are in fact seldom if ever expressed in females. Interestingly, however, small doses of testosterone in the water (Winge 1934; Haskins et al. 1961) or the food (Kavumpurath & Pandian 1993a) of female guppies usually result in the expression of the autosomal or X-linked color patterns. This not only provides a convenient experimental tool for those studying guppy color patterns and their inheritance but also demonstrates that the expression of many X-linked and autosomal color pattern genes is sex limited via epistatic effects with androgens, which may very well be mediated by Y-linked genes (also see Schlupp & Riesch, **chapter 5**).

23.4 Inheritance of color pattern elements

Early studies of the inheritance of color patterns by Schmidt (1920) and Winge (1921) showed strong patrilineal inheritance of large components of the color patterning. These studies proved to be the first descriptions of Y-linked inher-

itance in any species (Winge & Ditlevsen 1947). Although Winge is remembered as the "Father of Yeast Genetics" (Szybalski 2001), before he started his work on yeast he in fact made an enormous contribution to genetics in several systems, one of which was guppies. In addition to correctly identifying Y linkage as the mechanism for this new and peculiar pattern of inheritance (Winge 1922b, 1922a), he showed that while many color pattern genes are found only on the Y, some others recombine occasionally to the X, where their expression is sex limited to males (Winge 1927). Winge mapped over 20 genes to within two map units of the factor determining "maleness" (Winge & Ditlevsen 1947). Some of the patterns identified by Winge are illustrated in fig. 23.2.

Nowadays we know of at least 20 exclusively Y-linked color pattern alleles (i.e., located within the nonrecombining region close to the sex-determining region) and at least 28 that recombine between X and Y (i.e., within the large pseudoautosomal region of Y) (Winge & Ditlevsen 1947; Yamamoto 1975; Angus 1989a). There are, however, almost certainly considerably more sex-linked genes than

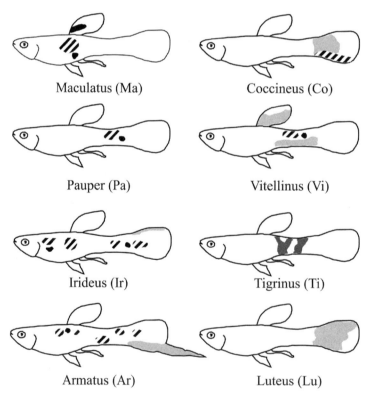

Maculatus (Ma)

Coccineus (Co)

Pauper (Pa)

Vitellinus (Vi)

Irideus (Ir)

Tigrinus (Ti)

Armatus (Ar)

Luteus (Lu)

Figure 23.2 The effects of some color genes in male guppies. The four color patterns on the left are examples of patterns found by Winge to be always Y linked. The four on the right are usually Y linked but can be found on the X chromosome (i.e., they sometimes recombine to X). Red spots (barred spots in the figure) may be orange or red, yellow spots (light gray in the figure) are always yellow, black spots are obligate black spots, and dark-gray spots are equivalent to facultatively expressed black spots (fuzzy black). Redrawn from Winge & Ditlevsen 1947.

these, and studies of particular color patterns of commercial interest to ornamental-guppy breeders continue to generate new insights into guppy color pattern genetics (Khoo et al. 1999a, 1999b; Phang et al. 1999).

As more powerful genomic tools become available (e.g., Tripathi et al. 2008; Tripathi et al. 2009b), we expect that the field will achieve both greater resolution and a more complete understanding of the allelic basis of color patterns. The genes that influence the presence of particular spots, including those identified by Winge, result in the predictable expression of one or more spots in a particular place on the body. Some of these genes, especially the Y-linked genes, result in the co-inheritance of several spots of different colors, which can occasionally become disassociated from one another (Winge & Ditlevsen 1947; Purdom 1993). This suggests that these "supergenes" may be tightly linked complexes that are almost exclusively co-inherited due to their proximity, yet are transcribed and thus expressed independently. It also suggests the potential for rare recombination events to create new long-lived associations of color spots.

In stark contrast to the sex-linked color pattern genes, there are only five documented autosomal alleles for color, and these all influence the background shade of the fish rather than the presence of particular color spots (Winge & Ditlevsen 1947; Yamamoto 1975). However, there is considerable scope for autosomal variation to influence the expression of color patterns in other ways, which will become more apparent when we consider the quantitative genetics of color patterns.

23.5 Quantitative genetic variation

Although we now know a considerable amount about the inheritance of particular discrete color pattern elements from evolutionary and domesticated breeding experiments, there is more to the genetics of color patterns that can be accessed only via the tools of quantitative genetics. Most quantitative genetic studies of guppy color patterns focus on the area of the male body covered by each color, the number of spots of a particular color, or overall properties of the color pattern such as contrast. These traits are influenced not only by the presence or absence of particular spots but by their size, intensity, and contrast as well. Quantitative genetic studies, therefore, estimate both the collective effects of the genes we studied in the previous section on Mendelian inheritance and the effects of other sources of variation (e.g., genetic and environmental) that modify the expression of these color pattern genes. Thereby, quantitative genetics deals with the more or less continuously distributed variation in the size of the particular color elements and allows for the partitioning of this variation so that we can resolve the importance of, for example, additive genetic, dominance, or maternal effects.

Traditional quantitative genetic studies can take the form of breeding designs or the study of natural pedigrees, in which the phenotypic resemblance among relatives provides an estimate of the relative importance of genes and environment in shaping variation in coloration. Given a particular selection regime, such estimates allow one to predict how the different color traits will evolve. The other way around, we can infer the genetic basis of a trait from the phenotypic change in coloration that is observed over the course of an artificial-selection experiment. This latter method in particular can provide important insights into how much variation is actually available to selection, because despite high univariate heritabilities, the genetic correlations among traits can often constrain a response (Lande & Arnold 1983; Falconer & Mackay 1996). Indeed, selection directly on color patterns in guppies has resulted in a rapid response in the amount of orange coloration (Houde 1992, 1994; Brooks & Couldridge 1999) and the amount of black (Brooks & Couldridge 1999). The speed of this response suggests very high levels of additive genetic variation in these measures.

23.5.1 Y-linked genetic variance

In accordance with what we know about the inheritance of discrete color patterns (see above), results from quantitative genetic studies of male coloration are typically consistent with a large Y-linked component of the variance in coloration (Houde 1992; Brooks & Endler 2001a; Karino & Haijima 2001; Hughes et al. 2005). For example, father-son regression analyses of males from the Paria River, Trinidad, showed that the variance in orange coloration is highly heritable (Houde 1992). In fact, when we assume autosomal inheritance (and thus double the slope of the regression; e.g., Falconer & Mackay 1996), the heritability estimates in two separate experiments for orange color area were 1.08 and 1.42 and thus exceeded the maximum heritability biologically possible. Likewise, father-son regression using feral guppies from Okinawa, Japan, resulted in a heritability estimate of 1.02 (Karino & Haijima 2001). These exceptionally high heritability estimates support the idea that much of the quantitative genetic variation underlying the total area of orange coloration is Y linked, at least in some populations.

Further support for the importance of Y linkage is provided by breeding designs in which the resemblance in coloration among full and/or half sibs is quantified. For example, Hughes et al. (2005) used a full-sib breeding design and showed that broad-sense heritability estimates were substantially inflated (>1) for several of their measures of orange, black, and silver coloration (they studied the inheritance of each color type on three different body regions) in males from the El Cedro and Guanopo rivers in Trinidad. Similarly, in a half-sib breeding design, the estimated variance attributable to sire was much higher than that attributed to dam for the areas of black, fuzzy black, orange, and total iridescence, as well as for overall mean brightness, contrast, mean chroma, and tail area, in males from Alligator Creek, Australia (where guppies are feral) (Brooks & Endler 2001a). All these patterns of inflated heritability estimates are consistent with a large proportion of the genetic variation in these traits being Y linked.

Importantly, however, none of the above approaches can tell us exactly how much of the variance in a given trait is autosomal and how much is Y linked, and hence, they do not allow for an explicit test for Y linkage. In a half-sib breeding design, for example, the sire variance component contains all the Y-linked variation but also one-half of the X-linked variance and one-fourth of the additive genetic variance, whereas the dam component contains one-half of the X-linked variance and one-fourth of the additive genetic variance, as well as one-eighth of the variance that is due to epistatic and dominance interactions among alleles and all the maternal effects (Falconer & Mackay 1996). The Y-linked variance component cannot, therefore, be formally partitioned from these other sources of variance.

Unlike these traditional methods, relatively new approaches in quantitative genetics that fit an animal model to complex pedigrees can, at least in theory, partition out other genetic components (Lynch & Walsh 1998). The simple patrilineal inheritance of the Y chromosome makes such an exercise particularly simple and appealing, at least once one has gone through the considerable trouble of establishing a multigeneration pedigree through which one is able to track the inheritance of each individual Y chromosome. We have recently completed exactly such an exercise in guppies from Alligator Creek (Brooks & Postma, unpublished data) and found evidence for significant Y-linked and non-Y-linked genetic variance across the six traits we studied (fig. 23.3). Interestingly, however, the relative amount of variance attributable to Y linkage varied substantially among traits, with Y linkage being most important in black, fuzzy black, and orange, whereas variation in body and tail size was predominantly autosomal or X linked.

A different but complementary approach to infer Y linkage of quantitative genetic variation in male coloration was taken by Tripathi et al. (2009b), who constructed a genetic linkage map using crosses between guppies from a high-predation (Cumaná, Venezuela) and a low-predation stream (Quaré River, Trinidad). They found significant quantitative trait loci (QTL) for a number of morphological and color traits, several of which mapped to the sex-

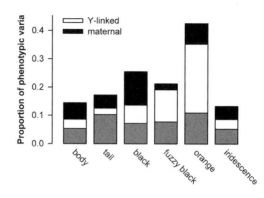

Figure 23.3 Partitioned sources of variance in the ornamental traits of male guppies. Using data from a multigeneration breeding experiment (Brooks & Postma, unpublished data), we partitioned the phenotypic variance in six ornamental traits into an autosomal, a Y-linked, a maternal, and a residual variance component. Note that the maternal component contains all variance attributable to the identity of the mother and may have either a genetic or an environmental basis.

determining locus on the Y chromosome. However, because of the reduced levels of recombination between the X and Y chromosomes, and the absence of Y-linked markers, it is difficult to resolve the actual number of Y-linked QTL and their precise location. Additionally, they found one marker associated with a pigment candidate gene that mapped to the sex-linked linkage group. However, in medaka (*Oryzias latipes*), mutations in this gene affect only the body color rather than any specific ornamental traits (Tripathi et al. 2008).

23.5.2 Non-Y-linked genetic variance

In addition to formally demonstrating a significant Y-linked component, the significant non-Y-linked variance in our pedigree-based study (Brooks & Postma, unpublished data) shows that much of the quantitative variation in these traits is attributable to autosomal or X-linked variation. This is corroborated by the significant dam variances for orange area and total spot number (as well as several other large but nonsignificant dam variances in other traits) in an earlier half-sib study (Brooks & Endler 2001a). Furthermore, several QTL for different color traits, as well as a number of candidate genes involved in pigment-cell differentiation, have been found to map to autosomal linkage groups (Tripathi et al. 2008; Tripathi et al. 2009b).

The substantial amount of autosomal genetic variation for quantitative color variation is in contrast to the predominantly Y-linked patterns of inheritance found for most discrete color patterns. Although this variation is likely to be due in part to the effects of the few X-linked and autosomal alleles that influence the presence or absence of particular spots and patterns (see above), it is also likely that there are additional segregating (non-Y-linked) alleles that affect the size and intensity of color spots, such as those that influence pteredine biosynthesis or carotenoid acquisition, over and above the Y-linked inheritance of their presence or absence.

23.5.3 Genetic covariances among traits

Given that ornamental traits are likely to share pathways associated with sex dependence, and given that many of the Mendelian genes involved influence more than one color, the different aspects of male coloration are unlikely to be phenotypically and genotypically independent. Indeed, there appears to be considerable pleiotropy between body size, tail size, and the area of each color in males. For example, there were strong genetic covariances attributable to sires in our earlier half-sib breeding design (Brooks & Endler 2001a). Moreover, there were significant patterns of genetic covariation among color traits in our recent pedigree-based study (Brooks & Postma, unpublished data). Interestingly, both genetic components of variation showed different patterns of covariation, suggesting very different sets of evolutionary forces acting on Y-linked and non-Y-linked variation. On the whole, it is clear that there is a considerable amount still to be learned about the quantitative genetic basis of variation in different male ornamental traits and their covariation in male guppies and how these are shaped by selection.

23.5.4 Genotype-environment interaction

Environmental factors such as the intensity of intrasexual competition for mates, availability of carotenoids and macronutrients, and the risk of predation might reasonably be expected to influence the expression of genetic variation for coloration. Although environmental variation among populations is associated with systematic differences in color patterns (see extensive discussion below as well as in Grether & Kolluru, **chapter 6**; Coleman, **chapter 7**; Kelley & Brown, **chapter 16**), the two studies testing for genotype-environment interactions (G × E) found little evidence that either food availability (Hughes et al. 2005) or the presence of females (Miller & Brooks 2005) interacted with genotype to alter color pattern expression.

23.6 Sex-dependent selection and sexually antagonistic genes

The selection operating on males and females often differs, sometimes dramatically. This sex-dependent selection fa-

vors the evolution of the sexes toward sex-specific phenotypic optima, thereby driving the evolution of sexual dimorphism in anatomy, physiology, behavior, and life history (Bonduriansky & Chenoweth 2009). The evolution of sexual dimorphism toward sex-specific optima is constrained, however, by the fact that under Mendelian inheritance an autosomal allele is expressed equally often in males and in females, and selection favors alleles that provide the greatest average benefit across the sexes (Rice 1992; Bonduriansky & Chenoweth 2009). Such alleles may either produce both males and females of intermediate quality or increase fitness when expressed in one sex but decrease fitness in the other. The latter are known as sexually antagonistic alleles. There is a growing appreciation that there can indeed be considerable sexually antagonistic genetic variation segregating within populations (Rice 1992; Chippindale et al. 2001; Fedorka & Mousseau 2004; Foerster et al. 2007; Prasad et al. 2007).

A sexually antagonistic allele is likely to be favored by selection if its positive effect on fitness in one sex outweighs its negative effect when expressed in the other sex, or if it tends to be expressed more often in the sex in which it is beneficial than in the other sex. Sex-specific selection and the resulting phenomenon of intralocus sexual conflict are thought, therefore, to be important to the evolution of sex limitation (Rhen 2000), sex linkage (Fisher 1930; Bull 1983; Rice 1984, 1996a), X-chromosome inactivation (Wu & Xu 2003; Engelstädter & Haig 2008), suppressed X-Y chromosome recombination (Rice 1996a), genomic imprinting (Iwasa & Pomiankowski 1999b; Day & Bonduriansky 2004), and meiotic drive (Wilkinson et al. 1998), all of which may bias the presence or expression of genes toward the sex in which they are favored by selection.

Fisher (1930) first recognized the significance of sexually antagonistic genes in his interpretation of Winge's (1922a, 1922b, 1927) findings of Y-linked color patterns in guppies. He realized that Y linkage presented the perfect solution to male-benefit sexually antagonistic genes because such genes, when inherited on the Y, are only ever carried by males. Females never bear the cost of their expression or of mechanisms that exist to limit their expression. Despite the importance of Fisher's (1930) insights on guppies to the idea of sexually antagonistic genes (although this term has been in existence less than 20 years), most work in this area has so far focused on the homogametic sex chromosome (X or Z) rather than on the heterogametic (Y or W) chromosome. This is probably due in large part to the fact that in many species with chromosomal sex determination, including mammals and *Drosophila*, the heterogametic chromosome has been eroded so that it represents only a small portion of the genome (Graves 1995; Rice 1996a). Nonetheless, in some taxa the heterogametic chromosome is in fact large and contains a large number of genes, and evidence is accumulating that many of these genes are sexually antagonistic in nature (Iyengar et al. 2002; Lindholm & Breden 2002; Lindholm et al. 2004; Ellegren & Parsch 2007).

Although one might argue that the large Y chromosome found in guppies and some other poeciliid fishes is rather exceptional, and that the strong Y linkage of male coloration cannot therefore be directly extrapolated to other taxa, it is important to realize that its simple patrilineal inheritance makes it much more straightforward to quantify the amount of Y linkage than X (or Z) linkage. Furthermore, selection for the evolution of Y (or W) linkage of sexually antagonistic alleles is expected to be much stronger than selection for X or Z linkage. These two factors together make Y linkage in guppies an exceptionally powerful model system for the study of the phenomenon of sex linkage and for the evolution of sex chromosomes in general.

23.6.1 Guppy color patterns as sexually antagonistic traits

As we will show below, it is difficult to imagine a clearer example of a sexually antagonistic trait than the color patterns of male guppies. Male guppies are probably unrivaled in their arduous and persistent courtship of females via their visual "sigmoid" display (Baerends et al. 1955). While some of these displays progress through a series of stereotyped steps all the way to the point of intromission, many displays are terminated by the female before that stage, and most displays in fact barely gain the attention of the female (Baerends et al. 1955; Houde 1987). Females therefore control the eventual outcome of each male courtship attempt, and such control is the primary avenue of precopulatory mate choice in guppies (see Rios-Cardenas & Morris, chapter 17).

A vast literature demonstrates that male courtship success (i.e., attractiveness) is associated with the male's color pattern (Endler 1983, 1988, 1995; Kodric-Brown 1985; Houde 1987; Houde & Endler 1990; Brooks & Endler 2001a). However, the particular color pattern components, alone or in combination with others, that are most attractive vary not only among populations (Houde & Endler 1990; Endler et al. 2001) but also with local conditions within populations (Endler 1987; Brooks 2002; Gamble et al. 2003). Furthermore, it has been found that the local rarity of a pattern also increases its attractiveness to females (Farr 1977; Hughes et al. 1999; Zajitschek & Brooks 2008), leading to negative frequency-dependent selection on color patterns. Zajitschek and Brooks (2008) showed experimentally that female preferences for particular com-

binations of male color patterns are in fact operating in parallel with preferences for rare patterns, suggesting two concurrent sources of sexual selection on male color patterns. Irrespective of these fascinating details and complexities, it is, however, beyond doubt that color patterns are essential to mate choice in guppies, with dull or indistinctive males experiencing much lower mating success than males with bright or distinctive color patterns. It is thus clear that these color patterns are under strong sexual selection via mate choice (Houde 1987, 1997) (Rios-Cardenas & Morris, chapter 17). However, as we will see below, such mating success comes at a cost.

One of the major reasons guppies are among the most iconic examples of the interplay between sexual and natural selection is the fascinating differences we find in coloration across the variety of streams in which they occur, particularly on the island of Trinidad. These streams differ in flow, productivity (Grether et al. 1999; Grether & Kolluru, chapter 6), availability of particular nutrients (Rodd et al. 2002), spectral transmission properties of the water (Endler 1991; Coleman, chapter 7), presence of competitors, and, especially, species of predators present (Endler 1978, 1980, 1982, 1983, 1995; Millar et al. 2006; Kelley & Brown, chapter 16).

Haskins and Haskins (1951) first noted an association between the predator fauna in a particular stream and the color pattern of male guppies in that stream. Their subsequent work with collaborators (Haskins et al. 1961), as well as a series of field and laboratory studies by Endler (1978, 1980, 1982, 1983), showed that selective predation of bright and distinctively colored individuals by large, visually hunting piscivores is driving the evolution of guppy color patterns. Differences among populations in predation have been implicated in the replicated divergence in a suite of life-history traits besides color patterns and, possibly, mating preferences (Reznick & Endler 1982; Reznick et al. 1996b; reviewed by Magurran 2005, 12–28; see also Pires et al., chapter 3; Johnson & Bagley, chapter 4). Despite the realization that factors like turbidity and productivity covary with predation (the more voracious predators such as the pike cichlid *Crenicichla frenata* occur in larger, lower-altitude streams that have higher nutrient loads and more turbid water), sifting through the evidence from a vast number of different localities, including translocations of prey and predators, led Millar et al. (2006) to conclude that the composition of the predatory-fish fauna remains the strongest but not the only factor influencing the mean level of male coloration in a given stream.

While net selection on a color pattern gene carried and expressed by a male is more likely to be positive, the same gene carried and expressed in a female is more likely to be negative, since it would attract the predation cost without the attractiveness benefit. Hence, the selection imposed by predators on male color patterns is perhaps the major feature that makes color pattern genes sexually antagonistic, making guppies an iconic example of the opposition between sexual selection favoring bright, conspicuous males and predator-induced natural selection favoring crypsis.

Given the strong Y linkage of many patterns, the study by Haskins et al. (1961) in which they examined the color patterns of male and female guppies caught below and above a large waterfall on the Aripo River is particularly noteworthy. Above this waterfall, fish experience relatively low predation, whereas below the waterfall, they experience much higher predation. Interestingly, females caught below the waterfall never expressed any color patterns when treated with testosterone. Females caught above the waterfall, on the other hand, did show such color patterns, suggesting the presence of X-linked or autosomal color pattern genes with sex-limited expression. Likewise, breeding experiments with a particular color pattern (*Sb*) that can be inherited on both the X and the Y chromosome showed that it was never X linked in high-predation downstream localities but that it was found on the X chromosome above the waterfall (Haskins et al. 1961).

Although it is clear that sexually antagonistic color pattern genes still have much to teach us about the evolution of suppressed X-Y recombination, this pattern provides a tantalizing glimpse of what may be the evolution of Y linkage in action, an important step in sex chromosome evolution (Rice 1987a, 1996a). On the whole, this and other available evidence suggests, therefore, that Fisher's (1930) prediction is correct and that the observed Y linkage, X linkage, and sex limitation of color pattern genes in guppies are due to their sexually antagonistic effects.

23.6.2 Other forms of sexual antagonism

Even in the absence of predation, however, color pattern genes are likely to be costly. For example, Winge (1934; Winge & Ditlevsen 1938) experimentally created XY females using a testosterone treatment and then crossed these to XY males to make YY males that were either homozygous or heterozygous for the Y-linked *Maculatus* gene (the patterns mentioned in this paragraph are illustrated in fig. 23.2). Later, Haskins and colleagues (Haskins et al. 1970) did the same for a variety of other Y-linked genes. They showed that in their homozygous form, the *Maculatus*, the *Armatus*, and the *Pauper* Y haplotypes were all lethal, suggesting the existence of recessive lethal alleles in association with each color pattern. Heterozygotes between each pair of these three patterns, as well as other patterns, proved

both viable and fertile, suggesting that the recessive lethal alleles associated with each pattern are different.

Another example comes from experiments by Farr (1980b, 1981), who used a variety of strains that had been kept in captivity for many years to each retain a particular color pattern, including *Maculatus* (started by Schmidt 1920) and *Pauper* (started by Winge 1927), as well as several others (some of which were started as recently as 1959). He showed that the longer a strain with a Y-linked color pattern had been kept in the laboratory, the more female biased the sex ratio was, with in the end only 30% of adults in the *Maculatus* strain being male. He considered a number of possible explanations and concluded that the evidence favored either a pleiotropic cost of the Y-linked color pattern or a genetic deterioration of the Y chromosomes over time due to the accumulation of deleterious alleles. In the latter case, reduced recombination on the Y chromosome together with selection to keep the color pattern intact could favor the accumulation of a significant genetic load. To date, however, we do not know the extent to which the deleterious recessives associated with *Maculatus*, *Armatus*, and *Pauper* (Winge & Ditlevsen 1938; Haskins et al. 1970) accumulated under strain domestication in the laboratory or if they were already present in the wild.

Finally, in a quantitative genetic study using wild-caught parents from Alligator Creek, Australia, Brooks (2000) found a strong negative genetic correlation between male attractiveness (from behavioral assays) and offspring survival to maturity. This pattern was driven by the survival of sons, as the number of sons at maturity was negatively genetically correlated with attractiveness. Moreover, the pattern continued beyond maturity with the longevity of adult males negatively correlated with attractiveness among families. This result could be a pleiotropic effect of the expression of attractive color patterns, possibly due to their costliness (see Kokko et al. 2002). However, it could also be due to the accumulation of deleterious alleles via genetic hitchhiking, with the most attractive Y-linked color pattern supergenes accumulating a heavier mutational load than less attractive supergenes because they tend to be favored early in a male's adult life due to increased mating success (Brooks 2000).

On the whole, these studies show that genetic variation in coloration comes with important benefits but also with some potentially severe costs, even in the absence of predation. As females are again most likely to carry only the costs, color genes can be considered sexually antagonistic, even when predation risk is low. This is particularly true if these costs are pleiotropic consequences of the expression of male color patterns rather than a consequence of tight linkage with deleterious alleles, although distinguishing between these two possibilities is beyond our current ability. Irrespective of the exact underlying mechanism, however, in addition to teaching us more about the nature of sexually antagonistic genes, guppies have considerable potential to provide insights into the evolution of reduced recombination and reduced physical size of Y chromosomes. The observations to date are consistent with, and constitute some of the best evidence for, the sexually antagonistic gene chain reaction hypothesis for the evolution of X-Y dimorphism (Rice 1987a, 1987b, 1994).

23.7 Inbreeding depression

The area and number of black and orange spots in male guppies have been shown to decrease under inbreeding in two out of three studies to date (Sheridan & Pomiankowski 1997b; van Oosterhout et al. 2003b; but not Mariette et al. 2006). Such inbreeding depression in guppy color patterns is especially interesting because of the large (but not exclusive) Y-linked component of variance in these traits. Y chromosomes are hemizygous, and so there is no opportunity for dominance effects, which are generally believed to be the main determinant of inbreeding depression. Inbreeding depression in Y-linked guppy color patterns may thus be due exclusively to the autosomal alleles that influence color pattern expression, including those that influence condition and resource acquisition (Grether 2000). Alternatively, inbreeding depression may be due, at least in part, to epistatic interactions between the Y chromosome and X-linked or autosomal genes (van Oosterhout et al. 2003b). Therefore, guppies make uniquely suitable subjects for valuable studies of the role of epistasis in inbreeding depression, a role that remains controversial (Charlesworth & Charlesworth 1987).

23.8 Why are guppy color patterns so polymorphic?

We began this chapter restating the assertion made by others (Haskins et al. 1970; Chenoweth & Blows 2006) that the genetic polymorphism in guppy color patterns remains one of the most extreme known in nature. The maintenance of such extreme polymorphism in traits under selection, however, is one of the most outstanding problems in evolutionary genetics. There is no shortage of hypothetical explanations for this polymorphism, including the usual suspects: negative frequency dependence, temporally fluctuating selection, spatial heterogeneity balanced with gene flow, mutation-selection balance, and antagonistic pleiotropy (Barton & Turelli 1989; Radwan 2008). These sce-

narios have been considered elsewhere, including by us (Brooks 2002), so we will not review them exhaustively here. Instead, to conclude, we highlight some of the most promising possibilities, paying particular attention to some exciting recent developments.

23.8.1 Negative frequency dependence

Negative frequency dependence is both the strongest theoretic explanation for the maintenance of genetic variance (Barton & Turelli 1989) and the best supported in guppies. The idea that predators could fail to detect and/or catch males with rare color patterns, perhaps due to the lack of a search image for such males, was suggested by Farr (1977) and Endler (1980), but it was only recently that this prediction was upheld with strong positive evidence. Exhaustive and careful release and recapture studies of marked males with known color patterns showed negative frequency-dependent survival, quite probably due to predation effects (Olendorf et al. 2006).

A body of evidence supports the existence of negative frequency-dependent mate choice, starting with Farr's (1977, 1980b) demonstration of female preference for males with rare or novel patterns. Later, Hughes et al. (1999) showed in a compelling experiment that males with color patterns that are unfamiliar to females enjoy such an advantage, and Eakley and Houde (2004) demonstrated that when it comes to second and subsequent matings, females discriminate against males that share the color pattern of previous mates. Finally, Zajitschek and Brooks (2008) experimentally showed that pattern rarity, rather than a lack of familiarity with the color pattern, confers increased mating success.

23.8.2 Temporal and spatial heterogeneity

Conditions such as predation, carotenoid availability, and light environment vary both among ponds or streams and over time, and much research has focused on the potential for such variation to maintain variation in male color patterns (Endler 1978, 1980, 1983; Grether et al. 1999; Rodd et al. 2002; Gamble et al. 2003). As a matter of fact, it is not even clear that it is reasonable to expect a single optimum color pattern in a given location (Endler 1980). While thus far there is very little empirical support for the idea that genotype-environment interaction could contribute to color pattern polymorphism (Hughes et al. 2005; Miller & Brooks 2005), there is no doubt that there is high gene flow among pools (Haskins et al. 1961; Shaw et al. 1992; Becher & Magurran 2000), and the preference for unfamiliar males with rare or novel patterns can only serve to facilitate such gene flow.

23.8.3 There may be less genetic variation than we think

At this stage it is beyond doubt that there is copious genetic variation in color patterns. Nevertheless, how much of this variation is actually exposed to selection? This is a problem that is presenting itself currently in a variety of species. For example, when one looks at the multivariate genetic variation in the direction of selection, there appears to be far less than in directions not under strong selection (Hine et al. 2004; Blows & Hoffmann 2005; Hunt et al. 2007; Van Homrigh et al. 2007). Brooks and Endler (2001a) showed that when the strong directional sexual selection gradients that they estimated were combined with the genetic variance-covariance matrix (**G**), there was very little predicted response in size or color traits. Further support comes from nonlinear analyses of selection in the same experiment, which showed that there were at least three ways to make an attractive guppy (three peaks on the fitness surface) (Blows et al. 2003). Without the capacity to shuffle color pattern genes via recombination, it appears that these three peaks will always be separate, and thus that there may always be more than one way to make an attractive guppy.

A conclusive test of this idea comes in the form of selection directly on attractiveness (Hall et al. 2004). When, instead of selecting some aspect of coloration, sexual selection was emulated by selecting on attractiveness itself, there was very little further evolution of attractiveness, suggesting the presence of little multivariate additive genetic variation in attractiveness (and thus coloration) in the direction of sexual selection.

These results, like many recent developments in the field, scratch the surface of a host of interesting new questions about the genes underlying ornamental traits and how they influence and are shaped by the selective forces operating in populations, including populations of guppies and other poeciliids. The guppy remains a unique, distinctive, and highly tractable genetic and ecological model species, and many exciting discoveries await, no doubt, as the genomic revolution reaches guppy labs.

Chapter 24 A primer of sex determination

Manfred Schartl, Delphine Galiana-Arnoux, Christina Schultheis, Astrid Böhne, and Jean-Nicolas Volff

24.1 Introduction

SEX DETERMINATION IS DEFINED as "the mechanism directing sex differentiation" rather than the process of sex differentiation itself, which is "the development of testes or ovaries from the undifferentiated gonad anlage" (Hayes 1998). Sex can be determined either by the genetic constitution of the individual (termed genetic sex determination, GSD) or by the environment (environmental sex determination, ESD). The most prominent environmental factor involved in ESD is temperature (temperature-dependent sex determination, TSD), but other triggers such as social factors may also play a role. Many different modes of GSD exist, but in the simplest case a single pair of chromosomes, called sex chromosomes, are involved. The development of males or females then depends on the dosage of the sex chromosomes. In the most common case there are two types of sex chromosomes, one of which is present only in one sex, while the other is present in both sexes, in either the heterozygous or the homozygous state. The underlying mechanism is that there is a dominance of factors that determine one of the sexes, and these accumulate on one of the sex chromosomes—or that there is even a single master sex-determining gene (male XY–female XX and male ZZ–female WZ mechanisms are most widespread). At the other extreme, sex-determining factors can be distributed over a few (oligofactorial, -genic) or many (polyfactorial, -genic) chromosomes. Another variation comes from the fact that sex determination mechanisms can act in different individuals, which are the male or the female (so-called gonochorists), or in the same individual, which then goes through a male and female phase (either consecutively or simultaneously) and which is known as a hermaphrodite.

Teleosts exhibit an enormous range of sex determination mechanisms, and poeciliids perfectly illustrate the diversity of GSD mechanisms observed in teleosts. Besides unisexuality (Schlupp & Riesch, **chapter 5**), a wide spectrum of sex-determining mechanisms has been described in poeciliids, including male and female heterogamety with or without an influence of autosomal genes, as well as more complicated situations such as multichromosomal and polyfactorial sex determination (fig. 24.1). Due to the presence of different mechanisms in closely related species, or even among populations within a species, poeciliids are very attractive models to study the evolution of sex determination.

This chapter gives an overview of our present knowledge about sex determination in poeciliids. Since poeciliids are well suited for genetic studies (see the other chapters in **part V** of this volume), a great deal of information has been amassed on the inheritance of genetic determinants of sex. This has made some poeciliids textbook examples of the various modes of GSD. In an era when molecular biology and genomics are applicable to a wider range of organisms than traditional mainstream models are, poeciliids are being used to isolate master sex-determining genes and to unravel the regulatory network that governs this developmental process. Such work will be of interest to more than just poeciliid biologists, since it is expected to contribute to the general understanding of the evolution of sex determination and the molecular mechanisms of sexual development, which, even in the classical models, are problems that have yet to be resolved.

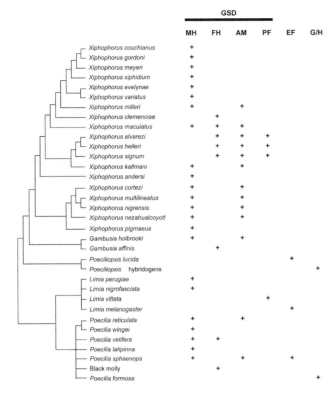

| | GSD | | | | | |
	MH	FH	AM	PF	EF	G/H
Xiphophorus couchianus	+					
Xiphophorus gordoni	+					
Xiphophorus meyeri	+					
Xiphophorus xiphidium	+					
Xiphophorus evelynae	+					
Xiphophorus variatus	+					
Xiphophorus milleri	+		+			
Xiphophorus clemenciae		+				
Xiphophorus maculatus	+	+	+			
Xiphophorus alvarezi		+	+	+		
Xiphophorus helleri		+	+	+		
Xiphophorus signum		+	+	+		
Xiphophorus kallmani	+		+			
Xiphophorus andersi	+					
Xiphophorus cortezi	+		+			
Xiphophorus multilineatus	+		+			
Xiphophorus nigrensis	+		+			
Xiphophorus nezahualcoyotl	+		+			
Xiphophorus pigmaeus	+					
Gambusia holbrooki	+		+			
Gambusia affinis		+				
Poeciliopsis lucida					+	
Poeciliopsis hybridogens						+
Limia perugiae	+					
Limia nigrofasciata	+					
Limia vittata				+		
Limia melanogaster					+	
Poecilia reticulata	+		+			
Poecilia wingei	+					
Poecilia velifera	+	+				
Poecilia latipinna	+					
Poecilia sphaenops	+		+	+		
Black molly		+				
Poecilia formosa						+

Figure 24.1 Overview of known or supposed sex determination systems in poeciliids and their phylogenetic relationships. For details on sex determination systems, see table 24.2 and text. The molecular tree is based on the analysis of mitochondrial sequences according to Meyer et al. 2006 and Hrbek et al. 2007. GSD, genetic sex determination; MH, male heterogamety; FH, female heterogamety; AM, autosomal modifiers; PF, polyfactorial determination; EF, environmental factors; G/H, gynogenesis/hybridogenesis.

24.2 Genetic sex determination in *Xiphophorus*

The genus *Xiphophorus* consists of four major subgroups: northern swordtails, southern swordtails, northern platyfish, and southern platyfish (Kallman & Kazianis 2006). The genus currently comprises 26 described species (see Rauchenberger et al. 1990; Kallman et al. 2004). Thanks to decades of intensive genetic studies, mainly by Klaus Kallman, this genus is the best-documented group among poeciliid fishes for studies of GSD. Simple male and female heterogametic systems together with more complicated situations involving multiple loci and chromosomes were found. Kallman (1984) concluded that the sex chromosomes of different *Xiphophorus* species with different sex determination mechanisms are derived from a common ancestral sex chromosome. This conclusion has yet to be confirmed by molecular markers. What we know is that the melanoma gene *xmrk* (Schartl & Meierjohann, **chapter 26**) is homologous in *X. maculatus*, *X. milleri*, *X. birchmanni*, *X. cortezi*, *X. montezumae*, *X. variatus*, *X. evelynae*, and *X. xiphidium* (Weis & Schartl 1998). In most of those species (where the sex determination system is known) the gene

is sex chromosomal and closely linked to the macromelanophore pattern locus *mdl*, with the exception of *X. cortezi* (Kallman & Atz 1966), where the macromelanophore locus and the associated *xmrk* are autosomal. These findings are consistent with a common origin of sex chromosomes. Alternatively, a strong selection pressure might exist to recruit (the same) pigmentation genes to sex chromosomes of different origin.

It should be noted that for some species an extensive data base is available, while for others only single experiments have been performed or single populations have been studied. Consequently, in some species the true situation may be more complicated, with different sex chromosomes and autosomal modifiers being present in addition to the simple picture emerging from a few laboratory crosses. No data exist for the sex determination systems in *X. birchmanni*, *X. continens*, *X. malinche*, *X. mayae*, *X. mixei*, *X. monticolus*, and *X. montezumae*.

24.2.1 *Xiphophorus maculatus*

Xiphophorus maculatus (southern platyfish) has three different genetically well-defined sex chromosomes: X, Y, and W (Kallman 1984). Possible female genotypes are XX, XW, and YW, whereas males are XY or YY. Males with two Y chromosomes—even when present in the same population (being identical in gene content)—are fully viable. Viable WW females have not been found in natural populations but can be generated under laboratory conditions (Kallman 1984). The W chromosome seems to be absent from some natural populations with simple XX-XY systems—for example, in the Río Jamapa (Veracruz, Mexico) population at the northern range of the species. However, an estimated 60% of *X. maculatus* populations have a sex determination scheme involving three sex chromosomes (Kallman 1965). Laboratory strains with exclusive female heterogamety (YW females and YY males) or male heterogamety (XX females and XY males) exist (Walter et al. 2006). Cases of atypical sex determination—for instance, XY females or XX males (less than 1% in the platyfish)—can be explained by the action of autosomal modifiers (Kallman 1984). Exogenous factors can also override GSD mechanisms when manipulated experimentally. For example, X-ray irradiation can "force" XY embryos to develop into females (Anders et al. 1969).

As in many other teleosts, the X and Y sex chromosomes recombine over a large portion of the whole linkage group, but recombination frequencies appear to be much lower around the *SD* region. This is evident from the clustering of markers around *SD*, making the sex chromosomes (linkage group 24) in platyfish the smallest of all linkage groups (Kazianis et al. 2004a), although it corresponds cytogeneti-

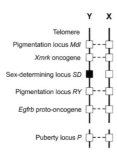

Figure 24.2 Gene loci and genes linked to the master sex-determining gene *SD* on the sex chromosomes of the southern platyfish *Xiphophorus maculatus*. The exact position of the *P*-locus relative to other loci is still unknown. *egfrb*, epidermal growth factor receptor b; *mdl*, macromelanophore-determining locus; *RY*, red-yellow pigmentation locus; *xmrk*, *Xiphophorus* melanoma receptor kinase.

cally to one of the larger chromosomes (Nanda et al. 2000; Schultheis et al. 2006).

Besides the master sex-determining locus *SD*, other genetic loci and genes have been mapped on the X and Y chromosomes (fig. 24.2) (Kallman 1975; Gutbrod & Schartl 1999). The *SD* locus is located between two genes, the melanoma-inducing oncogene *xmrk* and its proto-oncogene counterpart *egfrb*, which is the ancestral duplicate version of *xmrk* (Schartl & Meierjohann, **chapter 26**). Other linked loci have been mapped close to *SD* but have not yet been identified at the molecular level. One of these, the macromelanophore-determining locus *mdl*, controls polymorphic pigmentation patterns formed by large melanin-producing cells (macromelanophores), which are the cellular progenitors of melanoma in *Xiphophorus* (Schartl & Meierjohann, **chapter 26**). Another polymorphic locus linked to *SD* is the red-yellow locus *RY*, responsible for the red-yellow pigmentation patterns of the iris, skin, and fins in *Xiphophorus*.

Finally, the puberty locus *P* (P stands for pituitary, because this gland is critically involved in gonad maturation; Kallman 1989), which determines the timing of sexual maturity, resides in the same chromosomal region as *SD*, although its precise position has not yet been determined. In wild platyfish nine alleles (*P1–P9*) have been described so far, residing on either the X or the Y chromosome, or in some cases on both (Kallman 1989). *P1* determines very early puberty (*P1P1* males mature at 8 weeks) and small male size (standard length 21 mm), because males stop growing when they reach maturity. The *P5P2* combination codes for later maturity and large male size (more than 36 weeks for maturation and standard length more than 40 mm). Combinations of the other alleles of this series fall in between these extremes and lead to a continuum of male size classes in the southern platyfish. *P* is supposed to determine the timing of the first activation of the hypothalamus-pituitary-gonadal axis, possibly through regulating gonadotropin-releasing hormone and/or luteotrophic hormone–releasing hormone

(Halpern-Sebold et al. 1986; Kallman & Bao 1987). Evidence has been presented in a study on *X. nigrensis* and *X. multilineatus* that the *P*-locus is made up of a series of allelic copies of the melanocortin4 receptor gene (*mc4r*). Some copies may encode nonfunctional receptors that can dilute out the signal from the wild-type receptor (Lampert et al. 2010). It is hypothesized that a certain threshold of Mc4r signal has to be reached to trigger the onset of sexual maturation. Late *P* alleles contribute more nonfunctional receptors, and hence the threshold is reached later in life. It is, however, not totally clear whether the *P*-locus in all *Xiphophorus* species functions in the same way.

None of these loci has yet been identified on the W chromosome. Therefore, its relationship with the X and Y remains uncertain. Thus far, a single W/Y crossover event is known that transferred an *RY* allele to the W chromosome (Kallman 1984). It should be emphasized that a W (strongly female determining) chromosome has never been identified in other platyfish species.

Different models have been proposed to explain sex determination with three sex chromosomes (fig. 24.3). Kallman 1984 suggested that different alleles of a male-determining gene might be present on all three types of sex chromosomes. Only the Y-chromosomal allele would be active, whereas X- and W-chromosomal alleles would be suppressed by autosomal repressors, leading to a male phenotype only when the Y chromosome is present. In YW females, the W chromosome should itself carry a specific suppressor for the Y-chromosomal male-determining allele (Kallman 1984). Another theory involves a gene dosage mechanism based on gene copy number (Volff & Schartl

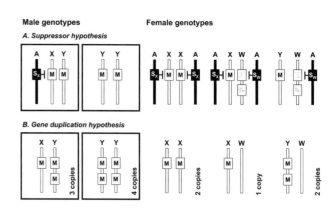

Figure 24.3 Models explaining sex determination with a male-determining gene and three types of sex chromosomes in *X. maculatus*. (A) Suppressor hypothesis (Kallman 1984). A male-determining gene (*M*) is present on all three types of sex chromosomes. The Y-chromosomal allele is active, while X- and W-chromosomal alleles are suppressed by autosomal repressors (S_X and S_W, respectively). In the YW female genotype, the Y-chromosomal allele of *M* is repressed by a W-chromosomal suppressor (S_Y). (B) Gene dosage hypothesis (Volff & Schartl 2001). Different types of sex chromosomes carry different copy numbers of a male-determining gene *M*. The sexual phenotype is determined by the total number of *M* genes provided by the different combinations of sex chromosomes: the presence of at least three copies leads to the male phenotype, and two copies or fewer, to the female phenotype.

2001). According to this model, the Y chromosome carries two copies of a male sex-determining gene; the X chromosome, only one; and the W chromosome, no copy. Depending on the combination of sex chromosomes and the total copy number of the male-determining gene, either the male or the female phenotype will be determined; at least three copies of the gene would lead to the male phenotype (YY four copies, XY three copies), and two copies or fewer would lead to the female phenotype (XX two copies, XW one copy, and WW no copy). A similar mechanism is also conceivable with a dosage-dependent female-determining gene with two copies on the W, one on the X, and no copy on the Y chromosome. In this case, females would possess at least two copies of the female-determining gene. This model is consistent with the frequent occurrence of gene duplications in the sex-determining region of X. maculatus sex chromosomes (Volff et al. 2003), as well as with a possible involvement of gene dosage at the top of the sex determination cascade as proposed for the avian *dmrt1* gene, which is present on the Z but absent from the W chromosome (Nanda et al. 1999).

24.2.2 Xiphophorus variatus, X. evelynae, and X. xiphidium

Like the southern platyfish X. maculatus, the three species of the northern platyfish group X. variatus, X. evelynae, and X. xiphidium have sex-linked macromelanophore (see Schartl & Meierjohann, **chapter 26**, table 26.1) and red-yellow pigmentation patterns. Based on the inheritance of pigmentation phenotypes, they all appear to have a simple sex determination mechanism with male heterogamety (Anders et al. 1973; Borowsky 1984; Kazianis et al. 2005). Although not yet noted from experimental crosses, the influence of rare autosomal modifiers cannot be excluded in these species or any other Xiphophorus species with apparently simple gonosomal sex determination.

24.2.3 Xiphophorus meyeri, X. gordoni, and X. couchianus

In their review, Kazianis et al. (2005) refer to unpublished work by Kallman that X. meyeri and X. gordoni have an XX/XY system. In X. couchianus, an XX/XY mechanism might also be operative (Gordon 1946).

24.2.4 Xiphophorus andersi

Xiphophorus andersi was determined to be XX/XY on the basis of the male-linked inheritance of P alleles for large and small male size in one laboratory stock derived from wild fish from the Río Atoyac (Veracruz, Mexico) (Kallman 1989).

24.2.5 Xiphophorus hellerii

Many laboratory stocks of X. hellerii exhibit sex ratio deviations (more males than females or the contrary). Based on extensive crosses with one stock from Belize (Bx) and one from the Río Lancetilla, Honduras (hIII), Peters (1964) explained the biased sex ratios by a polyfactorial sex-determining system. Such systems operate by male- (M) and female- (F) determining factors that may even occur as different alleles of different strength and that are distributed over various chromosomes, with no chromosome harboring the majority of these factors, or especially strong male- or female-determining alleles. Depending on the genotypic composition, the fish will be either male ($\Sigma M > \Sigma F$) or female ($\Sigma M < \Sigma F$). Peters was even able to explain in her crosses the occurrence of early-maturing, small males and late-maturing, large males. In early-maturing males the preponderance of M factors over F factors is high and guarantees the realization of the male phenotype much earlier during development than in the late-maturing ones. However, this explanation was called into question by Kallman (1984) because of the difficulties in reconciling it to what is known about the genetic and physiological control of puberty by the P-locus. According to the hypothesis of Peters, however, alleles of the puberty P-locus that determine early maturation could be M factor linked.

Intriguingly, other strains have been established in the laboratory that show completely even sex ratios over many generations. For two stocks, the Río Sarabia strain (Sara) (Oaxaca, Mexico) and a strain from Belize (Bel) different from that analyzed by Peters (1964), the use of pigmentation loci that code for either orange or green sword coloration revealed a female heterogametic system with sword coloration linked as two different alleles to the homogametic sex chromosome (Kazianis et al. 2005). Other strains show unbiased or only mildly biased sex ratios (Kallman 1984). This leads to the assumption that in X. hellerii there is a continuum of sex-determining mechanisms ranging from polyfactorial to monofactorial (= gonosomal) sex determination. A likely intermediate would be a situation with sex chromosomes that are especially sensitive to autosomal modifiers. Different chromosomal mechanisms might even operate in the same population or river system, as, for example, in the two strains from Belize mentioned above.

24.2.6 Xiphophorus signum

Female-to-male ratios in X. signum, a southern swordtail species, deviate greatly from 1:1 in different generations and pedigrees (Kallman 1984; Kazianis et al. 2005; M. Schartl, unpublished data). This may point to a similar situation as seen in X. hellerii, although populations or strains with stable gonosomal inheritance have yet to be detected.

24.2.7 *Xiphophorus nezahualcoyotl* and *X. milleri*

At first glance, the sex-determining system for *X. nezahualcoyotl* (previously designated *X. montezumae*; see Rauchenberger et al. 1990) and *X. milleri* appeared simple: a male heterogametic system. In fact, it is more complex, with two Y chromosomes: Y and Y′. The presence of Y invariably produces males. Y′ leads to XY′ females, probably due to the influence of an autosomal modifier that suppresses the male-determining locus on Y′ (Kallman 1984). Consequently, YY′ males can be produced. In both species the Y and Y′ chromosomes can be differentiated due to linked pigmentation loci. In *X. nezahualcoyotl* the Y′ does not harbor a macromelanophore pattern locus, while the strongly male-determining Y carries the macromelanophore pattern *Ss* (previously designated *Sp*; see Schartl & Meierjohann, **chapter 26**). In *X. milleri* the black pattern *Sv* is linked to Y′, and *Gn* to Y (Kallman 1984). In *X. nezahualcoyotl* the X carries the *s* allele of the *P*-locus, while the Y can have either P^s (*s* for small) or P^L (*L* for late) alleles (Kallman 1984).

24.2.8 *Xiphophorus nigrensis*

The sex-determining mechanism in *X. nigrensis* is also a sophisticated version of the XX-XY system. XY animals are all males. By contrast, for XX animals, sex determination depends on an autosomal modifier with two allelic forms, *A* and *a*. Animals with the [*XX, AA*] genotype will be female, but [*XX, aa*] individuals will develop as males. Most [*XX, Aa*] fish will be females, but males can be found at a frequency of about 5% (Kallman 1984). Hence, in the absence of the major Y-chromosomal male-determining gene, the allelic composition at an autosomal locus will determine the sex of individuals. The *A* allele might be necessary to activate a gene important for ovarian differentiation or to repress a testis differentiation gene. Reduction of its expression level in [*XX, Aa*] animals can occasionally lead to male development. The chromosomal location and molecular nature of the *A*-locus are unknown.

Interestingly, male size as determined by alleles of the *P*-locus is connected to sex determination. XX males do always belong to the smallest size class, while XY males can belong to either the small, intermediate, or large size class (Kallman 1984). The X harbors the P^s (*s* for small) allele. The Y can carry P^s, P^I (*I* for intermediate), or P^L (*L* for late) alleles (Kallman 1989).

24.2.9 *Xiphophorus multilineatus*

Xiphophorus multilineatus and *X. nigrensis* share many features and are closely related taxa (Rauchenberger et al. 1990; Meyer et al. 1994; Meyer et al. 2006). Until 1990 they were even regarded as populations (Río Coy and Río Choy, San Luis Potosi, Mexico) of the same species (Rosen 1979; Rauchenberger et al. 1990). In *X. multilineatus*, as in *X. nigrensis*, males occur in small, intermediate, and large sizes. We found that the early-maturing small males are either XX or XY and intermediate and large males are always XY, as in *X. nigrensis* (K. Lampert & M. Schartl, unpublished data), indicating a sex-determining system similar to that in *X. nigrensis*. Besides P^L and P^s, two different P^I alleles were found on the Y chromosome (Kallman 1989).

24.2.10 *Xiphophorus cortezi*

As none of the pigment pattern genes in *X. cortezi* are sex linked (Kallman 1971), the issue of sex determination is still not resolved. Based on crosses with other species of known sex chromosome constitution, and on some crosses with all-female offspring, Zander (1965) proposed an XX/XY system, which can easily be disrupted by autosomal factors. The production of all-female offspring was explained by postulating that the male has the exceptional XX constitution. The rare occurrence of XX males then is due to autosomal modifiers, which might interfere with the major male heterogametic system. Such modifier effects have not been seen in other studies (Kallman 1971). However, interspecific crosses for analyzing sex determination mechanisms can sometimes generate ambiguous results. Consequently, only a study of the inheritance of genetic markers that co-segregate with sex will be helpful in resolving these issues.

24.2.11 *Xiphophorus alvarezi*

A ZZ/ZW system has been inferred for *X. alvarezi* by correlating the inheritance of sword coloration with sex in a stock (DL) derived from the Río Dolores population (Guatemala) (Kallman & Bao 1987). However, in some populations, autosomal modifiers or a polygenic system are likely to act, as exemplified by a strain from the Río San Ramon (Guatemala) that shows strongly male-biased sex ratios (Kallman 1984).

24.2.12 *Xiphophorus kallmani*

Xiphophorus kallmani, a southern swordtail species (Meyer & Schartl 2003) endemic to the northeastern tributaries of Lake Catemaco (Veracruz, Mexico), was analyzed for sex-linked repetitive sequences. This indicated male heterogamety (Nanda et al. 1992). On the other hand, our laboratory stock shows a strong female bias (80% females,

20% males; M. Schartl, unpublished data). This points to a system of XY/XX sex determination with strong influence from autosomal modifiers.

24.2.13 *Xiphophorus clemenciae*

Based on a hybrid cross of female *X. clemenciae* with *X. milleri* (with a known XY system), Kazianis et al. (2005) concluded that *X. clemenciae* has female heterogamety. More studies are needed to find out whether there is purely a gonosomal mechanism or if autosomal modifiers also play an important role in natural populations of these fishes.

24.2.14 *Xiphophorus pygmaeus*

Based on the sex-linked inheritance of the yellow caudal fin pigmentation, an XX/XY sex-determining system was proposed for *X. pygmaeus* (Kallman 1989).

24.2.15 Spontaneous sex reversal in swordtails: an urban myth?

There are a few reports of sex reversal from the scientific literature and quite a number from fish hobbyist journals that report that functional female swordtails (in some cases *X. hellerii* but mostly ornamental hybrids with a southern swordtail phenotype) transformed into functional males. The described phenomenon is in stark contrast to data from scientists who have been working for decades with *Xiphophorus* fishes collected from the wild and who have raised thousands of swordtails in the laboratory (M. Gordon, G. Peters, K. Kallman, A. Anders, F. Anders, M. Schartl, R. Walter). This discrepancy has been discussed (see Kallman 1984). In general, the claim that a female had offspring (hence, was functional) before it changed sex is poorly documented. Essenberg (1992) claimed that two females gave rise to offspring before turning into functional males, which again produced offspring. For the three females reported by Lodi (1979), the unclear breeding history and maintenance in community tanks leave doubts about their functionality.

Most likely, the observed cases of female-to-male sex reversal in swordtails can be explained by a misinterpretation of two well-known phenomena in *X. hellerii*. Late-maturing males are phenotypically very similar to mature females for an extended period of their lives (except for the presence of the gravidity spot, which in ornamental fish is often invisible due to the intense body coloration). Such late-maturing males develop the typical male specific secondary sex characters (gonopodium, sword, intensification of pigmentation) only after puberty. A second phenomenon that can be mistaken as sex change is arrhenoidy, where females take

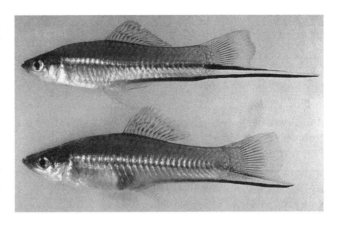

Figure 24.4 Arrhenoidy in *Xiphophorus kallmani*. Upper: normal male; lower: arrhenoid female developing the swordlike caudal appendix typical for males. Note that this phenotype can also be produced by androgen treatment of females.

on the external characteristics of the male. As pointed out by Atz (1964) and Peters (1964), a variety of conditions, most prominently old age, induce the development of male secondary sex characters (most obviously small swords) in females. Kallman reported that this phenomenon was seen in every stock of *X. hellerii* and also in *X. signum* and *X. alvarezi*, but not in any of the northern swordtail or platyfish species. We confirm this observation and add *X. kallmani* (fig. 24.4) and also species of the genus *Girardinus*, where such arrhenoid females have been found in laboratory stocks derived from wild fish (M. Schartl, unpublished data). The emergence of male secondary sex characters in females is obviously the result of a rise in androgen and a decrease in estrogen levels. The metabolic changes connected to aging and why particularly the southern swordtails show this phenomenon are unknown.

In conclusion, if—contrary to all our available reliable scientific information—true sex reversal during the lifetime of a swordtail female does indeed occur, it must be an extremely rare and exceptional phenomenon of no evolutionary consequence.

24.3 Genetic sex determination in *Poecilia*

24.3.1 *Poecilia reticulata*

In the guppy, many color patterns and other traits involved in sexual selection are located on the sex chromosomes (Lindholm & Breden 2002). Through analysis of body and fin color pattern inheritance, it has been concluded that guppies possess an XX-XY system. The X and Y chromosomes can recombine to a large extent (Winge 1922b, 1927). Using conventional cytogenetic techniques, it has not been possible to differentiate the sex chromosomes other than in a few ornamental strains where they differ in size (Nanda et al. 1990). Male heterogamety in ornamental guppies has been confirmed cytogenetically through fluorescent in situ hybridization (FISH) of repetitive DNA, by multilocus fingerprinting with simple repeat oligonucleotide probes, and by specific staining of the constitutive heterochromatin (Nanda et al. 1993), as well as through synaptonemal complex analysis and comparative genomic hybridization (Traut & Winking 2001).

Cases of atypical sex determination (XY females and XX males) indicate that autosomal loci can also interfere with the major sex chromosomal system. In XX males, autosomal male-determining factors are thought to drive male formation and to overwhelm the female-determining factors on both X chromosomes, while a strong female-determining locus might act in XY females (Winge 1930). XX males can mate with normal females and produce all-female offspring (Winge 1930; Nayudu 1979). One case has been described where a regular X chromosome acquired a male-determining function either through translocation of the male-specific region on the Y to the X, recruitment of a new male-determining gene from elsewhere in the genome, or mutation of an X-linked gene (Tripathi et al. 2009a). YY male guppies can be obtained from crosses between XY females and XY males, but only if the Y chromosomes originate from different populations and are not identical. Such YY males are fully viable and fertile, confirming that sex chromosomes in the guppy are not very differentiated. The presence of recessive lethal genes or the differential loss of essential genes on the Y chromosome of different strains might explain why YY males with two identical Y chromosomes are not viable, in contrast to males with two different Y chromosomes (Winge & Ditlevsen 1938).

Atypical sex determination depends on the strain analyzed, but there is so far no data to attribute this phenomenon either to a polymorphism of the Y-chromosomal sex-determining locus or to an autosomal modifier (Kallman 1984; Schröder et al. 1993). One strain ("ma-ze") established by O. Winge shows a high frequency of XY females that can be easily identified by the expression of a black spot on the dorsal fin encoded by the Y-chromosomal and tightly sex-determination-linked *maculatus* (Winge 1934) pigmentation gene. Interestingly, this pattern is expressed in females, while pigmentary loci in the pseudoautosomal region of the Y, and thus present also on the X, are never expressed in females.

Atypical sex determination is also found in natural populations, indicating that the phenomenon is not a laboratory artifact. As reported for many other teleost fish species, sex can be reversed in guppies through steroid hormone treatment (Dzwillo 1962; Takahashi 1975; Kavumpurath & Pandian 1993b).

The other recently described species of guppies, *P. wingei*, which includes the so-called Endler's livebearer, also has a male heterogametic sex determination system with a cytologically detectable Y chromosome in males (I. Nanda, S. Schories, & M. Schartl, unpublished data).

24.3.2 *Poecilia latipinna, P. velifera, P. sphenops,* and black mollies

Little is known about the genetics of sex determination in mollies. From laboratory strains, it has been concluded that *P. velifera* and *P. latipinna* have an XY mechanism and that the sex chromosomes of the two species are homologous (Schröder 1964). Conversely, for *P. velifera*, possibly from a different strain, female heterogamety was deduced from the presence of a sex-linked simple repetitive DNA sequence (Nanda et al. 1992; Nanda et al. 1993).

A ZZ-ZW mechanism has been proposed for the black molly (Haaf & Schmid 1984), which is an ornamental interspecific hybrid of unclear origin. Female heterogamety is supported cytogenetically through FISH of repetitive sequences and specific staining of constitutive heterochromatin (Nanda et al. 1993). The main problem with mollies is that so far all studies have been performed on ornamental fish and laboratory strains of partly uncertain origin; conclusions concerning wild populations should therefore be drawn with caution.

In the Río Jamapa (Veracruz, Mexico) population of *P. sphenops* a male was found with a conspicuous black spotting pattern. Generally, only the male offspring of this male showed the pigmentation phenotype (fig. 24.5). When unspotted females were treated with androgens, they remained unspotted, confirming that this is a true sex-linked trait rather than a sex hormone–dependent pattern like the pigmentation encoded on the pseudoautosomal part of the guppy X and Y chromosomes. Thus, we can conclude that the original male had an XX-XY system. One spotted female was obtained, which points to the presence of autosomal modifiers.

Figure 24.5 Y-linked inheritance of a black spotting gene in *Poecilia sphenops* (Río Jamapa population). Left: male; right: female.

24.3.3 Poecilia formosa

Poecilia formosa, the Amazon molly, is a gynogenetic species with clonal reproduction that originated from a hybridization event involving *Poecilia mexicana limantouri* and *Poecilia latipinna* (Avise et al. 1991; M. Schartl et al. 1995b; Lampert et al. 2007; Schlupp & Riesch, **chapter 5**). As a trigger for this form of reproduction, matings with males of another molly species living in the same habitat are necessary; gynogenesis is sperm dependent, even if the male partner generally does not contribute to the genome of the progeny. However, paternal contribution can occur occasionally and leads to stable triploid clones or clones with microchromosomes (Schlupp et al. 1998; Lamatsch et al. 2000a). The evolutionary significance of these phenomena might reside in their contribution to genetic variability in an otherwise clonal species.

In nature, in exceptionally rare cases, males of *P. formosa* occur (Hubbs & Hubbs 1932), and in the laboratory, fish with a complete male phenotype (including sperm production) can be obtained by androgen treatment (Turner & Steeves 1989; Schartl et al. 1991). This indicates that the genetic program for male differentiation is still present and largely intact in this all-female species. Interestingly, fish that have both forms of paternal contribution, triploidy and microchromosomes, can develop spontaneously as males (Lamatsch et al. 2000a). These can trigger gynogenetic offspring production when mated to normal *P. formosa* females but cannot produce offspring with females of gonochoristic *Poecilia* species, possibly due to the aneuploidy of their sperm cells.

So-called pseudomales of *P. formosa* have been found in nature and can be obtained in the laboratory (Schlupp et al. 1992). They are different from the above-described males in that they do not exhibit the full male phenotype. Instead, their phenotype better resembles the masculinization of southern swordtail arrhenoid females due to a hormonal imbalance created by an excess of androgens. The underlying trigger for formation of such pseudomales and its relevance for the gynogenetic reproductive system remain unclear.

24.4 Genetic sex determination in other poeciliid species

24.4.1 Genus *Gambusia*

Data on sex determination are scarce for the genus *Gambusia*, which comprises over 40 species. Different types of GSD have been found in two closely related species. In the eastern mosquitofish *G. holbrooki*, an XX-XY system has been deduced from the inheritance of male color patterns

(Regan 1961; Angus 1989b). The observation of females with a normally male-specific color phenotype suggests that atypical sex determination or sex chromosomal crossovers can occur in this species (Angus 1989b). In contrast, the western mosquitofish *G. affinis* has a ZZ-ZW female heterogametic mechanism, with heteromorphic sex chromosomes. Interestingly, the W chromosome in females is the largest of the karyotype and greatly exceeds the size of the corresponding Z (Black & Howell 1979; I. Nanda & M. Schartl, unpublished data).

24.4.2 Genus *Limia*

The genus *Limia* comprises more than 20 species, for which some data on sex determination have been obtained from the inheritance of different sex-linked traits. In *L. perugiae* (Erbelding-Denk et al. 1994), males typically fall into three size classes. Small males show sneaking behavior while the intermediate and large males engage in pronounced courtship display. It has been suggested that genes for small and large body size are linked to the Y chromosome, with autosomal recessive suppressors acting in intermediate-sized males. An XX/XY system has been proposed on the basis of Y-linked genes for small and large size that could explain the inheritance of male size classes in breeding experiments.

Based on interspecific crosses between *L. nigrofasciata* and *L. caudofasciata*, Breider (1935) concluded that *L. nigrofasciata* has an XX-XY mechanism. The highly biased sex ratios in broods of *L. vittata* resemble the situation in many *X. hellerii* strains and were taken as an indication of an analogous polyfactorial sex determination system (Kallman 1984). On the other hand, based on the morphology of the early differentiating gonad, a "juvenile" hermaphroditism, as described in the zebrafish (Takahashi 1977; Hsiao & Tsai 2003), was proposed for *L. vittata* (Breider 1935). Again, these results should be interpreted cautiously because they are derived from crossings of different species. In the event of differing sex determination systems (e.g., female vs. male heterogamety) with factors of different strength or the presence of autosomal modifiers, the gender of the interspecific hybrids will not be determined as simply as suggested by the presence of male and female (secondary) sex characters (for discussion, see Kallman 1984).

24.4.3 Genus *Poeciliopsis*

For one species, *P. lucida*, there is a report about the influence of temperature on sex determination in some but not all strains (see section 24.5). Otherwise, this genus is mainly of interest with respect to sex determination because several biotypes occur that reproduce by hybridogenesis or gynogenesis (Schlupp & Riesch, **chapter 5**). In northwestern Mexico hybrids exist that always have one chromosome complement of *P. monacha* combined with a haploid set of chromosomes (haplome) of either *P. lucida*, *P. occidentalis*, or *P. latidens*. They are all females, which during meiosis exclude the non-*monacha* genome and produce haploid eggs. These are then fertilized by sperm of males of the co-occurring "host" species. The resulting hybrid offspring again reproduce in the same way, leading to hemiclonal lineages with a nonrecombining *monacha* genome.

These all-female *Poeciliopsis* can be produced in the laboratory (Schultz 1973), and the clonal variation found in nature in the populations of the different hybridogens is also interpreted as being the result of repeated crosses leading to new clones (Wetherington et al. 1989a). In all cases the maternal (and asexually transmitted) genome is the *monacha* haplome (Schultz 1989). Considering gonosomal sex determination, this might lead to the conclusion that in *P. monacha* the female is the homogametic sex, or that only haplomes with a W can give rise to hybridogenetic lineages. Further crossings to a third species (e.g., *P. monacha–P. lucida* hybridogenetic females with *P. latidens* males) produced all-male offspring (Schultz 1977), while *P. monacha–P. latidens* × *P. latidens* crosses resulted in all-female offspring from some females but produced both sexes in others (Kallman 1984). At present the crossing experiments have elucidated neither the sex determination mechanism underlying all-female lineages nor that of the parental and host species. Such experiments also have the caveat that in interspecific hybrids the noncompatibility of differing sex determination mechanisms will lead to sex ratios that do not necessarily reflect the primary chromosomal sex-determining mechanism. Unfortunately, in *Poeciliopsis* species the phenotypic markers that were so useful in the analysis of the inheritance of sex chromosomes in the genus *Xiphophorus* and in the guppy are scarce. Molecular markers and linkage maps are required for analyzing these interesting breeding complexes.

Triploid hybrids (*P. monacha–P. 2lucida*, *P. 2monacha–P. lucida*, *P. monacha–P. viriosa–P. lucida*) exist. They reproduce like *P. formosa* by gynogenesis, using sperm of sympatric *Poeciliopsis* species to trigger the parthenogenetic development of the triploid eggs.

24.4.4 *Heterandria formosa*

Riehl (1991) claimed that hermaphroditism occurred in this species based on observations of an aquarium fish undergoing transition from female to male phenotype. Surprisingly, the gonad consisted of parts with fully developed ovarian structure, while in other areas mature sperm were clearly seen (Riehl 1991). This observation could be interpreted in two ways: either this was a case of simultaneous hermaphroditism, similar to the only other known case, *Kryptolebias marmoratus*, or this fish was a protogynous hermaphrodite in the transition phase. However, the report did not show by histology the mixed appearance of ovary and testes as known from *Kryptolebias*. Instead, the electron micrograph depicted only sperm in the arrangement found in mature spermatozeugmata. What were designated as Sertoli cells did not exhibit the typical morphology of this cell type, and other components of testes (Leydig cells, spermatids, spermatogonia, etc.) were not found. Hence, the picture is rather reminiscent of a normal ovary with sperm storage. The "gonopodium-like anal fin" of the analyzed fish had no features of the developing secondary male sex character seen in maturing males or in hormone-treated or arrhenoid females. Thus, the described phenomenon might be masculinization of adult females rather than a general mode of reproduction by true hermaphroditism. In our own laboratory stock of *H. formosa*, no sex change or even arrhenoidy has ever been observed in almost 20 years.

24.5 Environmental sex determination

While GSD triggers sexual development of the undifferentiated gonad anlage toward a testis or an ovary exclusively on the basis of the genotype (GSD), there are a number of organisms in which this trigger is supplied by the environment. The most striking examples are the turtles and crocodilians (Shoemaker & Crews 2009), where the incubation temperature of the embryos determines whether they will become male or female. It should be noted that ESD operates under natural conditions and that the signal from the environment is within the physiological range. For poeciliids certain physical parameters of the aqueous habitat, which are subject to change, could theoretically be such determinants. ESD has been proposed as the most basal and primitive mode of sex determination (see discussion by Johnston et al. 1995; Valenzuela & Lance 2004). There are also some reports on ESD in poeciliids. One study (Rubin 1985) claims an influence of pH on sex ratios in "*X. hellerii*," with low pH (6.2) producing exclusively males and high pH (7.8) a strong (>80%) bias toward females. Although such pH values can be found in some habitats, this study has to be interpreted with caution. All fish were ornamental fish from the aquarium trade, and only two broods per pH were raised. Thus, biased sex ratios could result just by chance, as such biased sex ratios are naturally seen in several *X. hellerii* strains (Kallman 1984). In addition, only the sex of adult fish was determined, making it impossible to exclude a pH-dependent selective mortality of one sex. Clearly, larger and more controlled studies are necessary to make a case for pH-dependent ESD.

Claims for TSD come from single reports of three poeciliid species: *Poecilia sphenops* (Baron et al. 2002), *Limia melanogaster* (Römer & Beisenherz 1996), and *Poeciliopsis lucida* (Sullivan & Schultz 1986). These cases have been critically analyzed by Ospina-Alvarez and Piferrer (2008). For *P. sphenops* (fish taken from a wild population in Oaxaca) a combination of pH variation and different physiological temperatures was tested. High pH (8.0) and low (22.6°C) or intermediate (26°C) temperature produced more males than any other combination (pH 6.2, 7.1, $t = 29.7$°C). However, at any combination the female sex ratio was always higher (M/F ranging from 0.1 to 0.7). Moreover, there is indication for GSD (by the presence of heteromorphic chromosomes) in this species (see section 24.3.2). For *L. melanogaster* and *P. lucida* the presence of GSD has not been investigated and was not helpful as an exclusion criterion. The temperature-induced bias in sex ratio again was unidirectional, with more males produced at high temperatures. For *P. lucida* the two (highly inbred) strains used even behaved differently. Only one responded consistently to temperature. In the other line certain females that were incubated at the same temperature produced variable offspring sex ratios due to unknown factors not expected under TSD (Schultz 1993). Certainly, much more work has to be done to establish the effect of temperature on the sex ratios of these and possibly other species of poeciliids.

In all cases discussed above, differential mortality of one sex at different temperature conditions was impossible to assess. Thus, so far we have no compelling evidence for true bidirectional TSD with an intermediate value where both sexes are produced in a 1:1 ratio in poeciliids. However, there is still the possibility that environmental influence can somehow override or modulate a GSD pathway. This would also offer a simple explanation for the male bias in the temperature experiments. Such a phenomenon has been documented in a number of teleost species, including the medaka and *Tilapia* (Baroiller & D'Cotta 2001; Nanda et al. 2003; Sato et al. 2005; Hattori et al. 2007). Here, although a firm GSD system exists, high temperatures can efficiently produce female to male sex reversal. It should also be noted that in those cases low temperature did not have the opposite effect on sex ratios, as in the true cases of TSD, for instance, in reptiles (Shoemaker & Crews 2009).

In general, the examples of true TSD in teleosts are scarce (Ospina-Alvarez & Piferrer 2008), and even the biological significance of environmental effects like temperature in overriding the primary GSD system is unclear. In addition, it is worth noting that the reproductive mode of viviparity seems to be incompatible with the requirements to develop TSD (Valenzuela et al. 2003; Ospina-Alvarez & Piferrer 2008).

24.6 Genomic organization of the sex-determining region of the platyfish *Xiphophorus maculatus*

Xiphophorus maculatus has been chosen as an anchor species for the comparative study of the molecular basis of GSD in poeciliids. The viability and fertility of both YY and WW genotypes indicate that degenerative gene loss, as observed for, instance, for the mammalian Y chromosome, did not occur for the Y and W chromosomes in *X. maculatus*. Accordingly, X and Y chromosomes have many loci in common in the sex-determining region. In addition, cytogenetic analyses demonstrated that the X and Y chromosomes are homomorphic (Nanda et al. 2000). Taken together, these observations support a relatively recent evolutionary origin for the sex chromosomes in the platyfish. However, a Y-specific accumulation of a particular repeat, which might represent an initial step of molecular differentiation, has been detected in one population of *X. maculatus* (Nanda et al. 2000).

The positional cloning of the master sex-determining gene of *X. maculatus* has been initiated. A bacterial artificial chromosome (BAC) genomic library of XY males has been constructed from the Jp 163A strain (genotype *X-mdl^{Sd} xmrk^{B}RY^{Dr} Y-mdl^{Sr}xmrk^{A}RY^{Ar}*, origin Río Jamapa, Veracruz, Mexico) (Froschauer et al. 2002). Sex chromosomal BAC clones containing *Xmrk*, *egfrb*, and other sex-linked markers have been isolated and used as start points for a chromosome walk. Megabase-sized contigs of BAC clones almost completely covering the sex-determining region of the X and Y chromosomes have been assembled. Initial sequencing and analysis of some BAC clones have revealed gene candidates, some of them apparently corresponding to pseudogenes (table 24.1, see online supplement at http://www.press.uchicago.edu/books/Evans) (Volff et al. 2003; Schultheis et al. 2006; Böhne et al. 2008). Remarkably, the sex-determining region of *X. maculatus* is very unstable and affected by many genomic rearrangements, particularly duplications, deletions, and transpositions (Volff et al. 2003). This high level of genomic plasticity might have important consequences for the evolution of sex determination and sex-linked traits in poeciliids, which are generally highly polymorphic.

Some sequences are pseudogenic duplicates of genes located elsewhere in the genome. This is the case for the postsynaptic protein gene *cript*, the mismatch repair gene *msh2*, the melanocortin receptor gene *mc4r*, the F-box protein gene *fbox11*, and the miRNA-processing RNase III gene *drosha* (Volff et al. 2003). Other genes such as the acetylcholine receptor gene *chrnd*, the transcription factor gene *irf3*, and two cadherin genes might be functional, but this remains to be demonstrated.

Comparative syntenic analysis with other model species (human, mouse, chicken, zebrafish, medaka, tetraodon) has revealed relationships with autosomes, but no synteny

Table 24.2 Types of sex determination in poeciliids

Species	Male heterogamety	Female heterogamety	Autosomal influence[a]	Polyfactorial	Gynogenesis	Environmental	Remarks
Gambusia affinis		+					
Gambusia holbrooki	+		(+)				
Limia melanogaster						Temperature?	
Limia nigrofasciata	+						
Limia perugiae	+						
Limia vittata				+			
Poecilia formosa					+		
Poecilia latipinna	+						
Poecilia reticulata	+		(+)				
Poecilia sphenops	+		(+)			pH? Temperature?	
Poecilia velifera	+	+					
Poecilia wingei	+						
Poeciliopsis lucida						Temperature?	
Xiphophorus alvarezi	+	+	+				
Xiphophorus andersi	+						
Xiphophorus clemenciae		+					
Xiphophorus continens	+						
Xiphophorus cortezi	+		(+)				
Xiphophorus couchianus	+						
Xiphophorus evelynae	+						
Xiphophorus gordoni	+						
Xiphophorus hellerii		+	+	+		pH?	
Xiphophorus kallmani	+		+				
Xiphophorus maculatus	+	+	(+)				W, X, Y
Xiphophorus meyeri	+						
Xiphophorus milleri	+		+				X, Y, Y′
Xiphophorus multilineatus	+		+				X, Y, A, a
Xiphophorus nezahualcoyotl	+		+				X, Y, Y′
Xiphophorus nigrensis	+		+				X, Y, A, a
Xiphophorus pygmaeus	+						
Xiphophorus signum				+			
Xiphophorus variatus	+						
Xiphophorus xiphidium	+						

[a]Parentheses indicate that these are very rare events.

was found with a sex chromosome of any other species, supporting an independent origin of sex chromosomes in different vertebrate lineages but also between different fish species. So far, all gene candidates identified in the *SD* region of the platyfish have been detected on both the X and the Y, supporting the idea that sex chromosomes are poorly differentiated and evolutionarily young. None of them has a known function in sex determination in fishes or other organisms, and so far no expression pattern compatible with a function in sexual development has been observed. Therefore, none of these genes represents an obvious candidate for the master sex-determining gene—even if we cannot eliminate the possibility that some of these genes are only functional on one type of sex chromosome.

24.7 Molecular markers and genomic organization of the sex chromosomes in the guppy

A second well-advanced model for molecular analysis in poeciliids is the guppy, *Poecilia reticulata*. A large number of polymorphic marker sequences have been generated from mapping crosses between wild guppies from Trinidad and Venezuela. From this data base a high-resolution genetic map for all 23 chromosomes of the guppy genome has been established, and the sex chromosomes have been assigned to linkage group 12 (Tripathi et al. 2009b). From conserved synteny to the medaka genome as an outgroup, it can be concluded that there is no homology to the sex chromosomal linkage group of *X. maculatus*, again confirming the typical independent origin of sex chromosomes, even in related taxa. The sex-determining region, harboring a Y-linked male-determining gene and the subset of strictly Y-linked pigmentation genes, was mapped distally to the telomere-near part of the linkage group (Tripathi et al. 2008, 2009a). The distance of the closest marker to the sex-determining locus is up to 2 cM. Thus, a positional cloning approach requires further and much closer markers.

24.8 Conclusions and perspectives

In contrast to birds and mammals, sex determination in fish is extremely variable and might involve the evolutionary turnover of sex chromosomes and master sex regula-

tors (Volff et al. 2007). The reasons behind this variability are still unclear but very interesting from the evolutionary point of view. Different types of sex determination and sex chromosomes are known in poeciliids (table 24.2, fig. 24.1). Thus, sex chromosome evolution appears to be rapid, and traits encoded by genes closely linked to the sex-determining region are highly polymorphic. Hence, poeciliids represent a prime model to analyze the evolutionary dynamics of the control of sexual development. To make full use of this model, a more precise knowledge of the genetics of sex determination and the molecular biology of the genes involved is required. Of particular interest will be not only the molecular identification of master genes but also the characterization of autosomal modifiers, which should provide new clues concerning the evolutionary transition between sex-determining systems.

With the exception of mammals and birds, most vertebrate lineages exhibit highly variable sex-determining mechanisms. Why are such systems so diverse, with frequent changes during evolution? We still have no clear answers to this question. This high variability might be adaptive to environmental changes involving temperature, pollutants, sex-manipulating parasites, or other sex ratio distorters. Comparative analysis in poeciliids and more generally in fish will certainly contribute to our understanding of the evolutionary dynamics of sex determination and to the identification of the evolutionary constraints driving this diversity.

Acknowledgments

We thank M. Niklaus-Ruiz for help in preparing the manuscript. Our work is supported by the Deutsche Forschungsgemeinschaft (Graduiertenkolleg 1048: Molecular Basis of Organ Development in Vertebrates) and DFG Forschungszentrum "Target Proteins" to M.S., and the Biofuture program of the German Bundesministerium für Bildung und Forschung (BMBF), the Association pour la Recherche contre le Cancer (ARC), the Institut National de la Recherche Agronomique (INRA), the Centre National de la Recherche Scientifique (CNRS), the Fondation pour la Recherche Médicale (FRM), and the Agence Nationale de la Recherche (ANR) to J.-N.V., C.S., and A.B. received fellowships from the CNRS and the French Ministère de l'Enseignement Supérieur et de la Recherche, respectively.

Chapter 25 Evolutionary genetics and molecular ecology of the major histocompatibility complex genes

Mark McMullan and Cock van Oosterhout

25.1 Introduction

THE MAJOR HISTOCOMPATIBILITY complex (MHC) is a large multigene family present in all jawed vertebrates (see Klein et al. 2000). It has a central role in immunocompetence and is implicated in mate choice and sexual selection (Apanius et al. 1997; Edwards & Hedrick 1998; Hughes & Yeager 1998; Penn & Potts 1999; Bernatchez & Landry 2003; Wegner et al. 2004; Sommer 2005). The histocompatibility genes were first studied in relation to transplantation compatibility in humans and mice (Dausset 1958) and then received attention in conservation biology with work on the cheetah (O'Brien et al. 1985). The MHC has a role in the detection of invading pathogens by recognition of small protein fragments called peptides or antigens.

The number of genes required to recognize antigens varies enormously between species. For instance, there are over 200 MHC loci (including pseudogenes) within the human MHC (called the human leukocyte antigen, or HLA) (MHC Sequencing Consortium 1999; Shiina et al. 2004) and just 19 within the chicken MHC (B locus) (Kelley et al. 2005). The number of MHC loci can also vary between individuals of the same species (Malaga-Trillo et al. 1998; Figueroa et al. 2001). Variation in the number of genetic loci within the MHC is predicted by various multigene models, including the accordion model of MHC evolution (Klein et al. 1993), the birth-and-death model, and concerted evolution (see review by Nei & Rooney 2005).

The MHC is characterized by extraordinary levels of polymorphism, as is illustrated, for example, by the human MHC HLA-B locus, which has over 1000 alleles (Robinson et al. 2003; IMGT/HLA Database, 2008, http://www.ebi.ac.uk/imgt/hla/stats.html). This level of polymorphism varies within and among populations (Wegner et al. 2003b; van Oosterhout et al. 2006b; Dionne et al. 2007), as well as between different MHC genes (Robinson et al. 2003) (IMGT/HLA Database, 2008). The highest level of polymorphism is observed within a specific region of the MHC known as the peptide-binding region (PBR) (Hughes & Yeager 1998). This region is responsible for the specificity of the MHC molecule with respect to bind particular peptides from different sources of antigens, including bacteria, viruses, and micro- and macroparasites (Bjorkman et al. 1987b; Stern et al. 1994)

Parasitism and host defense against pathogens are generally considered the primary mode of selection maintaining polymorphism at the MHC. The abundance and composition of pathogens within populations are proposed mechanisms for maintaining specific MHC alleles. Much of the work on this subject has focused on humans (Shiina et al. 2004) and mice (Penn & Potts 1999). Here, we focus on poeciliids, where the MHC has only recently been unraveled, as well as on other teleosts, such as other Cyprinodontiformes, Gasterosteiformes, and Salmoniformes.

In addition to natural selection by parasites, sexual selection may also play a role in maintaining MHC polymorphism (see Rios-Cardenas & Morris, **chapter 17**; and Evans & Pilastro, **chapter 18**). Sexual selection for "good genes" or increased genetic diversity has been shown to maintain MHC polymorphism in many species (Milinski 2006). Furthermore, the variation at the MHC may be an important signal used in mate choice, with individuals showing preference for genetically compatible mating partners, which may, for example, help to avoid inbreeding (reviewed by Penn & Potts 1999; Milinski 2006).

Sexual selection has been studied extensively in several poeciliid species (see, e.g., Rios-Cardenas & Morris, **chapter 17**). Mate-choice experiments in guppies (*Poecilia reticulata* Peters) (see Houde 1997) have primarily focused on visual cues, and relatively little research has tested the role of olfaction and chemical communication (Shohet & Watt 2004; but see Archard et al. 2008). Olfactory detection of MHC-compatible mates has been shown to occur in a number of other animal systems, including humans, mice, and sticklebacks (Penn & Potts 1998; Aeschlimann et al. 2003; Beauchamp & Yamazaki 2003). Given that poeciliids are suitable model organisms to use in studying the role of olfaction in mate choice, we will briefly discuss here the potential role of the MHC in sexual selection and the maintenance of MHC polymorphism.

25.2 The major histocompatibility complex

The MHC multigene family encodes a number of proteins responsible for the preparation, loading, and specific binding of antigen fragments (reviewed by Ting & Trowsdale 2002; Boss & Jensen 2003; Danchin et al. 2004b). The classical MHC genes are typically highly polymorphic and separated into the MHC class I and MHC class II subfamilies. Both classes of genes form part of the adaptive immune system. Nonclassical MHC genes display low levels of polymorphism and gene expression, and the DNA sequences of these genes commonly show little evidence of positive selection (Summers et al. 2009). The class III region contains a group of genes genetically and evolutionarily unrelated to class I or class II, although some of these genes have immune function, particularly in the innate immune response (Klein & Sato 1998). MHC class I and class II are linked in mammals and birds, but they occur unlinked on different chromosomes in teleost fishes (Sato et al. 2000; Stet et al. 2003). Consequently, some authors refer to the teleost MHC as the MH (see Ellis et al. 2006; Dixon 2008), although we prefer to use the more traditional abbreviation, MHC.

Class I and class II molecules differ in the type of antigens they present. Class I molecules are expressed on the surface of all nucleated somatic cells and present peptides originating from within the cell, such as viral peptides. Class II molecules are expressed primarily on antigen-presenting cells of the immune system, and they specialize in the presentation of extracellular pathogen peptides (e.g., bacteria and microparasites). These differences in subfamilies are echoed in the specific antigen-processing pathways

required to load the MHC molecule and in the type of T cell to which they present. Once degraded, antigen fragments are bound to the MHC glycoprotein and presented to T cells. T cells are responsible for the initiation of the adaptive immune response, which acts to target the pathogen specifically. T cells need to respond only to foreign antigens and not to the host's own tissue. To recognize "self" from "nonself," autoreactive thymocytes with T-cell receptors that interact with MHC molecules presenting self-peptides are eliminated during thymic selection in the developing embryo (Jameson et al. 1994; Jordan et al. 2001).

Much population genetic, immunogenetic, and evolutionary research on the MHC has focused on the second exon of classical MHC encoded molecules, as this transcribes a large proportion of the PBR. Positive selection is evident in only this part of the gene, and consequently, this exon shows the highest level of polymorphism of the entire MHC. Other exons of the MHC are relatively conserved, and the sequence variation in these exons is governed mainly by neutral evolutionary forces, such as mutation, genetic drift, and recombination (see review by Hughes & Yeager 1998).

25.3 Evolutionary genetics and molecular ecology of the MHC

25.3.1 Natural selection and the MHC

In the last three decades, molecular ecologists have studied neutral evolution using allozymes, minisatellites, microsatellites, and single-nucleotide polymorphisms (SNPs) (Hedrick 2005). Neutral theory asserts that the majority of new mutations are either deleterious and eliminated from a population by negative selection or are (nearly) neutral and subject to random genetic drift (Ford 2002; Garrigan & Hedrick 2003). The neutral polymorphism is maintained in a mutation-drift equilibrium, and this variation can be used to analyze the population's demography. Neutral marker loci are used to analyze genealogies, establish paternities, track population migration, measure gene flow, and estimate effective population size.

More recently, with the advent of economically viable sequencing and the identification of genes under positive selection, research has begun to focus on adaptive evolution. Adaptive evolution is driven by positive selection promoting novel (favorable) sequence variants. Positive selection can be inferred using several types of analyses of sequence data, and the d_N/d_S ratio is one commonly used method (Ford 2002; Garrigan & Hedrick 2003). The d_N/d_S value is essentially a ratio of the number of nonsynonymous mutations (i.e., amino acid–changing mutations) to synonymous ones (also known as silent substitutions). Synonymous mutations do not change the protein produced by the gene and are predicted to be selectively neutral. Consequently, synonymous mutations are accumulated and lost in a neutral fashion, and these mutations will occur in drift-mutation equilibrium (Crow & Kimura 1970; Hedrick 2005). In contrast, nonsynonymous mutations can alter protein function, and such mutations are thus usually subject to natural selection. Mutations that change the fitness of the organism can be either favorable (positively selected) or deleterious (negatively selected). In conserved genomic regions, nonsynonymous mutations tend to be purged by negative selection because changes to the protein structure are generally deleterious. Synonymous mutations, on the other hand, can persist at a relatively high frequency, and consequently, in conserved gene regions the d_N/d_S ratio will be below unity ($d_N/d_S < 1$). However, in gene regions where evolutionary change is favorable, positive selection is predicted to maintain nonsynonymous mutations, resulting in an elevated d_N/d_S ratio ($d_N/d_S \geq 1$). In the MHC, the protein region interacting with foreign peptides is transcribed from the codons of the PBR. The codons in this region are thought to accumulate nonsynonymous mutations, as these allow the recognition of novel parasites. By contrast, the areas outside the PBR are conserved, and nonsynonymous mutations are purged from these gene regions. Studies on the MHC have been facilitated by the fact that the PBR codons of the human MHC have been identified through X-ray crystallography (Bjorkman et al. 1987a, 1987b; Brown et al. 1993; Stern et al. 1994), and the MHC is highly conserved across vertebrates (but see Blais et al. 2007 for differences in PBR codons between humans and cichlid fish). This allows researchers working on the MHC of other vertebrate taxa a priori to identify the codons under positive and negative selection.

25.3.2 Balancing selection

Besides positive and negative selection, balancing selection plays a pivotal role in the evolution of the MHC. Balancing selection is thought to maintain polymorphism at a locus by one (or combination) of the following three processes: (1) heterozygote advantage (overdominant selection), (2) rare-allele advantage (negative frequency-dependent selection), or (3) selection varying in time or space (spatial or temporal variable selection). Genes that are evolving under balancing selection are distinct from neutral loci in that they show high levels of heterozygosity and many alleles at similar frequencies (hence the name "balancing" selection). Some alleles can be maintained over long periods of time, possibly exceeding speciation events, a phenomenon called trans-species polymorphism (reviewed by Klein et al. 1998). Furthermore, balancing selection can also affect the DNA substitution pattern in a way similar to positive selection, resulting in an elevated d_N/d_S ratio (see above).

1. *Overdominance selection*: Of the three hypotheses suggested to maintain a balanced polymorphism, the model of overdominant selection (heterozygote superiority) has received the most attention. The heterozygote advantage hypothesis relates directly to the MHC genotype and the role of specific MHC alleles present at the locus. Heterozygous individuals are thought to be able to recognize twice as many pathogens as homozygous individuals and, therefore, have a superior immune response and higher fitness (Doherty & Zinkernagel 1975). Landry et al. (2001) demonstrated selection for increased proportions of Atlantic salmon offspring that were heterozygous at the MHC class IIB locus. Similarly, Arkush et al. (2002) demonstrated increased survival of heterozygous MHC class IIB Chinook salmon after exposure to infectious hematopoietic necrosis virus (IHNV). The level of inbreeding (inbred or outbred) was not found to correlate with IHNV survival but was associated with increased resistance to another myxozoan pathogen. These findings highlight the importance of heterozygosity at MHC genes and throughout the genome as a means of pathogen defense (Spielman et al. 2004; Hale & Briskie 2007; van Oosterhout et al. 2007b). Penn (2002) performed a meta-analysis and found no clear evidence of MHC heterozygote advantage. Evidence is accumulating that an optimal, rather than a maximal, heterozygosity might be providing the highest resistance (see Nowak et al. 1992; Wegner et al. 2003a; Milinski 2006), which begs the question of how MHC polymorphism in populations is then maintained.

2. *Negative frequency-dependent selection*: Negative frequency-dependent selection is another model of balancing selection that can maintain polymorphisms in populations (Clarke & Kirby 1966; Kojima 1971). Whereas overdominant selection acts on MHC genotypes, the model of frequency-dependent selection assumes that each MHC allele recognizes particular parasites. Langefors et al. (2001) infected 4800 *Salmo salar* with a bacterium (*Aeromonas salmonicida*) and identified MHC alleles associated with either high vresistance or susceptibility. This shows that besides selection on heterozygous genotypes, parasite selection can also act on single MHC alleles, which may result in a dynamic coevolutionary arms race between the host immune system and parasite virulence genes (Lively & Dybdahl 2000; Carius et al. 2001). Parasites are expected to evolve to avoid detection by common MHC alleles. By contrast, the selection pressure on parasites to avoid recognition by rare MHC alleles is much lower (Slade & McCallum 1992). Consequently, rare MHC alleles are more likely to detect pathogens, and hence such alleles are predicted to confer a higher

fitness to the host than common alleles (Trachtenberg et al. 2003; Froeschke & Sommer 2005). Over time, parasite selection is expected to increase the frequency of rare alleles, as hosts carrying these rare alleles have a higher probability of avoiding parasitism. The coevolutionary arms race between the host immune system and parasite virulence results in a dynamic and balanced polymorphism in both host MHC and parasite virulence genes (Sommer 2005). We know of no long-term study demonstrating this cyclical pattern of MHC allele frequencies in fish (but for studies on other vertebrates, see Westerdahl et al. 2004). However, given the rapid generation time of poeciliids and the recent advances made by studies on their MHC and parasites, future studies on these fish offer promising insights into host-parasite coevolution and Red Queen dynamics (Bell 1982; Penn & Potts 1999).

3. *Spatial or temporal variable selection*: Spatial and/or temporal heterogeneity in selection pressure also can result in a balanced polymorphism at the MHC (Hedrick 2002). A nine-year study of the MHC class I variation of great red warblers found variation in allele frequencies over time (Westerdahl et al. 2004). The authors suggest that these fluctuations in MHC allele frequencies are due to temporal fluctuations in pathogen fauna, which is also consistent with frequency-dependent selection (see also Charbonnel & Pemberton 2005). A parasite of the Trinidadian guppy has also been implicated in temporal fluctuation in MHC allele frequency (Fraser et al. 2010b). Miller et al. (2001) sampled MHC polymorphism in 31 populations of sockeye salmon (*Oncorhynchus nerka*) within a river basin. They found evidence of balancing selection acting on the MHC in some but not all populations. The authors were able to separate population demography to rule out population bottlenecks in populations with low MHC diversity, showing that the patterns of allele frequencies in some populations were more consistent with directional selection (Landry & Bernatchez 2001).

These three models of balancing selection are not mutually exclusive but, rather, likely to act complementarily. De Boer et al. (2004) and Borghans et al. (2004) modeled negative frequency-dependent selection and overdominant selection and found that overdominance was unlikely to maintain the level of polymorphism observed at the MHC unless each MHC allele conferred a very similar fitness (e.g., symmetric overdominance). However, it is difficult to discriminate between different types of balancing selection in wild systems. For example, rare alleles that are predicted to be maintained under negative frequency-dependent selection are also more likely to be presented in heterozygote condition (Apanius et al. 1997).

These traditional models of balancing selection were developed to understand the evolution and population genetics of a single immune gene. However, the MHC is not a single gene but a multigene family, and these models ignore the potential role of linkage and epistatic gene-gene interactions. The human MHC genes are surrounded by linked genetic variation (SNPs) that is associated with more diseases than any other part of the human genome (de Bakker et al. 2006; Shiina et al. 2006). This suggests that this linked peri-MHC region is under strong selection and that it could play a potentially important role in the evolution of this multigene family. Van Oosterhout (2009) proposed a new theory of MHC evolution that incorporates the impact of selection on the region surrounding the MHC genes. The model is called associative balancing complex (ABC) evolution, and it proposes that selection acts on the deleterious mutations that are associated with the MHC genes (box 25.1).

Box 25.1 The ABC model of MHC evolution

The model of associative balancing complex (ABC) evolution (van Oosterhout 2009) proposes that recessive deleterious mutations accumulate near MHC genes. These mutations maintain polymorphism at the MHC and surrounding region by associative balancing selection. The recessive deleterious mutations can accumulate in the MHC because the effective rate of recombination in some areas of the MHC is very low (see, e.g., Stenzel et al. 2004; Gregersen et al. 2006). This reduces the efficiency of purifying selection in removing recessive deleterious mutations (Haddrill et al. 2007). Some of those mutations become fixed in all copies of a particular haploblock in a process analogous to Muller's ratchet (Muller 1932). ABC evolution is inspired by theoretical studies of self-incompatibility loci (S-loci) of plants (Uyenoyama 2003), and it offers a plausible explanation for the long-ranging linkage disequilibria and MHC-disease associations in well-researched organisms (humans, dogs, and mice) (van Oosterhout 2009). The new theory also elucidates a number of evolutionary properties of the MHC that are not well explained by the traditional theories, such as (1) the shape of the MHC genealogies and trans-species evolution, (2) the unexpectedly high levels of genetic differentiation of MHC genes under balancing selection, and (3) the high selection coefficients required to maintain MHC polymorphism in small, isolated populations (see van Oosterhout 2009).

25.4 Sexual selection, parasite selection, and inbreeding avoidance

According to the Red Queen hypothesis, the host's immune system evolves quickly in response to the rapid evolution of its parasites (Milinski 2006), and balancing selection by parasites can maintain MHC polymorphism as proposed by the various models (see above). However, mate choice can also result in rapid MHC evolution, or it may help to maintain polymorphism at these genes (Milinski 2006). The MHC is implicated in sexual selection at both pre- and postzygotic levels, for example, by means of the choice of mates with a specific MHC genotype, differential fertilization success, and the selective abortion of MHC-homozygous embryos (see Rios-Cardenas & Morris, chapter 17; Evans & Pilastro, chapter 18). Studies on a wide variety of animals support the role of the MHC in precopulatory sexual selection, including research on humans and rodents (reviewed by Penn & Potts 1999; Kavaliers et al. 2004), salmon (Landry et al. 2001; Neff et al. 2008a), and sticklebacks (reviewed by Milinski 2003). This work has focused in particular on the role of olfaction, which is thought to enable individuals to determine the compatibility and/or diversity of the MHC of prospective mating partners during courtship rituals. For example, research on sticklebacks showed that females use evolutionarily conserved structural features of MHC molecules to evaluate the diversity of the MHC of males (Milinski et al. 2005).

There has been considerable debate on the relative roles of MHC variation and genome-wide heterozygosity and their consequences for fitness (Arkush et al. 2002; Sommer 2005). Landry et al. (2001) were able to separate the effects of MHC heterozygosity and inbreeding avoidance (i.e., the relatedness measured using similarity at five neutral microsatellite loci). They sampled spawning Atlantic salmon and juvenile salmon after hatching and found that mates selected each other in order to increase genotypic differences in the MHC of juveniles and not to increase heterozygosity in the genome. This supports the role of the MHC in sexual selection for "good genes" (or genetic compatibility), although the authors acknowledge that the experiment cannot differentiate between active mate choice, differential fertilization success, and the higher survival of juveniles of heterozygous MHC genotypes (see also de Eyto et al. 2007; Neff et al. 2008a). Disentangling the relative importance of inbreeding avoidance, good-gene selection, and parasite resistance in MHC evolution is an important future challenge in MHC research. A particularly promising new avenue of research is offered by the comparison of MHC evolution in a pair of closely related poeciliids, one species being entirely clonal and the other sexually reproducing (see section 25.5.1 and Schlupp & Riesch, chapter 5).

There is increasing evidence that the outcome of sperm competition can depend on interactions between male and female genotypes (see Rios-Cardenas & Morris, chapter 17; Evans & Pilastro, chapter 18). Studies on organisms other than poeciliids suggest that the MHC may be implicated in nonrandom fertilization (see Simmons 2005 for review). However, these studies typically score paternity on newborn or adult offspring, making it difficult to disentangle the effect of sperm selection from differential embryo mortality. The only study that directly measured fertilization success in a fish (Arctic char, *Salvelinus alpinus*) showed that males heterozygous at the MHC had increased fertilization success compared with homozygotes and, in addition, that one MHC allele was associated with increased fertilization success (Skarstein et al. 2005).

25.5 The poeciliid MHC

The class I and II genes of the teleosts' MHC are unlinked and thought to reside on separate chromosomes, an arrangement that is markedly different from that of tetrapods (Sato et al. 2000; Stet et al. 2003; Dixon 2008). Free recombination between class I and II MHC regions in teleost fish is believed to be a derived characteristic, as the class I and II genes in cartilaginous fishes, the oldest class of extant vertebrates, are linked (Ohta et al. 2000). Class III genes are split over several different chromosomes in teleosts (Flajnik et al. 1999; Sambrook et al. 2002). The occurrence of separate linkage groups may have important implications for the evolution of teleost MHC, because with free recombination, selection can operate independently on the different gene classes. This suggests that teleost MHC is more versatile and perhaps less prone to accumulate mutations by genetic hitchhiking. Relatively little work has been done on MHC class I and III genes in poeciliids (but see Sato et al. 1995; Sato et al. 2000; Schaschl et al. 2008), possibly because class I genes are involved in the antigen presentation of more elusive intracellular parasites such as viruses. However, Figueroa et al. (2001) conduct a comprehensive survey of the MHC class I genes of 22 swordtail fishes. Sequences were found to fall within one of two ancient lineages, and the pattern of sequence similarity was consistent with the birth-and-death theory of multigene evolution (Nei & Rooney 2005). In this chapter we will focus on research on MHC class II loci, in particular, *DAB* and *DXB* genes and introns (box 25.2).

25.5.1 The poeciliid MHC *DAB* genes

The MHC multigene family and its large number of genes necessitate a clear classification of alleles and genes. Here,

Box 25.2 MHC introns

Introns are found throughout the eukaryote genome and have been proposed to fulfill a number of functions. Introns may increase the functional diversity of genes through alternative splicing, and they are thought to play a role in regulating the recombination rate (Duret 2001; Roy & Gilbert 2006). Microsatellites have been identified in a number of introns of the classical MHC of humans (Balas et al. 2005), chimps (Bak et al. 2006), Atlantic herring (Stet et al. 2008), Atlantic salmon (Stet et al. 2002), and sticklebacks (Reusch et al. 2004), and these repetitive elements are believed to affect protein expression and increase the recombination rate (Arnold et al. 2000; Reusch & Langefors 2005; Majumder et al. 2008). Recombination has been shown to occur both within and between MHC loci (Hughes et al. 1993; Richman et al. 2003; Reusch & Langefors 2005). The effects of recombination on genetic polymorphism are dependent on the type of selection acting on the gene (Reusch & Langefors 2005). Balancing selection is likely to preserve polymorphisms that are generated by recombination events, whereas polymorphisms are expected to be lost by random genetic drift at a neutrally evolving locus. Recombination is believed to play an important role in introducing polymorphism to the peptide-binding region (PBR) (Charlesworth 2006), providing balancing selection with novel variation above that of the background mutation rate. Evidence of these recombination events leading to the generation of PBR polymorphism may be present in the surrounding introns (Cereb et al. 1997; Bergstrom et al. 1998; Hughes & Yeager 1998; Bergstrom et al. 1999; Elsner et al. 2002; von Salomé et al. 2007). Besides the functional aspects of intron variation, the repetitive elements often observed within introns are potentially useful markers for population genetic studies. Microsatellites within the MHC region are receiving increased attention as proxy measures of the level of polymorphism at the MHC genes themselves (Stet et al. 2002; Santucci et al. 2007; Stet et al. 2008). The screening of microsatellite loci that are tightly linked to the MHC is a particularly time- and cost-effective method to study the role of MHC polymorphism in mate choice and parasite resistance. Future work on the poeciliid MHC would benefit from the development of such linked microsatellite markers.

we follow the nomenclature first proposed by Klein et al. (1990). The *DAB* loci in poeciliids (and many other fish species) are possibly the best-studied genes. *DAB* specifies a region of the MHC; *D* stands for "duo" (in reference to class II), *A* represents the family of genes, and *B* specifies the β heterodimer (class IIB). The species from which the MHC sequence was amplified is denoted by an abbreviation of the genus and species names; for example, *Mhc-Pore* specifies that the MHC sequence was obtained from the organism *Poecilia reticulata* (see Klein et al. 1990).

Sato et al. (1995) first described the MHC of the guppy using ornamental stocks and inbred wild fish to define the class I and class II regions of the MHC. The authors found very low variability at both classes of genes and predicted that guppies possess one MHC class II gene and one or two MHC class I genes. Van Oosterhout et al. (2006a) confirmed this finding, showing that ornamental-guppy lines possessed between one and three MHC class IIB alleles. However, they detected up to four alleles in individual wild-caught guppies, which implies that there are at least two *DAB* genes (Fraser et al. 2010a). Within populations, the authors detected 15–16 MHC alleles, demonstrating that these wild-guppy populations had a much higher level of MHC polymorphism than ornamental-guppy stocks. The disparity in allelic richness between wild and ornamental fish could not be explained by fixation of alleles by inbreeding in captivity nor by the presence of nonamplified sequences (i.e., null alleles). Rather, van Oosterhout et al. (2006a) suggested that during many generations in captivity, multigene evolution may have played an important role in shaping the organization and polymorphism of the MHC. In the ornamental lines, gene conversion may have fixed the same allele at duplicated MHC *DAB* genes. (For the role of gene conversion in teleost MHC, see, e.g., Reusch et al. 2004; Reusch & Langefors 2005.) Furthermore, van Oosterhout et al. (2006a) hypothesized that a reduced level of parasitism during more than 100 generations in captivity may have reduced the number of duplicated *DAB* genes in the ornamental strains, a hypothesis consistent with the accordion model of MHC evolution (see Klein et al. 1993).

The MHC has also been well studied in two other poeciliids, the Gila topminnow (*Poeciliopsis occidentalis occidentalis*) and the Yaqui topminnow (*Poeciliopsis occidentalis sonoriensis*). Work by Hedrick et al. is one of the first examples of the use of MHC variation in conservation of an endangered species (Hedrick & Parker 1998; Hedrick et al. 2001a; Hedrick et al. 2001b; see also section 25.6.3).

Schaschl et al. (2008) have analyzed the MHC class I and class IIB diversity of two closely related mollies. The Amazon molly (*Poecilia formosa*) is a clonal all-female poeciliid that lives in sympatry with the sexually reproducing sailfin molly (*Poecilia latipinna*) (see also Schlupp &

Riesch, **chapter 5**). Both neutral genetic and MHC class I and II diversity of the asexual molly were lower than in its sexual counterpart, although it still maintained a considerable level of MHC diversity. The Red Queen hypothesis suggests that the asexual molly should be unable to provide a "moving target" against rapidly evolving pathogens, because (meiotic) recombination cannot provide novel gene combinations in clonally reproducing organisms. However, previous research showed no difference in parasite prevalence between these sympatric species (Tobler & Schlupp 2005; Tobler et al. 2005). Comparison of MHC diversity between the asexually and sexually reproducing species offers a unique opportunity to disentangle the relative roles of parasites and sexual selection in the maintenance of MHC polymorphism.

25.5.2 The poeciliid MHC *DXB* genes

McConnell et al. (1998b) reported a novel MHC class IIB–like gene in *Xiphophorus maculatus* and *Xiphophorus hellerii*. This gene was labeled *DXB* because its family designation remains unknown (McConnell et al. 1998b). This locus is unlinked to the *DAB* locus and is expressed in *X. maculatus*, *Xiphophorus multilineatus*, and *Xiphophorus pygmaeus* (McConnell et al. 1998a; Roney et al. 2004). Genomic copies of this gene have also been identified in *X. hellerii* and *P. reticulata*, with features consistent with functional genes (McConnell et al. 1998a; McConnell et al. 1998b). *Xima-DXB* and *Pore-DXB* are thought to be orthologous, sharing 98% sequence similarity in exon 3 (i.e., *Xima-DXB**01 and *Pore-DXB**01). *DXB* has a sequence similarity of only 61%–63% in exon 3 with *DAB* alleles, and both genes are hypothesized to be the result of an ancient gene duplication event. However, unlike *DAB*, the *DXB* gene shows very little polymorphism and a weak signal of positive selection (Summers et al. 2009).

The exon-intron structure of the *DXB* is similar to amphibian, chicken, and human class IIB structure, which suggests that a *DXB*-like lineage is phylogenetically related to tetrapod MHC class IIB genes (McConnell et al. 1998a; McConnell et al. 1998b). Interestingly, Roney et al. (2004) identified two alternative splice patterns of the *DXB* gene (as well as one truncated version from a sample of *X. multilineatus* and *X. pygmaeus*). Alternative splicing is the transcription of only certain exons within a gene, and this allows a single gene to encode many distinct proteins (reviewed by Stetefeld & Ruegg 2005; Artamonova & Gelfand 2007). Alternative splicing may be an important feature of the MHC, potentially increasing its versatility, and in one human nonclassical class II chain molecule (*HLA-DM*) four alternative splice patterns have been detected (Modrek et al. 2001). Gene expression and population ge-

netics of the *DXB* gene in poeciliids have received relatively little attention. However, the interactions between *DXB* gene products and other MHC class II molecules (see Ting & Trowsdale 2002; Danchin et al. 2004b) and the relative conservation of this gene across phylogenetically distinct taxa suggest that it plays an important role in MHC evolution and its functioning (Jensen et al. 2008). As recently highlighted by Summers et al. (2009), research on fish, and in particular poeciliids, may help to unravel the enigmatic role of (nonclassical) MHC genes in other vertebrates.

The genomic organization of poeciliid MHC is relatively simple and is more similar to that of salmonids than to that of many other important evolutionary and ecological model species. Salmonids are believed to contain just one *DAB* locus (Langefors et al. 1998, 2000; Landry & Bernatchez 2001; Stet et al. 2002), which makes detecting putative associations between resistance and MHC alleles or genotypes easier than for species with many duplicated MHC genes. For example, the three-spined stickleback (*Gasterosteus aculeatus*) is thought to have 4 *DB* gene loci (*Gaac-DAB*, -*DBB*, -*DCB*, and -*DDB*) (Reusch & Langefors 2005), while cichlids may have up to 17 loci (Malaga-Trillo et al. 1998). Consequently, in these species associations between resistance and MHC can be detected only when analyzing multilocus genotypes. Although several studies tried to find the evolutionary or ecological cause for the high and variable number of *DB* loci, PCR amplifying of the alleles of a large number of loci is a technically challenging task. Indeed, the relative simplicity of poeciliid MHC makes this family an excellent model for future immunological, parasitological, and evolutionary studies. Furthermore, the apparent difference in genomic organization between ornamental and wild guppies makes this an interesting species in which to examine the role of different models of multigene evolution.

25.6 The poeciliid MHC, molecular evolution, and ecology

25.6.1 Parasite selection and the MHC in poeciliids

In natural populations, all genetic variation is affected by mutation and random genetic drift. In addition to these evolutionary forces, MHC genes are also affected by selection (Ford 2002; Garrigan & Hedrick 2003). To study the effects of selection, researchers tend to compare the genetic variation at MHC loci with that of neutral genetic markers (e.g., microsatellites). The signal of natural selection should be detectable only in the MHC, while population demographic forces (such as genetic drift, migration, and mutation) will be evident at both neutral and immunogenetic loci. This allows one to disentangle these demographic effects from the effects of selection on the MHC (van Oosterhout et al. 2006b).

Van Oosterhout et al. (2006a) assessed MHC class IIB diversity in two wild populations of guppies while also recording differences in parasite fauna. Populations inhabited the upper ($N_e \approx 100$) and lower ($N_e \approx 2400$) regions of the Aripo River in Trinidad. Both sites are characterized by distinctly different parasite faunas (see Cable, **chapter 8**); more specifically, *Gyrodactylus* burdens of individual fish were markedly different between the sites (van Oosterhout et al. 2006b). The upland population, with higher *Gyrodactylus* loads, was shown to maintain high levels of MHC diversity despite the effects of random genetic drift on this small population. Using computer simulations that incorporated the demographic estimates based on microsatellite variation, the authors estimated selection coefficients of $S \geq 0.2$ acting on the MHC in the more parasitized population.

In a subsequent study, van Oosterhout et al. (2007a) estimated the potential for parasite selection directly, using a mark-recapture experiment in the same upland Aripo guppy population in Trinidad. They marked almost 200 guppies and counted their gyrodactylid parasite load. Although individuals with a relatively high parasite burden were recaptured at a significantly reduced rate, the selection coefficient was lower ($S \leq 0.14$) than that required to explain the MHC variation observed in the previous study (van Oosterhout et al. 2006b). This suggests that other selective forces are operating on the MHC, such as sexual selection through female mate choice or selection on linked mutations that hitchhike with the MHC alleles (van Oosterhout 2009).

25.6.2 Sexual selection and the MHC in poeciliids

An impressive body of research on poeciliids has been dedicated to studying the role of female mate choice based on male color pattern and ornamentation (Rios-Cardenas & Morris, **chapter 17**). However, sexual selection based on other senses such as olfaction has received relatively little attention (Crow & Liley 1979; Meyer & Liley 1982; Griffiths & Magurran 1999; Shohet & Watt 2004; Archard et al. 2008). Nevertheless, studies on other fish species have indicated that multimodal interactions between visual and chemosensory cues potentially play a role in mate choice and sexual selection (e.g., three-spined sticklebacks, *Gasterosteus aculeatus*; McLennan 2003). Archard et al. (2008) showed that high concentrations of chemosensory stimulus were required to elicit a behavioral response in guppies. Such high concentrations can be attained only when the fish are in close proximity. This may arise when guppies shoal or when males display to females, which occurs within distances of a few centimeters (Endler 1995; Long

& Rosenqvist 1998). Shohet and Watt (2004) used an experimental flume to study the attraction of female guppies to conspecific chemosensory cues and found that females were attracted to cues from other females and that visually attractive males were chemically unattractive, and vice versa. The potential role of the MHC in this model and in other poeciliid model systems has, however, not been investigated. Future research into the MHC and sexual selection should probably utilize poeciliid species with little variation in male color patterns, since in these species olfaction may play a more prominent role than visual communication.

25.6.3 The MHC in conservation

MHC diversity is believed to play an important role in conservation genetics, particularly in captive populations bred for reintroduction programs (O'Brien & Evermann 1988; Hedrick et al. 2001b; Miller et al. 2001). Surprisingly, although the success rate of reintroduction programs is low, with only 11% attaining viable population sizes (Beck et al. 1994), few studies have investigated the efficacy of captive breeding regimes for the release of captive-bred vertebrates. Contemporary breeding strategies for endangered species typically aim to prevent loss of genetic variation, limit the adaptation to captive conditions, and maximize evolutionary potential (van Oosterhout et al. 2007b). Management of MHC diversity in captive gene pools has been suggested as an alternative breeding protocol (Hughes 1991; but see Miller & Hedrick 1991), although this suggestion has not been tested empirically. Indeed, whether the level of MHC variation affects the viability of populations and the success of reintroduced individuals is not yet clearly established (Gutierrez-Espeleta et al. 2001; Miller & Lambert 2004), although reduced immunocompetence (Hale & Briskie 2007) and increased risk of disease outbreaks (O'Brien & Evermann 1988; Spielman et al. 2004) have been noted in several bottlenecked and/or inbred populations.

Van Oosterhout et al. (2007b) tested the effects of breeding regimes on reintroduction success using wild-caught guppies from two populations in Trinidad. The authors bred fish for four generations in captivity and released the fish in a mesocosm in Trinidad. However, they did not examine whether particular MHC genotypes or alleles affected the survival rate of the reintroduced fish. The authors identified parasites as the main threat to survival in the wild, which suggests that MHC variation could play an important role. Guppies and other poeciliids offer an excellent model to test the effectiveness of breeding programs and, in particular, the importance of MHC diversity for viability of reintroduced populations.

Hedrick & Parker (1998) examined polymorphism at the MHC in the endangered Gila topminnow (*Poeciliopsis occidentalis occidentalis*). They examined four watersheds in the United States and found no MHC variation in the most severely bottlenecked population. Because the studied populations differed significantly in their MHC divergence, the authors suggested that further research was warranted to establish whether the watersheds should be considered for independent conservation and management. In a subsequent study, Hedrick et al. (2001b) used both microsatellites and the MHC to study the conservation genetics of the endangered Sonoran topminnow. The Sonoran topminnow (*Poeciliopsis occidentalis*) consisted of two putative subspecies, the Gila topminnow (*Poeciliopsis o. occidentalis*) and the Yaqui topminnow (*P. o. sonoriensis*). The authors used both sets of highly polymorphic markers to recommend that both subspecies be recognized as species (rather than subspecies) and identified as different evolutionarily significant units (ESUs) and management units (MUs). Hedrick et al. (2001a) also found that fish with homozygote MHC genotypes were marginally more susceptible to gyrodactylid parasite infections than heterozygous individuals, which could indicate that gene diversity of the MHC plays an important role in immunocompetence in Gila topminnows. However, Giese and Hedrick (2003) did not find an association between MHC genotype and resistance to a bacterial infection in the Gila topminnow.

25.7 Concluding remarks

Studies of the poeciliid MHC may unravel some of the most enigmatic questions in evolutionary biology, including the roles of parasite-driven selection and sexual selection in adaptive evolution. Poeciliid fishes are particularly well-suited models to examine those questions, given their often (relatively) simple MHC organization, their well-known parasite faunas, and their tractability in behavioral and field experiments. For example, contrasting the mother-embryo interactions between poeciliids with superfetation and those with ovoviviparity may help us understand the role of the MHC in embryo survival and MHC-mediated prenatal selection in newborns. Research on the MHC in poeciliids is also important for conservation biology, not only because these fishes are excellent models for examining captive breeding programs but also because several poeciliid species are vulnerable or endangered themselves.

Chapter 26 Oncogenetics

Manfred Schartl and Svenja Meierjohann

26.1 Introduction

PLATYFISH, SWORDTAILS, and their hybrids became the earliest animal models for studying genetic factors involved in the formation of tumors. Importantly, the *Xiphophorus* melanoma system was the first to prove the existence of oncogenes responsible for cancer. Already in the 1920s, approximately at the same time when it was noted that certain laboratory strains of mice have a higher incidence of spontaneous tumors than others (see Heston 1982), it was shown that hybridization of spotted individuals of the platyfish *Xiphophorus maculatus* with nonspotted southern green swordtails (*Xiphophorus hellerii*) results in offspring that develop pigmented tumors according to Mendelian laws (Gordon 1927; Häussler 1928; Kosswig 1928). Tumor formation was explained by the "inappropriate interaction" of the pigmentation locus from the platyfish with the *X. hellerii* genome (Kosswig 1929; Gordon 1931). The genetic components involved in this "inappropriate interaction" were recognized as tumor modifiers and later on defined as negatively acting tumor-inhibiting genes (Breider 1952). These discoveries were further developed into the current genetic model for tumor formation in *Xiphophorus* hybrids. In the original version the model included dominant differentiation genes converted to oncogenes by naturally occurring mutations and negatively acting tumor-suppressing genes (Anders 1967; Ahuja & Anders 1976). The clear-cut genetics of tumor formation and the strict hereditary etiology made it possible to define and map the dominant tumor gene, which on this basis could be isolated and functionally studied. Other genetic determinants of tumor formation have also been studied, as well as the genetics of the pigment patterns that predispose to tumors.

26.2 Genetics of melanoma formation in *Xiphophorus*

26.2.1 Macro- and micromelanophores

All poeciliid species have three different lineages of pigment cells: the reflecting iridophores, the yellow, orange, and red xantho-/erythrophores, and the dark brown to black melanophores. Importantly, the latter cell type occurs in two different forms in some genera. Micromelanophores have a maximal diameter of 100 μm and are present in all species. They are mostly dispersed on fins and body, thereby producing the typical grayish wild-type pigmentation. In addition, they can also concentrate in certain areas of the skin and extracutaneous tissues, leading to reticular or punctate patterns (Kallman 1975; Borowsky 1984). In these micromelanophore accumulations, cells are not clustered and do not overlap locally. Macromelanophores are much larger in size (300–500 μm in diameter). They occur in some species of *Poecilia* and *Girardinus* and in *Gambusia holbrooki*, *Limia vittata*, and *Phalloceros caudimaculatus* (M. Schartl, unpublished data), but most information on macromelanophores is available from their presence in some *Xiphophorus* species. Here, the local macromelanophore accumulations constitute heritable spot patterns (Kallman 1975; Borowsky 1984). The large cell size is probably due to endomitosis, as multiple nuclei are often observed in these cells (Gordon 1931; Anders et al. 1979).

Macromelanophores tend to overgrow each other and usually give rise to much more prominent pigment spots on the fish skin than micromelanophore spots. Macromelanophores are typical for poeciliids. Whether this cell type exists also in other fish and contributes to black spotting patterns (as, e.g., in some cichlids) is unknown.

The macromelanophore patterns of poeciliids are encoded by a series of codominant alleles (Kallman 1975) of the so-called *mdl* locus (macromelanophore-determining locus). The genetic information encoded by *mdl* determines macromelanophore development in different body compartments, the shape of the pattern (patches, spots, stripes, and their respective expansion), and onset of pattern formation. In the genus *Xiphophorus*, *mdl* alleles are present in some but not all species (table 26.1). Their presence does not follow a simple phylogenetic pattern, meaning that the *mdl*s of the extant species have an independent origin and show convergent evolution or that the *mdl* has been lost repeatedly in several species. The nomenclature for the different alleles follows the rules for naming genes in fish (http://zfin

.org/zf_info/nomen.html). The locus is written in italics, and the symbol for the respective encoded pattern is placed as a superscript. Notably, as long as molecular information on the *mdl*-encoded gene(s) is missing, patterns in different species, even if giving rise to the same or a highly similar phenotype, are given different names. One has to keep in mind that some of them might be homologous (Kallman & Atz 1966).

Expression of the *mdl* locus is enhanced in certain hybrids of *Xiphophorus*, giving rise to varying degrees of hyperpigmentation ranging from enlarged spot patterns to benign melanosis and to highly malignant melanoma.

26.2.2 The Gordon-Kosswig-Anders melanoma cross (spontaneous melanoma)

In the late 1920s, researchers found that certain hybrids of platyfish (*X. maculatus*) exhibiting macromelanophore patterns and swordtails without macromelanophores (*X. hellerii*) spontaneously develop malignant melanoma (Gordon

Table 26.1 Wild-type macromelanophore patterns in fish of the genus Xiphophorus

Macromelanophore pattern	*mdl* allele	Species	*Xmrk* associated	Remarks
Spotted dorsal	*Sd*	*X. maculatus*	Yes	
Spot-sided[a]	*Sp*		Yes	Similar to *Fl, Pu1, Se, Db2, Ss*
Striped	*Sr*		Yes	
Nigra	*Ni*		Yes	
Spotted belly	*Sb*		Yes	
Flecked	*Fl*	*X. xiphidium*	Yes	
Lineatus	*Li*	*X. variatus*	Yes	
Punctatus-1	*Pu1*		No	
Punctatus-2	*Pu2*		Yes	
Spotted ventral	*Sv*	*X. milleri*	Yes	This pattern was also designated *Sl* (Weis & Schartl 1998), but *Sv* is the priority name
Speckled	*Se*	*X. evelynae*	Yes	
Dabbed-1	*Db1*	*X. hellerii*	No	Spots arranged in longitudinal rows
Dabbed-2	*Db2*		No	Irregularly arranged spots
Caudal spotted	*Cs*	*X. birchmanni*	Yes	Very similar to *Sc*
Spotted caudal	*Sc*	*X. cortezi*	Yes	Very similar to *Cs*
Atromaculatus	*At*		?	
Spotted side	*Ss*	*X. nezahualcoyotl*	No	
Marmoratus	*Ma*	*X. montezumae*	Yes	
Spotted flanks	*Sf*		No	

[a]Kallman (1975) recognized many variants of the pattern, which fall into two subtypes: A (spots restricted to the flank below the dorsal fin and the peduncle) and B ("salt-and-pepper" appearance). Within the subtypes, different spotting phenotypes can be distinguished, which all are encoded by distinct genes at the *mdl* locus, indicated by a number following the pattern acronym (*Sp1, Sp2*, etc.).

1927; Häussler 1928; Kosswig 1928). It was suggested that the *mdl* of *X. maculatus* "interacted" with the *X. hellerii* genome differently than it interacted with the *X. maculatus* genome, thereby enabling melanoma development. The tumor-causing locus of *X. maculatus* was later defined more clearly and called *Tu* (for "tumor") (Anders 1967; Ahuja & Anders 1976). It was reasoned that the function of the *mdl* in determining development of the macromelanophore as a giant pigment cell type is equivalent to *Tu*, while the pattern formation function was assigned to a series of co-dominantly linked alleles that determine the compartment of appearance of macromelanophores. They were called R_{Co} (Anders & Anders 1978; Anders 1991).

The other, interacting genetic entity was termed *R* (for "regulator") or, more precisely, R_{Diff} (Ahuja & Anders 1976), also known as *Diff* (Vielkind 1976; Anders 1991) or *MelSev* (Morizot & Siciliano 1983). *R* was proposed to be present in the same fish that carries *Tu* (here *X. maculatus*). It functions by repressing *Tu* activity. The crossing partner was thought to either lack the *R*-locus in general or have a "nonfunctional" *R* (meaning not properly interacting with *Tu*). This genetic model is illustrated in fig. 26.1 in the so-called classical cross using a *X. maculatus* female with the macromelanophore pattern spotted dorsal (mdl^{Sd}, one to two spots of macromelanophores in the dorsal fin) and a *X. hellerii* male without an *mdl* allele. *Xiphophorus maculatus* is homozygous for the *Tu* locus (*Tu/Tu*) and the autosomal *R* locus (*R/R*), whereas *X. hellerii* is thought to be devoid of *Tu* and *R* (–/ –; –/–). Crossing these two parental fish results in offspring with one allele each of *Tu* and *R* (*Tu/–*; *R/–*). This genetic constitution becomes phenotypically apparent by a locally restricted and benign hyperproliferation of melanophores due to reduced activity of the remaining *Tu* allele, which in turn is due to the loss

of one *R* allele. Further crossing of these F_1 fish with *X. hellerii* leads to the following allele combinations in BC_1 (backcross generation number 1) fish: (–/–; –/–), (–/–; *R*/–), (*Tu/–*; *R*/–) or (*Tu/–*; –/–). Fish with the latter genotype spontaneously develop melanoma, as *Tu* repression is totally abolished. As a result, the macromelanophores proliferate without restriction.

The mechanism of action of *R* is still not known, as its molecular identity has not been uncovered yet. Because of this, instead of a tumor suppressor being present in *X. maculatus*, the crossing data can formally also be explained by *R* being a tumor-inducer present in *X. hellerii* that only in combination with the *Tu* allele exerts a dosage-dependent protumorigenic effect (Schartl 1995). However, in the meantime enough evidence has accumulated indicating that *Tu* function needs to be suppressed in the platyfish rather than being activated in the hybrids, most importantly because *Tu* encodes a highly malignant tumor gene (Schartl & Wellbrock 1998; Schartl et al. 1998); and see below).

26.2.3 The *Tu* locus encodes the *xmrk* gene

The tumor-determining gene of the *Tu* locus was identified as oncogenic receptor tyrosine kinase (RTK), termed *xmrk* (for *Xiphophorus* melanoma receptor kinase) (Wittbrodt et al. 1989). The gene *xmrk* arose from its proto-oncogene, the *Xiphophorus* co-orthologue of the epidermal growth factor receptor *egfrb*, due to a local gene duplication event.

In the healthy purebred *X. maculatus*, *xmrk* is expressed at very low levels in eyes, gills, and skin (Schartl et al. 1999). However, strong overexpression resulting in melanoma genesis occurs only in the black-pigment-cell lineage of hybrid fish. The proto-oncogene is expressed only at very low levels in the melanoma cells but is also expressed in some normal organs where *xmrk* expression is not detectable (Gomez et al. 2004). This difference in organ specificity and expression levels of the two duplicates (oncogene *xmrk* and its progenitor *egfrb*) points to a different transcriptional control. The new genomic environment created by the duplication event for the second copy is most likely responsible for this difference in expression of *xmrk* and *egfrb* (Adam et al. 1993; Volff et al. 2003).

It is assumed that *xmrk* expression in the parental *X. maculatus* is under the control of *R*. This control seems to be lost in *R*-deficient backcross hybrids. Indeed, (*Tu/–*; *R*/–) individuals display a moderate increase of *xmrk* expression and develop benign melanotic spots. (*Tu/–*; –/–) fish express much higher oncogene levels and here melanomas emerge in all individuals (Wittbrodt et al. 1989; Adam et al. 1991; Mäueler et al. 1993). Thus, melanoma development may correlate with the level of *xmrk* expression, and it is as-

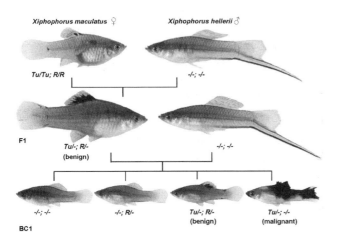

Figure 26.1 Gordon-Kosswig-Anders crossing procedure for producing fish with spontaneous melanoma or benign, premalignant pigment-cell lesion. The description is given in the text.

sumed that *R*, as a tumor suppressor, is involved (directly or indirectly) in transcriptional regulation of *xmrk*.

26.3 The *xmrk* oncogene

26.3.1 Structure of the Xmrk protein

Proteins in the EGFR family consist of an extracellular domain, which induces dimerization after ligand binding, a transmembrane domain, and a cytoplasmic domain harboring a tyrosine kinase activity and docking sites for intracellular adapter proteins. The *xmrk* oncogene differs from its progenitor proto-oncogene, *egfrb*, in that it has a number of nucleotides that accumulated after the gene duplication event. Altogether, these lead to 17 amino acid changes between the mature Xmrk and Egfrb proteins of *X. maculatus*. Nine of these are located in the extracellular domain. Here, the presence of intramolecular cysteine disulfide bridges is important for correct protein folding and thus functionality (Meierjohann et al. 2006a). At least two of the amino acid exchanges in the Xmrk receptor lead to the disruption of stabilizing intramolecular disulfide bonds. In one case, one of the two cysteines is exchanged for serine, and consequently one binding partner is missing. This results in a free binding availability of the other cysteine residue. If now two Xmrk molecules with unbound cysteines come closely together, a disulfide bridge is formed between the neighboring molecules (Gomez et al. 2001). The dimer structure generated by this mechanism mimics the structure of a ligand-bound wild-type receptor. This is sufficient for crossactivation of the intracellular kinase domains (Meierjohann et al. 2006a). In contrast to wild-type EGF receptors that only become activated if a ligand is present, Xmrk is permanently active.

26.3.2 Oncogenic signaling of Xmrk

As EGFR orthologue, Xmrk shares many characteristic biochemical features with the human EGFR. Because Xmrk is at the top of numerous signal transduction pathways, it is capable of inducing all necessary tumorigenic features of a cell such as proliferation, dedifferentiation, anti-apoptosis, invasion of foreign environments and survival therein, and migration. The specific steps are detailed below and are summarized in fig. 26.2.

The RAS/RAF/ERK1/2 cascade is a classical target signal transduction pathway of receptor tyrosine kinases in general, and via phosphorylation events its activation can affect the turnover of cytoskeletal proteins and the activity of metabolic enzymes and of nuclear transcription factors and many more (reviewed in Dhillon et al. 2007). The tran-

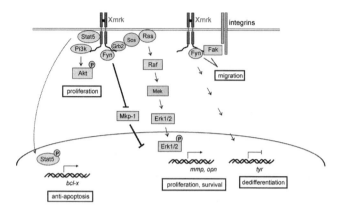

Figure 26.2 Signal transduction pathways induced by Xmrk. Grb2 is the adapter protein that directly docks to Xmrk. Sos is a guanine nucleotide exchange factor that activates Ras after binding to Grb2 and localizing to the membrane. Mek is the upstream kinase that phosphorylates and activates Erk1/2. All other proteins are explained in the text. mmp: matrix metalloprotease; tyr: tyrosinase; bcl-x: bcl2-like.

scription of many Xmrk-induced target genes is mediated via Erk1/2 (for an overview of known Xmrk target genes and their upstream activators, see table 26.2). Specific cellular roles have been attributed to most of these genes.

The secreted protein osteopontin (Opn) guarantees cellular survival in a three-dimensional collagen matrix. In the absence of Opn, melanocytes become apoptotic due to the presence of unligated integrins on their surface (Stupack et al. 2001). In vivo, this is considered a protection against dermal invasion and dissemination of epidermal cells. Production and secretion of Opn leads to its binding to integrins and prevents apoptosis (Geissinger et al. 2002). Interestingly, OPN plays a crucial role in human melanomas, too, where it is even discussed as a diagnostic serum marker (Zhou et al. 2005; Reiniger et al. 2007).

Matrix metalloproteases (MMP) are secreted proteins as well, and their most acknowledged function is the digestion of extracellular matrix during the process of cell migration (Sternlicht & Werb 2001; Folgueras et al. 2004). However, Xmrk-expressing melanocytes depend neither on Mmp activity nor on their upstream signaling pathway through Erk1/2 to be able to migrate (Meierjohann et al. 2006b; Meierjohann et al., 2010). They need the proteases for an efficient passage through the cell cycle, which gives the Ras/Raf/Erk1/2 pathway in general and Mmps in particular a pro-proliferative function. The exact reason for this is unknown, but as many MMPs are capable of providing active growth factors by cleaving their extracellular proforms, an outside-in signaling initiated by these growth factors is a possible explanation (Mook et al. 2004). ERK1/2 activation is also a hallmark of most human melanomas. It is the result of somatic mutations that give rise to oncogenic changes in its upstream activators RAS or B-RAF (reviewed in Fecher et al. 2008). The relevance of these findings is evi-

Table 26.2 List of known target genes induced by Xmrk (and responsible pathways)

Gene	Common function	Induced by MAPK	Induced by PI3K	Induced by STAT5
opn	Secreted protein, binds to integrins	X		
mmp-1, -3, -9, -13	Mostly secreted proteins, matrix proteases	X		
cyr61	Secreted, matrix-binding protein, acts angiogenically	X	X	
igfbp3	Secreted protein, binds to insulin-like growth factors	X	X	
Emp1	Membrane, adhesion molecule	X		
bcl-x	Intracellular protein, anti-apoptotic			X
egr-1	Nuclear, transcription regulator			X

Note: opn: osteopontin; *mmp:* matrix metalloprotease; *cyr61:* cysteine-rich angiogenic inducer 61; *igfbp3:* insulin-like growth factor binding protein 3; *emp1:* epithelial membrane protein 1; *bcl-x:* bcl2-like; *egr-1:* early growth response 1.

dent from the fact that MMPs are aberrantly expressed in human melanomas (Hofmann et al. 2000; Hofmann et al. 2005).

The small nonreceptor tyrosine kinase Fyn is a signal transduction protein that directly docks to phosphorylated tyrosine residues in the cytoplasmic domain of Xmrk (Wellbrock & Schartl 1999). It belongs to the family of Src kinases, which include a number of proto-oncogenes involved in multiple cellular events such as proliferation, survival, anchorage-independent growth, invasion, and angiogenesis (Mitra & Schlaepfer 2006). In the Xmrk melanoma model, Fyn is the only identified Src family kinase that becomes directly activated by Xmrk (Wellbrock et al. 1995). It can support the action of several pathways in the Xmrk melanoma model.

First, Fyn inhibits the Map kinase phosphatase 1 (Mkp-1), which usually dephosphorylates and thereby inactivates Erk1/2 (Wellbrock et al. 2002). As a consequence, Fyn is directly involved in mediating cell proliferation. In this context, differentiation of pigment cells is also blocked by the action of Fyn, as the active kinase can reduce the levels of the pigment-cell-specific transcription factor Mitf and of tyrosinase, the rate-determining enzyme in melanin synthesis (Wellbrock et al. 2002).

Second, Fyn interacts with focal adhesion kinase (Fak), which is a crucial determinant of the interaction strength between cell surface and substratum. By modulating Fak activity, Fyn can induce an increased turnover of focal adhesions, which are the contact sites between the cell membrane and the overlying extracellular matrix. This results in attenuated cell attachment and can finally result in migration (Meierjohann et al. 2006b). Human melanoma cells also express fair amounts of FYN, a fraction of which interacts with FAK protein. This suggests a function for both proteins in human melanoma signaling (Teutschbein et al. 2009).

Third, a complex of three proteins, namely Xmrk, Fyn,

and phosphoinositide-3 kinase (Pi3k), is present in Xmrk-expressing pigment cells (Wellbrock & Schartl 2000). Pi3k is a lipid kinase that catalyzes the conversion of phosphatidylinositol-(4,5)-bisphosphate to phosphatidylinositol-(3,4,5)-trisphosphate (PIP3) on the inner leaflet of the plasma membrane. PIP3 acts as second messenger and is capable of inducing several downstream pathway components, such as the serine/threonine kinase Akt. In the Xmrk melanoma model, activated Pi3k/Akt have no effect on cell migration but enable the entry into S phase and are therefore involved in mitogenic signaling (Wellbrock & Schartl 2000). Pi3k also plays an important role in human melanomas. Aberrant oncogenic RAS and inactivation of negative-feedback signals contribute to an elevated AKT activity in up to 67% of all human melanomas (Chin 2003; Stahl et al. 2004; Robertson 2005). Pi3k enhances the progression to more advanced aggressive tumors and is involved in survival of melanoma cells.

The "signal transducer and activator of transcription (STAT)" proteins are classical downstream targets of cytokine receptors and receptor tyrosine kinases. After phosphorylation by RTKs or membrane-recruited Janus kinases (JAK), they translocate to the nucleus and induce target gene transcription. Xmrk-expressing pigment cells exhibit a permanent nuclear Stat5 localization, which results in enhanced cell proliferation and anti-apoptotis (Baudler et al. 1999). The latter effect is most likely due to the induction of the anti-apoptotic gene *bcl-x* (Morcinek et al. 2002). In human melanomas, 60% of samples display activated STAT5, which contributes to interferon resistance and *BCL-2*-dependent cell survival (Wellbrock et al. 2005; Mirmohammadsadegh et al. 2006). STAT5 can also be activated by stimulating endogenous EGFR, which demonstrates the good comparability between the Xmrk-driven fish tumors and human melanomas (Mirmohammadsadegh et al. 2006).

26.3.3 *xmrk* transgenic fish

The high oncogenic potential of Xmrk is evident from transgenic experiments. When the oncogene is expressed under the control of a ubiquitously active promoter in embryos and larvae of another fish species, the medaka (*Oryzias latipes*), tumors are formed after short latency periods (Winkler et al. 1994; Dimitrijevic et al. 1998; Winnemoeller et al. 2005). In contrast to *Xiphophorus*, medaka are egg-laying fish, easily allowing genetic manipulation and the production of stable mutant lines. The medaka is a small aquarium fish species comparable and complementary to the well-known zebrafish (Wittbrodt et al. 2002). Remarkably, not only pigment-cell tumors but also a large variety of epithelial- and neural-tissue tumors develop in transient *xmrk*-transgenic medaka. However, expression of the proto-oncogene at similarly high levels does not lead to tumor formation (Dimitrijevic et al. 1998; Winnemoeller et al. 2005).

When the transgenic expression of *xmrk* is driven by a pigment-cell-specific promoter, tumors derive only from pigment cells (Schartl et al. 2010). Taking advantage of this fact, stable transgenic lines of medaka were produced, which now, similar to *Xiphophorus* hybrids, spontaneously develop melanoma.

In addition to the skin-derived melanomas, which are the overwhelming majority of tumors observed in *Xiphophorus*, medaka also display other types of pigment-cell tumors. On the one hand, melanoma develop from extracutaneous sites such as the meninges, the intestine, and the choroid layer of the eye. On the other hand, cells of the xantho-/erythrophore lineage give rise to exophytically growing yellow to orange pigment-cell tumors ("xantho-erythrophoroma"), which have a lower metastatic capacity than the melanomas (Schartl et al. 2010). The type of pigment-cell tumor that will develop depends on the genetic background of the transgenic strain, pointing to the action of strain-specific tumor modifier genes.

26.3.4 Control of *xmrk* expression

The promoter region of *xmrk* differs from that of its corresponding proto-oncogene *egfrb*. In the *Xiphophorus* model, promoter characteristics might play a major role in melanomagenesis and are thus of high interest. An upstream fragment (nucleotides −675/+34) from the *xmrk* promoter is responsible for high transcriptional activity. A GC-rich sequence identical to the human Sp1 binding site is located here, close to the transcription start site of *xmrk*. It was shown that GC-binding proteins are present in both melanoma and nonmelanoma cells; however, the promoter is active only in melanoma cells (Baudler et al. 1997). A po-

tential silencer element that prevents expression in nonmelanoma cells is therefore highly likely and might be linked to the repressing activity of *R*.

In addition to specific transcription factor binding sites, epigenetic mechanisms such as DNA methylation may regulate promoter activity. Classically, methylated cytosin residues in a number of concurrent CpG dinucleotides in the proximal promoter prevent transcription. The *xmrk* promoter is highly methylated in nonmelanoma tissue but unmethylated in melanoma tissue (Altschmied et al. 1997). This is consistent with overexpression in the pigment-cell lineage of hybrid genotypes.

26.4 Genetic organization of the melanoma locus

The clear-cut genetic model of the past and the ideas about the organization of the *Tu* locus are difficult to translate into the language of molecular genetics. Neither Mendelian genetics nor our molecular genetic knowledge nor the combination of both can explain all the phenomena observed.

Macromelanophore patterns and thus *mdl* alleles are present in *X. maculatus*, *X. xiphidium*, *X. milleri*, *X. variatus*, *X. evelynae*, *X. montezumae*, *X. nezahualcoyotl*, *X. hellerii*, *X. cortezi*, and *X. birchmanni*. Many of the patterns produce hyperpigmentation or even melanoma after appropriate crossings, but others certainly do not (table 26.1). Those *mdl* that are melanomagenic are always associated with the presence of the *xmrk* gene, while fish with *mdl*s that do not predispose to melanoma lack the *xmrk* gene. Thus, *xmrk* and *mdl* are separate genetic entities (Weis & Schartl 1998). The genes *mdl* and *xmrk* are closely linked, and *mdl* maps 5′ to *xmrk* in *X. maculatus* (Strain Jp163A) (Gutbrod & Schartl 1999). In this view the classical *Tu* locus would consist of *mdl*, which determines the development and physiology of the macromelanophore and the pattern phenotype, plus *xmrk*, which contributes only the capacity of neoplastic transformation to this specific cell lineage. Consequently, *mdl* is responsible for the body compartment and temporal onset of tumor formation in the hybrid melanoma.

The *xmrk* genes that are associated with different *mdl*s show sequence variations within and between species (S. Schories & M. Schartl, unpublished data). So far two allele classes have been identified in *X. maculatus* (fig. 26.3). Class A encodes a full-length EGF receptor, while class B has a deletion that includes one of the carboxyterminal exons (exon 26) and results in a protein with 35 amino acids missing from the cytoplasmic domain (Adam et al. 1991). The class B alleles are equally tumorigenic as the class A alleles. Further allele classes may exist; therefore, it is reasonable to assign the so far uncharacterized *xmrk* genes from

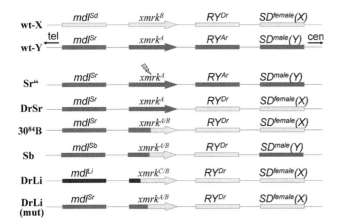

Figure 26.3 Genomic organization and allele structure of the wild-type (wt) X and Y chromosomal melanoma locus (*mdl* and *xmrk*) with linked red-yellow pigment pattern (RY) and sex-determining locus (SD) of *X. maculatus* (Río Jamapa, Jp163A) and different mutant chromosomes. Orientation toward the centromere (cen) and telomere (tel) is indicated. In the Sr'' mutant the flash indicates the so far unidentified mutation in *xmrk*. DrSr and 30⁸⁴B are X/Y sex chromosomal crossovers, Sb is obviously the result of two consecutive crossovers or a double crossover. For DrLi the crossover occurred between a *X. variatus* sex chromosome (indicated by black color) and the *X. maculatus* X chromosome. DrLi (mut) originated from a crossover of the DrLi chromosome with the Sr'' chromosome. The wt-X, Sr'', Sb, and DrLi chromosomes harbor melanomagenic loci, while the wt-Y, DrSr, 30⁸⁴B, and DrLi (mut) chromosomes do not lead to melanoma or even hyperpigmentation in hybrids. For more details about the genomic organization and location of chromosomal breakpoints, see Gutbrod and Schartl 1999.

other species to separate classes until information similar to what we have for the *X. maculatus* alleles is available (e.g., *xmrk^C* for the gene from *X. variatus*).

As a consequence, a specific melanomagenic locus is defined by the combination of the *mdl* and *xmrk* alleles. The famous spotted dorsal *Tu* locus from the classical cross would then be *Mdl^Sd*-*xmrk^B* (fig. 26.3). For convenience we propose adding the sex chromosomal location and collection site or laboratory strain information, respectively, in parentheses—in this case: (X, Jp163A).

The first full-length cDNA of *xmrk* was cloned from a melanoma-cell line (Wittbrodt et al. 1989). This cDNA from the PSM cell line (PSM, for platyfish swordtail hybrid melanoma; Wakamatsu 1981) has been used for many experiments. It represents a class A allele. The PSM cell line was established from ornamental fish that were obtained from the aquarium trade and that carried an *Sp*-type *mdl* allele; but the sex chromosomal origin is unknown.

Besides differences in the coding sequence, strong differences in the introns and 3′ and 5′ flanking regions of *xmrk* have also been noted (Volff et al. 2003; Schultheis et al. 2006). The whole chromosomal region that harbors *xmrk* (and probably also *mdl*) is a hot spot for transposon integration and other genomic alterations. Of note, the *mdl^Sr*-*xmrk^A* (Y, Jp163A) locus, which is not melanomagenic and does not even lead to hyperpigmentation in hybrid crosses,

carries a transposon (piggyBac-like element) insertion upstream of *xmrk* in the putative promoter (Volff et al. 2003). It is tempting to assume that the disruption of the promoter has irreversibly shut off *xmrk*, but supporting data about the transcriptional control of this allele are missing.

With respect to gene order, *mdl* is located 5′ to *xmrk*. Other known sex chromosomal loci from the *xmrk* region map 3′, including the red-yellow pattern locus *RY*, the puberty locus *P*, the sex-determining locus *SD*, and the proto-oncogene *egfrb* (Gutbrod & Schartl 1999). The distance from *xmrk* to *egfrb* is approximately 0.6 centimorgan. BAC contigs for the entire region on both sex chromosomes of *X. maculatus* (strain Jp163A) were established (Froschauer et al. 2002). Strategic sequencing led to the identification of multiple copies of a melanocortin receptor gene in this region. Those are discussed as candidates for the *mdl* encoded gene.

26.5 *xmrk* mutants

In the past, many *xmrk* and *mdl* mutations have been identified that affect macromelanophore pattern and melanoma phenotype (fig. 26.3; Kallman 1975, table 8). Loss-of-function mutants that produce neither macromelanophores nor melanomas have provided genetic evidence for the causative role of *xmrk* in melanoma development. In one such mutant, a disrupting insertion was found amid the *xmrk* gene, leading to its loss of function (Schartl et al. 1999). In another mutant, the entire *xmrk* was deleted (Wittbrodt et al. 1989).

Gain-of-function mutants are characterized by an enhanced number of macromelanophores in the nonhybrid fish. Generally the area of macromelanophore spots is enlarged, and sometimes the pattern is even different from the wild-type phenotype. In F₁ and backcross hybrids hyperpigmentation and the melanomas are more severe. Some of those mutations are due to *xmrk* intragenic sex chromosomal crossovers (Gutbrod & Schartl 1999). This has changed the combination of the 5′ upstream promoter region and other regulatory regions of the affected *xmrk* allele and also the association to *mdl* alleles.

A particular crossover occurred between the sex chromosomes of two species. One partner was the X chromosome of *X. variatus* that carried *mdl^Li*, encoding the *lineatus* macromelanophore pattern, which is associated with a weak allele of *xmrk*. In accordance with the above-outlined nomenclature, the *X. variatus xmrk* is designated as class C allele. The *mdl^Li*-*xmrk^C* wild-type locus produces only strong hyperpigmentation in hybrids. The other partner was *mdl^Sd*-*xmrk^B* (X, Jp163A) from the *X. maculatus* sex chromosome. In the resulting *DrLi* mutant the upstream region and the 5′ portion of the Mdl^{Li}-$xmrk^C$ is fused to the 3′ part of $xmrk^B$ (fig. 26.3). The macromelanophore pattern encoded by the recombinant locus is similar to neither lineatus (*Li*) nor spotted dorsal (*Sd*). No macromelanophores occur in the dorsal fin, but the entire flanks are covered with numerous irregular spots. In these fish, aggressive melanomas already develop in F_1 hybrids with *X. hellerii*. It has been proposed that the phenotype of the mutant is the result of the combination of mdl^{Li} and/or the promoter of $xmrk^C$ with the foreign $xmrk^B$ coding sequence. Whether other, more subtle mutations (e.g., affecting the *xmrk* sequence, the promoter, or the associated *mdl* allele) did also occur in *DrLi* or the other crossovers' mutant chromosomes (see fig. 26.3) is unknown, as well as the underlying molecular mechanisms leading to the altered phenotype.

Another mutation was induced by X-ray irradiation. The mutant locus designated *Sr″* arose from the mdl^{Sr}-$xmrk^A$ allele (Y, Jp163a) of *X. maculatus*, which is not melanomagenic (fig. 26.3). Sr″ fish, however, exhibit severe melanosis and even melanoma already in F_1 hybrids with *X. hellerii*. The mutation was mapped to a region within intron 1 or downstream thereof (Gutbrod & Schartl 1999), but its molecular nature is awaiting further analysis.

Other sex chromosomal crossovers did not change the macromelanophore pattern and the melanoma phenotype but instead placed the whole *mdl-xmrk* on the other sex chromosome. Interestingly, the $30^{84}B$ mutant is an intragenic sex chromosomal *xmrk* crossover, which occurred somewhere between the 3′ end of intron 1 and exon 15 (fig. 26.3). The mutant chromosome produces the wild-type mdl^{Sr} phenotype and is not melanomagenic. Hence, this further supports the reasoning that the promoter of the wild-type mdl^{Sr}-$xmrk^A$ allele is nonmelanomagenic due to a specific feature of its 5′ flanking region, for example, the insertion of the piggyBac-like transposon.

26.6 The *R* locus

The *R* locus is responsible for suppressing Xmrk-driven melanoma formation. The gene(s) encoded by the *R* locus has (have) not been identified yet. Early on, *R* was mapped in the classical cross to linkage group V, in proximity to an esterase isoenzyme (Ahuja et al. 1980; Morizot & Siciliano 1983; Förnzler et al. 1991). The current linkage map contains more than 20 loci (Kazianis et al. 1998; http://www .xiphophorus.org/mapping.htm). Of all genes identified here, the potential tumor suppressor *cdkn2ab*—formerly termed *cdkn2x*—shows the best genetic association in certain crosses with the *R*-mediated effect on melanoma formation. The human orthologue of this gene is mutated or epigenetically silenced in various human tumors, particularly in melanoma (Bennett 2008). This makes *cdkn2ab* a very attractive candidate for the tumor-modifying function of *R*. However, in *Xiphophorus* melanomas, the *cdkn2ab* promoter is basically unmethylated (meaning unrepressed), and RNA expression is even enhanced here in comparison to skin (Kazianis et al. 2000). In fact, a positive correlation between *xmrk* and *cdkn2ab* expression levels was noted (Butler et al. 2007). *CDKN2A* function as a tumor suppressor is compromised or even totally lost in those human tumors where the association between *CDKN2A* mutations and tumor development was noted. Hence, a high expression in tumors is unexpected and does not explain a potential tumor-modifying activity of this gene in *Xiphophorus*.

An amino acid comparison between the protein product of *cdkn2ab* of *X. maculatus* and *X. hellerii* revealed two differences between the two species: one conservative exchange at position 3 (Leu [*X. hellerii*], Val [*X. maculatus*]) and one nonconservative exchange at position 90 (Glu [*X. hellerii*], Lys [*X. maculatus*]) (Kazianis et al. 1999). There are numerous human *CDKN2A* mutations that are associated with melanoma development and that lead to decreased binding to CDK4 and CDK6 (Harland et al. 1997; Monzon et al. 1998; Becker et al. 2001). However, neither *Xiphophorus* Leu/Val3 nor Glu/Lys90 is predicted to be involved in CDK4/6 interaction. As yet, there is no experimental proof that *cdkn2ab* is the sole crucial *R* locus encoded gene.

26.7 Other melanoma models

26.7.1 Hybrid melanomas

Besides the classical Gordon-Kosswig-Anders cross, a number of backcrossing procedures have been established that result in fish with hyperpigmentation or melanoma (Atz 1962; Zander 1969, 1977). As in the classical cross, the penetrance of the tumor phenotypes is high, and the stereotypical development of the disease provides uniform and highly reproducible material for research (table 26.3).

26.7.2 Nonhybrid melanomas

Certain *Mdl-xmrk* combinations can lead to hyperpigmentation and even melanoma in the nonhybrid situation. They are found in wild fish of *X. variatus* with Mdl^{Pu2}-$xmrk^C$, *X. cortezi* with Mdl^{Sc}-$xmrk^D$, and *X. birchmanni* with Mdl^{Cs}-$xmrk^E$ (Borowsky 1973; A. Schartl et al. 1995; Fernandez & Morris 2008). These rare melanomas overexpress *xmrk* and are most likely induced by the same biochemical

Table 26.3 Xiphophorus crosses that lead to spontaneous or induced melanoma

Model	Crossing scheme[a]*	mdl/xmrk combination	Remarks
H001BC$_1$-B cross (hybrid 1)	*X. hellerii* (Sara) × (*X. maculatus* (Jp163B) × *X. hellerii* (Sara))	mdl^{Sp} - xmrk^B	UV melanoma model; MNU[b] melanoma model
H004BC$_1$-A cross (hybrid 4)	(*X. maculatus* (Jp163A) × *X. andersi* (and C)) × *X. andersi* (and C)	mdl^{Sd} - xmrk^B	
H005BC$_1$-B cross (hybrid 5)	*X. hellerii* (Sara) × (*X. maculatus* (Jp163A) × *X. hellerii* (Sara))	mdl^{Sd} - xmrk^B	Similar to Gordon-Kosswig-Anders cross[c]
H007BC$_1$-B cross (hybrid 7)	*X. couchianus* (Xc) × (*X. maculatus* (Jp163B) × *X. couchianus* (Xc))	mdl^{Sp} - xmrk^B	UV melanoma model
H008BC$_1$-A cross (hybrid 8)	(*X. maculatus* (Jp163A) × *X. couchianus* (Xc)) × *X. couchianus* (Xc)	mdl^{Sd} - xmrk^B	MNU tumor model
H009BC$_1$-A cross (hybrid 9)	(*X. nezahualcoyotl* (Ocampo) × *X. variatus* (Zarco)) × *X. nezahualcoyotl* (Ocampo)	mdl^{Pu2} - xmrk^C	
H010BC$_1$-A cross (hybrid 10)	(*X. maculatus* (Jp163B) × *X. andersi* (and C)) × *X. andersi* (and C)	mdl^{Sp} - xmrk^B	
H015BC$_1$-B cross (hybrid 15)	*X. hellerii* (Sara) × (*X. hellerii* (Sara) × *X. variatus* (Zarco))	mdl^{Pu2} - xmrk^C	

[a]Geographic origin or laboratory strain is given in parenthesis. For details see the *Xiphophorus* homepage (http://www.xiphophorus.org) and the *Xiphophorus* genetic stock center manual.
[b]MNU = *N-methyl-N*-nitrosourea.
[c]Note that other laboratories use *X. hellerii* from *Rio Lancetilla* as the recurrent parent for crossing and backcrossing and an F$_1$ female.

events as described above. Interestingly, they arise—as far as described—mostly in males carrying the macromelanophore pattern, indicating a possible hormone dependency.

26.7.3 Ultraviolet-induced melanomas

While the above-mentioned melanomas develop spontaneously, there are certain hybrids that display melanoma genesis in response to ultraviolet (UV) damage. Similar to the spontaneous melanoma model, UV-induced melanomas derive from fish with *xmrk* (reviewed in Mitchell & Nairn 2006). The *mdl* allele that is involved here (*mdl^{Sp}-xmrk^B* [X, Jp163B]) gives rise to *X. maculatus* fish with large spots located on the body (*Sp* for "spot-sided") instead of on the dorsal fins as in the classical cross. The backcrossing partner may be either *X. hellerii* (strain Río Sarabia) or *X. couchianus* (see table 26.3, hybrid 1 and hybrid 7). As in the Gordon-Kosswig-Anders cross, the presence of *xmrk* is essential for the generation of UV-induced melanoma. Only low spontaneous tumor rates are observed in this model in the BC$_1$ generation, but melanoma incidence rises to 40% after UV irradiation (Setlow et al. 1989).

UV-B damage is characterized by pyrimidine dimers in the DNA, which may result in mutations if the DNA damage systems are overstrained. The accumulation of these mutations is considered causative for melanoma development. More recent results hint at a similarly significant role for UV-A, which can photosensitize melanin, thereby leading to the formation of tumor-causing oxygen radicals (Setlow et al. 1993; Wood et al. 2006). These reactive oxygen species (ROS) can also alter genetic information, and UV-A-related DNA damage is represented by oxidative base alterations and, again, pyrimidine dimers (Ikehata et al. 2008). Similar events also happen in human melanocytes (Larsson et al. 2006; Placzek et al. 2007); thus, the *Xiphophorus* melanoma model supposedly reflects human UV-B- and UV-A-induced melanoma formation. The fact that this melanoma model requires further UV-induced mutations in addition to oncogenic *xmrk* points at the presence of tumor modifiers in the genetic background of the respective hybrids.

UV-induced DNA damage can be counteracted by different repair systems. In the case of oxidative base damage, base excision repair (BER) by cutting out the damaged base is activated. Although genetic hybridization can influence BER activity, a direct link between UV-induced melanoma and BER activity could not be demonstrated (Gimenez-Conti et al. 2001; Heater et al. 2007).

Photolyase-dependent repair of pyrimidine dimers, which is quite efficient in fishes (Regan et al. 1982), reduces the extent of melanoma formation (Setlow et al. 1993). In addition, pyrimidine dimers can be eliminated by the nucleotide excision repair (NER) system, which removes about 30 nucleotides, including the damaged ones, before the

DNA polymerase uses the undamaged second strand as a template for synthesis. Interestingly, a correlation between reduced NER and tumorigenesis was demonstrated in the *X. maculatus mdl^{Sp}–xmrk^B* (X, Jp163B)—*X. couchianus* cross (see table 26.3, hybrid 7) (Setlow et al. 1993). Altogether, the DNA damage repair system has a possible tumor modifier effect on UV-induced fish melanomas.

26.7.4 MNU-induced melanomas

Similar to UV, many carcinogens induce alteration of the DNA. *N*-methyl-*N*-nitrosourea (MNU) is a mutagen that acts through methylation of purines. If BC$_1$ fish from crossbreedings between *X. maculatus mdl^{Sp}-xmrk^B* (X, Jp163B) and *X. hellerii* (see table 26.3, hybrid 1) are treated with MNU, melanoma development is observed only in fish that possess the *Sp* pigment pattern. Among the *mdl^{Sp}-xmrk^B*-bearing offspring, a melanoma incidence of 37% is reached, while other tumor types occur only rarely (Kazianis et al. 2001b). This suggests the involvement of *xmrk* in MNU-induced melanoma development, similar to the observations for spontaneous and UV-induced melanoma. *Cdkn2ab*, however, does not seem to actively take part in this process (Rahn et al. 2009).

26.8 Other tumor models

Although *Xiphophorus* fish are generally considered melanoma models, other tumors are also observed here, particularly after exposure to carcinogenic agents. MNU treatment of BC$_1$ hybrids between *X. maculatus* (*mdl^{Sd}-xmrk^B*) and *X. couchianus* (see table 26.3, hybrid 8) mainly results in fibrosarcomas, neuroblastomas, schwannomas, and retinoblastomas (Kazianis et al. 2001a). A similar observation was made decades before by Schwab and colleagues, who treated a large number of different *Xiphophorus* backcross genotypes and obtained neuroblastomas, mainly in *X. variatus* × *X. hellerii* backcross hybrids with *mdl^{Li}-xmrk^C* (Schwab et al. 1979). Although the degree of tumor development was linked to the genetic background of the fish, detailed biochemical analyses of MNU-induced melanomas or neuroblastomas are not available to date.

In addition to tumors that show a certain predisposition depending on the *mdl-xmrk* genotype of the fish, a whole spectrum of tumors from different tissues, including fibrosarcomas, rhabdomyosarcomas, and hepatic carcinomas were induced in backcross segregants not containing *mdl-xmrk* (Schwab et al. 1979; Anders et al. 1984).

A high incidence of spontaneous formation of ocular tumors was noted in a hybrid line established from offspring of a cross between an ornamental albino swordtail female and a platyfish male carrying the spotted belly and spotted dorsal macromelanophore patterns (*mdl^{Sd}, mdl^{Sb}*) (Gordon 1947). The genetic determinant for the ocular tumor segregated independently from the melanoma locus. The histopathology of this tumor was not further specified, and obviously this line has been lost.

Xiphophorus has been used as a thyroid tumor model. In certain laboratory strains of *Xiphophorus montezumae*, a high incidence of spontaneous thyroid tumors was observed (Gorbman & Gordon 1951; Berg & Gorbman 1954). These tumors are of inhomogeneous phenotype, and they easily invade bone. The growing tumor also fills the visceral arches, thereby impairing gill function, which eventually leads to the death of the fish. Tumors occur more often in females and are fully formed only in fish of five months and older (Berg & Gorbman 1954).

In the Amazon molly, *Poecilia formosa* (see Schlupp and Riesch, **chapter 5**), massive tumor formation is observed if fish-derived thyroid cells are externally irradiated with UV and subsequently injected into the abdominal cavity of isogenic fish. Here, comparable to UV-induced *Xiphophorus* melanoma, the induction of DNA damage enzymes attenuates the tumor phenotype. If UV-treated cells are exposed to black light before injection, tumor prevalence is negligible (Setlow & Woodhead 2001). Black light induces the so-called "photoreactivating enzyme" (PR), which can convert the above-mentioned UV-induced pyrimidine dimers back to monomers. This demonstrates that the main reason for UV-induced tumors in *Poecilia formosa* is the generation of pyrimidine dimers.

Certain laboratory strains of *P. formosa* develop benign, mostly exophytic tumors of the skin that have a papillomatous growth and consist of varying degrees of cells of all pigment-cell lineages (chromatophoromas) (Schartl et al. 1997). The tumors develop preferentially in certain body compartments. The genetic situation leading to this tumor model can be conceptually compared with spontaneous melanoma formation in *Xiphophorus* hybrids, namely, dysregulation of a pigmentary locus in a foreign genetic background. In the laboratory, males of the ornamental *Poecilia* hybrid strain "black molly" are used for the gynogenetic propagation of *P. formosa*. Sperm of the host species triggers the parthenogenetic development of the ameiotic diploid eggs and usually do not contribute any genetic material. Exceptionally, however, the paternal DNA is not completely cleared from the inseminated oocyte and persists in the form of a microchromosome (M. Schartl et al. 1995a; Nanda et al. 2007). The observed chromatophoromas develop only in those strains of *P. formosa* that have a microchromosome with a pigmentation locus from the black molly, which themselves develop chromatophoromas only extremely rarely. If the black molly microchromosomes are in the genetic background of *P. formosa*, any trans-acting regulation will come from the *P. formosa* genome and not from the parental black molly genome. Such control might be insufficient and might allow local overproliferation of pigment-cell precursors. The clonality of the genomes of these tumor-bearing fish offers a unique possibility to analyze the contribution of epigenetic changes versus genomic alterations in the process of tumorigenesis.

26.9 Tumor modifiers

Genes that are not involved in the primary events of neoplastic transformation (like the classical oncogenes and tumor suppressor genes) but instead play a role in determining the course of the disease are called tumor modifiers. Such genes have become particularly evident in the *Xiphophorus* melanoma models. *The mdl* locus fulfills the criteria of a tumor modifier. It determines the onset of melanoma formation, the body compartment where the melanoma will arise, and to a certain extent the severity of the melanoma.

The *golden* (*g*) locus has been described as blocking melanocyte differentiation, affecting both the micromelanophore and macromelanophore lineages (Anders et al. 1979). When hybrids from *Xiphophorus* melanoma crosses that would on a wild-type pigmentation genetic background develop macromelanophore hyperpigmentation, melanosis, and malignant melanoma are homozygous for *golden*, they have only a few micromelanophores and are devoid of macromelanophores (Kallman & Brunetti 1983). In general, melanoma formation is suppressed. Sometimes the mutation becomes leaky, and single macromelanophores appear. In this case, localized hyperpigmentation or a melanoma develops, but the phenotype differs from the wild-type situation, in which multiple tumors develop simultaneously. Unfortunately, nothing is known about the mechanism of action and the molecular nature of the *golden* locus; even its location on the genetic map of *Xiphophorus* is unknown.

The recessive *albino* (*a*) mutation causes tyrosine-positive albinism. The locus is not mapped and its molecular nature is unknown. Crossing fish with conditioned *xmrk*-induced melanoma into an *a/a* background results in the appearance of gray melanoma due to the production of amorphous melanin by increased tyrosinase activity in the progression stage (Vielkind et al. 1971; Vielkind et al. 1977). The melanomas consist of lowly differentiated cells; thus, they grow faster and are generally more aggressive than the fully melanized tumors in the wild-type pigmented fish. The situation is reminiscent of observations in humans, where unpigmented melanomas are frequently more malignant than melanotic tumors.

26.10 Evolutionary origin of *xmrk*

The gene *xmrk* arose by a segmental gene duplication event of the *egfrb* gene, most probably already in the common ancestor of all *Xiphophorus* species (Weis & Schartl 1998). Although the validity of estimating divergence times from DNA sequence variations has been debated (Howell et al. 2008), at least a reasonable estimate can be reached using this method. This is particularly useful when calibrations are available from related groups of organisms (for a recent compilation, see Burridge et al. 2008). Using the fastest and slowest rates of all known calibrated molecular clocks in teleosts (including fast-clock lineages like the Great Lake cichlids) for the mitochondrial DNA sequence data set of the genus *Xiphophorus* (Meyer et al. 2006), we can estimate that the last common ancestor of all present-day *Xiphophorus* species lived between 1.1 and 11.6 million years ago. The gene duplication leading to *xmrk* has happened within this time frame.

Duplicated genes generally undergo one of three different fates. As only one of the two copies is needed and thus remains under stabilizing selection, the second can accumulate mutations, which might compromise its function and finally lead to degeneration (nonfunctionalization). Alternatively, mutations might allow a new function to be acquired (neofunctionalization). Finally, mutations in both copies might enable them to share the ancestral function (subfunctionalization). The latter two mechanisms lead to survival of both copies.

Xmrk and Egfrb differ by many amino acids. The difference between the ratio of synonymous to nonsynonymous substitutions for *xmrk* and the ratio of synonymous to nonsynonymous substitutions for its progenitor, *xegfrb*, and the other EGF receptors strongly suggest that *xmrk* evolves under purifying selection rather than as a pseudogene (Volff et al. 2003b). Both genes are differently expressed and vary in their biochemical properties, with *xmrk* encoding a constitutively active, oncogenic receptor and *egfrb* encoding a nononcogenic receptor with a presumably physiological function. This indicates neofunctionalization of *xmrk*. The only function of Xmrk known so far is melanoma formation. The gene *xmrk* has been lost from many species of the genus, and even those containing the gene are polymorphic for its presence, meaning that many individuals of a species do not possess *xmrk*. Thus, Xmrk may not have a relevant general physiological function. Moreover, melanoma formation by *xmrk* overexpression can even sometimes occur in wild nonhybrid fish (A. Schartl et al. 1995). It is counterintuitive to consider a potentially deleterious gene being kept in the genome for many millions of years.

Two hypotheses have been proposed to explain the evolutionary persistence of *xmrk*. The "hitchhiking" hypothesis considers the fact that *xmrk* is found so far exclusively tightly linked to *mdl* and thus associated with certain macromelanophore patterns. These patterns have been assigned a function in facilitating the finding of conspecific individuals for the purpose of shoaling, especially under murky water conditions (Franck et al. 2001). It has been shown that in the case of strongly positively selected alleles (like *mdl* for conspecific recognition) the frequency of closely linked alleles that are less positively or even negatively selected (as probably *xmrk*) can increase (Charlesworth 1994; Stephan et al. 2006). Thus, *mdl* would help to carry *xmrk* along. It is, however, unclear if this effect is strong enough to keep a gene functional over many millions of years.

Another hypothesis favors sexual selection. For mdl^{Sc}-$xmrk^D$ from *X. cortezi* it was found that in two of three investigated populations females prefer males with larger macromelanophore caudal-fin spots, especially those with strong hyperpigmentation or even melanoma (Fernandez & Morris 2008). If *xmrk* in the nonhybrid situation regulates the size of the macromelanophore spots (which is reasonable to assume for the mdl^{Sc}-$xmrk^D$ locus, because it is associated with nonhybrid melanoma), this would explain the maintenance of an oncogene by female preference in this species.

The situation described for *X. cortezi* may be exceptional because the penetrance of the macromelanophore pattern is much lower in females and the spots are much larger in males only in this species and very few other species. In most species the macromelanophore pattern is expressed in males and females, and there is no enhanced expression of the pattern phenotype in males. Thus, explanations other than sexual selection need to be found. Whatever positive function can be assigned to the *xmrk*-linked spotting patterns, they might also have negative features, for instance, making a fish more conspicuous to predators. This would explain why all species and populations with *mdl-xmrk* are polymorphic for the presence of that locus.

26.11 Evolution of *R*

As long as the molecular nature of *R* is unknown, we can only speculate about its evolution. When *xmrk* was generated by the gene duplication process, it may initially just have been a redundant copy of *egfrb* and might have fulfilled a redundant function. Thereafter, gross changes by insertion of DNA sequences into the 5′ flanking region from elsewhere in the genome (e.g., one multicopy locus, the so-called *D* locus) have considerably changed the promoter (Adam et al. 1993; Förnzler et al. 1996; Volff et al. 2003). It is tempting to speculate that through these sequences a

transcriptional rewiring of the duplicated gene occurred, and during this process control by *R* was acquired. A potential suppression of *xmrk* transcription by *R* might have allowed the evolution of the physiological proto-oncogene with an epidermal growth factor receptor function into an oncogenic receptor that escapes physiological control by a ligand. Thus, *R* may have preexisted as a transcriptional suppressor involved in regulation of one of the sequences inserted into the promoter region of *xmrk*. In this context it may be of note that *R* or *R*-like genes appear to be present in all *Xiphophorus* fish regardless of the presence of *xmrk*, even in those species where *xmrk* has been lost long ago (Schartl 2008).

26.12 The speciation gene hypothesis for the *xmrk*/*R* gene pair

It became textbook knowledge that *xmrk* and *R* are rare known examples for speciation genes. Dobzhansky and Müller proposed a genetic model for the evolution of postzygotic isolation mechanisms based on incompatible epistatic interactions between genes that have diverged during evolution of the two species that give rise to a hybrid. At first glance *xmrk* and *R* appear to be instrumental in preventing the production of viable offspring by different species, for example, *X. maculatus* and *X. hellerii*. These two species are indeed sympatric in many biotopes of the southeastern lowlands of Mexico. Of course, tumor-bearing hybrids would have a lower fitness or would not reproduce at all in those cases where the melanomas are fatal prior to sexual maturity. This is the case for the combinations of *xmrk* with some alleles of mdl^{Sp}, mdl^{Ni}, and mdl^{Sb} from *X. maculatus*, but certainly not for mdl^{Sd}, where many melanoma fish reach fertility. The allele combination mdl^{Sr}-$xmrk$ does not lead to tumors in hybrids at all and consequently has no effect on hybrid viability or fertility. The ineffectiveness of the *xmrk*/*R* gene pair in preventing successful hybridization is even more evident when the pigmentation polymorphisms are considered. There are sympatric species of *Xiphophorus* in which both partners lack *mdl-xmrk*. In all other cases *mdl-xmrk* loci are present in only a fraction of fish from the species, and *mdl-xmrk* homozygotes are extremely rare in nature (Gordon & Gordon 1957). Importantly, the natural *mdl-xmrk* allele combinations produce fatal malignant melanomas only in the second hybrid generation, while the F_1 fish generally show different degrees of hyperpigmentation, which are benign lesions. The Dobzhansky-Müller model postulates coevolution of both genes, but *xmrk* and *R* obviously do not coevolve, because xmrk is lacking in many individuals of a population or even in entire species, while *R* appears to be present ubiquitously. Another typical criterion of speciation genes, namely, an accelerated evolutionary rate, is also not met by *xmrk*. Because of these and other arguments the speciation gene hypothesis for the *xmrk/R* gene pair was refuted (Schartl 2008).

26.13 Perspectives

The *Xiphophorus* melanoma system gained its importance and value because it was possible to do genetics and to define the different genetic components that are involved in the development of the cancerous disease from the primary event of neoplastic transformation to the final stages of tumor progression. With the beginning of the era of molecular genetics, directed approaches toward the isolation of genes involved in tumorigenesis were successfully applied to *Xiphophorus*. However, the inability—due to the livebearing mode of reproduction—to produce transgenic fish and the relative difficulty of using the developmental biological methods that are used to study the egg-laying zebrafish and medaka have hampered the functional analysis of the genes of interest in *Xiphophorus* fish. Further biochemical analyses were possible by using the established analytical tools provided by mouse and human cell lines that were engineered to express the Xmrk oncogenic receptor. This has generated a wealth of information for a better understanding of human cancers—in particular, melanomas—and provided data on genes and proteins that can be applied to improve diagnosis and treatment (Meierjohann & Schartl 2006). The *xmrk* transgenic medaka fish, which, like the platyfish-swordtail hybrids, develop melanomas spontaneously, will allow functional studies and high-throughput approaches like mutagenesis and drug-testing screens. It can be expected that much future work on the molecular details will concentrate on these nonpoeciliid systems. However, the *Xiphophorus* system, with all its different tumor models, will remain a gold mine for cancer research. The action of the *R* locus, *mdl*, and the other tumor modifiers is recognizable only in the original hybrid crosses, and their molecular nature as well as their biological function can be elucidated only here. The availability of a fully sequenced *Xiphophorus* genome will certainly boost this research.

Acknowledgments

This chapter is dedicated to the memory of Steven Kazianis, who contributed so much to our knowledge and understanding of *Xiphophorus* genetics, in particular, the genetics of cancer. The work of the authors was supported by the Deutsche Forschungsgemeinschaft and the European Commission. We thank Monika Niklaus-Ruiz for help with preparing the manuscript.

Part VI

Conclusions

Chapter 27 Integrative commentary on *Ecology and Evolution of Poeciliid Fishes*

John A. Endler

THE SPECIES DIVERSITY, accessible biology, well-defined and accessible habitats, short generation time, and ease of laboratory culture make poeciliid fishes extraordinarily good subjects for research in both academic and applied biology. The chapters in this book show that nearly every major problem in biology has been addressed with poeciliids and that they have given major new insights in behavioral ecology, evolutionary biology, and genetics related to sexual reproduction. The chapter topics are so diverse that most readers (including me) will learn something new and interesting from this book. Unlike any other model system, poeciliids, and guppies (*Poecilia reticulata*) in particular, have a well-described and accessible ecology, allowing unusual power in understanding both the function of traits and the causes of their origin and evolutionary maintenance. This is impossible or very difficult in other model systems such as *Caenorhabditis*, *Drosophila*, zebrafish, or mice. The foreword, preface, and chapters extol the virtues of poeciliids, describe much interesting biology, and pose many unanswered questions, so I will not elaborate on them here. Instead, I will make some general remarks on poeciliids and identify some interesting and unanswered questions that are only hinted at in this volume.

The family has high diversity: more than 260 species in 22–28 genera, with 6 genera having 21–43 species each (e.g., see Stockwell & Henkanaththegedara, **chapter 12**). This means high potential power for using the comparative method to deduce the pattern, if not the causes, of evolution in many suites of traits, as well as the sequences of evolutionary events. In spite of this significant phylogenetic power, there are no large-scale (100+ species) comparative studies. This is at least partially due to insufficient sampling of species. A repeated theme of almost all the chapters is that too few species in too few genera have been studied to make generalizations about the family or the processes discussed, let alone to undertake serious comparative studies. In fact, most subjects are supported by work from only 3–10 species, or less than 5% of the total diversity. The potential for comparative studies is also hindered by insufficient species sampling; the best phylogenetic tree (Hrbek et al. 2007) includes only 48 species (18% of poeciliid diversity), although these species were well spread among the poeciliid lineages and they probably give us a good skeleton of the phylogenetic relationships within the family. A further complication, which plagues all phylogenetic and comparative work, is the possible presence of cryptic species, as suggested by the recent discovery of 21 new *Phalloceros* species (Stockwell & Henkanaththegedara, **chapter 12**) and many new *Xiphophorus* species (Schartl et al., **chapter 24**). Species sampling is a formidable gap in our knowledge and understanding, and obtaining more data on more species for more traits should have top priority because it can so easily lead to really significant new evolutionary (and biological) insights.

An indication of the power of phylogenetics to elucidate evolution can be seen if we step back and consider the Atherinomorphs, which include the Poeciliidae (within the order Cyprinodontiformes). This analysis (Mank & Avise 2006c) shows that viviparity evolved at least four times, once in Beloniformes and three times in Cyprinodontiformes, with only one (partial) reversal—in the poeciliid

Tomeurus. It would be interesting to know the conditions that favored viviparity in these clades and why reversals are so rare, as well as why poeciliids show the one possible reversal. The viviparous clades are more species rich than their oviparous sister clades but show no difference in extinction rates. This implies a tendency for increasingly viviparous species but is contradicted by the fact that only about 25% of Atherinomorph species are viviparous, and even fewer in the Actinopterygii as a whole. Perhaps the genetic, physiological, and morphological limitations of forming viviparous mechanisms are evolutionarily or developmentally difficult, or perhaps internal fertilization, which is also rare in fishes, must evolve first, making their joint evolution less likely (Mank & Avise 2006c). Interestingly, there is no evidence for differences in speciation in marine versus freshwater lineages, although freshwater lineages tend to have higher rates of extinction, as one would expect from the more geologically unstable nature of freshwater compared with marine habitats. If we had more complete data on more poeciliid species, we could ask similar questions: for example, the evolutionary steps to matrotrophy, what habitats favor it, what habitats and conditions favor particular degrees of matrotrophy, and whether certain traits and/or habitats favor speciation, extinction, and persistence.

Our understanding of the generation and maintenance of poeciliid diversity is rudimentary, and there are many unanswered questions. Rosenthal and García de León (**chapter 10**) suggest that the species diversity of poeciliids is low compared with African lake cichlids (500 species) and American *Anolis* lizards (400 species), and they ask why there are not more species. Are poeciliids actually less diverse than other fish, other aquatic groups (such as decapod Crustacea), or even terrestrial groups in the same landmass and same habitats? A different impression comes from a local perspective: poeciliids represent 35% of the Central American fish fauna, and only the cichlids are comparable in diversity there (Hrbek et al. 2007). *Anolis* have similar diversity: in Costa Rica, a Central American country with high habitat diversity, there are only 26 *Anolis* (now known as *Ctenonotus*, *Dacyloa*, and *Noropos*), where they make up 35% of the lizard fauna, and only 13% if we include all squamate reptiles, which are closely related (Savage 2002). According to *FishBase* (www.fishbase.org) there are 158 native freshwater fishes in Costa Rica, of which 13% are endemic. Poeciliids are the second most diverse family, composing 13.3% of the Costa Rican fauna. Cichlids are only slightly more diverse (15.2%), characins are similar with 12.7%, and the next most diverse Eleotridae provide only 7% of the fauna. The Costa Rican poeciliids are not significantly less diverse than cichlids and actually have the highest percentage of endemic species, 4.4%, compared with a little more than the 3.2% for characins, and much

more than the 1.3% for cichlids (there are no endemic eleotrids). There is no evidence for poeciliids having low diversity compared with other fish families in Central America. Poeciliids are much more diverse than other freshwater fish families in the Greater Antilles (when present): they are 38% of the fauna on Cuba and 52% on Hispaniola, but there are no natives on Puerto Rico and the smaller islands (*FishBase*). Cichlids and eleotrids form less than 1% of these faunas, which contain no characins. Endemism is even higher, with 66% of poeciliids endemic on Cuba and 84% on Hispaniola. This is in contrast to *Anolis*, where island and mainland populations are similar in diversity but have diverged with habitats in different ways (Pinto et al. 2008). Studies of poeciliids with the sophistication of the *Anolis* studies would be revealing, and the ability to do both laboratory and field experiments would make such studies even more powerful tools for understanding evolution and biodiversity than is possible with *Anolis*. Also, are some families more likely to speciate in lakes (such as cichlids) and others more likely to do so in river systems (poeciliids, characins, etc.)?

Poeciliids are a much smaller fraction of the South American fish fauna, yet they appear to have originated in South America and colonized Central America at least three times (Hrbek et al. 2007). Is this because they are excluded from most of the South American habitats by high predation except in shallower, more peripheral stream systems, which have fewer predator species? Are they also excluded by competition from the more diverse characins, cichlids, and catfish in South America? The relative effects of predation and competition are unknown, but both are likely to be important (Robinson & Wilson 1994). Putting them in a phylogenetic context and relating them to dispersal and invasion patterns would be even more interesting and should yield general new evolutionary insights. Another possible explanation for lower diversity in South America is that omnivory impedes dietary and other divergence because there is no need to diverge (Rosenthal & García de León, **chapter 10**). This might also explain the lower morphological diversity in Cuba and Hispaniola but is problematic for the diverse Central American poeciliid fauna. Even if omnivory does not impede divergence, perhaps the lower diversity in South America is due to fewer microhabitats as well as the less temporally and spatially stable differences among habitats and microhabitats in the smaller, shallower waters that poeciliids inhabit (as suggested in chapters 7 and 10). The fewer-habitats argument is supported by the lack of significant ecological radiation in the West Indies (no specialists like *Belonesox* or *Alfaro*), where there are much smaller river and lake systems than in Central America. Is species diversity related to the fraction of freshwater habitat area that is in the lowlands, given that fish habitat diversity and areas of each habitat type increase downstream? What other geographical properties favor diversity? What are the relative effects of habitat diversity, habitat geography, invasion, and evolutionary history? Were there fewer colonizations of the Greater Antilles than of Central America? Poeciliids would be particularly good models for addressing these questions.

Diversity and ubiquity may depend upon which taxa get to a landmass first and which can establish populations more quickly. Did poeciliids reach a higher diversity, endemism, and importance in Central America because their generalist ecology and viviparity allowed them to colonize the reemerging land more rapidly than the richer but more conservative fish clades in South America? Viviparity, matrotrophy, and superfetation may be an advantage in poorer and more fluctuating conditions, which may be characteristic of freshwater habitats on newly emerging landmasses (see chapters 2–4), as well as in extreme environments (Tobler & Plath, chapter 11), facilitating rapid invasion of these habitats. Moreover, viviparity may specifically allow poeciliids to invade newly evolving fish communities (Trexler et al., chapter 9). It is striking that the most divergent and ecologically specialized poeciliid genera, *Belonesox* (a pikelike predator) and *Alfaro* (a characinlike mid- and fast-water species feeding on objects that drop in the water), are found in Central America. Is this because their ancestors dispersed there earlier than other families and preempted these two niches? There is much evidence for competition, competitive release, and character displacement in fishes (Robinson & Wilson 1994), and competition could keep later arrivals out of relatively new freshwater habitats. Perhaps the relatively weaker radiation in morphology and ecology in the Greater Antilles is due to significant competition and predation from primarily marine fish families that range only a few miles upstream; most of the freshwater habitats in Caribbean islands are close to the sea, but only a small fraction of Central American freshwater habitats are so close to the sea. Of course, in South America even a smaller fraction of freshwater habitats are close to the sea, and poeciliids are mostly found near the coasts. What is the relationship between time since colonization of a previously empty landmass, colonization ability, and the presence of existing aquatic communities to speciation and to the degree of divergence? These questions also apply to recolonization after natural or human disturbance and to the future effects of climate change.

There are a host of questions about what makes colonization, establishment, persistence, divergence, and speciation more likely. Most of these could be addressed with poeciliids more easily than with other vertebrates. What is the effect of competition within the family or genus after establishment? In *Anolis* lizards there is some niche conservatism, and divergence in niches appears to occur more easily among distantly related sympatric species (Losos et al.

2003). Does this imply that multiple invasions are needed for strong divergence and specialization? It is suggestive that the two strongly divergent poeciliid genera (*Belonesox* and *Alfaro*) are found in Central America, which had several invasions from South America. It would be interesting to know if the smaller divergence in Cuba and Hispaniola is associated with fewer colonization events. Does matrotrophy and/or superfetation allow faster dispersal and faster establishment in new or well-established communities in similar habitats? Does the length of sperm storage, possible sperm nourishment by females, and the degree of polyandry in stored sperm (e.g., Greven, chapter 1; Evans & Pilastro, chapter 18) also affect establishment and initial growth of new populations in invasions of new landmasses as well as reinvasions after local catastrophes or seasonal fluctuations? Do these traits also encourage rapid colonization of new, as well as evolutionarily familiar, habitats and therefore divergence and even speciation among habitats? What are their effects relative to other fishes and decapod crustaceans with similar ecology but lacking these traits?

Do all these poeciliid-specific traits, which also favor higher genetic variation, also favor lower extinction rates? Which traits favor evolutionary persistence if not radiation over the range of degrees of environmental fluctuation and favor reinvasion after fluctuations or environmental change (see also Stockwell & Henkanaththegedara, chapter 12)? If the environment fluctuates so much that recolonization of streams is required each year, does this favor asexuality (Schlupp & Riesch, chapter 5) because asexual populations can colonize faster and expand faster than sexual ones? On the other hand, does asexuality result in poor persistence at intermediate levels of disturbance over geological time because of the loss of the capacity to evolve under slower or less frequent environmental change? Does ecological generalization allow faster dispersal but slower diversification and speciation? Are all these traits needed to invade extreme habitats (Tobler & Plath, chapter 11)?

Given that some poeciliids are invasive after human introductions and that species vary widely in expansion after human-induced colonization, poeciliids would be excellent for explicit studies of what favors colonization, establishment, and interaction with already-present species communities (Stockwell & Henkanaththegedara, chapter 12) and can be used to address experimentally the complex question of the relative importance of numbers of founders, frequency of repeated colonizations of the same place, genetic bottlenecks during colonization, or natural selection during establishment (Keller & Taylor 2008). A combination of population genetic and phylogenetic studies can reveal geographic and numerical patterns of previous colonizations and their effects on evolution (Olivieri 2009), and these ideas can be tested with known human-induced colonization and establishment events. A comparison among human-induced introductions of various poeciliid species, with various degrees of success, and colonizations of the Greater Antilles and of Central America would be very interesting.

The other side of invasion and colonization is species replacement and introgressive hybridization; this involves the invasion and recomposition of communities and genomes. These phenomena are intertwined with problems in conservation biology (Stockwell & Henkanaththegedara, **chapter 12**) as well as being fundamental to evolutionary biology (e.g., Lindholm et al. 2005). Poeciliids would be ideal to investigate these phenomena experimentally and in the field. This is especially important in a world of climate change and habitat destruction because habitat changes can change community structure, affect gene flow, and induce hybridization (e.g., Seehausen et al. 1997). Moreover, habitat changes can confuse taxonomy and hence conservation status, making it still more difficult to understand what has happened in the field. A case in point is Endler's livebearer (*Poecilia wingei*); see box 27.1.

Box 27.1 Endler's livebearer (*Poecilia wingei*): interactions between ecology, taxonomy, conservation biology, aquarists, and habitat disturbance

Endler's livebearer (*Poecilia wingei*) is also known as Endler's guppy, the Cumaná guppy, and the Campoma guppy and is a close relative of guppies (Alexander & Breden 2004). It is a good example of multiple problems arising from habitat disturbance and the multiple interactions between ecology, biology, conservation biology, taxonomy, and aquarists. I found these fish with guppies (*Poecilia reticulata*) in the Laguna de Los Patos, at Cumaná, in northeastern Venezuela in 1975 while looking at geographic variation of color patterns in guppies (Alexander & Breden 2004). At the time I thought that they were very different from guppies and gave them to Dr. Donn E. Rosen, of the American Museum of Natural History, to describe. Although he died before having a chance to describe them, he gave some live fish to Dr. Klaus Kallman, who then introduced them to aquarists, who in turn spread them around the world, under the name of "Endler's livebearer." Shortly after discovering them, I found a collection of guppies in the University of Michigan Museum of Zoology collected by Franklyn F. Bond in 1935–1937, also from the Laguna de Los Patos, and mixed in among many guppies in that museum bottle were the same fish I saw in 1975. These should have been the type specimens, but when Poeser described the fish as *P. wingei* (Poeser et al. 2005), he was unable to find the Laguna del Los Patos bottle of "guppies" in the museum. He also did not think that what was in the Laguna de Los Patos was the same fish that he found elsewhere in 2002, and so he used fish from Campoma (further east along the coast) as the type specimens. Given the strong interest by aquarists in these fish, but the striking lack of variation in aquarium fishes, I encouraged many groups of people to go back to my original site and find them again, and this has resulted in an interesting plot of the fraction of guppies in the Laguna de Los Patos as a function of time (box-fig. 27.1).

Several things have happened since F. F. Bond and I visited the site. When I visited, guppies (*P. reticulata*) were more common than *P. wingei*. First, as shown in box-figure 27.1, I sampled in the early stages of the decline of the frequency of *P. reticulata* versus *P. wingei*; the most recent visitors to the lagoon found no guppies at all (particular thanks to Armando Pou, who visited several times).

Second, I was interested in the genetics of color patterns, particularly the bright gold-bronze of this population, so I tried to hybridize them. After trying for more than a year with many pairs of individuals, I obtained three and gave up. In the meantime breeders had managed to hybridize the two entities, and photographs taken about five years after my first collection started to show clear introgression with guppies, and the distinction between the two species in aquaria stocks began to blur. This resulted in heated debates in the aquarium literature and Web sites (including some devoted to these

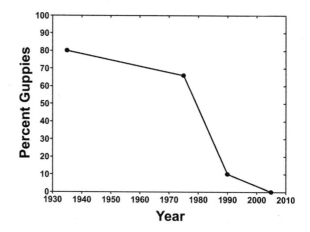

Box Figure 27.1 Decline of guppies (*Poecilia reticulata*) and increase in either *P. wingei* or a hybrid swarm of the two species.

There are large numbers of fundamental evolutionary, ecological, and behavioral questions that could be addressed with a much greater knowledge of the phylogeny and ecology of a much higher fraction of the family. There is probably a tight linkage between community ecology, divergence, speciation, and extirpation and extinction, yet this has not been studied in any animal group, and poeciliids would be an ideal group to investigate these joint processes. For example, if poeciliids invade an area with high predation, this might restrict them to peripheral habitats (as in South America) with at least two suites of consequences. Peripheral habitats (small, higher-gradient streams) are less productive, and this selects for generalists rather than specialists, which may lead to less speciation. Peripheral habitats and other habitats with low microhabitat diversity will lead to less natural-selection-induced divergence within and among stream systems and also less speciation, although random divergence might be higher. What happens when poeciliids invade low-predation areas? Would the reverse be true? Does this explain the differ-

---

fish) about just what Endler's livebearer was, as some people who had hybrids claimed that these were pure strains, etc. Clearly something happened in the aquarium stocks after 1975.

Third, it appears that something similar happened to the fish in the Laguna de Los Patos. The collectors who went there and brought back fishes found that the two entities hybridized readily in aquaria. It is possible that the declining proportion of guppies meant that the rare-male effect and other effects of asymmetrical abundances of two closely related species resulted in hybridization and introgressive hybridization in the lagoon, and collectors were now sampling a hybrid swarm. This caused even greater controversy among aquarists, but also caused problems in taxonomy. In fact, by the time Poeser visited the lagoon, he thought that *P. wingei* had been introduced there by aquarists and that the population was not suitable for being the type locality. The introgressive hybridization clearly has extended beyond appearances to molecular markers, as the fish vary in their similarity to guppies (Alexander & Breden 2004; Poeser et al. 2005; C. Dreyer & E. M. Willig, pers. comm., 2008). In any case, something interesting happened there in the 1980s.

Fourth, there has been increasing disturbance, fragmentation, and pollution in the Laguna de Los Patos. When I visited in 1975, the city dump was encroaching on one end of the lagoon, but the rest was intact and in reasonably good condition. Collectors who went back in the 1980s and 1990s found that the lagoon had become fragmented by development and that some parts were so polluted that no fish were present and others had many introduced fishes as well as the natives. A look at the lagoon on Google Earth in June 2009 confirmed the massive disturbance and fragmentation; I could not even recognize my original collecting site. It is a pity that the change in ecology, the change in the relative abundance of guppies, and the extent of hybridiza-

tion were not monitored over this period, because this might be a case similar to that of the human-induced hybridization of the cichlids described by Seehausen et al. (1997); as in that system, eutrophication, pollution, and concomitant changes in the visual and chemical environment may have caused or abetted the hybridization. It also stresses the importance of good long-term records (Reznick et al. 1994), which were not taken here given that the process was not initially (before 1980) obvious.

The same process may be at earlier stages in the other populations described and mapped by Poeser (Poeser et al. 2005). *P. wingei* is particularly vulnerable to human disturbance because all known populations are found below 250 meters in elevation, and these lowlands are particularly popular for human development, both urban and agricultural. Such disturbance not only endangers this interesting relative of guppies but also makes it difficult even to know exactly what the species is (compare Alexander & Breden 2004 with Poeser et al. 2005), further thwarting our understanding of the process of divergence (divergence at least to semispecies status), coexistence of closely related species or semispecies, and speciation itself. The situation is further complicated by aquarists and fish breeders, as well as introductions of aquarium fishes back into the wild.

Here we see a cycle of human disturbance possibly causing hybridization and replacement of two species by a hybrid swarm that confuses the species status in the affected population and confuses the taxonomic status of the entire entity, making conservation as well as ecological and evolutionary studies difficult. This problem is not limited to poeciliids. But it is a problem that perhaps could be best addressed experimentally and theoretically with poeciliids and may help us to conserve other species as well as understand conservation problems more deeply.

ence in diversity between Central America (colonization of new land areas) and the West Indies (already occupied by secondary marine-fish families)? What is the relationship between central and peripheral habitats in molding geographic patterns of community structure and speciation? It is ironic that the wonderfully accessible ecology of poeciliids that has allowed many first-class field experiments has not been utilized in ecological experiments addressing such questions as competition, predation, population dynamics, and community structure. Such studies would be valuable for ecology and would further illuminate all the questions asked with poeciliids as study animals.

Ecological speciation is finally being taken seriously (Schluter 2009), and some aspects of this process could be investigated experimentally in poeciliids in conjunction with sensory ecology (Coleman, **chapter 7**) and behavioral ecology (see **chapters 16–21**). For example, speciation might require strong divergence first, then sexual isolation. Although *Poecilia mexicana* shows significant morphological, physiological, and behavioral divergence between ordinary stream habitats and both sulfide-rich and cave habitats, there is no genetic incompatibility, and isolation appears to be achieved as a result of mortality of individuals traveling between habitats (Tobler et al. 2008a). Full speciation requires either postmating or premating isolation. It would be interesting to compare the degree of nonsexual and sexual divergence among the entire family in order to see whether divergence precedes or follows sexual isolation, and whether pre- or postmating isolation evolves first. Sexual communication may also be affected by environmental conditions (Endler 1992, 1995; Endler & Basolo 1998; Boughman 2002). For example, predation intensity may affect speciation probability and rates through its effect on sexual signal visibility. If there is higher predation, then males will evolve duller coloration, which may lead to fewer choice criteria and fewer ways to discriminate among males, which would lead to lower speciation rates. Alternatively, higher predation may lead to a change from primarily visual to olfactory and/or lateral-line signaling and female choice. This would either lead to more cryptic species (morphologically similar species) or perhaps have little effect on the speciation rate once the change was achieved. Changes to olfactory-based mate choice might affect olfactory-based foraging and allow the addition of new foods and hence divergence and possibly increased speciation, if other (nonpoeciliid) species did not prevent expansion into new niches. Higher predation may also lead to more cryptic (postcopulatory) female choice as another alternative to visual displays. This might lead to an unchanged speciation rate but much less morphological diversity. More predation may also lead to more sneaky mating and coercion (Magurran, **chapter 19**), more complex and longer gonopodia

(Langerhans, **chapter 21**), and various female means of counteracting the male strategies, and this could also increase the speciation rate (see Rosenthal & García de León, **chapter 10**). Of course, increased predation could also increase the extirpation rate, which would favor divergence among populations, or the extinction rate of new species, which would reduce species diversity. For all these reasons the relationship between predation, species diversity, and speciation rate is complex but could easily be investigated in a phylogenetic context in poeciliids.

Some species have discrete variation in color patterns (polymorphic) in both sexes, others show color pattern polymorphism only in males, others have monomorphic but different sexes (sexually dichromic), and others are monomorphic in both sexes (sexually monomorphic). There are many competing explanations for polymorphisms involving both predation and mate choice (Archer et al. 1987, Endler 1980, 1991; Brooks 2002; Hurtado-Gonzales & Uy 2009), and the reasons for these patterns, and their phylogenetic distributions, need further study. Interestingly, color polymorphism appears to be independent of the degree of sexual dichromatism (Endler 1983), and polymorphic species tend to be the only member of their genus in a single location, parapatric with all congeners, or syntopic with very few other congeners compared with the wide range of sympatry with congeners and other poeciliids in most monomorphic species. This should be investigated in more detail and may also be true for body shape and other visual or chemical cues. If generally true, this would suggest that speciation might be bi-stable, depending upon the degree of polymorphism in the basal part of the lineage. If a basal species is polymorphic, it may be harder to develop species recognition traits than if it were monomorphic (or less polymorphic), leading to a lower speciation rate. A lower speciation rate may lead to fewer syntopic species and hence encourage (or be permissive of) polymorphism, sending the system further in that direction. Alternatively, if a basal species is monomorphic, species recognition is easier, leading to a higher speciation rate and more syntopic species and favoring more monomorphism and monomorphic divergence. This might be difficult to test in a single family because it is tightly bound up with evolutionary history, and so many clades would be needed for a proper test. For example, is this why there are so many polymorphic species in the genus *Poecilia* (*reticulata, wingei, picta, parae*)? Or is this because these species live in peripheral habitats in the otherwise fish species rich South America? Why do *Xiphophorus maculatus* and *X. cortezi* have polymorphism (Fernandez & Morris 2008), whereas other congeners do not? The *Xiphophorus* polymorphisms may be partially linked to balancing selection between sexual selection and oncogenes (Fernandez & Morris 2008; see also Schartl &

Meierjohann, **chapter 26**), but the more extreme *Poecilia* polymorphisms are much more difficult to understand. In spite of these difficulties, it would be interesting to know if evolution tends to get trapped in monomorphism or polymorphism, with respect to visual, olfactory, sound, or lateral-line traits.

Sperm storage in females may also have an effect on speciation rates. If the sperm storage time is long (Evans & Pilastro, **chapter 18**), this may mean slower sperm turnover during storage. This may also be the case if more sperm were stored. Both would result in higher gene flow among populations and hence lower divergence and speciation rates. Moreover, the larger effective population sizes resulting from longer or larger sperm storage would also result in fewer random differences between source and both founder and exchanging populations, further inhibiting population divergence and speciation. On the other hand, these traits would result in greater founding effective population sizes, hence more genetic variation and faster expansion and adaptive divergence than in species with shorter and smaller sperm storage, and hence more speciation. This suggests that longer and larger sperm storage would favor more divergence and speciation after invasion of new habitats or landmasses but less divergence of speciation in situ. Shorter and smaller sperm storage may lead to more speciation in situ but also greater extinction. Greater extinction would cause greater species turnover, possibly resulting in competitive release of other species, which might inhibit the newly formed species from establishing and spreading. It would be interesting to test these ideas.

Because poeciliids are so experimentally tractable, they are particularly good for studies of development (**part I** of this volume), physiology, neurobiology, and neuroethology (e.g., Coleman, **chapter 7**). This means that we may be able to discover trade-offs between different traits that might be important to their ecology and evolution but have not yet been considered. For example, the presence of viviparity, matrotrophy, and superfetation immediately identifies trade-offs in the evolution of life-history traits (e.g., **chapters 2–6**), but there may be additional trade-offs. Frazier and Roth (2009) found that *Caenorhabditis elegans* nematodes can alter the maternal environment of their embryos adaptively in response to environmental stress (as may be the case for poeciliids), but because the mechanisms of salt tolerance interfere with hypoxia tolerance, mothers adapting for more saline conditions need more oxygen, and their offspring are more hypoxia sensitive. Do these kinds of trade-offs affect poeciliids, not only in maternal-offspring relationships but also more generally in the ability to disperse from freshwater across estuaries and the sea and invade extreme environments? Johnston (2006) reviewed fish muscle development and plasticity of response of muscle development to both active use and environmental conditions during embryonic and early growth stages. When early development is rapid, the environment can have significant effects on subsequent muscle mass and significantly affect all aspects of fitness. There are also trade-offs between fast-muscle mass, energy and nutrient supply and function in streamflow, predator escape, and courtship, and these are all affected by temperature and oxygen levels during embryo and larval development. Temperature and oxygen levels jointly affect the optimal fiber number. Johnston et al. (2009) found in zebrafish (*Danio rerio*) that there is an optimal temperature that results in more fast-muscle fibers; embryonic temperature affects both the intensity of muscle fiber production as well as the body length at which the transition between new fiber production and fiber growth occurs. Even within a short stretch of stream there is significant spatial variation in oxygen level, temperature, and flow rates, and these vary and covary with stream order.

Microhabitat preferences of populations and species should favor different sets of traits and solutions to all these physiological trade-offs, and the diversity of habitats and ease of experimental study make poeciliids particularly good for studies of physiology, plasticity and evolution. A good start has been made with plasticity in response to resource availability (Grether & Kolluru, **chapter 6**), but relating this to physiology and development would be particularly fruitful. Moreover, if physiological trade-offs of the kinds reviewed by Johnston (2006) and Grether and Kolluru (**chapter 6**) are commonplace, we should ask whether female poeciliids choose times and places for optimum development of the muscles of their broods, and whether such microhabitat choice is more important in high-flow environments, or in environments with frequent predator encounters, than in other environments. For example, do females spend more time in relatively deeper water to escape from male harassment (Magurran, **chapter 19**) or to provide better physiological conditions for offspring, or both? Can these trade-offs be used to predict what sorts of handicaps may be examined by females during mate choice? Moreover, do the muscle-environment trade-offs affect the evolution of size and shape in poeciliids, and do the existing sizes and shapes cause natural selection on the trade-offs themselves? Do poeciliids have physiological trade-offs that prevent them from invading niches and habitats held by other species from families that got there first? What is the general pattern of joint evolution of physiological traits, morphological traits, microhabitat choice, and the trade-offs induced by these factors? Do the patterns of multiple-trait evolution and constraints prevent speciation or encourage it? These questions are unexplored and poeciliids would be an ideal group to study them.

In spite of much research on the evolution of sex (Schartl et al., **chapter 24**), sex-linked color patterns (Brooks & Postma, **chapter 23**), and sexual selection (**chapters 17–21**), relatively little has been done with sexual size dimorphism and its consequences (see Endler 1983; Evans & Pilastro, **chapter 18**). Males are usually smaller than females (mean 0.7 male/female length), but there is a range of size ratios from 0.4 in *Poecilia scalprides* and *Poeciliopsis gracilis* to about 1.0 in *Poecilia petenensis* (Endler 1983). What ecological and behavioral factors are associated with these differences, and are they also associated with differences in gonopodium size and shape? What is the phylogenetic pattern of size dimorphism and how does it relate to changing mating systems, sneaky copulation, social-network structure, predation, and stream velocity? Is there sexual dimorphism in nonsexual traits such as those affecting life history, swimming, and foraging? In *Anolis* lizards Butler and Losos (2002) found that both males and females of many species repeatedly evolve morphology that matches their microhabitats, but that the sexes diverge in shape more than can be explained by sexual selection alone. Shape influences many different ecological factors, so these differences may have profound implications. It would be interesting to follow this up in poeciliids. Combinations of size dimorphism, gonopodium and other morphology, mating system, and morphology may covary in interesting ways that might be predicted from first principles and could be used to understand simultaneous evolution of multiple traits.

Poeciliids are one of the few taxonomic groups where a network of cause-and-effect relationships in natural selection on multiple traits has been worked out, for example, the multiple effects of predation risk on a range of color, life-history, and other traits (Endler 1995). In spite of wonderful work on the function of various poeciliid traits (reported throughout this volume) there is little work integrating function and selection on multiple traits with entirely different functions. One way to integrate genetically unrelated traits is sensory ecology (Coleman, **chapter 7**), because the interplay between signals, receivers, and habitats is inescapable. Sensory drive is a particularly promising approach to the integration of function, selection, and evolution of unrelated traits (Endler 1992; Endler & Basolo 1998). Sensory drive involves a hypothesized cycle of selective interactions between senses, signals, and preferences (Endler 1992) and includes known processes such as sensory exploitation and preexisting bias (review in Endler & Basolo 1998). Sensory drive has the following cycle of cause-and-effect relationships. New habitats or changing environments result in new sensory conditions that affect the efficacy of signal generation, transmission, reception, and perception (Endler 1993b) and therefore affect the ability to detect and discriminate among potential mating traits. This leads to the evolution of sexual traits with better properties in the new conditions. Mate preferences evolve in parallel by a correlated response to sexual-trait selection (review in Andersson 1994) but will also be driven directly by changed sensory conditions; traits perceived in new ways affect existing preferences and favor new traits and preferences. If signaling traits work best in certain environments, this favors choice of microhabitats with the beneficial sensory conditions. Specific sensory conditions favor sensory systems that work best in those conditions, leading to evolution of the senses and brain, with further evolutionary effects on preferences and chosen traits. If changed prey visibility or new prey species require different sensory processing, sensory evolution may affect mating preferences via sensory properties that evolved in the context of prey detection and discrimination (Rodd et al. 2002), biasing the direction of sexual-trait and preference evolution (Endler 1992). Studying sensory drive is valuable for understanding the evolution of multiple suites of traits and also has important implications for the origin of population divergence and the origin and maintenance of species because senses, choice, and traits are used in species recognition (Boughman 2002; Maan et al. 2006; Terai et al. 2006).

Each component of sensory drive has been demonstrated in guppies (*P. reticulata*): male color patterns evolve increased visual contrast under sexual selection over many generations (Endler 1980). Visual contrast changes with ambient light and depends upon eye properties (Endler 1991; Smith et al. 2002; White et al. 2003). Some (long-wavelength-sensitive) visual pigments are genetically polymorphic (Archer et al. 1987; Archer & Lythgoe 1990; Hoffmann et al. 2007; Weadick & Chang 2007), allowing color perception to evolve as well as inducing variable female perception and hence choices even if females use the same criteria (Archer et al. 1987; Endler 1991). Courtship timing (associated with specific light environments) results in higher male visual contrast than would be achieved at other times (Endler 1987, 1991). Female choice changes with visual backgrounds (Endler 1983) and with ambient-light spectra (Long & Houde 1989; Gamble et al. 2003). Female preferences are predictable from male visual contrast (Endler & Houde 1995). Female preferences may also be related to food choice (Rodd et al. 2002). Female choice is geographically correlated with male color patterns (Endler & Houde 1995). Artificial selection for color patterns results in changes in female preferences (Breden & Hornaday 1994; Houde 1994). Artificial selection for spectral sensitivity results in changes in the visual system (Endler et al. 2001). Many of these components have also been found in other taxa, but it is unusual to have all in a single experimentally tractable species. This means that sensory drive is likely to be important in guppies and provides a means of predicting the direction of evolution of male signals, perception, female preferences, and microhabitat choice—traits that are usually studied separately. This and other integrative approaches could easily be applied to other poeciliids and yield significant new insights about the evolution of multiple suites of genetically unrelated but functionally related traits.

Poeciliids provide extraordinary potential for studying the function, ecology, and evolution of traits and integrated organisms and relating them to environmental conditions. They have already shown their worth in helping us understand the balance between sexual selection and predation, how sexual selection works, how and why life-history traits and color patterns evolve, and other evolutionary mechanisms. Their greatest potential is still hardly touched, in the coevolution within and among suites of genetically unrelated traits and this coevolution relative to environmental parameters, the assembly of communities after invasion of new areas (both natural and man-made), the evolution of physiological traits (including both classical environmental physiology and sensory physiology), the evolution of behavior, and the function and evolution of both neuroethological and behavioral mechanisms. I really look forward to the third book on poeciliids, when many of these areas will have been explored.

Acknowledgments

I thank Armando Pou for many communications after his visits to La Laguna de Los Patos. I also thank the National Science Foundation (USA) for many years of support of my own and others' research on guppies. Finally, I thank all the authors for stimulating and interesting chapters and the editors for useful suggestions on this chapter.

References cited

Able, D. J. 1996. The contagion indicator hypothesis for parasite-mediated sexual selection. Proceedings of the National Academy of Sciences of the United States of America 92: 2229–2233.

Abney, M. A., and Rakocinski, C. 2004. Life history variation in Caribbean gambusia, *Gambusia puncticulata puncticulata* (Poeciliidae) from the Cayman Islands, British West Indies. Environmental Biology of Fishes 70: 67–79.

Abrahams, M. V. 1993. The trade-off between foraging and courting in male guppies. Animal Behaviour 45: 673–681.

Abrams, P. A. 2000. The evolution of predator-prey interactions: theory and evidence. Annual Review of Ecology and Systematics 31: 79–105.

Abrams, P. A. 2003. Can adaptive evolution or behaviour lead to diversification of traits determining a trade-off between foraging gain and predation risk? Evolutionary Ecology Research 5: 653–670.

Abrams, P. A., and Rowe, L. 1996. The effects of predation on the age and size of maturity of prey. Evolution 50: 1052–1061.

Adam, D., Mäueler, W., and Schartl, M. 1991. Transcriptional activation of the melanoma inducing *Xmrk* oncogene in *Xiphophorus*. Oncogene 6: 73–80.

Adam, D., Dimitrijevic, N., and Schartl, M. 1993. Tumor suppression in *Xiphophorus* by an accidentally acquired promoter. Science 259: 816–819.

Aeschlimann, P. B., Häberli, M. A., Reusch, T. B. H., Boehm, T., and Milinski, M. 2003. Female sticklebacks *Gasterosteus aculeatus* use self-reference to optimize MHC allele number during mate selection. Behavioral Ecology and Sociobiology 54: 119–126.

Agrawal, A. F. 2001. The evolutionary consequences of mate copying on male traits. Behavioral Ecology and Sociobiology 51: 33–40.

Agrillo, C., Dadda, M., and Bisazza, A. 2006. Sexual harassment influences group choice in female mosquitofish. Ethology 112: 592–598.

Agrillo, C., Dadda, M., and Bisazza, A. 2007. Quantity discrimination in female mosquitofish. Animal Cognition 10: 63–70.

Agrillo, C., Dadda, M., and Serena, G. 2008a. Choice of female groups by male mosquitofish (*Gambusia holbrooki*). Ethology 114: 479–488.

Agrillo, C., Dadda, M., Serena, G., and Bisazza, A. 2008b. Do fish count? Spontaneous discrimination of quantity in female mosquitofish. Animal Cognition 11: 495–503.

Agrillo, C., Dadda, M., and Bisazza, A. 2009a. Escape behaviour elicited by a visual stimulus: a comparison between lateralised and non-lateralised female topminnows. Laterality 14: 300–314.

Agrillo, C., Dadda, M., Serena, G., and Bisazza, A. 2009b. Use of number by fish. PloS ONE 4: e4786.

Ahuja, M. R., and Anders, F. 1976. A genetic concept of the origin of cancer, based in part upon studies of neoplasms. Progress in Experimental Tumor Research 20: 380–397.

Ahuja, M. R., Schwab, M., and Anders, F. 1980. Linkage between a regulatory locus for melanoma cell differentiation and an esterase locus in *Xiphophorus*. Journal of Heredity 71: 403–407.

Alemadi, S. D., and Jenkins, D. G. 2008. Behavioral constraints for the spread of the eastern mosquitofish, *Gambusia holbrooki* (Poeciliidae). Biological Invasions 10: 59–66.

Alexander, H. J., and Breden, F. 2004. Sexual isolation and extreme morphological divergence in the Cumaná guppy: a possible case of incipient speciation. Journal of Evolutionary Biology 17: 1238–1254.

Alexander, H. J., Taylor, J. S., Wu, S. S. T., and Breden, F. 2006. Parallel evolution and vicariance in the guppy (*Poecilia reticulata*) over multiple spatial and temporal scales. Evolution 60: 2352–2369.

Alkins-Koo, M. 2000. Reproductive timing of fishes in a tropical intermittent stream. Environmental Biology of Fishes 57: 49–66.

Allan, J. D. 1995. *Stream Ecology: Structure and Function of Running Waters*. New York: Chapman & Hall.

Allendorf, F. W. 1986. Genetic drift and the loss of alleles versus heterozygosity. Zoo Biology 5: 181–190.

Almaca, C. 1995. Freshwater fish and their conservation in Portugal. Biological Conservation 72: 125–127.

Alonzo, S. H., and Warner, R. R. 2000. Female choice, conflict between the sexes and the evolution of male alternative reproductive behaviours. Evolutionary Ecology Research 2: 149–170.

Althoff, D. M., and Pellmyr, O. 2002. Examining genetic structure in a bogus yucca moth: a sequential approach to phylogeography. Evolution 56: 1632–1643.

Altschmied, J., Ditzel, L., and Schartl, M. 1997. Hypomethylation of the *Xmrk* oncogene promoter in melanoma cells of *Xiphophorus*. Biological Chemistry 378: 1457–1466.

Alvarez del Villar, J. 1948. Descripción de una nueva especie de *Mollienisia* capturada en Baños del Azufre, Tabasco (Pisces,

Poeciliidae). Anales de la Escuela Nacional de Ciencias Biológicas 5: 275–281.

Amlacher, J., and Dugatkin, L. A. 2005. Preference for older over younger models during mate-choice copying in young guppies. Ethology Ecology and Evolution 17: 161–169.

Amourique, L. 1965. Origine de la substance dynamogène émise par *Lebistes reticulatus* femelle (Poisson, Poeciliidae, Cyprinodontiformes). Comptes Rendus Hebdomadaires des Séances de l'Academie des Sciences 260: 2334–2335.

Anders, A., and Anders, F. 1978. Etiology of cancer as studied in the platyfish-swordtail system. Biochimica et Biophysica Acta 516: 61–95.

Anders, A., Anders, F., and Rase, S. 1969. XY females caused by x-irradiation. Experientia 25: 871.

Anders, A., Anders, F., and Klinke, K. 1973. *Regulation of Gene Expression in the Gordon-Kosswig Melanoma System*. New York: Springer Verlag.

Anders, F. 1967. Tumour formation in platyfish-swordtail hybrids as a problem of gene regulation. Experientia 23: 1–10.

Anders, F. 1991. Contributions of the Gordon-Kosswig melanoma system to the present concept of neoplasia. Pigment Cell Research 3: 7–29.

Anders, F., Diehl, H., Schwab, M., and Anders, A. 1979. Contributions to an understanding of the cellular origin of melanoma in the Gordon-Kosswig Xiphophorine fish tumor system. Pigment Cell Research 4: 142–149.

Anders, F., Schartl, M., Barnekow, A., and Anders, A. 1984. *Xiphophorus* as an *in vivo* model for studies on normal and defective control of oncogenesis. Advances in Cancer Research 42: 191–275.

Anderson, E. J., McGillis, W. R., and Grosenbaugh, M. A. 2001. The boundary layer of swimming fish. Journal of Experimental Biology 204: 81–102.

Anderson, R. C., Searcy, W. A., and Nowicki, S. 2005. Partial song matching in an eastern population of song sparrows, *Melospiza melodia*. Animal Behaviour 69: 189–196.

Anderson, R. M., and Gordon, D. M. 1982. Processes influencing the distribution of parasite numbers within host populations with special emphasis on parasite-induced host mortalities. Parasitology 85: 373–398.

Andersson, M. 1994. *Sexual Selection*. Princeton, NJ: Princeton University Press.

Andersson, M., and Simmons, L. W. 2006. Sexual selection and mate choice. Trends in Ecology and Evolution 21: 296–302.

Andrew, R. J., and Rogers, L. J. 2002. *Comparative Vertebrate Lateralization*. Cambridge: Cambridge University Press.

Angus, R. A. 1989a. A genetic overview of poeciliid fishes. In G. K. Meffe and F. F. Snelson Jr. (eds.), *Ecology and Evolution of Livebearing Fishes (Poeciliidae)*, 51–68. Englewood Cliffs, NJ: Prentice Hall.

Angus, R. A. 1989b. Inheritance of melanistic pigmentation in the eastern mosquitofish. Journal of Heredity 80: 833–853.

Anteunis, A. 1959. Recherches sur la structure et le développement de l'ovaire et de l'oviducte chez *Lebistes reticulatus* (Téléostéen). Archives de Biologie 70: 783–807.

Apanius, V., Penn, D., Slev, P. R., Ruff, L. R., and Potts, W. K. 1997. The nature of selection on the major histocompatibility complex. Critical Reviews in Immunology 17: 179–224.

Archard, G. A., Cuthill, I. C., Partridge, J. C., and van Oosterhout, C. 2008. Female guppies (*Poecilia reticulata*) show no preference for conspecific chemosensory cues in the field or an artificial flow chamber. Behaviour 145: 1329–1346

Archdeacon, T. P. 2007. Effects of Asian tapeworm, mosquitofish, and food ration on Mohave tui chub growth and survival. Master's thesis, University of Arizona.

Archer, S. N. 1999. Light and photoreception: visual pigments and photoreception. In S. N. Archer, B. A. Djamgoz, and E. R. Loew (eds.), *Adaptive Mechanisms in the Ecology of Vision*, 3–23. England: Klumer Academic.

Archer, S. N., and Hirano, J. 1997. Opsin sequences of the rod visual pigments in two species of poeciliid fish. Journal of Fish Biology 51: 215–219.

Archer, S. N., and Lythgoe, J. N. 1990. The visual pigment basis for cone polymorphism in the guppy, *Poecilia reticulata*. Vision Research 30: 225–233.

Archer, S. N., Endler, J. A., Lythgoe, J. N., and Partridge, J. C. 1987. Visual pigment polymorphism in the guppy *Poecilia reticulata*. Vision Research 28: 1243–1252.

Arenas, M. I., Fraile, B., Demiguel, M. P., and Paniagua, R. 1995a. Cytoskeleton in Sertoli cells of the mosquito fish (*Gambusia affinis holbrooki*). Anatomical Record 241: 225–234.

Arenas, M. I., Fraile, B., Demiguel, M., and Paniagua, R. 1995b. Intermediate filaments in the testis of the teleost mosquito fish *Gambusia affinis holbrooki*: a light and electron-microscope immunocytochemical study and Western blotting analysis. Histochemical Journal 27: 329–337.

Arendt, J. D. 1997. Adaptive intrinsic growth rates: an integration across taxa. Quarterly Review of Biology 72: 149–177.

Arendt, J. D., and Reznick, D. N. 2005. Evolution of juvenile growth rates in female guppies (*Poecilia reticulata*): predator regime or resource level? Proceedings of the Royal Society of London Series B—Biological Sciences 272: 333–337.

Arias, A. L., and Reznick, D. 2000. Life history of *Phalloceros caudiomaculatus*: a novel variation on the theme of livebearing in the family Poeciliidae. Copeia 2000: 792–798.

Arkush, K. D., Giese, A. R., Mendonca, H. L., McBride, A. M., Marty, G. D., and Hedrick, P. W. 2002. Resistance to three pathogens in the endangered winter-run chinook salmon (*Oncorhynchus tshawytscha*): effects of inbreeding and major histocompatibility complex genotypes. Canadian Journal of Fisheries and Aquatic Sciences 59: 966–975.

Arndt, M., Parzefall, J., and Plath, M. 2004. Does sexual experience influence mate choice decisions in cave molly females (*Poecilia mexicana*, Poeciliidae, Teleostei)? Subterranean Biology 2: 53–58.

Arnegard, M. E., and Kondrashov, A. S. 2004. Sympatric speciation by sexual selection alone is unlikely. Evolution 58: 222–237.

Arnold, M. L. 2006. *Evolution through Genetic Exchange*. Oxford: Oxford University Press.

Arnold, R., Mäueler, W., Bassili, G., Lutz, M., Burke, L., Epplen, T. J., and Renkawitz, R. 2000. The insulator protein CTCF represses transcription on binding to the (gt)(22)(ga)(15) microsatellite in intron 2 of the *HLA-DRB1*0401* gene. Gene 253: 209–214.

Arnold, S. J., Pfrender, M. E., and Jones, A. G. 2001. The adaptive landscape as a conceptual bridge between micro- and macroevolution. Genetica 112–113: 9–32.

Arnott, G., and Elwood, R. W. 2008. Information gathering and decision making about resource value in animal contests. Animal Behaviour 76: 529–542.

Arnqvist, G. 1997. The evolution of animal genitalia: distinguishing between hypotheses by single species studies. Biological Journal of the Linnean Society 60: 365–379.

Arnqvist, G. 1998. Comparative evidence for the evolution of genitalia by sexual selection. Nature 393: 784–786.

Arnqvist, G., and Rowe, L. 2005. *Sexual Conflict*. Princeton, NJ: Princeton University Press.

Arnqvist, G., and Thornhill, R. 1998. Evolution of animal genitalia: patterns of phenotypic and genotypic variation and condition dependence of genital and non-genital morphology in water strider (Heteroptera: Gerridae: Insecta). Genetical Research 71: 193–212.

Artamonova, I. I., and Gelfand, M. S. 2007. Comparative genomics and evolution of alternative splicing: the pessimists' science. Chemical Reviews 107: 3407–3430.

Arthington, A. H. 1989. Diet of *Gambusia affinis holbrooki*, *Xiphophorus helleri*, *X. maculatus* and *Poecilia reticulata* (Pisces: Poeciliidae) in streams of Southeastern Queensland, Australia. Asian Fisheries Science 2: 193–212.

Arthington, A. H., and Lloyd, L. N. 1989. Introduced poeciliids in Australia and New Zealand. In G. K. Meffe and F. F. Snelson Jr. (eds.), *Ecology and Evolution of Livebearing Fishes (Poeciliidae)*, 333–348. Englewood Cliffs, NJ: Prentice Hall.

Arthington, A. H., and Mitchell, D. S. 1986. Aquatic invading species. In R. H. Groves and J. J. Burdon (eds.), *Ecology of Biological Invasions—an Australian Perspective*, 34–53. Canberra: Australian Academy of Science.

Aspbury, A. S. 2007. Sperm competition effects on sperm production and expenditure in sailfin mollies, *Poecilia latipinna*. Behavioral Ecology 18: 776–780.

Aspbury, A. S., and Basolo, A. L. 2002. Repeatable female preferences, mating order and mating success in the poeciliid fish, *Heterandria formosa*. Behavioral Ecology and Sociobiology 51: 238–244.

Aspbury, A. S., and Gabor, C. R. 2004a. Differential sperm priming by male sailfin mollies (*Poecilia latipinna*): effects of female and male size. Ethology 110: 193–202.

Aspbury, A. S., and Gabor, C. R. 2004b. Discriminating males alter sperm production between species. Proceedings of the National Acadamy of Sciences of the United States of America 101: 15970–15973.

Aspbury, A. S., Coyle, J. M., and Gabor, C. R. 2010a. Effect of predation on male mating behavior in a unisexual-bisexual mating system. Behaviour 147: 53–63.

Aspbury, A. S., Espinedo, C., and Gabor, C. R. 2010b. Lack of species discrimination based on chemical cues by male sailfin mollies, *Poecilia latipinna*. Evolutionary Ecology 24: 69–82.

Atz, J. W. 1962. Effects of hybridization on pigmentation in fishes of the genus *Xiphophorus*. Zoologica 47: 153–181.

Atz, J. W. 1964. Intersexuality in fishes. In A. J. Marshall and C. N. Armstrong (eds.), *Intersexuality in Vertebrates Including Man*, 145–232. New York: Academic Press.

Avise, J. C., Trexler, J. C., Travis, J., and Nelson, W. S. 1991. *Poecilia mexicana* is the recent female parent of the unisexual fish *P. formosa*. Evolution 46: 1530–1533.

Ayala, J. R., Rader, R. B., Belk, M. C., and Schaalje, G. B. 2007. Ground-truthing the impact of invasive species: spatio-temporal overlap between native least chub and introduced western mosquitofish. Biological Invasions 9: 857–869.

Baccetti, B., Nurrini, A. G., Collodel, G., Piomboni, B., Renieri, T., and Sensini, C. 1989. Localization of acrosomal enzymes in Arthropoda, Echinodermata and Vertebrata. Journal of Submicroscopical Cytology and Pathology 21: 385–389.

Baer, C. F. 1998a. Population structure in a south-eastern US freshwater fish, *Heterandria formosa*, II: gene flow and biogeography within the St. Johns River drainage. Heredity 81: 404–411.

Baer, C. F. 1998b. Species-wide population structure in a southeastern U.S. freshwater fish, *Heterandria formosa*: gene flow and biogeography. Evolution 52: 183–193.

Baerends, G. P., Brouwer, R., and Waterbolk, H. T. 1955. Ethological studies on *Lebistes reticulatus* (Peters), I: an analysis of the male courtship pattern. Behaviour 8: 249–334.

Bagarinao, T., and Vetter, R. D. 1989. Sulfide tolerance and detoxification in shallow water marine fishes. Marine Biology 103: 291–302.

Bagarinao, T., and Vetter, R. D. 1992. Sulfide-hemoglobin interactions in the sulfide-tolerant salt marsh resident, the California killifish *Fundulus parvipinnis*. Journal of Comparative Physiology A—Neuroethology, Sensory, Neural, and Behavioral Physiology 162: 614–624.

Bagarinao, T., and Vetter, R. D. 1993. Sulfide tolerance and adaptation in the California killifish, *Fundulus parvipinnis*, a salt marsh resident. Journal of Fish Biology 42: 729–748.

Bailey, R. J. 1933. The ovarian cycle in the viviparous teleost *X. helleri*. Biological Bulletin 64: 206–225.

Bak, E. J., Ishii, Y., Omatsu, T., Kyuwa, S., Tetsuya, T., Hayasaka, I., and Yoshikawa, Y. 2006. Identification and analysis of MHC class II *DRB-1* (*Patr-DRB1*) alleles in chimpanzees. Tissue Antigens 67: 134–142.

Bakke, T. A., Harris, P. D., and Cable, J. 2002. Host specificity dynamics: observations on gyrodactylid monogeneans. International Journal for Parasitology 32: 281–308.

Bakke, T. A., Cable, J., and Østbø, M. 2006. The ultrastructure of hypersymbionts on *Gyrodactylus salaris* (Monogenea) infecting Atlantic salmon (*Salmo salar*). Journal of Helminthology 80: 377–386.

Bakke, T. A., Cable, J., and Harris, P. D. 2007. The biology of gyrodactylid monogeneans: the "Russian-doll killers." Advances in Parasitology 64: 161–376.

Balas, A., Aviles, M. J., Alonso-Nieto, M., Zarapuz, L., Blanco, L., Garcia-Sanchez, F., and Vicario, J. L. 2005. *HLA-DQA1* introns 2 and 3 sequencing: *DQA1* sequencing-based typing and characterization of a highly polymorphic microsatellite at intron 3 of *DQA1*0505*. Human Immunology 66: 903–911.

Balsano, J. S., Rasch, E. M., and Monaco, P. J. 1989. The evolutionary ecology of *Poecilia formosa* and its triploid associate. In G. K. Meffe and F. F. Snelson Jr. (eds.), *Ecology and Evolution of Livebearing Fishes (Poeciliidae)*, 277–298. Englewood Cliffs, NJ: Prentice Hall.

Banet, A. I., and Reznick, D. N. 2008. Do placental species abort offspring? Testing an assumption of the Trexler-DeAngelis model. Functional Ecology 22: 323–331.

Barber, I., Hoare, D., and Krause, J. 2000. Effects of parasites on fish behaviour: a review and evolutionary perspective. Reviews in Fish Biology and Fisheries 10: 131–165.

Barney, R. L., and Anson, B. J. 1921. Seasonal abundance of the mosquito destroying top-minnow, *Gambusia affinis*, especially in relation to male frequency. Ecology 2: 53–69.

Baroiller, J. F., and D'Cotta, H. 2001. Environment and sex

determination in farmed fish. Comparative Biochemistry and Physiology C—Toxicology and Pharmacology 130: 399–409.

Baron, B., Buckle, F., and Espina, S. 2002. Environmental factors and sexual differentiation in *Poecilia sphenops* Valenciennes (Pisces: Poeciliidae). Aquaculture Research 33: 615–619.

Barrier, R. F. G., and Hicks, B. J. 1994. Behavioral interactions between back mudfish (*Neochanna diversus* Stokell, 1949: Galaxiidae) and mosquitofish (*Gambusia affinis* Baird and Girard, 1845). Ecology of Freshwater Fish 3: 93–99.

Barson, N. J., Cable, J., and van Oosterhout, C. 2009. Population genetic analysis of microsatellite variation of guppies (*Poecilia reticulata*) in Trinidad and Tobago: evidence for a dynamic source-sink metapopulation structure, founder events and population bottlenecks. Journal of Evolutionary Biology 22: 485–497.

Barth, K. A., Miklosi, A., Watkins, J., Bianco, I. H., Wilson, S. W., and Andrew, R. J. 2005. *fsi* zebrafish show concordant reversal of laterality of viscera, neuroanatomy, and a subset of behavioral responses. Current Biology 15: 844–850.

Barton, N. H. 1995. A general model for the evolution of recombination. Genetical Research 65: 123–144.

Barton, N. H., and Charlesworth, B. 1998. Why sex and recombination? Science 281: 1986–1990.

Barton, N. H., and Hewitt, G. M. 1985. Analysis of hybrid zones. Annual Review of Ecology and Systematics 16: 113–148.

Barton, N. H., and Turelli, M. 1989. Evolutionary quantitative genetics: how little do we know? Annual Review of Genetics 23: 337–370.

Bashey, F. 2006. Cross-generational environmental effects and the evolution of offspring size in the Trinidadian guppy *Poecilia reticulata*. Evolution 60: 348–361.

Bashey, F. 2008. Competition as a selective mechanism for larger offspring size in guppies. Oikos 117: 104–113.

Basolo, A. L. 1990. Female preference predates the evolution of the sword in swordtail fish. Science 250: 808–810.

Basolo, A. L. 1995a. A further examination of a pre-existing bias favouring a sword in the genus *Xiphophorus*. Animal Behaviour 50: 365–375.

Basolo, A. L. 1995b. Phylogenetic evidence for the role of a preexisting bias in sexual selection. Proceedings of the Royal Society of London Series B—Biological Sciences 259: 307–311.

Basolo, A. L. 1996. The phylogenetic distribution of a female preference. Systematic Biology 45: 290–307.

Basolo, A. L. 1998. Evolutionary change in a receiver bias: a comparison of female preference functions. Proceedings of the Royal Society of London Series B—Biological Sciences 265: 2223–2228.

Basolo, A. L. 2006. Genetic linkage and color polymorphism in the southern platyfish (*Xiphophorus maculatus*): a model system for studies of color pattern evolution. Zebrafish 3: 65–83.

Basolo, A. L., and Trainor, B. C. 2002. The conformation of a female preference for a composite male trait in green swordtails. Animal Behaviour 63: 469–474.

Basolo, A. L., and Wagner, W. E. 2004. Covariation between predation risk, body size and fin elaboration in the green swordtail, *Xiphophorus helleri*. Zoological Journal of the Linnean Society 83: 87–100.

Bateman, A. J. 1948. Intra-sexual selection in *Drosophila*. Heredity 2: 349–368.

Baudler, M., Duschl, J., Winkler, C., Schartl, M., and Altschmied, J. 1997. Activation of transcription of the melanoma inducing *Xmrk* oncogene by a GC box element. Journal of Biological Chemistry 272: 131–137.

Baudler, M., Schartl, M., and Altschmied, J. 1999. Specific activation of a STAT family member in *Xiphophorus* melanoma cells. Experimental Cell Research 249: 212–220.

Bauwens, D., and Thoen, C. 1981. Escape tactics and vulnerability to predation associated with reproduction in the lizard *Lacerta vivipara*. Journal of Animal Ecology 50: 733–743.

Bayley, M., Larsen, P. F., Baekgaard, H., and Baatrup, E. 2003. The effects of vinclozolin, an anti-androgenic fungicide, on male guppy secondary sex characters and reproductive success. Biology of Reproduction 69: 1951–1956.

Beamish, F. W. H. 1978. Swimming capacity. In W. S. Hoar and J. D. Randall (eds.), *Fish Physiology*, 101–187. New York: Academic Press.

Beauchamp, G. K., and Yamazaki, K. 2003. Chemical signalling in mice. Biochemical Society Transactions 31: 147–151.

Beaugrand, J. P., and Zayan, R. 1985. An experimental model of aggressive dominance in *Xiphophorus helleri* (Pisces, Poeciliidae). Behavioural Processes 10: 1–52.

Beaugrand, J. P., Caron, J., and Comeau, L. 1984. Social organization of small heterosexual groups of green swordtails (*Xiphophorus helleri*, Pisces, Poeciliidae) under conditions of captivity. Behaviour 91: 24–60.

Beaugrand, J. P., Goulet, C., and Payette, D. 1991. Outcome of dyadic conflict in male green swordtail fish, *Xiphophorus helleri*: effects of body size and prior dominance. Animal Behaviour 41: 417–424.

Becher, S. A., and Magurran, A. E. 2000. Gene flow in Trinidadian guppies. Journal of Fish Biology 56: 241–249.

Becher, S. A., and Magurran, A. E. 2004. Multiple mating and reproductive skew in Trinidadian guppies. Proceedings of the Royal Society of London Series B—Biological Sciences 271: 1009–1014.

Beck, B. B., Rapaport, L. G., Stanley Price, M. R., and Wilson, A. C. 1994. Reintroduction of captive-born animals. In P. J. S. Olney, G. M. Mace, and A. T. C. Feistner (eds.), *Creative Conservation: Interactive Management of Wild and Captive Animals*, 265–286. London: Chapman & Hall.

Becker, T. M., Rizos, H., Kefford, R. F., and Mann, G. J. 2001. Functional impairment of melanoma-associated p16(INK4a) mutants in melanoma cells despite retention of cyclin-dependent kinase 4 binding. Clinical Cancer Research 7: 3282–3288.

Beerli, P. 2006. Comparison of Bayesian and maximum-likelihood inference of population genetic parameters. Bioinformatics 22: 341–345.

Belk, M. C., and Lydeard, C. 1994. Effects of *Gambusia holbrooki* on a similar-sized, syntopic poeciliid, *Heterandria formosa*: competitor or predator? Copeia 1994: 296–302.

Bell, A. M. 2005. Behavioural differences between individuals and two populations of stickleback (*Gasterosteus aculeatus*). Journal of Evolutionary Biology 18: 464–473.

Bell, A. M. 2007. Future directions in behavioural syndromes research. Proceedings of the Royal Society of London Series B—Biological Sciences 274: 755–761.

Bell, G. 1982. *The Masterpiece of Nature: The Evolution and*

Genetics of Sexuality. Berkeley and Los Angeles, CA: University of California Press.

Bence, J. R. 1988. Indirect effects and biological control of mosquitoes by mosquitofish. Journal of Applied Ecology 25: 505–521.

Bennett, D. C. 2008. How to make a melanoma: what do we know of the primary clonal events? Pigment Cell and Melanoma Research 21: 27–38.

Bennett, W. A., and Beitinger, T. L. 1997. Temperature tolerance of the sheepshead minnow, *Cyprinodon variegatus.* Copeia 1997: 77–87.

Benson, K. E. 2007. Enhanced female brood patch size stimulates male courtship in *Xiphophorus helleri.* Copeia 2007: 212–217.

Benson, K. E., and Basolo, A. L. 2006. Male-male competition and the sword in male swordtails, *Xiphophorus helleri.* Animal Behaviour 71: 129–134.

Berg, O., and Gorbman, A. 1954. Iodine utilization by tumorous thyroid tissue of the swordtail *Xiphophorus montezumae.* Cancer Research 14: 232–236.

Berglund, A., Bisazza, A., and Pilastro, A. 1996. Armaments and ornaments: an evolutionary explanation of traits of dual utility. Biological Journal of the Linnean Society 58: 385–399.

Bergmann, M., Schindelmeiser, I., and Greven, H. 1984. The blood testis barrier in vertebrates having different testicular organisation. Cell and Tissue Research 238: 145–150.

Bergstrom, T. F., Josefsson, A., Erlich, H. A., and Gyllensten, U. 1998. Recent origin of *HLA-DRB1* alleles and implications for human evolution. Nature Genetics 18: 237–242.

Bergstrom, T. F., Erlandsson, R., Engkvist, H., Josefsson, A., Erlich, H. A., and Gyllensten, U. 1999. Phylogenetic history of hominoid *DRB* loci and alleles inferred from intron sequences. Immunological Reviews 167: 351–365.

Bernatchez, L., and Landry, C. 2003. MHC studies in nonmodel vertebrates: what have we learned about natural selection in 15 years? Journal of Evolutionary Biology 16: 363–377.

Berry, G. R. 1962. *The Classic Greek Dictionary: Greek-English and English-Greek.* Chicago: Follett Publishing Co.

Bertin, A., and Fairbairn, D. J. 2007. The form of sexual selection on male genitalia cannot be inferred from within-population variance and allometry: a case study in *Aquarius remigis.* Evolution 61: 825–837.

Beukeboom, L. W., and Vrijenhoek, R. C. 1998. Evolutionary genetics and ecology of sperm-dependent parthenogenesis. Journal of Evolutionary Biology 11: 755–782.

Bhat, A., and Magurran, A. E. 2006. Benefits of familiarity persist after prolonged isolation in guppies. Journal of Fish Biology 68: 759–766.

Bi, K., and Bogart, J. P. 2006. Identification of intergenomic recombinations in unisexual salamanders of the genus *Ambystoma* by genomic in situ hybridization (GISH). Cytogenetic and Genome Research 112: 307–312.

Bildsøe, M. 1988. Aggressive, sexual and foraging behaviour in *Poecilia velifera* (Pisces: Poeciliidae) during captivity. Ethology 79: 1–12.

Billard, R. 1986. Spermatogenesis and spermatology of some teleost fish species. Reproduction Nutrition Development 26: 877–920.

Billard, R., and Jalabert, B. 1973. Le glycogène au cours de la formation des spermatozoïdes et leur transit dans les tractus génitaux male et femelle chez le guppy (Poisson poeciliide).

Annales de Biologie Animale, Biochimie, Biophysique 13: 313–320.

Birkhead, T. R. 2000. Defining and demonstrating postcopulatory female choice—again. Evolution 54: 1057–1060.

Birkhead, T. R., and Pizzari, T. 2002. Postcopulatory sexual selection. Nature Review Genetics 3: 262–273.

Birkhead, T. R., Chaline, N., Biggins, J. D., Burke, T., and Pizzari, T. 2004. Nontransitivity of paternity in a bird. Evolution 58: 416–420.

Biro, P., and Stamps, J. 2008. Are animal personality traits linked to life-history productivity? Trends in Ecology and Evolution 23: 361–368.

Bisazza, A. 1993a. Male competition, female mate choice and sexual size dimorphism in poeciliid fishes. Marine Behaviour and Physiology 23: 257–286.

Bisazza, A. 1993b. Male competition, female mate choice and sexual size dimorphism in poeciliid fishes. In F. A. Huntingford and P. Torricelli (eds.), *Behavioural Ecology of Fishes*, 257–286. Chur, Switzerland: Harwood Academic Press.

Bisazza, A., and Dadda, M. 2005. Enhanced schooling performance in lateralized fishes. Proceedings of the Royal Society of London Series B—Biological Sciences 272: 1677–1681.

Bisazza, A., and Marin, G. 1991. Male size and female mate choice in the eastern mosquitofish (*Gambusia holbrooki*: Poeciliidae). Copeia 1991: 730–735.

Bisazza, A., and Marin, G. 1995. Sexual selection and sexual size dimorphism in the eastern mosquitofish *Gambusia holbrooki* (Pisces Poeciliidae). Ethology, Ecology and Evolution 7: 169–183.

Bisazza, A., and Pilastro, A. 1997. Small male mating advantage and reversed size dimorphism in poeciliid fishes. Journal of Fish Biology 50: 397–406.

Bisazza, A., and Pilastro, A. 2000. Variation of female preference for male coloration in the eastern mosquitofish *Gambusia holbrooki.* Behavior Genetics 30: 207–212.

Bisazza, A., and Vallortigara, G. 1996. Rotational bias in mosquitofish (*Gambusia holbrooki*): the role of laterality and sun-compass navigation. Laterality 1: 161–175.

Bisazza, A., Pilastro, A., Palazzi, R., and Marin, G. 1996. Sexual behaviour of immature male eastern mosquitofish: a way to measure intensity of intra-sexual selection? Journal of Fish Biology 48: 726–737.

Bisazza, A., Pignatti, R., and Vallortigara, G. 1997a. Detour tests reveal task- and stimulus-specific behavioural lateralization in mosquitofish (*Gambusia holbrooki*). Behavioural Brain Research 89: 237–242.

Bisazza, A., Pignatti, R., and Vallortigara, G. 1997b. Laterality in detour behaviour: interspecific variation in poeciliid fish. Animal Behaviour 54: 1273–1281.

Bisazza, A., Facchin, L., Pignatti, R., and Vallortigara, G. 1998a. Lateralization of detour behaviour in poeciliid fish: the effect of species, gender and sexual motivation. Behavioural Brain Research 91: 157–164.

Bisazza, A., Rogers, L. J., and Vallortigara, G. 1998b. The origins of cerebral asymmetry: a review of evidence of behavioural and brain lateralization in fishes, reptiles and amphibians. Neuroscience and Biobehavioral Reviews 22: 411–426.

Bisazza, A., De Santi, A., and Vallortigara, G. 1999. Laterality and cooperation: mosquitofish move closer to a predator when the companion is on their left side. Animal Behaviour 57: 1145–1149.

Bisazza, A., Cantalupo, C., Capocchiano, M., and Vallortigara, G. 2000a. Population lateralisation and social behaviour: a study with 16 species of fish. Laterality 5: 269–284.

Bisazza, A., Facchin, L., and Vallortigara, G. 2000b. Heritability of lateralization in fish: concordance of right-left asymmetry between parents and offspring. Neuropsychologia 38: 907–912.

Bisazza, A., Sovrano, V. A., and Vallortigara, G. 2001a. Consistency among different tasks of left-right asymmetries in lines of fish originally selected for opposite direction of lateralization in a detour task. Neuropsychologia 39: 1077–1085.

Bisazza, A., Vaccari, G., and Pilastro, A. 2001b. Female mate choice in a mating system dominated by male sexual coercion. Behavioral Ecology 12: 59–64.

Bisazza, A., Dadda, M., and Cantalupo, C. 2005. Further evidence for mirror-reversed laterality in lines of fish selected for leftward or rightward turning when facing a predator model. Behavioural Brain Research 156: 165–171.

Bisazza, A., Dadda, M., Facchin, L., and Vigo, F. 2007. Artificial selection on laterality in the teleost fish *Girardinus falcatus*. Behavioural Brain Research 178: 29–38.

Bisazza, A., Piffer, L., Serena, G., and Agrillo, C. 2010. Ontogeny of numerical abilities in fish. PloS ONE 5: e15516.

Bischoff, R. J., Gould, J. L., and Rubenstein, D. I. 1985. Tail size and female choice in the guppy (*Poecilia reticulata*). Behavioural Ecology and Sociobiology 17: 253–255.

Bjorkman, P. J., Saper, M. A., Samraoui, B., Bennett, W. S., Strominger, J. L., and Wiley, D. C. 1987a. Structure of the human class-I histocompatibility antigen, *HLA-A2*. Nature 329: 506–512.

Bjorkman, P. J., Saper, M. A., Samraoui, B., Bennett, W. S., Strominger, J. L., and Wiley, D. C. 1987b. The foreign antigen-binding site and T-cell recognition regions of class-I histocompatibility antigens. Nature 329: 512–518.

Black, D. A., and Howell, W. M. 1979. The North American mosquitofish, *Gambusia affinis*: a unique case in sex chromosome evolution. Copeia 1979: 509–513.

Blackburn, D. G. 1992. Convergent evolution of viviparity, matrotrophy, and specialization for fetal nutrition in reptiles and other vertebrates. American Zoologist 33: 313–321.

Blackburn, D. G. 1999. Viviparity and oviparity: evolution and reproductive strategies. In T. E. Knobil and J. D. Neill (eds.), *Encyclopedia of Reproduction*, 994–1003. New York: Academic Press.

Blackburn, D. G. 2005. Evolutionary origins of viviparity in fishes. In M. C. Uribe and H. J. Grier (eds.), *Viviparous Fishes*, 283–297. Homestead, FL: New Life Publications.

Blais, J., Rico, C., van Oosterhout, C., Cable, J., Turner, G. F., and Bernatchez, L. 2007. MHC adaptive divergence between closely related and sympatric african cichlids. PloS ONE 2: e734.

Blomberg, S. P., Garland, T., and Ives, A. R. 2003. Testing for phylogenetic signal in comparative data: behavioral traits are more labile. Evolution 57: 717–745.

Blount, J. D., Møller, A. P., and Houston, D. C. 2001. Antioxidants, showy males and sperm quality. Ecology Letters 4: 393–396.

Blows, M. W., and Hoffmann, A. A. 2005. A reassessment of genetic limits to evolutionary change. Ecology 86: 1371–1384.

Blows, M. W., Brooks, R., and Kraft, P. G. 2003. Exploring complex fitness surfaces: multiple ornamentation and polymorphism in male guppies. Evolution 57: 1622–1630.

Böhne, A., Schultheis, C., Zhou, Q., Froschauer, A., Schmidt, C., Selz, Y., Braasch, I., Ozouf-Costaz, C., Dettai, A., Ségurens, B., Couloux, A., Bernard-Samain, S., Chilmonczyk, S., Gannouni, A., Madani, K., Brunet, F., Galiana-Arnoux, D., Schartl, M., and Volff, J. N. 2008. Identification of new gene candidates on the sex chromosomes of the platyfish *Xiphophorus maculatus*. Cybium 32: suppl. 69–71.

Bolnick, D. I., and Fitzpatrick, B. M. 2007. Sympatric speciation: models and empirical evidence. Annual Review of Ecology, Evolution, and Systematics 38: 459.

Bolnick, D. I., and Near, T. J. 2005. Tempo of hybrid inviability in centrarchid fishes (Teleostei: Centrarchidae). Evolution 59: 1754–1767.

Bonduriansky, R. 2007. Sexual selection and allometry: a critical reappraisal of the evidence and ideas. Evolution 61: 838–849.

Bonduriansky, R., and Chenoweth, S. F. 2009. Intralocus sexual conflict. Trends in Ecology and Evolution 24: 280–288.

Bonduriansky, R., and Day, T. 2003. The evolution of static allometry in sexually selected traits. Evolution 57: 2450–2458.

Bonnie, K. E., and Earley, R. L. 2007. Expanding the scope for social information use. Animal Behaviour 74: 171–181.

Bono, L., Rios-Cardenas, O., and Morris, M. R. Forthcoming. Alternative life histories in *Xiphophorus multilineatus*: evidence for different ages at sexual maturity and growth responses in the wild.

Boogert, N. J., Reader, S. M., Hoppitt, W., and Laland, K. N. 2008. The origin and spread of innovations in starlings. Animal Behaviour 75: 1509–1518.

Boorman, E., and Parker, G. A. 1976. Sperm (ejaculate) competition in *Drosophila melanogaster* and the reproductive value of females to males in relation to female age and mating status. Ecological Entomology 1: 145–155.

Borghans, J. A. M., Beltman, J. B., and De Boer, R. J. 2004. MHC polymorphism under host-pathogen coevolution. Immunogenetics 55: 732–739.

Borowsky, R. 1973. Melanomas in *Xiphophorus variatus* (Pisces: Poeciliidae) in the absence of hybridization. Experientia 29: 1431–1433.

Borowsky, R. 1984. The evolutionary genetics of *Xiphophorus*. In B. J. Turner (ed.), *Evolutionary Genetics of Fishes*, 235–310. New York: Plenum Press.

Borowsky, R., and Kallman, K. D. 1976. Patterns of mating in natural populations of *Xiphophorus* (Pisces: Poeciliidae), I: *X. maculatus* from Belize and Mexico. Evolution 30: 693–706.

Borowsky, R., and Khouri, J. 1976. Patterns of mating in natural populations of *Xiphophorus*, II: *X. variatus* from Tamaulipas, Mexico. Copeia 1976: 727–734.

Boschetto, C., Gasparini, C., & Pilastro, A. 2010. Sperm number and velocity affect sperm competition success in the guppy (*Poecilia reticulata*). Behavioral Ecology and Sociobiology. Forthcoming.

Boss, J. M., and Jensen, P. E. 2003. Transcriptional regulation of the MHC class II antigen presentation pathway. Current Opinion in Immunology 15: 105–111.

Botham, M. S., Kerfoot, C. J., Louca, V., and Krause, J. 2006. The effects of different predator species on antipredator behavior in the Trinidadian guppy, *Poecilia reticulata*. Naturwissenschaften 93: 431–439.

Botham, M. S., Hayward, R. K., Morrell, L. J., Croft, D. P., Ward, J. R., Ramnarine, I., and Krause, J. 2008. Risk-sensitive antipredator behavior in the Trinidadian guppy, *Poecilia reticulata*. Ecology 89: 3174–3185.

Bouaïchi, A., Simpson, S. J., and Roessingh, P. 1996. The influence of environmental microstructure on the behavioural phase state and distribution of the desert locust *Schostocerca gregaria*. Physiological Entomology 21: 247–256.

Bouchard, J. T. J., and Loehlin, J. C. 2001. Genes, evolution, and personality. Behavior Genetics 31: 243–273.

Boughman, J. W. 2001. Divergent sexual selection enhances reproductive isolation in sticklebacks. Nature 411: 944–948.

Boughman, J. W. 2002. How sensory drive can promote speciation. Trends in Ecology and Evolution 17: 571–577.

Boughman, J. W., Rundle, H. D., and Schluter, D. 2005. Parallel evolution of sexual isolation in sticklebacks. Evolution 59: 361–373.

Boulding, E. G., and Hay, T. 2001. Genetic and demographic parameters determining population persistence. Heredity 86: 313–324.

Bowmaker, J. K., and Hunt, D. M. 1999. Molecular biology of photoreceptor visual sensitivity. In S. N. Archer, B. A. Djamgoz, and E. R. Loew (eds.), *Adaptive Mechanisms in the Ecology of Vision*, 439–465. Boston: Klumer Academic.

Bowmaker, J. K., Govardovskii, V. I., Shukolyukov, S. A., Zueva. L. V., Hunt, D. M., Sideleva, V. G., and Smirnova, O. G. 1994. Visual pigments and the photic environment: the Cottoid fish of Lake Baikal. Vision Research 34: 591–605.

Boyce, M. S. 1984. Restitution of r- and K-selection as a model of density-dependent natural selection. Annual Review of Ecology and Systematics 15: 427–447.

Boyd, R., and Richerson, P. J. 1985. *Culture and the Evolutionary Process*. Chicago: University of Chicago Press.

Boyd, R., and Richerson, R. 2002. Group beneficial norms can spread rapidly in structured populations. Journal of Theoretical Biology 215: 287–296.

Bozynski, C. C., and Liley, N. R. 2003. The effect of female presence on spermiation, and of male sexual activity on "ready" sperm in the male guppy. Animal Behaviour 65: 53–58.

Bradbury, J. W., and Vehrencamp, S. L. 1998. *Principles of Animal Communication*. Sunderland, MA: Sinauer Associates.

Bradford, D. F., Tabatabai, F., and Graber, D. M. 1993. Isolation of remaining populations of the native frog, *Rana mucosa*, by introduced fishes in Sequoia and King Canyon National Parks, California. Conservation Biology 7: 882–888.

Bradner, J., and McRobert, S. P. 2001. Background colouration influences body colour segregation in mollies. Journal of Fish Biology 59: 673–681.

Brassard, P., Rau, M. E., and Curtis, M. A. 1982. Parasite-induced susceptibility to predation in diplostomiasis. Parasitology 85: 495–501.

Brassil, C. E. 2006. Can environmental variation generate positive indirect effects in a model of shared predation? American Naturalist 167: 43–54.

Breden, F., and Hornaday, K. 1994. Test of indirect models of selection in the Trinidad guppy. Heredity 73: 291–297.

Breden, F., and Stoner, G. 1987. Male predation risk determines female preference in the Trinidad guppy. Nature 329: 831–833.

Breden, F., Novinger, D., and Schubert, A. 1995. The effect of experience on mate choice in the Trinidad guppy, *Poecilia reticulata*. Environmental Biology of Fishes 42: 323–328.

Breden, F., Ptacek, M. B., Rashed, M., Taphorn, D., and Figueiredo, C. A. 1999. Molecular phylogeny of the live-bearing fish genus *Poecilia* (Cyprinodontiformes: Poeciliidae). Molecular Phylogenetics and Evolution 12: 95–104.

Breder, C. M., and Coates, C. W. 1935. Sex recognition in the guppy, *Lebistes reticulatus* Peters. Zoologica 19: 187–207.

Breider, H. 1935. Geschlechtsbestimmung und -differenzierung bei *Limia nigrofasciata, caudofasciata, vittata* und deren Artbastarden. Zeitschrift für Induktive Abstammungs- und Vererbungslehre, Berlin 68: 265–299.

Breider, H. 1952. Über Melanosarkome, Melaninbildung und homologe Zellmechanismen. Strahlentherapie 88: 619–639.

Brendonck, L., Michels, E., De Meester, L., and Riddoch, B. 2002. Temporary pools are not "enemy-free." Hydrobiologia 486: 147–159.

Bretman, A., Wedell, N., and Tregenza, T. 2004. Molecular evidence of post-copulatory inbreeding avoidance in the field cricket *Gryllus bimaculatus*. Proceedings of the Royal Society of London Series B—Biological Sciences 271: 159–164.

Brett, B. L. H., and Grosse, D. J. 1982. A reproductive pheromone in the Mexican poeciliid fish, *Poecilia chica*. Copeia 1982: 219–223.

Brett, J. R. 1964. The respiratory metabolism and swimming performance of young sockeye salmon. Journal of the Fisheries Research Board of Canada 21: 1183–1226.

Brewster, J., and Houde, A. 2003. Are female guppies more likely to flee when approached by two males? Journal of Fish Biology 63: 1056–1059.

Briggs, S. E., Godin, J. G. J., and Dugatkin, L. A. 1996. Mate-choice copying under predation risk in the Trinidadian guppy (*Poecilia reticulata*). Behavioral Ecology 7: 151–157.

Bright, M., and Giere, O. 2005. Microbial symbiosis in annelida. Symbiosis 38: 1–45.

Britton, R. H., and Moser, M. E. 1982. Size specific predation by herons and its effect on the sex-ratio of natural populations of the mosquito fish *Gambusia affinis* Baird and Girard. Oecologia 53: 146–151.

Brock, R. E., and Kam, A. K. H. 1997. *Biological and Water Quality Characteristics of Anchialine Resources in Kaloko-Honokohau National Historic Park*. Honolulu: Cooperative National Park Resources Studies Unit, University of Hawaii at Manoa, Department of Botany.

Brockelman, W. Y. 1975. Competition, the fitness of offspring, and optimal clutch size. American Naturalist 109: 677–699.

Brockmann, H. J. 2001. The evolution of alternative strategies and tactics. Advances in the Study of Behavior 30: 1–51.

Bronikowski, A. M., Clark, M. E., Rodd, F. H., and Reznick, D. N. 2002. Population-dynamic consequences of predator-induced life history variation in the guppy (*Poecilia reticulata*). Ecology 83: 2194–2204.

Brönmark, C., and Edenhamn, P. 1994. Does the presence of fish affect the distribution of treefrogs (*Hyla arborea*)? Conservation Biology 8: 841–845.

Brooks, J. L., and Dodson, S. I. 1965. Predation, body size, and composition of plankton. Science 150: 28–35.

Brooks, R. 1996. Copying and the repeatability of mate choice. Behavioral Ecology and Sociobiology 39: 323–329.

Brooks, R. 2000. Negative genetic correlation between male sexual attractiveness and survival. Nature 406: 67–70.

Brooks, R. 2002. Variation in female mate choice within guppy populations: population divergence, multiple ornaments and the maintenance of polymorphism. Genetica 116: 343–358.

Brooks, R., and Caithness, N. 1995. Female choice in a feral guppy population: are there multiple cues? Animal Behaviour 50: 301–307.

Brooks, R., and Couldridge, V. 1999. Multiple sexual ornaments coevolve with multiple mating preferences. American Naturalist 154: 37–45.

Brooks, R., and Endler, J. A. 2001a. Direct and indirect sexual selection and quantitative genetics of male traits in guppies (Poecilia reticulata). Evolution 55: 1002–1015.

Brooks, R., and Endler, J. A. 2001b. Female guppies agree to differ: phenotypic and genetic variation in mate-choice behavior and the consequences for sexual selection. Evolution 55: 1644–1655.

Brown, C. 2005. Cerebral lateralisation, "social constraints," and coordinated anti-predator responses. Behavioral and Brain Sciences 28: 591–592.

Brown, C., and Braithwaite, V. A. 2004. Size matters: a test of boldness in eight populations of the poeciliid Brachyrhaphis episcopi. Animal Behaviour 68: 1325–1329.

Brown, C., and Braithwaite, V. A. 2005. Effects of predation pressure on the cognitive ability of the poeciliid Brachyrhaphis episcopi. Behavioral Ecology 16: 482–487.

Brown, C., and Laland, K. 2001. Social learning and life skills training for hatchery reared fish. Journal of Fish Biology 59: 471–493.

Brown, C., and Laland, K. N. 2002. Social learning of a novel avoidance task in the guppy: conformity and social release. Animal Behaviour 64: 41–47.

Brown, C., and Laland, K. N. 2003. Social learning in fishes: a review. Fish and Fisheries 4: 280–288.

Brown, C., and Warburton, K. 1999. Differences in timidity and escape responses between predator-naive and predator-sympatric rainbowfish populations. Ethology 105: 491–502.

Brown, C., Davidson, T., and Laland, K. 2003. Environmental enrichment and prior experience of live prey improve foraging behaviour in hatchery-reared Atlantic salmon. Journal of Fish Biology 63: 187–196.

Brown, C., Gardner, C., and Braithwaite, V. A. 2004. Population variation in lateralized eye use in the poeciliid Brachyrhaphis episcopi. Proceedings of the Royal Society of London Series B—Biological Sciences 271: S455–S457.

Brown, C., Gardner, C., and Braithwaite, V. A. 2005a. Differential stress responses in fish from areas of high- and low-predation pressure. Journal of Comparative Physiology 175: 305–312.

Brown, C., Jones, F., and Braithwaite, V. A. 2005b. In situ examination of boldness-shyness traits in the tropical poeciliid, Brachyrhaphis episcopi. Animal Behaviour 70: 1003–1009.

Brown, C., Laland, K., and Krause, J. 2006a. Fish Cognition and Behaviour. Oxford: Blackwell.

Brown, C., Laland, K., and Krause, J. 2006b. Fish cognition and behaviour. In C. Brown, K. Laland and J. Krause (eds.), Fish Cognition and Behavior, 1–8. Oxford: Blackwell.

Brown, C., Jones, F., and Braithwaite, V. A. 2007a. Correlation between boldness and body mass in natural populations of the poeciliid Brachyrhaphis episcopi. Journal of Fish Biology 71: 1590–1601.

Brown, C., Western, J., and Braithwaite, V. A. 2007b. The influence of early experience on, and inheritance of, cerebral lateralization. Animal Behaviour 74: 231–238.

Brown, G. E. 2002. The effects of reduced pH on chemical alarm signalling in ostariophysan fishes. Canadian Journal of Fisheries and Aquatic Sciences 59: 1331–1338.

Brown, G. E. 2003. Learning about danger: chemical alarm cues and local risk assessment in prey fishes. Fish and Fisheries 4: 227–234.

Brown, G. E., and Chivers, D. P. 2006. Learning about danger: chemical alarm cues and the assessment of predation risk by fishes. In C. Brown, K. Laland, and J. Krause (eds.), Fish Cognition and Behavior, 49–69. Oxford: Blackwell.

Brown, G. E., and Cowan, J. 2000. Foraging trade-offs and predator inspection in an ostariophysan fish: switching from chemical to visual cues. Behaviour 137: 181–195.

Brown, G. E., and Godin, J.-G. J. 1999a. Who dares, learns: chemical inspection behaviour and acquired predator recognition in a characin fish. Animal Behaviour 57: 475–481.

Brown, G. E., and Godin, J. G. J. 1999b. Chemical alarm signals in wild Trinidadian guppies (Poecilia reticulata). Canadian Journal of Zoology—Revue Canadienne de Zoologie 77: 562–570.

Brown, G. E., Chivers, D. P., and Smith, R. J. F. 1995. Fathead minnows avoid conspecific and heterospecific alarm pheromones in the feces of northern pike. Journal of Fish Biology 47: 387–393.

Brown, G. E., Paige, J. A., and Godin, J. G. J. 2000. Chemically mediated predator inspection behaviour in the absence of predator visual cues by a characin fish. Animal Behaviour 60: 315–321.

Brown, G. E., Golub, J. L., and Plata, D. L. 2001. Attack cone avoidance during predator inspection visits by wild finescale dace (Phoxinus neogaeus): The effects of predator diet. Journal of Chemical Ecology 27: 1657–1666.

Brown, G. E., Poirier, J.-F., and Adrian, J. C. J. 2004. Assessment of local predation risk: the role of subthreshold concentrations of chemical alarm cues. Behavioral Ecology 15: 810–815.

Brown, G. E., MacNaughton, C. J., Elvidge, C. K., Ramnarine, I., and Godin, J. G. J. 2009. Provenance and threat-sensitive predator avoidance patterns in wild-caught Trinidadian guppies. Behavioral Ecology and Sociobiology 63: 699–706.

Brown, J. H., Jardetzky, T. S., Gorga, J. C., Stern, L. J., Urban, R. G., Strominger, J. L., and Wiley, D. C. 1993. 3-Dimensional structure of the human class-II histocompatibility antigen HLA-DR1. Nature 364: 33–39.

Bruce, K. E., and White, W. G. 1995. Agonistic relationships and sexual behaviour patterns in male guppies, Poecilia reticulata. Animal Behaviour 50: 1009–1021.

Bryer, P. J., Mirza, R. S., and Chivers, D. P. 2001. Chemosensory assessment of predation risk by slimy sculpins (Cottus cognatus): responses to alarm, disturbance, and predator cues. Journal of Chemical Ecology 27: 533–546.

Bshary, R., Wickler, W., and Fricke, H. 2002. Fish cognition: a primate's eye view. Animal Cognition 5: 1–13.

Buchmann, K., Lindenstrøm, T., and Bresciani, J. 2003. Interactive associations between fish hosts and monogeneans. In G. F. Wiegertjes (ed.), Parasite Host Interactions, 161–184. Oxford: Bios Scientific Publishers.

Buckingham, J. N., Wong, B. B. M., and Rosenthal, G. G. 2007.

Shoaling decisions in female swordtails: how do fish gauge group size? Behaviour 144: 1333–1346.

Budaev, S. V. 1997. "Personality" in the guppy (*Poecilia reticulata*): a correlational study of exploratory behavior and social tendency. Journal of Comparative Psychology 111: 399–411.

Bull, J. J. 1983. *Evolution of Sex Determining Mechanisms*. Menlo Park, CA: Benjamin/Cummings Publishing Co.

Bumann, D., Krause, J., and Rubenstein, D. 1997. Mortality risk of spatial positions in animal groups: the danger of being in the front. Behaviour 134: 1063–1076.

Burgess, P., McMahon, S. M., and Price, D. J. 2005. Conservation of Trinidad populations of *Poecilia reticulata* Peters, 1859 (Poeciliidae). In M. C. Uribe and H. J. Grier (eds.), *Viviparous Fishes*, 426–434. Homestead, FL: New Life Publications.

Burley, N. 1980. Clutch overlap and clutch size: alternative and complementary reproductive tactics. American Naturalist 115: 223–246.

Burley, N. T., and Symanski, R. 1998. "A taste for the beautiful": latent aesthetic mate preferences for white crests in two species of Australian grassfinches. American Naturalist 152: 792–802.

Burns, J. G., and Rodd, F. H. 2008. Hastiness, brain size and predation regime affect the performance of wild guppies in a spatial memory task. Animal Behaviour 76: 911–922.

Burridge, C. P., Craw, D., Jack, D. C., King, T. M., and Waters, J. M. 2008. Does fish ecology predict dispersal across a river drainage divide? Evolution 62: 1484–1499.

Bussing, W. A. 2008. A new species of poeciliid fish, *Poeciliopsis santaelena*, from Peninsula Santa Elena, Area de Conservación Guanacaste, Costa Rica. Revista de Biologia Tropical 56: 829–838.

Butler, A. P., Trono, D., Della Coletta, L., Beard, R., Fraijo, R., Kazianis, S., and Nairn, R. S. 2007. Regulation of *CDKN2A/B* and retinoblastoma genes in *Xiphophorus melanoma*. Comparative Biochemistry and Physiology C—Toxicology and Pharmacology 145: 145–155.

Butler, M. A., and Losos, J. B. 2002. Multivariate sexual dimorphism, sexual selection, and adaptation in Greater Antillean *Anolis* lizards. Ecological Monographs 72: 541–559.

Butlin, R. 2002. The costs and benefits of sex: new insights from old asexual lineages. Nature Reviews Genetics 3: 311–317.

Butlin, R., Schon, I., and Martens, K. 1998. Asexual reproduction in nonmarine ostracods. Heredity 81: 473–480.

Byers, J. A., Wiseman, P. A., Jones, L., and Roffe, T. J. 2005. A large cost of female mate sampling in pronghorn. American Naturalist 166: 661–668.

Byrne, R., and Whiten, A. 1988. *Machiavellian Intelligence*. Oxford: Clarendon Press.

Byrne, R. W., and Bates, L. A. 2007. Sociality, evolution and cognition. Current Biology 17: R714–R723.

Cable, J., and Harris, P. D. 2002. Gyrodactylid developmental biology: historical review, current status and future trends. International Journal for Parasitology 32: 255–280.

Cable, J., and van Oosterhout, C. 2007a. The impact of parasites on the life history evolution of guppies (*Poecilia reticulata*): the effects of host size on parasite virulence. International Journal for Parasitology 37: 1449–1458.

Cable, J., and van Oosterhout, C. 2007b. The role of innate and acquired resistance in two natural populations of guppies (*Poecilia reticulata*) infected with the ectoparasite *Gyrodac-*
tylus turnbulli. Biological Journal of the Linnean Society 90: 647–655.

Cable, J., Scott, E. C. G., Tinsley, R. C., and Harris, P. D. 2002. Behavior favoring transmission in the viviparous monogenean *Gyrodactylus turnbulli*. Journal of Parasitology 88: 183–184.

Cable, J., van Oosterhout, C., Barson, N., and Harris, P. D. 2005. *Gyrodactylus pictae* n. sp. (Monogenea: Gyrodactylidae) from the Trinidadian swamp guppy *Poecilia picta* Regan, with a discussion on species of *Gyrodactylus* von Nordmann, 1832 and their poeciliid hosts. Systematic Parasitology 60: 159–164.

Calsbeek, R., and Sinervo, B. 2004. Within-clutch variation in offspring sex determined by differences in sire body size: cryptic mate choice in the wild. Journal of Evolutionary Biology 17: 464–470.

Camassa, M. M. 2001. Responses to light in epigean and hypogean populations of *Gambusia affinis* (Cyprinodontiformes: Poeciliidae). Environmental Biology of Fishes 62: 115–118.

Candolin, U. 2003. The use of multiple cues in mate choice. Biological Reviews 78: 575–595.

Candolin, U., and Reynolds, J. D. 2002. Adjustments of ejaculation rates in response to risk of sperm competition in a fish, the bitterling (*Rhodeus sericeus*). Proceedings of the Royal Society of London Series B—Biological Sciences 269: 1549–1553.

Cantalupo, C., Bisazza, A., and Vallortigara, G. 1995. Lateralization of predator-evasion response in a teleost fish (*Girardinus falcatus*). Neuropsychologia 33: 1637–1646.

Carius, H. J., Little, T. J., and Ebert, D. 2001. Genetic variation in a host-parasite association: potential for coevolution and frequency-dependent selection. Evolution 55: 1136–1145.

Carleton, K. L., and Kocher, T. D. 2001. Cone opsin genes in African cichlid fishes: tuning spectral sensitivity by differential gene expression. Molecular Biology and Evolution 18: 1540–1550.

Carpenter, S. R., and Kitchell, J. F. 1988. *The Trophic Cascade in Lakes*. Cambridge: Cambridge University Press.

Carrico, R., Blumberg, W., and Peisach, J. 1978. The reversible binding of oxygen to sulfhemoglobin. Journal of Biological Chemistry 253: 7212–7215.

Carter, A. J., and Wilson, R. S. 2006. Improving sneaky-sex in a low oxygen environment: reproductive and physiological responses of male mosquito fish to chronic hypoxia. Journal of Experimental Biology 209: 4878–4884.

Carvalho, G. R., Shaw, P. W., Magurran, A. E., and Seghers, B. H. 1991. Marked genetic divergence revealed by allozymes among populations of the guppy *Poecilia reticulata* (Poeciliidae), in Trinidad. Biological Journal of the Linnaean Society 42: 389–405.

Carveth, C. J., Widmer, A. M., and Bonar, S. A. 2006. Comparison of upper thermal tolerances of native and nonnative fish species in Arizona. Transactions of the American Fisheries Society 135: 1433–1440.

Casatti, L., Carvalho, F. R., Veronezi, J. L., and Lacerda, D. R. 2006a. Reproductive biology of the neotropical superfetaceous *Pamphorichthys hollandi* (Cyprinodontiformes: Poeciliidae). Ichthyological Exploration of Freshwaters 17: 59–64.

Casatti, L., Langeani, F., and Ferreira, C. P. 2006b. Effects of physical habitat degradation on the stream fish assemblage structure in a pasture region. Environmental Management 38: 974–982.

Caswell, H. 2001. *Matrix Population Models: Construction, Analysis, and Interpretation*. Sunderland, MA: Sinauer Associates.

Caswell, H., and Cohen, J. E. 1995. Red, white and blue—environmental variance spectra and coexistence in metapopulations. Journal of Theoretical Biology 176: 301–316.

Cavalli-Sforza, L. L., and Feldman, M. W. 1981. *Cultural Transmission and Evolution*. Princeton, NJ: Princeton University Press.

Cerdà, J., Reidenbach, S., Prätzel, S., and Franke, W. E. 1999. Cadherin-catenin complexes during zebrafish oogenesis: heterotypic junctions between oocytes and follicle cells. Biology of Reproduction 61: 692–704.

Cereb, N., Hughes, A. L., and Yang, S. Y. 1997. Locus-specific conservation of the HLA class I introns by intra-locus homogenization. Immunogenetics 47: 30–36.

Chambers, J. 1987. The cyprinodontiform gonopodium, with an atlas of the gonopodia of the fishes of the genus *Limia*. Journal of Fish Biology 30: 389–418.

Chambers, J. 1990. The gonopodia of the fishes of the tribe Cnesterodontini (Cyprinodontiformes, Poeciliidae). Journal of Fish Biology 36: 903–916.

Chambolle, P. 1973. Recherches sur les facteurs physiologiques de la reproduction chez les poissons "ovovivipares": analyse expérimentale sur *Gambusia* sp. Bulletin Biologique de la France et de la Belgique 107: 27–101.

Chandler, M., Chapman, L. J., and Chapman, C. A. 1995. Patchiness in the abundance of metacercariae parasitizing *Poecilia gillii* (Poeciliidae) isolated in pools of an intermittent tropical stream. Environmental Biology of Fishes 42: 313–321.

Chapman, B. B., Morrell, L. J., Benton, T. G., and Krause, J. 2008a. Early interactions with adults mediate the development of predator defenses in guppies. Behavioral Ecology 19: 87–93.

Chapman, B. B., Ward, A. J. W., and Krause, J. 2008b. Schooling and learning: early social environment predicts social learning ability in the guppy, *Poecilia reticulata*. Animal Behaviour 76: 923–929.

Chapman, F. A., FitzCoy, S. A., Thunberg, E. M., and Adams, C. M. 1997. United States of America trade in ornamental fish. Journal of the World Aquaculture Society 28: 1–10.

Chapman, L. J., and Chapman, C. A. 1993. Desiccation, flooding, and the behavior of *Poecilia gillii* (Pisces: Poeciliidae). Ichthyological Exploration of Freshwaters 4: 279–287.

Chapman, L. J., and Hulen, K. G. 2001. Implications of hypoxia for the brain size and gill morphometry of mormyrid fishes. Journal of Zoology 254: 461–472.

Chapman, L. J., and Kramer, D. L. 1991a. The consequences of flooding for the dispersal and fate of poeciliid fish in an intermittent tropical stream. Oecologia 87: 299–306.

Chapman, L. J., and Kramer, D. L. 1991b. Limnological observations of an intermittent tropical dry forest stream. Hydrobiologia 226: 153–166.

Chapman, L. J., Chapman, C. A., Brazeau, D. A., McLaughlin, B., and Jordan, M. 1999. Papyrus swamps, hypoxia, and faunal diversification: variation among populations of *Barbus neumayeri*. Journal of Fish Biology 54: 310–327.

Chapman, L. J., Galis, F., and Shinn, J. 2000. Phenotypic plasticity and the possible role of genetic assimilation: hypoxia-induced trade-offs in the morphological traits of an African cichlid. Ecology Letters 3: 387–393.

Charbonnel, N., and Pemberton, J. 2005. A long-term genetic survey of an ungulate population reveals balancing selection acting on MHC through spatial and temporal fluctuations in selection. Heredity 95: 377–388.

Charlesworth, B. 1980. *Evolution in Age Structured Populations*. Cambridge: Cambridge University Press.

Charlesworth, B. 1994. The effect of background selection against deleterious mutations on weakly selected, linked variants. Genetics Research 63: 213–227.

Charlesworth, B., and Charlesworth, D. 1998. Some evolutionary consequences of deleterious mutations. Genetica 103: 3–19.

Charlesworth, D. 2006. Balancing selection and its effects on sequences in nearby genome regions. PLoS Genetics 2: 379–384.

Charlesworth, D., and Charlesworth, B. 1987. Inbreeding depression and its evolutionary consequences. Annual Review of Ecology and Systematics 18: 237–268.

Chase, I. D., Bartolomeo, C., and Dugatkin, L. A. 1994. Aggressive interactions and inter-contest interval: how long do winners keep winning? Animal Behaviour 48: 393–400.

Chase, J. M., and Knight, T. M. 2003. Community genetics: toward a synthesis. Ecology 84: 580–582.

Chase, J. M., and Leibold, M. A. 2003. *Ecological Niches: Linking Classical and Contemporary Approaches*. Chicago: University of Chicago Press.

Chen, K., and Morris, J. 1972. Kinetics of oxidation of aqueous sulfide by O_2. Enviromental Science and Technology 6: 529–537.

Cheng, K. 1986. A purely geometric module in the rat's spatial representation. Cognition 23: 149–178.

Cheng, Y. Y. 2004. Sexual selection and the evolution of genitalia in the guppy (*Poecilia reticulata*). Master's thesis, University of Toronto.

Chenoweth, S. F., and Blows, M. W. 2006. Dissecting the complex genetic basis of mate choice. Nature Reviews Genetics 7: 681–692.

Cheong, R. T., Henrich, S., Farr, J. A., and Travis, J. 1984. Variation in fecundity and its relationship to body size in a population of the least killifish, *Heterandria formosa* (Pisces: Poeciliidae). Copeia 1984: 720–726.

Chervinski, J. 1984. Salinity tolerance of the guppy, *Poecilia reticulata* Peters. Journal of Fish Biology 24: 449–452.

Chesser, R. K., Smith, M. W., and Smith, M. H. 1984. Biochemical genetics of mosquitofish, III: incidence and significance of multiple insemination. Genetica 64: 77–81.

Chin, L. 2003. The genetics of malignant melanoma: lessons from mouse and man. Nature Reviews Cancer 3: 559–570.

Chinen, A., Hamaoka, T., Yamada, Y., and Kawamura, S. 2003. Gene duplication and spectral diversification of cone visual pigments in zebrafish. Genetics 163: 663–675.

Chippindale, A. K., Gibson, J. R., and Rice, W. R. 2001. Negative genetic correlation for adult fitness between sexes reveals ontogenetic conflict in *Drosophila*. Proceedings of the National Academy of Sciences of the United States of America 98: 1671–1675.

Chivers, D. P., and Smith, R. J. F. 1994. Fathead minnows, *Pimephales promelas*, acquire predator recognition when alarm substance is associated with the site of an unfamiliar fish. Animal Behaviour 48: 597–605.

Chivers, D. P., and Smith, R. J. F. 1995. Fathead minnows,

Pimephales promelas, learn to recognise chemical stimuli from high risk habitats by the presence of alarm substance. Behavioural Ecology 6: 155–158.

Chivers, D. P., and Smith, R. J. F. 1998. Chemical alarm signalling in aquatic predator-prey systems: a review and prospectus. Ecoscience 5: 338–352.

Chivers, D. P., Brown, G. E., and Smith, R. J. F. 1995. Familiarity and shoal cohesion in fathead minnows (*Pimephales promelas*): implications for antipredator behavior. Canadian Journal of Zoology—Revue Canadienne de Zoologie 73: 955–960.

Choudhury, A., and Dick, T. A. 2000. Richness and diversity of helminth communities in tropical freshwater fishes: empirical evidence. Journal of Biogeography 68: 935–956.

Chung, K. S. 2001. Critical thermal maxima and acclimation rate of the tropical guppy *Poecilia reticulata*. Hydrobiologia 462: 253–257.

Clark, A. B., and Ehlinger, T. J. 1987. Pattern and adaptation in individual behavioral differences. Animal Ecology 7: 1–47.

Clark, A. G., Begun, D. J., and Prout, T. 1999. Female × male interactions in *Drosophila* sperm competition. Science 283: 217–220.

Clark, E. 1950. A method for artificial insemination in viviparous fishes. Science 112: 722–723.

Clark, E., and Aronson, L. R. 1951. Sexual behaviour in the guppy, *Lebistes reticulatus* (Peters). Zoologica 36: 49–66.

Clark, E., Aronson, L. R., and Gordon, M. 1954. Mating behavior patterns in two sympatric species of xiphophorin fishes: their inheritance and significance in sexual isolation. Bulletin of the American Museum of Natural History 103: 135–226.

Clarke, B., and Kirby, D. R. S. 1966. Maintenance of histocompatibility complex polymorphisms. Nature 211: 999–1000.

Clayton, G. M., and Price, D. J. 1992. Interspecific and intraspecific variation in resistance to ichthyophthiriasis among poeciliid and goodeid fishes. Journal of Fish Biology 40: 445–453.

Clayton, G. M., and Price, D. J. 1994. Heterosis in resistance to *Ichthyophthirius multifiliis* infections in poeciliid fish. Journal of Fish Biology 44: 59–66.

Clayton, N. S., and Krebs, J. R. 1994. Memory for spatial and object-specific cues in food-storing and non-storing birds. Journal of Comparative Physiology A—Neuroethology, Sensory, Neural, and Behavioral Physiology 174: 371–379.

Clement, T. S., Parikh, V., Schrumpf, M., and Fernald, R. D. 2005. Behavioral coping strategies in a cichlid fish: the role of social status and acute stress response in direct and displaced aggression. Hormones and Behavior 47: 336–342.

Cline, J., and Richards, F. 1969. Oxygenation of hydrogen sulfide in seawater at constant salinity, temerature, and pH. Enviromental Science and Technology 3: 838–843.

Clutton-Brock, T. H. 1991. *The Evolution of Parental Care*. Princeton, NJ: Princeton University Press.

Clutton-Brock, T. H., and Harvey, P. H. 1980. Primates, brain and ecology. Journal of Zoology 190: 309–323.

Clutton-Brock, T. H., and Parker, G. A. 1995. Sexual coercion in animal societies. Animal Behaviour 49: 1345–1365.

Clutton-Brock, T. H., Albon, S. D., and Guinness, F. E. 1981. Parental investment in male and female offspring in polygynous mammals. Nature 289: 487–489.

Coleman, S. W. 2009. Taxonomic and sensory biases in the mate-choice literature: there are far too few studies of chemical and multimodal communication. Acta Ethologica 12: 45–48.

Coleman, S. W., and Rosenthal, G. G. 2006. Swordtail fry attend to chemical and visual cues in detecting predators and conspecifics. PloS ONE 1: e118.

Coleman, S. W., Harlin-Cognato, A., and Jones, A. G. 2009. Reproductive isolation, reproductive mode, and sexual selection: empirical tests of the viviparity-driven conflict hypothesis. American Naturalist 173: 291–303.

Collier, A. 1936. The mechanism of internal fertilization in *Gambusia*. Copeia 1936: 45–53.

Collyer, M. L., Novak, J. M., and Stockwell, C. A. 2005. Morphological divergence of native and recently established populations of White Sands pupfish (*Cyprinodon tularosa*). Copeia 2005: 1–11.

Colombo, L., Colombo Belvedere, P. C., Marconato, A., and Bentivegna, F. 1982. Pheromones in teleost fish. In C. J. J. Richter and H. J. T. Goos (eds.), *Proceedings of the International Symposium on Reproductive Physiology of Fish*, 84–94. Wageningen: Pudoc.

Combes, C. 2001. *Parasitism: The Ecology and Evolution of Intimate Interactions*. Chicago: University of Chicago Press.

Combes, C., and Nassi, H. 1997. Metacercarial dispersion and intracellular parasitism in a strigeid trematode. International Journal for Parasitology 7: 501–503.

Condon, C. H. L., and Wilson, R. S. 2006. Effect of thermal acclimation on female resistance to forced matings in the eastern mosquitofish. Animal Behaviour 72: 585–593.

Congdon, J. D., Dunham, A. E., Hopkins, W. A., Rowe, C. L., and Hinton, T. G. 2001. Resource allocation–based life histories: a conceptual basis for studies of ecological toxicology. Environmental Toxicology and Chemistry 20: 1698–1703.

Conover, D. O., and Schultz, E. T. 1995. Phenotypic similarity and the evolutionary significance of countergradient variation. Trends in Ecology and Evolution 10: 248–252.

Constantz, G. D. 1974. Reproductive effort in *Poeciliopsis occidentalis* (Poeciliidae). Southwestern Naturalist 19: 47–52.

Constantz, G. D. 1975. Behavioral ecology of mating in the male gila topminnow, *Poeciliopsis occidentalis* (Poeciliidae). Ecology 56: 966–973.

Constantz, G. D. 1979. Life history patterns of a livebearing fish in contrasting environments. Oecologia 40: 189–201.

Constantz, G. D. 1984. Sperm competition in poeciliid fishes. In R. L. Smith (ed.), *Sperm Competition and the Evolution of Animal Mating Systems*, 465–485. Orlando, FL: Academic Press.

Constantz, G. D. 1989. Reproductive biology of poeciliid fishes. In G. K. Meffe and F. F. Snelson Jr. (eds.), *Ecology and Evolution of Livebearing Fishes (Poeciliidae)*, 33–50. Englewood Cliffs, NJ: Prentice Hall.

Contreras, B. S., and Escalante, C. M. A. 1984. Distribution and known impacts of exotic fishes in Mexico. In W. R. Courtenay and J. R. Stauffer Jr. (eds.), *Distribution, Biology and Management of Exotic Fishes*, 102–130. Baltimore, MD: Johns Hopkins University Press.

Coolen, I., van Bergen, Y., Day, R. L., and Laland, K. N. 2003. Heterospecific use of public information by fish in a foraging context. Proceedings of the Royal Society of London Series B—Biological Sciences 270: 2413–2419.

Coolen, I., Ward, A. J. W., Hart, P. J. B., and Laland, K. N. 2005. Foraging nine-spined sticklebacks prefer to rely on public information over simpler social cues. Behavioral Ecology 16: 865–870.

Cooley, L. R., and Foighil, D. Ó. 2000. Phylogenetic analysis of the Sphaeriidae (Mollusca: Bivalvia) based on partial mitochondrial 16S rDNA gene sequences. Invertebrate Biology 119: 299–308.

Corander, J., Waldmann, P., Marttinen, P., and Sillanpaa, M. J. 2004. BAPS 2: enhanced possibilities for the analysis of genetic population structure. Bioinformatics 20: 2362–2369.

Cordero, C., and Eberhard, W. G. 2003. Female choice of sexually antagonistic male adaptations: a critical review of some current research. Journal of Evolutionary Biology 16: 1–6.

Costa, G. C., and Schlupp, I. 2010. Biogeography of the Amazon molly: ecological niche and range limits of an asexual hybrid species. Global Ecology and Biogeography 19: 442–451.

Courtenay, W. R., Jr., and Deacon, J. E. 1983. Fish introductions in the American southwest: a case study of Rogers Spring, Nevada. Southwestern Naturalist 28: 221–224.

Courtenay, W. R., Jr., and Meffe, G. K. 1989. Small fishes in strange places: a review of introduced poeciliids. In G. K. Meffe and F. F. Snelson Jr. (eds.), *Ecology and Evolution of Livebearing Fishes (Poeciliidae)*, 319–331. Englewood Cliffs, NJ: Prentice Hall.

CoussiKorbel, S., and Fragaszy, D. M. 1995. On the relation between social dynamics and social learning. Animal Behaviour 50: 1441–1453.

Couzin, I. D., and Krause, J. 2003. Selforganisation and collective behaviour of vertebrates. Advances in the Study of Behaviour 32: 1–67.

Couzin, I. D., Krause, J., James, R., Ruxton, G. D., and Franks, N. R. 2002. Collective memory and spatial sorting in animal groups. Journal of Theoretical Biology 218: 1–11.

Covich, A. 1981. Chemical refugia from predation for thin-shelled gastropods in a sulfide-enriched stream. Verhandlungen der Internationalen Vereinigung für Limnologie 21: 1632–1636.

Coyne, J. A., and Orr, H. A. 2004. *Speciation*. Sunderland, MA: Sinauer Associates.

Crandall, K. A., Bininda-Emonds, O. R. P., Mace, G. M., and Wayne, R. K. 2000. Considering evolutionary processes in conservation biology. Trends in Ecology and Evolution 15: 290–295.

Crapon de Caprona, M. D., and Ryan, M. J. 1990. Conspecific mate recognition in swordtails, *Xiphophorus nigrensis* and *X. pygmaeus* (Poeciliidae): olfactory and visual cues. Animal Behaviour 39: 290–296.

Creson, T. K., Woodruff, M. L., Ferslew, K. E., Rasch, E. M., and Monaco, P. J. 2003. Dose-response effects of chronic lithium regimens on spatial memory in the black molly fish. Pharmacology, Biochemistry, and Behavior 75: 35–47.

Crespi, B., and Semeniuk, C. 2004. Parent-offspring conflict in the evolution of vertebrate reproductive mode. American Naturalist 163: 635–653.

Cresswell, W. 1994. Flocking is an effective anti-predation strategy in redshanks, *Tringa tetanus*. Animal Behaviour 47: 433–442.

Crispo, E., Bentzen, P., Reznick, D. N., Kinnison, M. T., and Hendry, A. P. 2006. The relative influence of natural selection and geography on gene flow in guppies. Molecular Ecology 15: 49–62.

Crivelli, A. J. 1995. Are fish introductions a threat to endemic freshwater fishes in the northern Mediterranean region? Biological Conservation 72: 311–319.

Croft, D. P., Arrowsmith, B. J., Bielby, J., Skinner, K., White, E., Couzin, I. D., Magurran, A. E., Ramnarine, I., and Krause, J. 2003. Mechanisms underlying shoal composition in the Trinidadian guppy, *Poecilia reticulata*. Oikos 100: 429–438.

Croft, D. P., Arrowsmith, B. J., Webster, M., and Krause, J. 2004a. Intra-sexual preferences for familiar fish in male guppies. Journal of Fish Biology 64: 279–283.

Croft, D. P., Krause, J., and James, R. 2004b. Social networks in the guppy (*Poecilia reticulata*). Proceedings of the Royal Society of London Series B—Biological Sciences 271: S516–S519.

Croft, D. P., James, R., Ward, A. J. W., Botham, M. S., Mawdsley, D., and Krause, J. 2005. Assortative interactions and social networks in fish. Oecologia 143: 211–219.

Croft, D. P., James, R., Thomas, P. O. R., Hathaway, C., Mawdsley, D., Laland, K. N., and Krause, J. 2006a. Social structure and co-operative interactions in a wild population of guppies (*Poecilia reticulata*). Behavioral Ecology and Sociobiology 59: 644–650.

Croft, D. P., Morrell, L. J., Wade, A. S., Piyapong, C., Ioannou, C. C., Dyer, J. R. G., Chapman, B. B., Wong, Y., and Krause, J. 2006b. Predation risk as a driving force for sexual segregation: a cross-population comparison. American Naturalist 167: 867–878.

Croft, D. P., James, R., and Krause, J. 2008. *Exploring Animal Social Networks*. Princeton, NJ: Princeton University Press.

Crow, J. F., and Kimura, M. 1970. *An Introduction to Population Genetics Theory*. New York: Harper & Row.

Crow, R. T., and Liley, N. R. 1979. A sexual pheromone in the guppy, *Poecilia reticulata* (Peters). Canadian Journal of Zoology 57: 184–188.

Crozier, W. J., and Wolf, E. 1939. The flicker response contours for genetically related fishes, II. Journal of General Physiology 22: 463–484.

Csanyi, V. 1986. Ethological analysis of predator avoidance by the paradise fish (*Macropodus opercularis* L.), II: key stimuli in avoidance learning. Animal Learning and Behavior 14: 101–109.

Culumber, Z. W., Fisher, H. S., Tobler, M., Mateos, M., Sorenson, M. D., Barber, P. H., and Rosenthal, G. G. 2011. Replicated hybrid zones of *Xiphophorus* swordtails along an elevational gradient. *Molecular Ecology* 20: 342–356.

Cummings, M. E. 2007. Sensory trade-offs predict signal divergence in surfperch. Evolution 61: 530–545.

Cummings, M. E., and Partridge, J. C. 2001. Visual pigments and optical habitats of surfperch (Embiotocidae) in the California kelp forest. Journal of Comparative Physiology A—Neuroethology, Sensory, Neural, and Behavioral Physiology 187: 875–889.

Cunningham, E., and Janson, C. 2007. A socioecological perspective on primate cognition, past and present. Animal Cognition 10: 273–281.

Curtis, C., Bartholomew, T., Rose, F., and Dodgson, K. 1972. Detoxication of sodium ^{35}S-sulphide in the rat. Biochemical Pharmacology 21: 2313–2321.

Curtsinger, J. W. 1991. Sperm competition and the evolution of multiple mating. American Naturalist 138: 93–102.

Dadda, M., and Bisazza, A. 2006a. Does brain asymmetry allow efficient performance of simultaneous tasks? Animal Behaviour 72: 523–529.

Dadda, M., and Bisazza, A. 2006b. Lateralized female topminnows can forage and attend to a harassing male simultaneously. Behavioral Ecology 17: 358–363.

Dadda, M., Pilastro, A., and Bisazza, A. 2005. Male sexual harassment and female schooling behaviour in the eastern mosquitofish. Animal Behaviour 70: 473–471.

Dadda, M., Pilastro, A., and Bisazza, A. 2008. Innate responses to male sexual harassment in female mosquitofish. Behavioral Ecology and Sociobiology 63: 53–62.

Dadda, M., Zandonà, E., Agrillo, C., and Bisazza, A. 2009. The costs of hemispheric specialization in a fish. Proceedings of the Royal Society of London Series B—Biological Sciences 276: 4399–4407.

Dahlgren, B. T. 1979. The effects of population density on fecundity and fertility in the guppy, *Poecilia reticulata* (Peters). Journal of Fish Biology 15: 71–91.

Dale, S., and Slagsvold, T. 1996. Mate choice in female pied flycatchers. Behaviour 133: 903–944.

Dall, S. R. X., Giraldeau, L. A., Olsson, O., McNamara, J. M., and Stephens, D. W. 2005. Information and its use by animals in evolutionary ecology. Trends in Ecology and Evolution 20: 187–193.

Danchin, E., Giraldeau, L. A., Valone, T. J., and Wagner, R. H. 2004a. Public information: from nosy neighbors to cultural evolution. Science 305: 487–491.

Danchin, E., Vitiello, V., Vienne, A., Richard, O., Gouret, P., McDermott, M. F., and Pontarotti, P. 2004b. The major histocompatibility complex origin. Immunological Reviews 198: 216–232.

Danielsson, I. 2001. Antagonistic pre- and post-copulatory sexual selection on male body size in a water strider (*Gerris lacustris*). Proceedings of the Royal Society of London Series B—Biological Sciences 268: 77–81.

Darden, S. K., and Croft, D. P. 2008. Male harassment drives females to alter habitat use and leads to segregation of the sexes. Biology Letters 4: 449–451.

Darwin, C. 1859. *On the Origin of Species by Means of Natural Selection.* London: John Murray.

Darwin, C. 1871. *The Descent of Man, and Selection in Relation to Sex.* London: John Murray.

Dausset, J. 1958. Iso-leuko-antibodies. Acta Haematologica 20: 156–166.

Davis, S. K., Echelle, A. A., and Van den Bussche, R. A. 2006. Lack of cytonuclear genetic introgression despite long-term hybridization and backcrossing between two poeciliid fishes (*Gambusia heterochir* and *G. affinis*). Copeia 2006: 351–359.

Dawkins, R. 1976. *The Selfish Gene.* Oxford: Oxford University Press.

Dawkins, R., and Krebs, J. R. 1979. Arms races between and within species. Proceedings of the Royal Society of London Series B—Biological Sciences 205: 489–511.

Dawley, R. M. 1989. An introduction to unisexual vertebrates. In R. M. Dawley and J. P. Bogart (eds.), *Evolution and Ecology of Unisexual Vertebrates*, 1–19. Albany: New York State Museum.

Dawley, R. M., Rupprecht, J. D., and Schultz, R. J. 1997. Genome size of bisexual and unisexual *Poeciliopsis*. Journal of Heredity 88: 249–252.

Day, R. L., MacDonald, T., Brown, C., Laland, K. N., and Reader, S. M. 2001. Interactions between shoal size and conformity in guppy social foraging. Animal Behaviour 62: 917–925.

Day, T., and Bonduriansky, R. 2004. Intralocus sexual conflict can drive the evolution of genomic imprinting. Genetics 167: 1537–1546.

Day, T., Abrams, P. A., and Chase, J. M. 2002. The role of size-specific predation in the evolution and diversification of prey life histories. Evolution 56: 877–887.

Deacon, J. E., Hubbs, C., and Zahuranec, B. J. 1964. Some effects of introduced fishes on the native fish fauna of southern Nevada. Copeia 1964: 384–388.

DeAngelis, D. L., Trexler, J. C., and Loftus, W. F. 2005. Life history trade-offs and community dynamics of small fishes in a seasonally pulsed wetland. Canadian Journal of Fisheries and Aquatic Sciences 62: 781–790.

Deaton, R. 2008. Factors influencing male mating behaviour in *Gambusia affinis* (Baird and Girard) with a coercive mating system. Journal of Fish Biology 72: 1607–1622.

de Bakker, P. I. W., McVean, G., Sabeti, P. C., Miretti, M. M., Green, T., Marchini, J., Ke, X. Y., Monsuur, A. J., Whittaker, P., Delgado, M., Morrison, J., Richardson, A., Walsh, E. C., Gao, X. J., Galver, L., Hart, J., Hafler, D. A., Pericak-Vance, M., Todd, J. A., Daly, M. J., Trowsdale, J., Wijmenga, C., Vyse, T. J., Beck, S., Murray, S. S., Carrington, M., Gregory, S., Deloukas, P., and Rioux, J. D. 2006. A high-resolution HLA and SNP haplotype map for disease association studies in the extended human MHC. Nature Genetics 38: 1166–1172.

DeBlois, E. M., and Rose, G. A. 1996. Cross-shoal variability in the feeding habits of migrating Atlantic cod (*Gadus morhua*). Oecologia 108: 192–196.

De Boer, R. J., Borghans, J. A. M., van Boven, M., Kesmir, C., and Weissing, F. J. 2004. Heterozygote advantage fails to explain the high degree of polymorphism of the MHC. Immunogenetics 55: 725–731.

Deckel, A. W. 1995. Laterality of aggressive responses in *Anolis*. Journal of Experimental Zoology Part A: Ecological Genetics and Physiology 272: 194–200.

de Eyto, E., McGinnity, P., Consuegra, S., Coughlan, J., Tufto, J., Farrell, K., Megens, H. J., Jordan, W., Cross, T., and Stet, R. J. M. 2007. Natural selection acts on Atlantic salmon major histocompatibility (MH) variability in the wild. Proceedings of the Royal Society of London Series B—Biological Sciences 274: 861–869.

DeFelice, D., and Rasch, E. M. 1969. Chronology of spermatogenesis and spermiogenesis in poeciliid fishes. Journal of Experimental Zoology 171: 191–208.

DeMarais, A., and Oldis, D. 2005. Matrotrophic transfer of fluorescent microspheres in poeciliid fishes. Copeia 2005: 632–636.

DeMont, D. J., and Corkum, K. C. 1982. The life-cycle of *Octospiniferoides chandleri* Bullock, 1957 (Acanthocephala: Neochinorhynchidae) with some observations on parasite-induced photophilic behaviour in ostracods. Journal of Parasitology 68: 125–130.

Denk, A. G., Holzmann, A., Peters, A., Vermeirssen, E. L. M., and Kempenaers, B. 2005. Paternity in mallards: effects of sperm quality and female sperm selection for inbreeding avoidance. Behavioral Ecology 16: 825–833.

Dépêche, J. 1973. Infrastructure superficielle de la vésicule vitelline et du sac péricardique de l'embryon de *Poecilia reticulata* (Poison Téléostéen): application à l'étude du rôle des "cellules a chlorure" dans l'osmorégulation embryonnaire. Zeitschrift für Zellforschung 141: 235–253.

Dépêche, J. 1976. Acquisition et limites de l'autonomie tro-
phique embryonnaire au cours du développement du poisson
téléostéen vivipare *Poecilia reticulata*. Bulletin Biologique de
la France et de la Belgique 110: 45–97.

De Santi, A., Sovrano, V. A., Bisazza, A., and Vallortigara, G.
2001. Mosquitofish display differential left- and right-eye
use during mirror image scrutiny and predator inspection
responses. Animal Behaviour 61: 305–310.

Devlin, H., and Nagahama, Y. 2002. Sex determination and
sex differentiation in fish: an overview of genetic, physiological, and environmental influences. Aquaculture 208:
191–364.

DeWoody, Y. D., and DeWoody, J. A. 2005. On the estimation
of genome-wide heterozygosity using molecular markers.
Journal of Heredity 96: 85–88.

Dhillon, A. S., Hagan, S., Rath, O., and Kolch, W. 2007.
MAP kinase signalling pathways in cancer. Oncogene 26:
3279–3290.

Dieckmann, U., Doebeli, M., Metz, J. A. J., and Tautz, D. 2004.
Adaptive Speciation. Cambridge: Cambridge University Press.

DiIorio, P., Holsinger, K., Schlutz, R. J., and Hightower, L. 1996.
Quantitative evidence that both *Hsc70* and *Hsp70* contrib-
ute to thermal adaptation in hybrids of the livebearing fishes
Poeciliopsis. Cell Stress and Chaperones 1: 139–147.

Dildine, G. C. 1936. Studies on teleostean reproduction, I:
embryonic hermaphroditism in *Lebistes reticulatus*. Journal of
Morphology 60: 261–277.

Dill, L. M., Hedrick, A. V., and Fraser, A. 1999. Male mating
strategies under predation risk: do females call the shots?
Behavioral Ecology 10: 452–461.

Dimitrijevic, N., Winkler, C., Wellbrock, C., Gomez, A., Duschl, J.,
Altschmied, J., and Schartl, M. 1998. Activation of the *Xmrk*
proto-oncogene of *Xiphophorus* by overexpression and mutational alterations. Oncogene 16: 1681–1690.

Dingemanse, N. J., and Reale, D. 2005. Natural selection and
animal personality. Behaviour 142: 1159–1184.

Dingemanse, N. J., Wright, J., Kazem, A. J. N., Thomas, D. K.,
Hickling, R., and Dawnay, N. 2007. Behavioural syndromes
differ predictably between 12 populations of three-spined
stickleback. Journal of Animal Ecology 76: 1128–1138.

Dionne, M., Miller, K. M., Dodson, J. J., Caron, F., and Bernatchez, L. 2007. Clinal variation in MHC diversity with
temperature: evidence for the role of host-pathogen interaction on local adaptation in Atlantic salmon. Evolution 61:
2154–2164.

Dixon, B. 2008. Rene Stet's impact on the study of teleost major
histocompatibility genes: evolution from loci to populations.
Immunogenetics 60: 77–82.

Dobberfuhl, A. P., Ullmann, J. F. P., and Shumway, C. A. 2005.
Visual acuity, environmental complexity, and social organization in African cichlid fishes. Behavioral Neuroscience 119:
1648–1655.

Dobler, S., Mardulyn, P., Pasteels, J. M., and Rowell-Rahier, M.
1996. Host-plant switches and the evolution of chemical
defense and life history in the leaf beetle genus *Oreina*. Evolution 50: 2373–2386.

Dobzhansky, T. 1937. *Genetics and the Origin of Species*. New
York: Columbia University Press.

Doherty, P. C., and Zinkernagel, R. M. 1975. Enhanced immunological surveillance in mice heterozygous at the *H-2* gene
complex. Nature 256: 50–52.

Domenici, P. 2010. Escape responses in fish: kinematics, perfor-
mance, and behavior. In P. Domenici and B. G. Kapoor (eds.),
Fish Locomotion: An Etho-ecological Perspective, 123–170.
Enfield, NH: Science Publishers.

Dominey, W. J. 1981. Anti-predator functions of bluegill sunfish
nesting colonies. Nature 290: 586–587.

Dosen, L. D., and Montgomerie, R. 2004a. Female size influences mate preferences of male guppies. Ethology 110:
245–255.

Dosen, L. D., and Montgomerie, R. 2004b. Mate preferences
by male guppies (*Poecilia reticulata*) in relation to the risk of
sperm competition. Behavioral Ecology and Sociobiology 55:
266–271.

Douglas, M. E., and Vrijenhoek, R. C. 1983. Endangered
ichthyofauna of the southwest United States—an example
using the Gila topminnow, *Poeciliopsis occidentalis*. American
Zoologist 23: 966.

Doutrelant, C., McGregor, P. K., and Oliveira, R. F. 2001. The
effect of an audience on intrasexual communication in male
Siamese fighting fish, *Betta splendens*. Behavioral Ecology 12:
283–286.

Dove, A. D. M. 1998. A silent tragedy: parasites and the exotic
fishes of Australia. Proceedings of the Royal Society of
Queensland 107: 109–113.

Dove, A. D. M. 2000. Richness patterns in the parasite
communities of exotic poeciliid fishes. Parasitology 120:
609–623.

Dowling, D. K., and Simmons, L. W. 2009. Reactive oxygen
species as universal constraints in life-history evolution. Proceedings of the Royal Society of London Series B—Biological
Sciences 276: 1737–1745.

Downhower, J. F., Brown, L. P., and Matsui, M. L. 2000. Life
history variation in female *Gambusia hubbsi*. Environmental
Biology of Fishes 59: 415–428.

Downhower, J. F., Brown, L. P., and Matsui, M. L. 2002. Litter
overlap in *Gambusia hubbsi*: superfetation revisited. Environmental Biology of Fishes 65: 423–430.

Drent, R., and Daan, S. 1980. The prudent parent: energetic
adjustments in avian breeding. Ardea 68: 225–252.

Dreyer, C., Hoffmann, M., Lanz, C., Willing, E. M., Riester, M.,
Warthmann, N., Sprecher, A., Tripathi, N., Henz, S. R., and
Weigel, D. 2007. ESTs and EST-linked polymorphisms for ge-
netic mapping and phylogenetic reconstruction in the guppy,
Poecilia reticulata. BMC Genomics 8: 269.

Dries, L. A. 2003. Peering through the looking glass at a sexual
parasite: are Amazon mollies red queens? Evolution 57:
1387–1396.

Duckworth, R. A. 2009. The role of behavior in evolution: a
search for mechanism. Evolutionary Ecology 23: 513–531.

Dugatkin, L. A. 1992a. Sexual selection and imitation: females
copy the mate choice of others. American Naturalist 139:
1384–1389.

Dugatkin, L. A. 1992b. Tendency to inspect predators predicts
mortality risk in the guppy (*Poecilia reticulata*). Behavioral
Ecology 3: 124–127.

Dugatkin, L. A. 1996. Interface between culturally based
preferences and genetic preferences: female mate choice in
Poecilia reticulata. Proceedings of the National Academy of
Sciences of the United States of America 93: 2770–2773.

Dugatkin, L. A. 1997. *Cooperation among Animals: An Evolutionary Perspective*. Oxford: Oxford University Press.

Dugatkin, L. A. 2009. *Principles of Animal Behavior*. 2d ed. New York: Norton.

Dugatkin, L. A., and Alfieri, M. S. 2003. Boldness, behavioral inhibition, and learning. Ethology, Ecology, and Evolution 15: 43–49.

Dugatkin, L. A., and Druen, M. 2004. The social implications of winner and loser effects. Proceedings of the Royal Society of London Series B—Biological Sciences 271: S488–S489.

Dugatkin, L. A., and Dugatkin, A. D. 2007. Extrinsic effects, estimating opponents' RHP, and the structure of dominance hierarchies. Biology Letters 3: 614–616.

Dugatkin, L. A., and Godin, J. G. J. 1992a. Prey approaching predators: a cost-benefit perspective. Annales Zoologici Fennici 29: 233–252.

Dugatkin, L. A., and Godin, J. G. J. 1992b. Reversal of female mate choice by copying in the guppy (*Poecilia reticulata*). Proceedings of the Royal Society of London Series B—Biological Sciences 249: 179–184.

Dugatkin, L. A., and Godin, J. G. J. 1993. Female mate copying in the guppy (*Poecilia reticulata*): age-dependent effects. Behavioral Ecology 4: 289–292.

Dugatkin, L. A., and Godin, J. G. J. 1998. Effects of hunger on mate-choice copying in the guppy. Ethology 104: 194–202.

Dugatkin, L. A., and Wilson, D. S. 2000. Assortative interactions and the evolution of cooperation during predator inspection in guppies (*Poecilia reticulata*). Evolutionary Ecology Research 2: 761–767.

Dugatkin, L. A., Lucas, J. S., and Godin, J. G. J. 2002. Serial effects of mate-choice copying in the guppy (*Poecilia reticulata*). Ethology, Ecology, and Evolution 14: 45–52.

Dugatkin, L. A., Druen, M. W., and Godin, J. G. J. 2003. The disruption hypothesis does not explain mate-choice copying in the guppy (*Poecilia reticulata*). Ethology 109: 67–76.

Dukas, R. 2004. Causes and consequences of limited attention. Brain, Behavior and Evolution 63: 197–210.

Dulzetto, F. 1928. Osservazioni sulla vita sessuale di *Gambusia holbrooki*. Atti della Reale Accademia dei Lincei, Rendiconti 8: 96–101.

Duret, L. 2001. Why do genes have introns? Recombination might add a new piece to the puzzle. Trends in Genetics 17: 172–175.

Durgin, F. H. 1995. Texture density adaptation and the perceived numerosity and distribution of texture. Journal of Experimental Psychology: Human Perception and Performance 21: 149–169.

Dussault, G. V., and Kramer, D. L. 1981. Food and feeding behavior of the guppy, *Poecilia reticulata* (Pisces: Poeciliidae). Canadian Journal of Zoology 59: 684–701.

Dzieweczynski, T. L., Earley, R. L., Green, T. M., and Rowland, W. J. 2005. Audience effect is context dependent in Siamese fighting fish, *Betta splendens*. Behavioral Ecology 16: 1025–1030.

Dzikowski, R., Hulata, G., Harpaz, S., and Karplus, I. 2004. Inducible reproductive plasticity of the guppy *Poecilia reticulata* in response to predation cues. Journal of Experimental Zoology 301A: 776–782.

Dzwillo, M. 1962. Über künstliche Erzeugung funktioneller Männchen weiblichen Genotyps bei *Lebistes reticulatus*. Biologisches Zentralblatt 81: 575–584.

Eakley, A. L., and Houde, A. E. 2004. Possible role of female discrimination against "redundant" males in the evolution of colour pattern polymorphism in guppies. Proceedings of the Royal Society of London Series B—Biological Sciences 271: S299–S301.

Earley, R. L., and Dugatkin, L. A. 2002. Eavesdropping on visual cues in green swordtail (*Xiphophorus helleri*) fights: a case for networking. Proceedings of the Royal Society of London Series B—Biological Sciences 269: 943–952.

Earley, R. L., and Dugatkin, L. A. 2005. Fighting, mating and networking: pillars of poeciliid sociality. In P. K. McGregor (ed.), *Animal Communication Networks*, 84–113. Cambridge: Cambridge University Press.

Earley, R. L., Druen, M., and Dugatkin, L. A. 2005. Watching fights does not alter a bystander's response towards naive conspecifics in male green swordtail fish, *Xiphophorus helleri*. Animal Behaviour 69: 1139–1145.

Eberhard, W. G. 1985. *Sexual Selection and Animal Genitalia*. Cambridge, MA: Harvard University Press.

Eberhard, W. G. 1993. Evaluating models of sexual selection: genitalia as a test case. American Naturalist 142: 564–571.

Eberhard, W. G. 1996. *Female Control: Sexual Selection by Cryptic Female Choice*. Princeton, NJ: Princeton University Press.

Eberhard, W. G. 2009. Static allometry and animal genitalia. Evolution 63: 48–66.

Eberhard, W. G. 2010. Hypotheses to explain genitalic evolution: theory and evidence. In J. Leonard and A. Cordoba-Aguilar (eds.), *The Evolution of Primary Sexual Characters in Animals*. Oxford: Oxford University Press. Forthcoming.

Echelle, A. A. 1990. Nomenclature and non-Mendelian ("clonal") vertebrates. Systematic Zoology 39: 70–78.

Echelle, A. A., Wildrick, D. M., and Echelle, A. F. 1989. Allozyme studies of genetic variation in poeciliid fishes. In G. K. Meffe and F. F. Snelson Jr. (eds.), *Ecology and Evolution of Livebearing Fishes (Poeciliidae)*, 217–234. Englewood Cliffs, NJ: Prentice Hall.

Edwards, R. 1993. Entomological and mammalogical perspectives on genital differentiation. Trends in Ecology and Evolution 8: 406–409.

Edwards, S. V., and Hedrick, P. W. 1998. Evolution and ecology of MHC molecules: from genomics to sexual selection. Trends in Ecology and Evolution 13: 305–311.

Edwards, T. M., Miller, H. D., and Guillette, L. J. 2006. Water quality influences reproduction in female mosquitofish (*Gambusia holbrooki*) from eight Florida springs. Environmental Health Perspectives 114: 69–75.

Ehrlich, P. R. 1986. Which animal will invade? In H. A. Mooney and J. A. Drake (eds.), *Ecology of Biological Invasions of North American and Hawaii*, 79–95. New York: Springer.

Eigenmann, C. H. 1894. Notes on some South American fishes. Annals of the New York Academy of Sciences 7: 625–637.

Eigenmann, C. H. 1907. The poeciliid fishes of Rio Grande do Sul and the La Plata basin. Proceedings of the United States Natural History Museum 33: 425–433.

Eigenmann, C. H. 1909. Reports on the expedition to British Guiana of the Indiana University and the Carnegie Museum, 1908. Report no. 1: Some new genera and species of fishes from British Guiana. Annals of Carnegie Museum 6: 14–54.

Eldredge, L. G. 2000. Non-indigenous freshwater fishes, amphibians and crustaceans of the Pacific and Hawaiian Islands. In G. Sherley (ed.), *Invasive Species in the Pacific: A Technical*

Review and Draft Regional Strategy, 173–190. Samoa: South Pacific Regional Environment Program.

Ellegren, H., and Parsch, J. 2007. The evolution of sex-biased genes and sex-biased gene expression. Nature Reviews Genetics 8: 689–698.

Ellis, S. A., Bontrop, R. E., Antczak, D. F., Ballingall, K., Davies, C. J., Kaufman, J., Kennedy, L. J., Robinson, J., Smith, D. M., Stear, M. J., Stet, R. J. M., Waller, M. J., Walter, L., and Marsh, S. G. E. 2006. ISAG/IUIS-VIC Comparative MHC Nomenclature Committee report, 2005. Immunogenetics 57: 953–958.

Elsner, H. A., Rozas, J., and Blasczyk, R. 2002. The nature of introns 4–7 largely reflects the lineage specificity of *HLA-A* alleles. Immunogenetics 54: 447–462.

Endler, J. A. 1978. A predator's view of animal color patterns. Evolutionary Biology 11: 319–364.

Endler, J. A. 1980. Natural selection on color patterns in *Poecilia reticulata*. Evolution 34: 76–91.

Endler, J. A. 1982. Convergent and divergent effects of natural selection on color patterns in two fish faunas. Evolution 36: 178–188.

Endler, J. A. 1983. Natural and sexual selection on color patterns in poeciliid fishes. Environmental Biology of Fishes 9: 173–190.

Endler, J. A. 1987. Predation, light intensity and courtship behaviour in *Poecilia reticulata* (Pisces: Poeciliidae). Animal Behaviour 35: 1376–1385.

Endler, J. A. 1988. Sexual selection and predation risk in guppies. Nature 332: 593–594.

Endler, J. A. 1991. Variation in the appearance of guppy color patterns to guppies and their predators under different visual conditions. Vision Research 31: 587–608.

Endler, J. A. 1992. Signals, signal conditions, and the direction of evolution. American Naturalist 139: S125–S153.

Endler, J. A. 1993a. The color of light in forests and its implications. Ecological Monographs 63: 1–27.

Endler, J. A. 1993b. Some general comments on the evolution and design of animal communication systems. Philosophical Transactions of the Royal Society of London Series B—Biological Sciences 340: 215–225.

Endler, J. A. 1995. Multiple-trait coevolution and environmental gradients in guppies. Trends in Ecology and Evolution 10: 22–29.

Endler, J. A., and Basolo, A. L. 1998. Sensory ecology, receiver biases and sexual selection. Trends in Ecology and Evolution 13: 415–420.

Endler, J. A., and Houde, A. E. 1995. Geographic variation in female preferences for male traits in *Poecilia reticulata*. Evolution 49: 456–468.

Endler, J. A., Basolo, A. L., Glowacki, S., and Zerr, J. 2001. Variation in response to artificial selection for light sensitivity in guppies (*Poecilia reticulata*). American Naturalist 158: 36–48.

Engelstadter, J., and Haig, D. 2008. Sexual antagonism and the evolution of X chromosome inactivation. Evolution 62: 2097–2104.

Engeszer, R. E., Ryan, M. J., and Parichy, D. M. 2004. Learned social preference in zebrafish. Current Biology 14: 881–884.

Englund, R. A. 1999. The impacts of introduced poeciliid fish and Odonata on the endemic *Megalarion* (Odonata) damselflies of Oahu Island, Hawaii. Journal of Insect Conservation 3: 225–243.

Engqvist, L., and Reinhold, K. 2005. Pitfalls in experiments testing predictions from sperm competition theory. Journal of Evolutionary Biology 18: 116–123.

Enquist, M., and Leimar, O. 1983. Evolution of fighting behaviour: decision rules and assessment of relative strength. Journal of Theoretical Biology 102: 387–410.

Enquist, M., Leimar, O., Ljungberg, T., Mallner, Y., and Segerdahl, N. 1990. A test of the sequential assessment game: fighting in the cichlid fish *Nannacara anomala*. Animal Behaviour 40: 1–14.

Erbelding-Denk, C., Schröder, J. H., Schartl, M., Nanda, I., Schmid, M., and Epplen, J. T. 1994. Male polymorphism in *Limia perugiae* (Pisces: Poeciliidae). Behavior Genetics 24: 95–101.

Eschmeyer, W. N., and Fong, J. D. 2008. *Species of Fishes by Family/Subfamily*. http://research.calacademy.org/research/ichthyology/catalog/SpeciesByFamily.html (accessed 2009).

Essenberg, J. H. 1923. Sex differentiation in the viviparous teleost *Xiphophorus helleri* Heckel. Biological Bulletin 45: 46–97.

Evans, C. 1967. The toxicity of hydrogen sulphide and other sulphides. Quarterly Journal of Experimental Physiology 52: 231–248.

Evans, J. P. 2009. No evidence for sperm priming responses under varying sperm competition risk or intensity in guppies. Naturwissenschaften 96: 771–779.

Evans, J. P. 2010. Quantitative genetic evidence that males trade attractiveness for ejaculate quality in guppies. Proceedings of the Royal Society of London Series B–Biological Sciences 277: 3195–3201.

Evans, J. P. 2011. Patterns of genetic variation and covariation in ejaculate traits reveal potential evolutionary constraints in guppies. Heredity. Forthcoming.

Evans, J. P., and Kelley, J. L. 2008. Implications of multiple mating for offspring relatedness and shoaling behaviour in juvenile guppies. Biology Letters 4: 623–626.

Evans, J. P., and Magurran, A. E. 1999. Male mating behaviour and sperm production characteristics under varying sperm competition risk in guppies. Animal Behaviour 58: 1001–1006.

Evans, J. P., and Magurran, A. E. 2000. Multiple benefits of multiple mating in guppies. Proceedings of the National Academy of Sciences of the United States of America 97: 10074–10076.

Evans, J. P., and Magurran, A. E. 2001. Patterns of sperm precedence and predictors of paternity in the Trinidadian guppy. Proceedings of the Royal Society of London Series B—Biological Sciences 268: 719–724.

Evans, J. P., and Meisner, A. D. 2009. Copulatory structures: taxonomic overview and the potential for sexual selection. In B. G. M. Jamieson (ed.), *Reproductive Biology and Phylogeny of Fishes*, 138–180. Enfield, NH: Science Publishers.

Evans, J. P., and Rutstein, A. N. 2008. Postcopulatory sexual selection favours intrinsically good sperm competitors. Behavioral Ecology and Sociobiology 62: 1167–1173.

Evans, J. P., and Simmons, L. W. 2008. The genetic basis of traits regulating sperm competition and polyandry: can selection favour the evolution of good—and sexy—sperm? Genetica 134: 5–19.

Evans, J. P., Kelley, J. L., Ramnarine, I. W., and Pilastro, A. 2002a. Female behaviour mediates male courtship under predation risk in the guppy (*Poecilia reticulata*). Behavioral Ecology and Sociobiology 52: 496–502.

Evans, J. P., Pitcher, T. E., and Magurran, A. E. 2002b. The ontogeny of courtship, colour and sperm production in male guppies. Journal of Fish Biology 60: 495–498.

Evans, J. P., Pierotti, M., and Pilastro, A. 2003a. Male mating behavior and ejaculate expenditure under sperm competition risk in the eastern mosquitofish. Behavioral Ecology 14: 268–273.

Evans, J. P., Zane, L., Francescato, S., and Pilastro, A. 2003b. Directional postcopulatory sexual selection revealed by artificial insemination. Nature 421: 360–363.

Evans, J. P., Gasparini, C., and Pilastro, A. 2007. Female guppies shorten brood retention in response to predator cues. Behavioural Ecology and Sociobiology 61: 719–727.

Evans, J. P., Brooks, R. C., Zajitschek, S. R. K., and Griffith, S. C. 2008. Does genetic relatedness of mates influence competitive fertilization success in guppies? Evolution 62: 2929–2935.

Excoffier, L., Smouse, P. E., and Quattro, J. M. 1992. Analysis of molecular variance inferred from metric distances among DNA haplotypes: application to human mitochondrial restriction data. Genetics 131: 479–491.

Facchin, L., Bisazza, A., and Vallortigara, G. 1999. What causes lateralization of detour behavior in fish? Evidence for asymmetries in eye use. Behavioural Brain Research 103: 229–234.

Facchin, L., Argenton, F., and Bisazza, A. 2009. Lines of *Danio rerio* selected for opposite behavioural lateralization show differences in anatomical left-right asymmetries. Behavioural Brain Research 197: 157–165.

Fajen, A., and Breden, F. 1992. Mitochondrial DNA sequence variation among natural populations of the Trinidad guppy, *Poecilia reticulata*. Evolution 46: 1457–1465.

Falconer, D. S., and Mackay, T. F. C. 1996. *Introduction to Quantitative Genetics*. London: Longman Group.

Fares Alkahem, H., al-Ghanim, A. A., and Ahmad, Z. 2007. Studies on the feeding ecology of sailfin molly (*Poecilia latipinna*) dwelling in Wadi Haneefah stream, Riyadh. Pakistan Journal of Biological Science 10: 335–341.

Faria, P., van Oosterhout, C., and Cable, J. 2010. Optimal release strategies for captive-bred animals in reintroduction programs: the effects of prior parasite exposure and release protocol on host survival and infection rates. Biology Conservation 143: 35–41.

Farley, D. C., and Younce, L. C. 1977. Some effects of *Gambusia affinis* on selected non-target organisms in Fresno County rice fields. Proceedings and Papers of the Annual Conference of the California Mosquito Vector Control Association 45: 87–94.

Farr, J. A. 1975. The role of predation in the evolution of social behaviour of natural populations of the guppy, *Poecilia reticulata* (Pisces: Poeciliidae). Evolution 29: 151–158.

Farr, J. A. 1976. Social facilitation of male sexual behavior, intrasexual competition and sexual selection in the guppy, *Poecilia reticulata* (Pisces: Poeciliidae). Evolution 30: 707–717.

Farr, J. A. 1977. Male rarity or novelty, female choice behavior and sexual selection in the guppy *Poecilia reticulata* Peters (Pisces: Poeciliidae). Evolution 31: 162–168.

Farr, J. A. 1980a. The effects of sexual experience and female receptivity on courtship-rape decisions in male guppies, *Poecilia reticulata* (Pisces: Poeciliidae). Animal Behaviour 28: 1195–1201.

Farr, J. A. 1980b. Social behavior patterns as determinants of reproductive success in the guppy, *Poecilia reticulata* Peters (Pisces: Poeciliidae): an experimental study of the effects of intermal competition, female choice and sexual selection. Behaviour 74: 38–91.

Farr, J. A. 1981. Biased sex ratios in laboratory strains of guppies, *Poecilia reticulata*. Heredity 47: 237–248.

Farr, J. A. 1984. Premating behavior in the subgenus *Limia* (Pisces: Poeciliidae): sexual selection and the evolution of courtship. Zeitschrift für Tierpsychologie 65: 152–165.

Farr, J. A. 1989. Sexual selection and secondary sexual differentiation in poeciliids: determinants of male mating success and the evolution of female choice. In G. K. Meffe and F. F. Snelson Jr. (eds.), *Ecology and Evolution of Livebearing Fishes (Poeciliidae)*, 91–123. Englewood Cliffs, NJ: Prentice Hall.

Farr, J. A., and Travis, J. 1986. Fertility advertisement by female sailfin mollies, *Poecilia latipinna* (Pisces: Poeciliidae). Copeia 1986: 467–472.

Farr, J. A., Travis, J., and Trexler, J. C. 1986. Behavioral allometry and interdemic variation in sexual behavior of the sailfin molly, *Poecilia latipinna* (Pisces, Poeciliidae). Animal Behaviour 34: 497–509.

Fecher, L. A., Amaravadi, R. K., and Flaherty, K. T. 2008. The MAPK pathway in melanoma. Current Opinion in Oncology 20: 183–189.

Feder, J. L., Smith, M. L., Chesser, R. K., Godt, M. J. W., and Asbury, K. 1984. Biochemical genetics of mosquitofish, II: demographic differentiation of populations in a thermally altered reservoir. Copeia 1984: 108–119.

Feder, M. E., and Hofmann, G. E. 1999. Heat-shock proteins, molecular chaperones, and the stress response: evolutionary and ecological physiology. Annual Reviews of Physiology 61: 243–282.

Fedorka, K. M., and Mousseau, T. A. 2004. Female mating bias results in conflicting sex-specific offspring fitness. Nature 429: 65–67.

Feigenson, L., Carey, S., and Hauser, M. 2002a. The representations underlying infants' choice of more: object files versus analog magnitudes. Psychological Science 13: 150–156.

Feigenson, L., Carey, S., and Spelke, E. 2002b. Infants' discrimination of number vs. continuous extent. Cognitive Psychology 44: 33–66.

Feigenson, L., Dehaene, S., and Spelke, E. 2004. Core systems of number. Trends in Cognitive Sciences 8: 307–314.

Felsenstein, J. 1981. Skepticism towards Santa Rosalia, or why are there so few kinds of animals? Evolution 35: 124–138.

Felsenstein, J. 1985. Phylogenies and the comparative method. American Naturalist 125: 1–15.

Fernandez, A. A., and Morris, M. R. 2008. Mate choice for more melanin as a mechanism to maintain a functional oncogene. Proceedings of the National Academy of Sciences of the United States of America 105: 13503–13507.

Ferno, A., and Sjolander, S. 1973. Some imprinting experiments on sexual preferences for colour variants in the platyfish (*Xiphophorus maculatus*). Zeitschrift für Tierpsychologie 33: 417–423.

Ferrari, M. C. O., Rive, A. C., MacNaughton, C. J., Brown, G. E., and Chivers, D. P. 2008. Fixed vs. random temporal predictability of predation risk: an extension of the risk allocation hypothesis. Ethology 114: 238–244.

Ferrière, R., Dieckmann, U., and Couvet, D. 2004. *Evolutionary Conservation Biology*. Cambridge: Cambridge University Press.

Figueiredo, C. A. 2008. A new *Pamphorichthys* (Cyprinodontiformes: Poeciliidae: Poeciliini) from central Brazil. Zootaxa 2008: 59–68.

Figueroa, F., Mayer, W. E., Sato, A., Zaleska-Rutczynska, Z., Hess, B., Tichy, H., and Klein, J. 2001. MHC class I genes of swordtail fishes, *Xiphophorus*: variation in the number of loci and existence of ancient gene families. Immunogenetics 53: 695–708.

Fischer, K., Perlick, J., and Galetz, T. 2008. Residual reproductive value and male mating success: older males do better. Proceedings of the Royal Society of London Series B—Biological Sciences 275: 1517–1524.

Fish, F. E., and Lauder, G. V. 2006. Passive and active flow control by swimming fishes and mammals. Annual Review of Fluid Mechanics 38: 193–224.

Fisher, H. S., and Rosenthal, G. G. 2006a. Female swordtail fish use chemical cues to select well-fed mates. Animal Behaviour 72: 721–725.

Fisher, H. S., and Rosenthal, G. G. 2007. Male swordtails court with an audience in mind. Biology Letters 3: 5–7.

Fisher, H. S., and Rosenthal, G. G. 2006b. Hungry females show stronger mating preferences. Behavioral Ecology 17: 979–981.

Fisher, H. S., Wong, B. B. M., and Rosenthal, G. G. 2006. Alteration of the chemical environment disrupts communication in a freshwater fish. Proceedings of the Royal Society of London Series B—Biological Sciences 273: 1187–1193.

Fisher, H. S., Mascuch, S., and Rosenthal, G. G. 2009. Multivariate male traits misalign with multivariate female preferences in the swordtail fish, *Xiphophorus birchmanni*. Animal Behaviour 78: 265–269.

Fisher, J., and Hinde, R. A. 1949. The opening of milk bottles by birds. British Birds 42: 347–357.

Fisher, R. A. 1930. *The Genetical Theory of Natural Selection*. Oxford: Clarendon Press.

Fisher, R. A. 1958. *The Genetical Theory of Natural Selection*. 2d ed. New York: Dover.

Fishman, M. A. 1999. Predator inspection: closer approach as a way to improve assessment of potential threats. Journal of Theoretical Biology 196: 225–235.

Flajnik, M. F., Ohta, Y., Namikawa-Yamada, C., and Nonaka, M. 1999. Insight into the primordial MHC from studies in ectothermic vertebrates. Immunological Reviews 167: 59–67.

Foerster, K., Coulson, T., Sheldon, B. C., Pemberton, J. M., Clutton-Brock, T. H., and Kruuk, L. E. B. 2007. Sexually antagonistic genetic variation for fitness in red deer. Nature 447: 1107–1110.

Folgueras, A. R., Pendas, A. M., Sanchez, L. M., and Lopez-Otin, C. 2004. Matrix metalloproteinases in cancer: from new functions to improved inhibition strategies. International Journal of Developmental Biology 48: 411–424.

Font, W. F. 1997a. Distribution of helminth parasites of native and introduced stream fishes in Hawaii. Bishop Museum Occasional Papers 49: 56–62.

Font, W. F. 1997b. Improbable colonists: helminth parasites of freshwater fishes on an oceanic island. Micronesia 30: 105–115.

Font, W. F. 1998. Parasites in paradise: patterns of helminth distribution in Hawaiian stream fishes. Journal of Helminthology 72: 307–311.

Font, W. F. 2003. The global spread of parasites: what do Hawaiian streams tell us? BioScience 53: 1061–1067.

Font, W. F., and Tate, D. C. 1994. Helminth parasites of native Hawaiian freshwater fishes: an example of extreme ecological isolation. Journal of Parasitology 80: 682–688.

Fontaneto, D., Herniou, E. A., Boschetti, C., Caprioli, M., Melone, G., Ricci, C., and Barraclough, T. G. 2007. Independently evolving species in asexual bdelloid rotifers. PLoS Biology 5: 914–921.

Ford, M. J. 2002. Applications of selective neutrality tests to molecular ecology. Molecular Ecology 11: 1245–1262.

Förnzler, D., Wittbrodt, J., and Schartl, M. 1991. Analysis of an esterase linked to a locus involved in the regulation of the melanoma oncogene and isolation of polymorphic marker sequences in *Xiphophorus*. Biochemical Genetics 29: 509–524.

Förnzler, D., Altschmied, J., Nanda, I., Kolb, R., Baudler, M., Schmid, M., and Schartl, M. 1996. The *Xmrk* oncogene promoter is derived from a novel amplified locus of unusual organization. Genome Research 6: 102–113.

Fox, J. W. 2007. The dynamics of top-down and bottom-up effects in food webs of varying prey diversity, composition, and productivity. Oikos 116: 189–200.

Fragaszy, D. M., and Perry, S. 2003. *The Biology of Traditions*. Cambridge: Cambridge University Press.

Fraile, B., Saez, F. J., Vicentini, C. A., Demiguel, M. P., and Paniagua, R. 1992. The testicular cycle of *Gambusia affinis holbrooki* (Teleostei, Poeciliidae). Journal of Zoology 228: 115–126.

Fraile, B., Saez, F. J., Vicentini, C. A., Demiguel, M. P., and Paniagua, R. 1993. Effects of photoperiod on spermatogenesis in *Gambusia affinis holbrooki* (Teleostei: Poeciliidae) during the period of testicular quiescence. Journal of Zoology 230: 651–658.

Fraile, B., Saez, F. J., Vicentini, C. A., Gonzalez, A., Demiguel, M. P., and Paniagua, R. 1994. Effects of temperature and photoperiod on the *Gambusia affinis holbrooki* testis during spermatogenesis period. Copeia 1994: 216–221.

Franck, D., and Ribowski, A. 1987. Influence of prior agonistic experience on aggression measures in the male swordtail (*Xiphophorus helleri*). Behaviour 103: 217–240.

Franck, D., and Ribowski, A. 1989. Escalating fights for rank-order position between male swordtails (*Xiphophorus helleri*): effects of prior rank-order experience and information transfer. Behavioral Ecology and Sociobiology 24: 133–143.

Franck, D., Dikomey, M., and Schartl, M. 2001. Selection and the maintenance of a colour pattern polymorphism in the green swordtail (*Xiphophorus helleri*). Behaviour 138: 467–486.

Franck, D., Müller, A., and Rogmann, N. 2003. A colour and size dimorphism in the green swordtail (population Jalapa): female mate choice, male-male competition, and male mating strategies. Acta Ethologica 5: 75–79.

Frankham, R. 1995. Conservation genetics. Annual Review of Genetics 29: 305–327.

Franssen, C. M. 2008. The effect of heavy metal mine drainage on population size structure, reproduction, and condition of western mosquitofish, *Gambusia affinis*. Archives of Environmental Contamination and Toxicology 57: 145–156.

Franssen, C. M., Tobler, M., Riesch, R., García de León, F. J., Tiedemann, R., Schlupp, I., and Plath, M. 2008. Sperm production in an extremophile fish, the cave molly (*Poecilia mexicana*, Poeciliidae, Teleostei). Aquatic Ecology 42: 685–692.

Fraser, B. A., and Neff, B. D. 2010. Parasite mediated homogenizing selection at the MHC in guppies. Genetica 138: 273–278.

Fraser, B. A., Ramnarine, I. W., and Neff, B. D. 2010a. Selection at the MHC class IIB locus across guppy (*Poecilia reticulata*) populations. Heredity 104: 155–167.

Fraser, B. A., Ramnarine, I. W., and Neff, B. D. 2010b. Temporal variation at the MHC class IIB in wild populations of the guppy (*Poecilia reticulata*). Evolution 64: 2086–2096.

Fraser, D. F., Gilliam, J. F., Daley, M. J., Le, A. N., and Skalski, G. T. 2001. Explaining leptokurtic movement distributions: intrapopulation variation in boldness and exploration. American Naturalist 158: 124–135.

Fraser, D. F., Gilliam, J. F., Akkara, J. T., Albanese, B. W., and Snider, S. B. 2004. Night feeding by guppies under predator release: effects on growth and daytime courtship. Ecology 85: 312–319.

Fraser, E. A., and Renton, R. M. 1940. Observations on the breeding and development of the viviparous fish, *Heterandria formosa*. Quarterly Journal of Microscopical Science 81: 479–516.

Frazier, H. N., III, and Roth, M. B. 2009. Adaptive sugar provisioning controls survival of *C. elegans* embryos in adverse environments. Current Biology 19: 859–863.

Frick, J. E. 1998. Evidence of matrotrophy in the viviparous holothuroid *Synapta hydriformis*. Invertebrate Biology 117: 169–179.

Froeschke, G., and Sommer, S. 2005. MHC class II *DRB* variability and parasite load in the striped mouse (*Rhabdomys pumilio*) in the southern Kalahari. Molecular Biology and Evolution 22: 1254–1259.

Froese, R., and Pauly, D. 2008. *FishBase*. http://www.fishbase .org (accessed August 2008).

Froese, R., and Pauly, D. 2009. *FishBase*. http://www.fishbase .org (accessed 2009).

Froschauer, A., Korting, C., Katagiri, T., Aoki, T., Asakawa, S., Shimizu, N., Schartl, M., and Volff, J.-N. 2002. Construction and initial analysis of bacterial artificial chromosome (BAC) contigs from the sex-determining region of the platyfish *Xiphophorus maculatus*. Gene 295: 247–254.

Fuller, R. C., Fleishman, L. J., Leal, M., Travis, J., and Loew, E. 2003. Intraspecific variation in retinal cone distribution in the bluefin killifish, *Lucania goodei*. Journal of Comparative Physiology A—Neuroethology, Sensory, Neural, and Behavioral Physiology 189: 609–616.

Fuller, R. C., Carleton, K. L., Fadool, J. M., Spady, T. C., and Travis, J. 2004. Population variation in opsin expression in the bluefin killifish, *Lucania goodei*: a real-time PCR study. Journal of Comparative Physiology 190: 147–154.

Fuller, R. C., Carleton, K. L., Fadool, J. M., Spady, T. C., and Travis, J. 2005. Genetic and environmental variation in the visual properties of bluefin killifish, *Lucania goodei*. Journal of Evolutionary Biology 18: 516–523.

Gabor, C. R., and Aspbury, A. S. 2008. Non-repeatable mate choice by male sailfin mollies, *Poecilia latipinna*, in a unisexual-bisexual mating complex. Behavioral Ecology 19: 871–878.

Gabor, C. R., and Ryan, M. J. 2001. Geographical variation in reproductive character displacement in mate choice by male sailfin mollies. Proceedings of the Royal Society of London Series B—Biological Sciences 268: 1063–1070.

Gabor, C. R., Ryan, M. J., and Morizot, D. C. 2005. Character displacement in sailfin mollies, *Poecilia latipinna*: allozymes and behavior. Environmental Biology of Fishes 73: 75–88.

Gabor, C. R., Gonzalez, R., Parmley, M., and Aspbury, A. S. 2010. Variation in male sailfin molly preference for female size: does sympatry with sexual parasites drive preference for smaller conspecifics? Behavioral Ecology and Sociobiology 64: 783–792.

Gabriel, W., Lynch, M., and Bürger, R. 1993. Muller's ratchet and mutational meltdowns. Evolution 47: 1744–1757.

Gabriel, W., and Bürger, R. 2000. Fixation of clonal lineages under Muller's ratchet. Evolution 54: 1116–1125.

Gadgil, M., and Bossert, P. W. 1970. Life historical consequences of natural selection. American Naturalist 104: 1–24.

Galat, D. L., and Robertson, B. 1992. Response of endangered *Poeciliopsis occidentalis sonoriensis* in the Rio-Yaqui drainage, Arizona, to introduced *Gambusia affinis*. Environmental Biology of Fishes 33: 249–264.

Galef, B. G. 1998. Recent progress in studies of imitation and social learning in animals. Advances in Psychological Science 2: 275–299.

Galef, B. G., and Giraldeau, L. A. 2001. Social influences on foraging in vertebrates: causal mechanisms and adaptive functions. Animal Behaviour 61: 3–15.

Galef, B. G., and Laland, K. N. 2005. Social learning in animals: empirical studies and theoretical models. Bioscience 55: 489–499.

Galef, B. G., and Wigmore, S. W. 1983. Transfer of information concerning distant foods—a laboratory investigation of the information-centre hypothesis. Animal Behaviour 31: 748–756.

Gamble, S., Lindholm, A. K., Endler, J. A., and Brooks, R. 2003. Environmental variation and the maintenance of polymorphism: the effect of ambient light spectrum on mating behaviour and sexual selection in guppies. Ecology Letters 6: 463–472.

Gamradt, S. C., and Kats, L. B. 1996. Effect of introduced crayfish and mosquitofish on California newts. Conservation Biology 10: 1155–1162.

Gandolfi, G. 1969. A chemical sex attractant in the guppy *Poecilia reticulata* Peters (Pisces: Poeciliidae). Monitore Zoologico Italiano 3: 89–98.

Garcia, C., Rolan-Alvarez, E., and Sanchez, L. 1992. Alarm reaction and alert state in *Gambusia affinis* (Pisces, Poeciliidae) in response to chemical stimuli from injured conspecifics. Journal of Ethology 10: 41–46.

Garcia-Berthou, E. 1999. Food of introduced mosquitofish: ontogenetic diet shift and prey selection. Journal of Fish Biology 55: 135–147.

Garcia-Berthou, E., Alcaraz, C., Pou-Rovira, Q., Zamora, L., Coenders, G., and Feo, C. 2005. Introduction pathways and establishment rates of invasive aquatic species in Europe. Canadian Journal of Aquatic Sciences 62: 453–463.

Gardiner, D. M. 1978. Utilization of extracellular glucose by spermatozoa of two viviparous fishes. Comparative Biochemistry and Physiology 59A: 165–168.

Garrigan, D., and Hedrick, P. W. 2003. Perspective: detecting adaptive molecular polymorphism; lessons from the MHC. Evolution 57: 1707–1722.

Gasparini, C., Peretti, A., and Pilastro, A. 2009. Female presence influences sperm velocity in the guppy. Biology Letters 5: 792–794.

Gasparini, C., Marino, I. A. M., Boschetto, C., and Pilastro, A. 2010a. Effect of male age on sperm traits and sperm competition success in the guppy (Poecilia reticulata). Journal of Evolutionary Biology 23: 124–135.

Gasparini, C., Simmons, L. W., Beveridge, M., and Evans, J. P. 2010b. Sperm swimming velocity predicts competitive fertilization success in the green swordtail Xiphophorus helleri. Public Library of Science One 5: e12146.

Geiser, S. W. 1924. Sex-ratios and spermatogenesis in the topminnow, Gambusia holbrooki. Biological Bulletin 47: 175–207.

Geissinger, E., Weisser, C., Fischer, P., Schartl, M., and Wellbrock, C. 2002. Autocrine stimulation by osteopontin contributes to antiapoptotic signalling of melanocytes in dermal collagen. Cancer Research 62: 4820–4828.

Ghalambor, C. K., Reznick, D. N., and Walker, J. A. 2004. Constraints on adaptive evolution: the functional trade-off between reproduction and fast-start swimming performance in the Trinidadian guppy (Poecilia reticulata). American Naturalist 164: 38–50.

Ghalambor, C. K., McKay, J. K., Carroll, S. P., and Reznick, D. N. 2007. Adaptive versus non-adaptive phenotypic plasticity and the potential for contemporary adaptation in new environments. Functional Ecology 21: 394–407.

Ghedotti, M. J. 2000. Phylogenetic analysis and taxonomy of the poeciliid fishes (Teleostei: Cyprinodontiformes). Zoological Journal of the Linnean Society 130: 1–53.

Gheorghiu, C., Cable, J., Marcogliese, D. J., and Scott, M. E. 2007. Effects of waterborne zinc on reproduction, survival and morphometrics of Gyrodactylus turnbulli (Monogenea) on guppies (Poecilia reticulata). International Journal for Parasitology 37: 375–381.

Ghirlanda, S., and Vallortigara, G. 2004. The evolution of brain lateralization: a game-theoretical analysis of population structure. Proceedings of the Royal Society of London Series B—Biological Sciences 271: 853–857.

Gibson, D. I., Timofeeva, T. A., and Gerasev, P. A. 1996. A catalogue of the nominal species of the monogenean genus Dactylogyrus Diesing, 1850 and their host genera. Systematic Parasitology 35: 3–46.

Gibson, R. M., and Langen, T. A. 1996. How do animals choose their mates? Trends in Ecology and Evolution 11: 468–470.

Giese, A. R., and Hedrick, P. W. 2003. Genetic variation and resistance to a bacterial infection in the endangered Gila topminnow. Animal Conservation 6: 369–377.

Gill, H. S., Hambleton, S. J., and Morgan, D. L. 1999. Is the mosquitofish, Gambusia holbrooki (Poeciliidae), a major threat to the native freshwater fishes of south-western Australia? In B. Seret and J. Y. Sire (eds.), Proceedings of the 5th Indo-Pacific Fish Conference, Noumea, 393–403. Paris: Society of French Ichthyology.

Gilliam, J. F., Fraser, D. F., and Alkins-Koo, M. 1993. Structure of a tropical stream fish community: a role for biotic interactions. Ecology 74: 1856–1870.

Gimenez-Conti, I., Woodhead, A. D., Harshbarger, J. C., Kazianis, S., Setlow, R. B., Nairn, R. S., and Walter, R. B. 2001. A proposed classification scheme for Xiphophorus melanomas based on histopathologic analyses. Marine Biotechnology 3: 100–106.

Giraldeau, L. A., Valone, T. J., and Templeton, J. J. 2002. Potential disadvantages of using socially acquired information. Philosophical Transactions of the Royal Society of London Series B—Biological Sciences 357: 1559–1566.

Gladyshev, E. A., Meselson, M., and Arkhipova, I. R. 2008. Massive horizontal gene transfer in bdelloid rotifers. Science 320: 1210–1213.

Godin, J. G. J. 1995. Predation risk and alternative mating tactics in male Trinidadian guppies (Poecilia reticulata). Oecologia 103: 224–229.

Godin, J. G. J., and Briggs, S. E. 1996. Female mate choice under predation risk in the guppy. Animal Behaviour 51: 117–130.

Godin, J. G. J., and Davis, S. A. 1995. Who dares, benefits: predator approach behaviour in the guppy (Poecilia reticulata) deters predator pursuit. Proceedings of the Royal Society of London Series B—Biological Sciences 259: 193–200.

Godin, J. G. J., and Dugatkin, L. A. 1995. Variability and repeatability of female mating preference in the guppy. Animal Behaviour 49: 1427–1433.

Godin, J. G. J., and Dugatkin, L. A. 1996. Female mating preference for bold males in the guppy, Poecilia reticulata. Proceedings of the National Academy of Sciences of the United States of America 93: 10262–10267.

Godin, J. G. J., and McDonough, H. E. 2003. Predator preference for brightly colored males in the guppy: a viability cost for a sexually selected trait. Behavioral Ecology 14: 194–200.

Godin, J. G. J., Classon, L. J., and Abrahams, M. V. 1988. Group vigilance and shoal size in a small characin fish. Behaviour 104: 29–40.

Godin, J. G. J., Alfieri, M. S., Hoare, D. J., and Sadowski, J. A. 2003. Conspecific familiarity and shoaling preferences in a wild guppy population. Canadian Journal of Zoology 81: 1899–1904.

Godin, J. G. J., Herdman, E. J. E., and Dugatkin, L. A. 2005. Social influences on female mate choice in the guppy, Poecilia reticulata: generalized and repeatable trait-copying behaviour. Animal Behaviour 69: 999–1005.

Gomes, J. L., and Monteiro, L. R. 2008. Morphological divergence patterns among populations of Poecilia vivipara (Teleostei Poeciliidae): test of an ecomorphological paradigm. Biological Journal of the Linnean Society 93: 799–812.

Gomez, A., Wellbrock, C., Gutbrod, H., Dimitrijevic, N., and Schartl, M. 2001. Ligand-independent dimerization and activation of the oncogenic Xmrk receptor by two mutations in the extracellular domain. Journal of Biological Chemistry 276: 3333–3340.

Gomez, A., Volff, J.-N., Hornung, U., Schartl, M., and Wellbrock, C. 2004. Identification of a second egfr gene in Xiphophorus uncovers an expansion of the epidermal growth factor receptor family in fish. Molecular Biology and Evolution 21: 266–275.

Gomez-Pinilla, F., So, V., and Kesslak, J. P. 1998. Spatial learning and physical activitiy contribute to the induction of fibroblast growth factor: neural substrates for increased cognition associated with exercise. Neuroscience 85: 53–61.

Gomulkiewicz, R., and Holt, R. D. 1995. When does evolution by natural selection prevent extinction? Evolution 49: 201–207.

Gong, A. 1997. The effects of predator exposure on the female choice of guppies (*Poecilia reticulata*) from a high predation population. Behaviour 134: 373–389.

Gong, A., and Gibson, R. M. 1996. Reversal of a female preference after visual exposure to a predator in the guppy, *Poecilia reticulata*. Animal Behaviour 52: 1007–1015.

Gonzalez, R. J., Cooper, J., and Head, D. 2005. Physiological responses to hyper-saline waters in sailfin mollies (*Poecilia latipinna*). Comparative Biochemistry and Physiology A—Molecular and Integrative Physiology 142: 397–403.

Gonzalez-Voyer, A., Winberg, S., and Kolm, N. 2009. Social fishes and single mothers: brain evolution in African cichlids. Proceedings of the Royal Society of London Series B—Biological Sciences 276: 161–167.

Goodey, W., and Liley, N. R. 1986. The influence of early experience on escape behavior in the guppy (*Poecilia reticulata*). Canadian Journal of Zoology 64: 885–888.

Goodrich, H. B., Dee, J. E., Flynn, C. M., and Mercer, R. N. 1934. Germ cells and sex differentiation in *Lebistes reticulatus*. Biological Bulletin 67: 83–96.

Goodsell, J. A., and Kats, L. B. 1999. Effect of introduced mosquitofish on Pacific treefrogs and the role of alternative prey. Conservation Biology 13: 921–924.

Goodwin, T. W. 1984. *The Biochemistry of the Carotenoids*. London: Chapman & Hall.

Goodyear, C. P. 1973. Learned orientation in the predator avoidance behavior of mosquitofish, *Gambusia affinis*. Behaviour 45: 191–224.

Goodyear, C. P., and Ferguson, D. E. 1969. Sun-compass orientation in the mosquitofish, *Gambusia affinis*. Animal Behaviour 17: 636–640.

Gorbman, A., and Gordon, M. 1951. Spontaneous thyroidal tumors in the swordtail *Xiphophorus montezumae*. Cancer Research 11: 184–187.

Gordon, H., and Gordon, M. 1957. Maintenance of polymorphism by potentially injurious genes in eight natural populations of the platyfish, *Xiphophorus maculatus*. Journal of Genetics 55: 1–44.

Gordon, M. 1927. The genetics of viviparous top-minnow Platypoecilus: the inheritance of two kinds of melanophores. Genetics 12: 253–283.

Gordon, M. 1931. Hereditary basis of melanosis in hybrid fishes. American Journal of Cancer 15: 1495–1523.

Gordon, M. 1937. Heritable color variations in the Mexican swordtail-fish. Journal of Heredity 28: 223–230.

Gordon, M. 1946. Interchanging genetic mechanisms for sex determination in fishes under domestication. Journal of Heredity 37: 307–320.

Gordon, M. 1947. Genetics of ocular-tumor development in fishes (preliminary report). Journal of National Cancer Institute 7: 87–92.

Gordon, M. S., and Rosen, D. E. 1962. A cavernicolous form of the poeciliid fish *Poecilia sphenops* from Tabasco, México. Copeia 1962: 360–368.

Goren, M., and Ortal, R. 1999. Biogeography, diversity and conservation of the inland water fish communities in Israel. Biological Conservation 89: 1–9.

Gosling, S. D. 2001. From mice to men: what can we learn about personality from animal research? Psychological Bulletin 127: 45–86.

Gottlieb, G. 2002. Developmental-behavioral initiation of evolutionary change. Psychological Reviews 109: 211–218.

Graham, L. K. E., and Wilcox, L. W. 2000. The origin of alternation of generations in land plants: a focus on matrotrophy and hexose transport. Philosophical Transactions of the Royal Society of London Series B—Biological Sciences 355: 757–767.

Grand, T. C., and Dill, L. M. 1999. The effect of group size on the foraging behaviour of juvenile coho salmon: reduction of predation risk or increased competition? Animal Behaviour 58: 443–451.

Gravemeier, B., and Greven, H. 2006. The envelope of fully grown, unfertilized oocytes in *Heterandria formosa* (Poeciliidae) and *Xenotoca eiseni* (Goodeidae). Verhandlungen der Gesellschaft für Ichthyologie 5: 7–11.

Graves, J. A. M. 1995. The origin and function of the mammalian Y chromosome and the Y-borne genes—an evolving understanding. Bioessays 17: 311–321.

Gray, M. M., and Weeks, S. C. 2001. Niche breadth in clonal and sexual fish (*Poeciliopsis*): a test of the frozen niche variation model. Canadian Journal of Fisheries and Aquatic Sciences 58: 1313–1318.

Green, A. J. 1992. Positive allometry is likely with mate choice, competitive display and other functions. Animal Behaviour 43: 170–172.

Green, R. F., and Noakes, D. L. G. 1995. Is a little bit of sex as good as a lot? Journal of Theoretical Biology 174: 87–96.

Greene, J. M., and Brown, K. L. 1991. Demographic and genetic characteristics of multiply inseminated female mosquitofish (*Gambusia affinis*). Copeia 1991: 431–444.

Gregersen, J. W., Kranc, K. R., Ke, X. Y., Svendsen, P., Madsen, L. S., Thomsen, A. R., Cardon, L. R., Bell, J. I., and Fugger, L. 2006. Functional epistasis on a common MHC haplotype associated with multiple sclerosis. Nature 443: 574–577.

Grether, G. F. 2000. Carotenoid limitation and mate preference evolution: a test of the indicator hypothesis in guppies (*Poecilia reticulata*). Evolution 54: 1712–1724.

Grether, G. F. 2005. Environmental change, phenotypic plasticity, and genetic compensation. American Naturalist 166: E115–E123.

Grether, G. F., Hudon, J., and Millie, D. F. 1999. Carotenoid limitation of sexual coloration along an environmental gradient in guppies. Proceedings of the Royal Society of London Series B—Biological Sciences 266: 1317–1322.

Grether, G. F., Hudon, J., and Endler, J. A. 2001a. Carotenoid scarcity, synthetic pteridine pigments and the evolution of sexual coloration in guppies (*Poecilia reticulata*). Proceedings of the Royal Society of London Series B—Biological Sciences 268: 1245–1253.

Grether, G. F., Millie, D. F., Bryant, M. J., Reznick, D. N., and Mayea, W. 2001b. Rain forest canopy cover, resource availability, and life history evolution in guppies. Ecology 82: 1546–1559.

Grether, G. F., Kasahara, S., Kolluru, G. R., and Cooper, E. L.

2004. Sex-specific effects of carotenoid intake on the immunological response to allografts in guppies (*Poecilia reticulata*). Proceedings of the Royal Society of London Series B—Biological Sciences 271: 45–49.

Grether, G. F., Cummings, M. E., and Hudon, J. 2005a. Countergradient variation in the sexual coloration of guppies (*Poecilia reticulata*): drosopterin synthesis balances carotenoid availability. Evolution 59: 175–188.

Grether, G. F., Kolluru, G. R., Rodd, F. H., de la Cerda, J., and Shimazaki, K. 2005b. Carotenoid availability affects the development of a colour-based mate preference and the sensory bias to which it is genetically linked. Proceedings of the Royal Society of London Series B—Biological Sciences 272: 2181–2188.

Grether, G. F., Kolluru, G. R., Lin, K., Quiroz, M. A., Robertson, G., and Snyder, A. J. 2008. Maternal effects of carotenoid consumption in guppies (*Poecilia reticulata*). Functional Ecology 22: 294–302.

Greven, H. 2005. Structural and behavioral traits associated with sperm transfer in Poeciliinae. In M. C. Uribe and H. J. Grier (eds.), *Viviparous Fishes*, 145–163. Homestead, FL: New Life Publications.

Grier, H. J. 1981. Cellular organisation of the testis and spermatogenesis in fishes. American Zoologist 21: 345–357.

Grier, H. J., Burns, J. R., and Flores, J. A. 1981. Testis structure in three species of teleosts with tubular gonopodia. Copeia 1981: 797–801.

Grier, H. J., Uribe, M. C., Parenti, L. R., and DelaRosa-Cruz, G. 2005. Fecundity, the germinal epithelium, and folliculogenesis in viviparous fishes. In M. C. Uribe and H. J. Grier (eds.), *Viviparous Fishes*, 191–216. Homestead, FL: New Life Publications.

Grieshaber, M. K., and Völkel, S. 1998. Animal adaptations for tolerance and exploitation of poisonous sulfide. Annual Review of Physiology 60: 33–53.

Griffiths, S. W. 1996. Sex differences in the trade-off between feeding and mating in the guppy. Journal of Fish Biology 48: 891–898.

Griffiths, S. W. 2003. Learned recognition of conspecifics by fishes. Fish and Fisheries 4: 256–268.

Griffiths, S. W., and Magurran, A. E. 1997a. Familiarity in schooling fish: how long does it take to acquire? Animal Behaviour 53: 945–949.

Griffiths, S. W., and Magurran, A. E. 1997b. Schooling preferences for familiar fish vary with group size in a wild guppy population. Proceedings of the Royal Society of London Series B—Biological Sciences 264: 547–551.

Griffiths, S. W., and Magurran, A. E. 1998. Sex and schooling behaviour in the Trinidadian guppy. Animal Behaviour 56: 689–693.

Griffiths, S. W., and Magurran, A. E. 1999. Schooling decisions in gunnies (*Poecilia reticulata*) are based on familiarity rather than kin recognition by phenotype matching. Behavioral Ecology and Sociobiology 45: 437–443.

Griffiths, S. W., and Ward, A. J. W. 2006. Learned recognition of conspecifics. In C. Brown, K. Laland and J. Krause (eds.), *Fish Cognition and Behavior*, 139–157. Oxford: Blackwell.

Grimm, V., Berger, U., Bastiansen, F., Eliassen, S., Ginot, V., Giske, J., Goss-Custard, J., Grand, T., Heinz, S. K., Huse, G., Huth, A., Jepsen, J. U., Jorgensen, C., Mooij, W. M., Müller, B., Pe'er, G., Piou, C., Railsback, S. F., Robbins, A., Robbins, M. M., Rossmanith, E., Ruger, N., Strand, E., Souissi, S., Stillman, R. A., Vabo, R., Visser, U., and DeAngelis, D. L. 2006. A standard protocol for describing individual-based and agent-based models. Ecological Modelling 198: 115–126.

Gross, M. R. 1996. Alternative reproductive strategies and tactics: diversity within sexes. Trends in Ecology and Evolution 11: 92–98.

Grove, B. D., and Wourms, J. P. 1982. Embryonic nutrient absorption in the poeciliid *Heterandria formosa*. American Zoologist 22: 881.

Grove, B. D., and Wourms, J. P. 1983. Endocytosis of molecular tracers by embryos of the viviparous fish, *Heterandria formosa*. Journal of Cell Biology 97: A100.

Grove, B. D., and Wourms, J. P. 1991. The follicular placenta of the viviparous fish, *Heterandria formosa*, I: ultrastructure and development of the embryonic absorptive surface. Journal of Morphology 209: 265–284.

Grove, B. D., and Wourms, J. P. 1994. The follicular placenta of the viviparous fish, *Heterandria formosa*, II: ultrastructure and development of the follicular epithelium. Journal of Morphology 220: 167–184.

Grub, J. C. 1972. Differential predation by *Gambusia affinis* on the eggs of seven species of anuran amphibians. American Midland Naturalist 88: 102–108.

Guevara-Fiore, P., Skinner, A., and Watt, P. J. 2009. Do male guppies distinguish virgin females from recently mated ones? Animal Behaviour 77: 425–431.

Gumm, J. M., and Gabor, C. R. 2005. Asexuals looking for sex: conflict between species and mate-quality recognition in sailfin mollies (*Poecilia latipinna*). Behavioral Ecology and Sociobiology 58: 558–565.

Gupta, N. P., and Kumar, A. 2002. Lycopene therapy in idiopathic male infertility—a preliminary report. International Urology and Nephrology 34: 369–372.

Gutbrod, H., and Schartl, M. 1999. Intragenic sex-chromosomal crossovers of *Xmrk* oncogene alleles affect pigment pattern formation and the severity of melanoma in *Xiphophorus*. Genetics 151: 773–83.

Gutierrez-Espeleta, G. A., Hedrick, P. W., Kalinowski, S. T., Garrigan, D., and Boyce, W. M. 2001. Is the decline of desert bighorn sheep from infectious disease the result of low MHC variation? Heredity 86: 439–450.

Gutierrez-Rodriguez, C., Morris, M. R., Dubois, N. S., and de Queiroz, K. 2007. Genetic variation and phylogeography of the swordtail fish *Xiphophorus cortezi* (Cyprinodontiformes, Poeciliidae). Molecular Phylogenetics and Evolution 43: 111–123.

Gutierrez-Rodriguez, C., Shearer, A. E., Morris, M. R., and de Queiroz, K. 2008. Phylogeography and monophyly of the swordtail fish species *Xiphophorus birchmanni* (Cyprinodontiformes, Poeciliidae). Zoologica Scripta 37: 129–139.

Haaf, T., and Schmid, M. 1984. An early stage of ZW/ZZ sex chromosome differentiation in *Poecilia sphenops* var. melanistica (Poeciliidae, Cyprinodontiformes). Chromosoma 89: 37–41.

Haddrill, P. R., Halligan, D. L., Tomaras, D., and Charlesworth, B. 2007. Reduced efficacy of selection in regions of the *Drosophila* genome that lack crossing over. Genome Biology 8: R18.

Hager, M. C., and Helfman, G. S. 1991. Safety in numbers: shoal size choice by minnows under predatory threat. Behavioral Ecology and Sociobiology 29: 271–276.

Hain, T. J. A., and Neff, B. D. 2007. Multiple paternity and kin recognition mechanisms in a guppy population. Molecular Ecology 16: 3938–3946.

Hairston, N. G. 1991. *Ecological Experiments: Purpose, Design, and Execution.* Cambridge: Cambridge University Press.

Haldane, J. B. S. 1922. Sex ratio and unisexual sterility in hybrid animals. Journal of Genetics 12: 101–109.

Hale, K. A., and Briskie, J. V. 2007. Decreased immunocompetence in a severely bottlenecked population of an endemic New Zealand bird. Animal Conservation 10: 2–10.

Hall, D., and Suboski, M. D. 1995. Visual and olfactory stimuli in learned release of alarm reactions by zebra danio fish (*Brachydanio rerio*). Neurobiology of Learning and Memory 63: 229–240.

Hall, D. W., and Kirkpatrick, M. 2006. Reinforcement and sex linkage. Evolution 60: 908–921.

Hall, M., Lindholm, A. K., and Brooks, R. 2004. Direct selection on male attractiveness and female preference fails to produce a response. BMC Evolutionary Biology 4: 1471–2148.

Hallgren, S. L. E., Linderoth, M., and Olsen, K. H. 2006. Inhibition of cytochrome p450 brain aromatase reduces two male specific sexual behaviours in the male Endler guppy (*Poecilia reticulata*). General and Comparative Endocrinology 147: 323–328.

Halpern-Sebold, L. R., Schreibman, M. P., and Margolis-Nunno, H. 1986. Differences between early- and late-maturing genotypes of the platyfish (*Xiphophorus maculatus*) in the morphometry of their immunoreactive luteinizing hormone releasing hormone-containing cells: a developmental study. Journal of Experimental Zoology 240: 245–257.

Hamer, A. J., Lane, S. J., and Mahony, M. J. 2002. The role of introduced mosquitofish (*Gambusia holbrooki*) in excluding the native green and golden bell frog (*Litoria aurea*) from original habitats in south-eastern Australia. Oecologia 132: 445–452.

Hamilton, A. 2001. Phylogeny of *Limia* (Teleostei: Poeciliidae) based on NADH dehydrogenase subunit 2 sequences. Molecular Phylogenetics and Evolution 19: 277–289.

Hamilton, W. D. 1980. Sex versus non-sex versus parasite. Oikos 35: 282–290.

Hamilton, W. D., and Zuk, M. 1982. Heritable true fitness and bright birds: a role for parasites? Science 218: 384–387.

Hankison, S. J., and Morris, M. R. 2002. Sexual selection and species recognition in the pygmy swordtail, *Xiphophorus pygmaeus*: conflicting preferences. Behavioral Ecology and Sociobiology 51: 140–145.

Hankison, S. J., and Morris, M. R. 2003. Avoiding a compromise between sexual selection and species recognition: female swordtail fish assess multiple species-specific cues. Behavioral Ecology 14: 282–287.

Hankison, S. J., and Ptacek, M. B. 2008. Geographical variation of genetic and phenotypic traits in the Mexican sailfin mollies, *Poecilia velifera* and *P. petenensis*. Molecular Ecology 17: 2219–2233.

Hankison, S. J., Childress, M. J., Schmitter-Soto, J. J., and Ptacek, M. B. 2006. Morphological divergence within and between the Mexican sailfin mollies, *Poecilia velifera* and *Poecilia petenensis*. Journal of Fish Biology 68: 1610–1630.

Hannes, R. P., Franck, D., and Liemann, F. 1984. Effects of rank-order fights on whole-body and blood concentration of androgen and corticosteroids in the male swordfish (*Xiphophorus helleri*). Zeitschrift für Tierpsychologie 65: 53–65.

Harland, M., Meloni, R., Gruis, N., Pinney, E., Brookes, S., Spurr, N. K., Frischauf, A. M., Bataille, V., Peters, G., Cuzick, J., Selby, P., Bishop, D. T., and Bishop, J. N. 1997. Germline mutations of the *CDKN2* gene in UK melanoma families. Human Molecular Genetics 6: 2061–2067.

Harris, P. D. 1988. Changes in the site specificity of *Gyrodactylus turnbulli* Harris, 1986 (Monogenea) during infections of individual guppies (*Poecilia reticulata* Peters, 1859). Canadian Journal of Zoology 66: 2854–2857.

Harris, P. D., and Lyles, A. M. 1992. Infections of *Gyrodactylus bullatarudis* and *Gyrodactylus turnbulli* on guppies (*Poecilia reticulata*) in Trinidad. Journal of Parasitology 78: 912–914.

Harris, P. D., Shinn, A. P., Cable, J., Bakke, T. A., and Bron, J. 2008. GyroDb: gyrodactylid monogeneans on the web. Trends in Parasitology 24: 109–111.

Harvey, P. H., and Pagel, M. D. 1991. *The Comparative Method in Evolutionary Biology.* Oxford: Oxford University Press.

Haskins, C. P., and Haskins, E. F. 1949. The role of sexual selection as an isolating mechanism in three species of poeciliid fishes. Evolution 3: 160–169.

Haskins, C. P., and Haskins, E. F. 1951. The inheritance of certain colour patterns in wild populations of *Lebistes reticulatus* in Trinidad. Evolution 5: 216–225.

Haskins, C. P., Haskins, E. F., McLaughlin, J. J. A., and Hewitt, R. E. 1961. Polymorphism and population structure in *Lebistes reticulatus*, an ecological study. In W. F. Blair (ed.), *Vertebrate Speciation*, 320–395. Austin: University of Texas Press.

Haskins, C. P., Young, P., Hewitt, R. E., and Haskins, E. F. 1970. Stabilised heterozygosis of supergenes mediating certain Y-linked colour patterns in populations of *Lebistes reticulatus*. Heredity 25: 575–588.

Hatai, K., Chukanhom, K., Lawhavinit, O. A., Hanjavanit, C., Kunitsune, M., and Imai, S. 2001. Some biological characteristics of *Tetrahymena corlissi* isolated from guppy in Thailand. Fish Pathology 36: 195–199.

Hattori, R. S., Gould, R. J., Fujioka, T., Saito, T., Kurita, J., Strussmann, C. A., Yokota, M., and Watanabe, S. 2007. Temperature-dependent sex determination in Hd-rR medaka *Oryzias latipes*: gender sensitivity, thermal threshold, critical period, and *DMRT1* expression profile. Sexual Development 1: 138–146.

Haubruge, E., Petit, F., and Gage, M. J. G. 2000. Reduced sperm counts in guppies (*Poecilia reticulata*) following exposure to low levels of tributyltin and bisphenol A. Proceedings of the Royal Society of London Series B—Biological Sciences 267: 2333–2337.

Hauser, M. D., Tsao, F., Garcia, P., and Spelke, E. S. 2003. Evolutionary foundations of number: spontaneous representation of numerical magnitudes by cotton-top tamarins. Proceedings of the Royal Society of London Series B—Biological Sciences 270: 1441–1446.

Häussler, G. 1928. Über Melanombildungen bei Bastarden von *Xiphophorus maculatus* var. rubra. Klinische Wochenschrift 7: 1561–1562.

Hayes, T. B. 1998. Sex determination and primary sex differentiation in amphibians: genetic and developmental mechanisms. Journal of Experimental Zoology 281: 373–399.

Haynes, J. L. 1995. Standardized classification of poeciliid development for life-history studies. Copeia 1995: 147–154.

Head, M. L., and Brooks, R. 2006. Sexual coercion and the opportunity for sexual selection in guppies. Animal Behaviour 71: 515–522.

Head, M. L., Lindholm, A. K., and Brooks, R. 2007. Operational sex ratio and density do not affect directional selection on male sexual ornaments and behavior. Evolution 62: 135–144.

Healy, S. D., and Rowe, C. 2007. A critique of comparative studies of brain size. Proceedings of the Royal Society of London Series B—Biological Sciences 274: 453–464.

Heater, S. J., Rains, J. D., Wells, M. C., Guerrero, P. A., and Walter, R. B. 2007. Perturbation of DNA repair gene expression due to interspecies hybridization. Comparative Biochemistry and Physiology C—Toxicology and Pharmacology 145: 156–163.

Heath, D. D., and Blouw, D. M. 1998. Are maternal effects in fish adaptive or merely physiological side effects? In T. A. Mousseau and C. W. Fox (eds.), *Maternal Effects as Adaptations*, 178–201. Oxford: Oxford University Press.

Hedrick, P. W. 2002. Pathogen resistance and genetic variation at MHC loci. Evolution 56: 1902–1908.

Hedrick, P. W. 2005. *Genetics of Populations*. Sudbury, MA: Jones & Bartlett.

Hedrick, P. W., and Parker, K. M. 1998. MHC variation in the endangered Gila topminnow. Evolution 52: 194–199.

Hedrick, P. W., Kim, T. J., and Parker, K. M. 2001a. Parasite resistance and genetic variation in the endangered Gila topminnow. Animal Conservation 4: 103–109.

Hedrick, P. W., Parker, K. M., and Lee, R. N. 2001b. Using microsatellite and MHC variation to identify species, ESUs, and MUs in the endangered Sonoran topminnow. Molecular Ecology 10: 1399–1412.

Hedrick, P. W., Lee, R. N., and Hurt, C. R. 2006. The endangered Sonoran topminnow: examination of species and ESUs using three mtDNA genes. Conservation Genetics 7: 483–492.

Helfman, G. S. 1989. Threat-sensitive predator avoidance in damselfish-trumpetfish interactions. Behavioural Ecology and Sociobiology 24: 47–58.

Helfman, G. S., and Schultz, E. T. 1984. Social transmission of behavioural traditions in a coral reef fish. Animal Behaviour 32: 379–384.

Helleday, T. 2003. Pathways for mitotic homologous recombination in mammalian cells. Mutation Research 532: 103–115.

Hellriegel, B., and Reyer, H. U. 2000. Factors influencing the composition of mixed populations of a hemiclonal hybrid and its sexual host. Journal of Evolutionary Biology 13: 906–918.

Hendrickson, D. A., and Brooks, J. E. 1991. Transplanting short lived fishes in North American deserts: review, assessment and recommendations. In W. L. Minckley and J. E. Deacon (eds.), *Battle against Extinction*, 283–293. Tucson: University of Arizona Press.

Hendry, A. P., Wenburg, J. K., Bentzen, P., Volk, E. C., and Quinn, T. P. 2000. Rapid evolution of reproductive isolation in the wild: evidence from introduced salmon. Science 290: 516–518.

Hendry, A. P., Kelly, M. L., Kinnison, M. T., and Reznick, D. N. 2006. Parallel evolution of the sexes? Effects of predation and

habitat features on the size and shape of wild guppies. Journal of Evolutionary Biology 19: 741–754.

Henn, A. W. 1916. On various South American poeciliid fishes. Annals of the Carnegie Museum 10: 93–142.

Henrich, J., and Boyd, R. 1998. The evolution of conformist transmission and the emergence of between-group differences. Evolution and Human Behavior 19: 215–242.

Henrich, S. 1988. Variation in offspring size of the poeciliid fish *Heterandria formosa* in relation to fitness. Oikos 51: 13–18.

Hensor, E. M. A., Couzin, I. D., James, R., and Krause, J. 2005. Modelling density-dependent fish shoal distributions in the laboratory and field. Oikos 110: 344–352.

Herdman, E. J. E., Kelly, C. D., and Godin, J. G. J. 2004. Male mate choice in the guppy (*Poecilia reticulata*): do males prefer larger females as mates? Ethology 110: 97–111.

Hermer, L., and Spelke, E. S. 1994. A geometric process for spatial reorientation in young children. Nature 370: 57–59.

Hernandez-Martich, J. D., and Smith, M. H. 1997. Downstream gene flow and genetic structure of *Gambusia holbrooki* (eastern mosquitofish) populations. Heredity 79: 295–301.

Hester, F. J. 1964. Effects of food supply on fecundity in the female guppy *Lebistes reticulatus* (Peters). Journal of the Fisheries Research Board of Canada 21: 757–764.

Heston, W. E. 1982. Genetics: animal tumors. In F. F. Becker (ed.), *Cancer: A Comprehensive Treatise*, 47–71. New York: Plenum Press.

Heubel, K. U., and Plath, M. 2008. Influence of male harassment and female competition on female feeding behaviour in a sexual-asexual mating complex of mollies (*Poecilia mexicana, P. formosa*). Behavioral Ecology and Sociobiology 62: 1689–1699.

Heubel, K. U., and Schlupp, I. 2006. Turbidity affects association behaviour in male *Poecilia latipinna*. Journal of Fish Biology 68: 555–568.

Heubel, K. U., and Schlupp, I. 2008. Seasonal plasticity in male mating preferences in sailfin mollies. Behavioral Ecology 19: 1080–1086.

Heubel, K. U., Hornhardt, K., Ollmann, T., Parzefall, J., Ryan, M. J., and Schlupp, I. 2008. Geographic variation in female mate-copying in the species complex of a unisexual fish, *Poecilia formosa*. Behaviour 145: 1041–1064.

Heubel, K. U., Rankin, D. J., and Kokko, H. 2009. How to go extinct by mating too much: population consequences of male mate choice and efficiency in a sexual-asexual species complex. Oikos 118: 513–520.

Heuschele, J., and Candolin, U. 2007. An increase in pH boosts olfactory communication in sticklebacks. Biology Letters 3: 411–413.

Heuts, B. A. 1999. Lateralisation of trunk muscle volume, and lateralization of swimming turns of fish responding to external stimuli. Behavioral Processes 47: 113–124.

Hey, J., and Nielsen, R. 2004. Multilocus methods for estimating population sizes, migration rates and divergence time, with applications to the divergence of *Drosophila pseudoobscura* and *D. persimilis*. Genetics 167: 747–760.

Heyes, C. M. 1994. Social learning in animals: categories and mechanisms. Biological Reviews 69: 207–231.

Heyes, C. M., and Galef, B. G. 1996. *Social Learning in Animals: The Roots of Culture*. London: Academic Press.

Hibler, T. L., and Houde, A. E. 2006. The effect of visual obstructions on the sexual behaviour of guppies: the importance of privacy. Animal Behaviour 72: 959–964.

Higham, T. E., Hulsey, C. D., Rican, O., and Carroll, A. M. 2007. Feeding with speed: prey capture evolution in cichilds. Journal of Evolutionary Biology 20: 70–78.

Hildemann, W. H., and Wagner, E. D. 1954. Intraspecific sperm competition in *Lebistes reticulatus*. American Naturalist 88: 87–91.

Hill, S. E., and Ryan, M. J. 2006. The role of model female quality in the mate choice copying behaviour of sailfin mollies. Biology Letters 2: 203–205.

Hill, W. 1996. Effects of light. In R. J. Stevenson, M. L. Bothwell, and R. L. Lowe (eds.), *Algal Ecology: Freshwater Benthic Ecosystems*, 121–148. Aquatic Ecology Series. San Diego, CA: Academic Press.

Hill, W. L. 1986. Clutch overlap in American coots. Condor 88: 96–97.

Hine, E., Chenoweth, S. F., and Blows, M. W. 2004. Multivariate quantitative genetics and the lek paradox: genetic variance in male sexually selected traits of *Drosophila serrata* under field conditions. Evolution 58: 2754–2762.

Hoare, D. J., Couzin, I. D., Godin, J. G. J., and Krause, J. 2004. Context-dependent group size choice in fish. Animal Behaviour 67: 155–164.

Hoffman, E. A., Schueler, F. W., Jones, A. G., and Blouin, M. S. 2006. An analysis of selection on a colour polymorphism in the northern leopard frog. Molecular Ecology 15: 2627–2641.

Hoffmann, M., Tripathi, N., Henz, S. R., Lindholm, A. K., Weigel, D., Breden, F., and Dreyer, C. 2007. Opsin gene duplication and diversification in the guppy, a model for sexual selection. Proceedings of the Royal Society of London Series B—Biological Sciences 274: 33–42.

Hofmann, U. B., Westphal, J. R., Van Muijen, G. N., and Ruiter, D. J. 2000. Matrix metalloproteinases in human melanoma. Journal of Investigative Dermatology 115: 337–344.

Hofmann, U. B., Houben, R., Brocker, E. B., and Becker, J. C. 2005. Role of matrix metalloproteinases in melanoma cell invasion. Biochimie 87: 307–314.

Hogarth, P. J. 1968. Immunological aspects of foetal-maternal relations in lower vertebrates. Journal of Reproduction and Fertility Supplement 3: 15–27.

Hogarth, P. J. 1972a. Immune relations between mother and foetus in the viviparous poeciliid fish *Xiphophorus helleri* Haeckel, I: antigenicity of the foetus. Journal of Fish Biology 4: 265–269.

Hogarth, P. J. 1972b. Immune relations between mother and foetus in the viviparous poeciliid fish *Xiphophorus helleri* Haeckel, II: lack of status of the ovary as a favourable site for allograft survival. Journal of Fish Biology 4: 271–275.

Hogarth, P. J. 1973. Immune relations between mother and foetus in the viviparous poeciliid fish *Xiphophorus helleri* Haeckel, III: survival of embryos after ectopic transplantation. Journal of Fish Biology 5: 109–113.

Hogarth, P. J., and Sursham, C. M. 1972. Antigenicity of *Poecilia* sperm. Experientia 28: 463–464.

Hojesjo, J., Johnsson, J. I., Petersson, E., and Jarvi, T. 1998. The importance of being familiar: individual recognition and social behavior in sea trout (*Salmo trutta*). Behavioral Ecology 9: 445–451.

Holland, B., and Rice, W. R. 1999. Experimental removal of sexual selection reverses intersexual antagonistic coevolution and removes a reproductive load. Proceedings of the National Academy of Sciences of the United States of America 96: 5083–5088.

Holling, C. S. 1986. Resilience of ecosystems: local surprise and global change. In W. C. Clark and R. E. Munn (eds.), *Sustainable Development of the Biosphere*, 292–317. Cambridge: Cambridge University Press.

Hollis, K. L., Dumas, M. J., Singh, P., and Fackelman, P. 1995. Pavlovian conditioning of aggressive behavior in blue gourami fish (*Trichogaster trichopterus*): winners become winners and losers stay losers. Journal of Comparative Psychology 109: 125–133.

Hopper, A. F. 1943. The early embryology of *Platypoecilus maculatus*. Copeia 1943: 218–224.

Hopper, A. F. 1949. Development and regeneration of the anal fin of normal and castrate males and females of *Lebestes reticulatus*. Journal of Experimental Zoology 110: 299–319.

Hoppitt, W., and Laland, K. N. 2008. Social processes influencing learning in animals: a review of the evidence. Advances in the Study of Behaviour 38: 105–166.

Hori, M. 1993. Frequency-dependent natural selection in the handedness of scale-eating fish. Science 260: 216–219.

Horstkotte, J., and Plath, M. 2008. Divergent evolution of feeding substrate preferences in a phylogenetically young species flock of pupfish (*Cyprinodon* spp.). Naturwissenschaften 95: 1175–1180.

Horstkotte, J., and Strecker, U. 2005. Trophic differentiation in the phylogenetically young *Cyprinodon* species flock (Cyprinodontidae, Teleostei) from Laguna Chichancanab (Mexico). Biological Journal of the Linnean Society 85: 125–134.

Horstkotte, J., Riesch, R., Plath, M., and Jäger, P. 2010. Predation by three species of spiders on a cave fish in a Mexican sulfur cave. Bulletin of the British Arachnological Society 15: 55–58.

Horth, L. 2003. Melanic body colour and aggressive mating behaviour are correlated traits in male mosquitofish (*Gambusia holbrooki*). Proceedings of the Royal Society of London Series B—Biological Sciences 270: 1033–1040.

Horth, L. 2007. Sensory genes and mate choice: evidence that duplications, mutations, and adaptive evolution alter variation in mating cue genes and their receptors. Genomics 90: 159–175.

Hosken, D. J., and Snook, R. 2005. How important is sexual conflict? American Naturalist 165, Supplement 5: S1–S4.

Hosken, D. J., and Stockley, P. 2004. Sexual selection and genital evolution. Trends in Ecology and Evolution 19: 87–93.

Hosken, D. J., and Stockley, P. 2005. Sexual conflict. Current Biology 15: R535–R536.

Hosken, D. J., Taylor, M. L., Hoyle, K., Higgins, S., and Wedell, N. 2008. Attractive males have greater success in sperm competition. Current Biology 18: R553–R554.

Houde, A. E. 1987. Mate choice based upon naturally occurring color-pattern variation in a guppy population. Evolution 41: 1–10.

Houde, A. E. 1988. Genetic difference in female choice between two guppy populations. Animal Behaviour 36: 510–516.

Houde, A. E. 1992. Sex-linked heritability of a sexually selected character in a natural population of *Poecilia reticulata* (Pisces: Poeciliidae) (guppies). Heredity 69: 229–235.

Houde, A. E. 1994. Effect of artificial selection on male colour patterns on mating preference of female guppies. Proceedings of the Royal Society of London Series B—Biological Sciences 256: 125–130.

Houde, A. E. 1997. *Sex, Color, and Mate Choice in Guppies.* Princeton, NJ: Princeton University Press.

Houde, A. E., and Endler, J. A. 1990. Correlated evolution of female mating preferences and male color patterns in the guppy, *Poecilia reticulata.* Science 248: 1405–1408.

Houde, A. E., and Torio, A. J. 1992. Effect of parasitic infection on male color pattern and female choice in guppies. Behavioral Ecology 3: 346–351.

House, C. M., and Lewis, Z. 2007. Genital evolution: blurring the battle lines between the sexes. Current Biology 17: R1013–R1014.

Houston, A. I., Stephens, P. A., Boyd, I. L., Harding, K. C., and McNamara, J. M. 2007. Capital or income breeding? A theoretical model of female reproductive strategies. Behavioral Ecology 18: 241–250.

Howarth, F. G. 1993. High-stress subterranean habitats and evolutionary change in cave-inhabiting arthropods. American Naturalist 142: S65–S77.

Howe, D., and Denver, D. 2008. Muller's ratchet and compensatory mutation in *Caenorhabditis briggsae* mitochondrial genome evolution. BMC Evolutionary Biology 8: 62.

Howe, E., Howe, C., Lim, R., and Burchett, M. 1997. Impact of the introduced poeciliid *Gambusia holbrooki* (Girard, 1859) on the growth and reproduction of *Pseudomugil signifer* (Kner, 1865) in Australia. Marine and Freshwater Research 48: 425–433.

Howell, N., Howell, C., and Elson, J. L. 2008. Time dependency of molecular rate estimates for mtDNA: this is not the time for wishful thinking. Heredity 101: 107–108.

Howell, W. M., Black, D. A., and Bortone, S. A. 1980. Abnormal expression of secondary sex characters in a population of mosquitofish, *Gambusia affinis holbrooki*: evidence for environmentally-induced masculinization. Copeia 1980: 676–681.

Hrbek, T., and Meyer, A. 2003. Closing of the Tethys Sea and the phylogeny of Eurasian killifishes (Cyprinodontiformes: Cyprinodontidae). Journal of Evolutionary Biology 16: 17–36.

Hrbek, T., Seckinger, J., and Meyer, A. 2007. A phylogenetic and biogeographic perspective on the evolution of poeciliid fishes. Molecular Phylogenetics and Evolution 43: 986–998.

Hsiao, C. D., and Tsai, H. J. 2003. Transgenic zebrafish with fluorescent germ cell: a useful tool to visualize germ cell proliferation and juvenile hermaphroditism *in vivo*. Developmental Biology 262: 313–323.

Hsu, Y. Y., and Wolf, L. L. 1999. The winner and loser effect: integrating multiple experiences. Animal Behaviour 57: 903–910.

Hsu, Y. Y., and Wolf, L. L. 2001. The winner and loser effect: what fighting behaviours are influenced? Animal Behaviour 61: 777–786.

Hubbs, C. 1964. Interactions between bisexual fish species and its gynogenetic sexual parasite. Bulletin of the Texas Memorial Museum 8: 1–72.

Hubbs, C. 1971. Competition and isolation mechanisms in the *Gambusia affinis* ×*G. heterochir* hybrid swarm. Bulletin of the Texas Memorial Museum 19: 1–48.

Hubbs, C., and Brodrick, H. J. 1963. Current abundance of *Gambusia gaigei*, an endangered fish species. Southwestern Naturalist 8: 46–48.

Hubbs, C., and Dries, L. A. 2002. Geographic variation in interbrood interval in *Poecilia*. In M. D. L. Lozano-Vilano (ed.), *Libro jubilar en honor al Dr. Salvador Contreras Balderas*, 35–41. Monterrey: Universidad Autonoma de Nuevo Leon.

Hubbs, C., and Reynolds, R. A. 1957. Copulatory function of the modified pectoral fin of gambusiin fishes. American Naturalist 91: 333–335.

Hubbs, C., and Schlupp, I. 2008. Juvenile survival in a unisexual/sexual complex of mollies. Environmental Biology of Fishes 83: 327–330.

Hubbs, C. L. 1950. Studies of cyprinodont fishes, XX: a new subfamily from Guatemala, with ctenoid scales and a unilateral pectoral clasper. Miscellaneous Publications of the Museum of Zoology of the University of Michigan 78: 1–28.

Hubbs, C. L., and Hubbs, L. C. 1932. Apparent parthenogenesis in nature, in a form of fish of hybrid origin. Science 76: 628–630.

Hubbs, C. L., and Hubbs, L. C. 1945. Bilateral asymmetry and bilateral variation in fishes. Papers of the Michigan Academy of Science, Arts and Letters 30: 229–310.

Hubbs, C. L., and Springer, V. G. 1957. A revision of the *Gambusia nobilis* species group, with descriptions of three new species, and notes on their ecology, and evolution. Texas Journal of Science 9: 279–327.

Hudon, J., Grether, G. F., and Millie, D. F. 2003. Marginal differentiation between the sexual and general carotenoid pigmentation of guppies (*Poecilia reticulata*) and a possible visual explanation. Physiological and Biochemical Zoology 76: 776–790.

Hughes, A. L. 1985. Male size, mating success and mating strategy in the mosquitofish *Gambusia affinis* (Poeciliidae). Behavioral Ecology and Sociobiology 17: 271–278.

Hughes, A. L. 1991. MHC polymorphism and the design of captive breeding programs. Conservation Biology 5: 249–251.

Hughes, A. L., and Yeager, M. 1998. Natural selection at major histocompatibility complex loci of vertebrates. Annual Review of Genetics 32: 415–435.

Hughes, A. L., Hughes, M. K., and Watkins, D. I. 1993. Contrasting roles of interallelic recombination at the *HLA-A* and *HLA-B* loci. Genetics 133: 669–680.

Hughes, K. A., Du, L., Rodd, F. H., and Reznick, D. N. 1999. Familiarity leads to female mate preference for novel males in the guppy, *Poecilia reticulata*. Animal Behaviour 58: 907–916.

Hughes, K. A., Rodd, F. H., and Reznick, D. N. 2005. Genetic and environmental effects on secondary sex traits in guppies (*Poecilia reticulata*). Journal of Evolutionary Biology 18: 35–45.

Hugueny, B. 1989. West African rivers as biogeographic islands: species richness of fish communities. Oecologia 79: 236–243.

Hulsey, C. D., and García de León, F. 2005. Cichlid jaw mechanics: linking morphology to feeding specialization. Functional Ecology 19: 487–494.

Humphrey, N. K. 1976. The social function of intellect. In P. P. G. Bateson and R. A. Hinde (eds.), *Growing Points in Ethology*, 303–317. Cambridge: Cambridge University Press.

Humphries, J., and Miller, R. R. 1981. A remarkable species flock of pupfishes, genus *Cyprinodon*, from Yucatán, México. Copeia 1981: 52–64.

Hunt, J., Brooks, R., and Jennions, M. D. 2005. Female mate choice as a condition-dependent life-history trait. American Naturalist 166: 79–92.

Hunt, J., Blows, M. W., Zajitschek, F., Jennions, M. D., and Brooks, R. 2007. Reconciling strong stabilizing selection with the maintenance of genetic variation in a natural population of black field crickets (*Teleogryllus commodus*). Genetics 177: 875–880.

Hurd, P. L. 1997. Is signaling of fighting ability costlier for weaker individuals? Journal of Theoretical Biology 184: 83–88.

Hurk, R. v. d. 1974. Steroidogenesis in the testis and gonadotropic activity in the pituitary during postnatal development of the black molly (*Mollienisia latipinna*). Koninklijke Nederlaandse Akademie von Wetenschaapen 77: 193–200.

Hurlbert, S. H., and Mulla, M. S. 1981. Impacts of mosquitofish (*Gambusia affinis*) predation on plankton communities. Hydrobiologia 83: 125–151.

Hurlbert, S. H., Zedler, J., and Fairbanks, D. 1972. Ecosystem alteration by mosquitofish (*Gambusia affinis*) predation. Science 175: 639–641.

Hurt, C. R., and Hedrick, P. W. 2003. Initial stages of reproductive isolation in two species of the endangered Sonoran topminnow. Evolution 57: 2835–2841.

Hurt, C. R., Stears-Ellis, S., Hughes, K. A., and Hedrick, P. W. 2004. Mating behaviour in the endangered Sonoran topminnow: speciation in action. Animal Behaviour 67: 343–351.

Hurt, C. R., Farzin, M., and Hedrick, P. W. 2005. Premating, not postmating, barriers drive genetic dynamics in experimental hybrid populations of the endangered Sonoran topminnow. Genetics 171: 655–662.

Hurtado-Gonzales, J. L., and Uy, J. A. C. 2009. Alternative mating strategies may favour the persistence of a genetically based colour polymorphism in a pentamorphic fish. Animal Behaviour 77: 1187–1194.

Ikehata, H., Kawai, K., Komura, J., Sakatsume, K., Wang, L., Imai, M., Higashi, S., Nikaido, O., Yamamoto, K., Hieda, K., Watanabe, M., Kasai, H., and Ono, T. 2008. UVA1 genotoxicity is mediated not by oxidative damage but by cyclobutane pyrimidine dimers in normal mouse skin. Journal of Investigative Dermatology 128: 2289–2296.

Imai, S., Tsurimaki, S., Goto, E., Wakita, K., and Hatai, K. 2000. *Tetrahymena* infection in guppy, *Poecilia reticulata*. Fish Pathology 35: 67–72.

Ioannou, C. C., Tosh, C. R., Neville, L., and Krause, J. 2008. The confusion effect—from neural networks to reduced predation risk. Behavioral Ecology 19: 126–130.

Ip, Y. K., Kuah, S. S. L., and Chew, S. F. 2004. Strategies adopted by the mudskipper *Boleophthalmus boddaerti* to survive sulfide exposure in normoxia or hypoxia. Physiological and Biochemical Zoology 77: 824–837.

Irie-Sugimoto, N., Kobayashi, T., Sato, T., and Hasegawa, T. 2009. Relative quantity judgment by Asian elephants (*Elephas maximus*). Animal Cognition 12: 193–199.

Ishii, S. 1963. Some factors involved in the delivery of the young of the top-minnow, *Gambusia affinis*. Journal of the Faculty of Science, University of Tokyo 10: 181–187.

IUCN (International Union for Conservation of Nature) Red List of Threatened Species. 2008. http://www.iucnredlist.org/.

Ives, A. R., Midford, P. E., and Garland, T. J. 2007. Within-species variation and measurement error in phylogenetic comparative methods. Systematic Biology 56: 252–270.

Iwasa, Y., and Pomiankowski, A. 1994. The evolution of mate preferences for multiple sexual ornaments. Evolution 48: 853–867.

Iwasa, Y., and Pomiankowski, A. 1999a. Good parent and good genes models of handicap evolution. Journal of Theoretical Biology 200: 97–109.

Iwasa, Y., and Pomiankowski, A. 1999b. Sex specific X chromosome expression caused by genomic imprinting. Journal of Theoretical Biology 197: 487–495.

Iyengar, V. K., Reeve, H. K., and Eisner, T. 2002. Paternal inheritance of a female moth's mating preference. Nature 419: 830–832.

Jacobs, K. 1971. *Livebearing Aquarium Fishes*. New York: Macmillan.

Jalabert, B., and Billard, R. 1969. Étude ultrastructurale du sité de conservation des spermatozoides dans l'ovaire de *Poecilia reticulata* (Poisson, Téléostéen). Annales de Biologie Animale, Biochimie, Biophysique 2: 273–280.

Jameson, S. C., Hogquist, K. A., and Bevan, M. J. 1994. Specificity and flexibility in thymic selection. Nature 369: 750–752.

Jamieson, B. G. M. 1991. *Fish Evolution and Systematics: Evidence from Spermatozoa*. Cambridge: Cambridge University Press.

Jeffery, W. R. 2001. Cavefish as a model system in evolutionary developmental biology. Developmental Biology 231: 1–12.

Jeffery, W. R., Strickler, A. G., and Yamamoto, Y. 2003. To see or not to see: Evolution of eye degeneration in Mexican blind cavefish. Integrative and Comparative Biology 43: 531–541.

Jehle, R., Sztatecsny, M., Wolf, J. B. W., Whitlock, A., Hödl, W., and Burke, T. 2007. Genetic dissimilarity predicts paternity in the smooth newt (*Lissotriton vulgaris*). Biology Letters 3: 526–528.

Jelks, H. L., Walsh, S. J., Burkhead, N. M., Contreras-Balderas, S., Díaz-Pardo, E., Hendrickson, D. A., Lyons, J., Mandrak, N. E., McCormick, F., Nelson, J. S., Platania, S. P., Porter, B. A., Renaud, C. B., Schmitter-Soto, J. J., Taylor, E. B., and Warren, J. M. L. 2008. Conservation status of imperiled North American freshwater and diadromous fishes. Fisheries Management and Ecology 33: 372–407.

Jennions, M. D., and Kelly, C. D. 2002. Geographical variation in male genitalia in *Brachyrhaphis episcopi* (Poeciliidae): is it sexually or naturally selected? Oikos 97: 79–86.

Jennions, M. D., and Petrie, M. 1997. Variation in mate choice and mating preferences: a review of causes and consequences. Biological Review 72: 283–327.

Jennions, M. D., and Petrie, M. 2000. Why do females mate multiply? A review of the genetic benefits. Biological Review 75: 21–64.

Jennions, M. D., and Telford, S. R. 2002. Life-history phenotypes in populations of *Brachyrhaphis episcopi* (Poeciliidae) with different predator communities. Oecologia 132: 44–50.

Jennions, M. D., Hunt, J., Graham, R., and Brooks, R. 2004. No evidence for inbreeding avoidance through postcopulatory mechanisms in the black field cricket, *Teleogryllus commodus*. Evolution 58: 2472–2477.

Jennions, M. D., Wong, B. B. M., Cowling, A., and Donnelly, C. 2006. Life-history phenotypes in a live-bearing fish *Brachyrhaphis episcopi* living under different predator regimes: seasonal effects? Environmental Biology of Fishes 76: 211–219.

Jensen, L. F., Hansen, M. M., Mensberg, K. L., and Loeschcke, V. 2008. Spatially and temporally fluctuating selection at non-MHC immune genes: evidence from TAP polymorphism in populations of brown trout (*Salmo trutta*, L.). Heredity 100: 79–91.

Jewell, G., and McCourt, M. E. 2000. Pseudoneglect: a review and meta-analysis of performance factors in line bisection tasks. Neuropsychologia 38: 93–110.

Jirotkul, M. 1999. Population density influences male-male competition in guppies. Animal Behaviour 58: 1169–1175.

Jirotkul, M. 2000. Male trait distribution determined alternative mating tactics in guppies. Journal of Fish Biology 56: 1427–1434.

Johansen, P. H. 1985. Female pheromone and the behaviour of male guppies (*Poecilia reticulata*) in a temperature gradient. Canadian Journal of Zoology 63: 1211–1213.

Johansson, J., Turesson, H., and Persson, A. 2004. Active selection for large guppies, *Poecilia reticulata*, by the pike cichlid, *Crenicichla saxatilis*. Oikos 105: 595–605.

Johnen, P. 2006. Temperaturabhängige Geschlechtsbestimmung bei *Cnesterodon decemmaculatus* (Poeciliidae). In H. Greven and R. Riehl (eds.), *Biologie der Aquarienfische*, 39–43. Berlin: Tetra-Verlag.

Johnson, J. B. 2001a. Adaptive life-history evolution in the livebearing fish *Brachyrhaphis rhabdophora*: genetic basis for parallel divergence in age and size at maturity and a test of predator-induced plasticity. Evolution 55: 1486–1491.

Johnson, J. B. 2001b. Hierarchical organization of genetic variation in the Costa Rican livebearing fish *Brachyrhaphis rhabdophora* (Poeciliidae). Biological Journal of the Linnean Society 72: 519–527.

Johnson, J. B. 2002. Divergent life histories among populations of the fish *Brachyrhaphis rhabdophora*: detecting putative agents of selection by candidate model analysis. Oikos 96: 82–91.

Johnson, J. B., and Belk, M. C. 2001. Predation environment predicts divergent life-history phenotypes among populations of the livebearing fish *Brachyrhaphis rhabdophora*. Oecologia 126: 142–149.

Johnson, J. B., and Zúñiga-Vega, J. J. 2009. Differential mortality predicts life history evolution and shapes population demography in the livebearing fish *Brachyrhaphis rhabdophora*. Ecology 90: 2242–2252.

Johnson, J. E., and Hubbs, C. 1989. Status and conservation of poeciliid fishes. In G. K. Meffe and F. F. Snelson Jr. (eds.), *Ecology and Evolution of Livebearing Fishes (Poeciliidae)*, 301–317. Englewood Cliffs, NJ: Prentice Hall.

Johnsson, J. I. 2003. Group size influences foraging effort independent of predation risk: an experimental study on rainbow trout. Journal of Fish Biology 63: 863–870.

Johnsson, J. I., and Akerman, A. 1998. Watch and learn: preview of the fighting ability of opponents alters contest behaviour in rainbow trout. Animal Behaviour 56: 771–776.

Johnston, C. M., Barnett, M., and Sharpe, P. T. 1995. The molecular biology of temperature-dependent sex determination. Philosophical Transactions of the Royal Society of London Series B—Biological Sciences 350: 297–303 (discussion, 303–294).

Johnston, I. A. 2006. Environment and plasticity of myogenesis in teleost fish. Journal of Experimental Biology 209: 2249–2264.

Johnston, I. A., and Wilson, R. S. 2005. Temperature-induced developmental plasticity. In S. J. Warburton, W. W. Burggren, B. Pelster, C. L. Reiber, and J. Spicer (eds.), *Comparative Developmental Physiology: Contributions, Tools and Trends*, 124–138. Oxford: Oxford University Press.

Johnston, I. A., Lee, H.-T., Macqueen, D. J., Paranthaman, K., Kawashima, C., Anwar, A., Kinghorn, J. R., and Dalmay, T. 2009. Embryonic temperature affects muscle fibre recruitment in adult zebrafish: genomewide changes in gene and microRNA expression associated with the transition from hyperplastic to hypertrophic growth phenotypes. Journal of Experimental Biology 212: 1781–1793.

Johnstone, R. A. 1995. Honest advertisement of multiple qualities using multiple signals. Journal of Theoretical Biology 177: 87–94.

Johnstone, R. A. 1996. Multiple displays in animal communication: "Backup signals" and "multiple messages." Philosophical Transactions of the Royal Society of London Series B—Biological Sciences 351: 329–338.

Johnstone, R. A. 2001. Eavesdropping and animal conflict. Proceedings of the National Academy of Sciences of the United States of America 98: 9177–9180.

Jokela, J., Lively, C. M., Dybdahl, M. F., and Fox, J. A. 1997. Evidence for a cost of sex in the freshwater snail *Potamopyrgus antipodarum*. Ecology 78: 452–460.

Jollie, W. P., and Jollie, L. G. 1964a. The fine structure of the ovarian follicle of the ovoviviparous poeciliid fish, *Lebistes reticulatus*, I: Maturation of follicular epithelium. Journal of Morphology 114: 479–501.

Jollie, W. P., and Jollie, L. G. 1964b. The fine structure of the ovarian follicle of the ovoviviparous poeciliid fish, *Lebistes reticulatus*, II: Formation of follicular pseudoplacenta. Journal of Morphology 114: 503–526.

Jordan, M. S., Boesteanu, A., Reed, A. J., Petrone, A. L., Holenbeck, A. E., Lerman, M. A., Naji, A., and Caton, A. J. 2001. Thymic selection of CD4(+)CD25(+) regulatory T cells induced by an agonist self-peptide. Nature Immunology 2: 301–306.

Jordao, L. C. 2004. Disturbance chemical cues determine changes in spatial occupation by the convict cichlid *Archocentrus nigrofasciatus*. Behavioural Processes 67: 453–459.

Jordao, L. C., and Volpato, G. L. 2000. Chemical transfer of warning information in non-injured fish. Behaviour 137: 681–690.

Jørgensen, B. B. 1982. Ecology of the bacteria of the sulphur cycle with special reference to anoxic/oxic interface environments. Philosophical Transactions of the Royal Society of London Series B—Biological Sciences 298: 543–561.

Jørgensen, B. B. 1984. The microbial sulfur cycle. In W. Krumbein (ed.), *Microbial Geochemistry*, 91–124. Oxford: Blackwell.

Jørgensen, B. B., and Fenchel, T. 1974. The sulphur cycle of a marine sediment model system. Marine Biology 24: 189–201.

Joron, M., and Brakefield, P. M. 2003. Captivity masks inbreeding effects on male mating success in butterflies. Nature 424: 191–194.

Kadow, P. 1954. An analysis of sexual behavior and reproductive physiology in the guppy, *Lebistes reticulatus (Peters)*. PhD thesis, New York University.

Kahn, A. T., Mautz, B., and Jennions, M. D. 2010. Females prefer to associate with males with longer intromittent organs in mosquitofish. Biology Letters 6: 55–58.

Kallman, K. D. 1965. Genetics and geography of sex determination in the poeciliid fish, *Xiphophorus maculatus*. Zoologica 50: 151–190.

Kallman, K. D. 1971. Inheritance of melanophore patterns and sex determination in the Montezuma swordtail, *Xiphophorus montezumae cortezi* Rosen. Zoologica 56: 77–94.

Kallman, K. D. 1975. The platyfish, *Xiphophorus maculatus*. In R. C. King (ed.), *Handbook of Genetics*, 81–132. New York: Plenum Press.

Kallman, K. D. 1983. The sex-determining mechanism of the poeciliid fish, *Xiphophorus montezumae* Jordan and Snyder and the genetic control of the sexual maturation process and adult size. Copeia 3: 755–769.

Kallman, K. D. 1984. A new look at sex determination in poeciliid fishes. In B. J. Turner (ed.), *Evolutionary Genetics of Fishes*, 95–171. New York: Plenum Press.

Kallman, K. D. 1989. Genetic control of size at maturity in *Xiphophorus*. In G. K. Meffe and F. F. Snelson Jr. (eds.), *Ecology and Evolution of Livebearing Fishes (Poeciliidae)*, 163–184. Englewood Cliffs, NJ: Prentice Hall.

Kallman, K. D. 2005. Genetic and environmental factors controlling size in swordtails. In M. C. Uribe and H. J. Grier (eds.), *Viviparous Fishes*, 365–379. Homestead, FL: New Life Publications.

Kallman, K. D., and Atz, J. W. 1966. Gene and chromosome homology in fishes of the genus *Xiphophorus*. Zoologica 51: 107–135.

Kallman, K. D., and Bao, I. Y. 1987. Female heterogamety in the swordtail, *Xiphophorus alvarezi* Rosen (Pisces, Poeciliidae), with comments on a natural polymorphism affecting sword coloration. Journal of Experimental Zoology 243: 93–102.

Kallman, K. D., and Brunetti, V. 1983. Genetic basis of three mutant color varieties of *Xiphophorus maculatus*: the gray, gold and ghost platyfish. Copeia 1: 170–181.

Kallman, K. D., and Kazianis, S. 2006. The genus *Xiphophorus* in Mexico and Central America. Zebrafish 3: 271–285.

Kallman, K. D., Walter, R. B., Morizot, D. C., and Kazianis, S. 2004. Two new species of *Xiphophorus* (Poeciliidae) from the Isthmus of Tehuantepec, Oaxaca, Mexico, with a discussion of the distribution of the *X. clemenciae* clade. American Museum Novitates 3441: 1–34.

Karayucel, I., Orhan, A. K., and Karayucel, S. 2008. Effect of temperature on some reproductive parameters of gravid females and growth of newly hatched fry in guppy, *Poecilia reticulta* (Peters, 1860). Journal of Animal and Veterinary Advances 7: 1261–1266.

Karino, K., and Haijima, Y. 2001. Heritability of male secondary sexual traits in feral guppies in Japan. Journal of Ethology 19: 33–37.

Karino, K., and Haijima, Y. 2004. Algal-diet enhances sexual ornament, growth and reproduction in the guppy. Behaviour 141: 585–601.

Karino, K., and Kobayashi, M. 2005. Male alternative mating behaviour depending on tail length of the guppy, *Poecilia reticulata*. Behaviour 142: 191–202.

Karino, K., and Shinjo, S. 2007. Relationship between algal-foraging ability and expression of sexually selected traits in male guppies. Zoological Science 24: 571–576.

Karino, K., Utagawa, T., and Shinjo, S. 2005. Heritability of the algal-foraging ability: an indirect benefit of female mate preference for males' carotenoid-based coloration in the guppy, *Poecilia reticulata*. Behavioral Ecology and Sociobiology 59: 1–5.

Karino, K., Shinjo, S., and Sato, A. 2007. Algal-searching ability in laboratory experiments reflects orange spot coloration of the male guppy in the wild. Behaviour 144: 101–113.

Kats, L. B., and Dill, L. M. 1998. The scent of death: chemosensory assessment of predation risk by prey animals. Ecoscience 5: 361–394.

Katzir, G., and Camhi, J. M. 1993. Escape response of black mollies (*Poecilia sphenops*) to predatory dives of a pied kingfisher (*Ceryle rudis*). Copeia 1993: 549–553.

Kavaliers, M., Choleris, E., Agmo, A., and Pfaff, D. W. 2004. Olfactory-mediated parasite recognition and avoidance: linking genes to behavior. Hormones and Behavior 46: 272–283.

Kavumpurath, S., and Pandian, T. J. 1993a. Masculinization of *Poecilia reticulata* by dietary administration of synthetic or natural androgen to gravid females. Aquaculture 116: 83–89.

Kavumpurath, S., and Pandian, T. J. 1993b. Production of a YY female guppy, *Poecilia reticulata*, by endocrine sex reversal and progeny testing. Aquaculture 118: 183–189.

Kawai, M. 1965. Newly-acquirred pre-cultural behaviour of the natural troop of Japanese monkeys on Koshima islet. Primates 6: 1–30.

Kazianis, S. 2006. Historical, present, and future use of *Xiphophorus* fishes for research. Zebrafish 3: 9–10.

Kazianis, S., Gutbrod, H., Nairn, R. S., McEntire, B. B., Della Coletta, L., Walter, R. B., Borowsky, R. L., Woodhead, A. D., Setlow, R. B., Schartl, M., and Morizot, D. C. 1998. Localization of a *CDKN2* gene in linkage group V of *Xiphophorus* fishes defines it as a candidate for the DIFF tumor suppressor. Genes, Chromosomes, Cancer 22: 210–220.

Kazianis, S., Morizot, D. C., Della Coletta, L., Johnston, D. A., Woolcock, B., Vielkind, J. R., and Nairn, R. S. 1999. Comparative structure and characterization of a *CDKN2* gene in a *Xiphophorus* fish melanoma model. Oncogene 18: 5088–5099.

Kazianis, S., Della Coletta, L., Morizot, D. C., Johnston, D. A., Osterndorff, E. A., and Nairn, R. S. 2000. Overexpression of a fish *CDKN2* gene in a hereditary melanoma model. Carcinogenesis 21: 599–605.

Kazianis, S., Gimenez-Conti, I., Setlow, R. B., Woodhead, A. D., Harshbarger, J. C., Trono, D., Ledesma, M., Nairn, R. S., and Walter, R. B. 2001a. MNU induction of neoplasia in a platyfish model. Laboratory Investigation 81: 1191–1198.

Kazianis, S., Gimenez-Conti, I., Trono, D., Pedroza, A., Chovanec, L. B., Morizot, D. C., Nairn, R. S., and Walter, R. B. 2001b. Genetic analysis of neoplasia induced by *N*-nitroso-*N*-methylurea in *Xiphophorus* hybrid fish. Marine Biotechnology 3: S37–S43.

Kazianis, S., Khanolkar, V. A., Nairn, R. S., Rains, J. D., Trono, D., Garcia, R., Williams, E. L., and Walter, R. B. 2004a. Structural organization, mapping, characterization and evolutionary relationships of *CDKN2* gene family members in *Xiphophorus* fishes. Comparative Biochemistry and Physiology C–Toxicology and Pharmacology 138: 291–299.

Kazianis, S., Nairn, R. S., Walter, R. B., Johnston, D. A., Kumar, J., Trono, D., Della-Coletta, L., Gimenez-Conti, I., Rains, J. D., Williams, E. L., Pino, B. M., Mamerow, M. M., Kochan, K. J., Schartl, M., Vielkind, J. R., Volff, J.-N., Woolcock, B., and Morizot, D. C. 2004b. The genetic map of *Xiphophorus* fishes represented by 24 multipoint linkage groups. Zebrafish 1: 287–304.

Kazianis, S., Vielkind, J., Woolcock, B., Morizot, D. C., Wigler, M., Lucito, R., Nairn, R., Richards, J., Pedroza, A., Hollek, L., Hazlewood, L., Walter, R. B., and Kallman, K. D. 2005. Sex-determination in platyfishes and swordtails. In M. C. Uribe and H. J. Grier (eds.), *Viviparous Fishes*, 381–400. Homestead, FL: New Life Publications.

Keddy, P., and Weiher, E. 1999. Introduction: the scope and goals of research on assembly rules. In E. Weiher and P. Keddy (eds.), *Ecological Assembly Rules: Perspectives, Advances, Retreats*, 1–20. Cambridge: Cambridge University Press.

Keegan-Rogers, V. 1983. Differential reproductive success among clones of unisexual fish: genetic factors and rare female advantage. American Zoologist 23: 1022.

Keegan-Rogers, V. 1984. Unfamiliar female mating advantage among clones of unisexual fish (*Poeciliopsis*: Poeciliidae). Copeia 1: 169–174.

Keegan-Rogers, V., and Schultz, R. J. 1984. Differences in courtship aggression among 6 clones of unisexual fish. Animal Behaviour 32: 1040–1044.

Keegan-Rogers, V., and Schultz, R. J. 1988. Sexual selection among clones of unisexual fish (*Poeciliopsis*: Poeciliidae): genetic factors and rare-female advantage. American Naturalist 132: 846–868.

Keller, L., and Reeve, H. K. 1995. Why do females mate with multiple males? The sexually selected sperm hypothesis. Advances in the Study of Animal Behavior 24: 291–315.

Keller, S. R., and Taylor, D. R. 2008. History, chance and adaptation during biological invasion: separating stochastic phenotypic evolution from response to selection. Ecology Letters 11: 852–866.

Kelley, J., Walter, L., and Trowsdale, J. 2005. Comparative genomics of major histocompatibility complexes. Immunogenetics 56: 683–695.

Kelley, J. L., and Magurran, A. E. 2003. Learned predator recognition and antipredator responses in fishes. Fish and Fisheries 4: 216–226.

Kelley, J. L., and Magurran, A. E. 2006. Learned defences and counterdefences in predator-prey interactions. In C. Brown, K. Laland, and J. Krause (eds.), *Fish Cognition and Behavior*, 28–43. Oxford: Blackwell.

Kelley, J. L., Graves, J. A., and Magurran, A. E. 1999. Familiarity breeds contempt in guppies. Nature 401: 661.

Kelley, J. L., Evans, J. P., Ramnarine, I. W., and Magurran, A. E. 2003. Back to school: Can antipredator behaviour in guppies be enhanced through social learning? Animal Behaviour 65: 655–662.

Kelly, C. D., and Godin, J. G. J. 2001. Predation risk reduces male-male sexual competition in the Trinidadian guppy (*Poecilia reticulata*). Behavioral Ecology and Sociobiology 51: 95–100.

Kelly, C. D., Godin, J. G. J., and Wright, J. M. 1999. Geographical variation in multiple paternity within natural populations of the guppy (*Poecilia reticulata*). Proceedings of the Royal Society of London Series B—Biological Sciences 266: 2403–2408.

Kelly, C. D., Godin, J. G. J., and Abdallah, G. 2000. Geographical variation in the male intromittent organ of the Trinidadian guppy (*Poecilia reticulata*). Canadian Journal of Zoology 78: 1674–1680.

Kelly, J. M., Adrian, J. C., and Brown, G. E. 2006. Can the ratio of aromatic skeletons explain cross-species responses within evolutionarily conserved Ostariophysan alarm cues? Testing the purine-ratio hypothesis. Chemoecology 16: 93–96.

Kemp, D. J., Wiklund, C., and Gotthard, K. 2006. Life history effects upon contest behaviour: age as a predictor of territorial contest dynamics in two populations of the speckled wood butterfly, *Pararge aegeria* L. Ethology 112: 471–477.

Kemp, D. J., Reznick, D. N., and Grether, G. F. 2008. Ornamental evolution in Trinidadian guppies (*Poecilia reticulata*): insights from sensory processing–based analyses of entire colour patterns. Biological Journal of the Linnean Society 95: 734–747.

Kendal, J. R., Rendell, L., Pike, T. W., and Laland, K. N. 2009. Nine-spined sticklebacks deploy a hill-climbing social learning strategy. Behavioral Ecology 20: 238–244.

Kendal, R. L., Coolen, I., van Bergen, Y., and Laland, K. N. 2005. Trade-offs in the adaptive use of social and asocial learning. Advances in the Study of Behavior 35: 333–379.

Kennedy, C. E. J., Endler, J. A., Poynton, S. L., and McMinn, H. 1987. Parasite load predicts mate choice in guppies. Behavioral Ecology and Sociobiology 21: 291–295.

Kennedy, M., and Gray, R. D. 1994. Agonistic interactions and the distribution of foraging organisms: individual costs and social information. Ethology 96: 155–165.

Khoo, G., Lim, T. M., Chan, W. K., and Phang, V. P. E. 1999a. Genetic basis of the variegated tail pattern in the guppy, *Poecilia reticulata*. Zoological Science 16: 431–437.

Khoo, G., Lim, T. M., Chan, W. K., and Phang, V. P. E. 1999b. Sex-linkage of the black caudal-peduncle and red tail genes in the tuxedo strain of the guppy, *Poecilia reticulata*. Zoological Science 16: 629–638.

Kilian, A., Yaman, S., Von Fersen, L., and Gunturkun, O. 2003. A bottlenose dolphin discriminates visual stimuli differing in numerosity. Learning and Behavior 31: 133–142.

Kim, J. H., Hayward, C. J., Joh, S. J., and Heo, G. J. 2002. Parasitic infections in live freshwater tropical fishes imported to Korea. Diseases of Aquatic Organisms 52: 169–173.

Kime, N. M., Rand, A. S., Kapfer, M., and Ryan, M. J. 1998. Consistency of female choice in the tungara frog: a permissive preference for complex characters. Animal Behaviour 55: 641–649.

King, T. A., and Cable, J. 2007. Experimental infections of the monogenean *Gyrodactylus turnbulli* indicate that it is not a strict specialist. International Journal for Parasitology 37: 663–672.

King, T. A., van Oosterhout, C., and Cable, J. 2009. Experimental infections with the tropical monogenean, *Gyrodactylus bullatarudis*: potential invader or experimental fluke? Parasitology International 58: 249–254.

Kingman, J. F. C. 1982. The coalescent. Stochastic Processes and Their Applications 13: 235–248.

Kinnison, M. T., Hendry, A. P., and Stockwell, C. A. 2007. Contemporary evolution meets conservation biology, II: impediments to integration and application. Ecological Research 22: 947–954.

Kirkpatrick, M. 1985. Evolution of female choice and male parental investment in polygynous species: the demise of the "sexy son." American Naturalist 125: 788–810.

Kirkpatrick, M., and Dugatkin, L. A. 1994. Sexual selection and the evolutionary effects of copying mate choice. Behavioral Ecology and Sociobiology 34: 443–449.

Kirkpatrick, M., and Hall, D. W. 2004. Sexual selection and sex linkage. Evolution 58: 683–691.

Kirkpatrick, M., and Ryan, M. J. 1991. The evolution of mating preferences and the paradox of the lek. Nature 350: 33–38.

Kittell, M. M., Harvey, M. N., Balderas, S. C., and Ptacek., M. B. 2005. Wild-caught hybrids between sailfin and shortfin mollies (Poeciliidae, Poecilia): morphological and molecular verification. Hidrobiológica 15: 131–137.

Klein, J., and Sato, A. 1998. Birth of the major histocompatibility complex. Scandinavian Journal of Immunology 47: 199–209.

Klein, J., Bontrop, R. E., Dawkins, R. L., Erlich, H. A., Gyllensten, U. B., Heise, E. R., Jones, P. P., Parham, P., Wakeland, E. K., and Watkins, D. I. 1990. Nomenclature for the major histocompatibility complexes of different species: a proposal. Immunogenetics 31: 217–219.

Klein, J., Ono, H., Klein, D., and O'hUigin, C. 1993. The accordion model of Mhc evolution. Progress in Immunology 8: 137–143.

Klein, J., Sato, A., Nagl, S., and O'hUigin, C. 1998. Molecular trans-species polymorphism. Annual Review of Ecology and Systematics 29: 1–21.

Klein, J., Sato, A., and Mayer, W. E. 2000. Jaws and AIS. In M. Kasahara (ed.), Major Histocompatibility Complex: Evolution, Structure and Function, 3–26. Tokyo: Springer.

Klerks, P. L., and Lentz, S. A. 1998. Resistance to lead and zinc in the western mosquitofish Gambusia affinis inhabiting contaminated Bayou Trepagnier. Ecotoxicology 7: 11–17.

Kobayashi, H., and Iwamatsu, T. 2002. Fine structure of the storage micropocket of spermatozoa in the ovary of the guppy Poecilia reticulata. Zoological Science 19: 545–555.

Kodama, I., Yamanaka, A., Endo, K., and Koya, Y. 2008. Role of the yellow spot around the urogenital opening of female mosquitofish (Gambusia affinis) as a cue for copulation. Zoological Science 25: 1199–1204.

Kodric-Brown, A. 1985. Female preference and sexual selection for male coloration in the guppy (Poecilia reticulata). Behavioural Ecology and Sociobiology 17: 199–205.

Kodric-Brown, A. 1989. Dietary carotenoids and male mating success in the guppy: an environmental component to female choice. Behavioural Ecology and Sociobiology 25: 393–401.

Kodric-Brown, A. 1992. Male dominance can enhance mating success in guppies. Animal Behaviour 44: 165–167.

Kodric-Brown, A. 1993. Female choice of multiple male criteria in guppies: interacting effects of dominance, coloration and courtship. Behavioural Ecology and Sociobiology 32: 415–420.

Kodric-Brown, A. 1998. Sexual dichromatism and temporary color changes in the reproduction of fishes. American Zoologist 38: 70–81.

Kodric-Brown, A., and Nicoletto, P. F. 1997. Repeatability of female choice in the guppy: response to live and videotaped males. Animal Behaviour 54: 369–376.

Kodric-Brown, A., and Nicoletto, P. F. 2001. Age and experience affect female choice in the guppy (Poecilia reticulata). American Naturalist 157: 316–323.

Kodric-Brown, A., and Nicoletto, P. F. 2005. Courtship behavior, swimming performance, and microhabitat use of Trinidadian guppies. Environmental Biology of Fishes 73: 299–307.

Kojima, K. I. 1971. Is there a constant fitness value for a given genotype? No! Evolution 25: 281–285.

Kokko, H., and Johnstone, R. A. 2002. Why is mutual mate choice not the norm? Operational sex ratios, sex roles and the evolution of sexually dimorphic and monomorphic signalling. Philosophical Transactions of the Royal Society of London Series B—Biological Sciences 357: 319–330.

Kokko, H., Brooks, R., McNamara, J. M., and Houston, A. I. 2002. The sexual selection continuum. Proceedings of the Royal Society of London Series B—Biological Sciences 269: 1331–1340.

Kokko, H., Brooks, R., Jennions, M. D., and Morley, J. 2003. The evolution of mate choice and mating biases. Proceedings of the Royal Society of London Series B—Biological Sciences 270: 653–664.

Kokko, H., Heubel, K. U., and Rankin, D. J. 2008. How populations persist when asexuality requires sex: the spatial dynamics of coping with sperm parasites. Proceedings of the Royal Society of London Series B—Biological Sciences 275: 817–825.

Kolluru, G. R., and Grether, G. F. 2005. The effects of resource availability on alternative mating tactics in guppies (Poecilia reticulata). Behavioral Ecology 16: 294–300.

Kolluru, G. R., and Joyner, J. W. 1997. The influence of male body size and social environment on the mating behavior of Phallichthys quadripunctatus (Pisces: Poeciliidae). Ethology 103: 744–759.

Kolluru, G. R., and Reznick, D. N. 1996. Genetic and social control of male maturation in Phallichthys quadripunctatus (Pisces: Poeciliidae). Journal of Evolutionary Biology 9: 695–715.

Kolluru, G. R., Grether, G. F., South, S. H., Dunlop, E., Cardinali, A., Liu, L., and Carapiet, A. 2006. The effects of carotenoid and food availability on resistance to a naturally occurring parasite (Gyrodactylus turnbulli) in guppies (Poecilia reticulata). Biological Journal of the Linnean Society 89: 301–309.

Kolluru, G. R., Grether, G. F., and Contreras, H. 2007. Environmental and genetic influences on mating strategies along a replicated food availability gradient in guppies (Poecilia reticulata). Behavioral Ecology and Sociobiology 61: 689–701.

Kolluru, G. R., Grether, G. F., Dunlop, E., and South, S. H. 2009. Food availability and parasite infection influence mating tactics in guppies (Poecilia reticulata). Behavioral Ecology 20: 131–137.

Komers, P. E. 1997. Behavioural plasticity in variable environments. Canadian Journal of Zoology 75: 161–169.

Kondrashov, A. S. 1984. Deleterious mutations as an evolutionary factor, 1: the advantage of recombination. Genetical Research 44: 199–217.

Kondrashov, A. S. 1985. Deleterious mutations as an evolutionary factor, 2: facultative apomixes and selfing. Genetics 111: 635–653.

Kondrashov, A. S. 1988. Deleterious mutations and the evolution of sexual reproduction. Nature 336: 435–440.

Koolhaas, J. M., de Boer, S. F., Buwalda, B., and van Reenen, K. 2007. Individual variation in coping with stress: a multidimensional approach of ultimate and proximate mechanisms. Brain, Behavior, and Evolution 70: 218–229.

Körner, K. E., Schlupp, I., Plath, M., and Loew, E. R. 2006. Spectral sensitivity of mollies: comparing surface- and cave-dwelling Atlantic mollies, *Poecilia mexicana*. Journal of Fish Biology 69: 54–65.

Korniumshin, A. V., and Glaubrecht, M. 2003. Novel reproductive modes in freshwater clams: brooding and larval morphology in Southeast Asia taxa of *Corbicula* (Mollusca, Bivalvia, Corbiculidae). Acta Zoologica 84: 293–315.

Korsgaard, B. 1994. Proteins and amino acids in maternal-embryonic trophic relationships in viviparous teleost fishes. Israel Journal of Zoology 40: 417–429.

Korsgaard, B., and Weber, R. E. 1989. Maternal-fetal trophic and respiratory relationships in viviparous ectothermic vertebrates. Advances in Comparative and Environmental Physiology 5: 209–233.

Kosswig, C. 1928. Über Kreuzungen zwischen den Teleostiern *Xiphophorus helleri* und *Platypoecilus maculatus*. Zeitschrift für Induktive Abstammungs- und Vererbungslehre, Berlin 47: 150–158.

Kosswig, C. 1929. Das Gen in fremder Erbmasse. Züchter 1: 152–157.

Kotiaho, J. S., Simmons, L. W., Hunt, J., and Tomkins, J. L. 2003. Males influence maternal effects that promote sexual selection: a quantitative genetic experiment with dung beetles *Onthophagus taurus*. American Naturalist 161: 852–859.

Kotrschal, K., van Staaden, M. J., and Huber, R. 1998. Fish brains: evolution and environmental relationships. Reviews in Fish Biology and Fisheries 8: 373–408.

Koya, Y., and Kamiya, E. 2000. Environmental regulation of annual reproductive cycle in the mosquitofish, *Gambusia affinis*. Journal of Experimental Zoology 286: 204–211.

Koya, Y., Itazu, T., and Inoue, M. 1998. Annual reproductive cycle based on histological changes in the ovary of the female mosquitofish, *Gambusia affinis*, in central Japan. Ichthyological Research 45: 241–248.

Koya, Y., Inoue, M., Naruse, T., and Sawguchi, S. 2000. Dynamics of oocyte and embryonic development during ovarian cycle of the viviparous mosquitofish *Gambusia affinis*. Fisheries Science 66: 63–70.

Koya, Y., Fujita, A., Niki, F., Ishihara, E., and Miyama, H. 2003. Sex differentiation and pubertal development of gonads in the viviparous mosquitofish, *Gambusia affinis*. Zoological Science 20: 1231–1242.

Kozak, H. L., Cirino, L. A., and Ptacek, M. B. 2008. Female mating preferences for male morphological traits used in species and mate recognition in the Mexican sailfin mollies, *Poecilia velifera* and *Poecilia petenensis*. Behavioral Ecology 19: 463–474.

Kozlowski, J., and Uchmanski, J. 1987. Optimal individual growth and reproduction in perennial species with indeterminate growth. Evolutionary Ecology 1: 214–230.

Kraaijeveld-Smit, F. J. L., Ward, S. J., Temple-Smith, P. D., and Paetkau, D. 2002. Factors influencing paternity success in *Antechinus agilis*: last-male sperm precedence, timing of mat-

ing and genetic compatibility. Journal of Evolutionary Biology 15: 100–107.

Krakauer, D. C. 1995. Groups confuse predators by exploiting perceptual bottlenecks: a connectionist model of the confusion effect. Behavioral Ecology and Sociobiology 36: 421–429.

Kramer, D. L. 1983. The evolutionary ecology of respiratory modes in fishes: an analysis based on the costs of breathing. Environmental Biology of Fishes 9: 145–158.

Kramer, D. L., and Bryant, M. J. 1995. Intestine length in the fishes of a tropical stream, 2: relationships to diet—the long and short of a convoluted issue. Environmental Biology of Fishes 42: 129–141.

Kramer, D. L., and McClure, M. 1982. Aquatic surface respiration, a widespread adaptation to hypoxia in tropical freshwater fishes. Environmental Biology of Fishes 7: 47–55.

Kramer, D. L., and Mehegan, J. 1981. Aquatic surface respiration, an adaptive response to hypoxia in the guppy, *Poecilia reticulata* (Pisces, Poeciliidae). Environmental Biology of Fishes 6: 299–313.

Krause, J. 1993a. The relationship between foraging and shoal position in a mixed shoal of roach (*Rutilus rutilus*) and chub (*Leuciscus cephlus*): a field study. Oecologia 93: 356–359.

Krause, J. 1993b. Transmission of fright reaction between different species of fish. Behaviour 127: 37–48.

Krause, J., and Godin, J. G. 1995. Predator preferences for attacking particular prey group sizes: consequences for predator hunting success and prey predation risk. Animal Behaviour 50: 465–473.

Krause, J., and Godin, J. G. J. 1996a. Influence of parasitism on shoal choice in the banded killifish (*Fundulus diaphanus*, Teleostei, Cyprinodontidae). Ethology 102: 40–49.

Krause, J., and Godin, J. G. J. 1996b. Influence of prey foraging posture on flight behavior and predation risk: predators take advantage of unwary prey. Behavioral Ecology 7: 264–271.

Krause, J., and Ruxton, G. D. 2002. *Living in Groups*. Oxford: Oxford University Press.

Krause, J., Butlin, R., Peuhkuri, N., and Pritchard, V. L. 2000. The social organisation of fish shoals: a test of the predictive power of laboratory experiments for the field. Biological Reviews 75: 477–501.

Krause, J., Croft, D. P., and James, R. 2007. Social network theory in the behavioural sciences: potential applications. Behavioral Ecology and Sociobiology 62: 15–27.

Kristensen, T., Baatrup, E., and Bayley, M. 2005. 17α-ethinyl-estradiol reduces the competitive reproductive fitness of the male guppy (*Poecilia reticulata*). Biology of Reproduction 72: 150–156.

Krumholz, L. A. 1948. Reproduction in the western mosquitofish, *Gambusia affinis affinis* (Baird & Girard), and its use in mosquito control. Ecological Monographs 18: 1–43.

Kuckuck, C., and Greven, H. 1997. Notes on the mechanically stimulated discharge of spermiozeugmata in the guppy, *Poecilia reticulata*: a quantitative approach. Zeitschrift für Fischkunde 4: 73–88.

Kummer, H., and Goodall, J. 1985. Conditions of innovative behaviour in primates. Philosophical Transactions of the Royal Society of London Series B—Biological Sciences 308: 203–214.

Kuntz, A. 1914. Notes on the habits, morphology of the reproductive organs, and embryology of the viviparous fish *Gam*-

busia affinis. Bulletin of the United States Bureau of Fisheries 33: 177–190.

Kunz, Y. W. 1963. Die embryonale Harnblase von *Lebistes reticulatus.* Revue Suisse de Zoologie 71: 291–207.

Kunz, Y. W. 1971. Histological study of greatly enlarged pericardial sac in the embryo of the viviparous teleost *Lebistes reticulatus.* Revue Suisse de Zoologie 78: 187–207.

Kunz, Y. W. 2007. Review of development and aging in the eye of teleost fish. Neuroembryology and Aging 4: 31–60.

Künzler, R., and Bakker, T. C. M. 2001. Female preferences for single and combined traits in computer animated stickleback males. Behavioral Ecology 12: 681–685.

Lachlan, R. F., Crooks, L., and Laland, K. N. 1998. Who follows whom? Shoaling preferences and social learning of foraging information in guppies. Animal Behaviour 56: 181–190.

Lafleur, D. L., Lozano, G. A., and Sclafani, M. 1997. Female mate-choice copying in guppies, *Poecilia reticulata*: a re-evaluation. Animal Behaviour 54: 579–586.

Laha, M., and Mattingly, H. T. 2006. Identifying environmental conditions to promote species coexistence: an example with the native Barrens topminnow and invasive western mosquitofish. Biological Invasions 8: 719–725.

Laland, K. N. 1994. Sexual selection with a culturally transmitted mating preference. Theoretical Population Biology 45: 1–15.

Laland, K. N. 2004. Social learning strategies. Learning and Behavior 32: 4–14.

Laland, K. N., and Reader, S. M. 1999a. Foraging innovation in the guppy. Animal Behaviour 57: 331–340.

Laland, K. N., and Reader, S. M. 1999b. Foraging innovation is inversely related to competitive ability in male but not in female guppies. Behavioral Ecology 10: 270–274.

Laland, K. N., and van Bergen, Y. 2003. Experimental studies of innovation in the guppy. In S. M. Reader and K. N. Laland (eds.), *Animal Innovation,* 155–174. Oxford: Oxford University Press.

Laland, K. N., and Williams, K. 1997. Shoaling generates social learning of foraging information in guppies. Animal Behaviour 53: 1161–1169.

Laland, K. N., and Williams, K. 1998. Social transmission of maladaptive information in the guppy. Behavioral Ecology 9: 493–499.

Lamatsch, D. K., Nanda, I., Epplen, J. T., Schmid, M., and Schartl, M. 2000a. Unusual triploid males in a microchromosome-carrying clone of the Amazon molly, *Poecilia formosa.* Cytogenetic and Genome Research 91: 148–156.

Lamatsch, D. K., Steinlein, C., Schmid, M., and Schartl, M. 2000b. Noninvasive determination of genome size and ploidy level in fishes by flow cytometry: detection of triploid *Poecilia formosa.* Cytometry 39: 91–95.

Lamatsch, D. K., Nanda, I., Schlupp, I., Epplen, J. T., Schmid, M., and Schartl, M. 2004. Distribution and stability of supernumerary microchromosomes in natural populations of the Amazon molly, *Poecilia formosa.* Cytogenetic and Genome Research 106: 189–194.

Lambert, J. G. D. 1970a. The ovary of the guppy, *Poecilia reticulata*: the atretic follicle, a corpus atreticum or a corpus luteum preaovulationis. Zeitschrift für Zellforschung und Mikroskopische Anatomie 107: 54–67.

Lambert, J. G. D. 1970b. The ovary of the guppy *Poecilia*

reticulata: the granulosa cells as sites of steroid biosynthesis. General and Comparative Endocrinology 15: 464–476.

Lampert, K. P., Lamatsch, D. K., Epplen, J. T., and Schartl, M. 2005. Evidence for a monophyletic origin of triploid clones of the Amazon molly, *Poecilia formosa.* Evolution 59: 881–889.

Lampert, K. P., Lamatsch, D. K., Fischer, P., Epplen, J. T., Nanda, I., Schmid, M., and Schartl, M. 2007. Automictic reproduction in interspecific hybrids of poeciliid fish. Current Biology 17: 1948–1953.

Lampert, K. P., Lamatsch, D. K., Fischer, P., and Schartl, M. 2008. A tetraploid Amazon molly, *Poecilia formosa.* Journal of Heredity 99: 223–226.

Lampert, K. P., Schmidt, C., Fischer, P., Volff, J.-N., Hoffmann, C., Muck, J., Lohse, M. J., Ryan, M. J., and Schartl, M. 2010. Determination of onset of sexual maturation and mating behavior by melanocortin receptor 4 polymorphisms. Current Biology 20: 1729–1734.

Landau, H. G. 1951. On dominance relations and the structure of animal societies. Bulletin of Mathematical Biophysics 13: 245–262.

Lande, R. 1981. Models of speciation by sexual selection on polygenic traits. Proceedings of the National Acadamy of Sciences of the United States of America 78: 3721–3725.

Lande, R., and Arnold, S. J. 1983. The measurement of selection on correlated characters. Evolution 37: 1210–1226.

Landeau, L., and Terborgh, J. 1986. Oddity and the "confusion effect" in predation. Animal Behaviour 34: 1372–1380.

Landry, C., and Bernatchez, L. 2001. Comparative analysis of population structure across environments and geographical scales at major histocompatibility complex and microsatellite loci in Atlantic salmon (*Salmo salar*). Molecular Ecology 10: 2525–2539.

Landry, C., Garant, D., Duchesne, P., and Bernatchez, L. 2001. "Good genes as heterozygosity": the major histocompatibility complex and mate choice in Atlantic salmon (*Salmo salar*). Proceedings of the Royal Society of London Series B—Biological Sciences 268: 1279–1285.

Lane, J. E., Boutin, S., Gunn, M. R., Slate, J., and Coltman, D. W. 2007. Genetic relatedness of mates does not predict patterns of parentage in North American red squirrels. Animal Behaviour 74: 611–619.

Langecker, T. G. 2000. The effect of continuous darkness on cave ecology and cavernicolous evolution. In H. Wilkens, D. C. Culver, and W. F. Humphreys (eds.), *Subterranean Ecosystems,* 135–157. Ecosystems of the World 30. Amsterdam: Elsevier Science.

Langefors, A., Von Schantz, T., and Widegren, B. 1998. Allelic variation of Mhc class II in Atlantic salmon; a population genetic analysis. Heredity 80: 568–575.

Langefors, A., Lohm, J., von Schantz, T., and Grahn, M. 2000. Screening of Mhc variation in Atlantic salmon (*Salmo salar*): a comparison of restriction fragment length polymorphism (RFLP), denaturing gradient gel electrophoresis (DGGE) and sequencing. Molecular Ecology 9: 215–219.

Langefors, A., Lohm, J., Grahn, M., Andersen, O., and von Schantz, T. 2001. Association between major histocompatibility complex class IIB alleles and resistance to *Aeromonas salmonicida* in Atlantic salmon. Proceedings of the Royal Society of London Series B—Biological Sciences 268: 479–485.

Langen, T. A. 1999. How western scrub-jays (*Aphelocoma californica*) select a nut: effects of the number of options, variation in nut size, and social competition among foragers. Animal Cognition 2: 223–233.

Langerhans, R. B., and DeWitt, T. J. 2004. Shared and unique features of evolutionary diversification. American Naturalist 164: 335–349.

Langerhans, R. B., and Reznick, D. N. 2010. Ecology and evolution of swimming performance in fishes: predicting evolution with biomechanics. In P. Domenici and B. G. Kapoor (eds.), *Fish Locomotion: An Etho-ecological Perspective*, 200–248. Enfield, NH: Science Publishers.

Langerhans, R. B., Layman, C. A., Shokrollahi, A. M., and De-Witt, T. J. 2004. Predator-driven phenotypic diversification in *Gambusia affinis*. Evolution 58: 2305–2318.

Langerhans, R. B., Layman, C. A., and DeWitt, T. J. 2005. Male genital size reflects a tradeoff between attracting mates and avoiding predators in two live-bearing fish species. Proceedings of the National Academy of Sciences of the United States of America 102: 7618–7623.

Langerhans, R. B., Gifford, M. E., and Joseph, E. O. 2007. Ecological speciation in *Gambusia* fishes. Evolution 61: 2056–2074.

Langson, J. C. 1990. Major protozoan and metazoan parasitic diseases of Australian finfish. In Post-Graduate Committee in Veterinary Science, Fish Diseases: Refresher Course for Veterinarians, 233–255.. Sydney: University of Sydney.

Larsson, D. G. J., Kinnberg, K., Sturve, J., Stephensen, E., Skön, M., and Förlin, L. 2002. Studies of masculinization, detoxification, and oxidative stress responses in guppies (*Poecilia reticulata*) exposed to effluent from a pulp mill. Ecotoxicology and Environmental Safety 52: 13–20.

Larsson, P., Ollinger, K., and Rosdahl, I. 2006. Ultraviolet (UV)A- and UVB-induced redox alterations and activation of nuclear factor-kappaB in human melanocytes—protective effects of alpha-tocopherol. British Journal of Dermatology 155: 292–300.

Law, R. 1979. Optimal life histories under age-specific predation. American Naturalist 114: 399–417.

Lawler, S. P., Dritz, D., Strange, T., and Holyoak, M. 1999. Effects of introduced mosquitofish and bullfrogs on the threatened California red-legged frog. Conservation Biology 13: 613–622.

Lawton, B. R., Sevigny, L., Obergfell, C., Reznick, D., O'Neill, R. J., and O'Neill, M. J. 2005. Allelic expression of IGF2 in live-bearing, matrotrophic fishes. Development, Genes, and Evolution 215: 207–212.

Lazzari, M., Bettini, S., Ciani, F., and Franceschini, V. 2007. Light and transmission electron microscopy study of the peripheral olfactory organ of the guppy, *Poecilia reticulata* (Teleostei, Poecilidae). Microscopy Research and Technique 70: 782–789.

Leberg, P. L. 1990. Influence of genetic variability on population growth: implications for conservation. Journal of Fish Biology 37: 193–195.

Leberg, P. L. 1993. Strategies for population reintroduction: effects of genetic variability on population growth and size. Conservation Biology 7: 194–199.

Leberg, P. L., and Vrijenhoek, R. C. 1994. Variation among desert topminnows in their susceptibility to attack by exotic parasites. Conservation Biology 8: 419–424.

Ledesma, J. M., and McRobert, S. P. 2008. Shoaling in juvenile guppies: the effects of body size and shoal size. Behavioural Processes 77: 384–388.

Leips, J., and Travis, J. 1999. The comparative expression of life-history traits and its relationship to the numerical dynamics of four populations of the least killifish. Journal of Animal Ecology 68: 595–616.

Leips, J., Richardson, J. M. L., Rodd, F. H., and Travis, J. 2009. Adaptive maternal adjustments of offspring size in response to conspecific density in two populations of the least killifish, *Heterandria formosa*. Evolution 63: 1341–1347.

Lenormand, T. 2002. Gene flow and the limits to natural selection. Trends in Ecology and Evolution 17: 183–189.

Leo, P., and Greven, H. 1999. Beobachtungen zum Balz- und Paarungsverhalten männlicher Guppys (*Poecilia reticulata*) gegenüber rezeptiven und nichtrezeptiven Weibchen. In R. Riehl and H. Greven (eds.), *Fortpflanzungsbiologie der Aquarienfische (2)*, 133–147. Bornheim: Birgit Schmettkamp.

Leslie, J. F., and Vrijenhoek, R. C. 1977. Genetic analysis of natural populations of *Poeciliopsis monacha*. Journal of Heredity 68: 301–306.

Lessells, C. M. 2006. The evolutionary outcome of sexual conflict. Philosophical Transactions of the Royal Society of London Series B—Biological Sciences 361: 301–317.

Levine, J. S., and MacNichol, E. F. 1979. Visual pigments in teleost fishes: effects of habitat, microhabitat, and behavior on visual system evolution. Sensory Processes 3: 95–131.

Levins, R. 1968. *Evolution in Changing Environments: Some Theoretical Explorations*. Princeton, NJ: Princeton University Press.

Levsen, A. 2001. Transmission ecology and larval behaviour of *Camallanus cotti* (Nematoda, Camallanidae) under aquarium conditions. Aquarium Sciences and Conservation 3: 301–311.

Levsen, A., and Berland, B. 2002. The development and morphogenesis of *Camallanus cotti* Fujita, 1927 (Nematoda: Camallanidae), with notes on its phylogeny and definitive host range. Systematic Parasitology 53: 29–37.

Levsen, A., and Jakobsen, P. J. 2002. Selection pressure towards monoxeny in *Camallanus cotti* (Nematoda, Camallanidae) facing an intermediate host bottleneck situation. Parasitology 124: 625–629.

Lewis, S. E. M., Sterling, E. S. L., Young, I. S., and Thompson, W. 1997. Comparison of individual antioxidants of sperm and seminal plasma in fertile and infertile men. Fertility and Sterility 67: 142–147.

Lewis, W. 1970. Morphological adaptations of cyprinodontoids for inhabiting oxygen deficient waters. Copeia 1970: 319–326.

Licht, T. 1989. Discrimination between hungry and satiated predators: the response of guppies (*Poecilia reticulata*) from high and low predation sites. Ethology 82: 238–243.

Lieberman, E., Hauert, C., and Nowak, M. A. 2005. Evolutionary dynamics on graphs. Nature 433: 312–316.

Lighthill, M. J. 1970. Aquatic animal propulsion of high hydromechanical efficiency. Journal of Fluid Mechanics 44: 265–301.

Liley, N. R. 1966. Ethological isolating mechanisms in four sympatric species of poeciliid fishes. Behaviour Supplements 13: 1–197.

Liley, N. R. 1968. The endocrine control of reproductive behavior in the female guppy, *Poecilia reticulata* Peters. Animal Behaviour 16: 318–331.

Liley, N. R., and Seghers, B. H. 1975. Factors affecting the morphology and behaviour of guppies in Trinidad. In G. Baerends, C. Beer, and A. Manning (eds.), *Function and Evolution in Behaviour*, 92–118. Oxford: Clarendon Press.

Liley, N. R., and Wishlow, W. P. 1974. The interaction of endocrine and experiential factors in the regulation of sexual behaviour in the female guppy *Poecilia reticulata*. Behaviour 48: 185–214.

Lima, N. R. W. 1998. Genetic analysis of predatory efficiency in natural and laboratory made hybrids of *Poeciliopsis* (Pisces: Poeciliidae). Behaviour 135: 83–98.

Lima, N. R. W. 2005. Variations on maternal-embryonic relationship in two natural and six laboratory made hybrids of *Poeciliopsis monacha-lucida* (Pisces, Cyprinodontiformes). Brazilian Archives of Biology and Technology 48: 73–79.

Lima, N. R. W., Kobak, C. J., and Vrijenhoek, R. C. 1996. Evolution of sexual mimicry in sperm-dependent all-female forms of *Poeciliopsis* (Pisces: Poeciliidae). Journal of Evolutionary Biology 9: 185–203.

Lima, S. L. 1998. Stress and decision making under the risk of predation: recent developments from behavioral, reproductive, and ecological perspectives. Advances in the Study of Animal Behavior 27: 215–290.

Lima, S. L., and Bednekoff, P. A. 1999. Temporal variation in danger drives antipredator behavior: the predation risk allocation hypothesis. American Naturalist 153: 649–659.

Lima, S. L., and Dill, L. M. 1990. Behavioral decisions made under the risk of predation: a review and prospectus. Canadian Journal of Zoology 68: 619–640.

Lincoln, R., Boxshall, G. A., and Clark, P. 1998. *A Dictionary of Ecology, Evolution and Systematics*. 2d ed. Cambridge: Cambridge University Press.

Lindholm, A., and Breden, F. 2002. Sex chromosomes and sexual selection in poeciliid fishes. American Naturalist 160: S143–S224.

Lindholm, A. K., Brooks, R., and Breden, F. 2004. Extreme polymorphism in a Y-linked sexually selected trait. Heredity 92: 156–162.

Lindholm, A. K., Breden, F., Alexander, H. J., Chan, W. K., Thakurta, S. G., and Brooks, R. 2005. Invasion success and genetic diversity of introduced populations of guppies *Poecilia reticulata* in Australia. Molecular Ecology 14: 3671–3682.

Lindstrom, K., and Ranta, E. 1993. Social preferences by male guppies, *Poecilia reticulata*, based on shoal size and sex. Animal Behaviour 46: 1029–1031.

Ling, N. 2004. *Gambusia* in New Zealand: really bad or just misunderstood? New Zealand Journal of Marine and Freshwater Research 38: 473–480.

Lintermans, M. 2004. Human-assisted dispersal of alien freshwater fish in Australia. New Zealand Journal of Marine and Freshwater Research 38: 481–501.

Lively, C. M., and Dybdahl, M. F. 2000. Parasite adaptation to locally common host genotypes. Nature 405: 679–681.

Lively, C. M., Craddock, C., and Vrijenhoek, R. C. 1990. Red Queen Hypothesis supported by parasitism in sexual and clonal fish. Nature 344: 864–866.

Locatello, L., Rasotto, M. B., Evans, J. P., and Pilastro, A. 2006.

Colourful male guppies produce faster and more viable sperm. Journal of Evolutionary Biology 19: 1595–1602.

Lodi, E. 1979. Instances of sex inversion in the domesticated swordtail, *Xiphophorus helleri* Heckel (Pisces, Osteichthyes). Experientia 35: 1440–1441.

Lodi, E. 1981. Competition between palla and normal bearing spermatozoa of *Poecilia reticulata* (Pisces: Poeciliidae). Copeia 1981: 624–629.

Loewe, L., and Lamatsch, D. K. 2008. Quantifying the threat of extinction from Muller's ratchet in the diploid Amazon molly (*Poecilia formosa*). BMC Evolutionary Biology 8: 88.

Loftus, W. F. 2000. Accumulation and fate of mercury in an Everglades aquatic food web. PhD thesis, Florida International University.

Lomassese, S. S., Strambi, A., Charpin, P., Augier, R., Aouane, A., and Cayre, M. 2000. Influence of environmental stimulation on neurogenesis in the adult insect brain. Journal of Neurobiology 45: 162–171.

Lombardi, J. 1996. Postzygotic maternal influences and the maternal-embryonic relationship of viviparous fishes. American Zoologist 36: 106–115.

Long, K. D., and Houde, A. E. 1989. Orange spots as a visual cue for female mate choice in the guppy (*Poecilia reticulata*). Ethology 82: 316–324.

Long, K. D., and Rosenqvist, G. 1998. Changes in male guppy courting distance in response to a fluctuating light environment. Behavioral Ecology and Sociobiology 44: 77–83.

López, S. 1998. Acquired resistance affects male sexual display and female choice in guppies. Proceedings of the Royal Society of London Series B—Biological Sciences 265: 717–723.

López, S. 1999. Parasitized female guppies do not prefer showy males. Animal Behaviour 57: 1129–1134.

Lorier, E., and Berois, N. 1995. Reproduction and embrionary nurture in *Cnesterodon decemmaculatus* (Teleostei: Poeciliidae). Revista Brasileira de Biologia 55: 27–44.

Losos, J. B., Leal, M., Glor, R. E., de Queiroz, K., Hertz, P. E., Schettino, L. R., Lara, A. C., Jackman, T. R., and Larson, A. 2003. Niche lability in the evolution of a Caribbean lizard community. Nature 424: 542–545.

Lozano, G. A. 1994. Carotenoids, parasites, and sexual selection. Oikos 70: 309–311.

Lozano-Vilano, M. D., and Contreras-Balderas, S. 1999. *Cyprinodon bobmilleri*: New species of pupfish from Nuevo Leon, Mexico (Pisces: Cyprinodontidae). Copeia 1999: 382–387.

Lucinda, P. H. F. 2005a. Systematics and biogeography of the genus *Phalloptychus* Eigenmann, 1907 (Cyprinodontiformes: Poeciliidae: Poeciliinae). Neotropical Ichthyology 3: 373–382.

Lucinda, P. H. F. 2005b. Systematics of the genus *Cnesterodon* Garman, 1895 (Cyprinodontiformes: Poeciliidae: Poeciliinae). Neotropical Ichthyology 3: 259–270.

Lucinda, P. H. F. 2008. Systematics and biogeography of the genus *Phalloceros* Eigenmann, 1907 (Cyprinodontiformes: Poeciliidae: Poeciliinae), with the description of twenty-one new species. Neotropical Ichthyology 6: 113–158.

Lucinda, P. H. F., Rosa, R. D., and Reis, R. E. 2005. Systematics and biogeography of the genus *Phallotorynus* Henn, 1916 (Cyprinodontiformes: Poeciliidae: Poeciliinae), with description of three new species. Copeia 2005: 609–631.

Lucinda, P. H. F., Litz, T., and Recuero, R. 2006. *Cnesterodon holopteros* (Cyprinodontiformes: Poeciliidae: Poeciliinae), a new species from the Republic of Uruguay. Zootaxa 2006: 21–31.

Ludlow, A. M., and Magurran, A. E. 2006. Gametic isolation in guppies (*Poecilia reticulata*). Proceedings of the Royal Society of London Series B—Biological Sciences 273: 2477–2482.

Luo, J., Sanetra, M., Schartl, M., and Meyer, A. 2005. Strong reproductive skew among males in the multiply mated swordtail *Xiphophorus multilineatus* (Teleostei). Journal of Heredity 96: 346–355.

Luyten, P. H., and Liley, N. R. 1985. Geographic variation in the sexual behaviour of the guppy, *Poecilia reticulata* (Peters). Behaviour 95: 164–179.

Luyten, P. H., and Liley, N. R. 1991. Sexual selection and competitive mating success of male guppies (*Poecilia reticulata*) from four Trinidad populations. Behavioral Ecology and Sociobiology 28: 329–336.

Lydeard, C., and Belk, M. C. 1993. Management of indigenous fish species impacted by introduced mosquitofish: an experimental approach. Southwestern Naturalist 38: 370–373.

Lydeard, C., Wooten, M. C., and Meyer, A. 1995. Molecules, morphology, and area cladograms—a cladistic and biogeographic analysis of *Gambusia* (Teleostei, Poeciliidae). Systematic Biology 44: 221–236.

Lyles, A. M. 1990. Genetic variation and susceptibility to parasites: *Poecilia reticulata* infected with *Gyrodactylus turnbulli*. PhD thesis, Princeton University.

Lynch, M. 1996. A quantitative genetic perspective on conservation issues. In J. C. Avise and J. L. Hamrick (eds.), *Conservation Genetics: Case Histories from Nature*, 471–501. New York: Chapman & Hall.

Lynch, M., and Walsh, B. 1998. *Genetics and Analysis of Quantitative Traits*. Sunderland, MA: Sinauer Associates.

Lythgoe, J. N. 1979. *The Ecology of Vision*. Oxford: Clarendon Press.

Lythgoe, J. N., and Partridge, J. C. 1989. Visual pigments and the acquisition of visual information. Journal of Experimental Biology 146: 1–20.

Lythgoe, J. N., Muntz, W. R. A., Partridge, J. C., Shand, J., and Williams, D. M. 1993. The ecology of the visual pigments of snappers (Lutjanidae) on the Great Barrier Reef. Journal of Comparative Physiology A—Neuroethology, Sensory, Neural, and Behavioral Physiology 174: 461–467.

Maan, M. E., Hofker, K. D., van Alphen, J. J. M., and Seehausen, O. 2006. Sensory drive in cichlid speciation. American Naturalist 167: 947–954.

MacArthur, R. H. 1962. Some generalized theorems of natural selection. Proceedings of the National Academy of Sciences of the United States of America 48: 1893–1897.

MacArthur, R. H. 1972. *Geographical Ecology: Patterns in the Distribution of Species*. New York: Harper & Row.

MacArthur, R. H., and Wilson, E. O. 1967. *The Theory of Island Biogeography*. Princeton, NJ: Princeton University Press.

Macedonia, J. M., and Evans, C. S. 1993. Variation among mammalian alarm call systems and the problem of meaning in animal signals. Ethology 93: 177–197.

MacFarlane, R. B., and Bowers, M. J. 1995. Matrotrophic viviparity in the yellowtail rockfish *Sebastes flavidus*. Journal of Experimental Biology 198: 1197–1206.

Macías-Garcia, C., Saborío, E., and Berea, C. 1998. Does male

biased predation lead to male scarcity in viviparous fish? Journal of Fish Biology 53: 104–117.

Mack, P. D., Hammock, B. A., and Promislow, D. E. L. 2002. Sperm competitive ability and genetic relatedness in *Drosophila melanogaster*: similarity breeds contempt. Evolution 56: 1789–1795.

MacLaren, R. D., and Rowland, W. J. 2006. Female preference for male lateral projection area in the shortfin molly, *Poecilia mexicana*: evidence for a pre-existing bias in sexual selection. Ethology 112: 678–690.

Maddern, M. 2003. The distribution, biology and ecological impacts of three introduced freshwater teleosts in Western Australia. Honors thesis, Murdoch University, Western Australia.

Maddison, W. P., and Maddison, D. R. 2008. *Mesquite: A Modular System for Evolutionary Analysis*. http://mesquiteproject .org, version 2.5.

Madhavi, R., and Anderson, R. M. 1985. Variability in the susceptibility of the fish host, *Poecilia reticulata*, to infection with *Gyrodactylus bullatarudis* (Monogenea). Parasitology 91: 531–544.

Magellan, K., and Magurran, A. E. 2006. Habitat use mediates the conflict of interest between the sexes. Animal Behaviour 72: 75–81.

Magellan, K., and Magurran, A. E. 2007a. Behavioural profiles: individual consistency in male mating behaviour under varying sex ratios. Animal Behaviour 74: 1545–1550.

Magellan, K., and Magurran, A. E. 2007b. Mate choice, sexual coercion and gene flow in guppy populations. Journal of Fish Biology 71: 1864–1872.

Magellan, K., Pettersson, L. B., and Magurran, A. E. 2005. Quantifying male attractiveness and mating behaviour through phenotypic size manipulation in the Trinidadian guppy, *Poecilia reticulata*. Behavioral Ecology and Sociobiology 58: 366–374.

Maglio, V. J., and Rosen, D. E. 1969. Changing preference for substrate color by reproductively active mosquitofish, *Gambusia affinis* (Baird and Girard) (Poeciliidae, Atheriniformes). American Museum Novitates 2379: 1–37.

Magnhagen, C., and Borcherding, J. 2008. Risk-taking behaviour in foraging perch: does predation pressure influence age-specific boldness? Animal Behaviour 75: 509–517.

Magurran, A. E. 1989. Acquired recognition of predator odour in the European minnow (*Phoxinus phoxinus*). Ethology 82: 216–223.

Magurran, A. E. 1990. The inheritance and development of minnow antipredator behaviour. Animal Behaviour 39: 834–842.

Magurran, A. E. 2001. Sexual conflict and evolution in Trinidadian guppies. Genetica 112–113: 463–474.

Magurran, A. E. 2005. *Evolutionary Ecology: The Trinidadian Guppy*. Oxford: Oxford University Press.

Magurran, A. E., and Garcia, C. M. 2000. Sex differences in behaviour as an indirect consequence of mating system. Journal of Fish Biology 57: 839–857.

Magurran, A. E., and Higham, A. 1988. Information transfer across fish shoals under threat. Ethology 78: 153–158.

Magurran, A. E., and Nowak, M. A. 1991. Another battle of the sexes: the consequences of sexual asymmetry in mating costs and predation risk in the guppy, *Poecilia reticulata*. Proceedings of the Royal Society of London Series B—Biological Sciences 246: 31–38.

Magurran, A. E., and Phillip, D. A. T. 2001. Evolutionary implications of large-scale patterns in the ecology of Trinidadian guppies, *Poecilia reticulata*. Biological Journal of the Linnean Society 73: 1–9.

Magurran, A. E., and Ramnarine, I. W. 2004. Learned mate recognition and reproductive isolation in guppies. Animal Behaviour 67: 1077–1082.

Magurran, A. E., and Ramnarine, I. W. 2005. Evolution of mate discrimination in a fish. Current Biology 15: 867–868.

Magurran, A. E., and Seghers, B. H. 1990a. Population differences in the schooling behaviour of newborn guppies, *Poecilia reticulata*. Ethology 84: 334–342.

Magurran, A. E., and Seghers, B. H. 1990b. Risk sensitive courtship in the guppy (*Poecilia reticulata*). Behaviour 112: 194–201.

Magurran, A. E., and Seghers, B. H. 1991. Variation in schooling and aggression amongst guppy (*Poecilia reticulata*) populations in Trinidad. Behaviour 118: 214–234.

Magurran, A. E., and Seghers, B. H. 1994a. A cost of sexual harassment in the guppy, *Poecilia reticulata*. Proceedings of the Royal Society of London Series B—Biological Sciences 258: 89–92.

Magurran, A. E., and Seghers, B. H. 1994b. Predator inspection behavior covaries with schooling tendency amongst wild guppy, *Poecilia reticulata*, populations in Trinidad. Behaviour 128: 121–134.

Magurran, A. E., and Seghers, B. H. 1994c. Sexual conflict as a consequence of ecology: evidence from guppy, *Poecilia reticulata*, populations in Trinidad. Proceedings of the Royal Society of London Series B—Biological Sciences 255: 31–36.

Magurran, A. E., Seghers, B. H., Carvalho, G. R., and Shaw, P. W. 1992. Behavioural consequences of an artificial introduction of guppies (*Poecilia reticulata*) in N. Trinidad: evidence for the evolution of anti-predator behaviour in the wild. Proceedings of the Royal Society of London Series B—Biological Sciences 248: 117–122.

Magurran, A. E., Seghers, B. H., Shaw, P. W., and Carvalho, G. R. 1994. Schooling preferences for familiar fish in the guppy, *Poecilia reticulata*. Journal of Fish Biology 45: 401–406.

Magurran, A. E., Seghers, B. H., Shaw, P. W., and Carvalho, G. R. 1995. The behavioural diversity and evolution of guppy, *Poecilia reticulata*, populations in Trinidad. Advances in the Study of Animal Behavior 24: 155–202.

Major, P. F. 1978. Predator-prey interactions in two schooling fishes, *Caranx ignobilis* and *Stolephorus purpureus*. Animal Behaviour 26: 760–777.

Majumder, P., Gomez, J. A., Chadwick, B. P., and Boss, J. M. 2008. The insulator factor CTCF controls MHC class II gene expression and is required for the formation of long-distance chromatin interactions. Journal of Experimental Medicine 205: 785–798.

Malaga-Trillo, E., Zaleska-Rutczynska, Z., McAndrew, B., Vincek, V., Figueroa, F., Sultmann, H., and Klein, J. 1998. Linkage relationships and haplotype polymorphism among cichlid Mhc class II B loci. Genetics 149: 1527–1537.

Malo, A. F., Roldan, E. R. S., Garde, J., Soler, A. J., and Gomendio, M. 2005. Antlers honestly advertise sperm production and quality. Proceedings of the Royal Society of London Series B—Biological Sciences 272: 149–157.

Mandegar, M. A., and Otto, S. P. 2007. Mitotic recombination counteracts the benefits of genetic segregation. Proceedings of the Royal Society of London Series B—Biological Sciences 274: 1301–1307.

Mank, J. E., and Avise, J. C. 2006a. Comparative phylogenetic analysis of male alternative reproductive tactics in ray-finned fishes. Evolution 60: 1311–1316.

Mank, J. E., and Avise, J. C. 2006b. The evolution of reproductive and genomic diversity in ray-finned fishes: insights from phylogeny and comparative analysis. Journal of Fish Biology 69: 1–27.

Mank, J. E., and Avise, J. C. 2006c. Supertree analyses of the roles of viviparity and habitat in the evolution of atherinomorph fishes. Journal of Evolutionary Biology 19: 734–740.

Mansfield, S., and Mcardle, B. H. 1998. Dietary composition of *Gambusia affinis* (Family Poeciliidae) populations in the northern Waikato region of New Zealand. New Zealand Journal of Marine and Freshwater Research 32: 375–383.

Marcaillou, C., and A. Szöllosi. 1980. The "blood-testis" barrier in a nematode and a fish: a generalizable concept. Journal of Ultrastructure Research 70: 128–136.

Marchetti, M. P., and Nevitt, G. A. 2003. Effects of hatchery rearing on brain structures of rainbow trout, *Oncorhynchus mykiss*. Environmental Biology of Fishes 66: 9–14.

Marcus, J. M., and McCune, A. R. 1999. Ontogeny and phylogeny in the northern swordtail clade of *Xiphophorus*. Systematic Biology 48: 491–522.

Mariette, M., Kelley, J. L., Brooks, R., and Evans, J. P. 2006. The effects of inbreeding on male courtship behaviour and coloration in guppies. Ethology 112: 807–814.

Marler, C. A., Foran, C., and Ryan, M. J. 1997. The influence of experience on mating preferences of the gynogenetic Amazon molly. Animal Behaviour 53: 1035–1041.

Marler, P. 1957. Species distinctiveness in the communication signals of birds. Behaviour 11: 13–39.

Marsh-Matthews, E., and Deaton, R. 2006. Resources and offspring provisioning: a test of the Trexler-DeAngelis model for matrotrophy evolution. Ecology 87: 3014–3020.

Marsh-Matthews, E., Skierkowski, P., and DeMarais, A. 2001. Direct evidence for mother-to-embryo transfer of nutrients in the livebearing fish *Gambusia geiseri*. Copeia 2001: 1–6.

Marsh-Matthews, E., Brooks, M., Deaton, R., and Tan, H. 2005. Effects of maternal and embryo characteristics on postfertilization provisioning in fishes of the genus *Gambusia*. Oecologia 144: 12–24.

Marsh-Matthews, E., Deaton, R., and Brooks, M. 2010. Survey of matrotrophy in lecithotrophic poeciliids. In M. C. Uribe and H. J. Grier (eds.), *Viviparous Fishes II*. Homestead, FL: New Life Publications. Forthcoming.

Martin, C. H., and Johnsen, S. 2007. A field test of the Hamilton-Zuk hypothesis in the Trinidadian guppy (*Poecilia reticulata*). Behavioral Ecology and Sociobiology 61: 1897–1909.

Martin, S. B., Albert, J. S., and Leberg, P. L. 2010. The evolution of the poeciliid gonopodium: integrating morphological and behavioral traits. In M. C. Uribe and H. J. Grier (eds.), *Viviparous Fishes II*. Homestead, FL: New Life Publications. Forthcoming.

Mateos, M. 2005. Comparative phylogeography of livebearing fishes in the genera *Poeciliopsis* and *Poecilia* (Poeciliidae:

Cyprinodontiformes) in central Mexico. Journal of Biogeography 32: 775–780.

Mateos, M., and Vrijenhoek, R. C. 2002. Ancient versus reticulate origin of a hemiclonal lineage. Evolution 56: 985–992.

Mateos, M., and Vrijenhoek, R. C. 2005. Independent origins of allotriploidy in the fish genus *Poeciliopsis*. Journal of Heredity 96: 32–39.

Mateos, M., Sanjur, O. I., and Vrijenhoek, R. C. 2002. Historical biogeography of the livebearing fish genus *Poeciliopsis* (Poeciliidae: Cyprinodontiformes). Evolution 56: 972–984.

Mathis, A., and Smith, R. J. F. 1993. Fathead minnows, *Pimephales promelas*, learn to recognize northern pike, *Esox lucius*, as predators on the basis of chemical stimuli from minnows in the pike's diet. Animal Behaviour 46: 645–656.

Mathis, A., Chivers, D. P., and Smith, R. J. F. 1996. Cultural transmission of predator recognition in fishes: intraspecific and interspecific learning. Animal Behaviour 51: 185–201.

Matos, R. J., and Schlupp, I. 2005. Performing in front of an audience: signalers and the social environment. In P. K. McGregor (ed.), *Animal Communication Networks*, 13–37. Cambridge: Cambridge University Press.

Matthews, I. M. 1998. Mating behaviour and reproductive biology of the guppy, *Poecilia reticulata*. PhD thesis, University of St. Andrews.

Matthews, I. M., and Magurran, A. E. 2000. Evidence for sperm transfer during sneaky mating in wild Trinidadian guppies. Journal of Fish Biology 56: 1381–1386.

Matthews, I. M., Evans, J. P., and Magurran, A. E. 1997. Male display rate reveals ejaculate characteristics in the Trinidadian guppy, *Poecilia reticulata*. Proceedings of the Royal Society of London Series B—Biological Sciences 264: 695–700.

Mattingly, H. T., and Butler, M. J., IV. 1994. Laboratory predation on the Trinidadian guppy: implications for the size-selective predation hypothesis and guppy life history evolution. Oikos 69: 54–64.

Mäueler, W., Schartl, A., and Schartl, M. 1993. Different expression patterns of oncogenes and proto-oncogenes in hereditary and carcinogen-induced tumors of *Xiphophorus*. International Journal of Cancer 55: 288–296.

Maynard Smith, J. 1978. *The Evolution of Sex*. Cambridge: Cambridge University Press.

Maynard Smith, J. 1982. *Evolution and the Theory of Games*. Cambridge: Cambridge University Press.

Maynard Smith, J., and Price, G. R. 1973. The logic of animal conflict. Nature 246: 15–18.

Mayr, E. 1942. *Systematics and the Origin of Species*. New York: Columbia University Press.

Mayr, E. 1963. *Animal Species and Evolution*. Oxford: Oxford University Press.

Mayr, E. 1982. *The Growth of Biological Thought*. Cambridge, MA: Harvard University Press.

McAlister, W. H. 1958. The correlation of coloration with social rank in *Gambusia hurtadoi*. Ecology 39: 477–482.

McCauley, R. W., and Thomson, D. A. 1988. Thermoregulatory activity in the Tecopa pupfish, *Cyprinodon nevadensis amargosae*, an inhabitant of a thermal spring. Environmental Biology of Fishes 23: 135–139.

McConnell, T. J., Godwin, U. B., and Cuthbertson, B. J. 1998a. Expressed major histocompatibility complex class II loci in fishes. Immunological Reviews 166: 294–300.

McConnell, T. J., Godwin, U. B., Norton, S. F., Nairn, R. S.,

Kazianis, S., and Morizot, D. C. 1998b. Identification and mapping of two divergent, unlinked major histocompatibility complex class II B genes in *Xiphophorus* fishes. Genetics 149: 1921–1934.

McDowall, R. M. 1990. *New Zealand Freshwater Fishes: A Natural History and Guide*. Auckland: Heinemann Reed.

McGill, B. J., Enquist, B. J., Weiher, E., and Westoby, M. 2006. Rebuilding community ecology from functional traits. Trends in Ecology and Evolution 21: 178–185.

McGregor, P. K. 1993. Signalling in territorial systems: a context for individual identification, ranging and eavesdropping. Philosophical Transactions of the Royal Society of London Series B—Biological Sciences 340: 237–244.

McGregor, P. K., and Peake, T. M. 2000. Communication networks: social environments for receiver and signaller behaviour. Acta Ethologica 2: 71–81.

McGregor, P. K., Otter, K., and Peake, T. M. 2000. Communication networks: receiver and signaler perspectives. In Y. Espmark, T. Amundsen and G. Rosenqvist (eds.), *Animal Signals: Signalling and Signal Design in Animal Communication*, 405–416. Trondheim: Tapir Academic Press.

McGue, M., and Bouchard, T. J. 1998. Genetic and environmental influences on human behavioral differences. Annual Review of Neuroscience 21: 1–24.

McKay, F. E. 1971. Behavioral aspects of population dynamics in unisexual-bisexual *Poeciliopsis* (Pisces: Poeciliidae). Ecology 52: 778–790.

McKinnon, J. S., Mori, S., Blackman, B. K., David, L., Kingsley, D. M., Jamieson, L., Chou, J., and Schluter, D. 2004. Evidence for ecology's role in speciation. Nature 429: 294–298.

McKinsey, D. M., and Chapman, L. J. 1998. Dissolved oxygen and fish distribution in a Florida spring. Environmental Biology of Fishes 53: 211–223.

McLennan, D. A. 2003. The importance of olfactory signals in the gasterosteid mating system: sticklebacks go multimodal. Biological Journal of the Linnean Society 80: 555–572.

McLennan, D. A., and Ryan, M. J. 1997. Responses to conspecific and heterospecific olfactory cues in the swordtail *Xiphophorus cortezi*. Animal Behaviour 54: 1077–1088.

McLennan, D. A., and Ryan, M. J. 1999. Interspecific recognition and discrimination based upon olfactory cues in northern swordtails. Evolution 53: 880–888.

McLennan, D. A., and Ryan, M. J. 2008. Female swordtails, *Xiphophorus continens*, prefer the scent of heterospecific males. Animal Behaviour 75: 1731–1737.

McManus, M. G., and Travis, J. 1998. Effects of temperature and salinity on the life history of the sailfin molly (Pisces: Poeciliidae): lipid storage and reproductive allocation. Oecologia 114: 317–325.

McMinn, H. 1990. Effects of the nematode parasite *Camallanus cotti* on sexual and non-sexual behaviours in the guppy (*Poecilia reticulata*). American Zoologist 30: 245–249.

McNeely, D. L., and Wade, C. E. 2003. Relative abundance of the gynogen *Poecilia formosa* and its sexual host *Poecilia latipinna* (Teleostei: Poeciliidae) in some southern Texas habitats. Southwestern Naturalist 48: 451–453.

McRobert, S. P., and Bradner, J. 1998. The influence of body coloration on shoaling preferences in fish. Animal Behaviour 56: 611–615.

Medlen, A. B. 1951. Preliminary observations on the effects of temperature and light upon reproduction in *Gambusia affinis*. Copeia 1951: 148–152.

Meffe, G. K. 1984. Effects of abiotic disturbances on coexistence of predator-prey fish species. Ecology 65: 1525–1534.

Meffe, G. K. 1985a. Life history patterns of *Gambusia marshi* (Poeciliidae) from Cuatro Ciénegas, Mexico. Copeia 1985: 898–905.

Meffe, G. K. 1985b. Predation and species replacement in American southwestern fishes: a case study. Southwestern Naturalist 30: 173–187.

Meffe, G. K. 1991. Life-history changes in eastern mosquito-fish (*Gambusia holbrooki*) induced by thermal elevation. Canadian Journal of Fisheries and Aquatic Sciences 48: 60–66.

Meffe, G. K. 1992. Plasticity of life-history characters in eastern mosquitofish (*Gambusia holbrooki*, Poeciliidae) in response to thermal stress. Copeia 1992: 94–102.

Meffe, G. K., and Snelson, F. F., Jr. 1989a. An ecological overview of poeciliid fishes. In G. K. Meffe and F. F. Snelson, Jr. (eds.), *Ecology and Evolution of Livebearing Fishes (Poeciliidae)*, 13–31. Englewood Cliffs, NJ: Prentice Hall.

Meffe, G. K., and Snelson, F. F., Jr. 1989b. *Ecology and Evolution of Livebearing Fishes (Poeciliidae)*. Englewood Cliffs, NJ: Prentice Hall.

Meffe, G. K., and Vrijenhoek, R. C. 1981. Starvation stress and intraovarian cannibalism in livebearers (Atheriniformes: Poeciliidae). Copeia 1981: 702–705.

Meier, R., Kotrba, M., and Ferrar, P. 1999. Ovoviviparity and viviparity in the Diptera. Biological Reviews 74: 199–258.

Meierjohann, S., and Schartl, M. 2006. From Mendelian to molecular genetics: the *Xiphophorus* melanoma model. Trends in Genetics 22: 654–661.

Meierjohann, S., Müller, T., Schartl, M., and Bühner, M. 2006a. A structural model of the extracellular domain of the oncogenic EGFR variant Xmrk. Zebrafish 3: 359–369.

Meierjohann, S., Wende, E., Kraiss, A., Wellbrock, C., and Schartl, M. 2006b. The oncogenic epidermal growth factor receptor variant *Xiphophorus* melanoma receptor kinase induces motility in melanocytes by modulation of focal adhesions. Cancer Research 66: 3145–3152.

Meierjohann, S., Hufnagel, A., Wende, E., Kleinschmidt, M. A., Wolf, K., Friedl, P., Gaubatz, S., and Schartl, M. 2010. MMP13 mediates cell cycle progression in melanocytes and melanoma cells: in vitro studies of migration and proliferation. Molecular Cancer 9: 201.

Mendelson, T. C., Imhoff, V. E., and Venditti, J. J. 2007. The accumulation of reproductive barriers during speciation: postmating barriers in two behaviorally isolated species of darters (Percidae: *Etheostoma*). Evolution 61: 2596–2606.

Mendez, V., and Cordoba-Aguilar, A. 2004. Sexual selection and animal genitalia. Trends in Ecology and Evolution 19: 224–225.

Menzel, B. W., and Darnell, R. M. 1973. Systematics of *Poecilia mexicana* (Pisces: Poeciliidae) in northern Mexico. Copeia 2: 225–237.

Meredith, R. W., Pires, M. N., Reznick, D. N., and Springer, M. S. 2010. Molecular phylogenetic relationships and the evolution of the placenta in *Poecilia* (*Micropoecilia*) (Poeciliidae: Cyprinodontiformes). Molecular Phylogenetics and Evolution 55: 631–639.

Mertz, D. B. 1970. Notes on methods used in life-history studies. In J. Connell, D. B. Mertz, and W. W. Murdoch (eds.), *Readings in Ecology and Ecological Genetics*, 4–17. New York: Harper & Row.

Mesterton-Gibbons, M., and Sherrat, T. N. 2007. Social eavesdropping: a game-theoretic analysis. Bulletin of Mathematical Biology 69: 1255–1276.

Metcalf, C. J. E., and Pavard, S. 2007. Why evolutionary biologists should be demographers. Trends in Ecology and Evolution 22: 205–212.

Meyer, A., Morrissey, J. M., and Schartl, M. 1994. Recurrent origin of a sexually selected trait in *Xiphophorus* fishes inferred from a molecular phylogeny. Nature 368: 539–542.

Meyer, A., Salzburger, W., and Schartl, M. 2006. Hybrid origin of a swordtail species (Teleostei: *Xiphophorus clemenciae*) driven by sexual selection. Molecular Ecology 15: 721–730.

Meyer, J. H., and Liley, N. R. 1982. The control of production of a sexual pheromone in the female guppy, *Poecilia reticulata*. Canadian Journal of Zoology 60: 1505–1510.

Meyer, M. K. 1983. *Xiphophorus*-Hybriden aus Nord-Mexiko, mit einer Revision der Taxa *X. kosszanderi* und *X. roseni* (Osteichthyes, Poeciliidae). *Zoologische Abhandlungen* 38: 285–291.

Meyer, M. K., and Etzel, V. 2001. Additional notes on the genus *Brachyrhaphis* Regan, 1913, with description of a new species from Panama (Teleostei: Cyprinodontiformes: Poeciliidae). Zoologische Abhandlungen, Staatliches Museum für Tierkunde in Dresden 51: 33–39.

Meyer, M. K., and Schartl, M. 2003. *Xiphophorus kallmani* sp. n.—a new species of swordtail from Mexico (Teleostei, Cyprinodontiformes, Poeciliidae). Zoologische Abhandlungen, Staatliches Museum für Tierkunde in Dresden 53: 57–64.

Meyer, M. K., Schneider, K., Radda, A. C., Wilde, B., and Schartl, M. 2004. A new species of *Poecilia*, subgenus *Mollienesia*, from Upper Río Cahabón system, Guatemala, with remarks on the nomenclature of *Mollienesia petenensis* Günther, 1886 (Teleostei: Cyprinodontiformes: Poeciliidae). Zoologische Abhandlungen, Staatliches Museum für Tierkunde in Dresden 54: 145–154.

MHC Sequencing Consortium. 1999. Complete sequence and gene map of a human major histocompatibility complex. Nature 401: 921–923.

Michod, R. E. 1979. Evolution of life histories in response to age-specific mortality factors. American Naturalist 113: 531–550.

Midford, P. E., Garland, T., and Maddison, W. P. 2008. *PDAP Package of Mesquite*. http://mesquiteproject.org/pdap_mesquite, version PDAP:PDTREE.

Milinski, M. 1977. Experiments on the selection by predators against spatial oddity of their prey. Zeitschrift für Tierpsychologie 43: 311–325.

Milinski, M. 1987. Tit for Tat in sticklebacks and the evolution of cooperation. Nature 325: 433–435.

Milinski, M. 2003. The function of mate choice in sticklebacks: optimizing Mhc genetics. Journal of Fish Biology 63: 1–16.

Milinski, M. 2006. The major histocompatibility complex, sexual selection, and mate choice. Annual Review of Ecology Evolution and Systematics 37: 159–186.

Milinski, M., and Bakker, T. C. M. 1990. Female sticklebacks use male coloration in mate choice and hence avoid parasitised males. Nature 344: 330–333.

Milinski, M., Griffiths, S., Wegner, K. M., Reusch, T. B. H., Haas-Assenbaum, A., and Boehm, T. 2005. Mate choice decisions of stickleback females predictably modified by MHC peptide ligands. Proceedings of the National Academy of Sciences of the United States of America 102: 4414–4418.

Millar, N. P., Reznick, D. N., Kinnison, M. T., and Hendry, A. P. 2006. Disentangling the selective factors that act on male colour in wild guppies. Oikos 113: 1–12.

Miller, H. C., and Lambert, D. M. 2004. Genetic drift outweighs balancing selection in shaping post-bottleneck major histocompatibility complex variation in New Zealand robins (Petroicidae). Molecular Ecology 13: 3709–3721.

Miller, K. M., Kaukinen, K. H., Beacham, T. D., and Withler, R. E. 2001. Geographic heterogeneity in natural selection on an MHC locus in sockeye salmon. Genetica 111: 237–257.

Miller, L. K., and Brooks, R. 2005. The effects of genotype, age, and social environment on male ornamentation, mating behavior, and attractiveness. Evolution 59: 2414–2425.

Miller, P. S., and Hedrick, P. W. 1991. MHC polymorphism and the design of captive breeding programs: simple solutions are not the answer. Conservation Biology 5: 556–558.

Miller, R. R. 1960. Four new species of viviparous fishes, genus *Poeciliopsis*, from northwestern Mexico. Occasional Papers of the Museum of Zoology, University of Michigan 619: 1–11.

Miller, R. R. 1961. Man and the changing fish fauna of the American Southwest. Papers of the Michigan Academy Science, Arts, and Letters 46: 365–404.

Miller, R. R. 1975. Five new species of Mexican poeciliid fishes of the genera *Poecilia*, *Gambusia*, and *Poeciliopsis*. Occasional Papers of the Museum of Zoology, University of Michigan 672: 1–44.

Miller, R. R. 2005. *Freshwater Fishes of Mexico*. Chicago: University of Chicago Press.

Miller, R. R., and Schultz, R. J. 1959. All-female strains of the teleost fishes of the genus *Poeciliopsis*. Science 130: 1656–1657.

Mills, M. D., Rader, R. B., and Belk, M. C. 2004. Complex interactions between native and invasive fish: the simultaneous effects of multiple negative interactions. Oecologia 141: 713–721.

Milstead, E. 1980. Genetic differentiation among subpopulations of three *Gambusia* species (Pisces: Poeciliidae) in the Pecos River, Texas, and New Mexico. Master's thesis, Baylor University.

Minckley, W. L. 1969. Native Arizona fishes, part I: livebearers. Wildlife Views 16: 6–8.

Minckley, W. L. 1999. Ecological review and management recommendations for recovery of the endangered Gila topminnow. Great Basin Naturalist 59: 230–244.

Minckley, W. L., and Deacon, J. E. 1968. Southwestern fishes and the enigma of "endangered species." Science 159: 1424–1432.

Minckley, W. L., and Jensen, B. L. 1985. Replacement of Sonoran topminnow by Pecos *Gambusia* under hatchery conditions. Southwestern Naturalist 30: 465–466.

Minckley, W. L., and Meffe, G. K. 1987. Differential selection by flooding in stream-fish communities of the arid American Southwest. In W. J. Mathews and D. C. Heins (eds.), *Community and Evolutionary Ecology of North American Stream Fishes*, 93–104. Norman: University of Oklahoma Press.

Mirmohammadsadegh, A., Hassan, M., Bardenheuer, W.,

Marini, A., Gustrau, A., Nambiar, S., Tannapfel, A., Bojar, H., Ruzicka, T., and Hengge, U. R. 2006. STAT5 phosphorylation in malignant melanoma is important for survival and is mediated through SRC and JAK1 kinases. Journal of Investigative Dermatology 126: 2272–80.

Mirza, R. S., and Chivers, D. P. 2002. Behavioural responses to conspecific disturbance chemicals enhance survival of juvenile brook charr, *Salvelinus fontinalis*, during encounters with predators. Behaviour 139: 1099–1109.

Mirza, R. S., Scott, J. J., and Chivers, D. P. 2001. Differential responses of male and female red swordtails to chemical alarm cues. Journal of Fish Biology 59: 716–728.

Mitchell, D. L., and Nairn, R. S. 2006. Photocarcinogenesis in *Xiphophorus* hybrid models. Zebrafish 3: 311–323.

Mitra, S. K., and Schlaepfer, D. D. 2006. Integrin-regulated FAK-Src signaling in normal and cancer cells. Current Opinion in Cell Biology 18: 516–23.

Modrek, B., Resch, A., Grasso, C., and Lee, C. 2001. Genome-wide detection of alternative splicing in expressed sequences of human genes. Nucleic Acids Research 29: 2850–2859.

Møller, A. P., and Pomiankowski, A. 1993. Why have birds got multiple sexual ornaments? Behavioral Ecology and Sociobiology 32: 167–176.

Möller, D. 2001. Aspekte zur Populationsgenetik des eingeschlechtlichen Amazonenkärpflings *Poecilia formosa* (Girard 1859) unter Berücksichtigung der genetischen parentalen Arten, dem Breitflossenkärpfling *Poecilia latipinna* (LeSueur 1821) und dem Atlantikkärpfling *Poecilia mexicana* (Steindachner 1863). PhD thesis, University of Hamburg.

Monaco, P. J., Rasch, E. M., and Balsano, J. S. 1978. Cytological evidence for temporal differences during the asynchronous ovarian maturation of bisexual and unisexual fishes of the genus *Poecilia*. Journal of Fish Biology 13: 33–44.

Monaco, P. J., Rasch, E. M., and Balsano, J. S. 1983. The occurrence of superfetation in the Amazon molly, *Poecilia formosa*, and its related sexual species. Copeia 1983: 969–974.

Monaco, P. J., Swan, K. F., Rasch, E. M., and Musich, P. R. 1989. Characterization of a repetitive DNA in the unisexual fish *Poecilia formosa*, I: isolation and cloning of the Mbo I family. In G. K. Meffe and F. F. Snelson Jr. (eds.), *Ecology and Evolution of Livebearing Fishes (Poeciliidae)*, 123–132. Englewood Cliffs, NJ: Prentice Hall.

Monzon, J., Liu, L., Brill, H., Goldstein, A. M., Tucker, M. A., From, L., McLaughlin, J., Hogg, D., and Lassam, N. J. 1998. *CDKN2A* mutations in multiple primary melanomas. New England Journal of Medicine 338: 879–887.

Mook, O. R., Frederiks, W. M., and Van Noorden, C. J. 2004. The role of gelatinases in colorectal cancer progression and metastasis. Biochimica et Biophysica Acta 1705: 69–89.

Moore, J. S., Gow, J. L., Taylor, E. B., and Hendry, A. P. 2007. Quantifying the constraining influence of gene flow on adaptive divergence in the lake-stream threespine stickleback system. Evolution 61: 2015–2026.

Moore, W. S. 1977. An evaluation of narrow hybrid zones in vertebrates. Quarterly Review of Biology 52: 263–277.

Moore, W. S., and McKay, F. E. 1971. Coexistence in unisexual-bisexual breeding complexes of *Poeciliopsis* (Pisces: Poeciliidae). Ecology 52: 791–799.

Moravec, F., and Justine, J. L. 2006. *Camallanus cotti* (Nematoda: Camallanidae), an introduced parasite of fishes in New Caledonia. Folia Parasitologica 53: 287–296.

Morcinek, J. C., Weisser, C., Geissinger, E., Schartl, M., and Wellbrock, C. 2002. Activation of STAT5 triggers proliferation and contributes to anti-apoptotic signalling mediated by the oncogenic Xmrk kinase. Oncogene 21: 1668–1678.

Moretz, J. A. 2003. Aggression and RHP in the northern swordtail fish, *Xiphophorus cortezi*: the relationship between size and contest dynamics in male-male competition. Ethology 109: 995–1008.

Moretz, J. A. 2005. Aggression and fighting ability are correlated in the swordtail fish *Xiphophorus cortezi*: the advantage of being barless. Behavioral Ecology and Sociobiology 59: 51–57.

Moretz, J. A., and Morris, M. R. 2003. Evolutionarily labile responses to a signal of aggressive intent. Proceedings of the Royal Society of London Series B—Biological Sciences 270: 2271–2277.

Moretz, J. A., and Morris, M. R. 2006. Phylogenetic analysis of the evolution of a signal of aggressive intent in northern swordtail fishes. American Naturalist 168: 336–349.

Morgan, D. L., Gill, H. S., Maddern, M. G., and Beatty, S. J. 2004. Distribution and impacts of introduced freshwater fishes in Western Australia. New Zealand Journal of Marine and Freshwater Research 38: 511–523.

Moritz, C. 1994. Defining "evolutionarily significant units" for conservation. Trends in Ecology and Evolution 9: 373–375.

Morizot, D. C., and Siciliano, M. J. 1983. Linkage group V of platyfishes and swordtails of the genus *Xiphophorus* (Poeciliidae): linkage of loci for malate dehydrogenase-2 and esterase-1 and esterase-4 with a gene controlling the severity of hybrid melanomas. Journal of the National Cancer Institute 71: 809–813.

Morizot, D. C., Wright, D. A., and Sicilian, M. J. 1977. Three linked enzyme loci in fishes: implications in the evolution of vertebrate chromosomes. Genetics 86: 645–656.

Morrell, L. J., Hunt, K. L., Croft, D. P., and Krause, J. 2007. Diet, familiarity and shoaling decisions in guppies. Animal Behaviour 74: 311–319.

Morrell, L. J., Croft, D. P., Dyer, J. R. G., Chapman, B. B., Kelley, J. L., Laland, K. N., and Krause, J. 2008. Association patterns and foraging behaviour in natural and artificial guppy shoals. Animal Behaviour 76: 855–864.

Morris, M. R. 1998. Further examination of female preference for vertical bars in swordtails: preference for "no bars" in a species without bars. Journal of Fish Biology 53: 56–63.

Morris, M. R., and Casey, K. 1998. Female swordtail fish prefer symmetrical sexual signal. Animal Behaviour 55: 33–39.

Morris, M. R., and Ryan, M. J. 1992. Breeding cycles in natural populations of *Xiphophorus nigrensis*, *X. multilineatus*, and *X. pygmaeus*. Copeia 1992: 1074–1077.

Morris, M. R., Batra, P., and Ryan, M. J. 1992. Male-male competition and access to females in the swordtail *Xiphophorus nigrensis*. Copeia 1992: 980–986.

Morris, M. R., Gass, L., and Ryan, M. J. 1995. Assessment and individual recognition of opponents in the pygmy swordtails *Xiphophorus nigrensis* and *X. multilineatus*. Behavioral Ecology and Sociobiology 37: 303–310.

Morris, M. R., Wagner, W. E. J., and Ryan, M. J. 1996. A negative correlation between trait and mate preference in *Xiphophorus pygmaeus*. Animal Behaviour 52: 1193–1203.

Morris, M. R., Elias, J. A., and Moretz, J. A. 2001. Defining vertical bars in relation to female preference in the swordtail fish *Xiphophorus cortezi* (Cyprinodontiformes, Poeciliidae). Ethology 107: 827–837.

Morris, M. R., Nicoletto, P. F., and Hesselman, E. 2003. A polymorphism in female preference for a polymorphic male trait in the swordtail fish *Xiphophorus cortezi*. Animal Behaviour 65: 45–52.

Morris, M. R., Moretz, J. A., Farley, K., and Nicoletto, P. 2005. The role of sexual selection in the loss of sexually selected traits in the swordtail fish *Xiphophorus continens*. Animal Behaviour 69: 1415–1424.

Morris, M. R., Rios-Cardenas, O., and Tudor, M. S. 2006. Larger swordtail females prefer asymmetrical males. Biology Letters 2: 8–11.

Morris, M. R., Tudor, M. S., and Dubois, N. S. 2007. Sexually selected signal attracted females before deterring aggression in rival males. Animal Behaviour 74: 1189–1197.

Morris, M. R., Rios-Cardenas, O., and Darrah, A. 2008. Male mating tactics in the northern mountain swordtail fish (*Xiphophorus nezahualcoyotl*): coaxing and coercing females to mate. Ethology 114: 977–988.

Morris, M. R., Rios-Cardenas, O., and Brewer, J. 2010. Variation in mating preference within a wild population influences the mating success of alternative mating strategies. Animal Behaviour 79: 673–678.

Mossman, H. W. 1937. Comparative morphogenesis of the fetal membranes and accessory uterine structures. Carnegie Contributions to Embryology 26: 129–246.

Moyle, P. B. 1976. Fish introductions in California: history and impact on native fishes. Biological Conservation 9: 101–118.

Moyle, P. B. 2002. *Inland Fishes of California*. Rev. and exp. ed. Berkeley and Los Angeles: University of California Press.

Mueller, L. D., Guo, P., and Ayala, F. 1991. Density-dependent natural selection and trade-offs in life history traits. Science 253: 433–435.

Mueller, L. D. 1997. Theoretical and empirical examination of density-dependent selection. Annual Review of Ecology and Systematics 28: 269–288.

Muller, H. J. 1932. Some genetic aspects of sex. American Naturalist 66: 118–138.

Muller, H. J. 1964. The relation of recombination to mutational advance. Mutational Research 1: 2–9.

Munger, L., Cruz, A., and Applebaum, S. 2004. Mate choice copying in female humpback limia (*Limia nigrofasciata*, family Poeciliidae). Ethology 110: 563–573.

Murphy, H. T., VanDerWal, J., Lovett-Doust, L., and Lovett-Doust, J. 2006. Invasiveness in exotic plants: immigration and naturalization in an ecological continuum. In M. W. Cadotte, S. M. McMahon, and T. Fukami (eds.), *Conceptual Ecology and Invasion Biology: Reciprocal Approaches to Nature*, 65–105. Dordrecht: Springer.

Murphy, K. E., and Pitcher, T. J. 1997. Predator attack motivation influences the inspection behaviour of European minnows. Journal of Fish Biology 50: 407–417.

Myers, G. S. 1965. *Gambusia*, the fish destroyer. Australian Zoologist 13: 102.

Mylius, S. D., and Diekmann, O. 1995. On evolutionary stable life histories, optimization and the need to be specific about density dependence. Oikos 74: 218–224.

Nakamura, M., Nakamura, M., Kobayashi, T., Chang, X. T.,

and Nagahama, Y. 1998. Gonadal sex differentiation in teleost fish. Journal of Experimental Zoology 281: 362–372.

Nakamura, O., Tazumi, Y., Muro, T., Yashuhara, Y., and Wantanabe, T. 2004. Active uptake and transport of protein by the intestinal epithelial cells in embryo of viviparous fish *Neoditrema ransonneti* (Perciformes: Embiotocidae). Journal of Experimental Zoology 301: 38–48.

Nanda, I., Feichtinger, W., Schmid, M., Schröder, J. H., Zischler, H., and Epplen, J. T. 1990. Simple repetitive sequences are associated with differentiation of the sex chromosomes in the guppy fish. Journal of Molecular Evolution 30: 456–462.

Nanda, I., Schartl, M., Feichtinger, W., Epplen, J. T., and Schmid, M. 1992. Early stages of sex chromosome differentiation in fish as analysed by simple repetitive DNA sequences. Chromosoma 101: 301–310.

Nanda, I., Schartl, M., Epplen, J. T., Feichtinger, W., and Schmid, M. 1993. Primitive sex chromosomes in poeciliid fishes harbor simple repetitive DNA sequences. Journal of Experimental Zoology 265: 301–308.

Nanda, I., Schartl, M., Feichtinger, W., Schlupp, I., Parzefall, J., and Schmid, M. 1995. Chromosomal evidence for laboratory synthesis of a triploid hybrid between the gynogenetic teleost *Poecilia formosa* and its host species. Journal of Fish Biology 47: 619–623.

Nanda, I., Shan, Z., Schartl, M., Burt, D. W., Koehler, M., Nothwang, H., Grutzner, F., Paton, I. R., Windsor, D., Dunn, I., Engel, W., Staeheli, P., Mizuno, S., Haaf, T., and Schmid, M. 1999. 300 million years of conserved synteny between chicken Z and human chromosome 9. Nature Genetics 21: 258–259.

Nanda, I., Volff, J.-N., Weis, S., Korting, C., Froschauer, A., Schmid, M., and Schartl, M. 2000. Amplification of a long terminal repeat-like element on the Y chromosome of the platyfish, *Xiphophorus maculatus*. Chromosoma 109: 173–180.

Nanda, I., Hornung, U., Kondo, M., Schmid, M., and Schartl, M. 2003. Common spontaneous sex-reversed XX males of the medaka, *Oryzias latipes*. Genetics 163: 245–251.

Nanda, I., Schlupp, I., Lamatsch, D. K., Lampert, K. P., Schmid, M., and Schartl, M. 2007. Stable inheritance of host species–derived microchromosomes in the gynogenetic fish *Poecilia formosa*. Genetics 177: 917–926.

National Research Council. 1979. *Hydrogen Sulfide*. Baltimore, MD: University Park Press.

Nayudu, P. L. 1979. Genetic studies of melanic color patterns, and atypical sex determination in the guppy, *Poecilia reticulata*. Copeia 2: 225–231.

Neat, F. C., Taylor, A. C., and Huntingford, F. A. 1998. Proximate costs of fighting in male cichlid fish: the role of injuries and energy metabolism. Animal Behaviour 55: 875–882.

Neff, B. D., and Pitcher, T. E. 2002. Assessing the statistical power of genetic analyses to detect multiple mating in fishes. Journal of Fish Biology 61: 739–750.

Neff, B. D., and Wahl, L. M. 2004. Mechanisms of sperm competition: testing the fair raffle. Evolution 58: 1846–1851.

Neff, B. D., Garner, S. R., Heath, J. W., and Heath, D. 2008a. The MHC and non-random mating in a captive population of Chinook salmon. Heredity 101: 175–185.

Neff, B. D., Pitcher, T. E., and Ramnarine, I. W. 2008b. Inter-

population variation in multiple paternity and reproductive skew in the guppy. Molecular Ecology 17: 2975–2984.

Nei, M., and Rooney, A. P. 2005. Concerted and birth-and-death evolution of multigene families. Annual Review of Genetics 39: 121–152.

Neill, S. R. S. J., and Cullen, J. M. 1974. Experiments on whether schooling of prey affects hunting behaviour of cephalopods and fish predators. Journal of Zoology 172: 549–569.

Nelson, D., and Jannasch, H. 1983. Chemoautotrophic growth of marine *Beggiatoa* in sulfide-gradient cultures. Archives of Microbiology 136: 262–269.

Nelson, J. S., Crossman, E. J., Espinosa-Perez, H., Findley, L. T., Gilbert, C. R., Lea, R. N., and Williams, J. D. 2004. *Common and Scientific Names of Fishes from United States, Canada and Mexico*. 6th ed. American Fisheries Society, Special Publication 29. Bethesda, MD.

Nelson, J. S. 2006. *Fishes of the World*. 4th ed. New York: John Wiley & Sons.

Neubert, M. G., and Caswell, H. 2000. Density-dependent vital rates and their population dynamic consequences. Journal of Mathematical Biology 41: 103–121.

Neves, F. M., and Monteiro, L. R. 2003. Body shape and size divergence among populations of *Poecilia vivipara* in coastal lagoons of south-eastern Brazil. Journal of Fish Biology 63: 928–941.

Ng, P. K. L., Chou, L. M., and Lam, T. J. 1993. The status and impact of introduced freshwater animals in Singapore. Biological Conservation 64: 19–24.

Nicholls, P. 1975. The effect of sulphide on cytochrome aa3: isosteric and allosteric shifts of the reduced alpha-peak. Biochemica Biophysica Acta 1975: 24–35.

Nichols, J. D., Conley, W., Batt, B., and Tipton, A. R. 1976. Temporally dynamic reproductive strategies and the concept of r- and K-selection. American Naturalist 110: 995–1005.

Nicoletto, P. F. 1993. Female sexual response to condition-dependent ornaments in the guppy, *Poecilia reticulata*. Animal Behaviour 46: 441–450.

Nicoletto, P. F. 1995. Offspring quality and female choice in the guppy, *Poecilia reticulata*. Animal Behaviour 49: 377–387.

Nicoletto, P. F. 1996. The influence of water velocity on the display behavior of male guppies, *Poecilia reticulata*. Behavioral Ecology 7: 272–278.

Niemeitz, A., Kreutzfeldt, R., Schartl, M., Parzefall, J., and Schlupp, I. 2002. Male mating behaviour of a molly, *Poecilia latipunctata*: a third host for the sperm-dependent Amazon molly, *Poecilia formosa*. Acta Ethologica 5: 45–49.

Nishi, K. 1981. Circadian rhythm in the photosensitive development of the ovary in the mosquitofish, *Gambusia affinis affinis* (Baird et Girad). Bulletin of the Faculty of Fisheries, Hokkaido University 32: 211–220.

Noor, M. A. F. 2003. Evolutionary biology: genes to make new species. Nature 423: 699–700.

Noor, M. A. F., and Feder, J. L. 2006. Speciation genetics: evolving approaches. Nature Reviews Genetics 7: 851–861.

Nordell, S. E. 1998. The response of female guppies, *Poecilia reticulata*, to chemical stimuli from injured conspecifics. Environmental Biology of Fishes 51: 331–338.

Nordlie, F. G. 2006. Physicochemical environments and tolerances of cyprinodontoid fishes found in estuaries and salt marshes of eastern North America. Reviews in Fish Biology and Fisheries 16: 51–106.

Nordlie, F. G., Haney, D. C., and Walsh, S. J. 1992. Comparisons of salinity tolerances and osmotic regulatory capabilities in populations of sailfin molly (*Poecilia latipinna*) from brackish and fresh waters. Copeia 1992: 741–746.

Norris, C. E., Diiorio, P. J., Schultz, R. J., and Hightower, L. E. 1995. Variation in heat-shock proteins within tropical and desert species of poeciliid fishes. Molecular Biology and Evolution 12: 1048–1062.

Norton, S. F., Luczkovich, J. J., and Motta, P. J. 1995. The role of ecomorphological studies in the comparative biology of fishes. Environmental Biology of Fishes 44: 287–304.

Nosil, P., Funk, D. J., and Ortiz-Barrientos, D. 2009. Divergent selection and heterogeneous genomic divergence. Molecular Ecology 18: 375–402.

Nowak, M. A., Tarczyhornoch, K., and Austyn, J. M. 1992. The optimal number of major histocompatibility complex-molecules in an individual. Proceedings of the National Academy of Sciences of the United States of America 89: 10896–10899.

O'Boyle, M. W., Cunnington, R., Silk, T. J., Vaughan, D., Jackson, G., Syngeniotis, A., and Egan, G. F. 2005. Mathematically gifted male adolescents activate a unique brain network during mental rotation. Cognitive Brain Research 25: 583–587.

O'Brien, S. J., and Evermann, J. F. 1988. Interactive influence of infectious disease and genetic diversity in natural populations. Trends in Ecology and Evolution 3: 254–259.

O'Brien, S. J., Roelke, M. E., Marker, L., Newman, A., Winkler, C. A., Meltzer, D., Colly, L., Evermann, J. F., Bush, M., and Wildt, D. E. 1985. Genetic-basis for species vulnerability in the cheetah. Science 227: 1428–1434.

Odell, J. P. 2002. Evolution of physiological performance in the Trinidadian guppy (*Poecilia reticulata*: Peters). PhD thesis, University of California, Riverside.

Offen, N., Meyer, A., and Begemann, G. 2009. Identification of novel genes involved in the development of the sword and gonopodium in swordtail fish. Developmental Dynamics 238: 1674–1687.

Ohta, Y., Okamura, K., McKinney, E. C., Bartl, S., Hashimoto, K., and Flajnik, M. F. 2000. Primitive synteny of vertebrate major histocompatibility complex class I and class II genes. Proceedings of the National Academy of Sciences of the United States of America 97: 4712–4717.

Ohtsuki, H., Hauert, C., Lieberman, E., and Nowak, M. A. 2006. A simple rule for the evolution of cooperation on graphs and social networks. Nature 441: 502–505.

Ojanguren, A. F., and Magurran, A. E. 2004. Uncoupling the links between male mating tactics and female attractiveness. Proceedings of the Royal Society of London Series B—Biological Sciences 271: S427–S429.

Ojanguren, A. F., and Magurran, A. E. 2007. Male harassment reduces short-term female fitness in guppies. Behaviour 144: 503–514.

Ojanguren, A. F., Evans, J. P., and Magurran, A. E. 2005. Multiple mating influences offspring size in guppies. Journal of Fish Biology 67: 1184–1188.

Olendorf, R., Rodd, F. H., Punzalan, D., Houde, A. E., Hurt, C., Reznick, D. N., and Hughes, K. A. 2006. Frequency-dependent survival in natural guppy populations. Nature 441: 633–636.

Oliveira, R. F., McGregor, P. K., and Latruffe, C. 1998. Know thine enemy: fighting fish gather information from observing conspecific interactions. Proceedings of the Royal Society of London Series B—Biological Sciences 265: 1045–1049.

Oliveira, R. F., Lopes, M., Carneiro, L. A., and Canario, A. V. M. 2001. Watching fights raises fish hormone levels. Nature 409: 475.

Olivier, K. 2001. *The Ornamental Fish Market*. Globefish Research Programme, vol. 67. Rome: United Nations Food and Agriculture Organization.

Olivieri, I. 2009. Alternative mechanisms of range expansion are associated with different changes of evolutionary potential. Trends in Ecology and Evolution 24: 289–292.

Olsén, K. H., and Liley, N. R. 1993. The significance of olfaction and social cues in milt availability, sexual hormones status, and spawning behaviour of male rainbow trout (*Oncorhynchus mikiss*). General and Comparative Endocrinology 89: 107–118.

Olsén, K. H., Grahn, M., Lohm, J., and Langefors, A. 1998. MHC and kin discrimination in juvenile Arctic charr, *Salvelinus alpinus* (L.). Animal Behaviour 56: 319–327.

Olsson, M., Shine, R., and Madsen, T. 1996. Sperm selection by females. Nature 383: 585.

O'Neill, M. J., Ingram, R. S., Vrana, P. B., and Tilghman, M. 2000. Allelic expression of *IGF2* in marsupials and birds. Development Genes and Evolution 2210: 18–20.

O'Neill, M. J., Lawton, B. R., Mateos, M., Carone, D. M., Ferreri, G. C., Hrbek, T., Meredith, R. W., Reznick, D. N., and O'Neill, R. J. 2007. Ancient and continuing Darwinian selection on insulin-like growth factor II in placental fishes. Proceedings of the National Academy of Sciences of the United States of America 104: 12404–12409.

Orlando, E. F., Bass, D. E., Caltabiano, L. M., Davis, W. P., Gray, L. E., and Guillette, L. J. 2007. Altered development and reproduction in mosquitofish exposed to pulp and paper mill effluent in the Fenholloway River, Florida, USA. Aquatic Toxicology 84: 399–405.

Orr, H. A., Masly, J. P., and Presgraves, D. C. 2004. Speciation genes. Current Opinion in Genetics and Development 14: 675–679.

Ospina-Alvarez, N., and Piferrer, F. 2008. Temperature-dependent sex determination in fish revisited: prevalence, a single sex ratio response pattern, and possible effects of climate change. PloS ONE 3: e2837.

O'Steen, S., Cullum, A. J., and Bennett, A. F. 2002. Rapid evolution of escape ability in Trinidadian guppies (*Poecilia reticulata*). Evolution 56: 776–784.

Overstreet, R. M. 1997. Parasitological data as monitor of environmental health. Parasitologia 39: 169–175.

Pagel, M. 1994. Detecting correlated evolution on phylogenies: a general method for the comparative analysis of discrete characters. Proceedings of the Royal Society of London Series B—Biological Sciences 255: 37–45.

Pagel, M. 2000. Discrete. Version 4.0. A computer program distributed by the author.

Palan, P., and Naz, R. 1996. Changes in various antioxidant levels in human seminal plasma related to immunoinfertility. Archives of Andrology 36: 139–143.

Pandey, S. 1969. Effects of hypophysectomy on the testis and secondary sex characters of the adult Guppy *Poecilia reticulata* Peters. Canadian Journal of Zoology 47: 775–781.

Panhuis, T. M., Butlin, R., Zuk, M., and Tregenza, T. 2001. Sexual selection and speciation. Trends in Ecology and Evolution 16: 364–371.

Parenti, L. R. 1981. A phylogenetic and biogeographic analysis of cyprinodontiform fishes (Teleostei, Atherinomorpha). Bulletin of the American Museum of Natural History 168: 335–557.

Parenti, L. R. 1989. A phylogenetic revision of the phallostethid fishes (Atherinomorpha, Phallostethidae). Proceedings of the Californian Academy of Sciences 46: 243–277.

Parenti, L. R., and Grier, H. J. 2004. Evolution and phylogeny of gonad morphology in bony fishes. Integrative and Comparative Biology 44: 333–348.

Parenti, L. R., and Rauchenberger, M. 1989. Systematic overview of the poeciliines. In G. K. Meffe and F. F. Snelson Jr. (eds.), *Ecology and Evolution of Livebearing Fishes (Poeciliidae)*, 3–12. Englewood Cliffs, NJ: Prentice Hall.

Paris, F., Paaßen, U., and Blüm, V. 1998. Spermienspeicherung bei weiblichen Schwertträgern (*Xiphophorus helleri*). Verhandlungen der Gesellschaft für Ichthyologie 1: 157–165.

Park, C., Nagel, R., Blumberg, W., Peisach, J., and Maliozzo, R. 1986. Sulfhemoglobin: properties of partially sulfurated tetramers. Journal of Biological Chemistry 261: 8805–8810.

Park, G. A. S., Pappas, B. A., Murtha, S. M., and Ally, A. 1992. Enriched environment primes forebrain choline-acetyltransferase activity to respond to learning-experience. Neuroscience Letters 143: 259–262.

Parker, G. A. 1970. Sperm competition and its evolutionary consequences in the insects. Biological Reviews 45: 525–567.

Parker, G. A. 1974. Assessment strategy and the evolution of fighting behaviour. Journal of Theoretical Biology 47: 223–243.

Parker, G. A. 1979. Sexual selection and sexual conflict. In M. S. Blum and N. A. Blum (eds.), *Sexual Selection and Reproductive Competition in Insects*, 123–166. New York: Academic Press.

Parker, G. A. 1990. Sperm competition games: raffles and roles. Proceedings of the Royal Society of London Series B—Biological Sciences 242: 120–126.

Parker, G. A., and Begon, M. 1986. Optimal egg size and clutch size: effects of environment and maternal phenotype. American Naturalist 128: 573–592.

Parker, G. A., and Partridge, L. 1998. Sexual conflict and speciation. Philosophical Transactions of the Royal Society of London Series B—Biological Sciences 353: 261–274.

Parker, G. A., Ball, M. A., Stockley, P., and Gage, M. J. G. 1996. Sperm competition games: individual assessment of sperm competition intensity by group spawners. Proceedings of the Royal Society of London Series B—Biological Sciences 263: 1291–1297.

Parker, G. A., Ball, M. A., Stockley, P., and Gage, M. J. G. 1997. Sperm competition games: a prospective analysis of risk assessment. Proceedings of the Royal Society of London Series B—Biological Sciences 264: 1793–1802.

Parker, K. M., Sheffer, R. J., and Hedrick, P. W. 1999. Molecular variation and evolutionarily significant units in the endangered Gila topminnow. Conservation Biology 13: 108–116.

Partridge, L., and Sibly, R. 1991. Constraints in the evolution of life histories. Philosophical Transactions of the Royal Society of London Series B—Biological Sciences 332: 3–13.

Parzefall, J. 1969. Zur vergleichenden Ethologie verschiedener *Mollienesia*—Arten einschliesslich einer Höhlenform von *Mollienesia sphenops*. Behaviour 33: 1–37.

Parzefall, J. 1970. Morphologische Untersuchungen an einer Höhlenform von *Mollienesia sphenops* (Pisces, Poeciliidae). Zeitschrift für Morphologie der Tiere 68: 323–342.

Parzefall, J. 1973. Attraction and sexual cycle of poeciliids. In J. H. Schröder (ed.), *Genetics and Mutagenesis of Fish*, 177–183. New York: Springer Verlag.

Parzefall, J. 1974. Rückbildung aggressiver Verhaltensweisen bei einer Höhlenform von *Poecilia sphenops* (Pisces, Poeciliidae). Zeitschrift für Tierpsychologie 35: 66–84.

Parzefall, J. 1979. Genetics and biological significance of the aggressive behavior of *Poecilia sphenops* (Pisces, Poeciliidae): studies on hybrids of epigeous and hypogeous living populations. Zeitschrift für Tierpsychologie 50: 399–422.

Parzefall, J. 2001. A review of morphological and behavioural changes in the cave molly, *Poecilia mexicana*, from Tabasco, Mexico. Environmental Biology of Fishes 62: 263–275.

Parzefall, J., Kraus, C., Tobler, M., and Plath, M. 2007. Photophilic behaviour in surface- and cave-dwelling Atlantic mollies *Poecilia mexicana* (Poeciliidae). Journal of Fish Biology 71: 1225–1231.

Peake, T. M., and McGregor, P. K. 2004. Information and aggression in fishes. Learning and Behavior 32: 114–121.

Peden, A. E. 1972a. Differences in the external genitalia of female gambusiin fishes. Southwest Naturalist 17: 265–272.

Peden, A. E. 1972b. The function of gonopodial parts and behavioural pattern during copulation by *Gambusia* (Poeciliidae). Canadian Journal of Zoology 50: 955–968.

Peden, A. E. 1973. Variation in anal spot expression of Gambusiin females and its effect on male courtship. Copeia 1973: 250–263.

Peden, A. E. 1975. Differences in copulatory behavior as partial isolating mechanisms in poeciliid fish *Gambusia*. Canadian Journal of Zoology—Revue Canadienne de Zoologie 53: 1290–1296.

Penn, D., and Potts, W. K. 1998. Untrained mice discriminate MHC-determined odors. Physiology and Behavior 64: 235–243.

Penn, D. J., and Potts, W. K. 1999. The evolution of mating preferences and major histocompatibility complex genes. American Naturalist 153: 145–164.

Persson, L., Andersson, J., Wahlstrom, E., and Eklov, P. 1996. Size-specific interactions in lake systems: predator gape limitation and prey growth rate and mortality. Ecology 77: 900–911.

Peters, A. D., and Lively, C. M. 2007. Short- and long-term benefits and detriments to recombination under antagonistic coevolution. Journal of Evolutionary Biology 20: 1206–1217.

Peters, G. 1964. Vergleichende Untersuchungen an drei Subspecies von *Xiphophorus helleri* Heckel (Pisces). Zeitschrift für Zoologische Systematik und Evolutionsforschung 2: 185–271.

Peters, G., and Mäder, B. 1964. Morphologische Veränderungen der Gonadenausführgänge sich fortpflanzender Schwertträgerweibchen (*Xiphophorus helleri* Heckel). Zoologischer Anzeiger 173: 243–257.

Peters, N., and Peters, G. 1968. Zur genetischen Interpretation morphologischer Gesetzmässigkeiten der degenerative Evolution. Zeitschrift für Morphologie der Tiere 62: 211–244.

Peters, N., Peters, G., Parzefall, J., and Wilkens, H. 1973. Über degenerative und konstruktive Merkmale bei einer phylogenetisch jungen Höhlenform von *Poecilia sphenops* (Pisces, Poeciliidae). Internationale Revue der Gesamten Hydrobiologie 58: 417–436.

Pethiyagoda, R. 1991. *Freshwater Fishes of Sri Lanka*. Colombo: Wildlife Heritage Trust of Sri Lanka.

Pethiyagoda, R. 2006. Conservation of Sri Lankan freshwater fish. In C. N. B. Bambaradeniya (ed.), *The Fauna of Sri Lanka: Status of Taxonomy, Research and Conservation*, 102–112. Colombo: World Conservation Union and the Government of Sri Lanka.

Pettersson, L. B., Ramnarine, I. W., Becher, S. A., Mahabir, R., and Magurran, A. E. 2004. Sex ratio dynamics and fluctuating selection pressures in natural populations of the Trinidadian guppy, *Poecilia reticulata*. Behavioral Ecology and Sociobiology 55: 461–468.

Pfeiffer, W. 1977. The distribution of fright reaction and alarm substance cells in fishes. Copeia 1977: 653–665.

Pfeiffer, W. 1982. Chemical signals in communication. In T. J. Hara (ed.), *Chemoreception in Fishes*, 307–325. Amsterdam: Elsevier.

Pfennig, K. S. 1998. The evolution of mate choice and the potential for conflict between species and mate-quality recognition. Proceedings of the Royal Society of London Series B—Biological Sciences 256: 1743–1748.

Pfennig, K. S. 2000. Female spadefoot toads compromise on mate quality to ensure conspecific matings. Behavioral Ecology 11: 220–227.

Phang, V. P. E., Khoo, G., and Ang, S. P. 1999. Interaction between the autosomal recessive bar gene and the Y-linked snakeskin body (Ssb) pattern gene in the guppy, *Poecilia reticulata*. Zoological Science 16: 905–908.

Philippi, E. 1908. Fortpflanzungsgeschichte der viviparen Teleosteer *Glaridichthhys januarius* und *G. decem-maculatus* in ihrem Einfluß auf Lebensweise, makroskopische und mikroskopische Anatomie. Zoologische Jahrbücher, Abteilung Anatomie 27: 1–94.

Pianka, E. R. 1970. On r- and K-selection. American Naturalist 104: 592–597.

Pianka, E. R. 1979. Citation classic: on r- and K-selection. Current Contents 47: 10.

Pigliucci, M. 2001. *Phenotypic Plasticity: Beyond Nature and Nurture*. Baltimore, MD: Johns Hopkins University Press.

Pigliucci, M. 2007. Do we need an extended evolutionary synthesis? Evolution 61: 2743–2749.

Pilastro, A., and Bisazza, A. 1999. Insemination efficiency of two alternative male mating tactics in the guppy (*Poecilia reticulata*). Proceedings of the Royal Society of London Series B—Biological Sciences 266: 1887–1891.

Pilastro, A., Giacomello, E., and Bisazza, A. 1997. Sexual selection for small size in male mosquitofish (*Gambusia holbrooki*). Proceedings of the Royal Society of London Series B—Biological Sciences 264: 1125–1129.

Pilastro, A., Evans, J. P., Sartorelli, S., and Bisazza, A. 2002a. Male phenotype predicts insemination success in guppies. Proceedings of the Royal Society of London Series B—Biological Sciences 269: 1325–1330.

Pilastro, A., Scaggiante, M., and Rasotto, M. B. 2002b. Individual adjustment of sperm expenditure accords with sperm competition theory. Proceedings of the National Academy of Sciences of the United States of America 99: 9913–9915.

Pilastro, A., Benetton, S., and Bisazza, A. 2003. Female aggregation and male competition reduce costs of sexual harassment in the mosquitofish *Gambusia holbrooki*. Animal Behaviour 65: 1161–1167.

Pilastro, A., Simonato, M., Bisazza, A., and Evans, J. P. 2004. Cryptic female preference for colorful males in guppies. Evolution 58: 665–669.

Pilastro, A., Mandelli, M., Gasparini, C., Dadda, M., and Bisazza, A. 2007. Copulation duration, insemination efficiency and male attractiveness in guppies. Animal Behaviour 74: 321–328.

Pilastro, A., Gasparini, C., Boschetto, C., and Evans, J. P. 2008. Colorful male guppies do not provide females with fecundity benefits. Behavioral Ecology 19: 374–381.

Pimenta-Leibowitz, M., Ariav, R., and Zilberg, D. 2005. Environmental and physiological conditions affecting *Tetrahymena* sp. infection in guppies, *Poecilia reticulata* Peters. Journal of Fish Diseases 28: 539–547.

Pineda-López, R., Salgado-Maldonado, G., Soto-Galera, E., Hernández-Camacho, N., Orozco-Zamorano, A., Contreras-Robledo, S., Cabañas-Carranza, G., and Aguilar-Aguilar, R. 2005. Helminth parasites of viviparous fishes in Mexico. In M. C. Uribe and H. J. Grier (eds.), *Viviparous Fishes*, 437–456. Homestead, FL: New Life Publications.

Pinto, G., Mahler, D. L., Harmon, L. J., and Losos, J. B. 2008. Testing the island effect in adaptive radiation: rates and patterns of morphological diversification in Caribbean and mainland *Anolis* lizards. Proceedings of the Royal Society of London Series B—Biological Sciences 275: 2749–2757.

Pires, M. N. 2007. The evolution of placentas in poeciliid fishes. PhD thesis, University of California, Riverside.

Pires, M. N., McBride, K. E., and Reznick, D. N. 2007. Interpopulation variation in life-history traits of *Poeciliopsis prolifica*: Implications for the study of placental evolution. Journal of Experimental Zoology Part A—Ecological Genetics and Physiology 307A: 113–125.

Pires, M. N., Arendt, J. D., and Reznick, D. N. 2010. The evolution of placentas and superfetation in the fish genus *Poecilia* (Cyprinodontiformes: Poeciliidae: subgenera *Micropoecilia* and *Acanthophacelus*). Biological Journal of the Linnean Society 99: 784–796.

Pister, E. P. 1974. Desert fishes and their habitats. Transactions of the American Fisheries Society 103: 531–540.

Pitcher, T. E., and Evans, J. P. 2001. Male phenotype and sperm number in the guppy (*Poecilia reticulata*). Canadian Journal of Zoology 79: 1891–1896.

Pitcher, T. E., Neff, B. D., Rodd, F. H., and Rowe, L. 2003. Multiple mating and sequential mate choice in guppies: females trade up. Proceedings of the Royal Society of London Series B—Biological Sciences 270: 1623–1629.

Pitcher, T. E., Rodd, F. H., and Rowe, L. 2007. Sexual colouration and sperm traits in guppies. Journal of Fish Biology 70: 165–177.

Pitcher, T. E., Rodd, F. H., and Rowe, L. 2008. Female choice and the relatedness of mates in the guppy (*Poecilia reticulata*). Genetica 134: 137–146.

Pitcher, T. J. 1983. Heuristic definitions of schooling behaviour. Animal Behaviour 31: 611–613.

Pitcher, T. J., and Parrish, J. K. 1993. Functions of shoaling behaviour in teleosts. In T. J. Pitcher (ed.), *Behaviour of Teleost Fishes*, 363–439. London: Chapman & Hall.

Pitcher, T. J., Magurran, A. E., and Winfield, I. 1982. Fish in larger shoals find food faster. Behavioural Ecology and Sociobiology 10: 149–151.

Pitcher, T. J., Green, D. A., and Magurran, A. E. 1986. Dicing with death: predator inspection behaviour in minnow shoals. Journal of Fish Biology 28: 439–448.

Pitnick, S., and Brown, W. D. 2000. Criteria for demonstrating female sperm choice. Evolution 54: 1052–1056.

Piyapong, C., Butlin, R. K., Faria, J. J., Wang, J., and Krause, J. In press. Kin assortment in juvenile shoals in wild guppy populations. Heredity doi:10.1038/hdy.2010.115.

Pizzari, T., Froman, D. P., and Birkhead, T. R. 2002. Pre- and post-insemination episodes of sexual selection in the fowl, *Gallus g. domesticus*. Heredity 88: 112–116.

Placzek, M., Przybilla, B., Kerkmann, U., Gaube, S., and Gilbertz, K. P. 2007. Effect of ultraviolet (UV) A, UVB or ionizing radiation on the cell cycle of human melanoma cells. British Journal of Dermatology 156: 843–847.

Plath, M. 2004. Cave molly females (*Poecilia mexicana*) avoid parasitised males. Acta Ethologica 6: 47–51.

Plath, M. 2008. Male mating behavior and costs of sexual harassment for females in cavernicolous and extremophile populations of Atlantic mollies (*Poecilia mexicana*). Behaviour 145: 73–98.

Plath, M., and Schlupp, I. 2008. Parallel evolution leads to reduced shoaling behavior in two cave dwelling populations of Atlantic mollies (*Poecilia mexicana*, Poeciliidae, Teleostei). Environmental Biology of Fishes 82: 289–297.

Plath, M., and Tobler, M. 2010. The evolutionary ecology of the cave molly (*Poecilia mexicana*) from the Cueva del Azufre system. In E. Trajano, M. E. Bichuette, and B. G. Kapoor (eds.), *The Biology of Subterranean Fishes*, 283–332. Enfield, NH: Science Publishers.

Plath, M., Parzefall, J., and Schlupp, I. 2003. The role of sexual harassment in cave and surface dwelling populations of the Atlantic molly, *Poecilia mexicana* (Poeciliidae, Teleostei). Behavioral Ecology and Sociobiology 54: 303–309.

Plath, M., Parzefall, J., Körner, K. E., and Schlupp, I. 2004. Sexual selection in darkness? Female mating preferences in surface- and cave-dwelling Atlantic mollies, *Poecilia mexicana* (Poeciliidae, Teleostei). Behavioral Ecology and Sociobiology 55: 596–601.

Plath, M., Heubel, K. U., García de León, F., and Schlupp, I. 2005a. Cave molly females (*Poecilia mexicana*, Poeciliidae, Teleostei) like well-fed males. Behavioral Ecology and Sociobiology 58: 144–151.

Plath, M., Heubel, K. U., and Schlupp, I. 2005b. Field observations on male mating behavior in surface- and cave-dwelling Atlantic mollies (*Poecilia mexicana*, Poeciliidae). Zeitschrift für Fischkunde 7: 113–119.

Plath, M., Seggel, U., Burmeister, H., Heubel, K. U., and Schlupp, I. 2006. Choosy males from the underground: male mating preferences in surface- and cave-dwelling Atlantic mollies (*Poecilia mexicana*). Naturwissenschaften 93: 103–109.

Plath, M., Hauswaldt, J. S., Moll, K., Tobler, M., García de León, F. J., Schlupp, I., and Tiedemann, R. 2007a. Local adaptation and pronounced genetic differentiation in an extremophile fish, *Poecilia mexicana*, inhabiting a Mexican cave with toxic hydrogen sulphide. Molecular Ecology 16: 967–976.

Plath, M., Makowicz, A. M., Schlupp, I., and Tobler, M. 2007b. Sexual harassment in live-bearing fishes (Poeciliidae): comparing courting and noncourting species. Behavioral Ecology 18: 680–688.

Plath, M., Schlupp, I., Parzefall, J., and Riesch, R. 2007c. Female choice for large body size in the cave molly, *Poecilia mexicana* (Poeciliidae, Teleostei): influence of species- and sex-specific cues. Behaviour 144: 1147–1160.

Plath, M., Tobler, M., Riesch, R., García de León, F. J., Giere, O., and Schlupp, I. 2007d. Survival in an extreme habitat: the roles of behaviour and energy limitation. Naturwissenschaften 94: 991–996.

Plath, M., Blum, D., Schlupp, I., and Tiedemann, R. 2008a. Audience effect alters mating preferences in a livebearing fish, the Atlantic molly, *Poecilia mexicana*. Animal Behaviour 75: 21–29.

Plath, M., Blum, D., Tiedemann, R., and Schlupp, I. 2008b. A visual audience effect in a cavefish. Behaviour 145: 931–947.

Plath, M., Richter, S., Tiedemann, R., and Schlupp, I. 2008c. Male fish deceive competitors about mating preferences. Current Biology 18: 1138–1141.

Plath, M., Kromuszczynski, K., and Tiedemann, R. 2009. Audience effect alters male but not female mating preferences. Behavioral Ecology and Sociobiology 63: 381–390.

Plath, M., Hermann, B., Schröder, C., Riesch, R., Tobler, M., García de León, F. J., Schlupp, I., and Tiedemann, R. 2010a. Locally adapted fish populations maintain small-scale genetic differentiation despite perturbation by a catastrophic flood event. BMC Evolutionary Biology 2010: 256.

Plath, M., Riesch, R., Oranth, A., Dzienko, J., Karau, N., Schießl, A., Stadler, S., Wigh, A., Zimmer, C., Arias-Rodriguez, L., Schlupp, I., and Tobler, M. 2010b. Complementary effect of natural and sexual selection against immigrants maintains differentiation between locally adapted fish. Naturwissenschaften 97: 769–774.

Plaut, I. 2002. Does pregnancy affect swimming performance of female mosquitofish, *Gambusia affinis*? Functional Ecology 16: 290–295.

Poeser, F. N. 1998. The role of character displacement in the speciation of Central American members of the genus *Poecilia* (Poeciliidae). Italian Journal of Zoology 65: 145–147.

Poeser, F. N. 2002. *Poecilia kykesis* nom. nov., a new name for *Mollienesia petenensis* Gunther, 1866, and redescription, revalidation and the designation of a lectotype for *Poecilia petenensis* Gunther, 1866 (Teleostei: Poeciliidae). Contributions to Zoology 70: 243–246.

Poeser, F. N., Kempkes, M., and Isbrucker, I. J. H. 2005. Description of *Poecilia* (*Acanthophacelus*) *wingei* n. sp. from the Paria Peninsula, Venezuela, including notes on *Acanthophacelus Eigenmann*, 1907 and other subgenera of *Poecilia* Bloch and Schneider, 1801 (Teleostei, Cyprinodontiformes, Poeciliidae). Contributions to Zoology 74: 97–115.

Pollux, B. J. A., Pires, M. N., Banet, A. I., and Reznick, D. N. 2009. Evolution of placentas in the fish family poeciliidae: an empirical study of macroevolution. Annual Reviews of Ecology, Evolution, and Systematics 40: 271–289.

Pomiankowski, A., and Iwasa, Y. 1998. Runaway ornament diversity caused by Fisherian sexual selection. Proceedings of the National Academy of Sciences of the United States of America 95: 5106–5111.

Porter, M., and Crandall, K. A. 2003. Lost along the way: the significance of evolution in reverse. Trends in Ecology and Evolution 18: 541–547.

Posner, M. I., Rothbart, M. K., and Sheese, B. E. 2007. Attention genes. Developmental Science 10: 24–29.

Potter, H., and Kramer, C. R. 2000. Ultrastructural observations on sperm storage in the ovary on the platyfish, *Xiphophorus maculatus* (Teleostei: Poeciliidae): the role of the duct epithelium. Journal of Morphology 245: 110–129.

Potts, W. K. 1984. The chorus line hypothesis of manoeuvre coordination in avian flocks. Nature 309: 344–345.

Poulin, R. 2006. *Evolutionary Ecology of Parasites*. Princeton, NJ: Princeton University Press.

Poulson, T. L. 2001. Adaptations of cave fishes with some comparisons to deep-sea fishes. Environmental Biology of Fishes 62: 345–364.

Poulson, T. L., and Lavoie, K. H. 2000. The trophic basis of subterranean ecosystems. In H. Wilkens, D. C. Culver, and W. F. Humphries (eds.), *Subterranean Ecosystems*, 231–249. Ecosystems of the World 30. Amsterdam: Elsevier Science.

Pound, G. E., Cox, S. J., and Doncaster, C. P. 2004. The accumulation of deleterious mutations within the frozen niche variation hypothesis. Journal of Evolutionary Biology 17: 651–662.

Prasad, N. G., Bedhomme, S., Day, T., and Chippindale, A. K. 2007. An evolutionary cost of separate genders revealed by male-limited evolution. American Naturalist 169: 29–37.

Prenter, J., Taylor, P. W., and Elwood, R. W. 2008. Large body size for winning and large swords for winning quickly in swordtail males, *Xiphophorus helleri*. Animal Behaviour 75: 1981–1987.

Presgraves, D. C. 2008. Sex chromosomes and speciation in *Drosophila*. Trends in Genetics 24: 336–343.

Price, A. C., and Rodd, F. H. 2006. The effect of social environment on male-male competition in guppies (*Poecilia reticulata*). Ethology 112: 22–32.

Price, T. D., Qvarnström, A., and Irwin, D. E. 2003. The role of phenotypic plasticity in driving genetic evolution. Proceedings of the Royal Society of London Series B—Biological Sciences 270: 1433–1440.

Pritchard, J. K., Stephens, M., and Donnelly, P. 2000. Inference of population structure using multilocus genotype data. Genetics 155: 945–959.

Pritchard, V. L., Lawrence, J., Butlin, R. K., and Krause, J. 2001. Shoal choice in zebrafish, *Danio rerio*: the influence of shoal size and activity. Animal Behaviour 62: 1085–1088.

Pruett-Jones, S. G. 1992. Independent versus non-independent mate choice: do females copy each other? American Naturalist 140: 1000–1009.

Ptacek, M. B. 1998. Interspecific mate choice in sailfin and short-fin species of mollies. Animal Behaviour 56: 1145–1154.

Ptacek, M. B., and Breden, F. 1998. Phylogenetic relationships among the mollies (Poeciliidae: *Poecilia*: *Mollienesia* group) based on mitochondrial DNA sequences. Journal of Fish Biology 53: 64–81.

Ptacek, M. B., and Travis, J. 1996. Inter-population variation in male mating behaviours in the sailfin mollie, *Poecilia latipinna*. Animal Behaviour 52: 59–71.

Ptacek, M. B., and Travis, J. 1997. Mate choice in the sailfin molly, *Poecilia latipinna*. Evolution 51: 1217–1231.

Ptacek, M. B., and Travis, J. 1998. Hierarchical patterns of covariance between morphological and behavioural traits. Animal Behaviour 56: 1044–1048.

Purcell, K. M., Hitch, A. T., Klerks, P. L., and Leberg, P. L. 2008. Adaptation as a potential response to sea-level rise: a genetic basis for salinity tolerance in populations of a coastal marsh fish. Evolutionary Applications 1: 155–160.

Purdom, C. E. 1993. *Genetics and Fish Breeding*. London: Chapman & Hall.

Purser, G. L. 1937. Succession of broods of *Lebistes*. Nature 140: 155.

Pyke, G. H. 2005. A review of the biology of *Gambusia affinis* and *G. holbrooki*. Reviews in Fish Biology and Fisheries 15: 339–365.

Pyke, G. H. 2008. Plague minnow or mosquito fish? A review of the biology and impacts of introduced *Gambusia* species. Annual Review of Ecology Evolution and Systematics 39: 171–191.

Quattro, J. M., and Vrijenhoek, R. C. 1989. Fitness differences among remnant populations of endangered Sonoran topminnow. Science 245: 976–978.

Quattro, J. M., Avise, J. C., and Vrijenhoek, R. C. 1991. Molecular evidence for multiple origins of hybridogenetic fish clones (Poeciliidae: *Poeciliopsis*). Genetics 127: 391–398.

Quattro, J. M., Avise, J. C., and Vrijenhoek, R. C. 1992a. An ancient clonal lineage in the fish genus *Poeciliopsis atheriniformes* (Poeciliidae). Proceedings of the National Academy of Sciences of the United States of America 89: 348–352.

Quattro, J. M., Avise, J. C., and Vrijenhoek, R. C. 1992b. Mode of origin and sources of genotypic diversity in triploid gynogenetic fish clones (*Poeciliopsis*: Poeciliidae). Genetics 130: 621–628.

Quattro, J. M., Leberg, P. L., Douglas, M. E., and Vrijenhoek, R. C. 1996. Molecular evidence for a unique evolutionary lineage of endangered Sonoran Desert fish (genus *Poeciliopsis*). Conservation Biology 10: 128–135.

Radwan, J. 2008. Maintenance of genetic variation in sexual ornaments: a review of the mechanisms. Genetica 134: 113–127.

Rahn, J. J., Trono, D., Gimenez-Conti, I., Butler, A. P., and Nairn, R. S. 2009. Etiology of MNU-induced melanomas in *Xiphophorus* hybrids. Comparative Biochemistry and Physiology C—Toxicology and Pharmacology 149: 129–133.

Räsänen, K., and Hendry, A. P. 2008. Disentangling interactions between adaptive divergence and gene flow when ecology drives diversification. Ecology Letters 11: 624–636.

Rasch, E. M., and Balsano, J. S. 1989. Trihybrids related to the unisexual molly fish, *Poecilia formosa*. In R. M. Dawley and J. P. Bogart (eds.), *Evolution and Ecology of Unisexual Vertebrates*, 252–267. Albany: New York State Museum.

Rasch, E. M., Monaco, P. J., and Balsano, J. S. 1982. Cytomorphometric and autoradiographic evidence for functional apomixis in a gynogenetic fish, *Poecilia formosa*, and its related, triploid unisexuals. Histochemistry 73: 515–533.

Rauchenberger, M. 1989. Annotated list of species of the subfamily Poeciliinae. In G. K. Meffe and F. F. Snelson Jr. (eds.), *Ecology and Evolution of Livebearing Fishes (Poeciliidae)*, 359–368. Englewood Cliffs, NJ: Prentice Hall.

Rauchenberger, M., Kallman, K. D., and Morizot, D. C. 1990. Monophyly and geography of the Río Panuco basin swordtails (genus *Xiphophorus*) with descriptions of four new species. American Museum Novitates 2975: 1–41.

Reader, S. M., and Laland, K. N. 2000. Diffusion of foraging innovations in the guppy. Animal Behaviour 60: 175–180.

Reader, S. M., and Laland, K. N. 2003. *Animal Innovation.* Oxford: Oxford University Press.

Reader, S. M., Kendal, J. R., and Laland, K. N. 2003. Social learning of foraging sites and escape routes in wild Trinidadian guppies. Animal Behaviour 66: 729–739.

Real, L. A. 1990. Search theory and mate choice, I: models of single-sex discrimination. American Naturalist 136: 376–405.

Reale, D., Gallant, B. Y., Leblanc, M., and Festa-Bianchet, M. 2000. Consistency of temperament in bighorn ewes and correlates with behaviour and life history. Animal Behaviour 60: 589–597.

Reebs, S. G. 2000. Can a minority of informed leaders determine the foraging movements of a fish shoal? Animal Behaviour 59: 403–409.

Reebs, S. G. 2001. Influence of body size on leadership in shoals of golden shiners, *Notemigonus crysoleucas.* Behaviour 138: 797–809.

Reed, J. R. 1969. Alarm substances and fright reactions in some fishes from the southeastern United States. Transactions of the American Fisheries Society 4: 664–668.

Rees, J. T. 1979. Community development in freshwater microcosms. Hydrobiologia 63: 113–128.

Regan, C. T. 1906. On the fresh-water fishes of the island of Trinidad, based on the collection, notes and sketches made by Mr. Lechmere Guppy, Junr. Proceedings of the Zoological Society of London 1: 378–393.

Regan, C. T. 1913. A revision of the cyprinodont fishes of the subfamily Poeciliinae. Proceedings of the Zoological Society of London 2: 977–1018.

Regan, J. D. 1961. Melanism in the poeciliid fish, *Gambusia affinis* (Baird and Girard). American Midland Naturalist 65: 139–143.

Regan, J. D., Carrier, W. L., Samet, C., and Olla, B. L. 1982. Photoreactivation in two closely related marine fishes having different longevities. Mechanisms of Ageing and Development 18: 59–66.

Rehage, J. S., and Sih, A. 2004. Dispersal behavior, boldness, and the link to invasiveness: a comparison of four *Gambusia* species. Biological Invasions 6: 379–391.

Reiniger, I. W., Wolf, A., Welge-Lussen, U., Mueller, A. J., Kampik, A., and Schaller, U. C. 2007. Osteopontin as a serologic marker for metastatic uveal melanoma: results of a pilot study. American Journal of Ophthalmology 143: 705–707.

Rensch, B. 1956. Increase of learning capability with increase of brain size. American Naturalist 90: 81–95.

Reusch, T. B. H., and Langefors, A. 2005. Inter- and intralocus recombination drive MHC class IIB gene diversification in a teleost, the three-spined stickleback *Gasterosteus aculeatus.* Journal of Molecular Evolution 61: 531–541.

Reusch, T. B. H., Schaschl, H., and Wegner, K. M. 2004. Recent duplication and inter-locus gene conversion in major histocompatibility class II genes in a teleost, the three-spined stickleback. Immunogenetics 56: 427–437.

Reutter, K., Breipohl, W., and Bijvank, G. J. 1974. Taste bud types in fishes, II: scanning electron microscopical investigations of *Xiphophorus helleri* Heckel (Poeciliidae, Cyprinodontiformes, Teleostei). Cell and Tissue Research 153: 151–165.

Revkin, S. K., Piazza, M., Izard, V., Cohen, L., and Dehaene, S. 2008. Does subitizing reflect numerical estimation? Psychological Science 19: 607–614.

Reynolds, J. D. 1993. Should attractive individuals court more? Thoery and a test. American Naturalist 141: 914–927.

Reynolds, J. D., and Gross, M. R. 1992. Female mate preference enhances offspring growth and reproduction in a fish, *Poecilia reticulata.* Proceedings of the Royal Society of London Series B—Biological Sciences 264: 57–62.

Reynolds, J. D., Gross, M. R., and Coombs, M. J. 1993. Environmental conditions and male morphology determine alternative mating behaviour in Trinidadian guppies. Animal Behaviour 45: 145–152.

Reznick, D. N. 1981. Grandfather effects: the genetics of interpopulation differences in offspring size in the mosquito fish. Evolution 35: 941–953.

Reznick, D. N. 1983. The structure of guppy life histories: the tradeoff between growth and reproduction. Ecology 64: 862–873.

Reznick, D. N. 1989. Life-history evolution in guppies, 2: repeatability of field observations and the effects of season on life histories. Evolution 43: 1285–1297.

Reznick, D. N., and Braun, B. 1987. Fat cycling in the mosquitofish (*Gambusia affinis*): fat storage as a reproductive adaptation. Oecologia 73: 401–413.

Reznick, D. N., and Bryant, M. 2007. Comparative long-term mark-recapture studies of guppies (*Poecilia reticulata*): differences among high and low predation localities in growth and survival. Annales Zoologici Fennici 44: 152–160.

Reznick, D. N., and Bryga, H. 1987. Life-history evolution in guppies (*Poecilia reticulata*: Poeciliidae), I: phenotypic and genetic changes in an introduction experiment. Evolution 41: 1370–1385.

Reznick, D. N., and Bryga, H. 1996. Life-history evolution in guppies (*Poecilia reticulata*: Poeciliidae), V: genetic basis of parallelism in life histories. American Naturalist 147: 339–359.

Reznick, D. N., and Endler, J. A. 1982. The impact of predation on life history evolution in Trinidadian guppies (*Poecilia reticulata*). Evolution 36: 125–148.

Reznick, D. N., and Ghalambor, C. K. 2001. The population ecology of contemporary adaptations: what empirical studies reveal about the conditions that promote adaptive evolution. Genetica 112: 183–198.

Reznick, D. N., and Miles, D. B. 1989a. A review of life history patterns in poeciliid fishes. In G. K. Meffe and F. F. Snelson Jr. (eds.), *Ecology and Evolution of Livebearing Fishes (Poeciliidae),* 125–148. Englewood Cliffs, NJ: Prentice Hall.

Reznick, D. N., and Miles, D. B. 1989b. Appendix 3. In G. K. Meffe and F. F. Snelson Jr. (eds.), *Ecology and Evolution of Livebearing Fishes (Poeciliidae),* 374–377. Englewood Cliffs, NJ: Prentice Hall.

Reznick, D. [N.], and Travis, J. 1996. The empirical study of adaptation in natural populations. In M. R. Rose and G. V. Lauder (eds.), *Adaptation,* 243–289. San Diego: Academic Press.

Reznick, D. N., and Yang, A. P. 1993. The influence of fluctuating resources on life-history: patterns of allocation and plasticity in female guppies. Ecology 74: 2011–2019.

Reznick, D. N., Bryga, H., and Endler, J. A. 1990. Experimentally induced life-history evolution in a natural population. Nature 346: 357–359.

Reznick, D. N., Miles, D. B., and Winslow, S. 1992. Life history of *Poecilia picta* (Poeciliidae) from the island of Trinidad. Copeia 1992: 782–790.

Reznick, D. N., Meyer, A., and Frear, D. 1993. Life history of *Brachyrhaphis rhabdophora* (Pisces: Poeciliidae). Copeia 1993: 103–111.

Reznick, D. N., Baxter, R. J., and Endler, J. 1994. Long-term studies of tropical stream fish communities: the use of field notes and museum collections to reconstruct communities of the past. American Zoologist 34: 452–462.

Reznick, D. N., Callahan, H., and Llauredo, R. 1996a. Maternal effects on offspring quality in poeciliid fishes. American Zoologist 36: 147–156.

Reznick, D. N., Butler, M. J., Rodd, F. H., and Ross, P. 1996b. Life-history evolution in guppies (*Poecilia reticulata*: Poeciliidae), II: differential mortality as a mechanism for natural selection. Evolution 50: 1651–1660.

Reznick, D. N., Rodd, F. H., and Cardenas, M. 1996c. Life-history evolution in guppies (*Poecilia reticulata*: Poeciliidae), IV: parallelism in life-history phenotypes. American Naturalist 147: 319–338.

Reznick, D. N., Shaw, F. H., Rodd, F. H., and Shaw, R. G. 1997. Evaluation of the rate of evolution in natural populations of guppies (*Poecilia reticulata*). Science 275: 1934–1937.

Reznick, D. N., Butler, M. J., and Rodd, H. 2001. Life-history evolution in guppies, VII: the comparative ecology of high- and low-predation environments. American Naturalist 157: 126–140.

Reznick, D. N., Bryant, M. J., and Bashey, F. 2002a. r- and K-selection revisited: the role of population regulation in life-history evolution. Ecology 83: 1509–1520.

Reznick, D. N., Mateos, M., and Springer, M. S. 2002b. Independent origins and rapid evolution of the placenta in the fish genus *Poeciliopsis*. Science 298: 1018–1020.

Reznick, D. N., Nunney, L., and Rodd, H. 2004. Empirical evidence for rapid evolution. In R. Ferrière, U. Dieckmann, and D. Couvet (eds.), *Evolutionary Conservation Biology*, 101–118. Cambridge: Cambridge University Press.

Reznick, D. N., Hrbek, T., Caura, S., de Greef, J., and Roff, D. 2007a. Life history of *Xenodexia ctenolepis*: implications for life history evolution in the family Poeciliidae. Biological Journal of the Linnean Society 92: 77–85.

Reznick, D. N., Meredith, R., and Collette, B. B. 2007b. Independent evolution of complex life history adaptations in two families of fishes, live-bearing halfbeaks (Zenarchopteridae, Beloniformes) and Poeciliidae (Cyprinodontiformes). Evolution 61: 2570–2583.

Reznick, D. N., Ghalambor, C. K., and Crooks, K. 2008. Experimental studies of evolution in guppies: a model for understanding the evolutionary consequences of predator removal in natural communities. Molecular Ecology 17: 97–107.

Rhen, T. 2000. Sex-limited mutations and the evolution of sexual dimorphism. Evolution 54: 37–43.

Ribowski, A., and Franck, D. 1993. Demonstration of strength and concealment of weakness in escalating fights of male swordtails (*Xiphophorus helleri*). Ethology 93: 265–271.

Rice, W. R. 1984. Sex chromosomes and the evolution of sexual dimorphism. Evolution 38: 735–742.

Rice, W. R. 1987a. The accumulation of sexually antagonistic genes as a selective agent promoting the evolution of reduced recombination between primitive sex chromosomes. Evolution 41: 911–914.

Rice, W. R. 1987b. Genetic hitchiking and the evolution of reduced genetic activity of the Y sex chromosome. Genetics 116: 161–167.

Rice, W. R. 1992. Sexually antagonistic genes: experimental evidence. Science 256: 1436–1439.

Rice, W. R. 1994. Degeneration of a nonrecombining chromosome. Science 263: 230–232.

Rice, W. R. 1996a. Evolution of the Y sex chromosome in animals. Bioscience 46: 331–343.

Rice, W. R. 1996b. Sexually antagonistic male adaptation triggered by experimental arrest of female evolution. Nature 381: 232–234.

Rice, W. R., and Holland, B. 1997. The enemies within: intergenomic conflict, interlocus contest evolution (ICE), and the intraspecific Red Queen. Behavioural Ecology and Sociobiology 41: 1–10.

Richards, E. L., Van Oosterhout, C., and Cable, J. 2010. Gender-specific differences in shoaling affect the rate of parasite transmission in guppies. PloS ONE 5: e13285.

Richards, G. R., and Chubb, J. C. 1996. Host response to initial and challenge infections, following treatment, of *Gyrodactylus bullatarudis* and *G. turnbulli* (Monogenea) on the guppy (*Poecilia reticulata*). Parasitology Research 82: 242–247.

Richards, G. R., and Chubb, J. C. 1998. Longer-term population dynamics of *Gyrodactylus bullatarudis* and *G. turnbulli* (Monogenea) on adult guppies (*Poecilia reticulata*) in 50-L experimental arenas. Parasitology Research 84: 753–756.

Richards, G. R., Veltkamp, C. J., and Chubb, J. C. 2000. Differentiation of *Gyrodactylus bullatarudis* Turnbull, 1956 and *G. rasini* Lucky, 1973 (Monogenea) with reassignment of *Gyrodactylus bullatarudis* Turnbull, 1956 sensu Harris (1986) to *G. rasini*. Journal of Natural History 34: 341–353.

Richman, A. D., Herrera, L. G., Nash, D., and Schierup, M. H. 2003. Relative roles of mutation and recombination in generating allelic polymorphism at an MHC class II locus in *Peromyseus maniculatus*. Genetical Research 82: 89–99.

Ricklefs, R. E., and Wikelski, M. 2002. The physiology/life-history nexus. Trends in Ecology and Evolution 17: 462–468.

Riehl, R. 1991. Masculinization in a hermaphroditic female of the mosquitofish *Heterandria formosa*. Japanese Journal of Ichthyology 37: 374–380.

Riehl, R., and Greven, H. 1993. Fine-structure of egg envelopes in some viviparous goodeid fishes, with comments on the relation of envelope thinness to viviparity. Canadian Journal of Zoology 71: 91–97.

Riehl, R., and Greven, H. 2008. Tight and gap junctions in the follicle epithelium of vitellogenic oocytes in the least killifish, *Heterandria formosa* (Poeciliidae). Bulletin of Fish Biology 10: 93–96.

Riesch, R., Schlupp, I., and Plath, M. 2008. Female sperm limitation in natural populations of a sexual/asexual mating complex (*Poecilia latipinna*, *Poecilia formosa*). Biology Letters 4: 266–269.

Riesch, R., Plath, M., Schlupp, I., and Marsh-Matthews, E. 2009a. Matrotrophy in the cave molly: an unexpected provisioning strategy in an extreme environment. Evolutionary Ecology 24: 789–801.

Riesch, R., Tobler, M., Plath, M., and Schlupp, I. 2009b.

Offspring number in a livebearing fish (*Poecilia mexicana*, Poeciliidae): reduced fecundity and reduced plasticity in a population of cave mollies. Environmental Biology of Fishes 84: 89–94.

Riesch, R., Plath, M., García de León, F. J., and Schlupp, I. 2010. Convergent life-history shifts: toxic environments result in big babies in two clades of poeciliids. Naturwissenschaften 97: 133–141.

Riesch, R., Oranth, A., Dzienko, J., Karau, N., Schießl, A., Stadler, S., Wigh, A., Zimmer, C., Arias-Rodriguez, L., Schlupp, I., and Plath, M. 2010. Extreme habitats are not refuges: poeciliids suffer from increased aerial predation risk in sulfidic, southern Mexican habitats. Biological Journal of the Linnean Society 101: 417–426.

Rincon, P. A., Correas, A. M., Morcillo, F., Risueno, P., and Lobon-Cervia, J. 2002. Interaction between the introduced eastern mosquitofish and two autochthonous Spanish toothcarps. Journal of Fish Biology 61: 1560–1585.

Rios-Cardenas, O., Tudor, M. S., and Morris, M. R. 2007. Female preference variation has implications for the maintenance of an alternative mating strategy in a swordtail fish. Animal Behaviour 74: 633–640.

Rios-Cardenas, O., Darrah, A., and Morris, M. R. 2010. Female mimicry and an enhanced sexually selected trait: what does it take to fool a male? Behaviour 147: 1443–1460.

Ritchie, M. G. 2007. Sexual selection and speciation. Annual Review of Ecology Evolution and Systematics 38: 79–102.

Rivas, L. R. 1980. Eight new species of poeciliid fishes of the genus *Limia* from Hispaniola. Northeast Gulf Science 2: 98–112.

Rivera-Rivera, N. L., Martinez-Rivera, N., Torres-Vazquez, I., Serrano-Velez, J. L., Lauder, G. V., and Rosa-Molinar, E. 2010. A male poeciliid's sexually dimorphic body plan, behavior, and nervous system. Integrative and Comparative Biology 50: 1081–1090.

Robbins, L. W., Hartman, G. D., and Smith, M. H. 1987. Dispersal, reproductive strategies, and the maintenance of genetic variability in mosquitofish (*Gambusia affinis*). Copeia 1987: 156–164.

Robertson, G. P. 2005. Functional and therapeutic significance of Akt deregulation in malignant melanoma. Cancer and Metastasis Review 24: 273–285.

Robinson, B. W., and Wilson, D. S. 1994. Character release and displacement in fishes: a neglected literature. American Naturalist 144: 596–627.

Robinson, B. W., and Wilson, D. S. 1995. Experimentally induced morphological diversity in Trinidadian guppies (*Poecilia reticulata*). Copeia 1995: 294–305.

Robinson, D. M., and Morris, M. R. 2010. Unraveling the complexities of variation in female mate preference for vertical bars in the swordtail, *Xiphophorus cortezi*. Behavioral Ecology and Sociobiology 64: 1537–1545.

Robinson, D. M., Aspbury, A. S., and Gabor, C. R. 2008. Differential sperm expenditure by male sailfin mollies, *Poecilia latipinna*, in a unisexual-bisexual species complex and the influence of spermiation during mating. Behavioral Ecology and Sociobiology 62: 705–711.

Robinson, J., Waller, M. J., Parham, P., de Groot, N., Bontrop, R., Kennedy, L. J., Stoehr, P., and Marsh, S. G. E. 2003. IMGT/HLA and IMGT/MHC: sequence databases for the study of the major histocompatibility complex. Nucleic Acids Research 31: 311–314.

Rodd, F. H., and Reznick, D. N. 1991. Life history evolution in guppies, III: the impact of prawn predation on guppy life histories. Oikos 62: 13–19.

Rodd, F. H., and Sokolowski, M. B. 1995. Complex origins of variation in the sexual behaviour of male Trinidadian guppies, *Poecilia reticulata*: interactions between social environment, heredity, body size and age. Animal Behaviour 49: 1139–1159.

Rodd, F. H., Hughes, K. A., Grether, G. F., and Baril, C. T. 2002. A possible non-sexual origin of mate preference: are male guppies mimicking fruit? Proceedings of the Royal Society of London Series B—Biological Sciences 269: 475–481.

Rodriguez, F., Broglio, C., Duran, E., Gomez, A., and Salas, C. 2006. Neural mechanisms of learning in teleost fish. In C. Brown, K. Laland, and J. Krause (eds.), *Fish Cognition and Behavior*, 243–277. Oxford: Blackwell.

Rodriguez, R. L., Sullivan, L. E., and Cocroft, R. B. 2004. Vibrational communication and reproductive isolation in the *Enchenopa binotata* species complex of treehoppers (Hemiptera: Membracidae). Evolution 58: 571–578.

Roff, D. A. 1992. *The Evolution of Life Histories*. New York: Chapman & Hall.

Rogers, A. R. 1988. Does biology constrain culture? American Anthropologist 90: 819–831.

Rogers, L. 1996. Behavioral, structural and neurochemical asymmetries in the avian brain: a model system for studying visual development and processing. Neuroscience and Biobehavioral Reviews 20: 487–503.

Rogers, L. J. 1989. Laterality in animals. International Journal of Comparative Psychology 3: 5–25.

Rogers, L. J. 2000. Evolution of hemispheric specialization: advantages and disadvantages. Brain and Language 73: 236–253.

Rogers, L. J. 2002. Advantages and disadvantages of lateralization. In R. J. Andrew and L. J. Rogers (eds.), *Comparative Vertebrate Lateralization*, 126–153. Cambridge: Cambridge University Press.

Rogers, L. J., and Workman, L. 1989. Light exposure during incubation affects competitive behaviour in domestic chicks. Applied Animal Behaviour Science 23: 187–198.

Rogers, L. J., Zucca, P., and Vallortigara, G. 2004. Advantages of having a lateralized brain. Proceedings of the Royal Society of London Series B—Biological Sciences 271: S420–S422.

Rogers, S. M., and Bernatchez, L. 2007. The genetic architecture of ecological speciation and the association with signatures of selection in natural lake whitefish (*Coregonus* sp. Salmonidae). Molecular Biology and Evolution 24: 1423–1438.

Rogowski, D. L., and Stockwell, C. A. 2006a. Assessment of potential impacts of exotic species on populations of a threatened species, White Sands pupfish, *Cyprinodon tularosa*. Biological Invasions 8: 79–87.

Rogowski, D. L., and Stockwell, C. A. 2006b. Parasites and salinity: costly tradeoffs in a threatened species. Oecologia 146: 615–622.

Römer, U., and Beisenherz, W. 1996. Environmental determination of sex in *Apistogramma* (Cichlidae) and two other freshwater fishes (Teleostei). Journal of Fish Biology 48: 714–725.

Romero, A., and Green, S. M. 2005. The end of regressive evolution: examining and interpreting the evidence from cave fishes. Journal of Fish Biology 67: 3–32.

Roney, K. E., Cuthbertson, B. J., Godwin, U. B., Kazianis, S., Della Coletta, L., Rosenthal, G. G., Ryan, M. J., Schmidt, M., and McConnell, T. J. 2004. Alternative splicing of major histocompatibility complex class II *DXB* transcripts in *Xiphophorus* fishes. Immunogenetics 56: 462–466.

Rosa, R. S., and Costa, W. J. E. M. 1993. Systematic revision of the genus *Cnesterodon* (Cyprinodontiformes: Poeciliidae) with the description of two new species from Brazil. Copeia 1993: 696–708.

Rosales Lagarde, L., Boston, P. J., Campbell, A., and Stafford, K. W. 2006. Possible structural connection between Chichón Volcano and the sulfur-rich springs of Villa Luz Cave (a.k.a. Cueva de las Sardinas), southern Mexico. Association for Mexican Cave Studies Bulletin 19: 177–184.

Rosa-Molinar, E. 2005. Edwin S. Goodrich's theory of transposition revisited: the shift to a sexually dimorphic axial formulae and nervous system in a poeciliine fish. In C. M. Uribe and H. J. Grier (eds.), *Viviparous Fishes*, 59–70. Homestead, FL: New Life Publications.

Rosa-Molinar, E., Hendricks, S. E., Rodriguezsierra, J. F., and Fritzsch, B. 1994. Development of the anal fin appendicular support in the western mosquitofish, *Gambusia affinis affinis* (Baird and Girard, 1854): reinvestigation and reinterpretation. Acta Anatomica 151: 20–35.

Rosa-Molinar, E., Fritzsch, B., and Hendricks, S. E. 1996. Organizational-activational concept revisited: sexual differentiation in an atherinomorph teleost. Hormones and Behavior 30: 563–575.

Rosa-Molinar, E., Proskocil, B. J., Hendricks, S. E., and Fritzsch, B. 1998. A mechanism for anterior transposition of the anal fin and its appendicular support in the western mosquitofish, *Gambusia affinis affinis* [Baird and Girard, 1854]. Acta Anatomica 163: 75–91.

Rosen, D. E. 1979. Fishes from the uplands and intermontrane basins of Guatemala: revisionary studies and comparative geography. Bulletin of the American Museum of Natural History 162: 271–375.

Rosen, D. E., and Bailey, R. M. 1959. Middle-American poeciliid fishes of the genera *Carlhubbsia* and *Phallichthys*, with descriptions of two new species. Zoologica 44: 1–44.

Rosen, D. E., and Bailey, R. M. 1963. The poeciliid fishes (Cyprinodontiformes), their structure, zoogeography, and systematics. Bulletin of the American Museum of Natural History 126: 1–176.

Rosen, D. E., and Gordon, M. 1953. Functional anatomy and evolution of male genitalia in poeciliid fishes. Zoologica 38: 1–47.

Rosen, D. E., and Tucker, A. 1961. Evolution of secondary sexual characters and sexual behavior patterns in a family of viviparous fishes (Cyprinodontiformes: Poeciliidae). Copeia 1961: 201–212.

Rosengrave, P., Gemmell, N. J., Metcalf, V., McBride, K., and Montgomerie, R. 2008. A mechanism for cryptic female choice in chinook salmon. Behavioral Ecology 19: 1179–1185.

Rosenthal, G. G. 2000. The behavioral ecology of visual signaling in swordtails. PhD thesis, University of Texas.

Rosenthal, G. G. 2007. Spatiotemporal dimensions of visual signals in animal communication. Annual Review of Ecology, Evolution, and Systematics 38: 155–178.

Rosenthal, G. G., and García de León, F. J. 2006. Sexual behavior, genes, and evolution in *Xiphophorus*. Zebrafish 3: 85–90.

Rosenthal, G. G., and Evans, C. S. 1998. Female preference for swords in *Xiphophorus helleri* reflects a bias for large apparent size. Proceedings of the National Academy of Sciences of the United States of America 95: 4431–4436.

Rosenthal, G. G., Evans, C. S., and Miller, W. L. 1996. Female preference for dynamic traits in the green swordtail, *Xiphophorus helleri*. Animal Behaviour 51: 811–820.

Rosenthal, G. G., Martinez, T. Y. F., García de León, F. J., and Ryan, M. J. 2001. Shared preferences by predators and females for male ornaments in swordtails. American Naturalist 158: 146–154.

Rosenthal, G. G., Wagner, W. E., Jr., and Ryan, M. J. 2002. Secondary reduction of preference for the sword ornament in the pygmy swordtail *Xiphophorus nigrensis* (Pisces: Poeciliidae). Animal Behaviour 63: 37–45.

Rosenthal, G. G., De La Rosa Reyna, X. F., Kazianis, S., Stephens, M. J., Morizot, D. C., Ryan, M. J., and García de León, F. J. 2003. Dissolution of sexual signal complexes in a hybrid zone between the swordtails *Xiphophorus birchmanni* and *Xiphophorus malinche* (Poeciliidae). Copeia 2003: 299–307.

Rosenthal, H. L. 1952. Observations on reproduction of the poeciliid *Lebistes reticularis* (Peters). Biological Bulletin 102: 30–38.

Rothschild, L., and Mancinelli, R. 2001. Life in extreme environments. Nature 409: 1092–1101.

Rowe, L., and Day, T. 2006. Detecting sexual conflict and sexually antagonistic coevolution. Philosophical Transactions of the Royal Society of London Series B—Biological Sciences 361: 277–285.

Roy, S. W., and Gilbert, W. 2006. The evolution of spliceosomal introns: patterns, puzzles and progress. Nature Reviews Genetics 7: 211–221.

Rubin, D. A. 1985. Effect of pH on sex ratio in cichlids and a poeciliid (Teleostei). Copeia 1985: 233–235.

Ruehl, C. B., and DeWitt, T. J. 2005. Trophic plasticity and fine-grained resource variation in populations of western mosquitofish, *Gambusia affinis*. Evolutionary Ecology Research 7: 801–819.

Rugani, R., Regolin, L., and Vallortigara, G. 2008. Discrimination of small numerosities in young chicks. Journal of Experimental Psychology: Animal Behavior Processes 34: 388–399.

Rundle, H. D., and Nosil, P. 2005. Ecological speciation. Ecology Letters 8: 336–352.

Russell, S. T. 2003. Evolution of intrinsic post-zygotic reproductive isolation in fish. Annales Zoologici Fennici 40: 321–329.

Russell, S. T., and Magurran, A. E. 2006. Intrinsic reproductive isolation between Trinidadian populations of the guppy, *Poecilia reticulata*. Journal of Evolutionary Biology 19: 1294–1303.

Russell, S. T., Kelley, J. L., Graves, J. A., and Magurran, A. E. 2004. Kin structure and shoal composition dynamics in the guppy, *Poecilia reticulata*. Oikos 106: 520–526.

Russell, S. T., Ramnarine, I. W., Mahabir, R., and Magurran, A. E. 2006. Genetic detection of sperm from forced copulations between sympatric populations of *Poecilia reticulata* and *Poecilia picta*. Biological Journal of the Linnean Society 88: 397–402.

Ryan, M. J. 1985. *The Tungara Frog: A Study in Sexual Selection and Communication*. Chicago: University of Chicago Press.

Ryan, M. J. 1998. Receiver biases, sexual selection and the evolution of sex differences. Science 281: 1999–2003.

Ryan, M. J., and Causey, B. A. 1989. "Alternative" mating behaviour in the swordtails *Xiphophorus nigrensis* and *Xiphophorus pygmaeus* (Pisces: Poeciliidae). Behavioral Ecology and Sociobiology 24: 341–348.

Ryan, M. J., and Keddy-Hector, A. 1992. Directional patterns of female mate choice and the role of sensory biases. American Naturalist 139: S4–S35.

Ryan, M. J., and Rand, A. S. 1993. Species recognition and sexual selection as a unitary problem in animal communication. Evolution 47: 647–657.

Ryan, M. J., and Rand, A. S. 2003. Mate recognition in túngara frogs: a review of some studies of brain, behavior, and evolution. Acta Zoologica Sinica 49: 713–726.

Ryan, M. J., and Rosenthal, G. G. 2001. Variation and selection in swordtails. In L. A. Dugatkin (ed.), *Model Systems in Behavioral Ecology*, 133–148. Princeton, NJ: Princeton University Press..

Ryan, M. J., and Wagner, W. E. J. 1987. Asymmetries in mating preferences between species: female swordtails prefer heterospecific males. Science 236: 595–597.

Ryan, M. J., Hews, D. K., and Wagner, W. E. J. 1990. Sexual selection on alleles that determine body size in the swordtail *Xiphophorus nigrensis* (Pisces: Poeciliidae). Behavioral Ecology and Sociobiology 26: 231–237.

Ryan, M. J., Pease, C. M., and Morris, M. R. 1992. A genetic polymorphism in the swordtail *Xiphophorus nigrensis*: testing the prediction of equal fitnesses. American Naturalist 139: 21–31.

Ryan, M. J., Dries, L. A., Batra, P., and Hillis, D. M. 1996. Male mate preferences in a gynogenetic species complex of Amazon mollies. Animal Behaviour 52: 1225–1236.

Sakai, S., and Harada, Y. 2001. Why do large mothers produce large offspring? Theory and a test. American Naturalist 157: 348–359.

Salas, C., Rodriguez, F., Vargas, J., Duran, E., and Torres, B. 1996. Spatial learning and memory deficits after telencephalic ablation in goldfish trained in place and turn maze procedures. Behavioral Neuroscience 110: 965–980.

Salathe, M., Kouyos, R. D., and Bonhoeffer, S. 2008. The state of affairs in the kingdom of the Red Queen. Trends in Ecology and Evolution 23: 439–445.

Salgado-Maldonado, G. 2006. Checklist of helminth parasites of freshwater fishes from Mexico. Zootaxa 1324: 1–357.

Sambrook, Y. G., Russell, R., Umrania, Y., Edwards, Y. K. J., Campbell, R. D., Elgar, G., and Clark, M. S. 2002. Fugu orthologues of human major histocompatibility complex genes: a genome survey. Immunogenetics 54: 367–380.

Santos, F. C., Pacheco, J. M., and Lenaerts, T. 2006. Cooperation prevails when individuals adjust their social ties. PLoS Computational Biology 2: 1284–1291.

Santucci, F., Ibrahim, K. M., Bruzzone, A., and Hewit, G. M. 2007. Selection on MHC-linked microsatellite loci in sheep populations. Heredity 99: 340–348.

Sato, A., Figueroa, F., O'hUigin, C., Reznick, D. N., and Klein, J. 1995. Identification of major histocompatibility complex genes in the guppy, *Poecilia reticulata*. Immunogenetics 43: 38–49.

Sato, A., Figueroa, F., Murray, B. W., Malaga-Trillo, E., Zaleska-Rutczynska, Z., Sultmann, H., Toyosawa, S., Wedekind, C., Steck, N., and Klein, J. 2000. Nonlinkage of major histocompatibility complex class I and class II loci in bony fishes. Immunogenetics 51: 108–116.

Sato, T., Endo, T., Yamahira, K., Hamaguchi, S., and Sakaizumi, M. 2005. Induction of female-to-male sex reversal by high temperature treatment in medaka, *Oryzias latipes*. Zoological Science 22: 985–988.

Savage, J. M. 2002. *The Amphibians and Reptiles of Costa Rica*. Chicago: University of Chicago Press.

Schaefer, J. F., Heulett, S. T., and Farrell, T. M. 1994. Interactions between two poeciliid fishes (*Gambusia holbrooki* and *Heterandria formosa*) and their prey in a Florida marsh. Copeia 1994: 516–520.

Schaffer, W. M. 1974. Optimal reproductive effort in fluctuating environments. American Naturalist 108: 783–790.

Scharpf, C. 2008. *Checklist of Freshwater Fishes of North America, Including Subspecies and Undescribed Forms*. North American Native Fishes Association. http://www.nanfa.org/checklist.shtml (accessed 2008).

Schartl, A., Malitschek, B., Kazianis, S., Borowsky, R., and Schartl, M. 1995. Spontaneous melanoma formation in nonhybrid *Xiphophorus*. Cancer Research 55: 159–65.

Schartl, A., Hornung, U., Nanda, I., Wacker, R., Müller-Hermelink, H. K., Schlupp, I., Parzefall, J., Schmid, M., and Schartl, M. 1997. Susceptibility to the development of pigment cell tumors in a clone of the Amazon molly, *Poecilia formosa*, introduced through a microchromosome. Cancer Research 57: 2993–3000.

Schartl, M. 1995. Platyfish and swordtails: a genetic system for the analysis of molecular mechanisms in tumor formation. Trends in Genetics 11: 185–189.

Schartl, M. 2008. Evolution of *Xmrk*: an oncogene, but also a speciation gene? Bioessays 30: 822–832.

Schartl, M., and Wellbrock, C. 1998. Polygenic inheritance of melanoma in *Xiphophorus*. In T. A. Dragani (ed.), *Human Polygenic Diseases: Animal Models*, 167–187. Chur, Switzerland: Harwood Academic.

Schartl, M., Schlupp, I., Schartl, A., Meyer, M. K., Nanda, I., Schmid, M., Epplen, J. T., and Parzefall, J. 1991. On the stability of dispensable constituents of the eukaryotic genome: stability of coding sequences versus truly hypervariable sequences in a clonal vertebrate, the Amazon molly, *Poecilia formosa*. Proceedings of the National Academy of Sciences of the United States of America 88: 8759–8763.

Schartl, M., Nanda, I., Schlupp, I., Wilde, B., Epplen, J. T., Schmid, M., and Parzefall, J. 1995a. Incorporation of subgenomic amounts of DNA as compensation for mutational load in a gynogenetic fish. Nature 373: 68–71.

Schartl, M., Wilde, B., Schlupp, I., and Parzefall, J. 1995b. Evolutionary origin of a parthenoform, the Amazon molly *Poecilia formosa*, on the basis of a molecular genealogy. Evolution 49: 827–835

Schartl, M., Wilde, B., and Hornung, U. 1998. Triplet repeat variability in the signal peptide sequence of the *Xmrk* receptor tyrosine kinase gene in *Xiphophorus* fish. Gene 224: 17–21.

Schartl, M., Hornung, U., Gutbrod, H., Volff, J.-N., and Wittbrodt, J. 1999. Melanoma loss-of-function mutants in *Xiphophorus* caused by *Xmrk*-oncogene deletion and gene disruption by a transposable element. Genetics 153: 1385–1394.

Schartl, M., Wilde, B., Laisney, J. A., Takeda, S., Taniguchi, Y., and Meierjohann, S. 2010. A mutated EGFR is sufficient to induce malignant melanoma formation in a new transgenic tumor model. Journal of Investigative Dermatology 130: 249–258.

Schaschl, H., Tobler, M., Plath, M., Penn, D. J., and Schlupp, I. 2008. Polymorphic MHC loci in an asexual fish, the Amazon molly (Poecilia formosa; Poeciliidae). Molecular Ecology 17: 5220–5230.

Scheel, D. 1993. Profitability, encounter rates, and prey choice of African lions. Behavioral Ecology 4: 90–97.

Schelkle, B., Shinn, A. P., Peeler, E., and Cable, J. 2009. Treatment of gyrodactylid infections in fish. Diseases of Aquatic Organisms 86: 65–75.

Schenck, R. A., and Vrijenhoek, R. C. 1989. Coexistence among sexual and asexual Poeciliopsis: foraging behavior and microhabitat selection. In R. M. Dawley and J. P. Bogart (eds.), Evolution and Ecology of Unisexual Vertebrates, 39–48. Albany: New York State Museum.

Schindler, J. F., and Hamlett, W. C. 1993. Maternal-embryonic relations in viviparous teleosts. Journal of Experimental Zoology 266: 378–393.

Schlesinger, M. 1990. Heat shock proteins. Journal of Biological Chemistry 265: 12111–12114.

Schlupp, I. 2005. The evolutionary ecology of gynogenesis. Annual Review of Ecology, Evolution, and Systematics 36: 399–417.

Schlupp, I. 2009. Behaviour of fishes in the sexual/unisexual mating system of the Amazon molly (Poecilia formosa). Advances in the Study of Behavior 39: 153–183.

Schlupp, I., and Plath, M. 2005. Male mate choice and sperm allocation in a sexual/asexual mating complex of Poecilia (Poeciliidae, Teleostei). Biology Letters 1: 169–171.

Schlupp, I., and Ryan, M. J. 1996. Mixed-species shoals and the maintenance of a sexual-asexual mating system in mollies. Animal Behaviour 52: 885–890.

Schlupp, I., and Ryan, M. J. 1997. Male sailfin mollies (Poecilia latipinna) copy the mate choice of other males. Behavioral Ecology 8: 104–107.

Schlupp, I., Parzefall, J., and Schartl, M. 1991. Male mate choice in mixed bisexual/unisexual breeding complexes of Poecilia (Teleostei, Poeciliidae). Ethology 88: 215–222.

Schlupp, I., Parzefall, J., Epplen, J. T., Nanda, I., Schmid, M., and Schartl, M. 1992. Pseudomale behaviour and spontaneous masculinization in the all-female teleost Poecilia formosa (Teleostei: Poeciliidae). Behaviour 122: 88–104.

Schlupp, I., Marler, C., and Ryan, M. J. 1994. Benefit to male sailfin mollies of mating with heterospecific females. Science 263: 373–374.

Schlupp, I., Nanda, I., Döbler, M., Lamatsch, D. K., Epplen, J. T., Parzefall, J., Schmid, M., and Schartl, M. 1998. Dispensable and indispensable genes in an ameiotic fish, the Amazon molly Poecilia formosa. Cytogenetics and Cell Genetics 80: 193–198.

Schlupp, I., Waschulewski, M., and Ryan, M. J. 1999. Female preferences for naturally-occurring novel male traits. Behaviour 136: 519–527.

Schlupp, I., McKnab, R., and Ryan, M. J. 2001. Sexual harassment as a cost for molly females: bigger males cost less. Behaviour 138: 277–286.

Schlupp, I., Parzefall, J., and Schartl, M. 2002. Biogeography of the Amazon molly, Poecilia formosa. Journal of Biogeography 29: 1–6.

Schlupp, I., Riesch, R., and Tobler, M. 2007. Amazon mollies. Current Biology 17: R536–R537.

Schlupp, I., Taebel-Hellwig, A., and Tobler, M. 2010. Equal fecundity in asexual and sexual mollies (Poecilia). Environmental Biology of Fishes 88: 201–206.

Schlüter, A., Parzefall, J., and Schlupp, I. 1998. Female preference for symmetrical vertical bars in male sailfin mollies. Animal Behaviour 56: 147–153.

Schluter, D. 1996. Ecological causes of adaptive radiation. American Naturalist 148: S40–S64.

Schluter, D. 2000. The Ecology of Adaptive Radiation. Oxford: Oxford University Press.

Schluter, D. 2001. Ecology and the origin of species. Trends in Ecology and Evolution 16: 372–380.

Schluter, D. 2009. Evidence for ecological speciation and its alternative. Science 323: 737–741.

Schmidt, J. 1920. Racial investigations, IV: the genetic behavior of a secondary sexual character. Comptes Rendus des Travaux du Laboratorie Carlsberg, Serie Physiologique 14: 1–14.

Schneider, S., Roesslie, D., and Excoffier, L. 2000. Arlequin: a software for population genetics data analysis. Version 2.000. Genetics and Biometry Laboratory, Department of Anthropology, University of Geneva.

Schoenherr, A. A. 1977. Density dependent and density independent regulation of reproduction in the Gila topminnow, Poeciliopsis occidentalis (Baird and Girard). Ecology 58: 438–444.

Schoenherr, A. A. 1981. The role of competition in the replacement of native fishes by introduced species. In R. J. Naiman and D. L. Soltz (eds.), Fishes in North American Deserts, 173–203. New York: John Wiley & Sons.

Scholl, B. J., and Pylyshyn, Z. W. 1999. Tracking multiple items through occlusion: clues to visual objecthood. Cognitive Psychology 38: 259–290.

Schories, S., Lampert, K. P., Lamatsch, D. K., García de León, F. J., and Schartl, M. 2007. Analysis of a possible independent origin of triploid P. formosa outside of the Rio Purificacion river system. Frontiers in Zoology 4: 1–9.

Schrader, M., and Travis, J. 2005. Population differences in pre- and post-fertilization offspring provisioning in the least killifish, Heterandria formosa. Copeia 2005: 649–656.

Schrader, M., and Travis, J. 2009. Do embryos influence maternal investment? Evaluating maternal-fetal coadaptation and the potential for parent-offspring conflict in a placental fish. Evolution 63: 2805–2815.

Schreibman, M. P., Berkowitz, E. J., and van den Hurk, R. 1982. Histology and histochemistry of the testis and ovary of the platyfish, Xiphophorus maculatus, from birth to sexual maturity. Cell Tissue Research 224: 81–87.

Schröder, J. H. 1964. Genetische Untersuchungen an domestizierten Stämmen der Gattung Mollienesia (Poeciliidae). Zoologische Beiträge 10: 369–463.

Schröder, J. H., Bauer, J., and Schartl, M. 1993. Trends in Ichthyology. Oxford: Blackwell.

Schröder, J. H., Rodriguez, C. M., and Siegmund, R. 1996. Male size polymorphism in natural habitats of Poecilia (Limia) perugiae (Pisces: Poeciliidae) endemic to Hispaniola. Biologisches Zentralblatt 115: 315–327.

Schug, M. D., Downhower, J. F., Brown, L. P., Sears, D. B., and Fuerst, P. A. 1998. Isolation and genetic diversity of *Gambusia hubbsi* (mosquitofish) populations in blueholes on Andros Island, Bahamas. Heredity 80: 330–346.

Schultheis, C., Zhou, Q., Froschauer, A., Nanda, I., Selz, Y., Schmidt, C., Matschl, S., Wenning, M., Veith, A.-M., Naciri, M., Hanel, R., Braasch, I., Dettai, A., Böhne, A., Ozouf-Costaz, C., Chilmonczyk, S., Ségurens, B., Couloux, A., Bernard-Samain, S., Schmid, M., Schartl, M., and Volff, J.-N. 2006. Molecular analysis of the sex-determining region of the platyfish *Xiphophorus maculatus*. Zebrafish 3: 295–305.

Schultz, M. E., and Schultz, R. J. 1988. Differences in response to a chemical carcinogen within species and clones of the live-bearing fish *Poeciliopsis*. Carcinogenesis (Eynsham) 9: 1029–1032.

Schultz, R. J. 1967. Gynogenesis and triploidy in the viviparous fish *Poeciliopsis*. Science 157: 1564–1567.

Schultz, R. J. 1969. Hybridization, unisexuality, and polyploidy in the teleost *Poeciliopsis* (Poeciliidae) and other vertebrates. American Naturalist 103: 605–619.

Schultz, R. J. 1973. Unisexual fish: laboratory synthesis of a "species." Science 179: 180–181.

Schultz, R. J. 1977. Evolution and ecology of unisexual fishes. In M. K. Hecht, W. C. Steere, and B. Wallace (eds.), *Evolutionary Biology*, 277–331. New York: Plenum Press.

Schultz, R. J. 1989. Origins and relationships of unisexual poeciliids. In G. K. Meffe and F. F. Snelson Jr. (eds.), *Ecology and Evolution of Livebearing Fishes (Poeciliidae)*, 69–88. Englewood Cliffs, NJ: Prentice Hall.

Schultz, R. J. 1993. Genetic regulation of temperature-mediated sex ratios in the livebearing fish *Poeciliopsis lucida*. Copeia 4: 1148–1151.

Schultz, R. J., and Fielding, E. 1989. Fixed genotypes in variable environments. In R. M. Dawley and J. P. Bogart (eds.), *Evolution and Ecology of Unisexual Vertebrates*, 32–38. Albany: New York State Museum.

Schultz, R. J., and Miller, R. R. 1971. Species of the *Poecilia sphenops* complex (Pisces: Poeciliidae) in México. Copeia 1971: 282–290.

Schwab, M., Kollinger, G., Haas, J., Ahuja, M. R., Abdo, S., Anders, A., and Anders, F. 1979. Genetic basis of susceptibility for neuroblastoma following treatment with N-methyl-N-nitrosourea and X-rays in *Xiphophorus*. Cancer Research 39: 519–526.

Schwartz, A. K., and Hendry, A. P. 2007. A test for the parallel co-evolution of male colour and female preference in Trinidadian guppies (*Poecilia reticulata*). Evolutionary Ecology Research 9: 71–90.

Schwartz, M. L., and Dimock, R. V. J. 2001. Ultrastructural evidence for nutritional exchange between brooding unionid mussels and their glocidia larvae. Invertebrate Biology 120: 227–236.

Scoppettone, G. G. 1993. Interactions between native and non-native fishes of the Upper Muddy River, Nevada. Transactions of the American Fisheries Society 122: 599–608.

Scott, M. E. 1982. Reproductive potential of *Gyrodactylus bullatarudis* (Monogenea) on guppies (*Poecilia reticulata*). Parasitology 85: 217–236.

Scott, M. E. 1985. Experimental epidemiology of *Gyrodactylus bullatarudis* (Monogenea) on guppies (*Poecilia reticulata*): short- and long-term studies. In D. Rollinson and R. M. Anderson (eds.), *Ecology and Genetics of Host-Parasite Interactions*, 21–38. New York: Academic Press.

Scott, M. E., and Anderson, R. M. 1984. The population dynamics of *Gyrodactylus bullatarudis* (Monogenea) within laboratory populations of the fish host *Poecilia reticulata*. Parasitology 89: 159–194.

Scott, M. E., and Nokes, D. J. 1984. Temperature-dependent reproduction and survival of *Gyrodactylus bullatarudis* (Monogenea) on guppies (*Poecilia reticulata*). Parasitology 89: 221–227.

Scott, M. E., and Robinson, M. A. 1984. Challenge infections of *Gyrodactylus bullatarudis* (Monogenea) on guppies, *Poecilia reticulata* (Peters), following treatment. Journal of Fish Biology 24: 581–586.

Scotti, M. L., and Foster, S. A. 2007. Phenotypic plasticity and the ecotypic differentiation of aggressive behavior in threespine stickleback. Ethology 113: 190–198.

Scribner, K. T. 1993. Hybrid zone dynamics are influenced by genotype-specific variation in life-history traits—experimental evidence from hybridizing *Gambusia* species. Evolution 47: 632–646.

Scribner, K. T., and Avise, J. C. 1993. Cytonuclear genetic architecture in mosquitofish populations and the possible roles of introgressive hybridization. Molecular Ecology 2: 139–149.

Scribner, K. T., and Avise, J. C. 1994. Cytonuclear genetics of experimental fish hybrid zones inside Biosphere 2. Proceedings of the National Academy of Sciences of the United States of America 91: 5066–5069.

Scribner, K. T., Wooten, M. C., Smith, M. H., Kennedy, P. K., and Rhodes, O. E. 1992. Variation in life-history and genetic traits of Hawaiian mosquitofish populations. Journal of Evolutionary Biology 5: 267–288.

Scrimshaw, N. S. 1944a. Embryonic growth in the viviparous poeciliid, *Heterandria formosa*. Biological Bulletin 87: 37–51.

Scrimshaw, N. S. 1944b. Superfetation in poeciliid fishes. Copeia 1944: 180–183.

Scrimshaw, N. S. 1945. Embryonic development in poeciliid fishes. Biological Bulletin 88: 233–246.

Scrimshaw, N. S. 1946. Egg size in poeciliid fishes. Copeia 1946: 20–23.

Seal, W. P. 1911. Breeding habits of the viviparous fishes *Gambusia holbrooki* and *Heterandria formosa*. Proceedings of the Biological Society of Washington 24: 91–96.

Seale, A. 1917. The mosquito fish, *Gambusia affinis* (Baird and Girard), in the Philippine Islands. Philippines Journal of Science 12: 177–189.

Seehausen, O., Van Alphen, J. J. M., and Witte, F. 1997. Cichlid fish diversity threatened by eutrophication that curbs sexual selection. Science 277: 1808–1811.

Seehausen, O., Terai, Y., Magalhaes, I. S., Carleton, K. L., Mrosso, H. D. J., Miyagi, R., van der Sluijs, I., Schneider, M. V., Maan, M. E., Tachida, H., Imai, H., and Okada, N. 2008. Speciation through sensory drive in cichlid fish. Nature 455: 620–626.

Seghers, B. H. 1973. An analysis of geographic variation in the antipredator adaptations of the guppy, *Poecilia reticulata*. PhD thesis, University of British Columbia.

Seghers, B. H. 1974. Schooling behavior in the guppy (*Poecilia* reticulata): an evolutionary response to predation. Evolution 28: 486–489.

Sengün, A. 1949. Experimente zur sexuell-mechanischen Isolation, II. Revue de Faculté de Science Istanbul Series B 14: 114–128.

Setlow, R. B., and Woodhead, A. D. 2001. Three unique experimental fish stories: Poecilia (the past), Xiphophorus (the present), and medaka (the future). Marine Biotechnology 3: 17–23.

Setlow, R. B., Woodhead, A. D., and Grist, E. 1989. Animal model for ultraviolet radiation-induced melanoma: platyfish-swordtail hybrid. Proceedings of the National Academy of Sciences of the United States of America 86: 8922–8926.

Setlow, R. B., Grist, E., Thompson, K., and Woodhead, A. D. 1993. Wavelengths effective in induction of malignant melanoma. Proceedings of the National Academy of Sciences of the United States of America 90: 6666–6670.

Shaklee, A. B. 1963. Comparative studies of temperament: fear responses in different species of fish. Journal of Genetic Psychology 102: 295.

Shakunthala, K., and Reddy, S. R. 1977. Environmental restraints on food intake, growth and conversion efficiency of Gambusia affinis. Ceylon Journal of Science 12: 177–184.

Shapiro, A. M., and Porter, A. H. 1989. The lock-and-key hypothesis: evolutionary and biosystematic interpretation of insect genitalia. Annual Review of Entomology 34: 231–245.

Shariff, M., Richards, R. H., and Sommerville, C. 1980. The histopathology of acute and chronic infections of rainbow trout Salmo gairdneri Richardson with eye flukes, Diplostomum spp. Journal of Fish Diseases 3: 455–465.

Shaw, P. W., Carvalho, G. R., Seghers, B. H., and Magurran, A. E. 1992. Genetic consequences of an artificial introduction of guppies (Poecilia reticulata) in N. Trinidad. Proceedings of the Royal Society of London Series B—Biological Sciences 248: 111–116.

Shaw, P. W., Carvalho, G. R., Magurran, A. E., and Seghers, B. H. 1994. Factors affecting the distribution of genetic variability in the guppy, Poecilia reticulata. Journal of Fish Biology 45: 875–888.

Shaw, R. G. 1991. The comparison of quantitative genetic parameters between populations. Evolution 45: 143–151.

Sheffer, R. J., Hedrick, P. W., Minckley, W. L., and Velasco, A. L. 1997. Fitness in the endangered Gila topminnow. Conservation Biology 11: 162–171.

Sheldon, B. C. 2000. Differential allocation: tests, mechanisms and implications. Trends in Ecology and Evolution 15: 397–402.

Sheller, F. J., Fagan, W. F., and Unmack, P. J. 2006. Using survival analysis to study translocation success in the Gila topminnow (Poeciliopsis occidentalis). Ecological Applications 16: 1771–1784.

Sheridan, L., and Pomiankowski, A. 1997a. Female choice for spot asymmetry in the Trinidadian guppy. Animal Behaviour 54: 1523–1529.

Sheridan, L., and Pomiankowski, A. 1997b. Fluctuating asymmetry, spot asymmetry and inbreeding depression in the sexual coloration of male guppy fish. Heredity 79: 515–523.

Shiina, T., Inoko, H., and Kulski, J. K. 2004. An update of the HLA genomic region, locus information and disease associations: 2004. Tissue Antigens 64: 631–649.

Shiina, T., Ota, M., Shimizu, S., Katsuyama, Y., Hashimoto, N., Takasu, M., Anzai, T., Kulski, J. K., Kikkawa, E., Naruse, T., Kimura, N., Yanagiya, K., Watanabe, A., Hosomichi, K., Kohara, S., Iwamoto, C., Umehara, Y., Meyer, A., Wanner, V., Sano, K., Macquin, C., Ikeo, K., Tokunaga, K., Gojobori, T., Inoko, H., and Bahram, S. 2006. Rapid evolution of major histocompatibility complex class I genes in primates generates new disease alleles in humans via hitchhiking diversity. Genetics 173: 1555–1570.

Shine, R. 1980. "Costs" of reproduction in reptiles. Oecologia 46: 92–100.

Shine, R., and Downes, S. J. 1999. Can pregnant lizards adjust their offspring phenotypes to environmental conditions? Oecologia 119: 1–8.

Shine, R., and Olsson, M. 2003. When to be born? Prolonged pregnancy or incubation enhances locomotor performance in neonatal lizards (Scincidae). Journal of Evolutionary Biology 16: 823–832.

Shoemaker, C. M., and Crews, D. 2009. Analyzing the coordinated gene network underlying temperature-dependent sex determination in reptiles. Seminar in Cell and Developmental Biology 20: 293–303.

Shohet, A. J., and Watt, P. J. 2004. Female association preferences based on olfactory cues in the guppy, Poecilia reticulata. Behavioral Ecology and Sociobiology 55: 363–369.

Shoji, A., Yokoyama, J., and Kawata, M. 2007. Molecular phylogeny and genetic divergence of the introduced populations of Japanese guppies, Poecilia reticulata. Conservation Genetics 8: 261–271.

Shuster, S. M., and Wade, M. J. 2003. Mating Systems and Strategies. Princeton, NJ: Princeton University Press.

Sibly, R. M., and Calow, P. 1983. An integrated approach to life-cycle evolution using selective landscapes. Journal of Theoretical Biology 102: 527–547.

Sibly, R. M., and Calow, P. 1989. A life-cycle theory of responses to stress. Biological Journal of the Linnean Society 37: 101–116.

Siciliano, M. J. 1972. Evidence for a spontaneous ovarian cycle in fish of the genus Xiphophorus. Biological Bulletin 142: 480–488.

Sih, A. 1980. Optimal behaviour: can foragers balance two conflicting demands? Science 210: 1041–1043.

Sih, A., and Bell, A. M. 2004. Behavioral syndromes: an integrative overview. Quarterly Review of Biology 79: 241–277.

Sih, A., Englund, G., and Wooster, D. 1998. Emergent impacts of multiple predators on prey. Trends in Ecology and Evolution 13: 350–355.

Sih, A., Ziemba, R., and Harding, K. C. 2000. New insights on how temporal variation in predation risk shapes prey behavior. Trends in Ecology and Evolution 15: 3–4.

Sih, A., Bell, A., and Johnson, J. C. 2004. Behavioral syndromes: an ecological and evolutionary overview. Trends in Ecology and Evolution 19: 372–378.

Simanek, D. E. 1978. Genetic variability and population structure of Poecilia latipinna. Nature 276: 612–614.

Simmons, L. W. 2005. The evolution of polyandry: sperm competition, sperm selection and offspring viability. Annual Review of Ecology, Evolution and Systematics 36: 125–146.

Simmons, L. W., and Moore, A. J. 2009. Evolutionary quantitative genetics of sperm. In T. R. Birkhead, D. J. Hosken, and S. Pitnick (eds.), *Sperm Biology: An Evolutionary Perspective*, 405–434. Amsterdam: Elsevier.

Simmons, L. W., Beveridge, M., and Kennington, W. J. 2007. Polyandry in the wild: temporal changes in female mating frequency and sperm competition intensity in natural populations of the tettigoniid *Requena verticalis*. Molecular Ecology 16: 4613–4623.

Simmons, L. W., Beveridge, M., and Evans, J. P. 2008. Molecular evidence for multiple paternity in a feral population of green swordtails. Journal of Heredity 99: 610–615.

Simons, L. H., Hendrickson, D. A., and Papoulias, D. 1989. Recovery of the Gila topminnow: a success story? Conservation Biology 3: 11–15.

Sirot, L. K. 2003. The evolution of insect mating structures through sexual selection. Florida Entomologist 86: 124–133.

Skarstein, F., Folstad, I., Liljedal, S., and Grahn, M. 2005. MHC and fertilization success in the Arctic charr (*Salvelinus alpinus*). Behavioral Ecology and Sociobiology 57: 374–380.

Skinner, A. M. J., and Watt, P. J. 2007. Phenotypic correlates of spermatozoon quality in the guppy, *Poecilia reticulata*. Behavioral Ecology 18: 47–52.

Slade, R. W., and McCallum, H. I. 1992. Overdominant vs frequency-dependent selection at MHC loci. Genetics 132: 861–862.

Slotow, R., and Paxinos, E. 1997. Intraspecific competition influences food return–predation risk trade-off by white-crowned sparrows. Condor 99: 642–650.

Smith, B. R., and Blumstein, D. T. 2008. Fitness consequences of personality: a meta-analysis. Behavioral Ecology 19: 448–455.

Smith, C. C. 2007. Independent effects of male and female density on sexual harassment, female fitness, and male competition for mates in the western mosquitofish *Gambusia affinis*. Behavioral Ecology and Sociobiology 61: 1349–1358.

Smith, E. J., Partridge, J. C., Parsons, K. N., White, E. M., Cuthill, I. C., Bennett, A. T. D., and Church, S. C. 2002. Ultraviolet vision and mate choice in the guppy (*Poecilia reticulata*). Behavioural Ecology 13: 11–19.

Smith, M. E., and Belk, M. C. 2001. Risk assessment in western mosquitofish (*Gambusia affinis*): do multiple cues have additive effects? Behavioral Ecology and Sociobiology 51: 101–107.

Smith, M. H., Scribner, K. T., Hernandez, J. D., and Wooten, M. C. 1989. Demographic, spatial, and temporal genetic variation in *Gambusia*. In G. K. Meffe and F. F. Snelson Jr. (eds.), *Ecology and Evolution of Livebearing Fishes (Poeciliidae)*, 235–257. Englewood Cliffs, NJ: Prentice Hall.

Smith, R. J. F. 1977. Chemical communication as an adaptation: alarm substance of fish. In D. Müller-Schwarze and M. M. Mozell (eds.), *Chemical Signals in Vertebrates*, 303–320. New York: Plenum Press.

Smith, R. J. F. 1982. The adaptive significance of the alarm substance—the fright reaction system. In T. J. Hara (ed.), *Chemoreception in Fishes*, 327–342. Amsterdam: Elsevier.

Smith, R. J. F. 1992. Alarm signals in fishes. Reviews in Fish Biology and Fisheries 2: 33–63.

Smith, R. J. F. 1999. What good is smelly stuff in the skin? Cross function and cross taxa effects in fish "alarm substances." In R. E. Johnston, D. Müller-Schwarze, and P. W. Sorensen

(eds.), *Advances in Chemical Signals in Vertebrates*, 475–488. New York: Kluwer Academic.

Smith, R. L. 1984. *Sperm Competition and the Evolution of Animal Mating Systems*. Orlando, FL: Academic Press.

Smouse, P. E., Long, J. C., and Sokal, R. R. 1986. Multiple regression and correlation extensions of the Mantel test of matrix correspondence. Systematic Zoology 35: 627–632.

Smuts, B. B., and Smuts, R. W. 1993. Male aggression and sexual coercion of females in nonhuman primates and other mammals: evidence and theoretical implications. Advances in the Study of Behavior 22: 1–63.

Sneddon, L. U. 2003. The bold and the shy: individual differences in rainbow trout. Journal of Fish Biology 62: 971–975.

Sneddon, L. U., Braithwaite, V. A., and Gentle, M. J. 2003a. Do fish have nociceptors? Evidence for the evolution of a vertebrate sensory system. Proceedings of the Royal Society of London Series B—Biological Sciences 270: 1115–1121.

Sneddon, L. U., Braithwaite, V. A., and Gentle, M. J. 2003b. Novel object test: examining pain and fear in the rainbow trout. Journal of Pain 4: 431–440.

Snelson, F. F., Jr. 1989. Social and environmental control of life history traits in poeciliid fishes. In G. K. Meffe and F. F. Snelson Jr. (eds.), *Ecology and Evolution of Livebearing Fishes (Poeciliidae)*, 149–162. Englewood Cliffs, NJ: Prentice Hall.

Snelson, F. F., Jr., and Wetherington, J. D. 1980. Sex ratio in the sailfin molly, *Poecilia latipinna*. Evolution 34: 308–319.

Snelson, F. F., Jr., Wetherington, J. D., and Large, H. L. 1986. The relationship between interbrood interval and yolk loading in a generalized poeciliid fish, *Poecilia latipinna*. Copeia 1986: 295–304.

Snook, R. R. 2005. Sperm in competition: not playing by the numbers. Trends in Ecology and Evolution 20: 46–53.

Soltz, D. L., and Naiman, R. J. 1978. The natural history of native fishes in the Death Valley System. Natural History of Los Angeles County, Science Series 30: 1–76.

Sommer, S. 2005. The importance of immune gene variability (MHC) in evolutionary ecology and conservation. Frontiers in Zoology 2: 16–34.

Sontag, C., Wilson, D. S., and Wilcox, R. S. 2006. Social foraging in *Bufo americanus* tadpoles. Animal Behaviour 72: 1451–1456.

Sørensen, J. G., Kristensen, T. N., and Loeschcke, V. 2003. The evolutionary and ecological role of heat shock proteins. Ecology Letters 6: 1025–1037.

Soucy, S., and Travis, J. 2003. Multiple paternity and population genetic structure in natural populations of the poeciliid fish, *Heterandria formosa*. Journal of Evolutionary Biology 16: 1328–1336.

Sovrano, V. A., Bisazza, A., and Vallortigara, G. 2002. Modularity and spatial reorientation in a simple mind: encoding of geometric and nongeometric properties of a spatial environment by fish. Cognition 85: B51–B59.

Sovrano, V. A., Dadda, M., and Bisazza, A. 2005. Lateralized fish perform better than nonlateralized fish in spatial reorientation tasks. Behavioural Brain Research 163: 122–127.

Spady, T. C., Parry, J. W. L., Robinson, P. R., Hunt, D. M., Bowmaker, J. K., and Carleton, K. L. 2006. Evolution of the cichlid visual palette through ontogenetic subfunctionalization of the opsin gene arrays. Molecular Biology and Evolution 23: 1538–1547.

Spady, T. S., Seehausen, O., Loew, E. R., Jordan, R. C., Kocher, T. D., and Carleton, K. L. 2005. Adaptive molecular evolution in the opsin gene of rapidly speciating cichlid species. Molecular Biology and Evolution 22: 1412–1422.

Spielman, D., Brook, B. W., Briscoe, D. A., and Frankham, R. 2004. Does inbreeding and loss of genetic diversity decrease disease resistance? Conservation Genetics 5: 439–448.

Stacey, N. E., and Sorensen, P. 2002. Hormonal pheromones in fish. In D. W. Pfaff, A. P. Arnold, A. M. Etgen, S. E. Fartback, and F. Rubin (eds.), *Hormonal Pheromones in Fish*, 375–434. San Diego: Academic Press.

Stahl, J. M., Sharma, A., Cheung, M., Zimmerman, M., Cheng, J. Q., Bosenberg, M. W., Kester, M., Sandirasegarane, L., and Robertson, G. P. 2004. Deregulated Akt3 activity promotes development of malignant melanoma. Cancer Research 64: 7002–7010.

Stamps, J. A. 2007. Growth-mortality tradeoffs and "personality traits" in animals. Ecology Letters 10: 355–363.

Stanley, E. L., Kendal, R. L., Kendal, J. R., Grounds, S., and Laland, K. N. 2008. The effects of group size, rate of turnover and disruption to demonstration on the stability of foraging traditions in fish. Animal Behaviour 75: 565–572.

Stearns, S. C. 1978. Interpopulational differences in reproductive traits of *Neoheterandria tridentiger* (Pisces: Poeciliidae) in Panamá. Copeia 1: 188–190.

Stearns, S. C. 1983a. The evolution of life-history traits in mosquitofish since their introduction to Hawaii in 1905: rates of evolution, heritabilities, and developmental plasticity. American Zoologist 23: 65–75.

Stearns, S. C. 1983b. The genetic basis of differences in life-history traits among six populations of mosquitofish (*Gambusia affinis*) that shared ancestors in 1905. Evolution 37: 618–627.

Stearns, S. C. 1983c. A natural experiment in life-history evolution: field data on the introduction of mosquitofish (*Gambusia affinis*) to Hawaii. Evolution 37: 601–617.

Stearns, S. C. 1989. Trade-offs in life-history evolution. Functional Ecology 3: 259–268.

Stearns, S. C. 1992. *The Evolution of Life Histories*. Oxford: Oxford University Press.

Stearns, S. C. 2000. Life history evolution: successes, limitations, and prospects. Naturwissenschaften 87: 476–486.

Stenseth, N. C., Kirkendall, L. R., and Moran, N. 1985. On the evolution of pseudogamy. Evolution 39: 294–307.

Stenzel, A., Lu, T., Koch, W. A., Hampe, J., Guenther, S. M., De La Vega, F. M., Krawczak, M., and Schreiber, S. 2004. Patterns of linkage disequilibrium in the MHC region on human chromosome 6p. Human Genetics 114: 377–385.

Stephan, W., Song, Y. S., and Langley, C. H. 2006. The hitchhiking effect on linkage disequilibrium between linked neutral loci. Genetics 172: 2647–2663.

Stephens, D. W., and Krebs, J. R. 1986. *Foraging Theory*. Princeton, NJ: Princeton University Press.

Stern, L. J., Brown, J. H., Jardetzky, T. S., Gorga, J. C., Urban, R. G., Strominger, J. L., and Wiley, D. C. 1994. Crystal-structure of the human class II MHC protein HLA-DR1 complexed with an influenza virus peptide. Nature 368: 215–221.

Sternlicht, M. D., and Werb, Z. 2001. How matrix metalloproteinases regulate cell behavior. Annual Review of Cell and Developmental Biology 17: 463–516.

Stet, R. J. M., de Vries, B., Mudde, K., Hermsen, T., van Heerwaarden, J., Shum, B. P., and Grimholt, U. 2002. Unique haplotypes of co-segregating major histocompatibility class II A and class II B alleles in Atlantic salmon (*Salmo salar*) give rise to diverse class II genotypes. Immunogenetics 54: 320–331.

Stet, R. J. M., Kruiswijk, C. P., and Dixon, B. 2003. Major histocompatibility lineages and immune gene function in teleost fishes: the road not taken. Critical Reviews in Immunology 23: 441–471.

Stet, R. J. M., Mudde, K., Wynne, J. W., Nooijen, A., Dahlgren, T. G., Ruzzante, D. E., and Andre, C. 2008. Characterization of a major histocompatibility class II A gene (*Clha-DAA*) with an embedded microsatellite marker in Atlantic herring (*Clupea harengus* L.). Journal of Fish Biology 73: 367–381.

Stetefeld, J., and Ruegg, M. A. 2005. Structural and functional diversity generated by alternative mRNA splicing. Trends in Biochemical Sciences 30: 515–521.

Stewart, J. R., and Thompson, M. B. 2003. Evolutionary transformations of the fetal membranes of viviparous reptiles: a case study of two lineages. Journal of Experimental Zoology 299: 13–32.

Stöck, M., Lampert, K. P., Möller, D., Schlupp, I., and Schartl, M. 2010. Monophyletic origin of clonal lineages in an asexual fish (*Poecilia formosa*). Molecular Ecology 19: 5204–5215.

Stockley, P. 1997. No evidence of sperm selection by female common shrews. Proceedings of the Royal Society of London Series B—Biological Sciences 264: 1497–1500.

Stockley, P. 1999. Sperm selection and genetic incompatibility: does relatedness of mates affect male success in sperm competition? Proceedings of the Royal Society of London Series B—Biological Sciences 266: 1663–1669.

Stockwell, C. A., and Leberg, P. L. 2002. Ecological genetics and the translocation of native fishes: emerging experimental approaches. Western North American Naturalist 62: 32–38.

Stockwell, C. A., and Vinyard, G. L. 2000. Life history variation in recently established populations of western mosquitofish (*Gambusia affinis*). Western North American Naturalist 60: 273–280.

Stockwell, C. A., and Weeks, S. C. 1999. Translocations and rapid evolutionary responses in recently established populations of western mosquitofish (*Gambusia affinis*). Animal Conservation 2: 103–110.

Stockwell, C. A., Mulvey, M., and Vinyard, G. L. 1996. Translocations and the preservation of allelic diversity. Conservation Biology 10: 1133–1141.

Stockwell, C. A., Hendry, A. P., and Kinnison, M. T. 2003. Contemporary evolution meets conservation biology. Trends in Ecology and Evolution 18: 94–101.

Stoner, G., and Breden, F. 1988. Phenotypic differentiation in female preference related to geographic variation in male predation risk in the Trinidad guppy (*Poecilia reticulata*). Behavioral Ecology and Sociobiology 22: 285–291.

Strauss, R. E. 1990. Predation and life-history variation in *Poecilia reticulata* (Cyprinodontiformes, Poeciliidae). Environmental Biology of Fishes 27: 121–130.

Streelman, J. T., and Danley, P. D. 2003. The stages of vertebrate evolutionary radiation. Trends in Ecology and Evolution 18: 126–131.

Streelman, J. T., Peichel, C. L., and Parichy, D. M. 2007. Developmental genetics of adaptation in fishes: the case of novelty. Annual Review of Ecology Evolution and Systematics 38: 655–681.

Stumpp, M. 1975. Untersuchungen zur Morphologie und Biologie yon *Camallanus cotti* (Fujita, 1927). Parasitenkunde 46: 277–290.

Stupack, D. G., Puente, X. S., Boutsaboualoy, S., Storgard, C. M., and Cheresh, D. A. 2001. Apoptosis of adherent cells by recruitment of caspase-8 to unligated integrins. Journal of Cell Biology 155: 459–470.

Sugawara, T., Terai, Y., and Okada, N. 2000. Natural selection of the rhodopsin gene during the adaptive radiation of East African Great Lakes cichlid fishes. Molecular Biology and Evolution 19: 1807–1811.

Sugita, Y. 1980. Imitative choice behaviour in guppies. Japanese Psychological Research 22: 7–12.

Suk, H. Y., and Neff, B. D. 2009. Microsatellite genetic differentiation among populations of the Trinidadian guppy. Heredity 102: 425–434.

Sullivan, J. A., and Schultz, R. J. 1986. Genetic and environmental basis of variable sex ratios in laboratory strains of *Poeciliopsis lucida*. Evolution 40: 152–158.

Summers, K., and Crespi, B. 2005. Cadherins in maternal-foetal interactions: red queen with a green beard? Proceedings of the Royal Society of London Series B—Biological Sciences 272: 643–649.

Summers, K., Roney, K. E., da Silva, J., Capraro, G., Cuthbertson, B. J., Kazianis, S., Rosenthal, G. G., Ryan, M. J., and McConnell, T. J. 2009. Divergent patterns of selection on the *DAB* and *DXB* MHC class II loci in *Xiphophorus* fishes. Genetica 135: 379–390.

Sumner, I. T., Travis, J., and Johnson, C. D. 1994. Methods of female fertility advertisement and variation among males in responsiveness in the sailfin molly (*Poecilia latipinna*). Copeia 1994: 27–34.

Suomalainen, E., Saura, A., and Lokki, J. 1987. *Cytology and Evolution in Parthenogenesis*. Boca Raton: CRC Press.

Svendsen, J. C., Skov, J., Bildsøe, M., and Steffensen, J. F. 2003. Intra-school positional preference and reduced tail beat frequency in trailing positions in schooling roach under experimental conditions. Journal of Fish Biology 62: 834–846.

Swain, R., and Jones, S. M. 1997. Maternal-fetal transfer of ^{3}H-labelled leucine in the viviparous lizard *Nivescincus metallicus* (Scindidae: Lygosominae). Journal of Experimental Zoology 277: 139–145.

Swaney, W., Kendal, J., Capon, H., Brown, C., and Laland, K. N. 2001. Familiarity facilitates social learning of foraging behaviour in the guppy. Animal Behaviour 62: 591–598.

Szybalski, W. 2001. My road to Ojvind Winge, the father of yeast genetics. Genetics 158: 1–6.

Taborsky, M. 1994. Sneakers, satellites, and helpers: parasitic and cooperative behavior in fish reproduction. Advances in the Study of Behavior 23: 1–100.

Taborsky, M. 1997. Bourgeois and parasitic tactics: do we need collective, functional terms for alternative reproductive behaviours? Behavioral Ecology and Sociobiology 41: 361–362.

Taborsky, M. 2008. Alternative reproductive tactics in fish. In R. F. Oliveira, M. Taborsky, and H. J. Brockmann (eds.), *Alternative Reproductive Tactics: An Integrative Approach*, 251–299. New York: Cambridge University Press.

Takahashi, H. 1975. Functional masculinization of female guppies, *Poecilia reticulata*, influenced by methyltestosterone before birth. Bulletin of the Japanese Society of Scientific Fisheries 41: 499–506.

Takahashi, H. 1977. Juvenile hermaphroditism in the zebrafish, *Brachydanio rerio*. Bulletin of the Faculty of Fisheries, Hokkaido University 28: 57–65.

Takahashi, Y., and Ebrey, T. G. 2003. Molecular basis of spectral tuning in the newt short wavelength sensitive visual pigment. Biochemistry and Cell Biology—Biochimie et Biologie Cellulaire 42: 6025–6034.

Takeuchi, I. K. 1975. Electron microscopic study on erythrophores of the guppy, *Lebistes reticulatus* Peters. Annotationes Zoologicae Japonenses 48: 242–251.

Tatarenkov, A., Healey, C. I. M., Grether, G. F., and Avise, J. C. 2008. Pronounced reproductive skew in a natural population of green swordtails, *Xiphophorus helleri*. Molecular Ecology 17: 4522–4534.

Tavolga, W. N. 1949. Embryonic development of the platyfish (*Platypoecilus*), the swordtail (*Xiphophorus*), and their hybrids. American Museum of Natural History 94: 165–229.

Tavolga, W. N., and Rugh, R. 1947. Development of the platyfish, *Platypoecilus maculatus*. Zoologica Scripta 32: 1–15.

Taylor, P. D., Day, T., and Wild, G. 2007. Evolution of cooperation in a finite homogeneous graph. Nature 447: 469–472.

Temperton, V. M., and Hobbs, R. J. 2004. The search for ecological assembly rules and its relevance to restoration ecology. In V. M. Temperton, R. J. Hobbs, T. Nuttle, and S. Halle (eds.), *Assembly Rules and Restoration Ecology*, 34–54. Washington, DC: Island Press.

Templeton, C. N., and Shriner, W. M. 2004. Multiple selection pressures influence Trinidadian guppy (*Poecilia reticulata*) antipredator behaviour. Behavioral Ecology 15: 673–678.

Terai, Y., Seehausen, O., Sasaki, T., Takahashi, K., Mizoiri, S., Sugawara, T., Sato, T., Watanabe, M., Konijnendijk, N., Mrosso, H. D. J., Tachida, H., Imai, H., Shichida, Y., and Okada, N. 2006. Divergent selection on opsins drives incipient speciation in Lake Victoria cichlids. PLoS Biology 4: 2244–2251.

Teutschbein, J., Schartl, M., and Meierjohann, S. 2009. Interaction of *Xiphophorus* and murine Fyn with focal adhesion kinase. Comparative Biochemistry and Physiology C—Toxicology and Pharmacology 149: 168–174.

Theodorakis, C. W. 1989. Size segregation and the effects of oddity on predation risk in minnow shoals. Animal Behaviour 38: 496–502.

Thibault, R. E. 1974. Genetics of cannibalism in a viviparous fish and its relationship to population density. Nature 251: 138–140.

Thibault, R. E., and Schultz, R. J. 1978. Reproductive adaptations among viviparous fishes (Cyprinodontiformes: Poeciliidae). Evolution 32: 320–333.

Thiessen, D. D., and Sturdivant, S. K. 1977. Female pheromone in the black molly fish (*Mollienesia latipinna*): a possible metabolic correlate. Journal of Chemical Ecology 3: 207–217.

Thilakaratne, I. D. S. I. P., Rajapaksha, G., Hewakopara, A., Rajapakse, R. P. V. J., and Faizal, A. C. M. 2003. Parasitic infections in freshwater ornamental fish in Sri Lanka. Diseases of Aquatic Organisms 54: 157–162.

Thompson, M. B., and Speake, B. K. 2006. A review of the evolution of viviparity in lizards: structure, function, and physiology of the placenta. Journal of Comparative Physiology B—Biochemical, Systemic, and Environmental Physiology 176: 179–189.

Thompson, M. B., Stewart, J. R., and Speake, B. K. 2000. Comparison of nutrient transport across the placenta of lizards differing in placental complexity. Comparative Biochemistry and Physiology A—Molecular and Integrative Physiology 127: 469–479.

Thomson, J. D., Weiblen, G., Thomson, B. A., Alfaro, S., and Legendre, P. 1996. Untangling multiple factors in spatial distributions: lilies, gophers and rocks. Ecology 77: 1698–1715.

Thornhill, R. 1983. Cryptic female choice and its implications in the scorpionfly *Harpobittacus nigricepts*. American Naturalist 122: 765–788

Thorpe, R. S., Surget-Groba, Y., and Johansson, H. 2008. The relative importance of ecology and geographic isolation for speciation in anoles. Philosophical Transactions of the Royal Society of London Series B—Biological Sciences 363: 3071–3081.

Thuman, K. A., and Griffith, S. C. 2005. Genetic similarity and the nonrandom distribution of paternity in a genetically highly polyandrous shorebird. Animal Behaviour 69: 765–770.

Tiedemann, R., Moll, K., Paulus, K. B., and Schlupp, I. 2005. New microsatellite loci confirm hybrid origin, parthenogenetic inheritance, and mitotic gene conversion in the gynogenetic Amazon molly (*Poecilia formosa*). Molecular Ecology Notes 5: 586–589.

Timmerman, C. M., and Chapman, L. J. 2003. The effect of gestational state on oxygen consumption and response to hypoxia in the sailfin molly, *Poecilia latipinna*. Environmental Biology of Fishes 68: 293–299.

Timmerman, C. M., and Chapman, L. J. 2004. Hypoxia and interdemic variation in *Poecilia latipinna*. Journal of Fish Biology 65: 635–650.

Ting, C. T., Tsaur, S. C., and Wu, C. I. 2000. The phylogeny of closely related species as revealed by the genealogy of a speciation gene, *Odysseus*. Proceedings of the National Academy of Sciences of the United States of America 97: 5313–5316.

Ting, C. T., Takahashi, A., and Wu, C. I. 2001. Incipient speciation by sexual isolation in *Drosophila*: concurrent evolution at multiple loci. Proceedings of the National Academy of Sciences of the United States of America 98: 6709–6713.

Ting, J. P. Y., and Trowsdale, J. 2002. Genetic control of MHC class II expression. Cell 109: S21–S33.

Tobler, M. 2008. Divergence in trophic ecology characterizes colonization of extreme habitats. Biological Journal of the Linnean Society 95: 517–528.

Tobler, M. 2009. Does a predatory insect contribute to the divergence between cave- and surface-adapted fish populations? Biology Letters 5: 506–509.

Tobler, M., and Schlupp, I. 2005. Parasites in sexual and asexual mollies (*Poecilia*, Poeciliidae, Teleostei): a case for the Red Queen? Biology Letters 1: 166–168.

Tobler, M., and Schlupp, I. 2008a. Expanding the horizon: the Red Queen and potential alternatives. Canadian Journal of Zoology 86: 765–773.

Tobler, M., and Schlupp, I. 2008b. Influence of black spot disease on shoaling behaviour in female western mosquitofish, *Gambusia affinis* (Poeciliidae, Teleostei). Environmental Biology of Fishes 81: 29–34.

Tobler, M., and Schlupp, I. 2009. Threatened fishes of the world: *Poecilia latipunctata* Meek, 1904 (Poeciliidae). Environmental Biology of Fishes 85: 31–32.

Tobler, M., and Schlupp, I. 2010. Differential susceptibility to food stress in neonates of sexual and asexual mollies (*Poecilia, Poeciliidae*). Evolutionary Ecology 24: 39–47.

Tobler, M., Wahli, T., and Schlupp, I. 2005. Comparison of parasite communities in native and introduced populations of sexual and asexual mollies of the genus *Poecilia*. Journal of Fish Biology 67: 1072–1082.

Tobler, M., Plath, M., Burmeister, H., and Schlupp, I. 2006a. Black spots and female association preferences in a sexual/asexual mating complex (*Poecilia*, Poecilildae, Teleostei). Behavioral Ecology and Sociobiology 60: 159–165.

Tobler, M., Schlupp, I., Heubel, K. U., Riesch, R., García de León, F. J., Giere, O., and Plath, M. 2006b. Life on the edge: hydrogen sulfide and the fish communities of a Mexican cave and surrounding waters. Extremophiles 10: 577–585.

Tobler, M., Schlupp, I., García de León, F. J., Glaubrecht, M., and Plath, M. 2007a. Extreme habitats as refuge from parasite infections? Evidence from an extremophile fish. Acta Oecologica—International Journal of Ecology 31: 270–275.

Tobler, M., Schlupp, I., and Plath, M. 2007b. Predation of a cave fish (*Poecilia mexicana*, Poeciliidae) by a giant waterbug (*Belostoma*, Belostomatidae) in a Mexican sulphur cave. Ecological Entomology 32: 492–495.

Tobler, M., DeWitt, T. J., Schlupp, I., García de León, F. J., Herrmann, R., Feulner, P. G. D., Tiedemann, R., and Plath, M. 2008a. Toxic hydrogen sulfide and dark caves: phenotypic and genetic divergence across two abiotic environmental gradients in *Poecilia mexicana*. Evolution 62: 2643–2659.

Tobler, M., Franssen, C. M., and Plath, M. 2008b. Male-biased predation of a cave fish by a giant water bug. Naturwissenschaften 95: 775–779.

Tobler, M., Riesch, R., García de León, F. J., Schlupp, I., and Plath, M. 2008c. A new and morphologically distinct population of cavernicolous *Poecilia mexicana* (Poeciliidae: Teleostei). Environmental Biology of Fishes 82: 101–108.

Tobler, M., Riesch, R., García de León, F. J., Schlupp, I., and Plath, M. 2008d. Two endemic and endangered fishes, *Poecilia sulphuraria* (Alvarez, 1948) and *Gambusia eurystoma* Miller, 1975 (Poeciliidae, Teleostei) as only survivors in a small sulphidic habitat. Journal of Fish Biology 72: 523–533.

Tobler, M., Schlupp, I., and Plath, M. 2008e. Does divergence in female mate choice affect male size distributions in two cave fish populations? Biology Letters 4: 452–454.

Tobler, M., Riesch, R., Tobler, C. M., and Plath, M. 2009a. Compensatory behavior in response to sulfide-induced hypoxia affects time budgets, feeding efficiency, and predation risk. Evolutionary Ecology Research 11: 935–948.

Tobler, M., Riesch, R., Tobler, C. M., Schulz-Mirbach, T., and Plath, M. 2009b. Natural and sexual selection against immigrants maintains differentiation among micro-allopatric populations. Journal of Evolutionary Biology 22: 2298–2304.

Tobler, M., Coleman, S. W., Perkins, B. D., and Rosenthal, G. G.

2010. Reduced opsin gene expression in a cave-dwelling fish. Biology Letters 6: 98–101.

Toft, G., and Baatrup, E. 2001. Sexual characteristics are altered by 4-tert-octylphenol and 17 β-estradiol in the adult male guppy (*Poecilia reticulata*). Ecotoxicology and Environmental Safety 48: 76–84.

Toft, G., and Baatrup, E. 2003. Altered sexual characteristics in guppies (*Poecilia reticulata*) exposed to 17 beta-estradiol and 4-tert-octylphenol during sexual development. Ecotoxicology and Environmental Safety 56: 228–237.

Toft, G., Baatrup, E., and Guillette, L. J. 2004. Altered social behavior and sexual characteristics in mosquitofish (*Gambusia holbrooki*) living downstream of a paper mill. Aquatic Toxicology 70: 213–222.

Toolson, E. C. 1985. Uptake of leucine and water by *Centruroides sculpturatus* (Ewing) embryos (Scorpiones, Buthidae). Journal of Arachnology 13: 303–310.

Torchin, M. E., Lafferty, K. D., and Kuris, A. M. 2002. Parasites and marine invasions. Parasitology 124: 137–151.

Tosh, C. R., and Ruxton, G. D. 2006. Artificial neural network properties associated with wiring patterns in the visual projections of vertebrates and arthropods. American Naturalist 168: E38–E52.

Tosh, C. R., Jackson, A. L., and Ruxton, G. D. 2006. The confusion effect in predatory neural networks. American Naturalist 167: E52–E65.

Townsend, C. R., Begon, M. E., and Harper, J. L. 2003. *Essentials of Ecology*. 2d ed. Oxford: Blackwell.

Trachtenberg, E., Korber, B., Sollars, C., Kepler, T. B., Hraber, P. T., Hayes, E., Funkhouser, R., Fugate, M., Theiler, J., Hsu, Y. S., Kunstman, K., Wu, S., Phair, J., Erlich, H., and Wolinsky, S. 2003. Advantage of rare HLA supertype in HIV disease progression. Nature Medicine 9: 928–935.

Traut, W., and Winking, H. 2001. Meiotic chromosomes and stages of sex chromosome evolution in fish: zebrafish, platyfish and guppy. Chromosome Research 9: 659–672.

Travis, J. 1994. Size-dependent behavioral variation and its genetic control within and among populations. In C. R. B. Boake (ed.), *Quantitative Genetic Approaches to Animal Behavior*, 165–187. Chicago: University of Chicago Press.

Travis, J., and Woodward, B. D. 1989. Social context and courtship flexibility in male sailfin mollies, *Poecilia latipinna* (Pisces: Poeciliidae). Animal Behaviour 38: 1001–1011.

Travis, J., Farr, J. A., Henrich, S., and Cheong, R. T. 1987. Testing theories of clutch overlap with the reproductive ecology of *Heterandria formosa*. Ecology 68: 611–623.

Travis, J. T., Trexler, J. C., and Mulvey, M. 1990. Multiple paternity and its correlates in female *Poecilia latipinna* (Poeciliidae). Copeia 1990: 722–729.

Traxler, G. S., Richards, J., and McDonald, T. E. 1998. *Ichthyophthirius multifiliis* (Ich) epizootics in spawning sockeye salmon in British Columbia, Canada. Journal of Aquatic Animal Health 10: 143–151.

Tregenza, T., and Wedell, N. 2000. Genetic compatibility, mate choice and patterns of parentage: invited review. Molecular Ecology 9: 1013–1027.

Trendall, J. T. 1982. Covariation of life history traits in the mosquitofish, *Gambusia affinis*. American Naturalist 119: 774–783.

Trexler, J. C. 1985. Variation in the degree of viviparity in the sailfin molly, *Poecilia latipinna*. Copeia 1985: 999–1004.

Trexler, J. C. 1988. Hierarchical organization of genetic variation in the sailfin molly, *Poecilia latipinna* (Pisces: Poeciliidae). Evolution 42: 1006–1017.

Trexler, J. C. 1989. Phenotypic plasticity in poeciliid life histories. In G. K. Meffe and F. F. Snelson Jr. (eds.), *Ecology and Evolution of Livebearing Fishes (Poeciliidae)*, 201–213. Englewood Cliffs, NJ: Prentice Hall.

Trexler, J. C. 1997. Resource availability and plasticity in offspring provisioning: embryo nourishment in sailfin mollies. Ecology 78: 1370–1381.

Trexler, J. C., and DeAngelis, D. L. 2003. Resource allocation in offspring provisioning: an evaluation of the conditions favoring the evolution of matrotrophy. American Naturalist 162: 574–585.

Trexler, J. C., Tempe, R. C., and Travis, J. 1994. Size-selective predation of sailfin mollies by 2 species of heron. Oikos 69: 250–258.

Trexler, J. C., Travis, J., and Dinep, A. 1997. Variation among populations of the sailfin molly in the rate of concurrent multiple paternity and its implications for mating-system evolution. Behavioral Ecology and Sociobiology 40: 297–305.

Trexler, J. C., Loftus, W. F., Jordan, C. F., Chick, J. H., Kandl, K. L., McElroy, T. C., and Bass, O. L. 2001. Ecological scale and its implications for freshwater fishes in the Florida Everglades. In J. W. Porter and K. G. Porter (eds.), *The Everglades, Florida Bay, and Coral Reefs of the Florida Keys: An Ecosystem Sourcebook*, 153–181. Boca Raton: CRC.

Trezise, A. E. O., and Collin, S. P. 2005. Opsins: evolution in waiting. Current Biology 15: R794–R796.

Trick, L. M., and Pylyshyn, Z. W. 1994. Why are small and large numbers enumerated differently? A limited-capacity preattentive stage in vision. Psychological Review 101: 80–102.

Tripathi, N., Hoffmann, M., and Dreyer, C. 2008. Natural variation of male ornamental traits of the guppy, *Poecilia reticulata*. Zebrafish 5: 265–278.

Tripathi, N., Hoffmann, M., Weigel, D., and Dreyer, C. 2009a. Linkage analysis reveals independent origin of poeciliid sex chromosomes and a case of atypical sex inheritance in the guppy (*Poecilia reticulata*). Genetics 182: 365–374.

Tripathi, N., Hoffmann, M., Willing, E. M., Lanz, C., Weigel, D., and Dreyer, C. 2009b. Genetic linkage map of the guppy, *Poecilia reticulata*, and quantitative trait loci analysis of male size and colour variation. Proceedings of the Royal Society of London Series B—Biological Sciences 276: 2195–2208.

Trivers, R. L. 1974. Parent-offspring conflict. American Zoologist 14: 249–264.

Turcotte, M. M., Pires, M. N., Vrijenhoek, R. C., and Reznick, D. N. 2008. Pre- and post-fertilization maternal provisioning in livebearing fish species and their hybrids (Poeciliidae: *Poeciliopsis*). Functional Ecology 22: 1118–1124.

Turner, B. J., and Steeves, H. R. 1989. Induction of spermatogenesis in an all-female fish species by treatment with an exogenous androgen. In R. M. Dawley and J. P. Bogart (eds.), *Evolution and Ecology of Unisexual Vertebrates*, 113–122. Albany: New York State Museum.

Turner, C. L. 1937. Reproductive cycles and superfetation in poeciliid fishes. Biological Bulletin 72: 145–164.

Turner, C. L. 1938. The reproductive cycle of *Brachyrhaphis episcopi*, an oviparous poeciliid fish, in the natural tropical habitat. Biological Bulletin 75: 56–65.

Turner, C. L. 1940a. Adaptations for viviparity in jenynsiid fishes. Journal of Morphology 67: 291–297.

Turner, C. L. 1940b. Pseudoamnion, pseudochorion, and follicular pseudoplacenta in poeciliid fishes. Journal of Morphology 67: 59–89.

Turner, C. L. 1940c. Superfetation in viviparous cyprinodont fishes. Copeia 1940: 88–91.

Turner, C. L. 1941a. Morphogenesis of the gonopodium in *Gambusia affinis*. Journal of Morphology 69: 161–185.

Turner, C. L. 1941b. Regeneration of the gonopodium of *Gambusia* during morphogenesis. Journal of Experimental Zoology 87: 181–209.

Turner, C. L. 1942. Morphogenesis of the gonopodial suspensorium in *Gambusia affinis* and the induction of male suspensorial characters in the female by androgenic hormones. Journal of Experimental Zoology 91: 167–191.

Turner, G. F., and Pitcher, T. J. 1986. Attack abatement: a model for group protection by combined avoidance and dilution. American Naturalist 128: 228–240.

Turner, J. S., and Snelson, F. F., Jr. 1984. Population structure, reproduction and laboratory behavior of the introduced *Belonesox belizanus* (Poeciliidae) in Florida. Environmental Biology of Fishes 10: 89–100.

Uller, C., Jaeger, R., Guidry, G., and Martin, C. 2003. Salamanders (*Plethodon cinereus*) go for more: rudiments of number in an amphibian. Animal Cognition 6: 105–112.

Utne-Palm, A. C., and Hart, P. J. B. 2000. The effects of familiarity on competitive interactions between threespined sticklebacks. Oikos 91: 225–232.

Uyenoyama, M. K. 2003. Genealogy-dependent variation in viability among self-incompatibility genotypes. Theoretical Population Biology 63: 281–293.

Valenzuela, N., and Lance, V. 2004. *Temperature Dependent Sex Determination in Vertebrates*. Washington, DC: Smithsonian Books.

Valenzuela, N., Adams, D. C., and Janzen, F. J. 2003. Pattern does not equal process: exactly when is sex environmentally determined? American Naturalist 161: 676–683.

Valero, A., Garcia, C. M., and Magurran, A. E. 2008. Heterospecific harassment of native endangered fishes by invasive guppies in Mexico. Biology Letters 4: 149–152.

Valero, A., Magurran, A. E., and Garcia, C. M. 2009. Guppy males distinguish between familiar and unfamiliar females of a distantly related species. Animal Behaviour 78: 441–445.

Vallortigara, G. 2004. Visual cognition and representation in birds and primates. In L. J. Rogers and G. Kaplan (eds.), *Vertebrate Comparative Cognition: Are Primates Superior to Non-primates?* 57–94. Dordrecht: Kluwer Academic Publishers.

Vallortigara, G., and Bisazza, A. 2002. How ancient is brain lateralization? In R. J. Andrew and L. J. Rogers (eds.), *Comparative Vertebrate Lateralization*, 9–69. Cambridge: Cambridge University Press.

Vallortigara, G., Rogers, L. J., Bisazza, A., Lippolis, G., and Robins, A. 1998. Complementary right and left hemifield use for predatory and agonistic behaviour in toads. Neuroreport 9: 3341–3344.

Vallowe, H. H. 1953. Some physiological aspects of reproduction in *Xiphophorus maculatus*. Biological Bulletin 104: 240–249.

van Bergen, Y., Coolen, I., and Laland, K. N. 2004. Nine-spined sticklebacks exploit the most reliable source when public and private information conflict. Proceedings of the Royal Society of London Series B—Biological Sciences 271: 957–962.

van Damme, R., Bauwens, D., and Verheyen, R. F. 1989. Effect of relative clutch mass on sprint speed in the lizard *Lacerta vivipara*. Journal of Herpetology 23: 459–461.

Van Dine, D. L. 1907. The introduction of top-minnows (natural enemies of mosquitoes) into the Hawaiian Islands. Hawaiian Agricultural Experiment Station Press Bulletin 20: 1–10.

Van Dine, D. L. 1908. Mosquitoes. Annual Report of the Hawaii Agricultural Experiment Station for 1907, 38–39.

van Hazel, I., Santini, F., Müller, J., and Chang, B. S. W. 2006. Short-wavelength sensitive opsin (SWS1) as a new marker for vertebrate phylogenetics. BMC Evolutionary Biology 6: 1–15.

Van Homrigh, A., Higgie, M., McGuigan, K., and Blows, M. W. 2007. The depletion of genetic variance by sexual selection. Current Biology 17: 528–532.

van Oosterhout, C. 2008. The guppy as a conservative model: implications of parasitism and inbreeding for reintroduction success ([*errata corrige*] vol. 21, pg. 1579, 2007). Conservation Biology 22: 228.

van Oosterhout, C. 2009. A new theory of MHC evolution: beyond selection on the immune genes. Proceedings of the Royal Society of London Series B—Biological Sciences 276: 657–665.

van Oosterhout, C., Harris, P. D., and Cable, J. 2003a. Marked variation in parasite resistance between two wild populations of the Trinidadian guppy, *Poecilia reticulata* (Pisces: Poeciliidae). Biological Journal of the Linnean Society 79: 645–651.

van Oosterhout, C., Trigg, R. E., Carvalho, G. R., Magurran, A. E., Hauser, L., and Shaw, P. W. 2003b. Inbreeding depression and genetic load of sexually selected traits: how the guppy lost its spots. Journal of Evolutionary Biology 16: 273–281.

van Oosterhout, C., Joyce, D. A., and Cummings, S. M. 2006a. Evolution of MHC class IIB in the genome of wild and ornamental guppies, *Poecilia reticulata*. Heredity 97: 111–118.

van Oosterhout, C., Joyce, D. A., Cummings, S. M., Blais, J., Barson, N. J., Ramnarine, I. W., Mohammed, R. S., Persad, N., and Cable, J. 2006b. Balancing selection, random genetic drift, and genetic variation at the major histocompatibility complex in two wild populations of guppies (*Poecilia reticulata*). Evolution 60: 2562–2574.

van Oosterhout, C., Mohammed, R. S., Hansen, H., Archard, G. A., McMullan, M., Weese, D. J., and Cable, J. 2007a. Selection by parasites in spate conditions in wild Trinidadian guppies (*Poecilia reticulata*). International Journal for Parasitology 37: 805–812.

van Oosterhout, C., Smith, A. M., Hänfling, B., Ramnarine, I. W., Mohammed, R. S., and Cable, J. 2007b. The guppy as a conservation model: implications of parasitism and inbreeding for reintroduction success. Conservation Biology 21: 1573–1583.

van Oosterhout, C., Potter, R., Wright, H., and Cable, J. 2008. Gyro-scope: an individual-based computer model to forecast gyrodactylid infections on fish hosts. International Journal for Parasitology 38: 541–548.

Van Valen, L. 1973. A new evolutionary law. Evolutionary Theory 1: 1–30.

Vargas, J. P., Lopez, J. C., Salas, C., and Thinus-Blanc, C. 2004. Encoding of geometric and featural spatial information by goldfish (*Carassius auratus*). Journal of Comparative Psychology 118: 206–216.

Veith, A. M., Schäfer, M., Klüver, N., Schmidt, C., Schultheis, C., Schartl, M., Winkler, C., and Volff, J.-N. 2006. Tissue-specific expression of *dmrt* genes in embryos and adults of the platyfish *Xiphophorus maculatus*. Zebrafish 3: 325–337.

Veith, W. J. 1979. Reproduction in the livebearing teleost *Clinus superciliosos*. South African Journal of Zoology 14: 208–211.

Venkatesh, B., Tan, C. H., and Lam, T. J. 1992. Prostaglandin synthesis in vitro by ovarian follicles and extrafollicular tissue of the viviparous guppy (*Poecilia reticulata*) and its regulation. Journal of Experimental Zoology 262: 405–413.

Vielkind, J., Vielkind, U., and Anders, F. 1971. Electron microscopic studies on melanotic and amelanotic melanomas in Xiphophorin fish. Zeitschrift für Krebsforschung 75: 243–245.

Vielkind, U. 1976. Genetic control of cell differentiation in platyfish-swordtail melanomas. Journal of Experimental Zoology 196: 197–204.

Vielkind, U., Schlage, W., and Anders, F. 1977. Melanogenesis in genetically determined pigment cell tumors of platyfish and platyfish-swordtail hybrids: correlation between tyrosine activity and degree of malignancy. Zeitschrift für Krebsforschung und Klinische Onkologie 90: 285–299.

Viken, A., Fleming, I. A., and Rosenqvist, G. 2006. Premating avoidance of inbreeding absent in female guppies (*Poecilia reticulata*). Ethology 112: 716–723.

Vincent, A. G., and Font, W. F. 2003a. Host specificity and population structure of two exotic helminths, *Camallanus cotti* (Nematoda) and *Bothriocephalus acheilognathi* (Cestoda), parasitizing exotic fishes in Waianu Stream, O'ahu, Hawai'i. Journal of Parasitology 89: 540–544.

Vincent, A. G., and Font, W. F. 2003b. Seasonal and yearly population dynamics of two exotic helminths, *Camallanus cotti* (Nematoda) and *Bothriocephalus acheilognathi* (Cestoda), parasitizing exotic fishes in Waianu Stream, O'ahu, Hawaii. Journal of Parasitology 89: 756–760.

Vitousek, M. N., Mitchell, M. A., Woakes, A. J., Niemack, M. D., and Wikelski, M. 2007. High costs of female choice in a lekking lizard. PloS ONE 2: 557.

Vitousek, P. M. 1990. Biological invasions and ecosystem processes: towards an integration of population biology and ecosystem studies. Oikos 57: 7–13.

Volff, J.-N., and Schartl, M. 2001. Variability of genetic sex determination in poeciliid fishes. Genetica 111: 101–110.

Volff, J.-N., and Schartl, M. 2003. Evolution of signal transduction by gene and genome duplication in fish. Journal of Structural and Functional Genomics 3: 139–150.

Volff, J.-N., Korting, C., Froschauer, A., Zhou, Q. C., Wilde, B., Schultheis, C., Selz, Y., Sweeney, K., Duschl, J., Wichert, K., Altschmied, J., and Schartl, M. 2003. The *Xmrk* oncogene can escape nonfunctionalization in a highly unstable subtelomeric region of the genome of the fish *Xiphophorus*. Genomics 82: 470–479.

Volff, J.-N., Nanda, I., Schmid, M., and Schartl, M. 2007. Governing sex determination in fish: regulatory putsches and ephemeral dictators. Sexual Development 1: 85–99.

Vøllestad, L. A., Hindar, K., and Møller, A. P. 1999. A meta-analysis of fluctuating asymmetry in relation to heterozygosity. Heredity 83: 206–218.

von Frisch, K. 1938. Zur Psychologie des fischschwarmes. Naturwissenschaften 26: 601–606.

von Salomé, J., Gyllenstwn, U., and Bergstrom, T. F. 2007. Full-length sequence analysis of the *HLA-DRB1* locus suggests a recent origin of alleles. Immunogenetics 59: 261–271.

Vrijenhoek, R. C. 1979. Factors affecting clonal diversity and coexistence. American Zoologist 19: 787–798.

Vrijenhoek, R. C. 1984. Ecological differentiation among clones: the frozen niche variation model. In K. Wohrmann and V. Loeschcke (eds.), *Population Biology and Evolution*, 217–231. Berlin: Springer-Verlag.

Vrijenhoek, R. C. 1989. Genetic and ecological constraints on the origins and establishment of unisexual vertebrates. In R. M. Dawley and J. P. Bogart (eds.), *Evolution and Ecology of Unisexual Vertebrates*, 19–23. Albany: New York State Museum.

Vrijenhoek, R. C. 1994. Unisexual fish—model systems for studying ecology and evolution. Annual Review of Ecology and Systematics 25: 71–96.

Vrijenhoek, R. C. 1998. Clonal organisms and the benefits of sex. In G. R. Carvalho and L. Hauser (eds.), *Advances in Molecular Ecology*, 151–172. Amsterdam: IOS Press.

Vrijenhoek, R. C., and Pfeiler, E. 1997. Differential survival of sexual and asexual *Poeciliopsis* during environmental stress. Evolution 51: 1593–1600.

Vrijenhoek, R. C., and Schultz, R. J. 1974. Evolution of a trihybrid unisexual fish (*Poeciliopsis*, Poeciliidae). Evolution 28: 205–319.

Vrijenhoek, R. C., Douglas, M. E., and Meffe, G. K. 1985. Conservation genetics of endangered fish populations in Arizona. Science 229: 400–402.

Vrijenhoek, R. C., Dawley, R. M., Cole, C. J., and Bogart, J. P. 1989. A list of known unisexual vertebrates. In R. M. Dawley and J. P. Bogart (eds.), *Evolution and Ecology of Unisexual Vertebrates*, 19–23. Albany: New York State Museum.

Vukusic, P., and Sambles, J. R. 2003. Photonic structures in biology. Nature 424: 852–855.

Waddington, C. H. 1953. Genetic assimilation of an acquired character. Evolution 7: 118–126.

Wagner, W. E., and Basolo, A. L. 2008. Incidental sanctions and the evolution of direct benefits. Ethology 114: 521–539.

Wakamatsu, Y. 1981. Establishment of a cell line from the platyfish-swordtail hybrid melanoma. Cancer Research 41: 679–80.

Walker, J. A., Ghalambor, C. K., Griset, O. L., McKenney, D., and Reznick, D. N. 2005. Do faster starts increase the probability of evading predators? Functional Ecology 19: 808–815.

Wallace, R. A., and Selman, K. 1981. Cellular and dynamic aspects of oocyte growth in teleosts. American Zoologist 21: 325–343.

Walling, C. A., Royle, N. J., Metcalfe, N. B., and Lindström, J. 2007. Early nutritional conditions, growth trajectories and mate choice: does compensatory growth lead to a reduction in adult sexual attractiveness? Behavioral Ecology and Sociobiology 61: 1007–1014.

Walling, C. A., Royle, N. J., Lindström, J., and Metcalfe, N. B. 2008. Experience-induced preference for short-sworded males in the green swordtail, *Xiphophorus helleri*. Animal Behaviour 76: 271–276.

Walter, R. B., Rains, J. D., Russell, J. E., Guerra, T. M., Daniels, C., Johnston, D. A., Kumar, J., Wheeler, A., Kelnar, K., Khanolkar, V. A., Williams, E. L., Hornecker, J. L., Hollek, L., Mamerow, M. M., Pedroza, A., and Kazianis, S. 2004. A microsatellite genetic linkage map for *Xiphophorus*. Genetics 168: 363–372.

Walter, R. B., Hazlewood, L., and Kazianis, S. 2006. *The Xiphophorus Genetic Stock Center Manual.* San Marcos: Texas State University.

Walters, L. H., and Walters, V. 1965. Laboratory observations on a cavernicolous poeciliid from Tabasco, Mexico. Copeia 1965: 214–223.

Wang, J. L. 2004. Sibship reconstruction from genetic data with typing errors. Genetics 166: 1963–1979.

Warburton, B., Hubbs, C., and Hagen, D. W. 1957. Reproductive behavior of *Gambusia heterochir.* Copeia 1957: 299–300.

Warburton, K., and Lees, N. 1996. Species discrimination in guppies: learned responses to visual cues. Animal Behaviour 52: 371–378.

Ward, A. J. W., and Hart, P. J. B. 2003. The effects of kin and familiarity on interactions between fish. Fish and Fisheries 4: 348–358.

Ward, A. J. W., and Hart, P. J. B. 2005. Foraging benefits of shoaling with familiars may be exploited by outsiders. Animal Behaviour 69: 329–335.

Ward, A. J. W., Axford, S., and Krause, J. 2003. Cross-species familiarity in fish. Proceedings of the Royal Society of London Series B—Biological Sciences 270: 1157–1161.

Ward, A. J. W., Hart, P. J. B., and Krause, J. 2004. The effects of habitat- and diet-based cues on association preferences in three-spined sticklebacks. Behavioral Ecology 15: 925–929.

Ward, A. J. W., Holbrook, R. I., Krause, J., and Hart, P. J. B. 2005. Social recognition in sticklebacks: the role of direct experience and habitat cues. Behavioral Ecology and Sociobiology 57: 575–583.

Ward, A. J. W., Webster, M. M., and Hart, P. J. B. 2006. Intraspecific food competition in fishes. Fish and Fisheries 7: 1–31.

Ward, A. J. W., Webster, M. M., and Hart, P. J. B. 2007. Social recognition in wild fish populations. Proceedings of the Royal Society of London Series B—Biological Sciences 274: 1071–1077.

Ward, A. J. W., Sumpter, D. J. T., Couzin, L. D., Hart, P. J. B., and Krause, J. 2008. Quorum decision-making facilitates information transfer in fish shoals. Proceedings of the National Academy of Sciences of the United States of America 105: 6948–6953.

Ward, A. J. W., Webster, M. M., Currie, S., Magurran, A. E., and Krause, J. 2009. Species and population differences in social recognition between fishes: a role for ecology? Behavioural Ecology and Sociobiology 20: 511–516.

Ward, M. N., Churcher, A. M., Dick, K. J., Laver, C. R. J., Owens, G. L., Polack, M. D., Ward, P. R., Breden, F., and Taylor, J. S. 2008. The molecular basis of color vision in colorful fish: four long wave–sensitive (LWS) opsins in guppies (*Poecilia reticulata*) are defined by amino acid substitutions at key functional sites. BMC Evolutionary Biology 8: 210–233.

Warner, R. R. 1988. Traditionality of mating-site preferences in a coral reef fish. Nature 335: 719–721.

Weadick, C. J., and Chang, B. S. W. 2007. Long-wavelength sensitive visual pigments of the guppy (*Poecilia reticulata*): six opsins expressed in a single individual. BMC Evolutionary Biology 7 (Supplement 1): 11.

Webb, P. W., and Gerstner, C. L. 2000. Fish swimming behavior: predictions from physical principles. In P. Domenici and R. W. Blake (eds.), *Biomechanics in Animal Behavior,* 59–77. Oxford: BIOS Scientific Publishers.

Weber, A., Horst, R., Barbier, G., and Oesterhelt, C. 2007. Metabolism and metabolomics of eukaryotes living under extreme conditions. International Review of Cytology 256: 1–34.

Weber, J. M., and Kramer, D. L. 1983. Effects of hypoxia and surface access on growth, mortality, and behavior of juvenile guppies, *Poecilia reticulata.* Canadian Journal of Fisheries and Aquatic Sciences 40: 1583–1588.

Webster, M. M., and Laland, K. N. 2008. Social learning strategies and predation risk: minnows copy only when using private information would be costly. Proceedings of the Royal Society of London Series B—Biological Sciences 275: 2869–2876.

Webster, M. M., Adams, E. L., and Laland, K. N. 2008a. Diet-specific chemical cues influence association preferences and patch use in a shoaling fish. Animal Behaviour 76: 17–23.

Webster, M. M., Ward, A. J. W., and Hart, P. J. B. 2008b. Shoal and prey patch choice by co-occurring fish and prawns: inter-taxa use of socially transmitted cues. Proceedings of the Royal Society of London Series B—Biological Sciences 275: 203–208.

Webster, M. M., Goldsmith, J., Ward, A. J. W., and Hart, P. J. B. 2007. Habitat-specific chemical cues influence association preferences and shoal cohesion in fish. Behavioral Ecology and Sociobiology 62: 273–280.

Weeks, S. C. 1990. An Experimental Test of the Tangled Bank–Frozen Niche–Variation Models. PhD thesis, University of Maryland.

Weeks, S. C. 1993. Phenotypic plasticity of life-history traits in clonal and sexual fish (*Poeciliopsis*) at high and low densities. Oecologia 93: 307–314.

Weeks, S. C. 1995. Comparisons of life-history traits between clonal and sexual fish (*Poeciliopsis*: Poeciliidae) raised in monoculture and mixed treatments. Evolutionary Ecology 9: 258–274.

Weeks, S. C. 1996a. The hidden cost of reproduction: reduced food intake caused by spatial constraints in the body cavity. Oikos 75: 345–349.

Weeks, S. C. 1996b. A reevaluation of the Red Queen model for the maintenance of sex in a clonal-sexual fish complex (Poeciliidae: *Poeciliopsis*). Canadian Journal of Fisheries and Aquatic Sciences 53: 1157–1164.

Weeks, S. C., and Quattro, J. M. 1991. Life-history plasticity under resource stress in a clonal fish (Poeciliidae, *Poeciliopsis*). Journal of Fish Biology 39: 485–494.

Wegman, I., and Götting, K. J. 1971. Untersuchungen zur Dotterblikung in den Oocyten von *Xiphophorus helleri* (Heckel, 1848) (Teleosti: Poeciliidae). Zeitschrift für Zellforschung und Mikroskopische Anatomie 119: 405–433.

Wegner, K. M., Kalbe, M., Kurtz, J., Reusch, T. B. H., and Milinski, M. 2003a. Parasite selection for immunogenetic optimality. Science 301: 1343.

Wegner, K. M., Reusch, T. B. H., and Kalbe, M. 2003b. Multiple parasites are driving major histocompatibility complex polymorphism in the wild. Journal of Evolutionary Biology 16: 224–232.

Wegner, K. M., Kalbe, M., Schaschl, H., and Reusch, T. B. H. 2004. Parasites and individual major histocompatibility complex diversity—an optimal choice? Microbes and Infection 6: 1110–1116.

Weihs, D. 1993. Stability of aquatic animal locomotion. Contemporary Mathematics 141: 443–461.

Weir, B. S., and Cockerham, C. C. 1984. Estimating *F*-statistics for the analysis of population structure. Evolution 38: 1358–1370.

Weis, P. 1972. Hepatic ultrastructure in two species of normal fasted and gravid teleost fishes. American Journal of Anatomy 133: 317–332.

Weis, S., and Schartl, M. 1998. The macromelanophore locus and the melanoma oncogene *Xmrk* are separate genetic entities in the genome of *Xiphophorus*. Genetics 149: 1909–1920.

Weishaupt, E. 1925. Die Ontogenie der Genitalorgane von *Girardinus reticulatus*. Zeitschrift für wissenschaftliche Zoologie 126: 571–611.

Welcomme, R. L. 1988. International introductions of inland aquatic species. FAO Fisheries Technical Papers.

Wellborn, G. A., Skelly, D. K., and Werner, E. E. 1996. Mechanisms creating community structure across a freshwater habitat gradient. Annual Review of Ecology and Systematics 27: 337–363.

Wellbrock, C., and Schartl, M. 1999. Multiple binding sites in the growth factor receptor Xmrk mediate binding to p59fyn, GRB2 and Shc. European Journal of Biochemistry 260: 275–283.

Wellbrock, C., and Schartl, M. 2000. Activation of phosphatidylinositol 3-kinase by a complex of p59fyn and the receptor tyrosine kinase Xmrk is involved in malignant transformation of pigment cells. European Journal of Biochemistry 267: 3513–3522.

Wellbrock, C., Lammers, R., Ullrich, A., and Schartl, M. 1995. Association between the melanoma-inducing receptor tyrosine kinase Xmrk and src family tyrosine kinases in *Xiphophorus*. Oncogene 10: 2135–2143.

Wellbrock, C., Weisser, C., Geissinger, E., Troppmair, J., and Schartl, M. 2002. Activation of p59(Fyn) leads to melanocyte dedifferentiation by influencing MKP-1-regulated mitogen-activated protein kinase signaling. Journal of Biological Chemistry 277: 6443–6254.

Wellbrock, C., Weisser, C., Hassel, J. C., Fischer, P., Becker, J., Vetter, C. S., Behrmann, I., Kortylewski, M., Heinrich, P. C., and Schartl, M. 2005. STAT5 contributes to interferon resistance of melanoma cells. Current Biology 15: 1629–1639.

Werner, E. E., and Hall, D. J. 1988. Ontogenetic habitat shifts in bluegill: the foraging rate predation risk trade-off. Ecology 69: 1352–1366.

West, S. A., Lively, C. M., and Read, A. F. 1999. A pluralist approach to sex and recombination. Journal of Evolutionary Biology 12: 1003–1012.

West-Eberhard, M. J. 2003. *Developmental Plasticity and Evolution*. Oxford: Oxford University Press.

Westerdahl, H., Hansson, B., Bensch, S., and Hasselquist, D. 2004. Between-year variation of MHC allele frequencies in great reed warblers: selection or drift? Journal of Evolutionary Biology 17: 485–492.

Wetherington, J. D., Schenck, R. A., and Vrijenhoek, R. C. 1989a. The origins and ecological success of unisexual *Poeciliopsis*: The frozen niche variation model. In G. K. Meffe and F. F. Snelson Jr. (eds.), *Ecology and Evolution of*

Livebearing Fishes (Poeciliidae), 259–276. Englewood Cliffs, NJ: Prentice Hall.

Wetherington, J. D., Weeks, S. C., Kotorak, E., and Vrijenhoek, R. C. 1989b. Genotypic and environmental components of variation in growth and reproduction of fish hemiclones (*Poeciliopsis*: Poeciliidae). Evolution 43: 635–645.

White, E. M., Partridge, U. C., and Church, S. C. 2003. Ultraviolet dermal reflexion and mate choice in the guppy, *Poecilia reticulata*. Animal Behaviour 65: 693–700.

White, E. M., Church, S. C., Willoughby, L. J., Hudson, S. J., and Partridge, J. C. 2005. Spectral irradiance and foraging efficiency in the guppy, *Poecilia reticulata*. Animal Behaviour 69: 519–527.

Whiten, A., and Ham, R. 1992. On the nature and evolution of imitation in the animal kingdom: reappraisal of a century of research. Advances in the Study of Behavior 21: 239–283.

Wiernasz, D. C., and Kingsolver, J. G. 1992. Wing melanin pattern mediates species recognition in *Pteris occidentalis*. Animal Behaviour 43: 89–94.

Wilbur, H. M., Tinkle, D. W., and Collins, J. P. 1974. Environmental certainty, trophic level, and resource availability in life history evolution. American Naturalist 108: 805–816.

Wilcove, D. S., Rothstein, D., Dubow, J., Phillips, A., and Losos, E. 1998. Quantifying threats to imperilled species in the United States. Bioscience 48: 607–615.

Wilke, C. O. 2004. The speed of adaptation in large asexual populations. Genetics 167: 2045–2053.

Wilkens, H. 2007. Regressive evolution: ontogeny and genetics of cavefish eye rudimentation. Biological Journal of the Linnean Society 92: 287–296.

Wilkinson, G. S., Presgraves, D. C., and Crymes, L. 1998. Male eye span in stalk-eyed flies indicates genetic quality by meiotic drive suppression. Nature 391: 276–279.

Williams, G. C. 1975. *Sex and Evolution*. Princeton, NJ: Princeton University Press.

Williams, H., and Jones, A. 1994. *Parasitic Worms of Fish*. London: Taylor & Francis.

Williams, J. E., Johnson, J. E., Hendrickson, D. A., Contreras-Balderas, S., Williams, J. D., Navarro-Mendoza, M., McAllister, D. E., and Deacon, J. E. 1989. Fishes of North America endangered, threatened or of special concern. Fisheries Management and Ecology 14: 2–20.

Williford, A., Stay, B., and Bhattacharya, D. 2004. Evolution of a novel function: nutritive milk in the viviparous cockroach, *Diploptera punctata*. Evolution and Development 6: 67–77.

Wilson, D. S. 1998. Adaptive individual differences within single populations. Philosophical Transactions of the Royal Society of London Series B—Biological Sciences 353: 199–205.

Wilson, D. S., Coleman, K., Clark, A. B., and Biederman, L. 1993. An ecological study of a psychological trait. Journal of Comparative Psychology 107: 250–260.

Wilson, D. S., Clark, A. B., Coleman, K., and Dearstyne, T. 1994. Shyness and boldness in humans and other animals. Trends in Ecology and Evolution 9: 442–446.

Wilson, M. L., Hauser, M. D., and Wrangham, R. W. 2001. Does participation in intergroup conflict depend on numerical assessment, range location, or rank for wild chimpanzees? Animal Behaviour 61: 1203–1216.

Wilson, N., Tubman, S. C., Eady, P. E., and Robertson, G. W. 1997. Female genotype affects male success in sperm competition. Proceedings of the Royal Society of London Series B—Biological Sciences 264: 1491–1495.

Wilson, R. S. 2005. Temperature influences the coercive mating and swimming performance of male eastern mosquitofish. Animal Behaviour 70: 1387–1394.

Wilson, R. S., Condon, C. H. L., and Johnston, I. A. 2007a. Consequences of thermal acclimation for the mating behaviour and swimming performance of female mosquito fish. Philosophical Transactions of the Royal Society of London Series B—Biological Sciences 362: 2131–2139.

Wilson, R. S., Hammill, E., and Johnston, I. A. 2007b. Competition moderates the benefits of thermal acclimation to reproductive performance in male eastern mosquitofish. Proceedings of the Royal Society of London Series B—Biological Sciences 274: 1199–1204.

Winemiller, K. O. 1989. Development of dermal lip protuberances for aquatic surface respiration in South American characid fishes. Copeia 1989: 382–390.

Winemiller, K. O. 1993. Seasonality of reproduction by live-bearing fishes in tropical rain forest streams. Oecologia 95: 266–276.

Winemiller, K. O., Leslie, M., and Roche, R. 1990. Phenotypic variation in male guppies from natural inland populations: an additional test of Haskins sexual selection/predation hypothesis. Environmental Biology of Fishes 29: 179–191.

Winge, Ø. 1921. A peculiar mode of inheritance and its cytological explanation. Comptes Rendus des Travaux du Laboratoire Carlsberg 14: 1–9.

Winge, Ø. 1922a. A peculiar mode of inheritance and its cytological explanation. Journal of Genetics 12: 137–144.

Winge, Ø. 1922b. One-sided masculine and sex-linked inheritance in Lebistes reticulatus. Journal of Genetics 12: 145–162.

Winge, Ø. 1927. The location of eighteen genes in Lebistes reticulatus. Journal of Genetics 18: 1–43.

Winge, Ø. 1930. On the occurence of XX males in Lebistes with some remarks on Aida's so called "non-disjunctional males" in Aplocheilus. Journal of Genetics 23: 69–76.

Winge, Ø. 1934. The experimental alternation of sex chromosomes into autosomes and vice versa, as illustrated by Lebistes. Comptes Rendus des Travaux du Laboratoire Carlsberg 21: 1–49.

Winge, Ø. 1937. Succession of broods in Lebistes. Nature 140: 467.

Winge, Ø., and Ditlevsen, E. 1938. A lethal gene in the Y chromosome of Lebistes. Comptes Rendus des Travaux du Laboratoire Carlsberg 22: 203–211.

Winge, Ø., and Ditlevsen, E. 1947. Colour inheritance and sex determination in Lebistes. Heredity 1: 65–83.

Winkler, C., Wittbrodt, J., Lammers, R., Ullrich, A., and Schartl, M. 1994. Ligand-dependent tumor induction in medakafish embryos by Xmrk receptor tyrosine kinase transgene. Oncogene 9: 1517–1525.

Winkler, P. 1985. Persistent differences in thermal tolerance among acclimation groups of a warm spring population of Gambusia affinis determined under field and laboratory conditions. Copeia 1985: 456–461.

Winnemoeller, D., Wellbrock, C., and Schartl, M. 2005. Activating mutations in the extracellular domain of the melanoma inducing receptor Xmrk are tumorigenic in vivo. International Journal of Cancer 117: 723–729.

Wisenden, B. D., and Chivers, D. P. 2006. The role of public chemical information in antipredator behaviour. In F. Ladich, S. P. Collin, P. Moller, and B. G. Kapoor (eds.), Fish Communication, 259–278. Enfield, NH: Science Publishers.

Wisenden, B. D., Chivers, D. P., and Smith, R. J. F. 1995. Early warning in the predation sequence: a disturbance pheromone in Iowa darters (Etheostoma exile). Journal of Chemical Ecology 21: 1469–1480.

Wisenden, B. D., Pogatshnik, J., Gibson, D., Bonacci, L., Schumacher, A., and Willett, A. 2008. Sound the alarm: learned association of predation risk with novel auditory stimuli by fathead minnows (Pimephales promelas) and glowlight tetras (Hemigrammus erythrozonus) after single simultaneous pairings with conspecific chemical alarm cues. Environmental Biology of Fishes 81: 141–147.

Wittbrodt, J., Adam, D., Malitschek, B., Mäueler, W., Raulf, F., Telling, A., Robertson, S. M., and Schartl, M. 1989. Novel putative receptor tyrosine kinase encoded by the melanoma-inducing Tu locus in Xiphophorus. Nature 341: 415–421.

Wittbrodt, J., Shima, A., and Schartl, M. 2002. Medaka—a model organism from the Far East. Nature Reviews Genetics 3: 53–64.

Witte, K. 2006. Learning and mate choice. In C. Brown, K. Laland, and J. Krause (eds.), Fish Cognition and Behaviour, 70–95. Oxford: Blackwell.

Witte, K., and Massmann, R. 2003. Female sailfin mollies, Poecilia latipinna, remember males and copy the choice of others after 1 day. Animal Behaviour 65: 1151–1159.

Witte, K., and Ryan, M. J. 1998. Male body length influences mate-choice copying in the sailfin molly Poecilia latipinna. Behavioral Ecology 9: 534–539.

Witte, K., and Ryan, M. J. 2002. Mate choice copying in the sailfin molly, Poecilia latipinna, in the wild. Animal Behaviour 63: 943–949.

Witte, K., and Ueding, K. 2003. Sailfin molly females (Poecilia latipinna) copy the rejection of a male. Behavioral Ecology 14: 389–395.

Wolf, L. E. 1931. The history of the germ cells in the viviparous teleost Platypoecilus maculatus. Journal of Morphology and Physiology 52: 115–153.

Wong, B. B. M., and Candolin, U. 2005. How is female mate choice affected by male competition? Biological Reviews 80: 559–571.

Wong, B. B. M., and McCarthy, M. 2009. Prudent male mate choice under perceived sperm competition risk in the eastern mosquito fish. Behavioral Ecology 20: 278–282.

Wong, B. B. M., and Rosenthal, G. G. 2006. Female disdain for swords in a swordtail fish. American Naturalist 167: 136–140.

Wong, B. B. M., Fisher, H. S., and Rosenthal, G. G. 2005. Species recognition by male swordtails via chemical cues. Behavioural Ecology 16: 818–822.

Wood, S. R., Berwick, M., Ley, R. D., Walter, R. B., Setlow, R. B., and Timmins, G. S. 2006. UV causation of melanoma in Xiphophorus is dominated by melanin photosensitized oxidant production. Proceedings of the National Academy of Sciences of the United States of America 103: 4111–4115.

Woodhead, A. D. 1978. Ageing changes in the liver of two poeciliid fishes, the guppy Poecilia (Lebistes) reticulata and the Amazon molly, P. formosa. Experimental Gerontology 13: 37–45.

Woodhead, A. D. 1984. Aging changes in the heart of a poeciliid fish, the guppy *Poecilia reticulatus*. Experimental Gerontology 19: 383–391.

Woodhead, A. D., and Armstrong, N. 1985. Aspects of the mating behavior of male mollies *Poecilia* spp. Journal of Fish Biology 27: 593–602.

Woodhead, A. D., Setlow, R. B., and Hart, R. W. 1977. The development of thyroid neoplasia in old age in the Amazon molly, *Poecilia formosa*. Experimental Gerontology 12: 193–200.

Woodhead, A. D., Setlow, R. B., and Pond, V. 1984. The Amazon molly, *Poecilia formosa*, as a test animal in carcinogenicity studies: chronic exposures to physical agents. National Cancer Institute Monograph 65: 45–52.

Wooten, M. C., Scribner, K. T., and Smith, M. H. 1988. Genetic variability and systematics of *Gambusia* in the southeastern United States. Copeia 1988: 283–289.

Wourms, J. P. 1981. Viviparity: the maternal-fetal relationship in fishes. American Zoologist 21: 473–515.

Wourms, J. P., Grove, B. D., and Lombardi, J. 1988. The maternal-embryonic relationship in viviparous fishes. In W. S. Hoar and D. J. Randall (eds.), *Fish Physiology: The Physiology of Developing Fish*, 1–134. New York: Academic Press.

Wright, D., Nakamichi, R., Krause, J., and Butlin, R. K. 2006. QTL analysis of behavioral and morphological differentiation between wild and laboratory zebrafish (*Danio rerio*). Behavior Genetics 36: 271–284.

Wright, S. 1931. Evolution in Mendelian populations. Genetics 16: 97–159.

Wright, S. 1951. The genetical structure of populations. Annals of Eugenics 15: 323–354.

Wu, C. I., and Xu, E. Y. 2003. Sexual antagonism and X inactivation—the SAXI hypothesis. Trends in Genetics 19: 243–247.

Wu, W., Meijer, O. G., Lamoth, C., Uegaki, K., van Dieën, J. H., Wuisman, P., de Vries, J., and Beek, P. 2004. Gait coordination in pregnancy: transverse pevic and thoracic rotations and their relative phase. Clinical Biomechanics 19: 480–488.

Wyatt, T. D. 2003. *Pheromones and Animal Behavior: Communication by Smell and Taste*. Cambridge: Cambridge University Press.

Wyles, J. S., Kunkel, J. G., and Wilson, A. C. 1983. Birds, behaviour, and anatomical evolution. Proceedings of the National Academy of Sciences of the United States of America 80: 4394–4397.

Yamamoto, T. 1975. The medaka, *Oryzias latipes*, and the guppy, *Lebistes reticularis*. In R. C. King (ed.), *Handbook of Genetics*, 133–149. New York: Plenum Press.

Yan, H. Y. 1986. Reproductive strategies of the Clear Creek gambusia, *Gambusia heterochir*. PhD thesis, University of Texas at Austin.

Yasui, Y. 1997. A "good-sperm" model can explain the evolution of costly multiple mating by females. American Naturalist 149: 573–584.

Yokoyama, S. 2000. Molecular evolution of vertebrate visual pigments. Progress in Retinal and Eye Research 19: 385–419.

Yokoyama, S. 2002. Molecular evolution of color vision in vertebrates. Gene 300: 69–78.

Yokoyama, S., and Yokoyama, R. 1996. Adaptive evolution of photoreceptors and visual pigments in vertebrates. Annual Review of Ecology, Evolution, and Systematics 27: 543–67.

Yunker, W. K., Wein, D. E., and Wisenden, B. D. 1999. Conditioned alarm behavior in fathead minnows (*Pimephales promelas*) resulting from association of chemical alarm pheromone with a nonbiological visual stimulus. Journal of Chemical Ecology 25: 2677–2686.

Zajitschek, S. R. K., and Brooks, R. 2008. Distinguishing the effects of familiarity, relatedness and colour pattern rarity on attractiveness and measuring their effects on sexual selection in guppies (*Poecilia reticulata*). American Naturalist 172: 843–854.

Zajitschek, S. R. K., Lindholm, A. K., Evans, J. P., and Brooks, R. C. 2009. Experimental evidence that high levels of inbreeding depress sperm competitiveness. Journal of Evolutionary Biology 22: 1338–1345.

Zander, C. D. 1965. Die Geschlechtsbestimmung bei *Xiphophorus montezumae cortezi* Rosen (Pisces). Molecular and General Genetics 96: 128–141.

Zander, C. D. 1969. Über die Entstehung und Veränderung von Farbmustern in der Gattung *Xiphophorus* (Pisces), I: qualitative Veränderungen nach Artkreuzung. Mitteilungen Hamburg Zoologische Museum Institut 66: 241–271.

Zander, C. D. 1977. Über die Entstehung und Veränderung von Farbmustern in der Gattung *Xiphophorus* (Pisces), II: quantitative Untersuchungen zur Penetranz und Expressivität bei Artbastarden mit dem Gen Sd (= spotted dorsal). Biologisches Zentralblatt 96: 467–479.

Zane, L., Nelson, W. S., Jones, A. G., and Avise, J. C. 1999. Microsatellite assessment of multiple paternity in natural populations of a live-bearing fish, *Gambusia holbrooki*. Journal of Evolutionary Biology 12: 61–69.

Zauner, H., Begemann, G., Mari-Beffa, M., and Meyer, A. 2003. Differential regulation of *msx* genes in the development of the gonopodium, an intromittent organ, and of the "sword," a sexually selected trait of swordtail fishes (*Xiphophorus*). Evolution and Development 5: 466–477.

Zeh, D. W., and Zeh, J. A. 2000. Reproductive mode and speciation: the viviparity-driven conflict hypothesis. Bioessays 22: 938–946.

Zeh, J. A., and Zeh, D. W. 1997. The evolution of polyandry, II: post-copulatory defenses against genetic incompatibility. Proceedings of the Royal Society of London Series B—Biological Sciences 264: 69–75.

Zeh, J. A., and Zeh, D. W. 2003. Toward a new sexual selection paradigm: polyandry, conflict and incompatibility. Ethology 109: 929–950.

Zeiske, E. 1968. Prädispositionen bei *Mollienesia sphenops* (Pisces, Poeciliidae) für einen Übergang zum Leben in subterranen Gewässern. Zeitschrift für Vergleichende Physiologie 58: 190–222.

Zhou, Y., Dai, D. L., Martinka, M., Su, M., Zhang, Y., Campos, E. I., Dorocicz, I., Tang, L., Huntsman, D., Nelson, C., Ho, V., and Li, G. 2005. Osteopontin expression correlates with melanoma invasion. Journal of Investigative Dermatology 124: 1044–1052.

Zimmerer, E. J., and Kallman, K. D. 1989. Genetic basis for alternative reproductive tactics in the pygmy swordtail, *Xiphophorus nigrensis*. Evolution 43: 1298–1307.

Zuberbuhler, K., and Byrne, R. W. 2006. Social cognition. Current Biology 16: 786–790.

Zuk, M., Thornhill, R., Ligon, J. D., and Johnson, K. 1990. Parasites and mate choice in red jungle fowl. American Zoologist 30: 235–244.

Zúñiga-Vega, J. J., Reznick, D. N., and Johnson, J. B. 2007. Habitat predicts reproductive superfetation and body shape in the livebearing fish *Poeciliopsis turrubarensis*. Oikos 116: 995–1005.

Zupanc, G. 2006. Neuogenesis and neuronal regeneration in the adult fish brain. Journal of Comparative Physiology A—Neuroethology, Sensory, Neural, and Behavioral Physiology 192: 649–670.

Contributors

Justin C. Bagley Evolutionary Ecology Laboratories, Department of Biology, Brigham Young University, 109 WIDB, Provo, UT 84602, USA (justin.bagley@byu.edu)

Amanda I. Banet Department of Biology, University of California—Riverside, 900 University Ave., Riverside, CA 92521, USA (mandybanet@gmail.com)

Angelo Bisazza Dipartimento di Psicologia Generale, Università di Padova, Via Venezia 8, 35131 Padova, Italy (angelo.bisazza @unipd.it)

Astrid Böhne Institut de Génomique Fonctionnelle de Lyon, Université de Lyon, Université Lyon 1, CNRS, INRA, Ecole Normale Supérieure de Lyon, Lyon, France (astrid.bohne@ens-lyon.fr)

Felix Breden Department of Biological Sciences, Simon Fraser University, Burnaby, BC V5A 1S6, Canada (breden@sfu.ca)

Robert C. Brooks Evolution & Ecology Research Centre and School of Biological, Earth and Environmental Sciences, University of New South Wales, Sydney 2052, Australia (rob.brooks @unsw.edu.au)

Culum Brown Department of Biological Sciences, Macquarie University, Sydney 2109, New South Wales, Australia (cbrown@bio .mq.edu.au)

Joanne Cable School of Biosciences, Cardiff University, Cardiff CF10 3AX, UK (cablej@cardiff.ac.uk)

Seth William Coleman Biology Department, Gonzaga University, Spokane, WA 99208, USA (colemans@gonzaga.edu)

Darren P. Croft Centre for Research in Animal Behaviour, University of Exeter, School of Psychology, Washington Singer Labs, Perry Road, Exeter EX4 4QG, UK (D.P.Croft@exeter.ac.uk)

Donald L. DeAngelis U.S. Geological Survey, Biological Resources Division, Department of Biology, University of Miami, P.O. Box 249118, Coral Gables, FL 33124, USA (ddeangelis@bio .miami.edu)

Matt Druen Department of Biological Sciences, University of Louisville, Louisville, KY 40292, USA (mwdrue01@gwise.louisville .edu)

Lee A. Dugatkin Department of Biological Sciences, University of Louisville, Louisville, KY 40292, USA (lee.dugatkin@louisville .edu)

John A. Endler School of Life and Environmental Sciences, Deakin University at Waurn Ponds, Pigdons Road, Geelong, Victoria, 3217, Australia (john.endler@deakin.edu.au)

Jonathan P. Evans Centre for Evolutionary Biology, University of Western Australia, Nedlands 6009, Western Australia (jonathan .evans@uwa.edu.au)

Delphine Galiana-Arnoux Institut de Génomique Fonctionnelle de Lyon, Université de Lyon, Université Lyon 1, CNRS, INRA, Ecole Normale Supérieure de Lyon, Lyon, France (Delphine.Galiana-Arnoux@ens-lyon.fr)

Francisco J. García de León Laboratorio Genética para la Conservación, Centro de Investigaciones Biológicas del Noroeste, Mar Bermejo no. 195, Col. Playa Palo de Santa Rita, Apdo. Postal 128, La Paz, BCS 23090, México (fgarciadl@cibnor.mx)

Gregory F. Grether Department of Ecology and Evolutionary Biology, University of California, Los Angeles, CA 90095-1606, USA (ggrether@ucla.edu)

Hartmut Greven Institut für Zellbiologie und Morphologie der Heinrich-Heine-Universität, Universitätsstrasse 1, D-40225 Düsseldorf, Germany (grevenh@uni-duesseldorf.de)

Sujan M. Henkanaththegedara Biological Sciences, Environmental, and Conservation Sciences Graduate Program, Stevens Hall, NDSU Dept. 2715, P.O. Box 6050, North Dakota State University, Fargo, ND 58108-6050, USA (Henkanaththegedara .Maduranga@ndsu.edu)

Richard James Department of Physics, University of Bath, Bath BA2 7AY, UK (r.james@bath.ac.uk)

Jiang Jiang Department of Biology, University of Miami, P.O. Box 249118, Coral Gables, FL 33124, USA (jjiang@bio.miami .edu)

Jerald B. Johnson Evolutionary Ecology Laboratories, Department of Biology, Brigham Young University, 153 WIDB, Provo, UT 84602, USA (jerry.johnson@byu.edu)

Jennifer L. Kelley Centre for Evolutionary Biology, University of Western Australia, Nedlands 6009, Western Australia (jennifer.kelley@uwa.edu.au)

Gita R. Kolluru Biological Sciences Department, California Polytechnic State University, San Luis Obispo, CA 93407, USA (gkolluru@calpoly.edu)

Jens Krause Department of Biology and Ecology of Fishes, Leibniz-Institute of Freshwater Ecology and Inland Fisheries, Müggelseedamm 310, 12587 Berlin, Germany (j.krause@igb-berlin.de)

Kevin N. Laland School of Biology, University of St. Andrews, St. Andrews, Fife KY16 9TS, UK (knl1@st-andrews.ac.uk)

R. Brian Langerhans Department of Biology and W. M. Keck Center for Behavioral Biology, North Carolina State University, Campus Box 7617, Raleigh, NC 27695-7617, USA (langerhans@ncsu.edu)

Anna K. Lindholm Institute of Evolutionary Biology and Environmental Studies, University of Zürich, Winterthurerstrasse 190, Zürich 8057, Switzerland (anna.lindholm@ieu.uzh.ch)

Anne E. Magurran Gatty Marine Laboratory, School of Biology, University of St. Andrews, St. Andrews, Fife KY16 8LB, UK (aem1@st-and.ac.uk)

Edie Marsh-Matthews Sam Noble Oklahoma Museum of Natural History and Department of Zoology, University of Oklahoma, 2401 Chautauqua Ave., Norman, OK 73072, USA (emarsh@ou.edu)

Mark McMullan School of Environmental Sciences, University of East Anglia, Norwich NR4 7TJ, UK (mcmullano@googlemail.com)

Gary K. Meffe Department of Wildlife Ecology and Conservation, University of Florida, Gainesville, FL 32611-0430, USA (GMeffe@conbio.org)

Svenja Meierjohann Physiological Chemistry I, Biocenter, Am Hubland, University of Würzburg, 97074, Würzburg, Germany (svenja.meierjohann@biozentrum.uni-wuerzburg.de)

Molly R. Morris Department of Biological Sciences, Ohio University, Athens, OH 45701, USA (morrism@ohio.edu)

Andrea Pilastro Dipartimento di Biologia, Università di Padova, Via Venezia 8, 35131 Padova, Italy (andrea.pilastro@unipd.it)

Marcelo N. Pires Department of Biological Sciences, Saddleback College, 28000 Marguerite Pkwy., Mission Viejo, CA 92692, USA (mpires@saddleback.edu)

Martin Plath Department of Ecology and Evolution, Institute of Ecology, Evolution and Diversity, J. W. Goethe-University Frankfurt am Main, Siesmayerstrasse 70-72, D-60054 Frankfurt am Main, Germany (martin_plath@web.de)

Bart J. A. Pollux Department of Biology, University of California—Riverside, 900 University Ave., Riverside, CA 92521, USA (barthp@ucr.edu)

Erik Postma Institute of Evolutionary Biology and Environmental Studies, University of Zürich, Winterthurerstrasse 190, 8057 Zürich, Switzerland (e.postma@ieu.uzh.ch)

David N. Reznick Department of Biology, University of California—Riverside, 900 University Ave., Riverside, CA 92521, USA (david.reznick@ucr.edu)

Rüdiger Riesch Department of Biology and W. M. Keck Center for Behavioral Biology, North Carolina State University, 127 David Clark Labs, Raleigh, NC 27695-7617, USA (ruedigerriesch@web.de)

Oscar Rios-Cardenas Instituto de Ecología, A. C. Departamento de Biología Evolutiva, Km 2.5 Carretera Antigua a Coatepec no. 351, Congregación El Haya, Xalapa, Mexico (oscar.rios@inecol.edu.mx)

Gil G. Rosenthal Department of Biology, Texas A&M University, 3258 TAMU, College Station, TX 77843-3258, USA; Centro de Investigaciones Científicas de las Huastecas, Calnali, Hgo., Mexico (grosenthal@bio.tamu.edu)

Manfred Schartl Physiological Chemistry I, Biocenter, Am Hubland, University of Würzburg, 97074, Würzburg, Germany (phch1@biozentrum.uni-wuerzburg.de)

Ingo Schlupp Department of Zoology, University of Oklahoma, 730 Van Vleet Oval, Norman, OK 73019, USA (schlupp@ou.edu)

Christina Schultheis Institut de Génomique Fonctionnelle de Lyon, Université de Lyon, Université Lyon 1, CNRS, INRA, Ecole Normale Supérieure de Lyon, France (christina.schultheis@ens-lyon.fr)

Craig A. Stockwell Biological Sciences, Environmental, and Conservation Sciences Graduate Program, Stevens Hall, NDSU Dept. 2715, P.O. Box 6050, North Dakota State University, Fargo, ND 58108-6050, USA (craig.stockwell@ndsu.edu)

Michael Tobler Department of Zoology, Oklahoma State University, 501 Life Sciences West, Stillwater, OK 74078, USA (michi.tobler@okstate.edu)

Joel C. Trexler Department of Biological Sciences, Florida International University, Miami, FL 33181, USA (trexlerj@fiu.edu)

Cock van Oosterhout School of Environmental Sciences, University of East Anglia, Norwich NR4 7TJ, UK (c.van-oosterhoutpuea.ac.uk)

Jean-Nicolas Volff Institut de Génomique Fonctionnelle de Lyon, Université de Lyon, Université Lyon 1, CNRS, INRA, Ecole Normale Supérieure de Lyon, France (Jean-Nicolas.Volff@ens-lyon.fr)

Mike M. Webster School of Biology, University of St. Andrews, St. Andrews, Fife KY16 9TS, UK (mmw1@st-andrews.ac.uk)

Author index

Taxonomic index

Subject index